T0180522

Bioinformatic and Statistical Analysis
of Microbiome Data

Yinglin Xia • Jun Sun

Bioinformatic and Statistical Analysis of Microbiome Data

From Raw Sequences to Advanced Modeling with QIIME 2 and R

 Springer

Yinglin Xia
Department of Medicine
University of Illinois Chicago
Chicago, IL, USA

Jun Sun
Department of Medicine
University of Illinois Chicago
Chicago, IL, USA

ISBN 978-3-031-21393-9   ISBN 978-3-031-21391-5   (eBook)
https://doi.org/10.1007/978-3-031-21391-5

This Springer imprint is published by the registered company Springer Nature Switzerland AG
The registered company address is: Gewerbestrasse 11, 6330 Cham, Switzerland

To our late grandmothers:
Mrs. Zhong Yilan (钟宜兰), Mrs. Sun
Zhangqiong (孙章琼), and Mrs. Zhong
Yizhen (钟宜珍) for their deep, constant love
and support.

Preface

Over the past two decades, microbiome research has received much attention across diverse fields and has become a topic of great scientific and public interests.

The microbiome is invisible but exists within a body space or a particular environment, but it is essential for development, immunity, and nutrition and can change our health status and progression of multiple diseases. The human microbiome has been described as an "essential organ" of the human body, "the invisible organ," "the forgotten organ" and "the last human organ" under active research, which highlights the importance of the human microbiome in health.

Compared with other research fields, microbiome data are complicated and have several unique characteristics. Thus, choosing appropriate statistical test or method is a very important step for analysis of microbiome data. However, it is still a challenging task for those biomedical researchers without statistical background and for those biostatisticians who do not have research experiences in this field. An appropriate statistical test or method is chosen not only based on the assumptions and properties of the statistical methods but also based on sufficient knowledge of the type and unique characteristics of collected data and the objective of the study.

We have done pioneering work to establish the microbiome data statistics and analysis. In October 2018, we published *Statistical Analysis of Microbiome Data with R* (Springer Nature), where we described a framework of statistical analysis of microbiome data. As the first statistical analysis book in microbiome research, we have received positive feedback from the readers around the world and positive review by *Biometrical Journal* (Dr. Kim-Anh Lê Cao, Biometrical Journal, Vol. 61, 2019). In that book, we focused on statistical analysis of microbiome data and introduced bioinformatic analysis of microbiome data with one chapter. Bioinformatic analysis of microbiome data is also a very important topic in microbiome research. The quality of read counts generated by bioinformatic analysis will have a large impact on the quality of downstream statistical analysis of microbiome data. We believe that it is very important to describe the workflows from bioinformatic analysis to statistical analysis of microbiome data. Thus, since October 2018, right after publishing our first book on statistical analysis of microbiome data, we planned to write a new book for combined bioinformatic and statistical analysis of microbiome data. This idea was motivated by the readers of the 2018 book who asked us for providing more coverages of bioinformatic analysis.

Our current book was written under the framework to provide one of the most important workflows for microbiome data analysis: from 16S rRNA sequencing raw reads to statistical analysis. We aim to provide a comprehensive review on statistical theories and methods on microbiome data analysis and discuss the development of bioinformatic and biostatistical methods and models in microbiome research. Particularly, we aim to provide the step-by-step procedures to perform bioinformatic and statistical analysis of microbiome data.

Profiling of bacterial communities through bacterial 16S rRNA sequencing is one important approach for bioinformatic analysis of microbiome data, which this book focuses on. Another important approach for bioinformatic analysis of microbiome data is shotgun metagenomics. Shotgun metagenomics methods allow for direct whole-genome shotgun sequencing of the microbiome metagenomic DNA, which have significantly improved our understanding of microbial community composition in ecosystem (e.g., the human body).

The microbiome data are the collection of all microorganisms. Emerging evidence and needs have shown the importance of viruses, fungi, and other microbes. Although no standard approach as bacterial 16S rRNA sequencing is available for analysis of the viral community and profiling of viruses' communities is still very challenging, however, in current years, several viral metagenomic data analysis tools have been developed to characterize different features of viruses, including viral discovery, virome composition analysis, taxonomy classification, and functional annotation. Although the beginning interests started from the 16S rRNA sequencing approach, we hope to have opportunities to discuss the other microbial data analysis and particularly profiling and analysis of viruses' communities in the future.

Statistical tools for performing microbiome data analysis are now available in different languages and environments across different platforms, either in web-based or programming-based approaches. Obviously, R system and environment play a critical role in developing statistical methods and models for analyzing microbiome data. QIIME 2 is a bioinformatic analysis tool via wrapping other sequencing platforms and also provides basic statistical analysis. Because of its comprehensive features and documentation supporting, QIIME 2 is one of the most popular bioinformatic tools in analysis of microbiome data. Thus, in this book, we leverage the capabilities of R and QIIME 2 for bioinformatic and statistical analysis of microbiome data.

Our book with 18 chapters is organized in this way: in the beginning two chapters, we specially provide overview and introduction of QIIME 2 and R in analysis of microbiome data, respectively. Chapters 3 to 6 present bioinformatic analysis of microbiome data and mainly through QIIME 2.

Chapter 7 introduces the original Operational Taxonomic Unit (OTU) methods in numerical taxonomy and Chapter 8 describes a movement of moving beyond OTU methods that has arisen in microbiome research field. Chapters 9 to 18 present biostatistical analysis of microbiome data and mainly through R and also QIIME 2.

Chapter 3 describes the basic data processing in QIIME 2. Chapter 4 introduces how to build feature table and feature data from raw sequencing reads. Chapter 5 introduces assigning taxonomy and building phylogenetic tree. Chapter 6 introduces taxonomic classification of the representative sequences and how to cluster Operational Taxonomic Units (OTUs). Chapter 7 comprehensively describes the development of OTU methods in numerical taxonomy, which provides a theoretical background of the clustering-based OTU methods that are used in bioinformatic analysis of microbiome data. Chapter 8 describes a movement that moves beyond OTU methods arisen in microbiome research field, which provides a comprehensive review on bioinformatic analysis of microbiome data. Chapters 9 and 10 provide two basic statistical analyses of microbiome data: Chap. 9 introduces alpha diversity metrics and visualization, Chap. 10 introduces beta diversity metrics and ordination. Chapters 11 to 18 present more advanced statistical methods and models in microbiome research. Chapter 11 introduces nonparametric methods for multivariate analysis of variance in ecological and microbiome data and statistical testing of beta diversity in microbiome data. Chapter 12 discusses differential abundance analysis of microbiome data mainly through the metagenomeSeq package. Chapter 13 presents zero-inflated beta models for microbiome data. Chapter 14 introduces compositional data and specifically the newly developed models for compositional analysis of microbiome data. Chapter 15 introduces linear mixed-effects models and describes using them for analysis of longitudinal microbiome data. Chapter 16 describes generalized linear mixed models (GLMMs) including the brief history of generalized linear models and generalized nonlinear models, algorithms for fitting GLMMs, as well as statistical hypothesis testing and modeling in GLMMs. Chapter 17 specifically introduces the newly developed GLMMs for longitudinal microbiome data and adopting the GLMMs in other fields to analyze longitudinal microbiome data. Chapter 18 provides an overview of multivariate longitudinal microbiome data analysis and specifically introduces the newly developed non-parametric microbial interdependence test. The large P small N problem is also discussed in the last chapter of this book.

We hope the contents and organization of these chapters will provide a set of basic concepts of microbiome, a framework of bioinformatic and statistical analysis of microbiome data. We expect this book to be used by (1) graduate students who study bioinformatic and statistical analysis of microbiome data; (2) bioinformaticians and statisticians, working on microbiome projects, either for their own research or for their collaborative research for experimental design, grant application, and data analysis; and (3) researchers who investigate biomedical and biochemical projects with the microbiome, and multi-omics data analysis. The datasets and R and QIIME 2 commends used in this book are available from Springer's website or by requesting to the first author: Yinglin Xia at yinglin.xia2007@gmail.com.

Chicago, IL, USA   Yinglin Xia
Chicago, IL, USA   Jun Sun
May 2022

Acknowledgments

The authors wish to thank the editors and staff at Springer Nature for their feedback along the way of process to publishing. Very special thanks to Mrs. Merry Stuber, Senior Editor, New York, for her enthusiasm in supporting and guiding the present project from beginning to end. We also wish to thank Dr. Cherry Ma, Managing Editor, Dr. Yu Zhu, Senior Publisher, and Mrs. Emily Zhang, Editor, for their help in processing the book proposal review.

We thank the three anonymous reviewers for their positive reviews on this book proposal. Especially we thank the two anonymous reviewers for their very positive and constructive reviews on the first draft of this book. Their constructive feedback was helpful in improving our reversion of this book. Broadly, we greatly appreciate the developers of bioinformatic and statistical methods, models, and R packages and R system and environment in general. Without their great works, the book cannot be available in current breadth and scope. Our special thanks go to Dr. Joseph Paulson for sharing his research paper with us.

Our path to the current academic journey and success was paved by supports from families for generations. We all wish to express our deepest appreciation to our respective parents and parents-in-law – Xincui Wang, Qijia Xia, Xiao-Yun Fu, and Zong-Xiang Sun – and to our respective families – Yuxuan Xia and Jason Xia – for their love and support.

Yinglin's grandmother Zhong Yilan (钟宜兰) was very kind and generous. She was always willing to help others and encouraged him to help people who were in need. She took care of Yinglin's everyday life until he left his village to attend high school at county. Yinglin remember that his grandmother Zhong Yilan always brought him with her to visit their relatives before Yinglin attended school. Yinglin's grandmother from his mother's side Zhong Yizhen (钟宜珍) was very kind and skillful. When Yinglin was young, every year his grandmother Zhong Yizhen gave them delicious homemade foods, delicate hand-knitted straw hats, and hand-woven straw fan as gifts. She only had one daughter, Yinglin's mother. She lived with Yinglin's parents for several years at the late time of her life.

Jun's grandmother Sun Zhangqiong (孙章琼) was positive, warm-hearted, and open-minded. In a traditional society, women's role was narrowly defined around the family. However, she always encouraged her granddaughter to chase her dreams, to find her intellectual potential, and to work for the society. Her vision and value for life has a deep influence on Jun.

We would like to dedicate this book to our late grandmothers, Mrs. Zhong Yilan (钟宜兰), Mrs. Sun Zhangqiong (孙章琼), and Mrs. Zhong Yizhen (钟宜珍) for their deep, constant love and support. May this book honor their memory and legacy.

Finally, we would like to acknowledge the VA Merit Award 1101BX004824-01, the DOD grant W81XWH-20-1-0623 (BC191198), Crohn's & Colitis Foundation Senior Research Award (902766), the NIDDK/National Institutes of Health grant R01 DK105118, and R01DK114126 to Jun Sun. The study sponsors play no role in the study design, data collection, analysis, and interpretation of data. The contents do not represent the views of the United States Department of Veterans Affairs or the United States Government.

Contents

1	**Introduction to QIIME 2**....................................		1
	1.1	Overview of QIIME 2................................	1
	1.2	Core Concepts in QIIME 2...........................	3
		1.2.1 Artifacts..................................	3
		1.2.2 Visualizations.............................	4
		1.2.3 Semantic Type............................	4
		1.2.4 Plugins..................................	4
	1.3	Install QIIME 2....................................	5
	1.4	Store and Track Data...............................	7
	1.5	Extract Data from QIIME 2 Archives..................	8
	1.6	Summary...	8
	References...		8
2	**Introduction to R for Microbiome Data**.....................		11
	2.1	Some Useful R Functions............................	11
		2.1.1 Save Data into R Data Format................	12
		2.1.2 Read Back Data from R Data Format Files.......	14
		2.1.3 Use Download.File() to Download File from Website................................	15
	2.2	Some Useful R Packages for Microbiome Data...........	15
		2.2.1 Readr....................................	16
		2.2.2 Ggpubr...................................	17
		2.2.3 Tidyverse................................	23
	2.3	Specifically Designed R Packages for Microbiome Data...	24
		2.3.1 Phyloseq.................................	24
		2.3.2 Microbiome..............................	34
		2.3.3 ampvis2..................................	47
		2.3.4 curatedMetagenomicData....................	48
	2.4	Some R Packages for Analysis of Phylogenetics.........	48
		2.4.1 Ape......................................	49

2.4.2   Phytools   50
2.4.3   Castor   52
2.5   BIOM Format and Biomformat Package   54
2.6   Creating Analysis Microbiome Dataset   57
2.7   Summary   61
References   61

3   Basic Data Processing in QIIME 2   65
3.1   Importing Data into QIIME 2   65
3.1.1   Import FASTA Format Data   66
3.1.2   Import FASTQ Format Data   69
3.1.3   Import Feature Table   73
3.1.4   Import Phylogenetic Trees   75
3.2   Exporting Data from QIIME 2   76
3.2.1   Export Feature Table   76
3.2.2   Export Phylogenetic Trees   77
3.3   Extracting Data from QIIME 2 Archives   77
3.3.1   Extract Data Using the Qiime Tools Export
Command   78
3.3.2   Extract Data Using Unzip Program on macOS   78
3.4   Filtering Data in QIIME 2   79
3.4.1   Filter Feature Table   80
3.4.2   Taxonomy-Based Tables and Sequences Filtering   83
3.4.3   Filter Distance Matrices   86
3.5   Introducing QIIME 2 View   87
3.6   Communicating Between QIIME 2 and R   87
3.6.1   Export QIIME 2 Artifacts into R Using qiime2R
Package   87
3.6.2   Prepare Feature Table and Metadata in R
and Import into QIIME 2   90
3.7   Summary   92
References   92

4   Building Feature Table and Feature Representative
Sequences from Raw Reads   95
4.1   Analyzing Demultiplexed Paired-End FASTQ Data   96
4.1.1   Prepare Sample Metadata   96
4.1.2   Prepare Raw Sequence Data   99
4.1.3   Import Data Files as Qiime Zipped
Artifacts(.qza)   100
4.1.4   Examine and Visualize the Qualities
of the Sequence Reads   101
4.2   Analyzing Demultiplexed Paired-End FASTQ Data Using
DADA2 and q2-dada2 Plugin   103
4.2.1   Introduction to DADA2 and q2-dada2 Plugin   103

4.2.2   Denoise Sequences to Construct Feature Table
and Feature Data with q2-dada2 Plugin 104
4.2.3   Summarize the Feature Table and Feature Data
from q2-dada2 Plugin 108
4.3   Analyzing Multiplexed Paired-End FASTQ Data
Using q2-dada2 Plugin 108
4.3.1   Prepare Sample Metadata 109
4.3.2   Prepare Raw Sequence Data 109
4.3.3   Import Data Files as Qiime Zipped Artifacts(.qza) . . . 110
4.3.4   Demultiplexing Sequences 110
4.3.5   Summarize the Demultiplexing Results and
Examine Quality of the Reads 112
4.4   Analyzing Demultiplexed Paired-End FASTQ Data
Using Deblur and q2-deblur Plugin 113
4.4.1   Introduction to Deblur and q2-deblur Plugin 113
4.4.2   Process Initial Quality Filtering 114
4.4.3   Preliminary Works for Denoising with Deblur 114
4.4.4   Denoise Sequences with Deblur to Construct
Feature Table and Feature Data 116
4.4.5   Summarize the Feature Table and Feature Data
from Deblur 117
4.4.6   Remarks on DADA2 and Deblur 118
4.5   Summary 119
References 120

5   **Assigning Taxonomy, Building Phylogenetic Tree** 123
5.1   Assigning Taxonomy 123
5.1.1   Bioinformatics Tools and Reference Databases 123
5.1.2   QIIME 2-Formatted and Maintained Taxonomic
Reference Databases 125
5.1.3   DADA2-Formatted and Maintained Taxonomic
Reference Databases 128
5.1.4   Introduction to q2-Feature-Classifier 128
5.1.5   Assign Taxonomy Using the q2-Feature-Classifier . . . 132
5.1.6   Remarks on Taxonomic Classification 135
5.2   Building Phylogenetic Tree 135
5.2.1   Introduction to Phylogenetic Tree 135
5.2.2   Build a Phylogenetic Tree Using the Alignment
and Phylogeny Commands 136
5.2.3   Remarks on the Taxonomic and Phylogenetic
Trees 138
5.3   Summary 141
References 141

6   Clustering Sequences into OTUs 147
   6.1   Introduction to Clustering Sequences into OTUs 147
      6.1.1   Merge Reads 148
      6.1.2   Remove Non-biological Sequences 148
      6.1.3   Trim Reads Length 148
      6.1.4   Discard Low-Quality Reads 149
      6.1.5   Dereplicate Sequences 149
   6.2   Introduction to VSEARCH and q2-vsearch 149
   6.3   Closed-Reference Clustering 150
      6.3.1   Introduction 150
      6.3.2   Implement Cluster-Features-Closed-Reference 150
      6.3.3   Remarks on Closed-Reference Clustering 152
   6.4   *De Novo* Clustering 153
      6.4.1   Introduction 153
      6.4.2   Implement Cluster-Features-De-Novo 153
      6.4.3   Remarks on *De Novo* Clustering 154
   6.5   Open-Reference Clustering 155
      6.5.1   Introduction 155
      6.5.2   Implement Cluster-Features-Open-Reference 155
      6.5.3   Remarks on Open-Reference Clustering 156
   6.6   Summary 157
   References 157

7   OTU Methods in Numerical Taxonomy 161
   7.1   Brief History of Numerical Taxonomy 162
   7.2   Principles of Numerical Taxonomy 164
      7.2.1   Definitions of Characters and Taxa 165
      7.2.2   Sample Size Calculation: How Many Characters? 166
      7.2.3   Equal Weighting Characters 169
      7.2.4   Taxonomic Rank 169
   7.3   Phenetic Taxonomy: Philosophy of Numerical Taxonomy 170
      7.3.1   Phenetics: Numerical-Based Empiricism Versus
                  Aristotle's Essentialism 170
      7.3.2   Classification: Inductive Theory Versus Darwin'
                  Theory of Evolution 173
      7.3.3   Biological Classifications: Phenetic Approach
                  Versus Cladistic Approach 174
   7.4   Construction of Taxonomic Structure 176
      7.4.1   Defining the Operational Taxonomic Units 177
      7.4.2   Estimation of Taxonomic Resemblance 178
      7.4.3   Commonly Clustering-Based OTU Methods 192
   7.5   Statistical Hypothesis Testing of OTUs 208
      7.5.1   Hypothesis Testing on Similarity Coefficients
                  of OTUs 208
      7.5.2   Hypothesis Testing on Clustering OTUs 209

7.6 Some Characteristics of Clustering-Based OTU Methods 210
    7.6.1 Some Basic Questions in Numerical Taxonomy 211
    7.6.2 Characteristics of Numerical Taxonomy 212
7.7 Summary 218
References 219

8 Moving Beyond OTU Methods 227
8.1 Clustering-Based OTU Methods in Microbiome Study 228
    8.1.1 Common Clustering-Based OTU Methods 228
    8.1.2 Hierarchical Clustering OTU Methods 229
    8.1.3 Heuristic Clustering OTU Methods 230
    8.1.4 Limitations of Clustering-Based OTU Methods 231
    8.1.5 Purposes of Using OTUs in Microbiome Study 240
    8.1.6 Defining Species and Species-Level Analysis 241
8.2 Moving Toward Single-Nucleotide Resolution-Based OTU
    Methods 250
    8.2.1 Concept Shifting in Bioinformatic Analysis 250
    8.2.2 Single-Nucleotide Resolution Clustering-Based
           OTU Methods 252
8.3 Moving Beyond the OTU Methods 254
    8.3.1 Entropy-Based Methods: Oligotyping 255
    8.3.2 Denoising-Based Methods 256
8.4 Discussion on Moving Beyond OTU Methods 267
    8.4.1 Necessity of Targeting Single-Base Resolution 267
    8.4.2 Possibility of Moving Beyond Traditional OTU
           Methods 268
    8.4.3 Issues of Sub-OTU Methods 270
    8.4.4 Prediction of Sequence Similarity to Ecological
           Similarity 271
    8.4.5 Functional Analysis and Multi-omics Integration 271
8.5 Summary 273
References 273

9 Alpha Diversity 289
9.1 Abundance-Based Alpha Diversity Metrics 290
    9.1.1 Chao 1 Richness and Abundance-Based
           Coverage Estimator (ACE) 290
    9.1.2 Shannon Diversity 298
    9.1.3 Simpson Diversity 300
    9.1.4 Pielou's Evenness 304
9.2 Phylogenetic Alpha Diversity Metrics 305
    9.2.1 Phylogenetic Diversity 306
    9.2.2 Phylogenetic Entropy 306
    9.2.3 Phylogenetic Quadratic Entropy (PQE) 307
9.3 Exploring Alpha Diversity and Abundance 307
    9.3.1 Heatmap 308

	9.3.2	Boxplot	312
	9.3.3	Violin Plot	316
9.4	Statistical Hypothesis Testing of Alpha Diversity		318
	9.4.1	Summarize the Diversity Measures	319
	9.4.2	Plot Histogram of the Diversity Distributions	319
	9.4.3	Kruskal-Wallis Test	319
	9.4.4	Perform Multiple Comparisons	321
9.5	Alpha Diversity Analysis in QIIME 2		322
	9.5.1	Calculate Alpha Diversity Using Core-Metrics-Phylogenetic Method	322
	9.5.2	Calculate Alpha Diversity Using Alpha Method	324
	9.5.3	Calculate Alpha Diversity Using Alpha-Phylogenetic Method	326
	9.5.4	Test for Differences of Alpha Diversity Between Groups	327
	9.5.5	Alpha Rarefaction in QIIME 2	328
9.6	Summary		330
References			331

10 Beta Diversity Metrics and Ordination 335
10.1	Abundance-Based Beta Diversity Metrics		335
	10.1.1	Bray-Curtis Dissimilarity	337
	10.1.2	Jaccard Dissimilarity	339
	10.1.3	Sørensen Dissimilarity	340
10.2	Phylogenetic Beta Diversity Metrics		342
	10.2.1	Unweighted UniFrac	342
	10.2.2	Weighted UniFrac	343
	10.2.3	GUniFrac	344
	10.2.4	pldist	344
	10.2.5	Calculate (Un)Weighted UniFrac and GUniFrac Distances Using the GUniFrac Package	345
	10.2.6	Remarks on Rarefaction for Alpha and Beta Diversity Analysis	348
10.3	Ordination Methods		349
	10.3.1	Introduction to Ordination	349
	10.3.2	Ordination Plots in the ampvis2 Package	352
	10.3.3	Principal Component Analysis (PCA)	353
	10.3.4	Principal Coordinate Analysis (PCoA)	359
	10.3.5	Nonmetric Multidimensional Scaling (NMDS)	362
	10.3.6	Correspondence Analysis (CA)	366
	10.3.7	Detrended Correspondence Analysis (DCA)	371
	10.3.8	Redundancy Analysis (RDA)	374
	10.3.9	Canonical Correspondence Analysis (CCA)	377
10.4	Beta Diversity Metrics and Ordination in QIIME 2		381
	10.4.1	Calculate Beta Diversity Measures	381

        10.4.2   Explore Principal Coordinates (PCoA)
                 Using Emperor Plots   382
   10.5 Remarks on Ordination and Clustering   384
   10.6 Summary   388
   References   388

11 **Statistical Testing of Beta Diversity**   397
   11.1 Introduction to Nonparametric MANOVA Using
         Permutation Tests   397
   11.2 Analysis of Similarity (ANOSIM)   398
         11.2.1   Introduction of ANOSIM   399
         11.2.2   Perform ANOSIM Using the Vegan Package   401
         11.2.3   Remarks on ANOSIM   404
   11.3 Permutational MANOVA (PERMANOVA)   405
         11.3.1   Introduction to PERMANOVA   405
         11.3.2   Perform PERMANOVA Using the Vegan
                  Package   409
         11.3.3   Remarks on PERMANOVA   417
   11.4 Analysis of Multivariate Homogeneity of Group
         Dispersions   418
         11.4.1   Introduction to the Function betadisper()   418
         11.4.2   Implement the Function betadisper()   419
   11.5 Pairwise PERMANOVA   422
         11.5.1   Introduction to Pairwise PERMMANOVA   422
         11.5.2   Implement Pairwise PERMMANOVA Using the
                  RVAideMemoire Package   423
   11.6 Identify Core Microbial Taxa Using the Microbiome
         Package   426
   11.7 Statistical Testing of Beta Diversity in QIIME 2   428
         11.7.1   Significant Testing of Bray-Curtis Distance   429
         11.7.2   Significant Testing of Jaccard Distance   430
         11.7.3   Significant Testing of Unweighted UniFrac
                  Distance   430
         11.7.4   Significant Testing of Weighted UniFrac
                  Distance   431
   11.8 Summary   431
   References   431

12 **Differential Abundance Analysis of Microbiome Data**   435
   12.1 Zero-Inflated Gaussian (ZIG) and Zero-Inflated
         Log-Normal (ZILN) Mixture Models   436
         12.1.1   Total Sum Scaling (TSS)   437
         12.1.2   Cumulative Sum Scaling (CSS)   438
         12.1.3   ZIG and ZILN Models   438
   12.2 Implement ZILN via metagenomeSeq   440

12.3   Some Additional Statistical Tests in metagenomeSeq   453
        12.3.1   Log-Normal Permutation Test of Abundance   453
        12.3.2   Presence-Absence Testing of the Proportion/Odds . . .   454
        12.3.3   Discovery Odds Ratio Testing of the Proportion
                  of Observed Counts   455
        12.3.4   Perform Feature Correlations   455
12.4   Illustrate Some Useful Functions in metagenomeSeq   456
        12.4.1   Access the MRexperiment Object   456
        12.4.2   Subset the MRexperiment Object   457
        12.4.3   Filter the MRexperiment Object or Count Matrix   458
        12.4.4   Merge the MRexperiment Object   458
        12.4.5   Call the Normalized Counts Using the
                  cumNormMat() and MRcounts()   459
        12.4.6   Calculate the Normalization Factors Using the
                  calcNormFactors()   460
        12.4.7   Access the Library Sizes Using the libSize()   460
        12.4.8   Save the Normalized Counts Using
                  the exportMat()   461
        12.4.9   Save the Sample Statistics Using
                  the exportStats()   461
        12.4.10   Find Unique OTUs or Features   461
        12.4.11   Aggregate Taxa   462
        12.4.12   Aggregate Samples   463
12.5   Remarks on CSS Normalization, ZIG, and ZILN   465
12.6   Summary   466
References   467

13 Zero-Inflated Beta Models for Microbiome Data   469
13.1   Zero-Inflated Beta Modeling Microbiome Data   469
13.2   Zero-Inflated Beta Regression (ZIBSeq)   471
        13.2.1   Introduction to ZIBseq   471
        13.2.2   Implement ZIBseq   473
        13.2.3   Remarks on ZIBSeq   479
13.3   Zero-Inflated Beta-Binomial Model (ZIBB)   480
        13.3.1   Introduction to ZIBB   480
        13.3.2   Implement ZIBB   483
        13.3.3   Remarks on ZIBB   488
13.4   Summary   489
References   489

14 Compositional Analysis of Microbiome Data   491
14.1   Introduction to Compositional Data   492
        14.1.1   What Are Compositional Data?   492
        14.1.2   Microbiome Data Are Treated as Compositional   492
        14.1.3   Aitchison Simplex   493
        14.1.4   Challenges of Analyzing Compositional Data   494

        14.1.5   Fundamental Principles of Compositional Data
                  Analysis 495
        14.1.6   The Family of Log-Ratio Transformations 496
        14.1.7   Remarks on Log-Ratio Transformations 498
   14.2   ANOVA-Like Compositional Differential Abundance
       Analysis 500
        14.2.1   Introduction to ALDEx2 500
        14.2.2   Implement ALDEx2 Using R 505
        14.2.3   Remarks on ALDEx2 517
   14.3   Analysis of Composition of Microbiomes (ANCOM) 518
        14.3.1   Introduction to ANCOM 519
        14.3.2   Implement ANCOM Using QIIME 2 520
        14.3.3   Remarks on ANCOM 526
   14.4   Analysis of Composition of Microbiomes-Bias Correction
       (ANCOM-BC) 528
        14.4.1   Introduction to ANCOM-BC 528
        14.4.2   Implement ANCOM-BC Using the ANCOMBC
                  Package 532
        14.4.3   Remarks on ANCOM-BC 547
   14.5   Remarks on Compositional Data Analysis Approach 549
   14.6   Summary 551
   References 551

15   Linear Mixed-Effects Models for Longitudinal
**   Microbiome Data** 557
   15.1   Introduction to Linear Mixed-Effects Models (LMMs) 558
        15.1.1   Advantages and Disadvantages of LMMs 558
        15.1.2   Fixed and Random Effects 559
        15.1.3   Definition of LMMs 559
        15.1.4   Statistical Hypothesis Tests 560
        15.1.5   How to Fit LMMs 561
   15.2   Identifying the Significant Taxa Using the nlme Package 563
        15.2.1   Introduction to LMMs in Microbiome Research 563
        15.2.2   Longitudinal Microbiome Data Structure 564
        15.2.3   Fit LMMs Using the Read Counts
                  as the Outcome 565
   15.3   Modeling the Diversity Indices Using the lme4 and
       LmerTest Packages 570
        15.3.1   Introduction to the lme4 and lmerTest Packages 571
        15.3.2   Fit LMMs Using the Diversity Index
                  as the Outcome 572
   15.4   Implement LMMs Using QIIME 2 576
        15.4.1   Introduction to the QIIME Longitudinal
                  Linear-Mixed-Effects Command 576
        15.4.2   Fit LMMs in QIIME 2 577
        15.4.3   Perform Volatility Analysis 579

15.5   Remarks on LMMs 581
15.6   Summary 582
References 582

16   Introduction to Generalized Linear Mixed Models 587
16.1   Generalized Linear Models (GLMs) and Generalized
       Nonlinear Models (GNLMs) 588
16.2   Generalized Linear Mixed Models (GLMMs) 588
16.3   Model Estimation in GLMMs 590
16.4   Algorithms for Parameter Estimation in GLMMs 591
       16.4.1   Penalized Quasi-Likelihood-Based Methods
                Using Taylor-Series Linearization 592
       16.4.2   Likelihood-Based Methods Using Numerical
                Integration 593
       16.4.3   Markov Chain Monte Carlo-Based Integration 595
       16.4.4   IWLS and EM-IWLS Algorithms 596
16.5   Statistical Hypothesis Testing and Modeling in GLMMs 599
       16.5.1   Model Selection in Statistics 599
       16.5.2   Model Selection in Machine Learning 599
       16.5.3   Information Criteria for Model Selection 600
       16.5.4   Likelihood-Ratio Test 605
       16.5.5   Vuong Test 607
16.6   Summary 608
References 608

17   Generalized Linear Mixed Models for Longitudinal
     Microbiome Data 615
17.1   Generalized Linear Mixed Models (GLMMs)
       in Microbiome Research 616
       17.1.1   Data Transformation Versus Using GLMMs 617
       17.1.2   Model Selection in Microbiome Data 618
       17.1.3   Statistical Hypothesis Testing in
                Microbiome Data 619
17.2   Generalized Linear Mixed Modeling Using the
       glmmTMB Package 621
       17.2.1   Introduction to glmmTMB 621
       17.2.2   Implement GLMMs via glmmTMB 624
       17.2.3   Remarks on glmmTMB 646
17.3   Generalized Linear Mixed Modeling Using the
       GLMMadaptive Package 647
       17.3.1   Introduction to GLMMadaptive 647
       17.3.2   Implement GLMMs via GLMMadaptive 647
       17.3.3   Remarks on GLMMadaptive 659
17.4   Fast Zero-Inflated Negative Binomial Mixed Modeling
       (FZINBMM) 660

	17.4.1	Introduction to FZINBMM	660
	17.4.2	Implement FZINBMM Using the NBZIMM Package	664
	17.4.3	Remarks on FZINBMM	668
17.5	Remarks on Fitting GLMMs		669
17.6	Summary		670
References			671

18 Multivariate Longitudinal Microbiome Models 675
- 18.1 Overview of Multivariate Longitudinal Microbiome Analysis 675
  - 18.1.1 Multivariate Distance/Kernel-Based Longitudinal Models 676
  - 18.1.2 Multivariate Integration of Multi-omics Methods 677
  - 18.1.3 Univariate Analysis Versus Multivariate Analysis 678
- 18.2 Nonparametric Microbial Interdependence Test (NMIT) 678
  - 18.2.1 Introduction to NMIT 679
  - 18.2.2 Implement NMIT Using R 680
  - 18.2.3 Implement NMIT Using QIIME 2 687
  - 18.2.4 Remarks on NMIT 690
- 18.3 The Large P Small N Problem 692
- 18.4 Summary 693
- References 693

Correction to: Bioinformatic and Statistical Analysis of Microbiome Data C1

Index 697

The original version of this book was revised. A correction is available at https://doi.org/10.1007/978-3-031-21391-5_19

About the Authors

Yinglin Xia is a Research Professor in the Department of Medicine at the University of Illinois Chicago (UIC). He was a Research Assistant Professor in the Department of Biostatistics and Computational Biology at the University of Rochester (Rochester, NY) before joining AbbVie (North Chicago, IL) as a Clinical Statistician. He joined UIC as a Research Associate Professor in 2015. Dr. Xia has successfully applied his statistical study design and data analysis skills to clinical trials, medical statistics, biomedical sciences, and social and behavioral sciences. He has published more than 140 statistical methodology and research papers in peer-reviewed journals. He serves on the editorial boards of several scientific journals including as an Associate Editor of *Gut Microbes* and has served as a reviewer for over 100 scientific journals. Dr. Xia has published three books on statistical analysis of microbiome and metabolomics data. He is the lead author of *Statistical Analysis of Microbiome Data with R* (Springer Nature, 2018), which was the first statistics book in microbiome study, *Statistical Data Analysis of Microbiomes and Metabolomics* (American Chemical Society, 2022), and *An Integrated Analysis of Microbiomes and Metabolomics* (American Chemical Society, 2022).

Jun Sun is a tenured Professor of Medicine at the University of Illinois Chicago. She is an elected fellow of the American Gastroenterological Association (AGA) and American Physiological Society (APS). She chairs the AGA Microbiome and Microbial Therapy section (2020–2022). She is an internationally recognized expert on microbiome and human diseases, such as vitamin D receptor in inflammation, dysbiosis, and intestinal dysfunction in amyotrophic lateral sclerosis (ALS). Her lab is the first to discover that chronic effects and molecular mechanisms of *Salmonella* infection and development of colon cancer. Dr. Sun has published over 210 scientific articles in peer-reviewed journals and 8 books on microbiome. She is on the editorial boards of more than 10 peer-reviewed international scientific journals and serves on the study sections for the national and international research foundations. Dr. Sun is a believer of scientific art and artistic science. She enjoys writing her science papers in English and poems in Chinese. Her poetry collection book《让时间停留在这一刻》("Let Time Stay Still at This Moment") was published in 2018.

Chapter 1
Introduction to QIIME 2

Abstract This chapter describes the foundations of the QIIME 2 approach for bioinformatic and biostatistical analyses of microbiome data. It first provides an overview of QIIME 2, and then introduces the core concepts in QIIME 2 and the installation of QIIME 2. Next it introduces how to store, track, and extract data in QIIME 2.

Keywords QIIME · QIIME 2 · DADA2 · Deblur · Artifacts · Visualizations · Semantic type · Plugins · QIIME 2 archives · q2-data2 · q2-feature-table · q2-types

In this chapter, we provide the foundations of the QIIME 2 approach to bioinformatic and statistical analysis of microbiome data. We first provide an overview of QIIME 2 (Sect. 1.1). Then we introduce the core concepts in QIIME 2 (Sect. 1.2). Next, we introduce how to install QIIME 2 (Sect. 1.3). Following that, we introduce how to store and track data in QIIME 2 (Sect. 1.4) and extract data from QIIME 2 (Sect. 1.5). We complete this chapter by a brief summary (Sect. 1.6).

1.1 Overview of QIIME 2

QIIME (canonically pronounced *chime*: Quantitative Insights Into Microbial Ecology) is an open source microbiome bioinformatics platform designed for analyzing microbial ecological communities (J. G. Caporaso et al. 2010). It has been used in many microbiome studies including analysis of bacterial, archaeal, fungal, or viral sequence data. QIIME analysis generally starts with raw sequence data (in FASTA format) generated from any sequencing technology. QIIME scripts primarily wrap other software packages. It can analyze high-throughput data in a wide variety of ways. QIIME is implemented as a collection of command-line scripts designed to

---

Supplementary Information The online version contains supplementary material available at https://doi.org/10.1007/978-3-031-21391-5_1.

take users from raw sequence data and sample metadata through publication-quality graphics and statistics (Xia et al. 2018). QIIME 2 has succeeded QIIME 1 on January 1, 2018. QIIME 1 is no longer supported since end of 2017. QIIME 2 (Bolyen et al. 2019) is not only a complete redesign and rewritten version of the QIIME 1, but also a completely reengineered and rewritten system aiming to facilitate reproducible and modular analysis of microbiome data.

However, for the users who use QIIME 2 as an analytical platform of microbiome data, the following three functionalities are more important:

(1) QIIME 2 is a bioinformatic analysis tool via wrapping other sequencing plat-forms, such as DADA2 (Callahan et al. 2016) and Deblur (Amir et al. 2017) to perform analysis, data generation, sequence quality control, taxonomy assign-ment, phylogenetic insertion, and other functions. This topic is one core part of this book, which is substantially introduced in Chaps. 3, 4, 5, and 6.
(2) QIIME 2 is also able to perform basical statistical analysis of microbiome data, such as calculating and conducting alpha diversity and beta diversity analysis. This part of contents is introduced in Chaps. 9, 10, 11, 14, and 15, respectively.
(3) QIIME 2 can also be used to manage and visualize microbiome data and meta data, such as Sect. 3.5, and do emperor plots (see Sect. 10.4.2).

The QIIME 2 system architecture consists of three core components: the **frame-work**, the **interfaces**, and the **plugins**. **Interfaces** (q2cli, q2studio, and Artifact API) define how users interact with the system; **plugins** define all domain-specific func-tionality (such as q2-data2, q2-feature-table, and q2-types). Interfaces and plugins do not communicate directly with one another; it is the **framework** that mediates communication between plugins and interfaces and performs core functionality such as provenance tracking. For the details of how QIIME 2 system works, the interested reader can check the QIIME 2 website and Bolyen et al. (Bolyen et al. 2019).

QIIME is a python interface that combines many independent scripts for the analysis of microbiome data, allowing data to be analyzed all the way from raw sequence data to diversity indices and taxonomic breakdowns. QIIME 2 has some distinctive features that make it to be powerful and distinguish it from other open source software tools for microbiome data science.

First, QIIME 2 uniquely and powerfully wraps the data generation tools into plugins (software packages). Continuing the development of QIIME 1, QIIME 2 is developed based on a plugin architecture and wrapping other bioinformatics tools, providing its capabilities to analyze high-throughput data through various ways, such as barcode splitting, cleaning and filtering low-quality sequences, removing chimera sequences via USEARCH 6.1 (Edgar 2010), PyNAST alignment (J. Gregory Caporaso et al. 2009), taxonomic analysis, tree building, and clustering samples. Currently QIIME 2 has wrapped DADA2 and Deblur (the latest generation tools) for sequence quality control, taxonomy assignment, and phylogenetic inser-tion. Among a number of bioinformatics tools, QIIME and mothur (Schloss et al. 2009) were reviewed as the two outstanding pipelines (Nilakanta et al. 2014) because of their comprehensive features and support documentation. QIIME 2 is getting the most popular since its release of reengineered and rewritten system and currently has become the dominant tool for 16S microbiome data analysis.

Second, QIIME 2 provides basic statistical analysis, including alpha- and beta-diversify analyses, as well as supports qualitatively new functionality, such as microbiome paired sample, timeseries analysis, machine learning, compositional analysis, and gneiss analysis.

Third, QIIME 2 not only provides "upstream" processing steps (e.g., sequence demultiplexing and quality control) for generating data, but also provides many interactive visualization tools to facilitate exploratory analyses and provides publication-quality graphics for result reporting. The users can also through QIIME 2 View (https://view.qiime2.org) securely share and interact with results without installing QIIME 2.

Finally, QIIME 2 not only is a powerful marker gene analysis tool but also has the potential to serve as a multidimensional and powerful data science platform for multi-omic microbiome studies, for example, through equipping newly released plugins including q2-cscs, q2-metabolomics, q2-shogun, q2-metaphlan2, and q2-picrust2, and integrating other data types, such as metabolite or metatranscriptome profiles, allowing users to run MetaPhlAn2 through QIIME. QIIME 2 can perform initial analysis of metabolomics and shotgun metagenomics data. Thus, QIIME 2 can be rapidly adapted to analyze diverse microbiome features and multi-omics data.

1.2   Core Concepts in QIIME 2

In QIIME 2, an **Action** (i.e., methods, visualizers, and pipelines) creates a **Result**, and a Result can be either an **Artifact** or a **Visualization**. Each execution of an Action and all Results created by Actions are assigned with the version 4 universally unique identifiers (UUIDs). There are five core concepts in QIIME 2. We describe them below.

1.2.1   Artifacts

Artifacts Are Data Files In QIIME 2, **artifacts** are defined as data files. The term artifact is an object that is made in a similar way as in an archaeological artifact. An Artifact is data generated by one or more Actions; or formally artifacts are the way of existence for data produced by QIIME 2. An artifact contains data and metadata. A metadata is the dataset that is used to describe data type, format, and provenance (i.e., how it was generated). QIIME 2 typically uses the .qza file extension to store an artifact in a file. QIIME 2 does not work with simple data files (e.g., FASTA files), instead works with artifacts. There are two benefits for using artifacts instead of simple data files: (1) automatically tracks the type, format, and provenance of data and (2) enables researchers to focus on conducting the analyses, instead of paying much attention to the particular data format. Because QIIME 2 directly works with artifacts, thus, when we import data, we must first create a QIIME 2 artifact. Typically we start importing raw sequence data, although data can be imported in any step of analysis. In QIIME 2, data also can be exported from an artifact.

1.2.2   Visualizations

Visualizations Are also Data Files   Visualizations are another type of data generated by QIIME 2. QIIME 2 typically uses the .qzv file extension to store visualization files. Visualizations contain similar types of metadata as artifacts, including provenance information. Because in QIIME 2 both artifacts and visualizations include metadata and particularly have unique provenance information, so they can be archived or shared with collaborators. Both artifacts and visualizations files (generally with .qza and .qzv extensions, respectively) can be easily reviewed using the website (https://view.qiime2.org) without requiring a QIIME installation. The examples of visualizations are a statistical results table, an interactive visualization, static images, or really any combination of visual data representations. All these are **terminal outputs** of an analysis, which is contrast to artifacts. Thus, we cannot use visualizations as input to other analyses in QIIME 2.

1.2.3   Semantic Type

Every Artifact Has a Semantic Type Associated with It   In QIIME 2, all Artifacts are annotated with a semantic description of their type. Data types present how data is represented in memory, and file formats present how data is stored on disk, whereas **semantic types** differ from data type and file formats, conveying the meaning of the data. The semantic types are used to (1) identify artifacts for suitable inputs to an analysis, which can prevent incompatible artifacts from being used in the analysis. It effectively constrains multiple actions to only those which are semantically meaningful actions. For example, in QIIME 2, phylogenetic trees have two semantic types: Phylogeny[Rooted] and Phylogeny[Unrooted]. The beta-phylogenetic action needs a UniFrac distance matrix, which works only on Phylogeny[Rooted], while fasttree can only generate a Phylogeny[Unrooted]. Thus the semantic type system is used to determine that the output of fasttree should not be directly provided as input to beta-phylogenetic. (2) They help users avoid incorrectly using data for semantically incorrect analyses. For example, categorical data cannot be used for a quantitative analysis, vice versa. (3) Also they facilitate users in identifying relevant workflows to generate desired data or further explore data.

1.2.4   Plugins

Microbiome Analyses Are Implemented via Plugins   QIIME 2 implements microbiome analyses via plugins (software packages). Thus, at least one plugin needs to be installed to provide the specific analyses, such as the q2-demux plugin for demultiplexing raw sequence data, the q2-diversity plugin for alpha- or betadiversity analyses. The initial plugins for an end-to-end microbiome analysis pipeline were developed by the QIIME 2 team; other plugins for additional analyses are developed by third-party developers and volunteers.

Plugins Define Methods and Visualizers to Perform Analyses To perform microbiome analyses, QIIME 2 plugins must define **methods** and **visualizers**. The question is: how to define them? A **method** is defined to combine artifacts and parameters as input, which produces one or more artifacts as output. The resulted output artifacts could subsequently be used as input to other methods or visualizers. The outputs produced by methods can be intermediate or terminal. One example of method definition is as below: in the q2-feature-table plugin, the rarefy method is defined to use a feature table artifact and sampling depth as input, producing a rarefied feature table artifact as output. Then we could use the resulted rarefied feature table artifact to calculate alpha diversity by providing an alpha method in the q2-diversity plugin. A **visualizer** is similar to a **method** in the sense that it combines artifacts and parameters as input. However, a visualizer is distinctive to a method: it produces exactly one terminal **visualization** as output, i.e., the resulted visualizations cannot be used as input to other methods or visualizers.

1.3   Install QIIME 2

Currently QIIME 2 is available for macOS, Windows, and Linux users. All have three approaches of installation: a native conda installation, and Docker and VirtualBox. The native conda installation usually works well. Thus QIIME 2 generally recommends the method of native conda installation.

We can install QIIME 2 natively or use virtual machines. Below we introduce natively installing QIIME 2 (Core 2022.2 distribution). The readers can easily follow the instruction of QIIME 2 document to install QIIME 2 using virtual machines. The QIIME 2 Core 2022.2 distribution includes the QIIME 2 framework, q2cli (a QIIME 2 command-line interface), and the following plugins: q2-alignment, q2-composition, q2-cutadapt, q2-dada2, q2-deblur, q2-demux, q2-diversity, q2-diversity-lib, q2-emperor, q2-feature-classifier, q2-feature-table, q2-fragment-insertion, q2-gneiss, q2-longitudinal, q2-metadata, q2-phylogeny, q2-quality-control, q2-quality-filter, q2-sample-classifier, q2-taxa, q2-types, and q2-vsearch.

Install Miniconda and QIIME 2 on Mac The recommended way to natively install QIIME 2 is through installing Miniconda. Miniconda provides the conda environment and package manager. QIIME 2 works within a conda environment. To install Miniconda in Mac, the following steps are needed:

Follow the Miniconda instructions for downloading and installing Miniconda. QIIME 2 works with either Miniconda2 or Miniconda3 (i.e., Miniconda Python 2 or 3). You may choose either one. Please follow the directions provided in the Miniconda instructions; particularly ensure that you run conda init and your Miniconda installation is fully installed.

The latest version is QIIME 2 Core 2022.2 distribution (February, 2022). To install this version, we can take the following steps:

Step 1: Click the website link https://docs.conda.io/en/latest/miniconda.html to download Miniconda (64-bit bash installer).

You can choose either Python 3.7 or Python 2.7 for Mac OS X. A file called *Miniconda3-latest-MacOSX-x86_64.sh* will be shown in your Downloads folder.

Step 2: Open a Mac **Terminal** (Terminal is within Utilities folder nested in Applications folder). Type: *cd Downloads*

Step 3: Run the bash "shell" script to install Miniconda. In the terminal, type: *bash Miniconda3-latest-MacOSX-x86_64.sh*

Press the ENTER to scroll through the license, and accept all the default installations.

Step 4: Close the Terminal, and open a new Terminal. Type: *conda -V*

If you see the "conda 4.10.3" or later version information, then it indicates that you have successfully installed conda via miniconda on your Mac. To update Miniconda and check if you're running the latest version of conda, type:

conda update conda

Then, install wget, type:

conda install wget

Step 5: Install QIIME 2 within the conda environment.

Since we have Miniconda installed, a conda environment is created. Now we are ready for installing QIIME 2 within this conda environment. We can natively install the QIIME 2 Core 2022.2 distribution within the environment. Because there are many required dependencies that are not needed to be added to an existing environment, QIIME 2 highly recommends creating a new environment specifically for the QIIME 2 release being installed. Here, we name the environment qiime2-2022.2 to indicate this is QIIME 2 Core 2022.2 distribution. We choose macOS/OS X (64-bit) and in the **Terminal**, type:

```
wget https://data.qiime2.org/distro/core/qiime2-2022.2-py38-
osx-conda.yml
conda env create -n qiime2-2022.2 --file qiime2-2022.2-py38-osx-
conda.yml
# OPTIONAL CLEANUP
rm qiime2-2022.2-py38-osx-conda.yml
```

Step 6: Activate the conda environment.

To activate the QIIME 2 environment, type the environment's name (here, qiime2-2022.2) in the **Terminal**:

```
conda activate qiime2-2022.2
```

It you want to deactivate an environment, run `conda deactivate`.

Step 7: Test the installation.

To test the installation, activate the QIIME 2 environment and run:

```
qiime --help
```

If the help information comes out when running this command, suggesting that the installation was successful!

1.4   Store and Track Data

In QIIME 2, data are stored in a directory structure called an **Archive** with a single root directory (UUID) serving as the identity of the archive. Several sub-directories exist under the a single root directory (UUID), including (1) a data directory, (2) a metadata directory, (3) a provenance directory, and (4) a VERSION directory. The data directory contains only the data in a relevant format. For example, the fasta or fastq for sequence data, and newick for phylogenetic trees are stored in this data directory. All files in these archives are zipped for data delivery convenience and facilitating data sharing. For example, the extension .qza is for QIIME zipped artifact and the extension .qzv is for QIIME zipped visualization. We can use unzip, WinZip, 7Zip, or other common tools to unzip these standard zip files. The zipped files are easily shared by peers or collaborators, or submitted to journals.

QIIME 2 uses data provenance tracking to store all information about the series of Actions that led to a Result, packages, and Python dependencies and other environment information and the data itself. Microbiome data analyses usually generate many output files. Data-provenance tracking is used to avoid losing track of how each file was generated.

1.5   Extract Data from QIIME 2 Archives

QIIME 2 .qza and .qzv files are zip file containers with a defined internal directory structure. Depending on whether or not QIIME 2 and the q2cli command line interface are installed in your computer, you can use the qiime tools export command or use the standard decompression utilities such as unzip, WinZip, or 7zip to access these files. Any actual data files in the .qza artifact can be extracted directly using qiime tools export, which is basically just a wrapper for unzip. Alternatively, we can unzip the artifact directly using unzip -k file.qza from the files in the data folder.

1.6   Summary

In this introductory chapter of QIIME 2, we first provided an overview of QIIME 2. Then we described some core concepts in QIIME 2, including artifacts, visualizations, semantic type, and plugins. Next, we introduced how to install QIIME 2 step-by-step. Following that, storing, tracking data, and extracting data from QIIME 2 archives were introduced.

In Chap. 3 of this book, we will introduce some basic data processing procedures in QIIME 2. Chapter 4 will introduce building feature table and feature representative sequences from raw reads. Chapter 5 will introduce assigning taxonomy and building phylogenetic tree. Chapter 6 will introduce clustering sequences into OTUs. Before we move on to these chapters, let's introduce some general R functions, packages, and specifically designed R packages for microbiome data analysis that are used in this book (Chap. 2).

References

Amir, Amnon, Daniel McDonald, Jose A. Navas-Molina, Evguenia Kopylova, James T. Morton, Xu Zhenjiang Zech, Eric P. Kightley, Luke R. Thompson, Embriette R. Hyde, Antonio Gonzalez, and Rob Knight. 2017. Deblur rapidly resolves single-nucleotide community sequence patterns. *mSystems* 2 (2): e00191-16. https://doi.org/10.1128/mSystems.00191-16, https://pubmed.ncbi.nlm.nih.gov/28289731, https://www.ncbi.nlm.nih.gov/pmc/articles/PMC5340863/.

Bolyen, Evan, Jai Ram Rideout, Matthew R. Dillon, Nicholas A. Bokulich, Christian C. Abnet, Gabriel A. Al-Ghalith, Harriet Alexander, Eric J. Alm, Manimozhiyan Arumugam, Francesco Asnicar, Yang Bai, Jordan E. Bisanz, Kyle Bittinger, Asker Brejnrod, Colin J. Brislawn, C. Titus Brown, Benjamin J. Callahan, Andrés Mauricio Caraballo-Rodríguez, John Chase, Emily K. Cope, Ricardo Da Silva, Christian Diener, Pieter C. Dorrestein, Gavin M. Douglas, Daniel M. Durall, Claire Duvallet, Christian F. Edwardson, Madeleine Ernst, Mehrbod Estaki, Jennifer Fouquier, Julia M. Gauglitz, Sean M. Gibbons, Deanna L. Gibson, Antonio Gonzalez, Kestrel Gorlick, Jiarong Guo, Benjamin Hillmann, Susan Holmes, Hannes Holste, Curtis Huttenhower, Gavin A. Huttley, Stefan Janssen, Alan K. Jarmusch, Lingjing Jiang, Benjamin D. Kaehler, Kyo Bin Kang, Christopher R. Keefe, Paul Keim, Scott T. Kelley, Dan Knights,

Irina Koester, Tomasz Kosciolek, Jorden Kreps, Morgan G.I. Langille, Joslynn Lee, Ruth Ley, Yong-Xin Liu, Erikka Loftfield, Catherine Lozupone, Massoud Maher, Clarisse Marotz, Bryan D. Martin, Daniel McDonald, Lauren J. McIver, Alexey V. Melnik, Jessica L. Metcalf, Sydney C. Morgan, Jamie T. Morton, Ahmad Turan Naimey, Jose A. Navas-Molina, Louis Felix Nothias, Stephanie B. Orchanian, Talima Pearson, Samuel L. Peoples, Daniel Petras, Mary Lai Preuss, Elmar Pruesse, Lasse Buur Rasmussen, Adam Rivers, Michael S. Robeson 2nd, Patrick Rosenthal, Nicola Segata, Michael Shaffer, Arron Shiffer, Rashmi Sinha, Se Jin Song, John R. Spear, Austin D. Swafford, Luke R. Thompson, Pedro J. Torres, Pauline Trinh, Anupriya Tripathi, Peter J. Turnbaugh, Sabah Ul-Hasan, Justin J.J. van der Hooft, Fernando Vargas, Yoshiki Vázquez-Baeza, Emily Vogtmann, Max von Hippel, William Walters, Yunhu Wan, Mingxun Wang, Jonathan Warren, Kyle C. Weber, Charles H.D. Williamson, Amy D. Willis, Zhenjiang Zech Xu, Jesse R. Zaneveld, Yilong Zhang, Qiyun Zhu, Rob Knight, and J. Gregory Caporaso. 2019. Reproducible, interactive, scalable and extensible microbiome data science using QIIME 2. *Nature Biotechnology* 37 (8): 852–857. https://doi.org/10.1038/s41587-019-0209-9, https://pubmed.ncbi.nlm.nih.gov/31341288, https://www.ncbi.nlm.nih.gov/pmc/articles/PMC7015180/.

Callahan, Benjamin J., Paul J. McMurdie, Michael J. Rosen, Andrew W. Han, Jo A. Amy, and Johnson, and Susan P. Holmes. 2016. DADA2: High-resolution sample inference from Illumina amplicon data. *Nature Methods* 13 (7): 581–583. https://doi.org/10.1038/nmeth.3869.

Caporaso, J.G., J. Kuczynski, J. Stombaugh, K. Bittinger, et al. 2010. QIIME allows analysis of high-throughput community sequencing data. *Nature Methods* 7 (5): 335–336. https://doi.org/10.1038/nmeth.f.303.

Caporaso, J. Gregory, Kyle Bittinger, Frederic D. Bushman, Todd Z. DeSantis, Gary L. Andersen, and Rob Knight. 2009. PyNAST: A flexible tool for aligning sequences to a template alignment. *Bioinformatics* 26 (2): 266–267. https://doi.org/10.1093/bioinformatics/btp636.

Edgar, Robert C. 2010. Search and clustering orders of magnitude faster than BLAST. *Bioinformatics* 26 (19): 2460–2461. https://doi.org/10.1093/bioinformatics/btq461.

Nilakanta, Haema, Kimberly L. Drews, Suzanne Firrell, Mary A. Foulkes, and Kathleen A. Jablonski. 2014. A review of software for analyzing molecular sequences. *BMC Research Notes* 7 (1): 830. https://doi.org/10.1186/1756-0500-7-830.

Schloss, P.D., S.L. Westcott, T. Ryabin, J.R. Hall, M. Hartmann, E.B. Hollister, R.A. Lesniewski, B.B. Oakley, D.H. Parks, C.J. Robinson, J.W. Sahl, B. Stres, G.G. Thallinger, D.J. Van Horn, and C.F. Weber. 2009. Introducing mothur: open-source, platform-independent, community-supported software for describing and comparing microbial communities. *Applied and Environmental Microbiology* 75 (23): 7537–7541. https://doi.org/10.1128/aem.01541-09.

Xia, Yinglin, Jun Sun, and Ding-Geng Chen. 2018. Bioinformatic analysis of microbiome data. In *Statistical Analysis of Microbiome Data with R*, 1–27. Singapore: Springer Singapore.

Chapter 2
Introduction to R for Microbiome Data

Abstract This chapter first introduces some useful R functions and R packages for microbiome data. Then it illustrates some specifically designed R packages for microbiome data analysis (e.g., phyloseq and microbiome). Next it briefly describes three R packages for analysis of phylogenetics (ape, phytools, and castor). Following that it introduces the BIOM format and the biomformat package and illustrates creating a microbiome dataset for longitudinal data analysis.

Keywords saveRDS() · save() · save.image() · readRDS() · load() · download.file() · readr · ggpubr · Box plots · Violin plots · Density plots · Histogram plots · tidyverse · phyloseq package · phyloseq object · Microbiome package · CSV · Mothur · BIOM format · ampvis2 package · curatedMetagenomicData · Phylogenetics · Ape · Phytools · Castor · read.table() · read.delim() · read.csv() · read.csv2() · write.table() · read.csv()

In Chap. 1, we discussed the general features of QIIME 2. In Chap. 2, we first introduce some useful R functions (Sect. 2.1). Then, we introduce some useful R packages for microbiome data (Sect. 2.2). In Sect. 2.3, we illustrate some specifically designed R packages for microbiome data including the phyloseq and microbiome packages. In Sect. 2.4, we briefly describe three R packages for analysis of phylogenetics including ape, phytools, and castor. In Sects. 2.5 and 2.6, we introduce the BIOM format and the biomformat package and how to create a microbiome dataset for longitudinal data analysis. Finally, we summarize the main contents of this chapter in Sect. 2.7.

Supplementary Information The online version contains supplementary material available at https://doi.org/10.1007/978-3-031-21391-5_2.

2.1   Some Useful R Functions

This section illustrates some useful R functions. Most microbiome data analyses in this book will use R software (R version 4.1.2; 2021-11-01). Some general R functions, R packages, and specifically designed R packages for microbiome data analysis are very useful. Let's first briefly introduce these functions and packages before we move on to other chapters.

Example 2.1: Iris Data
The famous Iris data were collected by Edgar Anderson in 1935 (Anderson 1935) and first used by Fisher in 1936 (Fisher 1936) to illustrate a taxonomic problem. The dataset measures the variables in centimeters of sepal length and width, and petal length and width, respectively, for 50 flowers from each of 3 *Iris* species (*setosa*, *versicolor*, and *virginica*). The dataframe iris has 150 cases (rows) and 5 variables (columns) named Sepal.Length, Sepal.Width, Petal.Length, Petal.Width, and Species. The Iris dataset is wrapped in R package.

In this section, we use the Iris dataset to illustrate some useful R functions. In Chap. 4 (Introduction to R, RStudio, and ggplot2) of our previous book on microbiome data analysis (Xia et al. 2018), we introduced some basic R functions for data import and export, including **read.table()**, **read.delim()**, **read.csv()**, **read. csv2()**, and **write.table()**. Some of them such as **read.csv()** are used very frequently across most chapters in this book. Here, we introduce additional basic R functions that could be useful in your data analysis.

2.1.1   Save Data into R Data Format

Writing R data into txt, csv, or Excel file formats facilitates data analysis using other software, such as Excel. However, all these file formats have their own data structures rather than reserve the original R data structures, such as vectors, matrices, and data frames as well as column data types (numeric, character, or factor). The data files will be automatically compressed after the data are saved into R data formats (e.g., .rda files); thus, one benefit of saving data into R data formats can substantially reduce the size of converted files.

2.1.1.1   Use SaveRDS() to Save Single R Object

We can use saveRDS() to save single R object to the file with .rds format. The general syntax of this function is given:

```
saveRDS(obj, file = "name_file.rds")
```

where **obj** is the R object to save and **file** is used to specify the name of the file where the R object is saved to (the > below are prompts).

```
> setwd("~/Documents/QIIME2R/Ch2_IntroductionToR")
> data(iris)
> # Save a singel object to a file
> saveRDS(iris, file = "iris_saved.rds")
```

2.1.1.2 Use Save() to Save Multiple R Objects

We can use save() to save multiple R objects to the file with .RData format. The general syntax of this function is given:

```
save(obj, file = "name_file.RData")
> head(iris,3)
  Sepal.Length Sepal.Width Petal.Length Petal.Width Species
1     5.1         3.5          1.4         0.2 setosa
2     4.9         3.0          1.4         0.2 setosa
3     4.7         3.2          1.3         0.2 setosa

> # Save on object with RData format and factor
> save(iris, file = "iris.RData")
```

Next, let's save each column individually as vectors.

```
> # Save the iris data as vectors
> Sepal.Length<-iris$Sepal.Length
> Sepal.Width<-iris$Sepal.Width
> Petal.Length<-iris$Petal.Length
> Petal.Width<-iris$Petal.Width
> Species<-iris$Species

> class(Sepal.Length)
[1] "numeric"
> class(Sepal.Width)
[1] "numeric"
> class(Petal.Length)
[1] "numeric"
> class(Petal.Width)
[1] "numeric"
> class(Species)
[1] "factor"
```

We can use the **save()** function to save all these 4 vectors and 1 factor to an .rda file. The "file" is renamed as "iris_saved.rda."

```
> # Save to rda file
> save(Sepal.Length, Sepal.Width, Petal.Length, Petal.Width, Species,
file = "iris_saved.rda")
```

2.1.1.3   Use Save.Image() to Save Workspace Image

By default, **save.image()** stores the workspace to a file named .RData. We can also specify the file name for saving the work space, such as to save the work space as "myworkspace" using save.image(file = "myworkspace.RData"). Actually, you may notice that each time when you close R/RStudio, you are asked whether or not to save your workspace. If you click yes, then the next time when you start R, the saved workspace (with .RData) will be loaded.

2.1.2   Read Back Data from R Data Format Files

The saved R files can be read back to R workspace using either readRDS() or load() functions.

2.1.2.1   Use ReadRDS() to Read Back Single R Object

The general syntax of this function is given:

```
readRDS (file = "name_file.rds")
```

We can rename the restored object. Where the **file** is used to name the file where the R object is read from.

```
> # Restore a saved R object using a different name
> iris_retored <- readRDS("iris_saved.rds")
> head(iris_retored,3)
  Sepal.Length Sepal.Width Petal.Length Petal.Width Species
1      5.1         3.5          1.4         0.2 setosa
2      4.9         3.0          1.4         0.2 setosa
3      4.7         3.2          1.3         0.2 setosa
```

2.1.2.2   Use Load() to Read Back Multiple Objects from R File

The syntax of this function is given:

```
load("name_file.RData")
```

The **load()** function will automatically load the original object if the data was saved with **save()**. In other words, the original object cannot be renamed when it is read back.

```
> # Load the RData data
> load("iris.RData")
> head(iris,3)
Sepal.Length Sepal.Width Petal.Length Petal.Width Species
1     5.1        3.5         1.4         0.2 setosa
2     4.9        3.0         1.4         0.2 setosa
3     4.7        3.2         1.3         0.2 setosa


> # Load the rda data
> load(file = "iris_saved.rda")
> head(iris,3)
Sepal.Length Sepal.Width Petal.Length Petal.Width Species
1     5.1        3.5         1.4         0.2 setosa
2     4.9        3.0         1.4         0.2 setosa
3     4.7        3.2         1.3         0.2 setosa
```

2.1.3   Use Download.File() to Download File from Website

The R function **download.file()** can be used to download file from a website, such as a webpage, an R file, a tar.gz file, etc. The simplest syntax of this function is with two arguments: download.file(url, destfile). **url** is a character string (or longer vector) naming the URL of the file to be downloaded; **destfile** is a character string (or vector) with the name where the downloaded file is saved (path with a file name). For example, to directly download the QIIME 2 files: "feature-table.qza" from the following QIIME 2 website to R or RStudio, we call the download.file().

```
> download.file("https://data.qiime2.org/2022.2/tutorials/
exporting/feature-table.qza", "feature-table.qza")
```

The downloaded feature-table files can be read back to R using the qiime2R package (see Chap. 3).

```
> library(qiime2R)
> feature_tab<-read_qza("feature-table.qza")
> head(feature_tab)
```

2.2   Some Useful R Packages for Microbiome Data

There are very useful and often used R packages. In this section, we introduce and illustrate some of them.

2.2.1   Readr

Example 2.2: Iris Data, Example 2.1, Cont.
In this section, we use the famous Iris data to illustrate some useful R package. The package **readr** was developed by Hadley Wickham et al. (Wickham, Hester, and Francois 2018) for fast reading and writing delimited files or csv files (current version 1.4.0, March 2022). It contains the function **write_delim(), write_csv(),** and **write_tsv()** to easily export a data from R. We continue to use the Iris dataset to illustrate the **readr** package.

Installing and Loading readr and Data
```
> setwd("~/Documents/QIIME2R/Ch2_IntroductionToR")
# Installing
install.packages("readr")
# Loading
library("readr")

> # Load data
> data(iris)
> df <- iris
> head(df, 3)
  Sepal.Length Sepal.Width Petal.Length Petal.Width Species
1      5.1         3.5         1.4         0.2 setosa
2      4.9         3.0         1.4         0.2 setosa
3      4.7         3.2         1.3         0.2 setosa
```

Writing Data Using the Readr Functions
Depending on data format, we can choose the **write_delim(), write_csv(),** or **write_tsv()** to export a data table from R. Of which, the **write_delim()** function is a general function of writing data from R; **write_csv()** is used to write a comma (",") separated values; while **write_tsv()** is used to write a tab separated ("\t") values. The simplified general syntax of these functions is given as follows:

```
write_delim(df, file, delim = " ") # General function of writing data from
R
write_csv(df, file) # Write comma (",") separated value files
write_tsv(df, file) # Write tab ("\t") separated value files
```

where **df** is a data frame to be written; **file** is used to specify the file name to be saved; while **delim** is the delimiter that is used to separate values. It must be single character.

The following commands write the Iris dataset to a txt (i.e., tsv) file.

```
> # Write data from R to a txt (i.e., tsv) file
> write_tsv(iris, file = "iris.txt")
```

The built-in R iris dataset has been exported to a tab-separated (sep = "\t") file called iris.txt in the current working directory: setwd("~/Documents/QIIME2R/Ch2_IntroductionToR").

The following commands write the Iris dataset to a csv file.

```
> # Write data from R to a csv file
> write_csv(iris, file = "iris.csv")
```

2.2.2   Ggpubr

Example 2.3 Iris Data, Example 2.2, Cont.
The R package **ggpubr** written by Alboukadel Kassambara (alboukadel. kassambara@gmail.com) (Kassambara 2020-06-27), is a ggplot2-based package that provides some easy-to-use functions for creating and customizing publication ready plots (current version 0.4.0, March 2022). The **ggplots** package written by Hadley Wickham (2016) is an excellent and flexible visualization tool in R. However the default generated plots are not ready for publication. Furthermore, the customized ggplot syntax is opaque, which challenges the R beginners and researchers with no advanced R programming skills. The **ggpubr** package facilitates the creation of beautiful ggplot2-based graphs with some key features, including less opaque syntax than the wrapped **ggplot2** package, providing publication-ready plots, is able to automatically add p-values and significance levels to box, bar, and line plots, facilitating arranging and annotating multiple plots on the same page, and changing graphical parameters (e.g., colors and labels). Here, we illustrate some plots that are useful for exploration of microbiome data.

Installation and Loading
There are two ways to install the **ggpubr** package: One is from CRAN as follows:

```
install.packages("ggpubr")
```

Or we can type the codes to install the latest version from GitHub as follows:

```
if(!require(devtools)) install.packages("devtools")
devtools::install_github("kassambara/ggpubr")
```

After installation, we can set working directory, call the library, and load the dataset as below:

```
> setwd("~/Documents/QIIME2R/Ch2_IntroductionToR")
> library(ggpubr)
> data(iris)
> df <- iris
> head(df, 3)
```

```
  Sepal.Length Sepal.Width Petal.Length Petal.Width Species
1      5.1         3.5          1.4          0.2 setosa
2      4.9         3.0          1.4          0.2 setosa
3      4.7         3.2          1.3          0.2 setosa
```

2.2.2.1   Box Plots

The following commands create the box plots with jittered points (Fig. 2.1):

```
> # Figure 2.1
> # Use outline colors for groups: Species
> # Use custom color palette
> # Add jitter points and use different shapes for groups
> # The plus sign ("+") below appears in the console, meaning that
> # R is wanting to enter some additional information
> p <- ggboxplot(df, x = "Species", y = "Sepal.Length",
+       color = "Species", palette =c("#00AFBB", "#E7B800", "#FC4E07"),
+       add = "jitter", shape = "Species")
> p
```

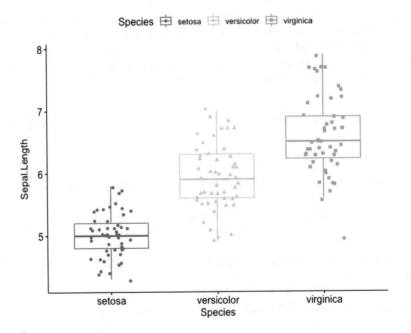

Fig. 2.1 Box plots with jittered points using Iris dataset

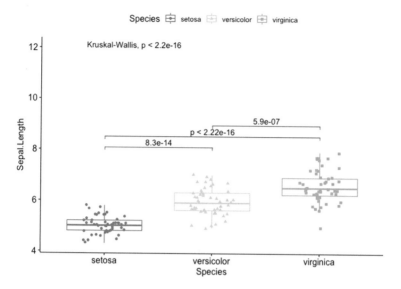

Fig. 2.2 Box plots with a global *p*-value and *p*-values for pairwise comparisons

The following commands add a global *p*-value and *p*-values for pairwise comparisons to the box plots with jittered points (Fig. 2.2):

```
> # Figure 2.2
> # Add p-values for comparing groups
> # Specify the pairwise group comparisons
> comps <- list ( c ("setosa", "versicolor"), c ("setosa", "virginica"), c
("versicolor", "virginica") )
> p + stat_compare_means (comparisons = comps) + # Add global p-value and
p-values for pairwise comparisons
+ stat_compare_means (label.y = 12)
```

The following plots show the significance levels for pairwise comparisons (Fig. 2.3):

```
> # Figure 2.3
> # Specify the comparisons of interest
> comps <- list ( c ("setosa", "versicolor"), c ("setosa", "virginica"), c
("versicolor", "virginica") )
> p + stat_compare_means (comparisons = comps, label = "p.signif") + #
Show the significance levels
+ stat_compare_means (label.y = 12)                    # Add global p-value
```

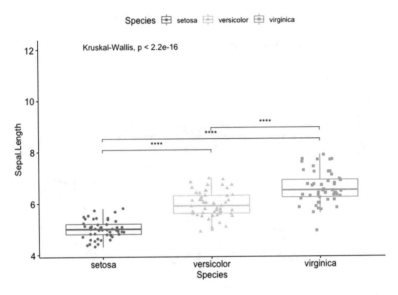

Fig. 2.3 Box plots with a global *p*-value and the significance levels for pairwise comparisons

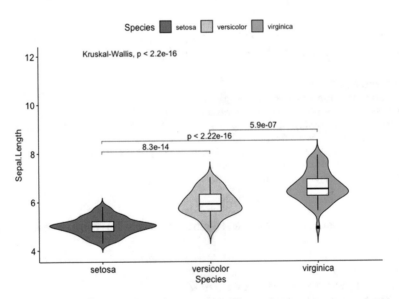

Fig. 2.4 Violin plots with box plots inside adding a global *p*-value and *p*-values for pairwise comparisons

2.2.2.2 Violin Plots

The following commands create the violin plots with box plots inside and *p*-values (Fig. 2.4):

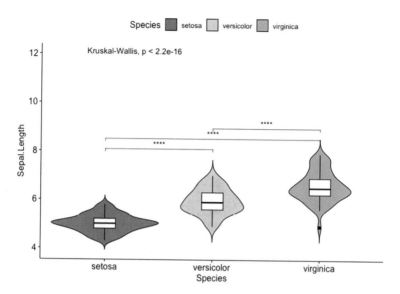

Fig. 2.5 Violin plots with box plots inside adding a global *p*-value and the significance levels for pairwise comparisons

```
> # Figure 2.4
> # Use outline colors for groups: Species
> # Use custom color palette
> # Add jitter points and use different shapes for groups
> # Add boxplot with white fill color
> # Specify the comparisons you want
> # Add p-values for comparing groups
> ggviolin(df, x = "Species", y = "Sepal.Length", fill = "Species",
+        palette = c("#00AFBB", "#E7B800", "#FC4E07"),
+        add = "boxplot", add.params = list(fill = "white")) +
+ stat_compare_means(comparisons = comps) + # Add pairwise comparisons
p-value
+ stat_compare_means(label.y = 12)
```

The following commands create the violin plots with box plots inside and label significance levels (Fig. 2.5):

```
> # Figure 2.5
> ggviolin(df, x = "Species", y = "Sepal.Length", fill = "Species",
+        palette = c("#00AFBB", "#E7B800", "#FC4E07"),
+        add = "boxplot", add.params = list(fill = "white")) +
+ stat_compare_means(comparisons = comps, label = "p.signif") + # Add
significance levels
+ stat_compare_means(label.y = 12) # Add global the p-value
```

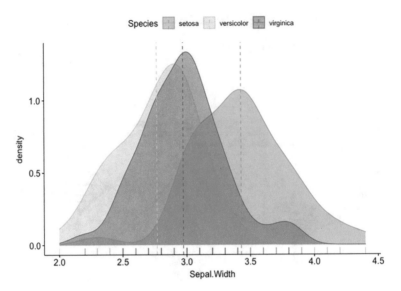

Fig. 2.6 Density plot of Sepal.Width with mean lines and marginal rug differentiating colors of species

2.2.2.3   Density Plots

The following commands plot the density of Sepal.Width with mean lines and marginal rug and use different colors to label the three species (Fig. 2.6):

```
> # Figure 2.6
> # Density plot with mean lines and marginal rug
> # Change outline and fill different colors for groups ("Species")
> # Use custom palette
> ggdensity(df, x = "Sepal.Width",
+       add = "mean", rug = TRUE,
+       color = "Species", fill = "Species",
+       palette = c("#00AFBB", "#E7B800", "#FC4E07"))
```

2.2.2.4   Histogram Plots

The following codes plot the histogram of Sepal.Width with mean lines and marginal rug and use different colors to label the three species (Fig. 2.7):

```
> # Figure 2.7
> # Change outline and fill different colors for groups ("Species")
> # Use custom color palette
> gghistogram(df, x = "Sepal.Width",
+       add = "mean", rug = TRUE,
```

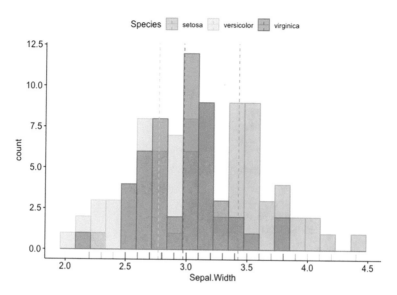

Fig. 2.7 Histogram plot of Sepal.Width with mean lines and marginal rug differentiating colors of species

```
+       color = "Species", fill = "Species",
+       bins = 20,
+       palette = c("#00AFBB", "#E7B800","#FC4E07"))
```

2.2.3   Tidyverse

The **tidyverse** package (a single "meta" package) was developed by Wickham et al. (Wickham et al. 2019) to collect R packages that share a high-level design philosophy and low-level grammar and data structures, enabling the R packages in the set work in harmony (current version 1.3.1, March 2022). The tidyverse is not the tools for statistical modelling or communication, instead its goal is to facilitate data import, tidying, manipulation, visualization, and programming. For example, it allows users to install all tidyverse packages with a single command. We can either install it from CRAN or via GitHub.

```
# Install from CRAN
> install.packages("tidyverse")

> # Or the development version from GitHub
> # install.packages("devtools")
> devtools::install_github("tidyverse/tidyverse")
```

```
> library(tidyverse)
```

When library(tidyverse) is called, the core tidyverse packages will be loaded:

```
── tidyverse 1.3.0 ──
✓ ggplot2 3.3.2 ✓ purrr 0.3.4
✓ tibble 3.0.3 ✓ dplyr 1.0.2
✓ tidyr 1.1.0 ✓ stringr 1.4.0
✓ readr 1.3.1 ✓ forcats 0.5.0
```

The conflicts with other previously loaded packages are also summarized:

```
── tidyverse_conflicts() ──
x dplyr::filter () masks stats::filter ()
x dplyr::lag() masks stats::lag()
```

The tidyverse consists of four programs: (1) import, (2) tidy, (3) understand (transformation, visualize, and model), and (4) communicate. The **programming** works out this way (Wickham et al. 2019): starts with data **import**→ tidy data→ data **transformation**→ understanding data: **visualization** and **modelling**→ **communication**.

2.3   Specifically Designed R Packages for Microbiome Data

In this section, we introduce some specifically designed R packages for microbiome data analysis.

2.3.1   *Phyloseq*

The **phyloseq** package was developed to handle and analyze the high-throughput microbiome census data. The package was written by McMurdie and Holmes (McMurdie and Holmes 2013) with contributions from Gregory Jordan and Scott Chamberlain (McMurdie and Holmes 2022) (current version 1.36.0, March 2022). The phyloseq classes and tools facilitate the import, storage, analysis, and graphical display of microbiome census data. The phyloseq package mainly wraps statistical functions from **vegan** package (F. Jari Oksanen et al. 2018) and **ggplot2** graphical package (Wickham 2016). Among the classes and tools, the phyloseq object and the operations on this object are very useful. The object and some functions of data operations have been adopted in other R microbiome packages for data processing and analysis such as in packages *curatedMetagenomicData* (Pasolli et al. 2017) and **microbiome** (Lahti 2017).

Example 2.4: Diet Swap Between Rural and Western Populations

O'Keefe et al. in 2015 (S.J.D. O'Keefe et al. 2015) reported a 2-week diet swap study between western (USA) and traditional (rural Africa) diets to investigate whether the higher rates of colon cancer are associated with consumption of higher animal protein and fat, and lower fiber. The data is available for download from Data Dryad (*S.J.D. O'Keefe* 2016).

The dataset contains a HITChip data matrix and a sample metadata. The HITChip data matrix is a csv file which contains 222 samples from this study, providing the absolute HITChip phylogenetic microarray signal estimate for 130 taxa (genus-like groups). The sample metadata is also a CSV file, containing various measurement variables, in which **SampleID** is the unique sample identified corresponding to samples in the HITChip data matrix; **subject** is the subject identifier (some subjects have multiple time points); **bmi** is the standard body-mass classification (defining underweight, bmi <18.5; lean, $18.5 \leq$ bmi <25; overweight, $25 \leq$ bmi <30; obese, $30 \leq$ bmi <35; severe obese, $35 \leq$ bmi <40; morbid obese, $40 \leq$ bmi <45; superobese, bmi >45); **sex** (male/female); **nationality** is African American (AAM) and Native African (AFR); **timepoint.total** refers to the time point in the overall dataset ranging from 1 to 6; **timepoint.group** is the time point (1/2) within the group and **group** the sample treatment group representing Dietary intervention (DI)/Home environment (HE)/Solid stool pre-colonoscopy (ED).

2.3.1.1   Create a phyloseq Object

Microbiome abundance and taxonomic datasets are generated by next-generation sequencing (NGS) techniques. Currently 16S rRNA gene sequencing and whole-metagenome shotgun sequencing are the two most commonly used sequencing approaches (Xia and Sun 2022).

After taking initial quality control steps to account for errors in the sequencing process, the data generated by microbiome community sequencing typically are organized into large matrices where usually the columns represent samples, and the rows contain observed counts of clustered sequences commonly known as operational taxonomic units (OTUs/ASVs) that represent types of bacteria or other features. We often call these tables as OTU/ASVs tables or feature tables. We will introduce how to build feature table from raw sequencing reads in Chap. 4, assign taxonomy, building phylogenetic tree in Chap. 5, and cluster OTUs in Chap. 6. For now, we focus on how to create a phyloseq object in the phyloseq package.

One important concept and procedure in **phyloseq** package is the **phyloseq object**. A full phyloseq object has four data components: otu_table, sample_data, tax_table, and phylo. At least two components are needed to analyze the data of phyloseq object. The phyloseq object is created via the `phyloseq()` function. When the phyloseq object is created, the validity and coherency between data components is checked by the phyloseq-class constructor, which is invoked internally by the importers.

Here, we use diet swap dataset to illustrate how to create a phyloseq object and some operations on phyloseq object. First, we need to install this package. Type the following commands in R (version "3.6" or later) or RStudio (the > below are prompts):

```
>if (!requireNamespace("BiocManager", quietly = TRUE))
> install.packages("BiocManager")
>BiocManager::install("phyloseq")


> packageVersion("phyloseq")
[1] '1.36.0'
```

In the directory "~/Documents/QIIME2R/Ch2_IntroductionToR", we store three datasets: (1) otu-table with Samples (rows) by OTUs (columns) format called "otu_table_dietswap.csv"; (2) taxa-table with OTUs(rows) by Taxa (columns) format called "tax_table_dietswap.csv"; and (3) sample metadata called "metadata_dietswap.csv", which is a data frame.

Step 1: Set the R working directory to the folder that these datasets are stored and load these datasets.

```
> setwd("~/Documents/QIIME2R/Ch2_IntroductionToR")
> otu_tab <- read.csv("otu_table_dietswap.csv", row.names = 1)
> tax_tab <- read.csv("tax_table_dietswap.csv", row.names = 1)
> meta_tab <- read.csv("metadata_dietswap.csv", row.names = 1)
```

There are two ways to create a phyloseq object: One is to use the **phyloseq()** function and another is to use the **merge_phyloseq** () function. The arguments of the **phyloseq** () are an otu_table and any unordered list of valid phyloseq components: sample_data, tax_table, phylo, or XStringSet. The tip labels of a phylo-object (tree) and the sequence names of an XStringSet object must match the OTU names of the otu_table. The **merge_phyloseq()** is most useful when we already created a phyloseq object and want to add a separately imported components to this already-created phyloseq object. The **merge_phyloseq()** can take any number of phyloseq objects and/or phyloseq components to combine them into one larger phyloseq object.

Step 2: Check and make sure that the datasets have been correctly loaded.

We can use the **head()** function to check whether we import the correct datasets including the sample IDs, data format, and taxonomy ranks.

```
> head(otu_tab,3)
> head(tax_tab,3)
> head(meta_tab,3)
```

Step 3: Check and make sure that the classes of datasets are correct.

One important thing when creating the phyloseq object is to make sure that both otu_table and tax_table components are matrices and sample_data is a data frame. The component of otu_table works on any numeric matrix. When we use the **phyloseq** () to create a phyloseq object, we must also specify if the taxa are rows or columns based on the loaded otu dataset. The component of tax_table works on any character matrix. The row names of tax_table must match the OTU names (taxa_names) of the otu_table if we combine these two tables with a phyloseq object. The component of sample_data works on any dataframe. The row names of sample_data must match the sample names in the otu_table if we combine them as a phyloseq object.

```
> class(otu_tab)
[1] "data.frame"
> class(tax_tab)
[1] "data.frame"
> class(meta_tab)
[1] "data.frame"
```

Since the above codes show that both otu_table and tax_table components are data frames, we need to convert them into matrices.

```
> otumat<-as.matrix(otu_tab)
> taxmat<-as.matrix(tax_tab)

> class(otumat)
[1] "matrix"
> class(taxmat)
[1] "matrix"
```

Step 4: Create phyloseq object using either phyloseq () or merge_phyloseq().

Now we can create a phyloseq object. We first use the **phyloseq()** function as below. It is important to make sure to specify whether taxa_are_rows = TRUE or FALSE based on the loaded otu matrix. Here, taxa_are_rows = FALSE.

```
> library(phyloseq)
> otu<-otu_table(otumat,taxa_are_rows = FALSE)
> tax<-tax_table(taxmat)
> sam<-sample_data(meta_tab)

> physeq = phyloseq(otu, tax, sam)
> physeq
phyloseq-class experiment-level object
otu_table() OTU Table: [ 37 taxa and 222 samples ]
sample_data() Sample Data: [ 222 samples by 7 sample variables ]
tax_table() Taxonomy Table: [ 37 taxa by 3 taxonomic ranks ]
```

The above created phyloseq object shows that OTU Table has 37 taxa and 222 samples, Sample Data has 222 samples by 7 sample variables, and Taxonomy Table has 37 taxa by 3 taxonomic ranks. One mistake that sometime happens is that the order or names of sample IDs in OTU Table and Sample Data are not exactly matched. We can use the **sample_names()** function to check.

```
> sample_names(otu)
 [1] "Sample-1" "Sample-2" "Sample-3" "Sample-4" "Sample-5"
 [6] "Sample-6" "Sample-7" "Sample-8" "Sample-9" "Sample-10"
[11] "Sample-11" "Sample-12" "Sample-13" "Sample-14" "Sample-15"
------
```

```
> sample_names(sam)
 [1] "Sample-1" "Sample-2" "Sample-3" "Sample-4" "Sample-5"
 [6] "Sample-6" "Sample-7" "Sample-8" "Sample-9" "Sample-10"
[11] "Sample-11" "Sample-12" "Sample-13" "Sample-14" "Sample-15"
------
```

Now let's illustrate how to use the **merge_phyloseq()** function to add the new data components to the phyloseq object we already have. Assume that we first use the **phyloseq()** to create a phyloseq object physeq1 by otu table and sample data.

```
> otu<-otu_table(otumat,taxa_are_rows = FALSE)
> tax<-tax_table(taxmat)
> sam<-sample_data(meta_tab)

> physeq1 = phyloseq(otu,sam)
> physeq1
phyloseq-class experiment-level object
otu_table() OTU Table:       [ 130 taxa and 222 samples ]
sample_data() Sample Data:        [ 222 samples by 7 sample variables ]
```

Now we can use the **merge_phyloseq()** function to merge the taxonomy data into the current phyloseq object physeq1:

```
> physeq2 = merge_phyloseq(physeq1,tax)
> physeq2
phyloseq-class experiment-level object
otu_table() OTU Table:       [ 37 taxa and 222 samples ]
sample_data() Sample Data:        [ 222 samples by 7 sample variables ]
tax_table() Taxonomy Table:        [ 37 taxa by 3 taxonomic ranks ]

> identical(physeq, physeq2)
[1] TRUE
```

The above prints show that the results using the functions **phyloseq()** and **merge_phyloseq()** are identical.

Step 5: Create phyloseq object with adding a random phylogenetic tree component.

We can use the **ape** package to create a random phylogenetic tree and add it to the dataset (current version 5.6.2, March 2022; see Sect. 2.4.1). For more information on phylogenetic tree, the readers are referred to Sect. 2.4 and Chap. 3 (Sect. 3.1.4). To ensure the tree be correctly created, make sure its tip labels match the names of OTU_table component.

```
> library("ape")
> random_tree = ape::rtree(ntaxa(physeq), rooted=TRUE, tip.
label=taxa_names(physeq))
> plot(random_tree)
```

Now merge the tree data to the phyloseq object we already have by using the **merge_phyloseq()**, or use the **phyloseq ()** to build it again from scratch. The results should be identical.

We first use the **merge_phyloseq()** to merge the tree data with the phyloseq object "physeq":

```
> physeq3 = merge_phyloseq(physeq, random_tree)
> physeq3
phyloseq-class experiment-level object
otu_table() OTU Table:        [ 37 taxa and 222 samples ]
sample_data() Sample Data:      [ 222 samples by 7 sample variables ]
tax_table() Taxonomy Table:       [ 37 taxa by 3 taxonomic ranks ]
phy_tree() Phylogenetic Tree:       [ 37 tips and 36 internal nodes ]
```

We then use the **phyloseq()** to build it again from scratch:

```
> library(phyloseq)
> physeq4 = phyloseq(otu, tax, sam, random_tree)
> physeq4
phyloseq-class experiment-level object
otu_table() OTU Table:        [ 37 taxa and 222 samples ]
sample_data() Sample Data:      [ 222 samples by 7 sample variables ]
tax_table() Taxonomy Table:       [ 37 taxa by 3 taxonomic ranks ]
phy_tree() Phylogenetic Tree:       [ 37 tips and 36 internal nodes ]
```

It is true that they are identical.

```
> identical(physeq3, physeq4)
[1] TRUE
```

2.3.1.2   Merge Samples in a phyloseq Object

Example 2.5: Diet Swap Between Rural and Western Populations, Example 2.4, Cont.

The **phyloseq** package supports two completely different categories of merging data objects: merging two data objects and merging data objects based upon a taxonomic or sample variable, i.e., merging the OTUs or samples in a phyloseq object. Merging OTU or sample indices based on variables in the data is useful to reduce noise or excess features in an analysis or graphic. It requires that the examples to be merged in a dataset are from the same environment and the OTUs to be merged are from the same taxonomic genera. The usual approach is to use a table-join, and hence the non-matching keys are omitted in the result, where keys are Sample IDs or Taxa IDs, respectively.

Here, we illustrate how to merge sample within a phyloseq object via the **merge_samples()** function. This function is a very useful tool to test an analysis effect by removing the individual effects between replicates or between samples from a particular explanatory variable. By performing the **merge_samples()** function, the abundance values of merged samples are summed.

First, remove empty samples, i.e., remove unobserved OTUs (sum 0 across all samples). In this case, all samples have sum of OTUs > 0, so the object physeq3a is identical to the object physeq3.

```
> physeq3a = physeq3
> physeq3a = prune_taxa(taxa_sums(physeq3) > 0, physeq3)
> physeq3a
phyloseq-class experiment-level object
otu_table()  OTU Table:      [ 37 taxa and 222 samples ]
sample_data() Sample Data:    [ 222 samples by 7 sample variables ]
tax_table()  Taxonomy Table:   [ 37 taxa by 3 taxonomic ranks ]
phy_tree()  Phylogenetic Tree:   [ 37 tips and 36 internal nodes ]
```

Then, add a new `sample_data` variable to the dataset.

```
> # Check the names of sample
> sample_variables(physeq3a)
[1] "subject"     "bmi"      "sex"      "nationality"
[5] "timepoint.total" "timepoint.group" "group"

> # Define obese variable
> obese = c("Obese")
> sample_data(physeq3a)$obese <- get_variable(physeq3a, "bmi") %in%
obese
> sample_data(physeq3a)$obese

> sampledata= sample_data(physeq3a)
> head(sampledata,3)
Sample Data: [3 samples by 8 sample variables]:
```

```
      subject bmi sex nationality timepoint.total timepoint.group
Sample-1    byn    Obese Male       AAM      4         1
Sample-2    nms    Lean Male        AFR      2         1
Sample-3    olt Overweight Male     AFR      2         1
 group obese
Sample-1    DI TRUE
Sample-2    HE FALSE
Sample-3    HE FALSE


print(sampledata)
> options(width=100)
> print(sampledata,3)
      subject     bmi    sex nationality timepoint.total timepoint.group
group obese
Sample-1    byn    Obese  Male      AAM         4            1  DI TRUE
Sample-2    nms    Lean   Male      AFR         2            1  HE FALSE
Sample-3    olt Overweight  Male    AFR            2         1  HE FALSE
Sample-4    pku    Obese Female     AFR         2            1  HE TRUE
Sample-5    qjy Overweight Female   AFR            2         1  HE FALSE
Sample-6    riv    Obese Female     AFR         2            1  HE TRUE
------


> # Merge obese variable
> mergedphyseq3a = merge_samples(physeq3a, "obese")
> mergedphyseq3a
phyloseq-class experiment-level object
otu_table()  OTU Table:      [ 37 taxa and 2 samples ]
sample_data() Sample Data:    [ 2 samples by 8 sample variables ]
tax_table()  Taxonomy Table:  [ 37 taxa by 3 taxonomic ranks ]
phy_tree()   Phylogenetic Tree: [ 37 tips and 36 internal nodes ]

> head(sample_names(sampledata))
[1] "Sample-1" "Sample-2" "Sample-3" "Sample-4" "Sample-5" "Sample-6"
> head(sample_names(physeq3a))
[1] "Sample-1" "Sample-2" "Sample-3" "Sample-4" "Sample-5" "Sample-6"
> sample_names(mergedphyseq3a)
[1] "FALSE" "TRUE"
```

The following commands define top 10 taxa:

```
> OTUnames10 = names(sort(taxa_sums(physeq3a), TRUE)[1:10])
> OTUnames10
 [1] "Akkermansia"   "Dialister"     "Bifidobacterium" "Collinsella"
"Enterococcus"
 [6] "Veillonella"   "Fusobacteria"  "Serratia"      "Campylobacter"
"Oceanospirillum"
```

The following commands create a phyloseq object for the top 10 taxa:

```
> physeq3a10 = prune_taxa(OTUnames10, physeq3a)
> physeq3a10
```

```
phyloseq-class experiment-level object
otu_table() OTU Table: [ 10 taxa and 222 samples ]
sample_data() Sample Data: [ 222 samples by 8 sample variables ]
tax_table() Taxonomy Table: [ 10 taxa by 3 taxonomic ranks ]
phy_tree() Phylogenetic Tree: [ 10 tips and 9 internal nodes ]
```

The following commands create a subset of phyloseq object for the top 10 taxa:

```
> mergedphyseq3a10 = prune_taxa(OTUnames10, mergedphyseq3a)
> mergedphyseq3a10
phyloseq-class experiment-level object
otu_table() OTU Table: [ 10 taxa and 2 samples ]
sample_data() Sample Data: [ 2 samples by 8 sample variables ]
tax_table() Taxonomy Table: [ 10 taxa by 3 taxonomic ranks ]
phy_tree() Phylogenetic Tree: [ 10 tips and 9 internal nodes ]
```

The following commands create a subset of sample for the obese subjects:

```
> obese_samples = sample_names(subset(sample_data(physeq3a),
bmi=="Obese"))
> print(obese_samples)
 [1] "Sample-1" "Sample-4" "Sample-6" "Sample-7" "Sample-12" "Sample-
14" "Sample-15"
 [8] "Sample-20" "Sample-22" "Sample-23" "Sample-28" "Sample-30"
"Sample-31" "Sample-36"
[15] "Sample-38" "Sample-40" "Sample-45" "Sample-47" "Sample-49"
"Sample-52" "Sample-55"
[22] "Sample-59" "Sample-61" "Sample-65" "Sample-67" "Sample-71"
"Sample-72" "Sample-73"
[29] "Sample-76" "Sample-78" "Sample-81" "Sample-85" "Sample-86"
"Sample-88" "Sample-92"
[36] "Sample-93" "Sample-95" "Sample-98" "Sample-99" "Sample-106"
"Sample-113" "Sample-115"
[43] "Sample-123" "Sample-125" "Sample-127" "Sample-134" "Sample-136"
"Sample-144" "Sample-146"
[50] "Sample-148" "Sample-150" "Sample-151" "Sample-152" "Sample-153"
"Sample-154" "Sample-155"
[57] "Sample-156" "Sample-160" "Sample-161" "Sample-162" "Sample-163"
"Sample-167" "Sample-168"
[64] "Sample-169" "Sample-170" "Sample-173" "Sample-174" "Sample-177"
"Sample-178" "Sample-182"
[71] "Sample-183" "Sample-184" "Sample-187" "Sample-188" "Sample-190"
"Sample-191" "Sample-193"
[78] "Sample-194" "Sample-196" "Sample-198" "Sample-200" "Sample-202"
"Sample-204" "Sample-206"
[85] "Sample-209" "Sample-211" "Sample-212" "Sample-217" "Sample-219"
"Sample-220"
```

The following commands find the OTU table for top 10 taxa in the obese subjects. We can see there are 10 taxa in 90 samples in the OTU table.

```
> otu_table(physeq3a10)[obese_samples, ]
OTU Table:        [10 taxa and 90 samples]
           taxa are columns
        Campylobacter   Serratia Fusobacteria Veillonella Akkermansia
Enterococcus Collinsella
Sample-1     302.8297 138.25918   337.9723   563.9701 1229.3921
269.0209   247.3910
Sample-4     313.0234 259.58027   350.2741   356.2538 16246.2164
373.2883 2987.0418
Sample-6     310.9208 655.30520   583.4619   538.3110 1092.0800
746.3891 1698.0409
------
```

We then sum up the reads over columns and rows for downstream analyses.

```
> colSums(otu_table(physeq3a10)[obese_samples, ]) ",]# otu_table has
sample by otu format
Bifidobacterium    Dialister Fusobacteria   Collinsella   Veillonella
Campylobacter
    484666       414137      51401       78701       55003       28298
  Enterococcus Oceanospirillum     Serratia   Akkermansia
    34466       33171       47480      436003
> rowSums(otu_table(physeq3a10)[obese_samples, ])
 Sample-1 Sample-4 Sample-6 Sample-7 Sample-12 Sample-14 Sample-15
Sample-20 Sample-22
 6420.736 53822.366 9784.471 23309.503 16467.630 63959.029 7991.240
9219.114 53930.128
------
```

2.3.1.3   Merge Two phyloseq Objects

In Sect. 2.3.1.1, we Illustrate how to create a phyloseq object using the function
merge_phyloseq (). The following commands show how to extract components
from an example dataset, and then build them back up to the original form using the
merge_phyloseq ().

```
> otu = otu_table(physeq3)
> tax = tax_table(physeq3)
> sam = sample_data(physeq3)
> tree = phy_tree(physeq3)
> otutax = phyloseq(otu, tax)
> otutax
phyloseq-class experiment-level object
otu_table() OTU Table: [ 37 taxa and 222 samples ]
tax_table() Taxonomy Table: [ 37 taxa by 3 taxonomic ranks ]
```

```
> physeq3b = merge_phyloseq(otutax, sam, tree)
> identical(physeq3b, physeq3)
[1] TRUE
```

The new `otutax` object only has the OTU table and taxonomy table. When we compare the physeq3b with the original physeq3 object, they are confirmed being identical, where the physeq3b is the object that is built up by the original physeq3 object by using the function `merge_phyloseq()`. The arguments to the `merge_phyloseq()` are a mixture of multi-component (`otutax`) and single-component objects.

2.3.2   Microbiome

The **microbiome** package was developed by Leo Lahti et al. (Lahti 2017). The package utilizes phyloseq object and wraps some functions of **vegan** (Jari Oksanen et al. 2019) and **phyloseq** (McMurdie and Holmes 2013) packages with less obscure syntax (current version 1.14.0, March 2022). It also uses tools from a number of other R extensions, including **ggplot2** (Wickham 2016), **dplyr** (Hadley Wickham et al. 2020), and **tidyr** (Wickham 2020). Here, we introduce some useful data processing functions in this package.

The microbiome package is built on the phyloseq objects and extends some functions of the phyloseq package in order to facilitate manipulation and processing microbiome datasets, including subsetting, aggregating, and filtering. To install microbiome package of Bioconductor development version in R, type the following R commands in R Console or RStudio Source editor panel.

```
> library(BiocManager)
> if (!requireNamespace("BiocManager", quietly=TRUE))
+ install.packages("BiocManager")
> BiocManager::install(version = "devel")
> BiocManager::install("microbiome")
```

2.3.2.1   Create a phyloseq Object

Example 2.6: Global Patterns Datasets
Caporaso et al. (2011)(Caporaso et al. 2011) published a study to compare the microbial communities from 25 environmental samples and three known "mock communities" – a total of 9 sample types – at a depth averaging 3.1 million reads per sample. This study demonstrated excellent consistency in taxonomic recovery and diversity patterns that were previously reported in many other published studies. It also demonstrated that 2000 Illumina single-end reads are sufficient to recapture the same relationships among samples that are observed with the full dataset.

These datasets have been used to illustrate the R programs in the other software including the phyloseq package. We store three data in the directory "~/Documents/ QIIME2R/Ch2_IntroductionToR": (1) "otu_table_GlobalPatterns.csv", (2) "tax_ table_GlobalPatterns.csv", and (3) "metadata_GlobalPatterns.csv".

The following commands load the package and example data in R.

```
> setwd("~/Documents/QIIME2R/Ch2_IntroductionToR")
> otu_tab <- read.csv("otu_table_GlobalPatterns.csv", row.names = 1)
> tax_tab <- read.csv("tax_table_GlobalPatterns.csv", row.names = 1)
> meta_tab <- read.csv("metadata_GlobalPatterns.csv", row.names = 1)

> otumat<-as.matrix(otu_tab)
> taxmat<-as.matrix(tax_tab)
> library(phyloseq)
> otu<-otu_table(otumat, taxa_are_rows = TRUE)
> tax<-tax_table(taxmat)
> sam<-sample_data(meta_tab)

> physeq = phyloseq(otu, tax, sam)
> physeq
phyloseq-class experiment-level object
otu_table() OTU Table: [ 19216 taxa and 26 samples ]
sample_data() Sample Data: [ 26 samples by 7 sample variables ]
tax_table() Taxonomy Table: [ 19216 taxa by 7 taxonomic ranks ]
```

The GlobalPatterns datasets are included in the phyloseq package. Below, we illustrate how to directly download these datasets from the phyloseq package.

```
> library(microbiome)
> library(phyloseq)
> library(knitr)
> data(GlobalPatterns)
> # Rename GlobalPatterns data (which is a phyloseq object)
> physeq <- GlobalPatterns
> physeq
phyloseq-class experiment-level object
otu_table() OTU Table: [ 19216 taxa and 26 samples ]
sample_data() Sample Data: [ 26 samples by 7 sample variables ]
tax_table() Taxonomy Table: [ 19216 taxa by 7 taxonomic ranks ]
phy_tree() Phylogenetic Tree: [ 19216 tips and 19215 internal nodes ]
```

The microbiome package has some very useful functions for processing microbiome data. We introduce them below.

2.3.2.2 Summarize the Contents of phyloseq Object

```
> summarize_phyloseq(physeq)
```

After performing the **summarize_phyloseq()** function, we obtain the information of the "physeq" object, including whether it is compositional, minimum, maximum, total, average, and median number of reads, sparsity, whether any OTU sum to 1, number of singletons, percent of OTUs that are singletons (i.e., exactly one read detected across all samples), and number of sample variables. In this case, we have 7 sample variables: X.SampleID, Primer, Final_Barcode, Barcode_truncated_plus_T, Barcode_full_length, SampleType, and Description.

2.3.2.3   Retrieve the Data Elements of phyloseq Object

A phyloseq object contains OTU table (taxa abundances), sample metadata, taxonomy table (mapping between OTUs and higher-level taxonomic classifications), and phylogenetic tree (relations between the taxa). Some of these are optional.

```
> physeq
phyloseq-class experiment-level object
otu_table()   OTU Table:      [ 19216 taxa and 26 samples ]
sample_data() Sample Data:    [ 26 samples by 7 sample variables ]
tax_table()   Taxonomy Table: [ 19216 taxa by 7 taxonomic ranks ]
phy_tree()    Phylogenetic Tree: [ 19216 tips and 19215 internal nodes ]
```

However, phyloseq is a type "S4" object. The information of object of type "S4" is not directly extractable via the basic R functions, such as **head()** to subset the data and **names()** to find the names of data. To extract the information, we need to find out names of slots in the object, and then process at the slot names. For example, we can subset the otu table and sample data names using the following R commands.

```
slotNames(physeq)
[1] "otu_table" "tax_table" "sam_data" "phy_tree" "refseq"
> head(physeq@otu_table,3)
OTU Table:        [3 taxa and 26 samples]
             taxa are rows
    CL3 CC1 SV1 M31Fcsw M11Fcsw M31Plmr M11Plmr F21Plmr M31Tong M11Tong
549322  0  0  0       0       0       0       0       0       0       0
522457  0  0  0       0       0       0       0       0       0       0
951     0  0  0       0       0       0       1       0       0       0
------
```

```
names(physeq@sam_data)
[1] "X.SampleID" "Primer" "Final_Barcode"
[4] "Barcode_truncated_plus_T" "Barcode_full_length" "SampleType"
[7] "Description"
```

The following functions in **microbiome** package facilitate this kind of processing and are very useful for downstream statistical analysis using the **microbiome** and other R packages, such as retrieving the data elements from phyloseq object.

First, let's use the **abundances()** function to retrieve absolute and relative taxonomic abundances from OTU table.

```
# Absolute abundances
> head(otu_abs<- abundances(physeq),3)
      CL3 CC1 SV1 M31Fcsw M11Fcsw M31Plmr M11Plmr F21Plmr M31Tong M11Tong
549322 0  0  0    0      0      0       0      0      0      0
522457 0  0  0    0      0      0       0      0      0      0
951    0  0  0    0      0      0       1      0      0      0
------
```

```
> # Relative abundances
> head(otu_rel <- abundances(physeq, "compositional"),3)
      CL3 CC1 SV1 M31Fcsw M11Fcsw M31Plmr  M11Plmr F21Plmr M31Tong M11Tong
549322 0  0  0    0      0      0 0.000e+00     0      0      0
522457 0  0  0    0      0      0 0.000e+00     0      0      0
951    0  0  0    0      0      0 2.305e-06     0      0      0
------
```

Then, we use the **readcount ()** function to retrieve the total read counts.

```
> read_tot <- readcount(physeq)
> head(read_tot)
   CL3     CC1    SV1 M31Fcsw M11Fcsw M31Plmr
864077 1135457 697509 1543451 2076476  718943
```

Next, we use the **tax_table()** function to retrieve the taxonomy table.

```
> #Taxonomy table
> tax <- tax_table(physeq)
> head((tax),3)
Taxonomy Table:    [3 taxa by 7 taxonomic ranks]:
    Kingdom  Phylum         Class         Order       Family
549322 "Archaea" "Crenarchaeota" "Thermoprotei" NA        NA
522457 "Archaea" "Crenarchaeota" "Thermoprotei" NA        NA
951   "Archaea" "Crenarchaeota" "Thermoprotei" "Sulfolobales"
"Sulfolobaceae"
    Genus     Species
549322 NA       NA
522457 NA       NA
951   "Sulfolobus" "Sulfolobusacidocaldarius"
```

Now, we use the **meta()** function to retrieve the metadata which is a data.frame.

```
> # Pick metadata as data.frame:
> meta <- meta(physeq)
> head((meta),3)
  X.SampleID Primer Final_Barcode Barcode_truncated_plus_T
CL3    CL3 ILBC_01     AACGCA              TGCGTT
CC1    CC1 ILBC_02     AACTCG              CGAGTT
```

```
SV1      SV1 ILBC_03     AACTGT              ACAGTT
  Barcode_full_length SampleType                    Description
CL3      CTAGCGTGCGT     Soil Calhoun South Carolina Pine soil, pH 4.9
CC1      CATCGACGAGT     Soil Cedar Creek Minnesota, grassland, pH 6.1
SV1      GTACGCACAGT     Soil Sevilleta new Mexico, desert scrub, pH 8.3
```

Finally, we can melt the phyloseq data as a data frame table for easier plotting and downstream statistical analysis.

```
> # Melt phyloseq data as a data frame table
> df <- psmelt (physeq)
> kable (head (df, 4))
|     |OTU   |Sample  | Abundance|X.SampleID |Primer  |Final_Barcode |
Barcode_truncated_plus_T |Barcode_full_length |SampleType     |
Description           |Kingdom |Phylum  |Class    |Order    |
Family   |Genus   |Species |
|:------|:------|:-------|---------:|:----------|:-------
|:------------|:----------------------|:-----------------
|:-------------|:--------------------------------------------
|:-------|:------------|:---------------|:------------
|:----------|:-------------|:-------|
|406582 |549656 |AQC4cm  | 1177685|AQC4cm  |ILBC_17 |ACAGCT   |AGCTGT
|CAAGCTAGCTG     |Freshwater (creek) |Allequash Creek, 3-4 cm depth
|Bacteria |Cyanobacteria |Chloroplast   |Stramenopiles |NA     |NA
|NA   |
|241435 |279599 |LMEpi24M |  914209|LMEpi24M  |ILBC_13 |ACACTG   |
CAGTGT          |CATGAACAGTG     |Freshwater   |Lake Mendota Minnesota,
24 meter epilimnion |Bacteria |Cyanobacteria |Nostocophycideae |
Nostocales  |Nostocaceae |Dolichospermum |NA   |
|406580 |549656 |AQC7cm  |  711043|AQC7cm  |ILBC_18 |ACAGTG   |CACTGT
|ATGAAGCACTG     |Freshwater (creek) |Allequash Creek, 6-7 cm depth
|Bacteria |Cyanobacteria |Chloroplast   |Stramenopiles |NA     |NA
|NA   |
|406574 |549656 |AQC1cm  |  554198|AQC1cm  |ILBC_16 |ACAGCA   |TGCTGT
|GACCACTGCTG     |Freshwater (creek) |Allequash Creek, 0-1cm depth
|Bacteria |Cyanobacteria |Chloroplast   |Stramenopiles |NA     |NA
|NA   |
```

2.3.2.4   Operate on Taxa

We can use the **ntaxa()** function to obtain the number of taxa and the most abundant taxa using the **top_taxa()** function.

```
> # Number of taxa
> num_tax <- ntaxa (physeq)
> num_tax
[1] 19216
```

```
> # Most abundant taxa
> top10tax <- top_taxa(physeq, n = 10)
> top10tax
 [1] "549656" "331820" "279599" "360229" "317182" "94166" "158660"
"329744" "550960" "189047"
```

Use the **rank_names** () function to get the names of taxa.

```
> ranks <- rank_names(physeq) # Taxonomic levels
> ranks
[1] "Kingdom" "Phylum" "Class" "Order" "Family" "Genus" "Species"
```

Use the **taxa** () function to get the names of taxa.

```
> taxa <- taxa(physeq) # Taxa names at the analyzed level
> head(taxa)
[1] "549322" "522457" "951" "244423" "586076" "246140"
```

Use the **subset_taxa**() function to subset taxa for analysis.

```
> physeq
phyloseq-class experiment-level object
otu_table() OTU Table: [ 19216 taxa and 26 samples ]
sample_data() Sample Data: [ 26 samples by 7 sample variables ]
tax_table() Taxonomy Table: [ 19216 taxa by 7 taxonomic ranks ]
phy_tree() Phylogenetic Tree: [ 19216 tips and 19215 internal nodes ]


> # Subset taxa
> sub_bac <- subset_taxa(physeq, Phylum == "Bacteroidetes")
> sub_bac
phyloseq-class experiment-level object
otu_table() OTU Table: [ 2382 taxa and 26 samples ]
sample_data() Sample Data: [ 26 samples by 7 sample variables ]
tax_table() Taxonomy Table: [ 2382 taxa by 7 taxonomic ranks ]
phy_tree() Phylogenetic Tree: [ 2382 tips and 2381 internal nodes ]
```

Use the **sample_sums** () function to obtain the total OTU abundance in each sample.
```
> # Total OTU abundance in each sample
> sam_sum <- sample_sums(physeq)
> head(sam_sum, 4)
 CL3 CC1 SV1 M31Fcsw
 864077 1135457 697509 1543451
```

We can prune (select) taxa using the **map_levels**() function. This function is used to map taxa between hierarchy levels. The syntax is map_levels(taxa = NULL, from, to, data), where **taxa** specifies taxa to convert, if NULL then considering all taxa in the tax.table; **from** specifies convert from taxonomic level; **to** specifies convert to taxonomic level; and data is either a phyloseq object or its taxonomy table-class.

```
> # List of Genera in the Bacteroideted Phylum
> #Prune (select) taxa:
> # List of Genera in the Bacteroideted Phylum
> tax_map <- map_levels("Bacteroidetes", "Phylum", "Genus", physeq)
> head(tax_map, 3)
$Bacteroidetes
 [1] "Balneola"        "Thermonema"
 [3] "Lutimonas"       "Kordia"
 [5] "Zhouia"          "Capnocytophaga"
------
```

```
> # Convert between taxonomic levels (here from Genus (Akkermansia)
> # to Phylum (Verrucomicrobia):
> tax_map2 <- map_levels("Akkermansia", "Genus", "Phylum", tax_table
(physeq))
> print(tax_map2)
[1] "Verrucomicrobia"
```

One useful function is the **prune_taxa()**, which is often used to prune(select) taxa with positive sum across samples.

```
> # Taxa with positive sum across samples
> prune <- prune_taxa(taxa_sums(physeq) > 0, physeq)
```

In microbiome data processing, we sometimes want to merge the desired taxa into "Other" category. The function **merge_taxa2()** is used to place certain OTUs or other groups into an "other" category. One syntax is given: merge_taxa2(x, taxa = NULL, pattern = NULL, name = "Merged"). The merge_taxa2 () function differs from phyloseq::merge_taxa () lies its last two arguments: here, we can specify the name of the new merged group, and we can specify a common pattern in the name to merge. Here, we merge all Bacteroides groups into a single group named Bacteroides.

```
> physeq2 <- merge_taxa2(physeq, pattern = "^Bacteroides", name =
"Bacteroides")
> head(taxa_bac <- tax_table(physeq2), 3)
Taxonomy Table:    [3 taxa by 7 taxonomic ranks]:
    Kingdom  Phylum         Class          Order          Family
549322 "Archaea" "Crenarchaeota" "Thermoprotei" NA          NA
522457 "Archaea" "Crenarchaeota" "Thermoprotei" NA          NA
951    "Archaea" "Crenarchaeota" "Thermoprotei" "Sulfolobales"
"Sulfolobaceae"
    Genus     Species
549322 NA        NA
522457 NA        NA
951    "Sulfolobus" "Sulfolobusacidocaldarius"
```

```
> head(taxa<- tax_table(physeq),3)
Taxonomy Table:   [3 taxa by 7 taxonomic ranks]:
    Kingdom Phylum       Class        Order      Family
549322 "Archaea" "Crenarchaeota" "Thermoprotei" NA      NA
522457 "Archaea" "Crenarchaeota" "Thermoprotei" NA      NA
951 "Archaea" "Crenarchaeota" "Thermoprotei" "Sulfolobales"
"Sulfolobaceae"
    Genus    Species
549322 NA     NA
522457 NA     NA
951 "Sulfolobus" "Sulfolobusacidocaldarius"
```

We can filter taxa based on user-specified function values (e.g., mean and variance).

```
> filt<- filter_taxa(physeq, function(x) mean(x) > 0.5, TRUE)
> filt2<- filter_taxa(physeq, function(x) var(x) > 0.001, TRUE)
```

We can pick a unique taxonomy level.

```
> # List unique phylum-level groups
> head(get_taxa_unique(physeq, "Phylum"))
[1] "Crenarchaeota" "Euryarchaeota" "Actinobacteria" "Spirochaetes"
"MVP-15"       "Proteobacteria"
```

We can also pick the taxa abundances for a given sample.

```
> # Pick the taxa abundances for a given sample
> sample1 <- sample_names(physeq)[[1]]
> sample1
[1] "CL3"
```

Below we pick abundances for a particular taxon (sample1).

```
> # Pick abundances for a particular taxon
> tax_abund <- abundances(physeq)[, sample1]
> head(tax_abund)
549322 522457   951 244423 586076 246140
   0    0    0    0    0    0
```

2.3.2.5   Aggregate Taxa to Higher Taxonomic Levels

One operation on taxa (aggregating taxa to higher taxonomic levels) is particularly useful. When the phylogenetic tree is missing, the **aggregate_taxa** () function can be used to aggregate taxa.

```
> physeq3 <- aggregate_taxa(physeq, "Phylum")
> head(Phylum <- tax_table(physeq3),3)
Taxonomy Table: [3 taxa by 3 taxonomic ranks]:
 Kingdom Phylum unique
Crenarchaeota "Archaea" "Crenarchaeota" "Crenarchaeota"
Euryarchaeota "Archaea" "Euryarchaeota" "Euryarchaeota"
ABY1_OD1 "Bacteria" "ABY1_OD1" "ABY1_OD1"
```

When the phylogenetic tree is available, we can use the functions **merge_samples()**, **merge_taxa()**, and **tax_glom()**.

2.3.2.6   Operate on Sample

We can use the **sample_names ()** function to retrieve sample names.

```
> head(sample_names(physeq))
[1] "CL3" "CC1" "SV1" "M31Fcsw" "M11Fcsw" "M31Plmr"
```

We can use the **subset_samples()** function to subset the metadata table by metadata fields (variables).

```
> physeq_sub <- subset_samples(physeq, SampleType == "Soil")
> physeq_sub
phyloseq-class experiment-level object
otu_table() OTU Table: [ 19216 taxa and 3 samples ]
sample_data() Sample Data: [ 3 samples by 7 sample variables ]
tax_table() Taxonomy Table: [ 19216 taxa by 7 taxonomic ranks ]
phy_tree() Phylogenetic Tree: [ 19216 tips and 19215 internal nodes ]
```

We also can subset a sample using sample names.

```
> # Check sample names for Soil and sample CL3 in this phyloseq object
> sub <- rownames(subset(meta(physeq), SampleType == "Soil" & X.
SampleID == "CL3"))
> head(sub,3)
[1] "CL3"
```

We also can pick the subset using sample names.

```
> # Pick the phyloseq subset with these sample names
> sub2 <- prune_samples(sub, physeq)
> sub2
phyloseq-class experiment-level object
otu_table() OTU Table: [ 19216 taxa and 1 samples ]
sample_data() Sample Data: [ 1 samples by 7 sample variables ]
tax_table() Taxonomy Table: [ 19216 taxa by 7 taxonomic ranks ]
phy_tree() Phylogenetic Tree: [ 19216 tips and 19215 internal nodes ]
```

The following commands check the total abundance in each sample.

```
> # Total OTU abundance in each sample
> sam_sum <- sample_sums(physeq)
> head(sam_sum)
   CL3     CC1     SV1 M31Fcsw M11Fcsw M31Plmr
 864077 1135457 697509 1543451 2076476 718943
```

Finally, for longitudinal data, we can use the **baseline**() function to pick samples at the baseline.

```
> # Pick samples at the baseline only.
> physeq0 <- baseline(physeq)
```

2.3.2.7   Operate on Metadata

For downstream statistical analysis, it is very important to familiarize metadata. First, let's use the **sample_variables** () function to check what sample variables are available in the metadata.

```
> sample_variables(physeq)
[1] "X.SampleID"        "Primer"
[3] "Final_Barcode"     "Barcode_truncated_plus_T"
[5] "Barcode_full_length"  "SampleType"
[7] "Description"
```

One frequently used operation in microbiome study is to calculate alpha and beta diversities and then assign these values as new fields to metadata, which will facilitate association (regression) analysis between the microbial taxa abundance and environmental/clinical factors. Below we first calculate Shannon diversity for samples and then assign the values to the sample data of "physeq" object.

```
> # Calculate diversity for samples
> shannon <- microbiome::alpha(physeq, index = "shannon")
> head(shannon,3)
 diversity_shannon
CL3       6.577
CC1       6.777
SV1       6.498
```

```
> # Assign the estimated Shannon diversity to sample metadata
> sample_data(physeq)$Shannon <- shannon
> head(sample_data(physeq)$Shannon,3)
 diversity_shannon
CL3       6.577
CC1       6.777
SV1       6.498
```

As recall, we have already retrieved sample data using the **meta()** above. Now we can directly assign Shannon diversity to the "meta" dataframe.

```
> meta <- meta(physeq)
> meta$Shannon <- shannon
> head(meta,3)
    X.SampleID Primer Final_Barcode Barcode_truncated_plus_T
CL3     CL3 ILBC_01     AACGCA          TGCGTT
CC1     CC1 ILBC_02     AACTCG          CGAGTT
SV1     SV1 ILBC_03     AACTGT          ACAGTT
    Barcode_full_length SampleType                Description
CL3     CTAGCGTGCGT     Soil  Calhoun South Carolina Pine soil, pH 4.9
CC1     CATCGACGAGT     Soil  Cedar Creek Minnesota, grassland, pH 6.1
SV1     GTACGCACAGT     Soil Sevilleta new Mexico, desert scrub, pH 8.3
    diversity_shannon
CL3         6.577
CC1         6.777
SV1         6.498
```

2.3.2.8   Data Transformations

The **microbiome** package provides a wrapper for transformation of standard sample/ OTU. One syntax is given: transform(x, transform = "identify", target = "OTU", shift = 0, scale = 1), where **x** is a phyloseq-class object; **transform** is used to specify the method of transformation to apply including "compositional" (i.e., relative abundance), "Z," "log10," "log10p," "hellinger," "identity," "clr," or any method from the vegan::decostand function. The **target** argument is used to specify whether the transformation applies for "sample" or "OTU," but this option does not affect the log transformation. The **shift** argument is used to specify a constant to shift the baseline abundance (in transform = "shift"), and the **scale** is used to scale constant for the abundance values (in transform = "scale").

One often used data transformation is to transform absolute abundances to relative scale of abundances. Below we print the first three observations of absolute abundances to see what they look like.

```
# Transform to relative abundances
> head(otu_table(physeq),3)
OTU Table:      [3 taxa and 26 samples]
         taxa are rows
     CL3 CC1 SV1 M31Fcsw M11Fcsw M31Plmr M11Plmr F21Plmr M31Tong M11Tong
549322 0 0 0     0       0       0       0       0       0       0
522457 0 0 0     0       0       0       0       0       0       0
951    0 0 0     0       0       0       1       0       0       0
------
```

Now we transform the absolute abundances to relative abundances. By applying the transformation of "compositional," abundances are returned as relative abundances in [0, 1], i.e., convert to percentages by multiplying with a factor of 100.

```
> physeq_comp <- microbiome::transform(physeq, "compositional")
> head(otu_table(physeq_comp),3)
OTU Table:        [3 taxa and 26 samples]
             taxa are rows
    CL3 CC1 SV1 M31Fcsw M11Fcsw M31Plmr  M11Plmr F21Plmr M31Tong M11Tong
549322 0 0 0    0     0      0 0.000e+00     0      0      0
522457 0 0 0    0     0      0 0.000e+00     0      0      0
951    0 0 0    0     0      0 2.305e-06     0      0      0
------
```

The Hellinger transformation is square root of the relative abundance but instead given at the scale [0,1]. The following commands perform the Hellinger transformation.

```
> physeq_hellinger <- microbiome::transform(physeq, "hellinger")
> head(otu_table(physeq_hellinger),3)
OTU Table:        [3 taxa and 26 samples]
             taxa are rows
    CL3 CC1 SV1 M31Fcsw M11Fcsw M31Plmr  M11Plmr F21Plmr M31Tong M11Tong
549322 0 0 0    0     0      0 0.000000      0      0      0
522457 0 0 0    0     0      0 0.000000      0      0      0
951    0 0 0    0     0      0 0.001518      0      0      0
------
```

The syntaxes of Log10 and Log10p transforms are transform(x, 'log10') and transform(x, 'log10p'), respectively. The log10p transformation always implements $\log10(1 + x)$. The log10 transformation is applied as $\log10(1 + x)$ if the data contains zeros. The following commands perform log10 transformation.

```
> physeq_log10 <- microbiome::transform(physeq, "log10")
> head(otu_table(physeq_log10),3)
OTU Table:        [3 taxa and 26 samples]
             taxa are rows
    CL3 CC1 SV1 M31Fcsw M11Fcsw M31Plmr  M11Plmr F21Plmr M31Tong M11Tong
549322 0 0 0    0     0      0 0.000       0      0      0
522457 0 0 0    0     0      0 0.000       0      0      0
951    0 0 0    0     0      0 0.301       0      0      0
------
```

For compositional approach of microbiome data analysis, the central log-ratio (CLR) transformation is often used. Since log zero is undefined, CLR transformation usually adds a pseudocount to avoid zeros, such as 0.5, 1, or a pseudocount of min (relative abundance)/2 to exact zero relative abundance entries in OTU table before taking logs. Some software provides an alternative method to impute the zero-inflated unobserved values. For example, the **zCompositions** R package has options to choose a multiplicative Kaplan-Meier smoothing spline (KMSS) replacement (by implementing the **multKM()** function), multiplicative lognormal replacement (via the **multLN ()** function), or multiplicative simple replacement (via the **multRepl()** function). By implementing these functions, in practice, at least draw 1000 (setting n.draws = 1000).

```
> physeq_clr <- microbiome::transform(physeq, "clr")
> head(otu_table(physeq_clr),3)
OTU Table:       [3 taxa and 26 samples]
        taxa are rows
     CL3    CC1   SV1 M31Fcsw M11Fcsw M31Plmr M11Plmr F21Plmr M31Tong
549322 -1.503 -1.629 -1.247 -0.416 -0.3503 -0.6268 -0.9298 -0.6795
-0.3541
522457 -1.503 -1.629 -1.247 -0.416 -0.3503 -0.6268 -0.9298 -0.6795
-0.3541
951 -1.503 -1.629 -1.247 -0.416 -0.3503 -0.6268 1.5438 -0.6795 -0.3541
------
```

Other common transformations are also available in the microbiome package including "Z" and "shift."

2.3.2.9   Export phyloseq Data into CSV Files

First, we find the slot names by implementing the slotNames() function to phyloseq object "physeq."

```
> slotNames(physeq)
[1] "otu_table" "tax_table" "sam_data" "phy_tree" "refseq"
```

Then, we define otu_table, tax_table, and sam_data as data frames.

```
> otu_df = as.data.frame(physeq@otu_table)
> tax_df = as.data.frame(physeq@tax_table)
> sam_df = as.data.frame(physeq@sam_data)
```

Finally, we export otu_table, tax_table, and sam_data using the readr package.

```
> library(readr)
> # Write otu_table data to a csv file
> write_csv(otu_df, file = "otu_tab_GlobalPatterns.csv")
> # Write tax_table data to a csv file
> write_csv(tax_df, file = "tax_tab_GlobalPatterns.csv")
> # Write sam_table data to a csv file
> write_csv(sam_df, file = "sam_tab_GlobalPatterns.csv")
```

2.3.2.10 Import CSV, Mothur, and BIOM Format Files into a phyloseq Object

We can use the **read_phyloseq()** function to read the otu_table, taxonomy_table, and metadata_table with .csv, mothur, and BIOM formats. The syntax is given:

```
read_phyloseq(otu.file = NULL, taxonomy.file = NULL,
metadata.file = NULL, type = c("simple", "mothur", "biom"), sep = ",")
```

where **otu.file** is used to specify the file containing the OTU table (for mothur this is the file with the .shared extension); **taxonomy.file** is used to specify the file containing the taxonomy table (for mothur this is typically the consensus taxonomy file with the .taxonomy extension); **metadata.file** is used to specify the file containing samples by variables metadata; **type** is used to specify the input data type: either "mothur" or "simple" or "biom" type; and **sep** is for CSV file separator. The **read_phyloseq()** function returns a phyloseq-class object. Here, we illustrate read CSV files into a phyloseq-class object.

```
> library(microbiome)
> # Read CSV files into R
> physeq_csv <- read_phyloseq(otu.file = "otu_table_GlobalPatterns.
csv",
+               taxonomy.file = "tax_table_GlobalPatterns.csv",
+               metadata.file = "metadata_GlobalPatterns.csv",
+               type= "simple")


> physeq_csv
phyloseq-class experiment-level object
otu_table() OTU Table:      [ 19216 taxa and 26 samples ]
sample_data() Sample Data:  [ 26 samples by 7 sample variables ]
tax_table() Taxonomy Table: [ 19216 taxa by 8 taxonomic ranks ]
```

2.3.3 ampvis2

The R package **ampvis2** was developed by Andersen et al. to analyze and visualize 16S rRNA amplicon data (Andersen et al. 2018) (current version 2.7.17, March 2022). Like the **phyloseq** and **microbiome** packages, ampvis2 mainly wraps statistical functions from **vegan** package and **ggplot2** graphical package. The attractiveness of ampvis2 is that its syntax of functions is very simple, and their plots such as ordination have very visualized effects. We will introduce ampvis2 to calculate alpha diversity (in Chap. 9), to calculate beta diversity, and to perform ordination plots (in Chap. 10).

2.3.4   curatedMetagenomicData

Shotgun metagenomic sequencing has provided amount of data for studying the taxonomic composition and functional potential of the human microbiome. However, various challenges prevent researchers from taking full advantage of these resources. The Bioconductor package **curatedMetagenomicData** (Pasolli et al. 2017) was developed to overcome these challenges and help the more shotgun data available publicly for using to perform hypothesis testing and meta-analysis (current version 3.0.10, March, 2022). The curatedMetagenomicData data package was developed for distribution through the Bioconductor (Huber et al. 2015) ExperimentHub platform. The database of curatedMetagenomicData so far provides thousands of uniformly processed and manually annotated human microbiome profiles (Pasolli et al. 2017), including bacterial, fungal, archaeal, and viral taxonomic abundances and gene marker presence and absence from MetaPhlAn2 (Truong et al. 2015), in addition to quantitative metabolic functional profiles and standardized metadata, including coverage and abundance of metabolic pathways and gene families abundance from HUMAnN2 (Abubucker et al. 2012). Metagenomic data with matched health and socio-demographic data are provided as Bioconductor **ExpressionSet** objects, with options for automatic conversion of taxonomic profiles to **phyloseq** or **metagenomeSeq** objects for microbiome-specific analyses. In this book, we will not implement this package; the interested readers are referred to (Lucas Schiffer and Waldron 2021).

2.4   Some R Packages for Analysis of Phylogenetics

Evolutionary analysis (analyses of phylogenetic trees) provides insights of phylogenetics and evolution that facilitate comparative biology and microbiome study. In this section, we introduce three phylogenetic R packages: **ape**, **phytools**, and **castor**. Each package provides various functions for writing, reading, plotting, and manipulating, and even inferring phylogenetic trees. However, we focus on phylogenetic tree generation, writing, and reading and provide simple tree plotting. For building phylogenetic tree in QIIME 2, the reader is referred to Chap. 5 (Sect. 5.2).

Example 2.7: Diet Swap Between Rural and Western Populations, Example 2.4, Cont.

In Sect. 2.3.1.1, we have created a phyloseq object using Example 2.4 (Diet swap between rural and western populations). Here we continue to use this dataset to illustrate the analyses of phylogenetics.

```
> physeq
phyloseq-class experiment-level object
otu_table() OTU Table: [ 37 taxa and 222 samples ]
sample_data() Sample Data: [ 222 samples by 7 sample variables ]
tax_table() Taxonomy Table: [ 37 taxa by 3 taxonomic ranks ]
```

2.4.1   Ape

The R package **ape** stands for Analyses of Phylogenetics and Evolution and is a modern and useful software for the study of biodiversity and evolution (Paradis and Schliep 2018)(current version 5.6.2, March 2022).

The ape package provides functions for reading, writing, plotting, and manipulating phylogenetic trees, conducting analyses of comparative data in a phylogenetic framework, ancestral character, diversification and macroevolution, computing distances from DNA sequences, reading and writing nucleotide sequences, as well as performing other functionality such as Mantel's test, generalized skyline plots. Here, we illustrate how to create and write a random tree using the function **rtree()**. This function generates a tree by splitting randomly the edges until the specified number of tips is obtained. One syntax of the rtree() is given:

```
rtree(n, rooted = TRUE, tip.label = NULL, br = runif, equiprob = FALSE)
```

where the argument **n** is an integer giving the number of tips in the tree; **rooted** is a logical value that is used to indicate whether the tree should be rooted (the default); **tip.label** is a character vector that is used to specify the tip labels with the tips "t1," "t2," ..., being given if not specified. The argument **br** is used to specify an R function used to generate the branch lengths; use NULL to simulate only a topology. The argument **equiprob** (new since ape version 5.4.1) is a logical value that is used for specifying whether topologies are generated in equal frequencies. If equiprob = FALSE is specified, then the unbalanced topologies are generated in higher proportions than the balanced ones. The function rtree() returns an object of class "phylo" (Fig. 2.8).

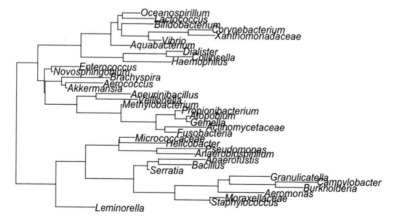

Fig. 2.8 A random phylogenetic tree generated by the ape package using the Diet swap between rural and western populations

Step 1: Generate random rooted and unrooted trees.

```
library("ape")
> # Generate a random rooted tree
random_tree = rtree(ntaxa(physeq), rooted=TRUE, tip.
label=taxa_names(physeq))
> # Generate a random unrooted tree
> random_unrooted_tree = rtree(ntaxa(physeq), rooted=FALSE, tip.
label=taxa_names(physeq))
```

Step 2: Plot the random trees.

```
> # Figure 2.8
> # Plot this random tree
plot(random_tree)
```

Step 3: Write the phylogenetic trees.

The generated tree can be easily written to and read back from files. The following commands write a rooted and an unrooted phylogenetic tree using the function write.tree():

```
> # Write the rooted random tree
> write.tree(random_tree, "Rooted_tree_dietswap.tre")
> # Write the unrooted random tree
> write.tree(random_unrooted_tree, "Unrooted_tree_dietswap.tre")
```

Step 4: Read back the phylogenetic trees.

The following commands read back a rooted and an unrooted phylogenetic tree using the function readLines ():

```
> cat(readLines("Rooted_tree_dietswap.tre"))
> cat(readLines("Unrooted_tree_dietswap.tre"))
```

2.4.2 Phytools

The R package **phytools** (Liam J. Revell 2022; Liam J Revell 2012) developed by Liam J. Revell is phylogenetic tools for comparative biology (and other things) (current Version 1.0–1, January, 2022). It provides functions in phylogenetic comparative biology and numerous methods for visualizing, manipulating, reading or writing, and even inferring phylogenetic trees.

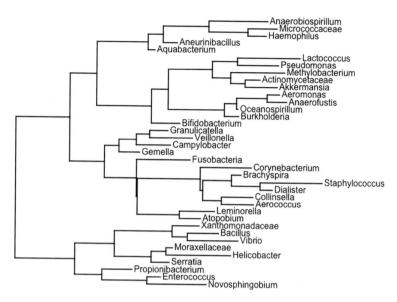

Fig. 2.9 The rooted phylogenetic tree plotted by the plotTree () function via the phytools package using dietswap data

Step 1: Generate random rooted and unrooted trees.

In Sect. 2.4.1, we have generated a random tree named as "random_tree" using the following commands (Fig. 2.9).

```
> # Generate a random rooted tree
> random_tree = ape::rtree(ntaxa(physeq), rooted=TRUE, tip.
label=taxa_names(physeq))
```

Step 2: Write the phylogenetic trees

The following commands write out this tree using the function writeNexus() from the **phytools** package.

This function writes a tree to file in Nexus format. It is redundant with ape::write. nexus(). The syntax is writeNexus(tree, "file = ""), where the argument **tree** is the object of class "phylo" or "multiPhylo";

file is the file name for output.

```
> library(phytools)
> library(ape)
> library(maps)
> writeNexus(random_tree, "Rooted_tree_dietswap.nex")
```

Step 3: Read back the phylogenetic trees.

The following commands read trees from files in the directory.

```
> cat(readLines("Rooted_tree_dietswap.nex"), sep = "\n")
```

Step 4: Plot the phylogenetic trees.

We can use the **plotTree()** function to plot rooted phylogenetic tree (a rooted phylogram).

```
> Figure 2.9
> plotTree(random_tree)
```

2.4.3 Castor

The class "phylo" in phylogenetic tree is the most important data unit. It was introduced in ape in August 2002. The class "phylo" with the tree topology encoded in the member variable `tree.edge`. Most existing phylogenetic packages have been designed using the "phylo" format for much smaller trees containing at most a few thousand tips, and thus scale poorly to large datasets (Louca and Doebeli 2017). Unlike these existing phylogenetic packages, the R package **castor** (Louca and Doebeli 2017) was developed to compare phylogenetics on large phylogenetic trees ($>$10,000 tips)(current version 1.7.2, March, 2022). It provides efficient implementations of common phylogenetic functions, focusing on analysis of trait evolution on fixed trees. This package provides four kinds of functions: (1) efficient tree manipulation functions including pruning, rerooting, and calculation of most-recent common ancestors; (2) calculating distances from the tree root and pairwise distance matrices; (3) calculating phylogenetic signal and mean trait depth (trait conservatism); and (4) providing efficient ancestral state reconstruction and hidden character prediction of discrete characters on phylogenetic trees using maximum likelihood and maximum parsimony methods. Below we introduce how to write and read a tree in Newick (parenthetic) format using this package. For more information about Newick format, see Chap. 3 (Sect. 3.1.4).

To install the latest version of this package, enter the following commands in R or RStudio.

```
> install.packages("castor")
> library(castor)
```

Step 1: Write the phylogenetic trees.

The **write_tree()** function writes a phylogenetic tree to a file or a string, in Newick (parenthetic) format. If the tree is unrooted, it is first rooted internally at the first node. The syntax of this function is given:

```
write_tree (tree, file = "", append = FALSE, digits = 10, quoting =
0, include_edge_labels = FALSE,
include_edge_numbers = FALSE)
```

The arguments: **tree** is a tree of class "phylo"; **file is** an optional path to a file, to which the tree should be written. The file may be overwritten without warning. If left empty (default), then a string is returned representing the Newick format tree; otherwise, the tree is directly written to the file. The argument **append** is a logical value that is used for specifying whether the tree should be appended at the end of the file, rather than replacing the entire file (if it exists). **digits** is an integer that is used to specify the number of significant digits for writing edge lengths; **quoting** is an integer, specifying whether and how to quote tip/node/edge names with 0, no quoting at all; 1, always use single quotes; 2, always use double quotes; −1, only quote when needed and prefer single quotes if possible; −2, only quote when needed and prefer double quotes if possible. The arguments **include_edge_labels** and **include_edge_numbers** are the logical values; they are used for specifying whether to include edge labels (if available) in the output tree, inside square brackets, edge numbers (if available) in the output tree, inside curly braces, respectively. These two arguments are an extension (Matsen et al. 2012) to the standard Newick format. This function **write_tree()** is comparable to (but typically much faster than) the ape function **write.tree()**.

```
# Obtain a string representation of the tree in Newick format
> Newick_string = write_tree(random_tree)
> Newick_string
```

Step 2: Read back a tree from a string or file.

The **read_tree()** function loads a tree from a string or file in Newick (parenthetic) format. The syntax is given:

```
read_tree( string = "", file = "", edge_order = "cladewise",
include_edge_lengths = TRUE, look_for_edge_labels = FALSE,
look_for_edge_numbers = FALSE, include_node_labels = TRUE,
underscores_as_blanks = FALSE, check_label_uniqueness = FALSE,
interpret_quotes = FALSE, trim_white = TRUE)
```

where **string** is a character containing a single tree in Newick format; **file** is character that is used to specify a path to an input text file containing a single tree in Newick format. The argument **edge_order** is either "cladewise" or "pruningwise," specifying the order in which edges should be listed in the returned tree. For other arguments of read_tree(), the readers are referred to the manual of the castor package.

```
> # Re-parse tree from string
> parsed_tree = read_tree(Newick_string)
> parsed_tree
Phylogenetic tree with 37 tips and 36 internal nodes.
Tip labels:
 Anaerobiospirillum, Micrococcaceae, Haemophilus, Aneurinibacillus,
 Aquabacterium, Lactococcus, ...
Rooted; includes branch lengths.
```

2.5   BIOM Format and Biomformat Package

The Biological Observation Matrix (BIOM) format(McDonald et al. 2012) or the BIOM file format (canonically pronounced *biome*) was designed to be a general-use format for representing biological sample by observation contingency tables (i.e., OTU tables, metagenome tables, or a set of genomes). BIOM is a recognized standard for the Earth Microbiome Project and is a Genomics Standards Consortium supported project. See more information on BIOM format in Chap. 3 (Sect. 3.1.3).

The **biomformat** package (McMurdie and Paulson 2021) is a utility package. Its use depends on other packages. It provides I/O functionality and functions to facilitate data processing from BIOM format files, such as providing tools to access data in "data.frame," "matrix," and "Matrix" classes.

To install this package, start R (version 4.1) and enter:

```
if (!require("BiocManager", quietly = TRUE))
install.packages("BiocManager")
BiocManager::install("biomformat")

> library("biomformat");
packageVersion("biomformat")
[1] '1.20.0'
```

Step 1: Read BIOM format file into R.

The following commands import BIOM formats of otu-table into R using the function **read_biom** ().

```
> setwd("~/Documents/QIIME2R/Ch2_IntroductionToR")
> otu_tab_biom <- "seq_table_L7.biom"
> biom_obj <- read_biom(otu_tab_biom)
> biom_obj
```

```
biom object.
type: OTU table
matrix_type: sparse
637 rows and 82 columns
```

The contents of BIOM data are briefly summarized above.

Step 2: Access BIOM data.

The biomformat package includes accessor functions. We can use them to access the data with BIOM format files.

Access Core Observation Data

The core "observation" data is stored in either sparse or dense matrices with the BIOM format. The sparse matrix support is carried with R via the Matrix package. We can use the **biom_data()** function to access the BIOM format matrices.

```
> head(biom_data(biom_obj),3)
3 x 82 sparse Matrix of class "dgCMatrix"
  [[ suppressing 82 column names 'Sun071.PG1', 'Sun027.BF2', 'Sun003.BA3'
... ]]

D_0__Archaea;D_1__Thaumarchaeota;D_2__Nitrososphaeria;
D_3__Nitrososphaerales;D_4__Nitrososphaeraceae;Other;Other .
D_0__Bacteria;D_1__Acidobacteria;D_2__Acidobacteriia;
D_3__Solibacterales;D_4__Solibacteraceae (Subgroup 3);
D_5__Bryobacter;Other .
D_0__Bacteria;D_1__Acidobacteria;D_2__Acidobacteriia;
D_3__Solibacterales;D_4__Solibacteraceae (Subgroup 3);Other;Other
```

Obviously the accessed results from the fancier sparse Matrix of class are not easy to read. We can use the **as()** function to easily coerce the data to the simple, standard "matrix" class.

```
> head(as(biom_data(biom_obj), "matrix"),3)
                                            Sun071.PG1 Sun027.BF2
Sun003.BA3
D_0__Archaea;D_1__Thaumarchaeota;D_2__Nitrososphaeria;
D_3__Nitrososphaerales;D_4__Nitrososphaeraceae;Other;Other
0      0     0
D_0__Bacteria;D_1__Acidobacteria;D_2__Acidobacteriia;
D_3__Solibacterales;D_4__Solibacteraceae (Subgroup 3);
D_5__Bryobacter;Other      0     0     0
D_0__Bacteria;D_1__Acidobacteria;D_2__Acidobacteriia;
D_3__Solibacterales;D_4__Solibacteraceae (Subgroup 3);Other;Other
0      0     0
```

Access Observation Metadata

Observation metadata is metadata associated with the individual units being counted/recorded in a sample. For microbiome census data, it is often a taxonomic classification and something about a particular OTU/species. Here, the observations are counts of OTUs and the metadata is taxonomic classification, if present. In this case, the absence of observation metadata is reported. We can access observation metadata using the **observation_metadata** () function.

```
> observation_metadata(biom_obj)
NULL
```

Access Sample Metadata

Sample metadata describe the properties of the samples. Similarly, we can access sample metadata using the **sample_metadata()** function.

```
> sample_metadata(biom_obj)
NULL
```

Step 3: Write BIOM format data.

The following commands write the biom objects to a temporary file using the **write_biom** () function and then read it back using the read_biom () function and store it as variable infile. Next it checks whether these two objects are identical using the identical() function.

```
> outfile_biom = tempfile()
> write_biom(biom_obj, outfile_biom )
> infile = read_biom(outfile_biom)
> identical(biom_obj, infile)
[1] FALSE
```

Step 4: Perform visualization and analysis on BIOM data.

Since the BIOM format files can be accessed, we can visualize and analyze these data using other R functions such as plots.

```
> plot(biom_data(biom_obj))
> boxplot(as(biom_data(biom_obj), "vector"))
> heatmap(as(biom_data(biom_obj), "matrix"))
```

2.6 Creating Analysis Microbiome Dataset

Example 2.8: Vaginal Microbiome Study

DiGiulio et al. (2015) conducted a longitudinal vaginal microbiome study to investigate the bacterial taxonomic composition for pregnant and postpartum women. This case-control study included 49 pregnant women, 15 of whom delivered preterm. The discovery data of 3767 specimens from 40 women were collected prospectively and weekly during gestation and monthly after delivery from the vagina, distal gut, saliva, and tooth/gum. The final dataset for pregnant women who delivered at term and preterm contained a total of 1271 taxa from 3432 specimens. Three datasets were downloaded and named as "TemporalSpatialOTU.csv," "TemporalSpatialTaxonomy.csv," and "TemporalSpatialClinicalMerged.csv," respectively. The OTU data are available at species level. The taxonomy data are ranked at 7 levels: Kingdom, Phylum, Class, Order, Family, Genus, and Species. The available clinical information included race, weeks/days at the collection of samples, way of delivery, and household income level. The clinical data of 9 pregnant women in validated dataset is not available.

The study found that those women who with microbiome profiles had high abundances of *Lactobacillus* were less likely to experience preterm births, defined as delivery before gestational weeks. The study also found that those women who had high amounts of *Prevotella* had a higher occurrence of preterm births.

These datasets have been used by others to illustrate their proposed models including Zhang et al. (Zhang et al. 2018). In this section, we illustrate how to create a microbiome dataset that is suitable for longitudinal data analysis. We illustrate the dataset derivation by the following 6 steps.

Step 1: Load raw microbiome datasets.

We first use **read.csv()** function to load the three raw pregnancy microbiome datasets including OTU, taxonomy, and clinical data.

```
> setwd("~/Documents/QIIME2R/Ch2_IntroductionToR")
> otu_tab <- read.csv("TemporalSpatialOTU.csv", row.names = 1, check.
names = F)
> tax_tab <- read.csv("TemporalSpatialTaxonomy.csv", row.names = 1)
> sam_tab <- read.csv("TemporalSpatialClinicalMerged.csv", row.names
= 1)
```

Step 2: Prune taxa using the gsub() function, and string pattern matching and replacement.

The goal of pruning taxa is to make data more readable. The original names of taxonomy are split by ":" and "-". We want to replace them with "_".

```
> head(tax_tab,3)
    Kingdom Phylum Class Order Family Genus Species
4330849 Bacteria P:SR1 P:SR1 P:SR1 P:SR1 P:SR1 P:SR1
4400869 Bacteria P:SR1 P:SR1 P:SR1 P:SR1 P:SR1 P:SR1
4457086 Bacteria P:GN02 C:BD1-5 C:BD1-5 C:BD1-5 C:BD1-5
```

We can use the functions **apply()** and **gsub()** as well as the string pattern matching and replacement techniques to replace ":" and "-" with "_" (Xia et al. 2018).

```
> tax_tab <- apply(tax_tab, 2, function(x) gsub(':|-', '_', x))
> head(tax_tab,3)
    Kingdom Phylum Class Order Family Genus Species
4330849 "Bacteria" "P_SR1" "P_SR1" "P_SR1" "P_SR1" "P_SR1" "P_SR1"
4400869 "Bacteria" "P_SR1" "P_SR1" "P_SR1" "P_SR1" "P_SR1" "P_SR1"
4457086 "Bacteria" "P_GN02" "C_BD1_5" "C_BD1_5" "C_BD1_5" "C_BD1_5"
"C_BD1_5"
```

Step 3: Use the phyloseq and Tidyverse packages to filter and subset taxonomy and sample data.

Combining the **phyloseq** and **tidyverse** packages is an effective way to filter and subset microbiome data. We first create a **phyloseq** object using the **phyloseq** package, and then process the filtering and subsetting with **tidyverse** package.

```
> library(phyloseq)
> OTU <- otu_table(otu_tab, taxa_are_rows = FALSE)
> TAX <- tax_table(as.matrix(tax_tab))
> SAM <- sample_data(sam_tab)


> table(SAM$Preg, SAM$BodySite)
      Saliva Stool Tooth_Gum Vaginal_Swab
 FALSE   167   152     147        166
 TRUE    740   580     719        761


> table(SAM$Outcome, SAM$BodySite)
            Saliva Stool Tooth_Gum Vaginal_Swab
 Marginal     108    92     110         111
 Preterm       82    42      82          82
 Term         668   557     623         684
 VeryPreterm   49    41      51          50


> # Genus level
> library(tidyverse)
> # Keep only samples collected during pregnancy
```

```
> # Keep only samples in vaginal swab
> genus <- phyloseq(OTU, TAX, SAM) %>%
+ tax_glom(taxrank = 'Genus') %>% # Combine taxa at genus level
+ subset_samples(Outcome != 'Marginal') %>%# Same as in the original
paper
+ subset_samples(BodySite == 'Vaginal_Swab') %>% # Look only at vaginal
samples
+ subset_samples(GDColl <= GDDel) # Remove samples collected after
delivery
```

```
> genus
phyloseq-class experiment-level object
otu_table()  OTU Table:   [ 269 taxa and 678 samples ]
sample_data() Sample Data: [ 678 samples by 64 sample variables ]
tax_table()  Taxonomy Table: [ 269 taxa by 7 taxonomic ranks ]
```

From the above phyloseq object, we can see there are 678 samples with 64 sample variables. The original paper analyzed the abundance data at species level and divided the samples into 5 Vaginal Community State Types (CSTs) and only analyzed samples with a *Lactobacillus*-poor vaginal community state type (CST 4). Here, for the illustration of data derivation, we averaged the abundance data at genus level and did not use the complete same criteria for sample filtering.

Step 4: Use the Tidyverse and dplyr packages to finalize analysis dataset.

There are 64 sample variables in genus sample data. We can check the variable names using colnames(sample_data(genus)) or head(sample_data(genus)).

```
> colnames(sample_data(genus))
 [1] "SampleID"
 [2] "BarcodeSequence"
 [3] "LinkerPrimerSequence"
 [4] "BodySite"
 [5] "SubjectID"
------
```

The following procedure is used to create or rename the variables. This is easy to do using **tidyverse** package combining with the **mutate()** function.

```
> library(tidyverse)
> # Final metadata
> meta_tab <- data.frame(sample_data(genus)) %>%
+ filter(GDColl >= 0) %>%
+ mutate(Weeks = GWColl) %>%
+ mutate(N = NumReads) %>% # Total reads: some models also use
LibrarySize
+ mutate(Subject = SubjectID) %>%
+ mutate(Preterm = (Outcome != 'Term')) %>%
+ mutate(Delivery = 1) %>%
+ mutate(TrimColl = factor(TrimColl))
```

```
> dim(meta_tab)
[1] 678 68
```

Finally, for ease of processing, we want to remove the variables that we do not use in the analysis. Although the **tidyverse** package contains the **dplyr** package, sometimes they are masked. In this case, we explicitly call the **select()** function of **dplyr** using the function **dplyr::select()**.

```
> # Obtain final meta dataset using the setect() function from the dplyr
package
> head(meta_genus<-dplyr::select(meta_tab,Subject,BodySite, GDColl,
GWColl,GDDel,GWDel,
+ Weeks,N,NumReads,TrimColl,Race,Ethnicity,LibrarySize,PrePreg,
Preg,Term,Preterm,Outcome),3)
  Subject BodySite GDColl GWColl GDDel GWDel Weeks N NumReads TrimColl
1 10003 Vaginal_Swab  198   29   260   38  29 2341  2341      3
2 10003 Vaginal_Swab  205   30   260   38  30 1136  1136      3
3 10003 Vaginal_Swab  212   31   260   38  31 2344  2344      3
        Race   Ethnicity LibrarySize PrePreg Preg Term Preterm Outcome
1 American Indian Non-hispanic     2347  FALSE TRUE TRUE  FALSE   Term
2 American Indian Non-hispanic     1144  FALSE TRUE TRUE  FALSE   Term
3 American Indian Non-hispanic     2356  FALSE TRUE TRUE  FALSE   Term

> colnames(meta_genus)
 [1] "Subject"   "BodySite"  "GDColl"   "GWColl"   "GDDel"
 [6] "GWDel"     "Weeks"     "N"        "NumReads" "TrimColl"
[11] "Race"      "Ethnicity" "LibrarySize" "PrePreg"  "Preg"
[16] "Term"      "Preterm"   "Outcome"
> dim(meta_genus)
[1] 678 18
> dim(otu_genus)
[1] 678 269
> dim(meta_genus)
[1] 678 18
> tax_genus<-as.data.frame(tax_table(genus))
> dim(tax_genus)
[1] 269  7
```

Step 5: Use the Readr package to write the final analysis datasets to csv and tsv files.

```
> # Write otu_genus data to a csv file
> write_csv(otu_genus, file = "otu_genus.csv")


> # Write tax_genus data to a csv file
> write_csv(tax_genus, file = "tax_genus.csv")
```

```
> # Write meta_genus data to a csv file
> write_tsv(meta_genus, file = "meta_genus.csv")
```

```
> # Write meta_genus data to a tsv file
> write_tsv(meta_genus, file = "meta_genus.tsv")
```

For some readers, this example may be too complicated at the beginning chapter. Do not worry, you can skip these materials. When you read further in the later chapters, you can go back to look at these programs again; you will find that they are very useful.

2.7 Summary

Chapter 2 described and introduced some useful R functions, R packages, specifically designed R packages for microbiome data, and some R packages for analysis of phylogenetics, as well as BIOM format and biomformat package. The R programming language was created by statisticians Ross Ihaka and Robert Gentleman for statistical computing and graphics supported by the R Core Team and the R Foundation for Statistical Computing. R first appeared in August 1993 and the stable version 4.1.2 was released on November 1, 2021. R is available as open-source free software and is getting popular in academic research for statistical analysis and computing, and specifically is a main tool for development of statistical methods and models for microbiome and other omics studies. This chapter provided 6 sections on the R functions and packages for microbiome data analysis, including (1) useful R functions for saving, reading, and downloading data; (2) R utility and data visualization packages; (3) specifically designed R packages for microbiome data with focusing on the phyloseq and microbiome packages; (4) some R packages for analysis of phylogenetics; (5) BIOM format and biomformat package; and (6) creating analysis microbiome dataset. In Chap. 3, we will introduce some basic data processing operations in QIIME 2.

References

Abubucker, Sahar, Nicola Segata, Johannes Goll, Alyxandria M. Schubert, Jacques Izard, Brandi L. Cantarel, Beltran Rodriguez-Mueller, Jeremy Zucker, Mathangi Thiagarajan, Bernard Henrissat, Owen White, Scott T. Kelley, Barbara Methé, Patrick D. Schloss, Dirk Gevers, Makedonka Mitreva, and Curtis Huttenhower. 2012. Metabolic reconstruction for metagenomic data and its application to the human microbiome. *PLoS Computational Biology* 8 (6): e1002358. https://doi.org/10.1371/journal.pcbi.1002358.
Andersen, Kasper S., Rasmus H. Kirkegaard, Søren M. Karst, and Mads Albertsen. 2018. ampvis2: An R package to analyse and visualise 16S rRNA amplicon data. *BioRxiv* 299537.
Anderson, Edgar. 1935. The irises of the Gaspe peninsula. *Bull. Am. Iris Soc.* 59: 2–5.

Caporaso, J. Gregory, Christian L. Lauber, William A. Walters, Donna Berg-Lyons, Catherine
    A. Lozupone, Peter J. Turnbaugh, Noah Fierer, and Rob Knight. 2011. Global patterns of 16S
    rRNA diversity at a depth of millions of sequences per sample. *Proceedings of the National
    Academy of Sciences* 108 (Supplement 1): 4516–4522.
DiGiulio, D.B., B.J. Callahan, P.J. McMurdie, E.K. Costello, D.J. Lyell, A. Robaczewska,
    C.L. Sun, D.S. Goltsman, R.J. Wong, G. Shaw, D.K. Stevenson, S.P. Holmes, and
    D.A. Relman. 2015. Temporal and spatial variation of the human microbiota during pregnancy.
    Proceedings of the National Academy of Sciences of the United States of America 112 (35):
    11060–11065. https://doi.org/10.1073/pnas.1502875112.
Fisher, Ronald A. 1936. The use of multiple measurements in taxonomic problems. *Annals of
    Eugenics* 7 (2): 179–188.
Ordination methods, diversity analysis and other functions for community and vegetation ecolo-
    gists. http://CRAN.R-project.org/package=vegan
Huber, Wolfgang, Vincent J. Carey, Robert Gentleman, Simon Anders, Marc Carlson, Benilton
    S. Carvalho, Hector Corrada Bravo, Sean Davis, Laurent Gatto, Thomas Girke, Raphael
    Gottardo, Florian Hahne, Kasper D. Hansen, Rafael A. Irizarry, Michael Lawrence, Michael
    I. Love, James MacDonald, Valerie Obenchain, Andrzej K. Oleś, Hervé Pagès, Alejandro
    Reyes, Paul Shannon, Gordon K. Smyth, Dan Tenenbaum, Levi Waldron, and Martin Morgan.
    2015. Orchestrating high-throughput genomic analysis with Bioconductor. *Nature Methods*
    12 (2): 115–121. https://doi.org/10.1038/nmeth.3252. https://pubmed.ncbi.nlm.nih.gov/25633
    503; https://www.ncbi.nlm.nih.gov/pmc/articles/PMC4509590/.
Jari Oksanen, F.G.B., Michael Friendly, Roeland Kindt, Pierre Legendre, Dan McGlinn, Peter
    R. Minchin, R.B. O'Hara, Gavin L. Simpson, Peter Solymos, M. Henry, and H. Stevens. 2018.
    Vegan: Community ecology package. *R Package Version* 2 (6).
Jari Oksanen, F., Guillaume Blanchet, Michael Friendly, Roeland Kindt, Pierre Legendre, Dan
    McGlinn, Peter R. Minchin, R.B. O'Hara, Gavin L. Simpson, Peter Solymos, M. Henry
    H. Stevens, Eduard Szoecs, and Helene Wagner. 2019. *Vegan: Community ecology package.*
    R Package Version 2.
Kassambara, Alboukadel. 2020 June 27. ggpubr: 'ggplot2' based publication ready plots. https://
    cran.r-project.org/web/packages/ggpubr/index.html
Lahti, Leo, Sudarshan Shetty, et al. 2017. *Tools for microbiome analysis*. R. Version 1.9.95.
Louca, Stilianos, and Michael Doebeli. 2017. Efficient comparative phylogenetics on large trees.
    Bioinformatics 34 (6): 1053–1055. https://doi.org/10.1093/bioinformatics/btx701.
Matsen, Frederick A., Noah G. Hoffman, Aaron Gallagher, and Alexandros Stamatakis. 2012. A
    format for phylogenetic placements. *PLoS One* 7 (2): e31009. https://doi.org/10.1371/journal.
    pone.0031009.
McDonald, Daniel, Jose C. Clemente, Justin Kuczynski, Jai Ram Rideout, Jesse Stombaugh, Doug
    Wendel, Andreas Wilke, Susan Huse, John Hufnagle, Folker Meyer, Rob Knight, and
    J. Gregory Caporaso. 2012. The biological observation matrix (BIOM) format or: How I learned
    to stop worrying and love the ome-ome. *GigaScience* 1 (1). https://doi.org/10.1186/2047-217x-
    1-7.
McMurdie, Paul J., and Susan Holmes. 2013. phyloseq: An R package for reproducible interactive
    analysis and graphics of microbiome census data, *PLOS ONE*. (8, 4): e61217. https://doi.org/10.
    1371/journal.pone.0061217.
———. 2022. "Handling and analysis of high-throughput microbiome census data." Accessed
    March 4, 2022. https://www.bioconductor.org/packages/release/bioc/html/phyloseq.html
McMurdie, Paul J., and Joseph N. Paulson. 2021. *Biomformat: An interface package for the BIOM
    file format*. R/Bioconductor Package Version 1.23.0. Last Modified October 26, 2021. Accessed
    5 Mar 2022.
O'Keefe, Stephen J.D., Jia V. Li, Leo Lahti, Ou Junhai, Franck Carbonero, Khaled Mohammed,
    Joram M. Posma, James Kinross, Elaine Wahl, Elizabeth Ruder, Kishore Vipperla, Vasudevan
    Naidoo, Lungile Mtshali, Sebastian Tims, Philippe G.B. Puylaert, James DeLany, Alyssa
    Krasinskas, Ann C. Benefiel, Hatem O. Kaseb, Keith Newton, Jeremy K. Nicholson, Willem

M. de Vos, H. Rex Gaskins, and Erwin G. Zoetendal. 2015. Fat, fibre and cancer risk in African Americans and rural Africans. *Nature Communications* 6 (1): 6342. https://doi.org/10.1038/ncomms7342.

O'Keefe, Stephen J. D. et al. 2016. *Data from: Fat, fibre and cancer risk in African Americans and rural Africans, Dryad*, Dataset. Stephen J. D. et al. O'Keefe: Dryad.

Paradis, Emmanuel, and Klaus Schliep. 2018. Ape 5.0: An environment for modern phylogenetics and evolutionary analyses in R. *Bioinformatics* 35 (3): 526–528. https://doi.org/10.1093/bioinformatics/bty633.

Pasolli, Edoardo, Lucas Schiffer, Paolo Manghi, Audrey Renson, Valerie Obenchain, Duy Tin Truong, Francesco Beghini, Faizan Malik, Marcel Ramos, Jennifer B. Dowd, Curtis Huttenhower, Martin Morgan, Nicola Segata, and Levi Waldron. 2017. Accessible, curated metagenomic data through ExperimentHub. *Nature Methods* 14 (11): 1023–1024. https://doi.org/10.1038/nmeth.4468.

Revell, Liam J. 2012. Phytools: An R package for phylogenetic comparative biology (and other things). *Methods in Ecology and Evolution* 2: 217–223.

Revell, Liam J. 2022. Phytools: phylogenetic tools for comparative biology (and other things). https://cran.r-project.org/web/packages/phytools/index.html.

Schiffer, Lucas, and Levi Waldron. 2021. *curatedMetagenomicData*. Last Modified 22 December 2021. Accessed 6 Mar 2022. https://bioconductor.org/packages/release/data/experiment/vignettes/curatedMetagenomicData/inst/doc/curatedMetagenomicData.html

Truong, Duy Tin, Eric A. Franzosa, Timothy L. Tickle, Matthias Scholz, George Weingart, Edoardo Pasolli, Adrian Tett, Curtis Huttenhower, and Nicola Segata. 2015. MetaPhlAn2 for enhanced metagenomic taxonomic profiling. *Nature Methods* 12 (10): 902–903. https://doi.org/10.1038/nmeth.3589.

Wickham, Hadley. 2016. *ggplot2: Elegant graphics for data analysis*. New York: Springer-Verlag.

———. 2020. Tidyr: Tidy messy data https://CRAN.R-project.org/package=tidyr

Wickham, Hadley, Jim Hester, and Romain Francois. 2018. *readr: Read rectangular text data*. R Package Version 1.3.1. https://CRAN.R-project.org/package=readr.

Wickham, Hadley, Mara Averick, Jennifer Bryan, Winston Chang, Lucy McGowan, Romain François, Garrett Grolemund, Alex Hayes, Lionel Henry, Jim Hester, Max Kuhn, Thomas Pedersen, Evan Miller, Stephan Bache, Kirill Müller, Jeroen Ooms, David Robinson, Dana Seidel, Vitalie Spinu, and Hiroaki Yutani. 2019. Welcome to the tidyverse. *Journal of Open Source Software* 4: 1686. https://doi.org/10.21105/joss.01686.

Wickham, Hadley, Romain François, Lionel Henry, and Kirill Müller. 2020. *dplyr: A grammar of data manipulation. A fast, consistent tool for working with data frame like objects, both in memory and out of memory*. R Package Version 1.0.7.

Xia, Yinglin, and Jun Sun. 2022. *An integrated analysis of microbiomes and metabolomics*. American Chemical Society.

Xia, Yinglin, Jun Sun, and Ding-Geng Chen. 2018. Introduction to R, RStudio and ggplot2. In *Statistical analysis of microbiome data with R*, 77–127. Springer.

Zhang, Xinyan, Yu-Fang Pei, Lei Zhang, Boyi Guo, Amanda H. Pendegraft, Wenzhuo Zhuang, and Nengjun Yi. 2018. Negative binomial mixed models for analyzing longitudinal microbiome data. *Frontiers in Microbiology* 9: 1683–1683. https://doi.org/10.3389/fmicb.2018.01683. https://pubmed.ncbi.nlm.nih.gov/30093893; https://www.ncbi.nlm.nih.gov/pmc/articles/PMC6070621/.

Chapter 3
Basic Data Processing in QIIME 2

Abstract This chapter presents some basic data processing in QIIME 2. First it introduces importing and exporting data. Then it introduces extracting data from QIIME 2 archives. Next, it describes how to filter data, review data in QIIME 2, as well as how to communicate between QIIME 2 and R.

Keywords FASTA · FASTQ · Feature table · Phylogenetic trees · Filter · QIIME 2 View · qiime2R Package · .qza file · Demultiplexed · .tsv file · Newick tree format

In the last two chapters, we provided an overview of QIIME 2 and R for microbiome data analysis. Starting with this chapter and until Chap. 6, we will focus on bioinformatic analysis of microbiome data using QIIME 2. In this chapter, we introduce some basic data processing in QIIME 2. We first introduce importing and exporting data in Sects. 3.1 and 3.2, respectively. We then introduce how to extract data from QIIME 2 archives (Sect. 3.3). Next, we describe how to filter data in QIIME 2 (Sect. 3.4). In Sect. 3.5, we introduce reviewing data in QIIME 2. Section 3.6 focuses on communicating between QIIME 2 and R. We complete this chapter with a brief summary (Sect. 3.7).

3.1 Importing Data into QIIME 2

QIIME 2 stores input data in artifacts (i.e., .qza files). Thus in order to use a QIIME 2 action, except for some metadata, all data must be imported as a QIIME 2 artifact.

QIIME 2 uses **the plugin qiime tools import** to import data. In QIIME 2, there are dozens of format types. Different data format types need different importing methods to import them into QIIME. You can use **qiime tools import --show-**

Supplementary Information The online version contains supplementary material available at https://doi.org/10.1007/978-3-031-21391-5_3.

importable-formats to check all the available import formats and **qiime tools import --show-importable-types** to check all available import types, respectively.

Currently either the QIIME 2 command-line interface (q2cli), or QIIME 2 Studio (q2studio), or Artifact API can be used to import input data. Depending on the task you want to implement, importing can be performed at any step in your analysis although importing typically starts with your raw sequence (e.g., FASTA or FASTQ) data. For "downstream" statistical analyses, typically importing starts with a feature table in either .biom or .csv format.

QIIME 2 supports importing many types of input data. Type the following command in the terminal to check which formats of input data are importable:

```
source activate qiime2-2022.2
qiime tools import \
 --show-importable-formats
```

Type the following command to check which QIIME 2 types you can use to import these formats:

```
qiime tools import \
 --show-importable-types
```

Currently no detailed documentations are available from QIIME 2 to tell us which QIIME 2 data types need what data formats although the information is indicated in the names of these formats and types. The most commonly used data formats are FASTA (sequences without quality information), FASTQ (sequence data with sequence quality information), feature table data, and phylogenetic trees.

3.1.1 Import FASTA Format Data

FASTA and FASTQ are the two basic and ubiquitous text-based formats for storing nucleotide and protein sequences. Common FASTA/Q file manipulations or processing include converting, searching, filtering, deduplication, splitting, shuffling, and sampling (Shen et al. 2016).

FASTA sequence file format or briefly **FASTA format** was originally invented by William Pearson in the FASTA software package (DNA and protein sequence alignment) (Lipman and Pearson 1985; Pearson and Lipman 1988). Nowadays FASTA format almost becomes a universal standard format in bioinformatics. The FASTA format represents either nucleotide sequences or amino acid (protein) sequences, in which nucleotides or amino acids are represented using single-letter codes.

There is no standard filename extension for FASTA file although each extension has its respective meaning (Wikipedia 2021). For example, **fasta**, or **fa**, means generic FASTA, which represents any generic fasta file; **fna** means FASTA nucleic acid, which is used generically to specify nucleic acids; **ffn** means FASTA nucleotide of gene regions, which contains coding regions for a genome; **faa** means

FASTA amino acid, i.e., containing amino acid sequences; and **frn** means FASTA non-coding RNA, i.e., containing non-coding RNA regions for a genome, in DNA alphabet e.g., tRNA and rRNA. One typical FASTA format file used in QIIME 1 and currently supported by QIIME 2 is called the post-split libraries FASTA file format. We cite an example of this format as follows (QIIME 2022):

```
>PC.634_1 FLP3FBN01ELBSX orig_bc=ACAGAGTCGGCT new_bc=ACAGAGTCGGCT
bc_diffs=0
CTGGGCCGTGTCTCAGTCCCAATGTGGCCGTTTACCCTCTCAGGCCGGCTACGCATCATCG
CCTTGGTGGGCCGTT
```

The sequence in FASTA format consists of exactly two lines per record: header (label line or description line) and sequence. They are distinguished by a greater-than (">") symbol in the first column.

The label line is separated by spaces and has five fields. From left to right, they are (1) the ID with the format <sample-id>_<seq-id> (e.g., PC.634_1), <sample-id> is used to identify the sample the sequence belongs to, and <seq-id> is used to identify the sequence *within* its sample; (2) the unique sequence id (e.g., FLP3FBN01ELBSX); (3) the original barcode (e.g., orig_bc=ACAGAGTCGGCT); (4) the new barcode after error-correction (e.g., new_bc=ACAGAGTCGGCT); and (5) the number of positions that differs between the original and new barcode (e.g. bc_diffs=0). A(**A**denine), C(**C**ytosine), G(**G**uanine), and T(**T**hymine) represent the four nucleobases in the nucleic acid of DNA in the letters G–C–A–T.

Each sequence must span exactly one line and cannot be split across multiple lines. The ID in each header must follow the format. The sequences in this data format are without quality information.

A feature sequence data with a FASTA format including DNA, RNA, or protein sequences could be aligned or unaligned. The purpose of aligning sequences is to identify regions of similarity that may be due to a consequence of functional, structural, or evolutionary relationships between the sequences. Aligned sequences of nucleotide or amino acid residues are typically represented as rows within a matrix. In order to align the columns to each other, gaps in a column (typically a dash "-") are inserted between the residues so that identical or similar characters are aligned in successive columns (Edgar 2004). Thus, all aligned sequences result in exactly the same length.

When importing FASTA format files, QIIME 2 specifies type as "**FeatureData [Sequence]**" for unaligned sequences and type as "**FeatureData [AlignedSequence]**" for aligned sequences. Here, we show how to import unaligned and aligned sequences into QIIME 2, respectively.

Example 3.1: VDR Fasta Data File

The SequencesVDR fasta data file was from the study of Vitamin D Receptor(VDR) and the murine intestinal microbiome (Jin et al. 2015). This study investigates whether VDR status regulates the composition and functions of the intestinal bacterial community. Here, we use this "SequencesVDR.fna" file to illustrate FASTA format data importation.

We take three steps to import this FASTA data into QIIME 2.

Step 1: Create a directory to store the fasta.gz files.

First, we need to create a directory folder to store the sequences data files (here, QIIME2R-Bioinformatics/Ch3). We can create the folder directly in computer or via the terminal of Mac: mkdir QIIME2R-Bioinformatics/Ch3. Then in the terminal, type source activate qiime2-2022.2 (depending on your QIIME 2 version) to activate QIIME 2 environment, and type cd QIIME2R-Bioinformatics/Ch3 to direct the QIIME 2 command to this folder.

Step 2: Store the fasta.gz files in this created directory.

We save the data files "SequencesVDR.fna" in the directory "QIIME2R-Bioinformatics/Ch3."

Step 3: Import the data into QIIME 2 artifacts (i.e., qza files) using "qiime tools import" command.

As we described in Chap. 1, all input data to QIIME 2 is in form of QIIME 2 artifacts, containing information about the type of data and the source of the data. Thus, we first need to import these sequence data files into a QIIME 2 artifact. For unaligned sequences, the semantic type of QIIME 2 artifact is FeatureData [Sequence]. We name the output file as "SequencesVDR.qza" in "output-path." The following commands can be used to import unaligned sequences into QIIME 2:

```
qiime tools import \
 --input-path SequencesVDR.fna \
 --output-path SequencesVDR.qza \
 --type 'FeatureData[Sequence]'
```

Imported SequencesVDR.fna as DNASequencesDirectoryFormat to SequencesVDR.qza

In above commands, "qiime tools import" defines the action, "input-path" specifies the data file path, and "output-path" specifies output data file path. We can see that SequencesVDR.fna was imported to SequencesVDR.qza as DNASequencesDirectoryFormat.

Example 3.2: Aligned Fasta Data File

The following aligned sequences were downloaded from QIIME 2 website. We extract two sequences from AlignedSequencesQiime2.fna (open using SeqKit software) to see what the aligned sequences look like.

```
>New.CleanUp.ReferenceOTU998 M2.Index.L_12921
-CTGGGCCGTATCTCAGTC-CCAATGTGGCCGGTCGCCCT---------CTCAGGCCGGC
TACCCGTCAAGGCC-TTGGTGGG-CCACTA-CCC-C-ACCAACAAGCTGATAGGCCGCGA
-G-ACGATCC-CTGACCGCA------------AAAA------G----------C-TTT-
-------CCAACAAC-CC-------GG--A---TG--CCCGG-G-AAA-----------
---CTG-AATAT-T--CGG-GA-TTA--------------C--CAC-C-T---GTTTCC
--AAG---T--GCT--A--T-ACC-A--AAG-TCA-AG--GG-------CA-CG-TT-C-
```

```
--C--TCA-CG-TG----------------TTACT-C---ACCCGTT-CGCCA-CT---
-------------------------------------
>New.CleanUp.ReferenceOTU999 M2.Ring.R_1432
-CTGGGCCGTATCTCAGTC-CCAATGTGGCCGGTCACCCT---------CTCAGGCCGGC
TACCCGTCGCCGCC-TTGGTAGG-CCACTA-CCC-C-ACCAACAAGCTGATAGGCCGCGA
-G-TCCATCC-ACAACCGCC-----------GGAG-----------------C-TTT-
-------CCAACCCC-CA-------CC--A---TG--CAGCA-G-GAG-----------
---CA--CATAT-C--CAG-TA-TTA--------------G--CAC-C-A---GTTTCC
--TAG---C--GTT--A--T-CCC-A--AAG-TTG-TG--GG-------CA-GG-TT-A-
--C--TCA-CG-TG----------------TTACT-C---ACCCG-----------
-------------------------------------
```

Now we use this fasta data file to illustrate importing the aligned sequences into QIIME 2. For aligned sequences, the semantic type of QIIME 2 artifact is FeatureData[AlignedSequence]. We can use the following commands.

```
qiime tools import \
 --input-path AlignedSequencesQiime2.fna \
 --output-path AlignedSequencesQiime2.qza \
 --type 'FeatureData[AlignedSequence]'
```

Imported AlignedSequencesQiime2.fna as AlignedDNASequencesDirectoryFormat to AlignedSequencesQiime2.qza

3.1.2 Import FASTQ Format Data

FASTQ sequence file format or briefly **FASTQ format** was originally developed at the Wellcome Trust Sanger Institute (Cock et al. 2010) as a simple extension to the FASTA format to store each nucleotide in a sequence and its corresponding quality score. For sequence file with FASTQ format, both the sequence letter and quality score are each encoded with a single ASCII character. In the field of DNA sequencing, the FASTQ file format has emerged as de facto standard format for storing the output of high-throughput sequencing instruments such as the Illumina Genome Analyzer and data exchange between tools (Cock et al. 2010).

A FASTQ file typically has four lines per sequence:

- Line 1 is the @title and optional description, begins with a "@" character and is followed by a sequence identifier and an *optional* description. This is a free format field with no length limit and allows including arbitrary annotation or comments.
- Line 2 is sequence line(s): the raw sequence letters (like in the FASTA format).
- Line 3 is +optional repeat of title line: signaling the end of the sequence lines and the start of the quality string. It begins with a "+" character and may include the same sequence identifier (and any description) again.

- Line 4 is quality line(s): encodes the quality values for the sequence in Line 2, and must contain the same number of symbols as letters in the sequence. They use a subset of the ASCII printable characters (at most ASCII 33–126 inclusive) with a simple offset mapping and the "@" marker character (ASCII 64) may be anywhere in the quality string.

A FASTQ file containing a single sequence might look like this:

```
@M00967:43:000000000-A3JHG:1:1101:18327:1699 1:N:0:188
NACGGAGGATGCGAGCGTTATCCGGATTTATTGGGTTTAAAGGGTGCGTAGGCGGCCTG
CCAAGTCAGCGGTAAAATTGCGGGGCTCAACCCCGTACAGCCGTTGA
AACTGCCGGGCTCGAGTGGGCGAGAAGTATGCGGAATGCGTG
GTGTAGCGGTGAAATGCATAGATATCACGCAGAACCCCGATTGCGAA
GGCAGCATACCGGCGCCCTACTGACGCTGAGGCACGAAAGTGCGGGGATCAAACAG
+

#>>AABABBFFFGGGGGGGGGGGGGGGGGGHHHHHHHGGGHHHHHGHGGGGGGGGHGGGGGGHHHHH
HHHHHGGGGGHHHHGHGGGGGGGHHBGHGDGGGGGHHHGGGGHHHHHHHGGGGGHG@DHHGHE
GGGGGGBFGGEGGGGGGGG.DFEFFFFFFFDCFFFFFFFFFFFFFFFFFFFFFFFFFFFDFDFFFEFFCFF?F
DFFFFFFFFAFFFFFFFFFFFFBDDFFFFFEFADFFFFFBAFFFA?EFFFBFF
```

The Earth Microbiome Project (EMP) founded in 2010 is a systematic effort to characterize global microbial taxonomic and functional diversity on this for planet earth (Thompson et al. 2017; Gilbert et al. 2010, 2014). "EMP protocol" has two fastq formats: multiplexed single-end and paired-end. In QIIME 2 terminology, the *single-end reads* refers to forward or reverse reads in isolation; the *paired-end reads* refers to forward and reverse reads that have not yet been joined; and the *joined reads* refers to forward and reverse reads that have already been joined (or merged).

"EMP Protocol" Multiplexed Single-End fastq
Single-end "Earth Microbiome Project (EMP) protocol" formatted reads total have two fastq.gz files: one contains the single-end reads, and another contains the associated barcode reads. The corresponding association between a sequence read and its barcode read is defined by the order of the records in these two files.

"Earth Microbiome Project (EMP) Protocol" Multiplexed Paired-End fastq
EMP paired-end formatted reads have three fastq.gz files total: one contains the forward sequence reads, another contains the reverse sequence reads, and a third contains the associated barcode reads.

The Illumina 1.8 FASTQ format was created and maintained by the Institute for Integrative Genome Biology UC Riverside. Each entry in a FASTQ file consists of four lines: Sequence identifier, Sequence, Quality score identifier line (consisting of a +), Quality score. An example of a valid entry is as follows:

```
@HWI-ST279:211:C0BFTACXX:3:1101:3469:2181 1:N:0:ACTTGA
GAACTATGCCTGATCAGGTTGAAGTCAGGGGAAACCCTGATGGAG
GACCGA + CCCFFFFFHHHHHJJJJJIIIJJJHJJJJJJJIJJJJIIIJJJIJJJJJJJ
```

Casava 1.8 Single-End Demultiplexed fastq

This fastq data file has one fastq.gz file for each sample in the study which contains the single-end reads for that sample. The file name includes the sample identifier, which looks like: L2S357_15_L001_R1_001.fastq.gz. The underscore-separated fields in this file name by order are the sample identifier, the barcode sequence or a barcode identifier, the lane number, the direction of the read (i.e., only R1, because these are single-end reads), and the set number.

Casava 1.8 Paired-End Demultiplexed fastq

This fastq format has two fastq.gz files for each sample in the study, each containing the forward or reverse reads for that sample. The file name includes the sample identifier. The forward and reverse read file names for a single sample might look like:

```
L2S357_15_L001_R1_001.fastq.gz and L2S357_15_L001_R2_001.fastq.gz,
respectively.
```

The underscore-separated fields in this file name are the sample identifier, the barcode sequence or a barcode identifier, the lane number, the direction of the read (i.e., R1 or R2), and the set number.

If the data do not have either EMP or Casava format, the data need to be manually imported into QIIME 2. First you need to create a "manifest" text file and then use the qiime tools import command. The specifications are different in the EMP or Casava import commands. The manifest file is a tab-separated (i.e., .tsv) text file: the first column defines the Sample ID, while the second (and optional third) column is the absolute file path to the forward (and optional reverse) reads. There are four variants of manifest FASTQ data in QIIME 2, including:

(1) **singleEndFastqManifestPhred33V2;**
(2) *singleEndFastqManifestPhred64V2;*
(3) **pairedEndFastqManifestPhred33V2;** and
(4) **pairedEndFastqManifestPhred64V2.**

In the format names, "Phred" indicates the PHRED software. This software reads DNA sequencing trace files, calls bases, and assigns a quality value to each base called (Ewing et al. 1998; Ewing and Green 1998), which defines the PHRED quality score of a base call in terms of the estimated probability of error. To hold these quality scores, PHRED introduced a new file format called the QUAL format. This is FASTA-like format, holding PHRED scores as space separated plain text integers and supplement a corresponding FASTA file with the associated sequences (Cock et al. 2010).

Phred33 means PHRED scores with an ASCII offset of 33, which is associated with Sanger FASTQ format. To be easily readable and editable by human, Sanger restricted the ASCII printable characters to 32–126 (decimal). Since ASCII 32 is the space character, Sanger FASTQ files use ASCII 33–126 to encode PHRED qualities from 0 to 93, which sets PHRED ASCII offset of 33.

Table 3.1 FASTQ data formats and the importing functions

Data formats	Command with data type
"EMP protocol" multiplexed single-end fastq	Implement command "qiime tools import" with specifying data type as " EMPSingleEndSequences"
"EMP protocol" multiplexed paired-end fastq	Implement command "qiime tools import" with specifying data type as "EMPPairedEndSequences"
Casava 1.8 single-end demultiplexed fastq	Implement command "qiime tools import" with specifying data type as "'SampleData[SequencesWithQuality]'" and input-format as "CasavaOneEightSingleLanePerSampleDirFmt"
Casava 1.8 paired-end demultiplexed fastq	Implement command "qiime tools import" with specifying data type as "'SampleData [PairedEndSequencesWithQuality]'" and input-format as "CasavaOneEightSingleLanePerSampleDirFmt"
SingleEndFastqManifestPhred33V2	Implement command "qiime tools import" with specifying data type as "'SampleData[SequencesWithQuality]'" and input-format as "SingleEndFastqManifestPhred33V2"
PairedEndFastqManifestPhred64V2	Implement command "qiime tools import" with specifying data type as "'SampleData [PairedEndSequencesWithQuality]'" and input-format as "PairedEndFastqManifestPhred64V2"

Phred64 means PHRED scores with an ASCII offset of 64, which is associated with Illumina 1.3+ FASTQ format. The Illumina FASTQ format encodes PHRED scores with an ASCII offset of 64, which can hold PHRED scores from 0 to 62 (ASCII 64–126) (Cock et al. 2010).

The encoded quality scores of PHRED 64 are different from PHRED 33; however, the encoded quality scores of PHRED 64 will be converted to those of PHRED 33 during importing.

Different types of FASTQ data need different functions to import. Table 3.1 summarizes FASTQ data formats and the importing functions in QIIME 2.

Example 3.3: "EMP Protocol" Multiplexed Single-End fastq Sequences Data File

We downloaded the example data "Moving Pictures" from QIIME 2 website including the single-end reads ("sequences.fastq") and its associated barcode reads ("barcodes.fastq") to illustrate this importation.

We take the following three steps to import FASTQ data into QIIME 2.

Step 1: Create a directory to store these two fastq.gz files.

Here, we create a directory called "QIIME2RCh3EMPSingleEndSequences." By typing the following command in a terminal, mkdir QIIME2RCh3EMPSingleEndSequences, we create a directory "QIIME2RCh3EMPSingleEndSequences" for Ch3 (the name suggests that the data is "EMP protocol" multiplexed single-end fastq, you can choose any name for the directory) to store the data file.

Step 2: Store the two fastq.gz files in this created directory.

We save the two data files "sequences.fastq" and "barcodes.fastq" in the directory "QIIME2RCh3EMPSingleEndSequences."

Step 3: Import the data into QIIME 2 artifacts (i.e., qza files) using "qiime tools import" command.

For "EMP protocol" multiplexed single-end fastq, the semantic type of QIIME 2 artifact is EMPSingleEndSequences, which contains sequences that are multiplexed, meaning that the sequences have not yet been assigned to samples and hence we need to include both sequences.fastq.gz file and barcodes.fastq.gz file, where it contains the barcode read associated with each sequence in sequences.fastq.gz.

With both two files "sequences.fastq.gz" and "barcodes.fastq.gz" stored in the directory "QIIME2RCh3EMPSingleEndSequences," now you can import these data into QIIME 2 artifacts (i.e., qza files). In the terminal, first type source activate qiime2-2022.2 to activate QIIME 2, and then type the following commands.

```
qiime tools import \
 --type EMPSingleEndSequences \
 --input-path QIIME2RCh3EMPSingleEndSequences\
 --output-path QIIME2RCh3EMPSingleEndSequences.qza
```

```
Imported QIIME2RCh3EMPSingleEndSequences as EMPSingleEndDirFmt to
QIIME2RCh3EMPSingleEndSequences.qza
```

In above commands, "qiime tools import" defines the action, "type" specifies the data type (in this case, the data type is "EMPSingleEndSequences"), "input-path" specifies the data file path, and "output-path" specifies output data file path. We can see that the data "QIIME2RCh3EMPSingleEndSequences.qza" are stored in QIIME 2 artifacts as format:"EMPSingleEndDirFmt".

Similarly, you can import "EMP protocol" multiplexed paired-end fastq, Casava 1.8 single-end demultiplexed fastq, and Casava 1.8 paired-end demultiplexed fastq files.

3.1.3   Import Feature Table

In Chap. 2 (Sect. 2.5), we have briefly introduced that the BIOM (Biological Observation Matrix) format is designed to be a general-use format for representing biological sample by counts of observation contingency tables (McDonald et al. 2012), and is a recognized standard for the Earth Microbiome Project and Genomics Standards Consortium candidate project.

Currently the BIOM file format has three versions: versions 1.0.0, 2.0.0, and 2.1.0. Here, we briefly introduce format specifications for version 1.0.0 and 2.1.0 and how to import pre-processed feature tables with BIOM format into QIIME 2. BIOM v1.0.0 format is based on **JSON** (JavaScript Object Notation) to provide the overall structure for the format (biom-format.org 2020a). BIOM v2.1.0 format is based on HDF5® Enterprise Support to provide the overall structure for the format (biom-format.org 2020b).

The BIOM format is generally used in various omics. For example, in marker-gene surveys, OTU or AVS tables primarily use this format; in metagenomics, metagenome tables also use this format; in genome data, a set of genomes uses this format too. Currently many projects support the BIOM format including QIIME 2, Mothur, phyloseq, MG-RAST, PICRUSt, MEGAN, VAMPS, metagenomeSeq, Phinch, RDP Classifier, USEARCH, PhyloToAST, EBI Metagenomics, GCModeller, and MetaPhlAn 2. The phyloseq package includes BIOM format examples with the four main types of biom files. The `import_biom()` function can be used to simultaneously import an associated phylogenetic tree file and reference sequence file (e.g., fasta).

Example 3.4: BIOM Sequences Data File with Version 1.0 .0 BIOM Format

The Seq_tableQTRT1.biom is the BIOM sequences data file with version 1.0 .0 BIOM format. The data was from the study of tRNA queuosine(Q)-modifications on the gut microbiome in breast cancers (Zhang et al. 2020). This study investigates how the enzyme queuine tRNA ribosyltransferase catalytic subunit 1 (QTRT1) affects tumorigenesis.

To import this file into QIIME 2, we first store it in the folder QIIME2R-Bioinformatics/Ch3. Then we type cd QIIME2R-Bioinformatics/Ch3 and the following commands in the terminal.

```
qiime tools import \
 --input-path Seq_tableQTRT1.biom \
 --type 'FeatureTable[Frequency]' \
 --input-format BIOMV100Format \
 --output-path Seq_tableQTRT1.qza
```

Imported Seq_tableQTRT1.biom as BIOMV100Format to Seq_tableQTRT1.qza

Example 3.5: BIOM Sequences Data File with Version 2.1.0 BIOM Format

The data "feature-table-v210.biom" was downloaded from the QIIME 2 website and renamed as "FeatureTablev210.biom," which was stored in the folder QIIME2R-Bioinformatics/Ch3. We type cd QIIME2R-Bioinformatics/Ch3 and the following commands in the terminal to import it into QIIME 2.

```
qiime tools import \
 --input-path FeatureTablev210.biom \
```

```
--type 'FeatureTable[Frequency]' \
--input-format BIOMV210Format \
--output-path FeatureTablev2.qza
```

Imported FeatureTablev210.biom as BIOMV210Format to FeatureTablev2.qza

3.1.4   Import Phylogenetic Trees

The Newick (parenthetic) tree format was introduced in the package **castor** in Sect. 2.4.3 of Chap. 2.

The Newick (parenthetic) tree format standard was adopted on June 26, 1986, by James Archie, William H. E. Day, Joseph Felsenstein, Wayne Maddison, Christopher Meacham, F. James Rohlf, and David Swofford, in an informal committee meeting in Durham, New Hampshire, and the second meeting in 1986, which was at Newick's restaurant in Dover, New Hampshire, US. This is the reason that the name of Newick came from. The adopted format represents a generalization of the format developed by Christopher Meacham in 1984 for the first tree-drawing programs in Felsenstein's PHYLogeny Inference Package (**PHYLIP**) (Felsenstein 1981, 2021).

The Newick format defines a tree by creating a minimal representation of nodes and their relationships to each other, which stores spanning-trees with weighted edges and node names in a minimal file format. Gary Olsen in 1990 provided an interpretation of the "Newick's 8:45" tree format standard (Olsen 1990). Newick formatted files are useful for representing phylogenetic trees and taxonomies.

A **phylogenetic tree** (a.k.a. **phylogeny** or **evolutionary tree**) is a branching diagram or a tree that represents evolutionary relationships among various biological species or other organisms based on similarities and differences in their physical or genetic characteristics (Felsenstein 2004). Phylogenetic trees may be rooted or unrooted. In a rooted phylogenetic tree, each node (called a taxonomic unit) has descendants to represent the inferred most recent common ancestor of those descendants, and in some trees the edge lengths may be interpreted as time estimates, whereas unrooted trees illustrate only the relatedness of the leaf nodes without assuming and do not require the ancestral root to be known or inferred (NIH 2002).

Example 3.6: Unrooted and Rooted Phylogenetic Trees, Example 2.7, Cont.
In Chap. 2, Example 2.7, we generated two tree data based on Dietswap study via the **ape** package:

Unrooted_tree_dietswap.tre and Rooted_tree_dietswap.tre. Here, we rename them as UnrootedTreeDietswap.tre and RootedTreeDietswap.tre, respectively, and use them to illustrate the importation of phylogenetic trees into QIIME 2. The following command can be used to import unrooted tree.

```
source activate qiime2-2022.2
cd QIIME2R-Bioinformatics/Ch3
```

```
qiime tools import \
 --input-path UnrootedTreeDietswap.tre \
 --output-path UnrootedTreeDietswap.qza \
 --type 'Phylogeny[Unrooted]'
```

Imported UnrootedTreeDietswap.tre as NewickDirectoryFormat to UnrootedTreeDietswap.qza

If you have a rooted tree, you can use --type 'Phylogeny[Rooted]' instead. The following command can be used to import rooted tree.

```
qiime tools import \
 --input-path RootedTreeDietswap.tre \
 --output-path RootedTreeDietswap.qza \
 --type 'Phylogeny[Rooted]'
```

Imported RootedTreeDietswap.tre as NewickDirectoryFormat to RootedTreeDietswap.qza

3.2   Exporting Data from QIIME 2

With QIIME 2 installed, you can export data from a QIIME 2 artifact to statistically analyze the data in R or using a different microbiome analysis software. This can be achieved using the **qiime tools export** command. Below we illustrate how to export feature table and phylogenetic tree.

3.2.1   Export Feature Table

The **qiime tools export** command takes a QIIME 2 artifact (.qza) file and an output directory as input. The data in the artifact will be exported to one or more files depending on the specific artifact. A FeatureTable[Frequency] artifact will be exported as a BIOM v2.1.0 formatted file.

Example 3.7: Exporting Feature Table, Example 3.5, Cont.
In Example 3.5, we imported a FeatureTablev210.biom as BIOMV210Format to FeatureTablev2.qza. Now we use the following command to export this feature table.qza data file to ExportedFeatureTable directory.

```
qiime tools export \
 --input-path FeatureTablev2.qza \
 --output-path ExportedFeatureTable
```

Exported FeatureTablev2.qza as BIOMV210DirFmt to directory ExportedFeatureTable

3.2.2 Export Phylogenetic Trees

Example 3.8: Exporting Phylogenetic Tree, Example 3.6, Cont.
Both UnrootedTreeDietswap.qza and RootedTreeDietswap.qza generated in Example 2.7 were stored in the directory folder QIIME2R-Bioinformatics/Ch3. We can export the unrooted tree data into the directory folder "ExportedTreeUnrooted" via the following command.

```
qiime tools export \
 --input-path UnrootedTreeDietswap.qza \
 --output-path ExportedTreeUnrooted
```

Exported UnrootedTreeDietswap.qza as NewickDirectoryFormat to directory ExportedTree

We can export the rooted tree data into the directory folder "ExportedTreeRooted" via the following command.

```
qiime tools export \
 --input-path RootedTreeDietswap.qza \
 --output-path ExportedTreeRooted
```

Exported RootedTreeDietswap.qza as NewickDirectoryFormat to directory ExportedTree

3.3 Extracting Data from QIIME 2 Archives

In Chap. 1, we have introduced that QIIME 2 .qza and .qzv files are zip file archives or containers with a defined internal directory structure. The data files stored in the file archives can be either exported or extracted; however, do not confuse "extract" and "export." In QIIME 2, extracting and exporting are two different data processing operations. Extracting an artifact differs from exporting an artifact. Exporting an artifact will only place the data files in the output directory; whereas extracting will not only place the data files, but also provide QIIME 2's metadata about an artifact, including the artifact's provenance in plain-text formats in the output directory. The output directory must already exist; otherwise must be created before extracting.

There are two ways to extract the data from the archives: one is to use the **qiime tools export** command if QIIME 2 and the **q2cli** command line interface are installed; another is to use standard decompression utilities such as unzip, WinZip, or 7zip when QIIME 2 is not installed. We illustrate these two ways to extract data below, respectively.

3.3.1 Extract Data Using the Qiime Tools Export Command

To extract QIIME 2 artifacts using **qiime tools extract** command, we first need to create an output directory such as "ExtractedFeatureTable," then call **qiime tools extract** command and specify input-path with file name (in this case, "FeatureTableMiSeq_SOP.qza") and just created output-path "ExtractedFeatureTable."

Example 3.9: FeatureTableMiSeq_SOP
The original sequencing data was downloaded from the published paper by Schloss et al. (2012) entitled "Stabilization of the murine gut microbiome following weaning." We generated the Feature Table using bioinformatic tool **data2** software through **QIIME 2** in Chap. 4. The objective of this study was to investigate the development and stabilization of the murine microbiome. The 360 fecal samples were collected from 12 mice (6 female and 6 male) longitudinally over the first year of life at 35 time points. Two mock community samples were added in the analysis for estimating the error rate. The raw sequence data are demultiplexed paired-end 16S rRNA gene reads generated using highly overlapping Illumina's MiSeq 2x250 amplicon sequencing platform from the V4 region of the 16S gene. The mouse gut dataset has been successfully used for testing new protocols and workflows of microbiome data analysis and new tool for integrative microbiome analysis (Buza et al. 2019; Westcott and Schloss 2015; Callahan et al. 2016). We use this dataset here and other chapters of this book to illustrate bioinformatic analysis using QIIME 2 and statistical analysis using R. More information on this MiSeq_SOP dataset, see Example 4.1: (MiSeq_SOP: One sample demultiplexed paired-end FASTQ data).

```
mkdir ExtractedFeatureTable
qiime tools extract \
   --input-path FeatureTableMiSeq_SOP.qza \
   --output-path ExtractedFeatureTable
Extracted FeatureTableMiSeq_SOP.qza to directory ExtractedFeatureTable/46eef13e-a20c-43f2-a7cf-944d36a8ebac
```

In the above commands, we first make a directory "ExtractedFeatureTable" by the command: mkdir ExtractedFeatureTable. Then use the command: qiime tools extract to extract the data file "FeatureTableMiSeq_SOP.qza" to the created directory "ExtractedFeatureTable." The output directory contain a new directory whose name is the artifact's UUID (in this case, 46eef13e-a20c-43f2-a7cf-944d36a8ebac). You can check that all artifact data and metadata are stored in this directory.

3.3.2 Extract Data Using Unzip Program on macOS

Above "FeatureTableMiSeq_SOP.qza" artifact also can be extracted using unzip program as below:

```
unzip FeatureTableMiSeq_SOP.qza
Archive: FeatureTableMiSeq_SOP.qza
 inflating: 46eef13e-a20c-43f2-a7cf-944d36a8ebac/metadata.yaml
 inflating: 46eef13e-a20c-43f2-a7cf-944d36a8ebac/checksums.md5
 inflating: 46eef13e-a20c-43f2-a7cf-944d36a8ebac/VERSION
 inflating: 46eef13e-a20c-43f2-a7cf-944d36a8ebac/provenance/
metadata.yaml
 inflating: 46eef13e-a20c-43f2-a7cf-944d36a8ebac/provenance/
citations.bib
 inflating: 46eef13e-a20c-43f2-a7cf-944d36a8ebac/provenance/VERSION
 inflating: 46eef13e-a20c-43f2-a7cf-944d36a8ebac/provenance/
artifacts/18ca53e7-d11f-4b48-9a33-72562f66084c/metadata.yaml
 inflating: 46eef13e-a20c-43f2-a7cf-944d36a8ebac/provenance/
artifacts/18ca53e7-d11f-4b48-9a33-72562f66084c/citations.bib
 inflating: 46eef13e-a20c-43f2-a7cf-944d36a8ebac/provenance/
artifacts/18ca53e7-d11f-4b48-9a33-72562f66084c/VERSION
 inflating: 46eef13e-a20c-43f2-a7cf-944d36a8ebac/provenance/
artifacts/18ca53e7-d11f-4b48-9a33-72562f66084c/action/action.yaml
 inflating: 46eef13e-a20c-43f2-a7cf-944d36a8ebac/provenance/action/
action.yaml
 inflating: 46eef13e-a20c-43f2-a7cf-944d36a8ebac/data/feature-table.
biom
```

The above unzip action created a new directory. The name of that directory is the UUID of the artifact being unzipped: 46eef13e-a20c-43f2-a7cf-944d36a8ebac. We can achieve a similar thing on Windows or Linux.

3.4   Filtering Data in QIIME 2

In this section, we will introduce how to filter feature tables, sequences, and distance matrices in QIIME 2.

Example 3.10: FeatureTableMiSeq_SOP. Example 3.9, Cont.

The data that are used to illustrate the filtering functionality in QIIME 2 are FeatureTableMiSeq_SOP.qza (feature table), TaxonomyMiSeq_SOP.qza (taxonomy data), SampleMetadataMiSeq_SOP.tsv (sample metadata), "BrayCurtisDistanceMatrixMiSeq_SOP.qza" (distance matrix), and "sequences. qza" (sequence data).

First, we create a directory for working on.

```
mkdir QIIME2R-Bioinformatics/Ch3/Filtering
cd QIIME2R-Bioinformatics/Ch3/Filtering
```

Then, we put all above data into the directory just created.

3.4.1   Filter Feature Table

Filtering feature tables include filtering (i.e., removing) samples and features from a feature table. Feature tables consist of the sample axis and the feature axis. The filtering operations are generally applicable to these two axes. The filter-samples method is used to filter sample axis, whereas the filter-features method is used to filter the feature axis. Both methods are implemented in the **q2-feature-table** plugin. We can also use the filter-table method in the **q2-taxa** plugin to perform the taxonomy-based filtering: filter features from a feature table.

3.4.1.1   Total-Frequency-Based Filtering

As the name suggested, total-frequency-based filtering filters samples or features based on the frequencies that samples or features are represented in the feature table. Two usual situations are (1) filter samples when total frequency is an outlier detected in the distribution of sample frequencies; (2) set up a cut-off point or minimum total frequency and then use it as a criterion to remove samples with a total frequency less than this cut-off point.

We can use the **--p-max-frequency** command to filter samples and features based on the maximum total frequency. We can also combine the commands **--p-min-frequency** and **--p-max-frequency** to filter samples and features based on lower and upper limits of total frequency.

The following commands filter (i.e., remove) samples with a total frequency less than 1500 from FeatureTableMiSeq_SOP.qza.

```
qiime feature-table filter-samples \
  --i-table FeatureTableMiSeq_SOP.qza \
  --p-min-frequency 1500 \
  --o-filtered-table SampleFrequencyFilteredFeatureTableMiSeq_SOP.
qza
```

Saved FeatureTable[Frequency] to: SampleFrequencyFilteredFeatureTableMiSeq_SOP.qza

The following commands remove all features with a total abundance (summed across all samples) of less than 10 from FeatureTableMiSeq_SOP.qza.

```
qiime feature-table filter-features \
  --i-table FeatureTableMiSeq_SOP.qza \
  --p-min-frequency 10 \
  --o-filtered-table FeatureFrequencyFilteredTable.qza
```

Saved FeatureTable[Frequency] to: FeatureFrequencyFilteredTable.qza

3.4.1.2   Contingency-Based Filtering

Those features that present in only one or a few samples may not represent real biological diversity but rather PCR or sequencing errors (such as PCR chimeras). Contingency-based filtering is designed to filter samples or features from a table contingent on the number of each other they contain. The following commands remove the features from FeatureTableMiSeq_SOP.qza that are not contained in at least 2 samples.

```
qiime feature-table filter-features \
  --i-table FeatureTableMiSeq_SOP.qza \
  --p-min-samples 2 \
  --o-filtered-table SampleContingencyFilteredTable.qza
```

Saved FeatureTable[Frequency] to: SampleContingencyFilteredTable.qza

The following commands remove samples from FeatureTableMiSeq_SOP.qza that contain less or equal to 10 features.

```
qiime feature-table filter-samples \
  --i-table FeatureTableMiSeq_SOP.qza \
  --p-min-features 10 \
  --o-filtered-table FeatureContingencyFilteredTable.qza
```

Saved FeatureTable[Frequency] to: FeatureContingencyFilteredTable.qza

Similar as the total-frequency-based filtering methods, contingency-based filtering methods can use the --p-max-features and --p-max-samples parameters to filter contingent on the maximum number of features or samples. They also can optionally be used in combination with --p-min-features and --p-min-samples.

3.4.1.3   Identifier-Based Filtering

When we want to keep the specific samples or features for analysis, we can define a user-specified list of samples or features based on their identifiers (IDs) in a QIIME 2 metadata file and then use the identifier-based filtering to retain these samples or features. Since IDs will be used to identify samples or features, then a QIIME 2 metadata file that contains the IDs in the first column is required. The metadata file is used as input with the --m-metadata-file parameter.

We can use either already existed metadata file or create a new one containing the IDs of the samples to filter by. To illustrate how to remove samples from a feature table using the identifier-based filtering method, below we create a simple QIIME 2 metadata file called SamplesToKeep.tsv that consists of two sample IDs to keep.

```
echo SampleID > SamplesToKeep.tsv
echo F3D0 >> SamplesToKeep.tsv
echo F3D9 >> SamplesToKeep.tsv
```

The following commands use the identifier-based filtering method to retain these two samples from FeatureTableMiSeq_SOP.qza.

```
qiime feature-table filter-samples \
  --i-table FeatureTableMiSeq_SOP.qza \
  --m-metadata-file SamplesToKeep.tsv \
  --o-filtered-table IdFilteredTable.qza
```

Saved FeatureTable[Frequency] to: IdFilteredTable.qza

After running the filter-samples method with the parameter --m-metadata-file SamplesToKeep.tsv, only the F3DO and F3D9 samples are retained in the IdFilteredTable.qza file.

3.4.1.4 Metadata-Based Filtering

Similar to identifier-based filtering, metadata-based filtering uses metadata search criteria to filter the feature table to keep the samples that the user wants to retain. This is achieved in QIIME 2 by combining the --p-where parameter and the --m-meta-data-file parameter. The following commands filter FeatureTableMiSeq_SOP.qza to contain only samples from Male mice.

```
qiime feature-table filter-samples\
  --i-table FeatureTableMiSeq_SOP.qza\
  --m-metadata-file SampleMetadataMiSeq_SOP.tsv\
  --p-where "Sex='Male'"\
  --o-filtered-table MaleFilteredTable.qza
```

Saved FeatureTable[Frequency] to: MaleFilteredTable.qza

We can also use multiple values in a single metadata column to filter samples. As in other programs, such as SAS, the **IN** clause can be used to specify those values. In this example, Time variable has two values Early and Later. The following commands can be used to retain both Early and Later samples. Please note that because Early and Later samples are all the samples for this dataset, so the command actually will not filter out any samples. Here, we just use this dataset to illustrate the program.

```
qiime feature-table filter-samples \
  --i-table FeatureTableMiSeq_SOP.qza \
  --m-metadata-file SampleMetadataMiSeq_SOP.tsv \
  --p-where "Time IN ('Early', 'Later')" \
  --o-filtered-table TimeFilteredTable.qza
```

Saved FeatureTable[Frequency] to: TimeFilteredTable.qza

Like in other programs, the keywords **AND** and **OR** can be used in **--p-where** parameter to evaluate both of the expressions or either of the expressions. The following commands are used to retain only those Early and Female samples.

```
qiime feature-table filter-samples \
  --i-table FeatureTableMiSeq_SOP.qza \
  --m-metadata-file SampleMetadataMiSeq_SOP.tsv \
  --p-where "Time='Early' AND Sex='Female'" \
  --o-filtered-table EarlyFemaleFilteredTable.qza
```

Saved FeatureTable[Frequency] to: EarlyFemaleFilteredTable.qza

The following commands use **OR** keyword syntax to retain samples.

```
qiime feature-table filter-samples \
  --i-table FeatureTableMiSeq_SOP.qza \
  --m-metadata-file SampleMetadataMiSeq_SOP.tsv \
  --p-where "Time='Early' OR Sex='Female'" \
  --o-filtered-table EarlyORFemaleFilteredTable.qza
```

Saved FeatureTable[Frequency] to: EarlyORFemaleFilteredTable.qza

Specifying Time='Early', Later samples would not be in the resulting table, but both Female and Male would retain in the resulting table; specifying Sex='Female', Male samples would not be in the resulting table, but both Early and Later samples would retain in the resulting table. Thus, actually evaluating **OR** syntax in this case would retain all of the samples. Here we just use it to illustrate the **OR** syntax.

The following commands will retain only the Early and Male samples in SampleMetadataMiSeq_SOP.tsv.

```
qiime feature-table filter-samples \
  --i-table FeatureTableMiSeq_SOP.qza \
  --m-metadata-file SampleMetadataMiSeq_SOP.tsv \
  --p-where "Time='Early' AND NOT Sex='Female'" \
  --o-filtered-table EarlyNonFemaleFilteredTable.qza
```

Saved FeatureTable[Frequency] to: EarlyNonFemaleFilteredTable.qza

3.4.2   Taxonomy-Based Tables and Sequences Filtering

The **filter-table** method in QIIME 2's **q2-taxa** plugin is designed to facilitate the process of taxonomy-based filtering, which is one of the most common types of feature-metadata-based filtering. The specific taxa can be retained or removed from a table using **--p-include** or **p-exclude** parameters, respectively.

3.4.2.1 Filter Tables Based on Taxonomy

Search terms in the **--p-mode** parameter by default are case insensitive. Thus, in the following commands, **--p-exclude** parameter would result in removing all features annotated as mitochondria and Mitochondria from the table.

```
qiime taxa filter-table \
   --i-table FeatureTableMiSeq_SOP.qza \
   --i-taxonomy TaxonomyMiSeq_SOP.qza \
   --p-exclude mitochondria \
   --o-filtered-table FeatureTableMiSeq_SOPNoMitochondria.qza
```

Saved FeatureTable[Frequency] to: FeatureTableMiSeq_SOPNoMitochondria.qza

Removing features can be done using more than one search term via listing a comma-separated search terms.

For example, the following commands will remove all features that contain either mitochondria or Rhodobacteraceae in their taxonomic annotation table.

```
qiime taxa filter-table \
   --i-table FeatureTableMiSeq_SOP.qza \
   --i-taxonomy TaxonomyMiSeq_SOP.qza \
   --p-exclude mitochondria,Rhodobacteraceae\
   --o-filtered-table
FeatureTableMiSeq_SOPNoMitochondriaNoRhodobacteraceae.qza
```

Saved FeatureTable[Frequency] to: FeatureTableMiSeq_SOPNoMitochondriaNoRhodobacteraceae.qza

The **--p-include** parameter is used to filter a table for retaining only specific features. For example, the following commands include **p__** in **--p-include** parameter to retain only features that contain a phylum-level annotation.

```
qiime taxa filter-table \
   --i-table FeatureTableMiSeq_SOP.qza \
   --i-taxonomy TaxonomyMiSeq_SOP.qza \
   --p-include p__\
   --o-filtered-table FeatureTableMiSeq_SOPWithPhyla.qza
```

Saved FeatureTable[Frequency] to: FeatureTableMiSeq_SOPWithPhyla.qza

The **--p-include** and **--p-exclude** parameters can be used combinedly. For example, the following commands use the **--p-include** parameter to retain all features that contain a phylum-level annotation(**p__**), and use **--p-exclude** parameter to exclude all features that contain either mitochondria or Rhodobacteraceae in their taxonomic annotation.

```
qiime taxa filter-table \
   --i-table FeatureTableMiSeq_SOP.qza \
   --i-taxonomy TaxonomyMiSeq_SOP.qza \
   --p-include p__ \
   --p-exclude mitochondria,Rhodobacteraceae\
   --o-filtered-table FeatureTableMiSeq_SOPWithPhylaButNo
MitochondriaNoRhodobacteraceae.qza
```

Saved FeatureTable[Frequency] to:
FeatureTableMiSeq_SOPWithPhylaButNoMitochondriaNoRhodobacteraceae.qza

By default, QIIME 2 matches the term(s) provided for **--p-include** or **--p-exclude** if they are contained in a taxonomic annotation.

However, sometimes we want to match the terms only if they are the complete taxonomic annotation. The parameter **--p-mode exact** (to indicate the search should require an exact match) is designed to achieve this goal. Since the search is an exact match, the search terms are case sensitive when searching with -p-mode exact. Thus, the search term mitochondria would not return the same results as the search term Mitochondria.

The following commands remove mitochondrial and chloroplast sequences with an exact match.

```
qiime taxa filter-table \
   --i-table FeatureTableMiSeq_SOP.qza \
   --i-taxonomy TaxonomyMiSeq_SOP.qza \
   --p-include p__ \
   --p-exclude mitochondria,chloroplast \
   --o-filtered-table table-with-phyla-no-mitochondria-no-
chloroplast.qza
```

Saved FeatureTable[Frequency] to: table-with-phyla-no-mitochondria-no-chloroplast.qza

In QIIME 2, we can also use **qiime feature-table filter-features** with the **--p-where** parameter to achieve the taxonomy-based filtering of tables. The qiime feature-table filter-features supports more complex filtering query than the **qiime taxa filter-table** filtering.

3.4.2.2 Filter Sequences Based on Taxonomy

The **filter-seqs** method in QIIME 2's **q2-taxa** plugin is designed to filter **FeatureData[Sequence]** based on a feature's taxonomic annotation. The **filter-seqs** method has very similar functionality that provided in **qiime taxa filter-**

table. Below, the **filter-seqs** method is used to retain all features that contain a phylum-level annotation, but exclude all features that contain either mitochondria or Rhodobacteraceae in their taxonomic annotation.

```
qiime taxa filter-table \
   --i-table FeatureTableMiSeq_SOP.qza \
   --i-taxonomy TaxonomyMiSeq_SOP.qza \
   --p-include p__ \
   --p-exclude mitochondria,Rhodobacteraceae\
   --o-filtered-table
SequencesMiSeq_SOPWithPhylaButNoMitochondriaNoRhodo
bacteraceae.qza
```

Saved FeatureTable[Frequency] to:
SequencesMiSeq_SOPWithPhylaButNoMitochondriaNoRhodobacteraceae.qza

For other filtering-sequences methods, we refer the reader to the **q2-feature-table** and **q2-quality-control** plugins. The q2-feature-table plugin also has a **filter-seqs** method, which can be used to remove sequences based on various criteria, including which features are present within a feature table. The q2-quality-control plugin has an **exclude-seqs** action, which can be used for filtering sequences based on alignment to a set of reference sequences or primers.

3.4.3   Filter Distance Matrices

The **q2-diversity** plugin provides the **filter-distance-matrix** method to filter (i.e., remove) samples from a distance matrix. It works the same way as filtering feature tables by identifiers or sample metadata.

3.4.3.1   Filtering Distance Matrix Based on Identifiers

The following commands filter the Bray-Curtis distance matrix to retain the two samples specified in SamplesToKeep.tsv above.

```
qiime diversity filter-distance-matrix \
   --i-distance-matrix BrayCurtisDistanceMatrixMiSeq_SOP.qza \
   --m-metadata-file SamplesToKeep.tsv \
   --o-filtered-distance-matrix
IdentifierFilteredBrayCurtisDistanceMtrix.qza
```

Saved DistanceMatrix to: IdentifierFilteredBrayCurtisDistanceMtrix.qza

3.4.3.2   Filter Distance Matrix Based on Sample Metadata

The following commands filter the Bray-Curtis distance matrix to retain only samples from Female mice.

```
qiime diversity filter-distance-matrix \
    --i-distance-matrix BrayCurtisDistanceMatrixMiSeq_SOP.qza \
    --m-metadata-file SampleMetadataMiSeq_SOP.tsv \
    --p-where "Sex='Female'" \
    --o-filtered-distance-matrix
FemaleFilteredBrayCurtisDistanceMatrix.qza
```

Saved DistanceMatrix to: FemaleFilteredBrayCurtisDistanceMatrix.qza

3.5   Introducing QIIME 2 View

QIIME 2 View (https://view.qiime2.org) is designed to allow the user to use the browser to directly open and read .qza and .qzv files that are archived on the user's computer. Thus, it facilitates sharing the visualizations generated in QIIME 2 with a collaborator who can explore the results interactively without having QIIME 2 installed. To use QIIME 2 View, simply open it with `qiime tools view` or https://view.qiime2.org/ and then drag the .qza and .qzv files to the area of QIIME 2 View.

3.6   Communicating Between QIIME 2 and R

To use QIIME 2 and R integratively, some communicating tools to link them have been developed. Here, we first introduce the **qiime2R** package and then describe how to prepare a feature table and metadata table in R and import them into QIIME 2.

3.6.1   Export QIIME 2 Artifacts into R Using qiime2R Package

As we reviewed in Chap. 1 and so far covered in this chapter, QIIME 2 artifact is a crucial and novel concept in QIIME 2. As a method for storing the inputs and outputs for QIIME 2 as well as associated metadata and provenance information about how the object was formed, QIIME 2 artifact file in reality is a compressed directory with an intuitive structure, which has the extension of .qza. Thus QIIME 2 artifact facilitates the data storage and delivery. Although QIIME 2 equips the export tool to export QIIME 2 artifact such as exporting feature table and sequences from the artifact, however, it does not mean it is easy to import to R for the R users.

The **qiime2R** package was developed for importing QIIME 2 artifacts directly into R (current version 0.99.6, March 2022). The package has two important usages: (1) the `read_qza()` function and (2) the `qza_to_phyloseq()` wrapper. By using the `read_qza()` function, the artifact can be easily obtained into R without discarding any of the associated data. The `qza_to_phyloseq()` wrapper can be used to generate a **phyloseq** object, which is very useful when you use the **phyloseq** package to further analyze data. We briefly introduce these two functions below.

To use this package, we first install this package by entering the following commands in R or RStudio:

```
install.packages("remotes")
remotes::install_github("jbisanz/qiime2R")
```

Example 3.11: FeatureTableMiSeq_SOP, Example 3.9, Cont.
We continue to use the data from Example 3.9 to illustrate the **qiime2R** package.

3.6.1.1   Read a .qza File

To read a .qza file, we first call library qiime2R:

```
> setwd("~/Documents/QIIME2R/Ch3_DataProcessing ")
> library(qiime2R)
```

Then, use the **read_qza( )** to read the file:

```
> feature_tab<-read_qza("FeatureTableMiSeq_SOP.qza")
> names(feature_tab)
[1] "uuid" "type" "format" "contents" "version"
[6] "data" "provenance"
```

3.6.1.2   Create a phyloseq Object

A **phyloseq** object consists of at least two out of four files: (1) feature, (2) taxonomy, (3) tree, and (4) metadata. The four QIIME 2 files, (1) FeatureTableMiSeq_SOP.qza, (2) TaxonomyMiSeq_SOP.qza, (3) RootedTreeMiSeq_SOP.qza, and (4) SampleMetadataMiSeq_SOP.tsv, have been saved in R source file directory

"~/Documents/QIIME2R/Ch3_DataProcessing"). Given the files are available, we now call the function **qza_to_phyloseq()** to build a phyloseq object as below:

```
> library(phyloseq)
> phyloseqObj<-qza_to_phyloseq(features="FeatureTableMiSeq_SOP.
qza", taxonomy = "TaxonomyMiSeq_SOP.qza", tree =
"RootedTreeMiSeq_SOP.qza", metadata="SampleMetadataMiSeq_SOP.tsv")
> phyloseqObj
phyloseq-class experiment-level object
otu_table() OTU Table:     [ 392 taxa and 360 samples ]
sample_data() Sample Data:    [ 360 samples by 11 sample variables ]
tax_table() Taxonomy Table:    [ 392 taxa by 7 taxonomic ranks ]
phy_tree() Phylogenetic Tree:    [ 392 tips and 389 internal nodes ]


> otu<-otu_table(phyloseqObj)
> head(otu,3)
OTU Table:        [3 taxa and 360 samples]
            taxa are rows
                F3D0 F3D1 F3D11 F3D125 F3D13 F3D141 F3D142 F3D143 F3D144
F3D145
b14d7992a4619e3524cad64f88ff8aa8  0   0    0     0     0    0      0      0      0     0
528ba5bd8a07c70f82636810d4a7743b  0   0    0     2     0    0      0      0      0     0
------


> tax<-tax_table(phyloseqObj)
> head(tax,3)
Taxonomy Table:     [3 taxa by 7 taxonomic ranks] :
                 Kingdom    Phylum       Class          Order
b14d7992a4619e3524cad64f88ff8aa8 "Bacteria" "Proteobacteria"
"Alphaproteobacteria" "Rhizobiales"
528ba5bd8a07c70f82636810d4a7743b "Bacteria" "Proteobacteria"
"Alphaproteobacteria" "Rhodobacterales"
5e13b5d5c72d5fb765a27828562246bb "Bacteria" "Proteobacteria"
"Alphaproteobacteria" "Rickettsiales"
                 Family      Genus      Species
------
> sam<-sample_data(phyloseqObj)
> head(sam,3)
Sample Data:      [3 samples by 11 sample variables] :
   BarcodeSequence ForwardPrimerSequence ReversePrimerSequence
ForwardRead
F3D0      <NA>        <NA>          <NA> F3D0_S188_L001_R1_001.fastq.gz
F3D1      <NA>        <NA>          <NA> F3D1_S189_L001_R1_001.fastq.gz
F3D11     <NA>        <NA>          <NA> F3D11_S198_L001_R1_001.fastq.gz
          ReverseRead Group  Sex Time DayID DPW    Description
F3D0  F3D0_S188_L001_R2_001.fastq.gz F3D0 Female Early D000
0 QIIME2RAnalysisSet
F3D1  F3D1_S189_L001_R2_001.fastq.gz F3D1 Female Early D001
1 QIIME2RAnalysisSet
F3D11 F3D11_S198_L001_R2_001.fastq.gz F3D11 Female Early D011
11 QIIME2RAnalysisSet
```

```
> tree<-phy_tree(phyloseqObj)
> head(tree,3)
$edge
     [,1] [,2]
 [1,]  393 394
 [2,]  394 395
 [3,]  395   1
 ------
$edge.length
 [1] 0.013504585 0.063435583 0.028701321 0.046779347 0.017936212
0.431774093 0.018533412 0.000000005
 [9] 0.000000005 0.000000005 0.095598550 0.081652745 0.000000005
0.004416175 0.000000005 0.042783284
[17] 0.038235871 0.046480155 0.004419571 0.000000005 0.021835292
0.076448202 0.162150745 0.022725035
 ------
$Nnode
[1] 389
```

3.6.2   *Prepare Feature Table and Metadata in R and Import into QIIME 2*

When using QIIME 2 to analyze microbiome data, probably most artifacts already have been generated from a count table. However, when using R for data analysis, an artifact may be not available. In this section, we demonstrate how to generate an artifact from a count table and then import this artifact into QIIME 2. We also demonstrate how to import metadata with an appropriate format into QIIME 2.

Example 3.12: QTRT1 Data, Example 3.4, Cont.
In Example 3.4, we used the sequences data from QTRT1 (Zhang et al. 2020) to demonstrate how to import BIOM sequences data file into QIIME 2. Here, we use this dataset to illustrate how to first generate feature table and metadata table and then import an artifact and metadata into QIIME 2.

Step 1: Generate feature table in R or RStudio.

```
> setwd("~/QIIME2R-Bioinformatics/Ch3")
> otu_tab <- read.csv("otu_table_genus_QTRT1.csv", check.names =
FALSE)
> meta_tab <- read.csv("metadata_QTRT1.csv", check.names = FALSE)
> head(otu_tab,3)

> # Remove rownames
> otu <- cbind(rownames(otu_tab), otu_tab[,2:41])
> head(otu,3)
```

```
> dim(otu)
[1] 586 41

> # Qiime 2 needs a featureid column
> colnames(otu)[1] <- "featureid"
> colnames(otu)[1]
[1] "featureid"

> # Remove rowname
> rownames(otu) <- NULL
> head(otu,3)

> write.table(otu, "feature_table_genus_QTRT1.txt", sep = "\t", col.
names=TRUE, row.names=FALSE, quote = FALSE)
```

Step 2: Generate metadata table in R or RStudio.

```
> head(meta_tab,3)
   SampleID Group  Time    Group4
1 Sun071.PG1  KO  Post    KO_POST
2 Sun027.BF2  WT  Before  WT_BEFORE
3 Sun066.PF1  WT  Post    WT_POST
> write.table(meta_tab, "metadata_QTRT1.txt", sep = "\t", col.
names=TRUE, row.names=FALSE, quote = FALSE)
```

Now we exit R and continue to process in QIIME 2. We need make sure that QIIME 2 and R have the same directory (in this case, "QIIME2R-Bioinformatics/Ch3") because "feature_table_genus_QTRT1.txt" and "metadata_QTRT1.txt" are written into this directory folder.

Step 3: Convert feature table into OTU table with biom2.0 format.

```
# Make sure to activate conda (QIIME 2) environment
source activate qiime2-2022.2
cd QIIME2R-Bioinformatics/Ch3
# Convert the feature_table_genus_QTRT1 dataset to biom2.0
biom convert -i feature_table_genus_QTRT1.txt -o
feature_table_genus_QTRT1.hdf5 --table-type="OTU table" --to-hdf5
```

Step 4: Import biom2.0 format OTU table into qiime 2.

```
# Import the biom2.0 format into qiime2
# This makes an artifact to be used.
qiime tools import --input-path feature_table_genus_QTRT1.hdf5 --type
```

```
FeatureTable[Frequency] --input-format BIOMV210Format --output-path
feature_table_genus_QTRT1.qza
```

Imported feature_table_genus_QTRT1.hdf5 as BIOMV210Format to feature_table_genus_QTRT1.qza

After both feature table and metadata are imported into QIIME 2, we can use them to analyze in QIIME 2.

3.7 Summary

This chapter demonstrated some basic data processing procedures in QIIME 2 with real microbiome datasets. First, importing FASTA and FASTQ format data as well as importing feature table and phylogenetic trees were described and illustrated. Then, BIOM format and Newick tree format were described and exporting feature table and exporting rooted and unrooted phylogenetic trees were illustrated. Next, two ways of extracting data from QIIME 2 archives, using the QIIME tools export command and using Unzip program on macOS, were illustrated. Followed that various filtering data methods including filtering feature table, taxonomy-based tables, and sequences filtering as well as filtering distance matrices were demonstrated. QIIME 2 View was also introduced. Finally, two ways of communicating between QIIME 2 and R were introduced and illustrated: exporting QIIME 2 artifacts into R using qiime2R package and preparing feature table and metadata in R and then importing them into QIIME 2. In Chap. 4, we will introduce building feature table and feature representative sequences from raw reads in QIIME 2.

References

biom-format.org. 2020a. *The biom file format: Version 1.0.* The BIOM Format Development Team. Last modified 05 Nov 2020. Accessed 8 March 2022. http://biom-format.org/documentation/format_versions/biom-1.0.html.
———. 2020b. *The biom file format: Version 2.1.* The BIOM Format Development Team. Last modified 05 Nov 2020. Accessed 8 March 2022. http://biom-format.org/documentation/format_versions/biom-2.1.html.
Buza, Teresia M., Triza Tonui, Francesca Stomeo, Christian Tiambo, Robab Katani, Megan Schilling, Beatus Lyimo, Paul Gwakisa, Isabella M. Cattadori, Joram Buza, and Vivek Kapur. 2019. iMAP: An integrated bioinformatics and visualization pipeline for microbiome data analysis. *BMC Bioinformatics* 20 (1): 374. https://doi.org/10.1186/s12859-019-2965-4.
Callahan, Ben J., Kris Sankaran, Julia A. Fukuyama, Paul J. McMurdie, and Susan P. Holmes. 2016. Bioconductor workflow for microbiome data analysis: From raw reads to community analyses. *F1000Research* 5: 1492–1492. https://doi.org/10.12688/f1000research.8986.2. https://www.ncbi.nlm.nih.gov/pubmed/27508062. https://www.ncbi.nlm.nih.gov/pmc/articles/PMC4955027/.
Cock, Peter J.A., Christopher J. Fields, Naohisa Goto, Michael L. Heuer, and Peter M. Rice. 2010. The Sanger FASTQ file format for sequences with quality scores, and the Solexa/Illumina

FASTQ variants. *Nucleic Acids Research* 38 (6): 1767–1771. https://doi.org/10.1093/nar/gkp1137. https://www.ncbi.nlm.nih.gov/pubmed/20015970. https://www.ncbi.nlm.nih.gov/pmc/articles/PMC2847217/.

Edgar, Robert C. 2004. MUSCLE: A multiple sequence alignment method with reduced time and space complexity. *BMC Bioinformatics* 5 (1): 113. https://doi.org/10.1186/1471-2105-5-113.

Ewing, B., and P. Green. 1998. Base-calling of automated sequencer traces using phred. II. Error probabilities. *Genome Research* 8 (3): 186–194.

Ewing, B., L. Hillier, M.C. Wendl, and P. Green. 1998. Base-calling of automated sequencer traces using phred. I. Accuracy assessment. *Genome Research* 8 (3): 175–185. https://doi.org/10.1101/gr.8.3.175.

Felsenstein, Joseph. 1981. Evolutionary trees from DNA sequences: A maximum likelihood approach. *Journal of Molecular Evolution* 17 (6): 368–376.

———. 2004. *Inferring phylogenies*. Sunderland: Sinauer Associates, Inc.

———. 2021. *The Newick tree format*. Accessed January 17. https://evolution.genetics.washington.edu/phylip/newicktree.html.

Gilbert, J.A., F. Meyer, D. Antonopoulos, P. Balaji, C.T. Brown, C.T. Brown, N. Desai, J.A. Eisen, D. Evers, D. Field, W. Feng, D. Huson, J. Jansson, R. Knight, J. Knight, E. Kolker, K. Konstantindis, J. Kostka, N. Kyrpides, R. Mackelprang, A. McHardy, C. Quince, J. Raes, A. Sczyrba, A. Shade, and R. Stevens. 2010. Meeting report: The terabase metagenomics workshop and the vision of an Earth microbiome project. *Standards in Genomic Sciences* 3 (3): 243–248. https://doi.org/10.4056/sigs.1433550.

Gilbert, Jack A., Janet K. Jansson, and Rob Knight. 2014. The Earth Microbiome project: Successes and aspirations. *BMC Biology* 12 (1): 69. https://doi.org/10.1186/s12915-014-0069-1.

Jin, Dapeng, Wu Shaoping, Yong-guo Zhang, Lu Rong, Yinglin Xia, Hui Dong, and Jun Sun. 2015. Lack of Vitamin D receptor causes dysbiosis and changes the functions of the murine intestinal microbiome. *Clinical Therapeutics* 37 (5): 996–1009.e7. https://doi.org/10.1016/j.clinthera.2015.04.004. https://www.sciencedirect.com/science/article/pii/S0149291815002283.

Lipman, D.J., and W.R. Pearson. 1985. Rapid and sensitive protein similarity searches. *Science* 227 (4693): 1435–1441. https://doi.org/10.1126/science.2983426. https://science.sciencemag.org/content/sci/227/4693/1435.full.pdf.

McDonald, Daniel, Jose C. Clemente, Justin Kuczynski, Jai Ram Rideout, Jesse Stombaugh, Doug Wendel, Andreas Wilke, Susan Huse, John Hufnagle, Folker Meyer, Rob Knight, and J. Gregory Caporaso. 2012. The Biological Observation Matrix (BIOM) format or: How I learned to stop worrying and love the ome-ome. *GigaScience* 1 (1): 7. https://doi.org/10.1186/2047-217X-1-7.

NIH. 2002. *"Tree" facts: Rooted versus unrooted trees*. Last modified revised 15 July 2002. https://www.ncbi.nlm.nih.gov/Class/NAWBIS/Modules/Phylogenetics/phylo9.html.

Olsen, Gary. 1990. *Interpretation of "Newick's 8:45" tree format*. Accessed 17 Jan. https://evolution.genetics.washington.edu/phylip/newick_doc.html.

Pearson, W.R., and D.J. Lipman. 1988. Improved tools for biological sequence comparison. *Proceedings of the National Academy of Sciences of the United States of America* 85 (8): 2444–2448. https://doi.org/10.1073/pnas.85.8.2444. https://www.ncbi.nlm.nih.gov/pubmed/3162770. https://www.ncbi.nlm.nih.gov/pmc/articles/PMC280013/.

QIIME. 2022. *Post- split_libraries FASTA File Overview*. QIIME.org. Accessed 8 Mar 2022. http://qiime.org/documentation/file_formats.html#post-split-libraries-fasta-file-overview.

Schloss, Patrick D., Alyxandria M. Schubert, Joseph P. Zackular, Kathryn D. Iverson, Vincent B. Young, and Joseph F. Petrosino. 2012. Stabilization of the murine gut microbiome following weaning. *Gut Microbes* 3 (4): 383–393. https://doi.org/10.4161/gmic.21008. https://www.ncbi.nlm.nih.gov/pubmed/22688727. https://www.ncbi.nlm.nih.gov/pmc/articles/PMC3463496/.

Shen, Wei, Shuai Le, Yan Li, and Hu. Fuquan. 2016. SeqKit: A cross-platform and ultrafast toolkit for FASTA/Q file manipulation. *PLoS One* 11 (10): e0163962–e0163962. https://doi.org/10.1371/journal.pone.0163962. https://pubmed.ncbi.nlm.nih.gov/27706213. https://www.ncbi.nlm.nih.gov/pmc/articles/PMC5051824/.

Thompson, Luke R., Jon G. Sanders, . . ., Janet K. Jansson, Jack A. Gilbert, Rob Knight, and The
    Earth Microbiome Project Consortium. 2017. *A communal catalogue reveals Earth's multiscale
    microbial diversity. Nature* 551: 457. https://doi.org/10.1038/nature24621. https://www.nature.
    com/articles/nature24621#supplementary-information.
Westcott, Sarah L., and Patrick D. Schloss. 2015. De novo clustering methods outperform
    reference-based methods for assigning 16S rRNA gene sequences to operational taxonomic
    units. *PeerJ* 3: e1487–e1487. https://doi.org/10.7717/peerj.1487. https://www.ncbi.nlm.nih.
    gov/pubmed/26664811. https://www.ncbi.nlm.nih.gov/pmc/articles/PMC4675110/.
Wikipedia. 2021. *FASTA format.* From Wikipedia, the free encyclopedia. Last modified 16 Nov
    2021. Accessed 8 Mar 2022. https://en.wikipedia.org/wiki/FASTA_format.
Zhang, J., R. Lu, Y. Zhang, Ż. Matuszek, W. Zhang, Y. Xia, T. Pan, and J. Sun. 2020. tRNA
    queuosine modification enzyme modulates the growth and microbiome recruitment to breast
    tumors. *Cancers (Basel)* 12 (3). https://doi.org/10.3390/cancers12030628.

Chapter 4
Building Feature Table and Feature Representative Sequences from Raw Reads

Abstract Bioinformatic techniques have advanced to correct sequencing errors to determine real biological sequences at single nucleotide resolution by generating amplicon sequence variants (ASVs) or sub-OTUs. QIIME 2 has warped the two most widely used denoising packages DADA2 and Deblur to generate ASVs and sub-OTUs with 100% identities to clinical variation. This chapter describes and illustrates their uses to generate ASVs or sub-OTUs. First, it introduces how to analyze the demultiplexed paired-end FASTQ data. Then it introduces using DADA2 and q2-dada2 plugin to analyze demultiplexed paired-end FASTQ data and the multiplexed paired-end FASTQ data. Next, it introduces using Deblur and q2-deblur plugin to analyze demultiplexed paired-end FASTQ data.

Keywords Demultiplexed paired-end FASTQ data · Sample metadata · Raw sequence data · Qiime zipped artifacts (.qza) · Multiplexed paired-end FASTQ data · Quality of the reads · q2-deblur plugin · list.files() · seqkit · Keemei

Traditionally, sequence reads are clustered into operational taxonomic units (OTUs) at a defined identity threshold to avoid sequencing errors generating spurious taxonomic units. However, with the bioinformatic technique advancement, recently several bioinformatic software can correct sequencing errors to determine real biological sequences at single nucleotide resolution by generating amplicon sequence variants (ASVs) or sub-OTUs. Both ASVs and sub-OTUs are 100% OTUs and supposedly have 100% identities to clinical variation. The two most widely used denoising packages **DADA2** (Callahan et al. 2016b) and **Deblur** (Amir et al. 2017) have been warped into QIIME 2. In this chapter, we describe and illustrate their uses to generate ASVs or sub-OTUs. First, we introduce how to analyze the demultiplexed paired-end FASTQ data (Sect. 4.1), and then we introduce DADA2 and q2-dada2 plugin and how to use them to analyze demultiplexed

---

Supplementary Information The online version contains supplementary material available at https://doi.org/10.1007/978-3-031-21391-5_4.

paired-end FASTQ data (Sect. 4.2) and analyze the multiplexed paired-end FASTQ data using q2-dada2 plugin (Sect. 4.3), respectively. Next, we introduce Deblur and q2-deblur plugin and how to use them to analyze demultiplexed paired-end FASTQ data (Sect. 4.4). Finally, we briefly summarize in this chapter (Sect. 4.5).

4.1   Analyzing Demultiplexed Paired-End FASTQ Data

Example 4.1: MiSeq_SOP: One Sample Demultiplexed Paired-End FASTQ Data

The sample paired-end fastq data and metadata we analyze here are from a published paper by Schloss et al. (2012) entitled "Stabilization of the murine gut microbiome following weaning." We have introduced this data in Example 3.9. Here and in other chapters of this book, we use this dataset to illustrate bioinformatic analysis using QIIME 2 and statistical analysis using R. We downloaded the data from http://www.mothur.org/MiSeqDevelopmentData/StabilityNoMetaG.tar. You can download, unzip, and save this dataset into the directory on your computer, and then you can use the following R function **list.files()** or software **seqkit** to open the raw sequence data.

```
# Update  to the directory that contains the fastq files
> path <- "./MiSeq_SOP"
> list.files(path)
```

Here we extract some rows of them below to give you some idea of these raw sequence data.

```
[1] "F3D0_S188_L001_R1_001.fastq.gz"  "F3D0_S188_L001_R2_001.fastq.gz"
[3] "F3D1_S189_L001_R1_001.fastq.gz"  "F3D1_S189_L001_R2_001.fastq.gz"
[5] "F3D11_S198_L001_R1_001.fastq.gz" "F3D11_S198_L001_R2_001.fastq.gz"
[7] "F3D125_S206_L001_R1_001.fastq.gz" "F3D125_S206_L001_R2_001.fastq.gz"
[9] "F3D13_S199_L001_R1_001.fastq.gz" "F3D13_S199_L001_R2_001.fastq.gz"
[11] "F3D141_S207_L001_R1_001.fastq.gz" "F3D141_S207_L001_R2_001.fastq.gz"
------

[713] "M6D65_S174_L001_R1_001.fastq.gz" "M6D65_S174_L001_R2_001.fastq.gz"
[715] "M6D7_S163_L001_R1_001.fastq.gz"  "M6D7_S163_L001_R2_001.fastq.gz"
[717] "M6D8_S164_L001_R1_001.fastq.gz"  "M6D8_S164_L001_R2_001.fastq.gz"
[719] "M6D9_S165_L001_R1_001.fastq.gz"  "M6D9_S165_L001_R2_001.fastq.gz"
[721] "Mock_S280_L001_R1_001.fastq.gz"  "Mock_S280_L001_R2_001.fastq.gz"
[723] "Mock2_S366_L001_R1_001.fastq.gz" "Mock2_S366_L001_R2_001.fastq.gz"
```

4.1.1   Prepare Sample Metadata

Sample metadata contains important biological information in microbiome analysis. They are created typically to collect data information on technical details (i.e., the DNA barcodes), descriptions of the experiment design and samples (i.e., the group, subject, time point, and body site that the sample belongs to). There are no specific restrictions on what types of sample metadata should be used and no enforced "metadata standards" in QIIME 2; however, QIIME 2 does provide some general formatting requirements when creating metadata. For example, although data with any file extensions can be used in QIIME 2, QIIME 2 prefers that sample metadata is stored in a tab-separated values (TSV) file rather than other formats such as the common used comma-separated values (CSV) format. The reason that QIIME 2 uses TSV instead of CSV is that CSV needs to use escape commas which often causes difficulties.

TSV files are simple text files used to store tabular structure data, such as database table or spreadsheet data. The format is supported by many spreadsheet programs and databases. Thus, we can easily use a spreadsheet program such as Microsoft Excel or Google Sheets to edit and export our metadata files. This is also QIIME 2's recommendation. For TSV files, each row in the table is one line of the text file; each column value of a row is separated from the next by a tab character. Thus, the TSV format belongs to a type of the more general delimiter-separated values format; and is an alternative to the CSV format. The interested reader can check the documents of QIIME 2 for details. Here, we just briefly describe formatting requirements for QIIME 2 sample metadata files.

- **Identifier Column.** The identifier (ID) column is the first column in the metadata file, which defines the sample IDs for sample metadata. The ID column name (i.e., *ID header*) must be one of the following case-insensitive values, and they are not allowed to be used for naming other IDs or columns in the file: id, sampleid, sample id, sample-id, featureid, feature id, and feature-id.
- **IDs.** IDs may consist of any unicode characters excepting starting with the pound sign (#). One file needs at least one ID. IDs must be unique, but cannot be empty, and cannot use any of the reserved ID column names listed above.
- **Identifiers.** Identifiers should be less or equal to 36 characters, and contain only ASCII alphanumeric characters (i.e., in the range of [a-z], [A-Z], or [0-9]), the period (.) character, or the dash (-) character.
- **Column Types.** QIIME 2 currently supports both *categorical* and *numeric* metadata columns. By default, if one column consists only of numbers or missing data, then QIIME 2 will treat the type of metadata column as *numeric*. Otherwise, if the column contains any non-numeric values, QIIME 2 will treat the column as *categorical*. Both categorical columns and numeric columns support missing data (i.e., empty cells).

The SampleMetadataMiSeq_SOP.tsv was prepared based on SampleMetadata. xlsx from the published paper. We can prepare this sample metadata by taking the following steps.

Step 1: Collect the study design and sample information using an Excel sheet.
Step 2: Upload the Excel sheet into Google sheet.
    First go to Google Drive homepage and log in using your credentials. In the Google Drive homepage, click **New** → select **Google Sheets**→click **File**→**Import**→in Import file screen, click **Upload**→ then you can either drag the SampleMetadata excel sheet file to the box, or click **Select a file from your**

device→then open the file. In the Import file screen, default option is "Replace spreadsheet"; just choose it and click **Import data**. Then the SampleMetadata. xlsx was uploaded into Google Sheet.

Step 3: Install open source Google sheets add-on Keemei program.

QIIME 2 needs sample metadata spreadsheet correctly formatted. You can set this file as a Google sheet and then use the **Keemei** (canonically pronounced *key may*) program (Rideout et al. 2016) for Google sheets to check whether the file is correctly formatted. **Keemei** is an open source Google Sheets add-on for **cloud-based** validating tabular bioinformatics file formats, including QIIME 2 metadata files. To install **Keemei**, first log in to your free Google account, then you have two options to install **Keemei** to your Google sheets. (1) Go to https://keemei. qiime2.org/ webpage, click the Chrome Web Store, and then click INSTALL and follow the direction to install. (2) From within a Google Sheet: click and search for **Keemei**. Once **Keemei** is installed, you can use it to validate the SampleMetadata.xlsx file.

Step 4: Check whether the sample metadata spreadsheet is correctly formatted using Keemeil.

To validate the SampleMetadata.xlsx, click **File**→**Make a copy**, and name it as Copy of SampleMetadata. Now you can start to validate the sample metadata with **Keemei**. To validate this active sheet, click **Add-ons**→**Validate QIIME 2 metadata file**. When you see **Keemeil** validation report says "Good news! Sheet metadata is a valid QIIME 2 metadata file," then your spreadsheet is correctly formatted for QIIME 2. In this case, the SampleMetadata spreadsheet passes **Keemei** validation. Now, click **File** again and choose **Download** as **Table-separated values (.tsv, current sheet)**. You now can save and rename it as what you want. In this case, we name it as SampleMetadataMiSeq_SOP.tsv.

Once this spreadsheet is correctly formatted for QIIME 2, the file is ready for use. If cells come out with red, it suggests that these cells have errors; if cells come out with yellow, which suggests that these cells have warnings. Then a sidebar summaries the validation report and lists invalid cells. Locate the cells with errors and warnings, fix all the invalid cells, and revalidate until all cells are valid. To clear the validation status on the active sheet, by clicking **Add-ons➔Keemei➔Clear validation status**, the cell background colors will reset to white and notes will be cleared.

Step 5: Further inspect the sample metadata in QIIME 2.

To further inspect the sample metadata in QIIME 2, create a working directory (here, QIIME2R-Bioinformatics) and put the SampleMetadataMiSeq_SOP.tsv in this working directory.

```
source activate qiime2-2022.2
mkdir QIIME2R-Bioinformatics
cd QIIME2R-Bioinformatics
```

Fig. 4.1 Inspection of sample metadata for the mouse gut microbiome study

```
              COLUMN NAME   TYPE
          =====================   ===========
         BarcodeSequence   categorical
    ForwardPrimerSequence   categorical
    ReversePrimerSequence   categorical
             ForwardRead   categorical
             ReverseRead   categorical
                   Group   categorical
                     Sex   categorical
                    Time   categorical
                   DayID   categorical
                     DPW   numeric
             Description   categorical
          =====================   ===========
                    IDS:   360
                COLUMNS:   11
```

Then type the following commands in terminal (Fig. 4.1).

```
# Figure 4.1
qiime tools inspect-metadata SampleMetadataMiSeq_SOP.tsv
```

4.1.2   Prepare Raw Sequence Data

Different sequencing platforms (e.g., Illumina vs. Ion Torrent) or different sequencing approaches (e.g., single-end vs. paired-end) will provide us different structured raw data. In addition, any pre-processing steps such as joined paired ends and barcodes in fastq header performed by sequencing centers also will result in different structured raw data.

The downloaded paired-end raw sequence data in Example 4.1 was demultiplexed consisting of both `forward.fastq.gz` and `reverse.fastq.gz`. We first create a sub-directory called "MiSeq_SOP" within the directory QIIME2R-Bioinformatics and then store the sequence file there.

```
mkdir QIIME2R-Bioinformatics/MiSeq_SOP
```

Now the files are ready for analysis. Currently two approaches are available in QIIME 2 to construct a feature table from raw reads: either using data2 or deblur plugins. We illustrate their uses respectively as below.

4.1.3   Import Data Files as Qiime Zipped Artifacts(.qza)

QIIME 2 works with artifacts (.qza). We must first import the FASTQ files as a QIIME artifact using the import command **qiime tools import**. As we described in Chap. 3, if the data do not have either EMP or Casava format, the data need to be manually imported into QIIME 2. First you need to create a tab-separated (i.e., .tsv) "manifest" text file. In this case, we created a manifest file called ManifestMiSeq_SOP.tsv as the same way we created the SampleMetadataMiSeq_SOP.tsv and stored it in the same working directory: QIIME2R-Bioinformatics.

One important thing we emphasize here is: the first column in manifest file defines the Sample ID, while the second and third columns are the absolute file path to the forward and reverse reads, respectively. The names of sample-id (e.g., F3D0 in first row) must be same as names in the parts of sequences (e.g., F3D0 in first row for that sequences) for each sample.

```
sample-id forward-absolute-filepath reverse-absolute-filepath
F3D0 $PWD/MiSeq_SOP/F3D0_S188_L001_R1_001.fastq.gz $PWD/MiSeq_SOP/
F3D0_S188_L001_R2_001.fastq.gz
F3D1 $PWD/MiSeq_SOP/F3D1_S189_L001_R1_001.fastq.gz $PWD/MiSeq_SOP/
F3D1_S189_L001_R2_001.fastq.gz
F3D2 $PWD/MiSeq_SOP/F3D2_S190_L001_R1_001.fastq.gz $PWD/MiSeq_SOP/
F3D2_S190_L001_R2_001.fastq.gz
```

$PWD/MiSeq_SOP/ is absolute-file path which links the sample names to sequence information for each sample. $PWD is a bash variable for the full path of the current working directory (here, QIIME2R-Bioinformatics). In this case, ManifestMiSeq_SOP.tsv was stored in the directory QIIME2R-Bioinformatcs, and raw sequences were stored in the sub-directory "MiSeq_SOP." The absolute path $PWD/MiSeq_SOP/ links them. In the following commands, we put the ManifestMiSeq_SOP.tsv in the input-path option position.

```
source activate qiime2-2022.2
cd QIIME2R-Bioinformatics
qiime tools import\
 --type 'SampleData[PairedEndSequencesWithQuality]'\
 --input-path ManifestMiSeq_SOP.tsv\
 --output-path PairedEndDemuxMiSeq_SOP.qza \
 --input-format PairedEndFastqManifestPhred33V2
```

```
Imported ManifestMiSeq_SOP.tsv as PairedEndFastqManifestPhred33V2 to
PairedEndDemuxMiSeq_SOP.qza
```

QIIME 2 uses SampleData[PairedEndSequencesWithQuality] to indicate the sequence data for each sample are paired forward/reverse FASTQ files. So we specify the data type as 'SampleData[PairedEndSequencesWithQuality]'. We specify

input data format as PairedEndFastqManifestPhred33V2 and name the output artifact as PairedEndDemuxMiSeq_SOP.qza. This file will contain a copy of each of the
sequence data files, which will enhance research reproducibility.

You can check this qiime zipped artifact (.qza) using qiime tools peek command.

```
Qiime tools peek PairedEndDemuxMiSeq_SOP.qza
UUID:         4eade1ee-ce58-459b-9dc7-7e474383528b
Type:         SampleData[PairedEndSequencesWithQuality]
Data format:  SingleLanePerSamplePairedEndFastqDirFmt
```

4.1.4   Examine and Visualize the Qualities of the Sequence Reads

To generate visualizations of the sequence qualities, you can run the command:

```
# Figures 4.2 and 4.3
qiime demux summarize \
  --p-n 10000 \
  --i-data PairedEndDemuxMiSeq_SOP.qza \
  --o-visualization PairedEndDemuxMiSeq_SOP.qzv


        Saved Visualization to: PairedEndDemuxMiSeq_SOP.qzv
```

Or, the following command can be used since 10000 random sampling is default
in QIIME 2:

```
# Figures 4.2 and 4.3
qiime demux summarize \
  --i-data PairedEndDemuxMiSeq_SOP.qza \
  --o-visualization PairedEndDemuxMiSeq_SOP.qzv


        Saved Visualization to: PairedEndDemuxMiSeq_SOP.qzv
```

To review the visualization of the PairedEndDemuxMiSeq_SOP.qzv file, you can
navigate to QIIME2 viewer in browser. In other words, copy over the
PairedEndDemuxMiSeq_SOP.qzv output to your computer, and open this file in
www.view.qiime2.org. The following plot is from the interactive quality plot.

Figures 4.2 and 4.3 show the quality profile across a sample of 10,000 for forward
and reverse reads, respectively.

Illumina sequencing data generally show a trend of decreasing average quality
towards the end of sequencing reads. In this case, the forward reads and the reverse
reads display different patterns of quality: the forward reads maintain high quality
over through, whereas the quality of the reverse reads drops significantly at around

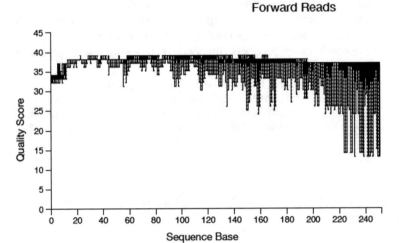

Fig. 4.2 Quality score box plots sampled from 10,000 random forward reads

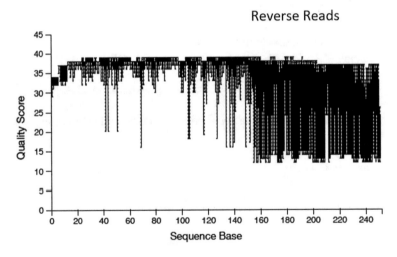

Fig. 4.3 Quality score box plots sampled from 10,000 random reverse reads

the position 160. Based on the different quality information for forward and reverse reads, we differentially truncate the forward reads at position 240, and the reverse reads at position 160 as did by Callahan et al. (2016a).

We also notice that both quality plots have slightly lower quality scores at the beginning of each read, which are caused by the homogeneity of the primer sequences. It is difficult for Illumina sequencer to properly identify clusters of DNA molecules (Fadrosh et al. 2014). Thus, the primer sequences must be removed at the denoising stage. Typically, the first 10 nucleotides of each read will be trimmed based on empirical observations across many Illumina datasets because these base positions are particularly likely to contain pathological errors (Callahan et al. 2016a).

4.2 Analyzing Demultiplexed Paired-End FASTQ Data Using DADA2 and q2-dada2 Plugin

The 16S rRNA marker gene sequencing approach has several advantages (Xia et al. 2018) including its own unique structure that contains both conserved and variable regions and its presence in all known Bacteria and Archaea species. This sequencing approach also has the advantages compared to shotgun metagenomic sequencing: low cost and avoid of problems with sequencing non-microbial DNA from host contamination. However, the 16S rRNA marker gene sequencing approach has the issue of sequencing errors, which makes it difficult to distinguish biologically real nucleotide differences in 16S sequences from sequencing artifacts. For example, traditionally the OTU method clusters sequence reads into Operational Taxonomic Units (OTUs). This method is used by most of the available pipelines (Caporaso et al. 2010; Schloss 2020; Mysara et al. 2017; Kumar et al. 2011; Hildebrand et al. 2014). However, the OTU method has several fundamental weaknesses such as clustering sequences with a fixed 3% dissimilarity threshold might avoid fine-scale variation among sequences (Rosen et al. 2012); therefore it often eliminates biological information present in the data. OTUs are not species; thus their construction is not necessitated by amplicon errors (Callahan et al. 2016a). In Chap. 6, we illustrate how to cluster OTUs via QIIME 2.

The methods for processing and analysis of 16S marker gene sequencing data continue to improve. **DADA2**(DADA: Divisive Amplicon Denoising Algorithm) (Callahan et al. 2016b) and **Deblur** (Amir et al. 2017) methods have been developed and recognized as one major advance towards quality control measures through denoising sequences to better discriminate between true sequence diversity and sequencing errors. Performing quality control of the sequences typically is performed prior to taxonomic classification. The goal is to identify the poor-quality reads and residual contamination in the dataset.

In this chapter, we illustrate how to conduct quality controlling sequences or denoising and QC filtering to generate feature table and feature data.

4.2.1 Introduction to DADA2 and q2-dada2 Plugin

DADA2 was proposed to use "amplicon sequence variants" (ASVs) to replace OTUs as the standard unit of marker-gene analysis and reporting (Callahan et al. 2017). **DADA2** uses an error-modeling approach for denoising and clustering amplicons and outputs exact sequence variants (or ASVs). DADA2 aims to overcome the fundamental weaknesses of traditional OTU methods, and to improve the performance of newly proposed bioinformatic sequence denoising approaches including UPARSE, MED, mothur (average linkage), and QIIME (uclust) OTU methods (Callahan et al. 2016a). It was demonstrated that DADA2 methods are more accurate compared to these four methods (Callahan et al. 2016a). It was shown (Callahan et al. 2016b) that DADA2 exactly infers sample sequences and resolves differences of as little as one nucleotide. To model and correct Illumina-sequenced amplicon errors, the software package DADA2 was developed. The R package DADA2 can implement the full amplicon workflow from filtering, dereplication, sample inference, chimera identification to merging the paired-end reads (Callahan et al. 2016a). For the details on algorithm and the development of DADA2, the reader is referred to Sect. 8.3.2.4.

Here, we illustrate how to implement the DADA2 methods via the DADA2 plugin in QIIME 2. The DADA2 plugin has several methods to denoise reads, including (1) denoise paired-end, which requires unmerged, paired-end reads (i.e., both forward and reverse); and (2) denoise single-end, which accepts either single-end or unmerged paired-end data. When the unmerged paired-end data are provided, only the forward reads will be used and the reverse reads will be ignored.

Implementing **DADA2** and **Deblur** methods will generate two QIIME 2 artifacts: a `FeatureTable[Frequency]` and a `FeatureData[Sequence]`. The `FeatureTable[Frequency]` artifact contains counts (frequencies) of each unique sequence in each sample in the dataset, and the `FeatureData[Sequence]` artifact maps feature identifiers in the FeatureTable to the sequences they represent. In QIIME 1, they were called Biom table and rep_set fasta file, respectively.

DADA2 and **Deblur** are currently the two denoising methods available in QIIME 2. We apply the **DADA2** approach to the mouse gut microbiome data (Example 4.1: MiSeq_SOP).

4.2.2   Denoise Sequences to Construct Feature Table and Feature Data with q2-dada2 Plugin

A feature is a species or an OTU in the context of microbiome sequencing or a gene in the RNA-Seq context. Specially in **DADA2** and **Deblur**, features are ASVs or sub-OTUs, respectively. In QIIME 2, denoising Illumina sequence via **DADA2** is an alternative option to OTU clustering as defined sequence-identify cut-off (e.g., 97%). The **qiime dada2 denoise-paired** method performs merging and denoising paired-end reads to denoise paired-end sequences, dereplicate them, and filter chimeras.

The **dada2 denoise-paired** method requires one paired-end demultiplexed sequences (an artifact SampleData[PairedEndSequencesWithQuality]) as input data and four parameters that are used in quality filtering: `--p-trim-left-f`, `--p-trim-left-r`, `--p-trunc-len-f`, and `--p-trunc-len-r`.

- The parameters **--p-trim-left-f** and **--p-trim-left-r** (optional) are used to trim the 5′ end of the input sequences, which will be the bases that were sequenced in the first cycles. When primers are present in the input sequence files, DADA2 requires removing the primers from the data to prevent false positive detection of chimeras as result of degeneracy in the primers before denoising DADA2 remove the setted length of the primer sequences. The parameter **--p-trim-left-f** is used to specify an integer for the position at which forward read sequences should be trimmed due to low quality. Default is 0 for not trimming the forward read sequences. The parameter **--p-trim-left-r** is used to specify an integer for the position at which reverse read sequences should be trimmed due to low quality. Default is 0.

- The parameter **--p-trunc-len-f** indicates the position at which the forward sequence will be truncated and parameter **--p-trunc-len-r** indicates the position at which the reverse read will be truncated. They are used to truncate the 3′ end of the of the input sequences, which will be the bases that were sequenced in the last cycles. In above Interactive Quality Plot tab in the visualization of PairedEndDemuxMiSeq_SOP.qzv file that was generated by **qiime demux summarize** command, there are quality scores for each reads. To determine what values to pass for these two parameters (**--p-trunc-len-f and --p-trunc-len-r**), you should review the Interactive Quality Plot tab. Specifying 0, no truncation or length filtering will be performed.

- The parameter **--p-max-ee** (--p-max-ee-f and --p-max-ee-r for forward reads and reverse reads, respectively) (optional) controls the maximum number of expected errors in a sequence before it is discarded (default is 2). The default 2 is used to enforce a maximum of 2 expected errors per-read (Edgar and Flyvbjerg 2015), which combines the trimming parameter with standard filtering parameters, and is considered as a better filter than simply averaging quality scores (Edgar and Flyvbjerg 2015). **DADA2** trims and filters paired reads jointly, i.e., both reads must pass the filter for the pair to pass.

- The parameter **--p-truncac-q** (optional) is used to truncate the sequence after the first position that has a quality score equal to or less than the provided value (default is 2).

- The parameter **--p-pooling-method** (optional) is used to specify pool samples for denoising. By default, samples are denoised independently ("independent"). If it is specified as ("pseudo"), the pseudo-pooling method is used to approximate pooling of samples.

- The parameter **--p-chimera-method** (optional) is used to specify the method ("none," "pooled," "consensus") to remove chimeras. Specifying "none," no chimera is removed; specifying "pooled," all reads are pooled prior to chimera detection, while by default ("consensus"), chimeras are detected in samples individually, and sequences are removed if chimeras are found in a sufficient fraction of samples.

- The parameter **--p-n-threads** (optional) is used to specify the number of threads to use for multithreaded. Processing with 1 as default and 0 for using all available cores.
- The parameter **--p-n-reads-learn** (optional) is used to specify the number of reads to use when training the error model. By default, 1,000,000 is used with smaller numbers for a shorter run time but a less reliable error model.
- The reader can refer to the QIIME 2 documentation for other input parameters.

We implement the following commands to denoise the sequences with **DADA2** based on quality score visualizations.

```
cd QIIME2R-Bioinformatics
qiime dada2 denoise-paired \
 --i-demultiplexed-seqs PairedEndDemuxMiSeq_SOP.qza \
 --p-trim-left-f 10 \
 --p-trim-left-r 10 \
 --p-trunc-len-f 240 \
 --p-trunc-len-r 160 \
 --p-n-threads 4 \
 --o-table FeatureTableMiSeq_SOP \
 --o-representative-sequences RepSeqsMiSeq_SOP.qza \
 --o-denoising-stats DenoisingStatsMiSeq_SOP.qza \
 --verbose
```

Running external command line application(s). This may print messages to stdout and/or stderr.

The command(s) being run are below. These commands cannot be manually re-run as they will depend on temporary files that no longer exist.

```
Command: run_dada_paired.R /var/folders/80/
b4jv62j553b9g3s7l5vxbcg40000gn/T/tmpxqoc0vjc/forward /var/folders/
80/b4jv62j553b9g3s7l5vxbcg40000gn/T/tmpxqoc0vjc/reverse /var/
folders/80/b4jv62j553b9g3s7l5vxbcg40000gn/T/tmpxqoc0vjc/output.
tsv.biom /var/folders/80/b4jv62j553b9g3s7l5vxbcg40000gn/T/
tmpxqoc0vjc/track.tsv /var/folders/80/
b4jv62j553b9g3s7l5vxbcg40000gn/T/tmpxqoc0vjc/filt_f /var/folders/
80/b4jv62j553b9g3s7l5vxbcg40000gn/T/tmpxqoc0vjc/filt_r 240 160 10 10
2.0 2.0 2 12 independent consensus 1.0 4 1000000

R version 4.1.2 (2021-11-01)
Loading required package: Rcpp
DADA2: 1.22.0 / Rcpp: 1.0.8 / RcppParallel: 5.1.5
1) Filtering
.......................................................................
....................................................................
....................................................................
....................................................................
....................................................................
..........................................................
2) Learning Error Rates
```

```
232305520 total bases in 1010024 reads from 94 samples will be used for
learning the error rates.
151503600 total bases in 1010024 reads from 94 samples will be used for
learning the error rates.
3) Denoise samples
.................................................................
................................................................
................................................................
................................................................
...............................................................
..........................................................
............
...............................................................
................................................................
.............................................................
...........................................................
........................................................
.....................................................
..............................................................
4) Remove chimeras (method = consensus)
6) Write output
```

```
Saved FeatureTable[Frequency] to: FeatureTableMiSeq_SOP.qza
Saved FeatureData[Sequence] to: RepSeqsMiSeq_SOP.qza
Saved SampleData[DADA2Stats] to: DenoisingStatsMiSeq_SOP.qza
```

In above commands, we use the parameter **--p-n-threads 4** to allow the program to perform parallel computations on 4 threads. If your datasets are very large, **DADA2** may be slow. You may need increase the number of threads. The option **--verbose** is used to display the DADA2 progress in the terminal as shown above. The printed information in the terminal with the **--verbose** option shows 6 stages of denoising process.

The denoising process outputs three artifacts: (1) a FeatureTable[Frequency] file via **--o-table** (we named it as FeatureTableMiSeq_SOP.qza), (2) a FeatureData [Sequence] via **--o-representative-sequences**, which is representative sequence file (we named it as RepSeqsMiSeq_SOP.qza), and (3) an artifact via **--o-denoising-stats** (DADA 2 Stats , we named it as DenoisingStatsMiSeq_SOP.qza). All these three output file names are required. The feature table file is the Biological Observation Matrix(BIOM) format file. The representative sequence file contains the denoised sequences, while the table file maps each of the sequences onto their denoised parent sequence.

The produced feature table by **DADA2** method is a higher-resolution analogue of the common "OTU table"; however, the count reads are called amplicon sequence variants (ASVs) instead of OTUs, which are thought resolving variants that differ by as little as one nucleotide (Callahan et al. 2016b).

4.2.3 Summarize the Feature Table and Feature Data from q2-dada2 Plugin

After successfully denoising sequences for each sample and generating feature table and representative sequences, we can summarize the denoised data using the **qiime feature table** command.

```
qiime feature-table summarize \
  --i-table FeatureTableMiSeq_SOP.qza \
  --o-visualization FeatureTableMiSeq_SOP.qzv \
  --m-sample-metadata-file SampleMetadataMiSeq_SOP.tsv

    Saved Visualization to: FeatureTableMiSeq_SOP.qzv


qiime feature-table tabulate-seqs \
  --i-data RepSeqsMiSeq_SOP.qza \
  --o-visualization RepSeqsMiSeq_SOP.qzv

    Saved Visualization to: RepSeqsMiSeq_SOP.qzv
```

The two produced visualization files (.qzv) by the above commands can be explored via the QIIME2 viewer. The "interactive Sample Detail" tab provides detailed information about the denoised sequence counts, such as the number of sequence per sample. We can explore to determine how rarefaction depths (subsampling) will impact your data. For example, we may check which samples have lowest sequencing depths to be dropped.

4.3 Analyzing Multiplexed Paired-End FASTQ Data Using q2-dada2 Plugin

To illustrate the bioinformatic workflow of QIIME 2 and multiplexed paired-end fastq data using QIIME 2, in this section, we demonstrate the steps of analyzing demultiplexed paired-end fastq data using Atacama soil microbiome data.

Example 4.2: Atacama Soil Microbiome
The data used here was originally from the study of "Significant Impacts of Increasing Aridity on the Arid Soil Microbiome" (Neilson et al. 2017), which analyzes the soil samples from the Atacama Desert in northern Chile. This desert is one of the most arid locations on Earth, where some areas receive less than a millimeter of rain per decade. We downloaded the data from the QIIME 2 website and use the data to illustrate how to denoise sequences to construct a feature table and the associated feature sequences using **Deblur** along with the importing, demultiplexing, and some other preliminary works.

4.3.1 Prepare Sample Metadata

To store the data, we first create a subdirectory within the QIIME2R-Bioinformatics directory.

```
cd QIIME2R-Bioinformatics
mkdir Atacama
cd Atacama
```

The sample metadata is available as a Google Sheet from QIIME 2 website. We download and save as SampleMetadataAtacama to the directory Atacama. Since the sample metadata is from QIIME 2, of course it has passed the validation of Keemei program and ready for use. We can further inspect the sample metadata in QIIME 2 by typing the following commands in terminal (Fig. 4.1).

```
#Figure 4.1 qiime tools inspect-metadata SampleMetadataAtacama.tsv
```

We can also create a visualization of this sample metadata (a qiime zipped visualization (.qzv file)) and navigate to QIIME2 viewer in browser to view this visualization.

```
qiime metadata tabulate\
 --m-input-file SampleMetadataAtacama.tsv\
 --o-visualization TabulatedSampleMetadataAtacama.qzv

    Saved Visualization to: TabulatedSampleMetadataAtacama.qzv
```

4.3.2 Prepare Raw Sequence Data

To store the raw sequence data, we create a sub-directory within the directory Atacama.

```
mkdir EmpPairedEndSequences
```

The paired-end raw sequence data consist of three `fastq` format files: `forward.fastq.gz`, `reverse.fastq.gz`, and `barcodes.fastq.gz`. They represent forward reads, reverse reads, and barcodes in sequencing run, respectively. Here, we use a 10% subsample data downloaded from QIIME 2 website. We move these three fastq files into the EmpPairedEndSequences working directory we just created and now the files are ready for use. The FASTQ data have the specific format of EMP (`EMPPairedEndSequences`).

The sequences with the format of EMPPairedEndSequences in QIIME 2 artifacts are multiplexed, suggesting that the sequences have not yet been assigned to samples, and hence to process this kind of sequences, both sequences.fastq.gz and barcodes.fastq.gz files are needed, in which the barcodes. fastq.gz contains the barcode read associated with each sequence in sequences.fastq. gz.

4.3.3   Import Data Files as Qiime Zipped Artifacts(.qza)

The data format used here is called EMPPairedEndSequences. We import the sequences into an artifact using the **qiime tools import** commands, which creates an artifact of the data.

```
qiime tools import \
 --type EMPPairedEndSequences \
 --input-path EmpPairedEndSequences \
 --output-path EmpPairedEndSequencesAtacama.qza
```

```
Imported EmpPairedEndSequences as EMPPairedEndDirFmt to
EmpPairedEndSequencesAtacama.qza
```

We can check this qiime zipped artifact (.qza) using **qiime tools peek** command.

```
Qiime tools peek EmpPairedEndSequencesAtacama.qza
UUID:         6b7ffd2d-1698-40b6-9d67-2be09f6c5a08
Type:         EMPPairedEndSequences
Data format: EMPPairedEndDirFmt
```

4.3.4   Demultiplexing Sequences

The next-generation sequencing instruments are able to analyze multiple samples in a single lane/run through *multiplexing* these samples. These samples are typically appended a unique barcode (a.k.a. index or tag) sequence to one or both ends of each sequence to identify their originals. Detecting these barcode sequences and mapping them back to the samples they belong to is called *demultiplexing* sequences.

In QIIME 2, two plugins are available for demultiplexing sequences: **q2-demux** and **q2-cutadapt**. However, depending on the type of raw sequence data either EMP Single End, EMP Paired End, Multiplexed Single End Barcode, or Multiplexed Paired End Barcode, usually only one demultiplexing action available in **q2-demux** or **q2-cutadapt** for the data. In the case, the barcodes have already been removed

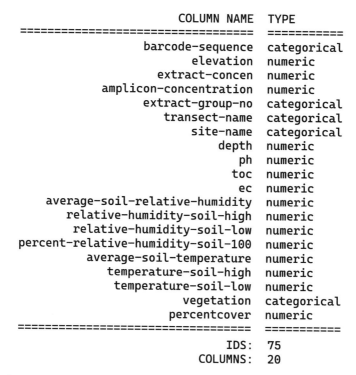

```
                           COLUMN NAME  TYPE
==================================== ===========
                    barcode-sequence  categorical
                           elevation  numeric
                      extract-concen  numeric
               amplicon-concentration  numeric
                    extract-group-no  categorical
                       transect-name  categorical
                           site-name  categorical
                               depth  numeric
                                  ph  numeric
                                 toc  numeric
                                  ec  numeric
         average-soil-relative-humidity  numeric
           relative-humidity-soil-high  numeric
            relative-humidity-soil-low  numeric
     percent-relative-humidity-soil-100  numeric
              average-soil-temperature  numeric
               temperature-soil-high  numeric
                temperature-soil-low  numeric
                          vegetation  categorical
                        percentcover  numeric
==================================== ===========
                                IDS:  75
                            COLUMNS:  20
```

Fig. 4.4 Inspection of sample metadata for the Atacama soil microbiome study

from the reads and are in a separate file, then **q2-demux** can be used; while if the barcodes are still in the sequences, then **q2-cutadapt** can be used instead.

Since demultiplexing sequences are to detect the barcode sequences and to map them back to the samples they belong to, thus, the sample metadata file is required. To *demultiplex*, we must specify which column in the sample metadata file contains the per-sample barcodes. As shown in Fig. 4.4, in this case, that column name is **barcode-sequence**. Additionally, in this dataset, the barcode reads are the reverse complement of those included in the sample metadata file, so we also need to include the **--p-rev-comp-mapping-barcodes** parameter.

```
qiime demux emp-paired \
 --m-barcodes-file SampleMetadataAtacama.tsv \
 --m-barcodes-column barcode-sequence \
 --p-rev-comp-mapping-barcodes \
 --i-seqs EmpPairedEndSequencesAtacama.qza \
 --o-per-sample-sequences DemuxAtacama.qza \
 --o-error-correction-details DemuxDetailsAtacama.qza

 Saved SampleData[PairedEndSequencesWithQuality] to:
 DemuxAtacama.qza
 Saved ErrorCorrectionDetails to: DemuxDetailsAtacama.qza
```

4.3.5 Summarize the Demultiplexing Results and Examine Quality of the Reads

After demultiplexing, we can use the following commands to create a visualization.

```
qiime demux summarize \
--i-data DemuxAtacama.qza \
--o-visualization DemuxAtacama.qzv
```

        Saved Visualization to: DemuxAtacama.qzv

Similarly as we review the visualization of the sample metadata, to view this visualization we can navigate to QIIME2 viewer in browser or copy over the .qzv output to the computer, and open DemuxAtacama.qzv in www.view.qiime2.org. We can view how many sequences were obtained per sample. Click "**Overview**"; it shows the demultiplexed sequence counts summary (minimum, median, mean, maximum, total) and the per-sample sequence counts (total samples, sample name, and sequence count). The demultiplexed sequence counts summary is displayed as a frequency plot, which is downloadable as a .pdf file. The per-sample sequence counts can be downloaded as a .csv format file. In the "**Interactive Quality Plot**," we can hover over a specific position to check how many reads are at least that long. These are the reads that were sampled for computing sequence quality. We can click at any position on the plots of Forward Reads and Reverse Reads to check the quality score for that position. For example, for Forward Reads, the 50th percentile (median) quality score at position 100 is 38. We can download both forward and reverse parametric seven-number summaries (2nd, 9th, 25th, 50th (Median), 75th, 91st, and 98th percentiles) as a .csv format file. The "Interactive Quality Plot" also includes a demultiplexed sequence length summary (Figs. 4.5 and 4.6).

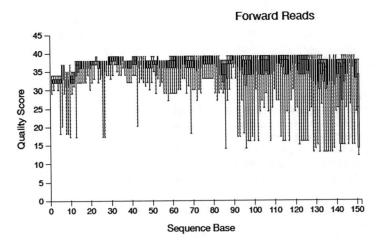

Fig. 4.5 Quality score box plots sampled from 10,000 random forward reads for Atcama soil study

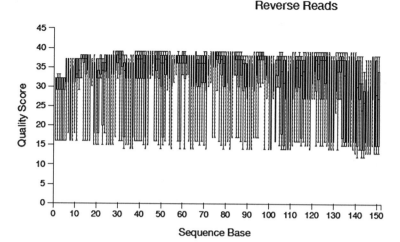

Fig. 4.6 Quality score box plots sampled from 10,000 random reverse reads for Atcama soil study

4.4 Analyzing Demultiplexed Paired-End FASTQ Data Using Deblur and q2-deblur Plugin

In mouse gut microbiome example (Example 4.1), we show how to conduct denoising and QC filtering sequences to generate feature table and feature data using **DADA2** method. Here, we illustrate the **Deblur**, another denoising method currently available in QIIME 2.

4.4.1 Introduction to Deblur and q2-deblur Plugin

Deblur (Amir et al. 2017) was developed by taking a sub-operational-taxonomic-unit (sub-OTU or sOTU) approach, aiming to identify exact sequences, i.e., obtain putative error-free sequences or single-nucleotide resolution in amplicon studies such as from Illumina MiSeq and HiSeq sequencing platforms. To obtain single-nucleotide resolution, Deblur employs a sample-by-sample approach combined with a greedy algorithm, which compares sequence-to-sequence Hamming distances within a sample to an upper-bound error profile (Amir et al. 2017). Deblur uses an upper error rate bound and a constant probability of indels and the mean read error rate together, this algorithm enabling removing predicted error-derived reads from neighboring sequences when the sequences are aligned together into "sub-OTUs" (Amir et al. 2017). Like ASVs, the sub-OTUs are considered representing the true biological sequences present in the data. The benefits of employing a sample-by-sample approach are that Deblur reduces both memory requirements and computational demand (Amir et al. 2017; Nearing et al. 2018). For the details on algorithm and the development of Deblur, the reader is referred to Sect. 8.3.2.6.

4.4.2   Process Initial Quality Filtering

To obtain high-quality sequence variant data, **Deblur** uses sequence error profiles to relate erroneous sequence reads to the true biological sequence from which they belong to. To achieve this goal, an initial quality filtering approach should be taken based on quality scores, which was recommended by Bokulich et al. in 2013 (Bokulich et al. 2013). To perform sequence quality control, QIIME 2 has wrapped **Deblur** quality filtering method in the **q2-deblur** plugin as implemented in **q2-quality-filter**.

Below we process an initial quality-filter with default settings prior to using **Deblur**. As **DADA2** method, **Deblur** also has an option to truncate reads to a constant length prior to denoising with the parameter **–p-trim-length**. The truncating parameter is optional. We can specify the parameter **–p-trim-length -1** to disable truncation in Deblur.

```
cd QIIME2R-Bioinfromatics/Atacama
qiime quality-filter q-score\
 --i-demux DemuxAtacama.qza\
 --o-filtered-sequences DemuxFilteredAtacama.qza\
 --o-filter-stats DemuxFilterStatsAtacama.qza

    Saved SampleData[SequencesWithQuality] to:
    DemuxFilteredAtacama.qza
    Saved QualityFilterStats to: DemuxFilterStatsAtacama.qza
```

4.4.3   Preliminary Works for Denoising with Deblur

The above artifact DemuxAtacama.qza is a demultplexed EmpPairedEndSequencesAtacama.qza file obtained in Sect. 4.3.4. It is SampleData[PairedEndSequencesWithQuality] file. After implementing qiime quality-filter q-score, this artifact was saved as DemuxFilteredAtacama.qza into the QIIME2R-Bioinfromatics/Atacama directory folder. Two denoise-methods are available in the deblur plugin to denoise sequences: (1) denoise-16S for denoising 16S sequences and (2) denoise-other for denoising other types of sequences. Some preliminary works need to be done before denoising with Deblur.

Step 1: Join reads.

Deblur needs the paired-end sequences jointed before using it to denoise sequences. Because the paired-end sequences have not been jointed, we here join the paired-end reads in QIIME 2 using the **q2-vsearch** plugin.

```
qiime vsearch join-pairs \
 --i-demultiplexed-seqs DemuxAtacama.qza\
 --o-joined-sequences DemuxJoinedAtacama.qza
```

```
      Saved SampleData[JoinedSequencesWithQuality] to:
      DemuxJoinedAtacama.qza
```

Step 2: Generate a summary of the joined data (in this case, DemuxJoinedAtacama.qza).

```
qiime demux summarize \
 --i-data DemuxJoinedAtacama.qza \
 --o-visualization DemuxJoinedAtacama.qzv
```

```
      Saved Visualization to: DemuxJoinedAtacama.qzv
```

Like other demultiplexing sequences, it is crucial to view the summary of joined data with read quality via the QIIME2 viewer. This summary provides several particularly useful information including how long the joined reads are and how many reads were used to estimate the quality score distribution at a specific position. The quality score plots show that most sequences are at least 250 bases long, which provides the important information to specify the long value to trim the sequences (Fig. 4.7).

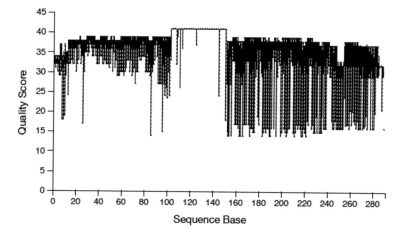

Fig. 4.7 Quality score box plots sampled from 10,000 random reads for Atcama soil study

Step 3: Conduct sequence quality control to the sequences using quality-filter q-score-joined

The quality-filter q-score-joined method is identical to quality-filter q-score, except that it operated on joined reads.

```
qiime quality-filter q-score \
--i-demux DemuxJoinedAtacama.qza \
--o-filtered-sequences DemuxJoinedFilteredAtacama.qza \
--o-filter-stats DemuxJoinedFilterStatsAtacama.qza

Saved SampleData[JoinedSequencesWithQuality] to:
DemuxJoinedFilteredAtacama.qza
Saved QualityFilterStats to: DemuxJoinedFilterStatsAtacama.qza
```

4.4.4 Denoise Sequences with Deblur to Construct Feature Table and Feature Data

The two denoise-sequences methods in the **deblur** plugin, **denoise-16S** and **denoise-other**, are used in different ways. When using **denoise-16S**, an initial positive filtering step will be performed to discard those reads that have less than 60% identity similarity to sequences from the 85% OTU GreenGenes database. Otherwise **Deblur** recommends using the **denoise-other** method. For example, when you apply **Deblur** to 18S data, you need to specify a reference composed of 18S sequences so that you can filter out sequences which do not appear to be 18S.

The **qiime deblur denoise-16S** performs sequence quality control for Illumina data using a 16S reference as a positive filter. Currently QIIME 2 only supports forward reads and uses the 88% OTUs from Greengenes 13_8 as the specific reference. Thus, this method is only limited to be used for a 16S amplicon protocol on an Illumina platform.

Deblur can start here to denoise the sequences, which will provide us additional quality control and similarly much higher-quality results as DADA2 did. The crucial action here is to specify a sequence length value for **--p-trim-length** parameter based on reviewing quality score plots (250 in this case). All the sequences will be trimmed to this length, and any sequences which are not at least this long will be discarded.

Here we use the **denoise-16S** method. In the following commands, the required inputs (demultiplexed sequences) are one of artifacts: (1) SampleData[SequencesWithQuality], (2) SampleData[PairedEndSequencesWithQuality], or (3) SampleData[JoinedSequencesWithQuality](DemuxJoinedFilteredAtacama. qza), which will be denoised. The parameter **--p-trim-length** is used to specify sequence trim length, which is also required. Specifying −1 to disable trimming.

The parameters **--p-sample-stats** or **--p-no-sample-stats** are used to gather or not gather stats per sample. The default is false. Here we want to gather the stats.

Three outputs need to be required, including **--o-table,** an artifact of FeatureTable [Frequency], which is the resulting denoised feature table; **--o-representative-sequences**, an artifact of FeatureData[Sequence], which is the resulting feature sequences; and **--o-stats**, an artifact of per-sample stats.

```
qiime deblur denoise-16S \
  --i-demultiplexed-seqs DemuxJoinedFilteredAtacama.qza \
  --p-trim-length 250 \
  --p-sample-stats \
  --o-table FeatureTableAtacama.qza \
  --o-representative-sequences RepSeqsAtacama.qza \
  --o-stats StatsAtacama.qza
```

```
Saved FeatureTable[Frequency] to: FeatureTableAtacama.qza
Saved FeatureData[Sequence] to: RepSeqsAtacama.qza
Saved DeblurStats to: StatsAtacama.qza
```

In above commands, we did not specify other parameters, instead used the defaults: (1) **--p-left-trim-len** with range(0, None) is used to trim sequence from the 5′ end. The default value of 0 will disable this trim. (2) **--p-mean-error** is used to specify the mean per nucleotide error for original sequence estimate. The default value is 0.005. (3) **--p-indel-prob** is used to specify the insertion/deletion (indel) probability (same for N indels). The default value is 0.01. (4) **--p-indel-max** is used to specify the maximum number of insertion/deletions. The default value is 3. (5) **--p-min-reads** is used to specify to retain only features appearing at least min-reads across all samples in the resulting feature table. The default value is 10. (6) **--p-min-size** is used to specify to discard all features with an abundance less than min-size in each sample. The default value is 2. (7) **--p-jobs-to-start** is used to specify the number of jobs to start (if to run in parallel). The default value is 1. And (8) **--p-hashed-feature-ids / --p-no-hashed-feature-ids** is used to specify whether or not hash the feature IDs. The default is true.

4.4.5   Summarize the Feature Table and Feature Data from Deblur

As in **DADA2** approach, now we can summarize the feature table and feature data and review them via the QIIME2 viewer.

```
qiime feature-table summarize \
  --i-table FeatureTableAtacama.qza \
```

```
--o-visualization FeatureTableAtacama.qzv \
--m-sample-metadata-file SampleMetadataAtacama.tsv
```

Saved Visualization to: FeatureTableAtacama.qzv

```
qiime feature-table tabulate-seqs \
 --i-data RepSeqsAtacama.qza \
 --o-visualization RepSeqsAtacama.qzv
```

Saved Visualization to: RepSeqsAtacama.qzv

```
qiime metadata tabulate \
 --m-input-file DemuxJoinedFilterStatsAtacama.qza \
 --o-visualization DemuxJoinedFilterStatsAtacama.qzv
```

Saved Visualization to: DemuxJoinedFilterStatsAtacama.qzv

4.4.6   Remarks on DADA2 and Deblur

DADA2 and **Deblur** share some common strengths over traditional denoising and QC filtering sequences methods, including:

- Both **DADA2** and **Deblur** bioinformatic sequence denoising approaches were proposed to correct sequencing errors to improve taxonomic resolution. Similar to traditional OTU method, both pipelines are self-contained, performing 16S rRNA gene sequencing data from raw sequences (i.e., FASTQ). However, both methods have advanced the sequencing from traditionally binning sequences into 97% OTUs to effectively using the sequences themselves as the unique identifier for a taxon (also referred to 100% OTU).
- Both methods perform quality filtering, denoising, and chimera removal and require little data preparation, and don't need to perform any quality screening prior to running them.
- Both DADA2 and Deblur methods have their own advantages and disadvantages. They also share some common properties such as obtaining single-nucleotide resolution. However, we have no intention to compare DADA2 and Deblur methods.
- Deblur runs its denoising process sample-by-sample, which helps lower Deblur's computational requirements; however, it also reduces its ability to correct multi-run.
- Another major difference between Deblur and DADA2 is that the built-in Deblur function in QIIME 2 uses a positive filter. That is, they use different strategies to

handle errors: DADA2 corrects/changes errors or "erroneous" sequences to match the sequence from which they're inferred to have arisen. In contrast, Deblur removes those error sequences. By using the default setting Deblur discards reads that do not reach the threshold to any sequence in the 88% representative sequences Greengenes database, which results in the remaining total numbers of sequences and features after denoising using Deblur are less than the total numbers of sequences and features after denoising using DADA2.

- Deblur is able to handle processed data whereas DADA2 can only handle raw data. This difference highlights its importance of Deblur because many publicly available datasets are already processed.
- Deblur was evaluated having poor performance than DADA2 and open-reference OTU clustering to detect low abundant taxa in the extreme dataset at 97% identity (Nearing et al. 2018).
- However, our main goal here is to illustrate these two methods via QIIME 2 and introduce their theories accordingly as necessary to better understand their approaches.

4.5 Summary

In this chapter, we described two methods of building feature table and feature representative sequences data from raw reads, DADA2 and Deblur, illustrated their use in QIIME 2 plugins using two real 16S rRNA microbiome datasets. First, we introduced some general procedures to analyze demultiplexed paired-end FASTQ data including preparation of sample metadata and raw sequence data, importation of data files as Qiime Zipped Artifacts (.qza), as well as examination and visualization of the qualities of the sequence reads. Second, we introduced DADA2 and q2-dada2 plugin and illustrated how to use them to analyze demultiplexed paired-end FASTQ data and specifically to denoise sequences to construct feature table and feature data with q2-dada2 plugin as well as how to summarize the feature table and feature data from q2-dada2 plugin. Third, we illustrated how to analyze multiplexed paired-end FASTQ data using q2-dada2 plugin. Fourth, we introduced Deblur and q2-deblur plugin and illustrated how to use them to analyze demultiplexed paired-end FASTQ Data such as processing initial quality filtering, conducting preliminary works for denoising with Deblur, and particularly denoising sequences with Deblur to construct feature table and feature data and summarize them. Chapter 5 will introduce how to assign taxonomy and build phylogenetic tree. Chapter 6 will introduce the traditional OTUs clustering methods.

References

Amir, Amnon, Daniel McDonald, Jose A. Navas-Molina, Evguenia Kopylova, James T. Morton, Xu Zhenjiang Zech, Eric P. Kightley, Luke R. Thompson, Embriette R. Hyde, Antonio Gonzalez, and Rob Knight. 2017. Deblur rapidly resolves single-nucleotide community sequence patterns. *mSystems* 2 (2): e00191–e00116. https://doi.org/10.1128/mSystems. 00191-16. https://www.ncbi.nlm.nih.gov/pubmed/28289731. https://www.ncbi.nlm.nih.gov/ pmc/articles/PMC5340863/.

Bokulich, Nicholas A., Sathish Subramanian, Jeremiah J. Faith, Dirk Gevers, Jeffrey I. Gordon, Rob Knight, David A. Mills, and J. Gregory Caporaso. 2013. Quality-filtering vastly improves diversity estimates from Illumina amplicon sequencing. *Nature Methods* 10 (1): 57–59. https:// doi.org/10.1038/nmeth.2276.

Callahan, B.J., P.J. McMurdie, M.J. Rosen, A.W. Han, A.J. Johnson, and S.P. Holmes. 2016a. DADA2: High-resolution sample inference from Illumina amplicon data. *Nature Methods* 13 (7): 581–583. https://doi.org/10.1038/nmeth.3869. https://www.ncbi.nlm.nih.gov/ pubmed/27214047. https://www.ncbi.nlm.nih.gov/pmc/articles/PMC4927377/.

Callahan, Ben J., Kris Sankaran, Julia A. Fukuyama, Paul J. McMurdie, and Susan P. Holmes. 2016b. Bioconductor Workflow for Microbiome Data Analysis: From raw reads to community analyses. *F1000Research* 5: 1492–1492. https://doi.org/10.12688/f1000research.8986.2. https://www.ncbi.nlm.nih.gov/pubmed/27508062. https://www.ncbi.nlm.nih.gov/pmc/articles/ PMC4955027/.

Callahan, Benjamin, Paul McMurdie, and Susan Holmes. 2017. Exact sequence variants should replace operational taxonomic units in marker-gene data analysis. *The ISME Journal* 11: 2639.

Caporaso, J.G., J. Kuczynski, J. Stombaugh, K. Bittinger, F.D. Bushman, E.K. Costello, N. Fierer, A.G. Peña, J.K. Goodrich, J.I. Gordon, G.A. Huttley, S.T. Kelley, D. Knights, J.E. Koenig, R.E. Ley, C.A. Lozupone, D. McDonald, B.D. Muegge, M. Pirrung, J. Reeder, J.R. Sevinsky, P.J. Turnbaugh, W.A. Walters, J. Widmann, T. Yatsunenko, J. Zaneveld, and R. Knight. 2010. QIIME allows analysis of high-throughput community sequencing data. *Nature Methods* 7 (5): 335–336. https://doi.org/10.1038/nmeth.f.303.

Edgar, Robert C., and Henrik Flyvbjerg. 2015. Error filtering, pair assembly and error correction for next-generation sequencing reads. *Bioinformatics* 31 (21): 3476–3482. https://doi.org/10.1093/ bioinformatics/btv401.

Fadrosh, Douglas W., Bing Ma, Pawel Gajer, Naomi Sengamalay, Sandra Ott, Rebecca M. Brotman, and Jacques Ravel. 2014. An improved dual-indexing approach for multiplexed 16S rRNA gene sequencing on the Illumina MiSeq platform. *Microbiome* 2 (1): 6–6. https://doi. org/10.1186/2049-2618-2-6. https://www.ncbi.nlm.nih.gov/pubmed/24558975. https://www. ncbi.nlm.nih.gov/pmc/articles/PMC3940169/.

Hildebrand, Falk, Raul Tadeo, Anita Yvonne Voigt, Peer Bork, and Jeroen Raes. 2014. LotuS: An efficient and user-friendly OTU processing pipeline. *Microbiome* 2 (1): 30. https://doi.org/10. 1186/2049-2618-2-30.

Kumar, Surendra, Tor Carlsen, Bjørn-Helge Mevik, Pål Enger, Rakel Blaalid, Kamran Shalchian-Tabrizi, and Håvard Kauserud. 2011. CLOTU: An online pipeline for processing and clustering of 454 amplicon reads into OTUs followed by taxonomic annotation. *BMC Bioinformatics* 12 (1): 182. https://doi.org/10.1186/1471-2105-12-182.

Mysara, Mohamed, Mercy Njima, Natalie Leys, Jeroen Raes, and Pieter Monsieurs. 2017. From reads to operational taxonomic units: An ensemble processing pipeline for MiSeq amplicon sequencing data. *GigaScience* 6 (2). https://doi.org/10.1093/gigascience/giw017.

Nearing, Jacob T., Gavin M. Douglas, André M. Comeau, and Morgan G.I. Langille. 2018. Denoising the Denoisers: An independent evaluation of microbiome sequence error-correction approaches. *PeerJ* 6: e5364–e5364. https://doi.org/10.7717/peerj.5364. https://www.ncbi.nlm. nih.gov/pubmed/30123705. https://www.ncbi.nlm.nih.gov/pmc/articles/PMC6087418/.

Neilson, Julia W., Katy Califf, Cesar Cardona, Audrey Copeland, Will Van Treuren, Karen L. Josephson, Rob Knight, Jack A. Gilbert, Jay Quade, J. Gregory, and Caporaso. 2017.

Significant impacts of increasing aridity on the arid soil microbiome. *MSystems* 2 (3): e00195–e00116.

Rideout, Jai Ram, John H. Chase, Evan Bolyen, Gail Ackermann, Antonio González, Rob Knight, and J. Gregory Caporaso. 2016. Keemei: Cloud-based validation of tabular bioinformatics file formats in Google Sheets. *GigaScience* 5 (1). https://doi.org/10.1186/s13742-016-0133-6.

Rosen, Michael J., Benjamin J. Callahan, Daniel S. Fisher, and Susan P. Holmes. 2012. Denoising PCR-amplified metagenome data. *BMC Bioinformatics* 13: 283–283. https://doi.org/10.1186/1471-2105-13-283. https://www.ncbi.nlm.nih.gov/pubmed/23113967. https://www.ncbi.nlm.nih.gov/pmc/PMC3563472/.

Schloss, Patrick D. 2020. Reintroducing mothur: 10 years later. *Applied and Environmental Microbiology* 86 (2): e02343–e02319.

Schloss, Patrick D., Alyxandria M. Schubert, Joseph P. Zackular, Kathryn D. Iverson, Vincent B. Young, and Joseph F. Petrosino. 2012. Stabilization of the murine gut microbiome following weaning. *Gut Microbes* 3 (4): 383–393. https://doi.org/10.4161/gmic.21008. https://www.ncbi.nlm.nih.gov/pubmed/22688727. https://www.ncbi.nlm.nih.gov/pmc/articles/PMC3463496/.

Xia, Yinglin, Jun Sun, and Ding-Geng Chen. 2018. Bioinformatic analysis of microbiome data. In *Statistical analysis of microbiome data with R*, 1–27. Singapore: Springer.

Chapter 5
Assigning Taxonomy, Building Phylogenetic Tree

Abstract Taxonomic classification of the representative sequences and clustering of OTUs (Operational Taxonomic Units) are two core steps in bioinformatic analysis of microbiome data. Chapter 4 described and illustrated how to generate feature table and feature data (i.e., representative sequences). This chapter describes and illustrates two more core bioinformatic analyses: assigning taxonomy and building phylogenetic tree.

Keywords Reference databases · q2-feature-classifier · Taxonomic classification · RDP · Greengenes · SILVA · UNITE · NCBI · EzBioCloud · GTDB (Genome Taxonomy Database) · HITdb · nifHdada2 · pr2database · MAFFT · FastTree

Taxonomic classification of the representative sequences and clustering of OTUs (Operational Taxonomic Units) are two core steps in bioinformatic analysis of microbiome data. In Chap. 4, we described and illustrated how to generate feature table and feature data (i.e., representative sequences), which is one crucial component for microbiome study because it provides the most wanted data for downstream analysis. In Chap. 5, we first describe and illustrate other core bioinformatic analyses: assign taxonomy (Sect. 5.1) and build phylogenetic tree (Sect. 5.2). Then we briefly summarize the materials covered in this chapter (Sect. 5.3).

5.1 Assigning Taxonomy

Taxonomic assignment is a crucial step in bioinformatic analysis of microbiome data, while reference databases are essential component in the analysis of microbiomes because they are used to transform sequences into readable taxonomy (e.g., bacterial) names.

Supplementary Information The online version contains supplementary material available at https://doi.org/10.1007/978-3-031-21391-5_5.

5.1.1   Bioinformatics Tools and Reference Databases

Various bioinformatics tools are available for analysis of 16S rRNA gene amplicon sequencing data (Plummer et al. 2015; Nilakanta et al. 2014). Among these software, the most widely used are QIIME (Caporaso et al. 2010) and its extension of QIIME 2 (Bolyen et al. 2019), and DADA2 (Callahan 2021). In Chap. 4, we used DADA2 via QIIME 2 to generate feature frequency table and its representative sequence data table which contains the denoised sequences.

Taxonomic classification is performed after the sequences pass the filtering process, which is typically searched against a known reference taxonomy classifier at a pre-determined threshold. Most classifiers including the Ribosomal Database Project (RDP) (Cole et al. 2005), Greengenes (DeSantis et al. 2006), SILVA 16S rRNA gene database (Yilmaz et al. 2014), the UNITE database (Kõljalg et al. 2013), the National Center for Biotechnology Information (NCBI) (Federhen 2011), and EzBioCloud (Yoon et al. 2017) are publicly available. When performing taxonomic classification, we should use the frequently updated databases to avoid mapping the sequences to obsolete taxonomy names.

Currently the most often used reference databases for 16S taxonomy assignment are Silva (Quast et al. 2013), RDP (Cole et al. 2005; Maidak et al. 2000), NCBI (Federhen 2011; Geer et al. 2010), and Greengenes (McDonald et al. 2012).

SILVA (from Latin silva, forest) (Pruesse et al. 2007) is a comprehensive online resource providing quality controlled databases of aligned rRNA sequences data for all three domains of life (Bacteria, Archaea, and Eukarya) (Pruesse et al. 2012). The SILVA database is based on phylogenies for small subunit rRNAs (16S and 18S), and manually curates its taxonomic rank assignment (Yilmaz et al. 2014). Other pipelines, such as QIIME 2, DADA2, and mothur (Schloss et al. 2009), use the SILVA 16S rRNA gene reference database (Plummer et al. 2015).

Like SILVA, **RDP** database (Cole et al. 2005) also provides the aligned and annotated rRNA gene sequences for all three domains of life (Bacteria, Archaea, and Eukarya) and a phylogenetically consistent taxonomic framework for these data. The RDP database obtains bacterial rRNA sequences from the International Nucleotide Sequence Databases (INSD: GenBank/EMBL/DDBJ) on a monthly basis (Nakamura et al. 2012). As a Bayesian classifier, the RDP Classifier can rapidly and accurately classify bacterial 16S rRNA sequences into the new higher-order with the majority of classifications (98%) having high estimated confidence ($\geq$95%) and high accuracy (98%) (Wang et al. 2007). RDP provides taxonomic assignments from domain to genus (Wang et al. 2007), and thus these collected 16S sequences in RDP database have not all been assigned to species level taxonomic names.

The **NCBI** Taxonomy project began in 1991. NCBI taxonomy database (Federhen 2011; Geer et al. 2010) is the standard nomenclature and classification repository for the International Nucleotide Sequence Database Collaboration (INSDC), comprising the GenBank (Benson et al. 2013), the European Nucleotide Archive (ENA) (EMBL) (Leinonen et al. 2011), and the DNA DataBank of Japan (DDBJ) (Mashima et al. 2017) databases (Federhen 2011). It contains all organism names and taxonomic lineages for each of the sequences associated with submissions to the NCBI global database and is manually curated to maintain a phylogenetic taxonomy for the source organisms represented in the sequence databases (Federhen 2011).

Greengenes (McDonald et al. 2012; DeSantis et al. 2006) is a chimera-check 16S rRNA gene database that has Bacteria and Archaea sequences. It provides chimera screening, standard alignment, and taxonomic classification using multiple published taxonomies. Most of the sequences in Greengenes are retrieved from NCBI database (DeSantis et al. 2006). Greengenes taxonomic classification has been improved by explicit ranks for ecological and evolutionary analyses of bacteria and archaea (McDonald et al. 2012): (1) In Greengenes a "taxonomy to tree" approach has been used for transferring group names from an existing taxonomy to a tree topology, and applied to the Greengenes, NCBI, and cyanoDB (Cyanobacteria only) taxonomies to a de novo tree sequences (McDonald et al. 2012). Reference phylogenies provide the crucial information for a taxonomic framework in interpreting marker gene and metagenomic surveys, and help to reveal novel species remarkably. (2) Explicit rank information provided by the NCBI taxonomy has been incorporated to group names for better user orientation and classification consistency and hence significantly improved the classification of the sequences in the merged taxonomy (McDonald et al. 2012). In summary, Greengenes is a dedicated full-length 16S rRNA gene database providing a curated taxonomy based on de novo tree inference. The database is used for closed-reference OTU clustering. Reads are clustered against this reference database. It has been wrapped in QIIME 1 and QIIME 2.

However, as reviewed in Chap. 4, the OTU method clustering sequences with a fixed 97% similarity threshold might avoid fine-scale variation among sequences (Rosen et al. 2012). OTUs are not species, and hence OTU method often eliminates biological information of the data, and the construction of OTUs is not necessitated by amplicon errors (Callahan et al. 2016). Thus, *DADA2 uses* an alternative error-modeling approach for denoising and clustering amplicons. This may be the reason that *the source of GreenGenes database will no longer be maintained in DADA2 because it is deprecated (see Table 5.2)*. For clustering sequences into OTUs, the reader is referred to Chap. 6.

There are a variety of databases available; QIIME 2 and DADA2 have formatted and maintained most often used taxonomic reference databases.

5.1.2 QIIME 2-Formatted and Maintained Taxonomic Reference Databases

We summarize the QIIME 2-formatted and maintained taxonomic reference databases into Table 5.1 (Bolyen et al. 2019). When the reader uses these databases in Table 5.1, please check the updating from QIIME 2 and cite QIIME 2 and the original databases.

Table 5.1 QIIME 2-formatted and maintained taxonomic reference databases

Category	Database name	Description
Taxonomy classifiers for use with q2-feature-classifier	Naive Bayes classifiers (Bokulich et al. 2018, 2021) trained on: Silva 138 99% OTUs full-length sequences (MD5: b8609f23e9b17bd4a1321a8971303310) (Quast et al. 2012; Yilmaz et al. 2013) Silva 138 99% OTUs from 515F/806R region of sequences (MD5: e05afad0fe87542704be96ff48382d4) (Quast et al. 2012; Yilmaz et al. 2013) Greengenes 13_8 99% OTUs full-length sequences (MD5: 6bbc9b3f29b51d663063a7979dd95f1) (McDonald et al. 2012) Greengenes 13_8 99% OTUs from 515F/806R region of sequences (MD5: 9e82e8969303b3a86ac941ceafeeac86) (McDonald et al. 2012)	Pre-trained classifiers can be used with q2-feature-classifier However, QIIME 2 warns that currently using pre-trained classifiers presents a security risk and this security risk will be addressed in a future version of q2-feature-classifier Taxonomic classifiers have best performance when they are trained based on the specific sample preparation and sequencing parameters of the study (e.g., the primers that were used for amplification and the length of your sequence reads) Therefore QIIME 2 recommends in general the instructions in "Training feature classifiers with q2-feature-classifier" should be followed when the users train their own taxonomic classifiers QIIME 2 notes that these classifiers were trained using scikit-learn 0.24.1, and therefore can only be used with scikit-learn 0.24.1 Using the pretrained-classifiers that were published with the release of QIIME 2 if the errors related to scikit-learn version mismatches are observed
Weighted Taxonomic Classifiers	Weighted pre-trained classifiers (Kaehler et al. 2019): Weighted Silva 138 99% OTUs full-length sequences (MD5: 48965bb0a9e63c411452a460d92cfc04) Weighted Greengenes 13_8 99% OTUs full-length sequences (MD5: 2baf87fce174c5f6c22a4c40866f1f1fe) Weighted Greengenes 13_8 99% OTUs from 515F/80 6R region of sequences (MD5: 8fb808c4af1c7526a2bdfaafa764e21f)	Trained with weights to take into account the fact that not all species are equally likely to be observed Provide superior classification precision if the V4 sample comes from any of the 14 QIIME 2 tested habitat types They might still help even if the sample doesn't come from one of those habitats Training with weights specific to the habitat should help even more Weights for a range of habitats are available from https://github.com/BenKaehler/readytowear

Marker gene reference databases	Greengenes (16S rRNA) (DeSantis et al. 2006; McDonald et al. 2012): 13_8 (most recent) 13_5 12_10 February 4th, 2011 Silva (16S/18S rRNA) (Bokulich et al. 2021): Silva 138 SSURef NR99 full-length sequences (MD5: de8886bb2c059b1e87522255d271f3010) (Quast et al. 2012; Yilmaz et al. 2013) Silva 138 SSURef NR99 full-length taxonomy (MD5: f12d5b78bf4b1519721fe52803581c3d) (Quast et al. 2012; Yilmaz et al. 2013) Silva 138 SSURef NR99 515F/806R region sequences (MD5: a914837bc3f8964b156a9653e2420d22) Silva 138 SSURef NR99 515F/806R region taxonomy (MD5: e2c40ae4c60cbf75e24312bb24652f2c) (Quast et al. 2012; Yilmaz et al. 2013) UNITE (fungal ITS) (Põlme et al. 2020): All releases are available for download at https://unite.ut.ee/repository.php	Formatted for use with QIIME 1 and QIIME 2 Need to import them into artifacts if using these databases with QIIME 2 Silva (16S/18S rRNA): QIIME is compatible SILVA releases (up to release 132) The pre-formatted SILVA 138 release reference sequence and taxonomy files provided here by QIIME were processed using RESCRIPt (https://github.com/bokulich-lab/RESCRIPt) and q2-feature-classifier (https://github.com/qiime2/q2-feature-classifier/) UNITE (fungal ITS): Find more information about UNITE at https://unite.ut.ee/
SEPP reference databases	SEPP references (SEPP-Refs project): Silva 128 SEPP reference database (MD5: 7879792a6f42c532531de986f5c4de) Greengenes 13_8 SEPP reference database (MD5: 9ed215415b52c362e25cb0a8a46e1076)	These databases: Are intended for use with q2-fragment-insertion Are constructed directly from the SEPP-Refs project

5.1.3 DADA2-Formatted and Maintained Taxonomic Reference Databases

DADA2 formatted 16S rRNA gene sequences for both bacteria and archaea (Alishum 2019). DADA2 collated and formatted two combined bacterial and archaeal 16S rRNA gene sequence databases (RefSeq+RDP and Genome Taxonomy Database (GTDB)) and used various sources for assigning taxonomy. DADA2 categorizes the 16S databases **into Maintained and Contributed.** DADA2 maintains reference fastas for the three most common 16S databases: Silva, RDP, and GreenGenes. It also maintains the General Fasta releases of the UNITE project for ITS taxonomic assignment. DADA2 also makes formatted versions of other databases available as "contributed."

DADA2 created the dada2-compatible training fastas from the Silva NR99 and taxonomy files, the RDP trainset 16 and release 11.5 database, and the GreenGenes 13.8 OTUs clustered at 97%.

We summarize the DADA2-formatted and maintained taxonomic reference databases into Table 5.2 (Callahan 2021). When the reader uses these databases in Table 5.2, please check the updating from DADA2 and cite DADA2 and the original databases.

5.1.4 Introduction to q2-Feature-Classifier

We may choose to train our classifiers using a suitable method, such as using **q2-feature-classifier** protocol that is available in QIIME 2 (Bokulich et al. 2018). The **q2-feature-classifier** is a QIIME 2 plugin for taxonomy classification of marker-gene sequences. It contains several novel machine-learning and alignment-based methods including a scikit-learn naïve Bayes machine-learning classifier, and alignment-based taxonomy consensus methods based on VSEARCH, and BLAST + for classification of bacterial 16S rRNA and fungal ITS (internal transcribed spacer) marker-gene amplicon sequence data, which were evaluated as match or outperform the species-level accuracy of other commonly used methods designed for classification of marker gene sequences (Bokulich et al. 2018).

The **q2-sample-classifier** plugin employs scikit-learn (Pedregosa et al. 2011) for supervised learning (SL) to classify sequence and feature selection algorithms. The classify-sklearn method is a pre-fitted sklearn-based taxonomy classifier for implementing scikit-learn machine learning algorithms, while maintaining an easy-to-use interface tightly integrated with the Python language with several distinctive features including (1) it is distributed under the BSD (Berkeley Source Distribution) license; thus it has low restriction and requirement for using the distribution of many free and open source software; (2) it incorporates compiled code for efficiency; (3) it depends only on numpy and scipy to facilitate easy distribution; and (4) it focuses on imperative programming (Pedregosa et al. 2011).

Table 5.2 DADA2-formatted and maintained taxonomic reference databases

Category	Database name	Description
Maintained databases	Silva (16S/18S rRNA) (Bokulich et al. 2021): • Silva version 138.1 - UPDATED Mar 10, 2021 • Silva version 132 • Silva version 128 • Silva version 123	Like Silva version 138, the DADA2-formatted reference fastas are optimized for classification of Bacteria and Archaea, and are not suitable for classifying Eukaryotes
	RDP (Cole et al. 2005): • RDP trainset 18 • RDP trainset 16 • RDP trainset 14	
	UNITE (fungal ITS) (Põlme et al. 2020): • UNITE (use the General Fasta releases)	
	Greengenes (16S rRNA) (DeSantis et al. 2006; McDonald et al. 2012; Callahan 2016): • GreenGenes version 13.8	*DADA2 will no longer maintain the source GreenGenes database because it is deprecated*

(continued)

Table 5.2 (continued)

Category	Database name	Description
Contributed databases	RefSeq + RDP (NCBI RefSeq 16S rRNA database supplemented by RDP) (Alishum 2019): • Reference files formatted for assignTaxonomy • Reference files formatted for assignSpecies	DADA2 compiled this database on May 14, 2018, from predominantly the NCBI RefSeq 16S rRNA database (https://www.ncbi.nlm.nih.gov/refseq/targetedloci/16S_process/) and was supplemented with extra sequences from the RDP database (https://rdp.cme.msu.edu/misc/resources.jsp) This database contains 14676 bacterial and 660 archaea full 16S rRNA gene sequences
	GTDB (Genome Taxonomy Database) (Alishum 2019): • GTDB Version 202: Genome Taxonomy Database ○ Version 86 for assignTaxonomy and assignSpecies	DADA2 downloaded this database from (http://gtdb.ecogenomic.org/downloads) on November 20, 2018 DADA2 formatted GTDB this reference sequence set which contains 20486 bacteria and 1073 archaea full 16S rRNA gene sequences
	Human InTestinal 16S rRNA (Diener 2016; Ritari et al. 2015): • HitDB version 1	HITdb is a reference taxonomy for Human **Intestinal** 16S rRNA genes as described in Ritari et al. (2015) **HITdb v1.00 for Dada2 is converted version to be used with dada2 HITdb is specific for intestinal samples; thus it might lead to arbitrarily wrong results for non-intestinal samples**
	RDP fungi LSU (Czaplicki 2017): • RDP fungi LSU trainset 11	**RDP LSU taxonomic training data was formatted for DADA2 (trainingset 11)**
	SILVA v128 and v132 dada2 formatted 18s "train sets" (Morien and Parfrey 2018):	These are species-level taxonomy classification training sets for the assignTaxonomy () function in DADA2 The v132 and v128 training sets include every Eukaryotic organism from SILVA's v132 and v128 databases, respectively, clustered at 99% similarity It also includes corrected species labels for the Blastocystis clade, and includes 37 Entamoeba

• Silva Eukaryotic 18S, v132 & v128	sequences sourced from GenBank not present in the original v128 db The v128 training set is modified specifically to allow for better species-level assignments for those two clades in mammalian gut microbiome studies
nifHdada2: v1.1.0 (Moynihan 2020): • nifH ARB, version 1	This is the new reference sequences added to database
pr2database(https://github.com/pr2data base/pr2database/releases): • PR2 version 4.7.2+.	The provided latest PR2 version 4.14.0 is a single SSU database that contains sequences for: 18S rRNA from nuclear and nucleomorph, 16S rRNA from plastid, apicoplast, chromatophore, Mitochondrion, as well as 16S rRNA from a small selection of bacteria **DADA2 note:** PR2 has different taxLevels than the DADA2 default. When assigning taxonomy against PR2, use the following: assignTaxonomy(..., taxLevels = c ("Kingdom", "Supergroup", "Division", "Class", "Order", "Family", "Genus", "Species")). There are many contributors and references for this database (https://github.com/pr2database/pr2database/releases)

The current improvement of methods for sequencing is capable to differentiate single nucleotide base: Amplicon Sequence Variants (ASVs), or sub-OTUs, which is 100% OTU. More researchers now in microbiome field including the developers of QIIME 2 recommend working with ASVs or sub-OTUs to assign taxonomy to the sequence variants, especially in 16S/18S/ITS amplicon sequencing. Thus, QIIME 2 workflow by default does not include a typical OTU picking step. Here, we follow this default direction; go directly into taxonomy assignment after using DADA2 to quality filter dataset. We will take the denoised sequences (RepSeqsMiSeq_SOP. qza) from Chap. 4 after taking denoising step, and assign taxonomy to each sequence (phylum→ class→. . .genus→species). A trained classifier is required for this step. We can either use a reference set to train a naïve Bayes classifier and save as a QIIME 2 artifact for latter re-use, which avoids re-training the classifier between runs and saves overall running time or download a pretrained classifier.

We can use the **qiime feature-classifier fit-classifier-naïve-bayes** command to train a naïve Bayes classifier. If we want to use a pretrained classifier, there is a pre-trained naïve Bayes classifier artifact available in QIIME 2. This classifier was trained against Greengenes (13_8 reversion) trimmed to contain only the V4 hyper-variable region and pre-clustered at 99% sequence identity (McDonald et al. 2012). We can check the QIIME 2 website (https://docs.qiime2.org/) to look for other available pre-trained artifacts.

5.1.5   Assign Taxonomy Using the q2-Feature-Classifier

Example 5.1: RepSeqsMiSeq_SOP.qza
We assign taxonomy based on the denoised sequence "RepSeqsMiSeq_SOP.qza" artifact using **the q2-feature-classifier**.

Once we obtain an appropriate classifier artifact, we can use the **qiime feature-classifier** command to generate the taxonomic classification results. Here, we use a pretrained classifier from GreenGenes database with 99% OTUs. We download this classifier gg-13-8-99-515-806-nb-classifier.qza from the QIIME 2 website (https://docs.qiime2.org/2022.2/data-resources/). To compare the denoised sequences (RepSeqsMiSeq_SOP.qza) to the GreenGenes reference database to assign taxonomy based on pairwise identity of rRNA sequences, place this classifier gg-13-8-99-515-806-nb-classifier.qza from the download in the working directory.

We need to take a few steps to assign taxonomy to the sequences, as shown below.

Step 1: Import reference data files as Qiime 2 Zipped Artifacts (.qza).

In this case, the downloaded gg-13-8-99-515-806-nb-classifier.qza is already Qiime 2 zipped artifacts (.qza), so we can skip importing it as an artifact (.qza). Otherwise if the downloaded data are zipped (.gz) files or other text files, then we need **qiime tools import** command to import it as an artifact (.qza).

Step 2: Assign taxonomy using QIIME 2 feature-classifier plugin.

Please note that the scikit-learn version used to generate the reference artifact should match the current version of scikit-learn installed. Otherwise the **qiime feature-classifier** will not work and a plugin error message from feature-classifier will be generated. We specify the **classify-sklearn** method in QIIME 2 **feature-classifier** plugin to assign taxonomy to the representative sequences RepSeqsMiSeq_SOP.qza and save the classified taxonomy files as artifacts and name as TaxonomyMiSeq_SOP.qza.

```
source activate qiime2-2022.2
mkdir QIIME2R-Bioinformatics
cd QIIME2R-Bioinformatics
qiime feature-classifier classify-sklearn \
 --i-classifier gg-13-8-99-515-806-nb-classifier.qza\
 --i-reads RepSeqsMiSeq_SOP.qza \
 --o-classification TaxonomyMiSeq_SOP.qza
```

```
Saved FeatureData[Taxonomy] to: TaxonomyMiSeq_SOP.qza
```

Step 3: Generate a visualization of the taxonomy artifact.

```
qiime metadata tabulate \
 --m-input-file TaxonomyMiSeq_SOP.qza \
 --o-visualization TaxonomyMiSeq_SOP.qzv
```

```
Saved Visualization to:
```

Now we can review the visualization of the classified sequences in the QIIME2 viewer.

Step 4: Visualize taxonomic classifications.

Since we now have three datasets: feature table, taxonomy, and sample metadata, we can use **qiime taxa barplot** command to create a bar plot to explore the distribution of taxonomy for each sample. Figure 5.1 can be reproduced using the following QIIME 2 commands.

```
# Figure 5.1:
qiime taxa barplot \
  --i-table FeatureTableMiSeq_SOP.qza \
  --i-taxonomy TaxonomyMiSeq_SOP.qza \
  --m-metadata-file SampleMetadataMiSeq_SOP.tsv \
 --o-visualization TaxaBarPlotsMiSeq_SOP.qzv
```

```
Saved Visualization to: TaxaBarPlotsMiSeq_SOP.qzv
```

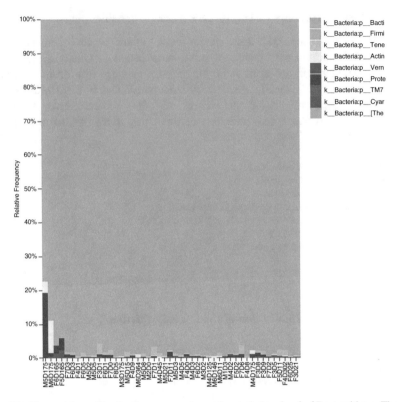

Fig. 5.1 Taxonomic profiles for the mouse gut samples at phylum level of Bacteroidetes. The bar plot was generated by first choosing Taxonomic Level(L2:phylum) and then by sorting samples by the taxonomic abundance(k_Bacteria;p_Bacteroidetes)

We can review the visualization of the taxa bar plot in the QIIME2 viewer or use the command **qiime tools view** to review TaxaBarPlotsMiSeq_SOP.qzv. The generated bars can be aggregated at the desired taxonomic level, and the abundance can be sorted by a specific taxonomic group. By providing sample metadata file, we can also sort the abundance by metadata groupings. We can also interactively change color schemes, and save plots and legends in vector graphic format.

Step 5: Create a BIOM table with taxonomy annotations(optional).

The .biom format files are often used in microbiome studies. The .biom files consist of two kinds of information: one is feature table [frequency]; another is taxonomy. Thus, we first export the data as a .biom file using **qiime tools export** command as below.

```
qiime tools export \
  --input-path FeatureTableMiSeq_SOP.qza \
  --output-path ExportedFeatureTableMiSeq_SOP
```

```
Exported FeatureTableMiSeq_SOP.qza as BIOMV210DirFmt to directory
ExportedFeatureTableMiSeq_SOP
```

Then we export taxonomy information as below.

```
qiime tools export \
 --input-path TaxonomyMiSeq_SOP.qza \
 --output-path ExportedFeatureTableMiSeq_SOP
```

```
Exported TaxonomyMiSeq_SOP.qza as TSVTaxonomyDirectoryFormat to
directory ExportedFeatureTableMiSeq_SOP
```

5.1.6 Remarks on Taxonomic Classification

A number of bioinformatics tools and reference databases are available for analysis of 16S rRNA amplicon microbiome data. It was shown that taxonomy assignment often obtains different results (Balvočiūtė and Huson 2017) and especially at genus level (Sierra et al. 2020) when using different reference databases. However, currently there is no general criterion to guide for choosing appropriate bioinformatics tools and reference databases for analysis of microbiome data; and especially there are no defined criteria for data curation and validation of annotations (Sierra et al. 2020). Thus, the annotated results may be inaccurate and irreproducible, making it difficult to compare data across studies.

5.2 Building Phylogenetic Tree

5.2.1 Introduction to Phylogenetic Tree

Microbiome data are encoded as a phylogenetic tree, which relates all the microbial species, containing the evolution information of the species. Thus, a phylogenetic tree is useful for incorporating biological structure (Xia 2020). Thus, one central method in computational biology is to infer evolutionary relationships or phylogenies from families of related DNA or protein sequences (Price et al. 2010). The FastTree method developed by Price et al. (2009) is to compute large minimum evolution trees with profiles instead of a distance matrix (Price et al. 2009).

Tree construction is optional; however, a phylogenetic tree has two primary applications:

1. Phylogenetic tree measures are used in computing phylogenetically based alpha diversity metrics such as unweighted uniFrac (Lozupone and Knight 2005), weighted uniFrac (Lozupone et al. 2007), Faith's Phylogenetic Diversity (PD) (Faith 1992), or generalized UniFrac distance (Chen et al. 2012). For

example, QIIME supports several phylogenetic diversity metrics. In QIIME 2 we can calculate alpha diversities and output core metrics, which include Faith PD, unweighted uniFrac, and weighted uniFrac distance measures. To generate these metrics, except for providing the FeatureTable[Frequency] artifact, a rooted phylogenetic tree that relates the features to one another is needed.

2. Phylogenetic tree-based association analyses also need to provide the information of phylogenetic tree. For example, a phylogenetic tree data are utilized in the general framework for association analysis of taxa (Tang et al. 2017), a predictive method based on a generalized mixed-models framework (Xiao et al. 2018) and a phylogenetic tree-based microbiome association test (Kim et al. 2019). For building phylogenetic tree in R, the reader is referred to Chap. 2 (Sect. 2.4).

5.2.2 Build a Phylogenetic Tree Using the Alignment and Phylogeny Commands

A phylogenetic tree is built in QIIME 2 via four steps: (1) multiple sequence alignment, (2) masking, (3) tree building, and (4) rooting.

Example 5.2: RepSeqsMiSeq_SOP.qza, Example 5.1 cont.
We can build a phylogenetic tree based on the denoised sequences "RepSeqsMiSeq_SOP.qza" artifact using **align-to-tree-mafft-fasttree** pipeline from the **q2-phylogeny** plugin in the four-step process as below.

Step 1: Conduct a multiple sequence alignment using MAFFT.

MAFFT is a multiple sequence alignment program based on fast Fourier transform in evolutionary analyses of biological sequences (Katoh and Standley 2013; Katoh et al. 2002). MAFFT includes various alignment strategies as its options: progressive methods, iterative refinement methods, and structural alignment methods for RNAs. MAFFT is a similarity-based multiple sequence alignment (MSA) method, while taking evolutionary information into account because evolutionary information is useful even for similarity-based methods (Katoh and Standley 2013). QIIME wraps MAFFT's multiple sequence alignment in the **qiime alignment mafft** command.

```
source activate qiime2-2022.2
cd QIIME2R-Bioinformatics
qiime alignment mafft \
  --i-sequences RepSeqsMiSeq_SOP.qza \
  --o-alignment AlignedRepSeqsMiSeq_SOP.qza
```

Saved FeatureData[AlignedSequence] to: AlignedRepSeqsMiSeq_SOP.qza

Above MAFFT commands aligned the denoised sequences in the FeatureData [Sequence] (in this case, RepSeqsMiSeq_SOP.qza) and created a FeatureData

[AlignedSequence] QIIME 2 artifact (we named as AlignedRepSeqsMiSeq_SOP. qza).

Step 2: Mask the alignment.

Highly variable positions could add noise to a resulting phylogenetic tree. The purpose of masking (i.e., filtering) the alignment is to remove these highly variable positions (highly gapped columns) from an alignment so that the sequences contain enough conservation to provide meaningful information. Below, we mask the uninformative positions via **qiime alignment mask** command.

QIIME 2 uses 40% (the default) minimum conservation as meaningful information to reproduce the mask presented in Lane (1991) via the parameter **--p-min-conservation**. Providing a value of 0.4 (the default), only the column that contains at least one character that is present in at least 40% of the sequences will be retained. Another default parameter used here is **--p-max-gap-frequency**, which is value of 1, retaining all columns regardless of gap character frequency. If a value of 0 is chosen, then retain only those columns without gap characters.

```
qiime alignment mask \
  --i-alignment AlignedRepSeqsMiSeq_SOP.qza \
  --o-masked-alignment MaskedAlignedRepSeqsMiSeq_SOP.qza
```

```
Saved FeatureData[AlignedSequence] to:
MaskedAlignedRepSeqsMiSeq_SOP.qza
```

Step 3: Create the tree using the Fasttree program.

FastTree (Price et al. 2009, 2010) is a bioinformatic tool for inferring phylogenies for alignments. So far two versions of FastTree have been released.

Utilizing the "minimum-evolution" principle, FastTree 1 (Price et al. 2009) tries to find a topology that minimizes the amount of evolution, or the sum of the branch lengths. While via using a heuristic variant of neighbor joining method (Saitou and Nei 1987; Studier and Keppler 1988), FastTree 1 quickly finds a starting tree and using nearest-neighbor interchanges (NNIs) refines the topology (Price et al. 2010). FastTree 2 has improved its topological accuracy (the proportion of the splits in the true trees that are recovered) and 100–1,000 times faster compared to FastTree 1 and outperforms other methods including PhyML 3's approach with default settings (NNI search) (Guindon et al. 2009, 2010), standard implementation of maximum-likelihood NNIs, minimum-evolution and parsimony methods, although not as accurate as the maximum-likelihood (ML) methods that use subtree-pruning-regrafting (SPR) moves (Price et al. 2010). The topological accuracy and outperformances are achieved by FastTree 2 mostly because FastTree 2 (1) adds minimum-evolution SPRs, (2) adds maximum likelihood NNIs, (3) uses heuristics to restrict the search for better trees, and (4) estimates a rate of evolution for each site (Price et al. 2010).

QIIME builds a phylogenetic tree based on FastTree (Price et al. 2010) via the **qiime phylogeney fasttree** command.

```
qiime phylogeny fasttree \
  --i-alignment MaskedAlignedRepSeqsMiSeq_SOP.qza \
  --o-tree UnrootedTreeMiSeq_SOP.qza
```

```
Saved Phylogeny[Unrooted] to: UnrootedTreeMiSeq_SOP.qza
```

Step 4: Root the tree using the longest root.

By processing the FastTree method, an unrooted phylogenetic tree is generated from the masked alignment. However, some downstream analyses require a rooted tree; thus we use the longest branch to root the tree at the midpoint of the two leaves that are the furthest from one another (called "midrooting"), producing the rooted tree artifact file "RootedTreeMiSeq_SOP.qza" that can be used as input to generate phylogenetic-diversity measures.

```
qiime phylogeny midpoint-root \
  --i-tree UnrootedTreeMiSeq_SOP.qza \
  --o-rooted-tree RootedTreeMiSeq_SOP.qza
```

```
Saved Phylogeny[Rooted] to: RootedTreeMiSeq_SOP.qza
```

5.2.3   Remarks on the Taxonomic and Phylogenetic Trees

Taxonomy and phylogeny are two concepts involved in the classification of organisms. Taxonomy stems from ancient Greek taxis, meaning "arrangement," and nomia, meaning "method." Taxonomy is a field of classification, identification, and naming of biological organisms based on their shared characteristics of similarities and dissimilarities (Xia et al. 2018).

Classification of organisms was first introduced by the Swedish botanist Carl Linnaeus (1707–1778) (known as the father of taxonomy). He developed a system for categorization of organisms, known as Linnaean taxonomy and binomial nomenclature for categorizing and naming organisms.

Linnaeus and others ranked all living organisms into seven biological groups or levels of classification in the taxonomic hierarchy: kingdom, phylum, class, order, family, genus, and species. There are no domains in their classifications. The classification of domain was first proposed by Woese et al. in 1977 (Woese and Fox 1977; Woese et al. 1990). They added a level called "domain" above the level of kingdom. The three domains of life are Archaea, Bacteria, and Eukarya, and the five major kingdoms are monera, protista, fungi, plantae, and animalia. Thus, we can classify all living organisms into eight major hierarchical levels, from domain (the most general) to species (the most specific): domain, kingdom, phylum, class, order, family, genus, and species.

A phylogenetic tree (also called phylogeny or evolutionary tree) (Felsenstein 2004) is a branching diagram or a tree showing the evolutionary relationship of a species or a group of species with a common ancestor based upon similarities and differences in their physical or genetic characteristics.

Various ways have been developed to graphically represent the phylogenetic trees (Letunic and Bork 2006). Both taxonomy and phylogeny are important for classification of organisms. Phylogeny is important in building taxonomy. Researchers have attempted to synthesize phylogeny and taxonomy into a comprehensive tree of life (Hinchliff et al. 2015). However, taxonomy and phylogeny as well as taxonomic tree and phylogenetic tree are different. The key difference between these two pairs of concept lies in the fact that **taxonomy/**taxonomic tree **involves naming and classifying organisms while phylogeny/**phylogenetic tree **involves the evolution of the species or groups of species.** Taxonomy **focuses on** naming and classifying organisms, and hence does not reveal anything about the shared evolutionary history of organisms. In contrast, phylogeny **focuses on** evolutionary relationships of organisms and hence reveals the shared evolutionary history.

Additionally, as shown in above illustrating examples, their reconstructions are also different. For amplicon sequencing, the taxonomic tree is typically reconstructed when the phylogeny is not available but taxonomic annotations are available. The taxonomic tree is reconstructed from lineages extracted from regularly updated databases such as from NCBI (Federhen 2011; Geer et al. 2010) and represents the alignment from domain to species rank; as discovery of new species continues, assignment of new taxa in the taxonomic hierarchy will never end. Thus, taxonomic trees are highly polyatomic. In contrast, the phylogenetic tree is reconstructed based on the sequence divergence of taxa (of the marker-gene) (Price et al. 2010) and it encodes the common evolutionary history of the taxa. In other words, phylogenetic trees are re**constructed** usually based on morphological or genetic homology to reveal the evolutionary relationships of species via comparison of anatomical traits and to reveal the ancestral genes (identify descent from an ancestral gene) via analysis of genetic differences of species. Thus, unlike taxonomic trees, phylogenetic trees are hypothetic (Dubois et al. 2021; Felsenstein 2004).

Phylogenetic classification has two main advantages over the Linnaean classification. First, phylogenetic classification reveals the evolutionary history of the organism: the important underlying biological processes that are responsible for the diversity of organisms. Second, phylogenetic classification does not attempt to "rank" organisms and hence avoids the misleading of considering different groupings with the same rank are equivalent and comparable. Actually, comparing to phylogenetic tree, taxonomic tree ignores the granular differences of the taxa belonging to the same rank. However, the advantage of phylogenies is that they do not capture similarities between taxa in terms of abundance profiles (Bichat et al. 2020).

Human microbiome is very complicated with existing genetic and evolutionary relationships among species. To understand the complexity of the human microbiome, it is important to recognize the genetic and evolutionary relationships between species. Microbiome data are encoded as the taxonomic and phylogenetic trees. Both taxonomic and phylogenetic trees play important roles in microbiome studies. As two unique features in the microbiome data, these two trees are usually used for different measures and are required by different strategies of statistical analysis.

Integrating the information of the taxonomic and phylogenetic trees into statistical analysis will increase statistical power in statistical hypothesis testing. For example, it has been believed that reference phylogenies can prove the crucial information into a taxonomic framework for interpretation of marker gene and metagenomic surveys to speed revealing novel species (McDonald et al. 2012), and leveraging the phylogenetic tree of the taxa can increase statistical power while controlling the False Discovery Rate (FDR) (Sankaran and Holmes 2014; Xiao et al. 2017).

However, the premise that the phylogenetic (or taxonomic) tree is the relevant hierarchical structure to incorporate in differential studies has been questioned in recent study; and it was showed that incorporating phylogenetic information in microbiome differential abundance analyses has no effect on detection power and FDR control (Bichat et al. 2020). Instead in this study (Bichat et al. 2020) a correlation-tree was proposed and advocated for use. A correlation-tree is a clustering tree built based on the abundance profiles of taxa across samples, in which taxa with highly correlated abundances are very close in the tree. The correlation tree is built involving three logical steps: (1) computing the pairwise correlation matrix using the Spearman correlation and excluding samples where both taxa are absent; (2) using the transformation to change the correlation matrix into a dissimilarity matrix; and (3) creating the correlation tree using hierarchical clustering with Ward linkage on this dissimilarity matrix. Branch lengths correspond to the dissimilarity cost of merging two subtrees.

The correlation tree was considered being better than the phylogenetic tree for a proxy of biological functions and increasing the detection power while with better FDR control (Bichat et al. 2020). However, it needs for evaluation by other studies and/or deserves to be further discussed and assessed whether above arguments are true and whether or not the correlation tree is more important than the phylogenetic tree in differential abundance analysis.

As reviewed the history of numerical taxonomy, we learn that early in 1950s the "taxonomic importance" was criticized by Cain (Cain 1958). Cain was not opposed to phylogeny per se, instead thought that we ought to incorporate phylogenetic information into classification when it is available and reliable. However, he recognized it was very difficult to do and in some cases classification should be purely phenotypic (Vernon 1988).

5.3 Summary

This chapter covered two important topics in bioinformatic analysis of microbiome data: assigning taxonomy and building phylogenetic tree. For assigning taxonomy, first, various bioinformatics tools and reference databases were reviewed, and then specifically both QIIME 2 and DADA2 formatted and maintained taxonomic reference databases were summarized in tables. Next, the q2-feature-classifier was introduced and how to assign taxonomy using the q2-feature-classifier was illustrated, followed by a brief remark on taxonomic classification. For building phylogenetic tree, phylogenetic tree was first introduced and then how to build a phylogenetic tree using the alignment and phylogeny commands was illustrated. Finally, comprehensive remarks on the taxonomic and phylogenetic trees were provided. Chapter 6 will introduce how to cluster sequences into OTUs.

References

Alishum, Ali. 2019. *DADA2 formatted 16S rRNA gene sequences for both bacteria & archaea (Version 1)* [Data set]. Zenodo. Accessed August 12. https://doi.org/10.5281/zenodo.2541239.

Balvočiūtė, Monika, and Daniel H. Huson. 2017. SILVA, RDP, greengenes, NCBI and OTT – How do these taxonomies compare? *BMC Genomics* 18 (2): 114. https://doi.org/10.1186/s12864-017-3501-4.

Benson, D.A., M. Cavanaugh, K. Clark, I. Karsch-Mizrachi, D.J. Lipman, J. Ostell, and E.W. Sayers. 2013. GenBank. *Nucleic Acids Research* 41 (Database issue): D36–D42. https://doi.org/10.1093/nar/gks1195.

Bichat, Antoine, Jonathan Plassais, Christophe Ambroise, and Mahendra Mariadassou. 2020. Incorporating phylogenetic information in microbiome differential abundance studies has no effect on detection power and FDR control. *Frontiers in Microbiology* 11 (649). https://doi.org/10.3389/fmicb.2020.00649. https://www.frontiersin.org/article/10.3389/fmicb.2020.00649.

Bokulich, Nicholas A., Benjamin D. Kaehler, Jai Ram Rideout, Matthew Dillon, Evan Bolyen, Rob Knight, Gavin A. Huttley, and J. Gregory Caporaso. 2018. Optimizing taxonomic classification of marker-gene amplicon sequences with QIIME 2's q2-feature-classifier plugin. *Microbiome* 6 (1): 90. https://doi.org/10.1186/s40168-018-0470-z.

Bokulich, Nicholas, Mike Robeson, Matthew Dillon, Michal Ziemski, Ben Kaehler, and Devon O'Rourke. 2021. *bokulich-lab/RESCRIPt: 2021.8.0.dev0 (2021.8.0.dev0).* Zenodo. Accessed 12 Aug. 2021

Bokulish, Nicolas, Matthew Dillon, Evan Bolyen, Benjamin Kaehler, Gavin Huttley, and J. Caporaso. 2018. q2-sample-classifier: Machine-learning tools for microbiome classification and regression. *Journal of Open Source Software* 3: 934. https://doi.org/10.21105/joss.00934.

Bolyen, Evan, Jai Ram Rideout, Matthew R. Dillon, Nicholas A. Bokulich, Christian C. Abnet, Gabriel A. Al-Ghalith, Harriet Alexander, Eric J. Alm, Manimozhiyan Arumugam, Francesco Asnicar, Yang Bai, Jordan E. Bisanz, Kyle Bittinger, Asker Brejnrod, Colin J. Brislawn, C. Titus Brown, Benjamin J. Callahan, Andrés Mauricio Caraballo-Rodríguez, John Chase, Emily K. Cope, Ricardo Da Silva, Christian Diener, Pieter C. Dorrestein, Gavin M. Douglas, Daniel M. Durall, Claire Duvallet, Christian F. Edwardson, Madeleine Ernst, Mehrbod Estaki, Jennifer Fouquier, Julia M. Gauglitz, Sean M. Gibbons, Deanna L. Gibson, Antonio Gonzalez, Kestrel Gorlick, Jiarong Guo, Benjamin Hillmann, Susan Holmes, Hannes Holste, Curtis Huttenhower, Gavin A. Huttley, Stefan Janssen, Alan K. Jarmusch, Lingjing Jiang, Benjamin

D. Kaehler, Kyo Bin Kang, Christopher R. Keefe, Paul Keim, Scott T. Kelley, Dan Knights, Irina Koester, Tomasz Kosciolek, Jorden Kreps, Morgan G.I. Langille, Joslynn Lee, Ruth Ley, Yong-Xin Liu, Erikka Loftfield, Catherine Lozupone, Massoud Maher, Clarisse Marotz, Bryan D. Martin, Daniel McDonald, Lauren J. McIver, Alexey V. Melnik, Jessica L. Metcalf, Sydney C. Morgan, Jamie T. Morton, Ahmad Turan Naimey, Jose A. Navas-Molina, Louis Felix Nothias, Stephanie B. Orchanian, Talima Pearson, Samuel L. Peoples, Daniel Petras, Mary Lai Preuss, Elmar Pruesse, Lasse Buur Rasmussen, Adam Rivers, Michael S. Robeson 2nd, Patrick Rosenthal, Nicola Segata, Michael Shaffer, Arron Shiffer, Rashmi Sinha, Se Jin Song, John R. Spear, Austin D. Swafford, Luke R. Thompson, Pedro J. Torres, Pauline Trinh, Anupriya Tripathi, Peter J. Turnbaugh, Sabah Ul-Hasan, Justin J.J. van der Hooft, Fernando Vargas, Yoshiki Vázquez-Baeza, Emily Vogtmann, Max von Hippel, William Walters, Yunhu Wan, Mingxun Wang, Jonathan Warren, Kyle C. Weber, Charles H.D. Williamson, Amy D. Willis, Zhenjiang Zech Xu, Jesse R. Zaneveld, Yilong Zhang, Qiyun Zhu, Rob Knight, and J. Gregory Caporaso. 2019. Reproducible, interactive, scalable and extensible microbiome data science using QIIME 2. *Nature Biotechnology* 37 (8): 852–857. https://doi.org/10.1038/ s41587-019-0209-9. https://pubmed.ncbi.nlm.nih.gov/31341288. https://www.ncbi.nlm.nih. gov/pmc/articles/PMC7015180/.

Cain, A.J. 1958. Chromosomes and their taxonomic importance. *Proceedings of the Linnean Society of London* 169: 125–128.

Callahan, Benjamin. 2016. *The RDP and GreenGenes taxonomic training sets formatted for DADA2* [Data set]. Zenodo. Accessed 13 Aug.

———. 2021. *DADA2 pipeline tutorial (1.16)*. https://benjjneb.github.io/dada2/tutorial.html. Accessed 25 Jan 2021.

Callahan, B.J., P.J. McMurdie, M.J. Rosen, A.W. Han, A.J. Johnson, and S.P. Holmes. 2016. DADA2: High-resolution sample inference from Illumina amplicon data. *Nature Methods* 13 (7): 581–583.

Caporaso, J. Gregory, Justin Kuczynski, Jesse Stombaugh, Kyle Bittinger, Frederic D. Bushman, Elizabeth K. Costello, Noah Fierer, Antonio Gonzalez Peña, Julia K. Goodrich, Jeffrey I. Gordon, Gavin A. Huttley, Scott T. Kelley, Dan Knights, Jeremy E. Koenig, Ruth E. Ley, Catherine A. Lozupone, Daniel McDonald, Brian D. Muegge, Meg Pirrung, Jens Reeder, Joel R. Sevinsky, Peter J. Turnbaugh, William A. Walters, Jeremy Widmann, Tanya Yatsunenko, Jesse Zaneveld, and Rob Knight. 2010. QIIME allows analysis of high-throughput community sequencing data. *Nature Methods* 7: 335. https://doi.org/10.1038/nmeth.f.303. https://www. nature.com/articles/nmeth.f.303#supplementary-information.

Chen, Jun, Kyle Bittinger, Emily S. Charlson, Christian Hoffmann, James Lewis, Gary D. Wu, Ronald G. Collman, Frederic D. Bushman, and Hongzhe Li. 2012. Associating microbiome composition with environmental covariates using generalized UniFrac distances. *Bioinformatics (Oxford, England)* 28 (16): 2106–2113. https://doi.org/10.1093/bioinformatics/bts342. https:// pubmed.ncbi.nlm.nih.gov/22711789. https://www.ncbi.nlm.nih.gov/pmc/articles/PMC34133 90/.

Cole, J.R., B. Chai, R.J. Farris, Q. Wang, S.A. Kulam, D.M. McGarrell, G.M. Garrity, and J.M. Tiedje. 2005. The Ribosomal Database Project (RDP-II): Sequences and tools for high-throughput rRNA analysis. *Nucleic Acids Research* 33 (Database issue): D294–D296. https:// doi.org/10.1093/nar/gki038. https://www.ncbi.nlm.nih.gov/pubmed/15608200. https://www. ncbi.nlm.nih.gov/pmc/articles/PMC539992/.

Czaplicki, Lauren. 2017. *RDP LSU taxonomic training data formatted for DADA2 (trainingset 11)* [Data set]. Zenodo. Accessed 13 Aug.

DeSantis, T.Z., P. Hugenholtz, N. Larsen, M. Rojas, E.L. Brodie, K. Keller, T. Huber, D. Dalevi, P. Hu, and G.L. Andersen. 2006. Greengenes, a chimera-checked 16S rRNA gene database and workbench compatible with ARB. *Applied and Environmental Microbiology* 72 (7): 5069–5072. https://doi.org/10.1128/AEM.03006-05. https://journals.asm.org/doi/abs/10.1128/ AEM.03006-05 %X A 16S rRNA gene database (http://greengenes.lbl.gov) addresses limita-tions of public repositories by providing chimera screening, standard alignment, and taxonomic classification using multiple published taxonomies. It was found that there is incongruent taxonomic nomenclature among curators even at the phylum level. Putative chimeras were identified in 3% of environmental sequences and in 0.2% of records derived from isolates. Environmental sequences were classified into 100 phylum-level lineages in the Archaea and Bacteria.

Diener, Christian 2016. *HITdb v1.00 for Dada2* [Data set]. Zenodo. Accessed 13 Aug.

Dubois, Alain, Annemarie Ohler, and R alexander Pyron. 2021. New concepts and methods for phylogenetic taxonomy and nomenclature in zoology, exemplified by a new ranked cladonomy of recent amphibians (Lissamphibia). *Megataxa* 5 (1): 1–738.

Faith, Daniel P. 1992. Conservation evaluation and phylogenetic diversity. *Biological Conservation* 61 (1): 1–10. https://doi.org/10.1016/0006-3207(92)91201-3. http://www.sciencedirect.com/science/article/pii/0006320792912013.

Federhen, Scott. 2011. The NCBI taxonomy database. *Nucleic Acids Research* 40 (D1): D136–D143. https://doi.org/10.1093/nar/gkr1178.

Felsenstein, Joseph. 2004. *Inferring phylogenies*. Sunderland: Sinauer Associates, Inc.

Geer, L.Y., A. Marchler-Bauer, R.C. Geer, L. Han, J. He, S. He, C. Liu, W. Shi, and S.H. Bryant. 2010. The NCBI BioSystems database. *Nucleic Acids Research* 38 (Database issue): D492–D496. https://doi.org/10.1093/nar/gkp858.

Guindon, Stéphane, Frédéric Delsuc, Jean-François Dufayard, and Olivier Gascuel. 2009. Estimating maximum likelihood phylogenies with PhyML. In *Bioinformatics for DNA sequence analysis*, 113–137. Springer.

Guindon, Stéphane, Jean-François Dufayard, Vincent Lefort, Maria Anisimova, Wim Hordijk, and Olivier Gascuel. 2010. New algorithms and methods to estimate maximum-likelihood phylogenies: Assessing the performance of PhyML 3.0. *Systematic Biology* 59 (3): 307–321. https://doi.org/10.1093/sysbio/syq010.

Hinchliff, Cody E., Stephen A. Smith, James F. Allman, J. Gordon Burleigh, Ruchi Chaudhary, Lyndon M. Coghill, Keith A. Crandall, Jiabin Deng, Bryan T. Drew, Romina Gazis, Karl Gude, David S. Hibbett, Laura A. Katz, H. Dail Laughinghouse, Emily Jane McTavish, Peter E. Midford, Christopher L. Owen, Richard H. Ree, Jonathan A. Rees, Douglas E. Soltis, Tiffani Williams, and Karen A. Cranston. 2015. Synthesis of phylogeny and taxonomy into a comprehensive tree of life. *Proceedings of the National Academy of Sciences* 112 (41): 12764–12769. https://doi.org/10.1073/pnas.1423041112. https://www.pnas.org/content/pnas/112/41/12764. full.pdf.

Kaehler, Benjamin D., Nicholas A. Bokulich, Daniel McDonald, J. Rob Knight, Gregory Caporaso, and Gavin A. Huttley. 2019. Species abundance information improves sequence taxonomy classification accuracy. *Nature Communications* 10 (1): 4643. https://doi.org/10.1038/s41467-019-12669-6.

Katoh, Kazutaka, and Daron M. Standley. 2013. MAFFT multiple sequence alignment software version 7: Improvements in performance and usability. *Molecular Biology and Evolution* 30 (4): 772–780. https://doi.org/10.1093/molbev/mst010.

Katoh, Kazutaka, Kazuharu Misawa, Kei-ichi Kuma, and Takashi Miyata. 2002. MAFFT: A novel method for rapid multiple sequence alignment based on fast Fourier transform. *Nucleic Acids Research* 30 (14): 3059–3066. https://doi.org/10.1093/nar/gkf436.

Kim, Kang Jin, Jaehyun Park, Sang-Chul Park, and Sungho Won. 2019. Phylogenetic tree-based microbiome association test. *Bioinformatics*. https://doi.org/10.1093/bioinformatics/btz686.

Kõljalg, Urmas, R. Henrik Nilsson, Kessy Abarenkov, Leho Tedersoo, Andy F.S. Taylor, Mohammad Bahram, Scott T. Bates, Thomas D. Bruns, Johan Bengtsson-Palme, Tony M. Callaghan, Brian Douglas, Tiia Drenkhan, Ursula Eberhardt, Margarita Dueñas, Tine Grebenc, Gareth W. Griffith, Martin Hartmann, Paul M. Kirk, Petr Kohout, Ellen Larsson, Björn D. Lindahl, Robert Lücking, María P. Martín, P. Brandon Matheny, Nhu H. Nguyen, Tuula Niskanen, Jane Oja, Kabir G. Peay, Ursula Peintner, Marko Peterson, Kadri Põldmaa, Lauri Saag, Irja Saar, Arthur Schüßler, James A. Scott, Carolina Senés, Matthew E. Smith, D. Ave Suija, M. Lee Taylor, Teresa Telleria, Michael Weiss, and Karl-Henrik Larsson. 2013.

Towards a unified paradigm for sequence-based identification of fungi. *Molecular Ecology* 22 (21): 5271–5277. https://doi.org/10.1111/mec.12481. https://onlinelibrary.wiley.com/doi/abs/10.1111/mec.12481.

Lane, D.J. 1991. 16S/23S rRNA sequencing. In *Nucleic acid techniques in bacterial systematics*, 115–175. New York: Wiley.

Leinonen, Rasko, Ruth Akhtar, Ewan Birney, Lawrence Bower, Ana Cerdeno-Tárraga, Ying Cheng, Iain Cleland, Nadeem Faruque, Neil Goodgame, Richard Gibson, Gemma Hoad, Mikyung Jang, Nima Pakseresht, Sheila Plaister, Rajesh Radhakrishnan, Kethi Reddy, Siamak Sobhany, Petra Ten Hoopen, Robert Vaughan, Vadim Zalunin, and Guy Cochrane. 2011. The European Nucleotide Archive. *Nucleic Acids Research* 39 (Database issue): D28–D31. https://doi.org/10.1093/nar/gkq967. https://pubmed.ncbi.nlm.nih.gov/20972220. https://www.ncbi.nlm.nih.gov/pmc/articles/PMC3013801/.

Letunic, Ivica, and Peer Bork. 2006. Interactive Tree Of Life (iTOL): An online tool for phylogenetic tree display and annotation. *Bioinformatics* 23 (1): 127–128. https://doi.org/10.1093/bioinformatics/btl529.

Lozupone, Catherine, and Rob Knight. 2005. UniFrac: A new phylogenetic method for comparing microbial communities. *Applied and Environmental Microbiology* 71 (12): 8228–8235. https://doi.org/10.1128/AEM.71.12.8228-8235.2005. https://www.ncbi.nlm.nih.gov/pubmed/16332807. https://www.ncbi.nlm.nih.gov/pmc/articles/PMC1317376/.

Lozupone, Catherine A., Micah Hamady, Scott T. Kelley, and Rob Knight. 2007. Quantitative and qualitative beta diversity measures lead to different insights into factors that structure microbial communities. *Applied and Environmental Microbiology* 73 (5): 1576–1585. https://doi.org/10.1128/AEM.01996-06. https://www.ncbi.nlm.nih.gov/pubmed/17220268. https://www.ncbi.nlm.nih.gov/pmc/articles/PMC1828774/.

Maidak, Bonnie L., James R. Cole, Timothy G. Lilburn, Charles T. Parker Jr, Paul R. Saxman, Jason M. Stredwick, George M. Garrity, Bing Li, Gary J. Olsen, Sakti Pramanik, Thomas M. Schmidt, and James M. Tiedje. 2000. The RDP (Ribosomal Database Project) continues. *Nucleic Acids Research* 28 (1): 173–174. https://doi.org/10.1093/nar/28.1.173.

Mashima, Jun, Yuichi Kodama, Takatomo Fujisawa, Toshiaki Katayama, Yoshihiro Okuda, Eli Kaminuma, Osamu Ogasawara, Kousaku Okubo, Yasukazu Nakamura, and Toshihisa Takagi. 2017. DNA Data Bank of Japan. *Nucleic Acids Research* 45 (D1): D25–D31. https://doi.org/10.1093/nar/gkw1001. https://pubmed.ncbi.nlm.nih.gov/27924010. https://www.ncbi.nlm.nih.gov/pmc/articles/PMC5210514/.

McDonald, Daniel, Morgan N. Price, Julia Goodrich, Eric P. Nawrocki, Todd Z. DeSantis, Alexander Probst, Gary L. Andersen, Rob Knight, and Philip Hugenholtz. 2012. An improved Greengenes taxonomy with explicit ranks for ecological and evolutionary analyses of bacteria and archaea. *The ISME Journal* 6 (3): 610–618. https://doi.org/10.1038/ismej.2011.139.

Morien, Evan, and Laura W. Parfrey. 2018. *SILVA v128 and v132 dada2 formatted 18s 'train sets' (1.0)* [Data set]. Zenodo. Accessed 13 Aug.

Moynihan, M.A. 2020. nifHdada2: v1.1.0 (v1.1.0). Zenodo. Accessed 13 Aug.

Nakamura, Yasukazu, Guy Cochrane, Ilene Karsch-Mizrachi, and on behalf of the International Nucleotide Sequence Database Collaboration. 2012. The International Nucleotide Sequence Database Collaboration. *Nucleic Acids Research* 41 (D1): D21–D24. https://doi.org/10.1093/nar/gks1084.

Nilakanta, Haema, Kimberly L. Drews, Suzanne Firrell, Mary A. Foulkes, and Kathleen A. Jablonski. 2014. A review of software for analyzing molecular sequences. *BMC Research Notes* 7: 830–830. https://doi.org/10.1186/1756-0500-7-830. https://pubmed.ncbi.nlm.nih.gov/25421430. https://www.ncbi.nlm.nih.gov/pmc/articles/PMC4258797/.

Pedregosa, Fabian, Gaël Varoquaux, Alexandre Gramfort, Vincent Michel, Bertrand Thirion, Olivier Grisel, Mathieu Blondel, Peter Prettenhofer, Ron Weiss, Vincent Dubourg, Jake Vanderplas, Alexandre Passos, David Cournapeau, Matthieu Brucher, Matthieu Perrot, and Édouard Duchesnay. 2011. Scikit-learn: Machine learning in Python. *Journal of Machine Learning Research* 12: 2825–2830.

Plummer, E., J. Twin, D.M. Bulach, S.M. Garland, and S.N. Tabrizi. 2015. A comparison of three bioinformatics pipelines for the analysis of preterm gut microbiota using 16S rRNA gene sequencing data. *Journal of Proteomics and Bioinformatics* 8: 283–291. https://doi.org/10. 4172/jpb.1000381.

Põlme, Sergei, Kessy Abarenkov, Rolf Henrik Nilsson, Björn Lindahl, Karina Clemmensen, Håvard Kauserud, Nhu Nguyen, Rasmus Kjøller, Scott Bates, Petr Baldrian, Tobias Frøslev, Kristjan Adojaan, Alfredo Vizzini, Ave Suija, Donald Pfister, Hans-Otto Baral, Helle Järv, Hugo Madrid, and Jenni Nordén. 2020. FungalTraits: A user-friendly traits database of fungi and fungus-like stramenopiles. *Fungal Diversity* 105: 1–16. https://doi.org/10.1007/s13225-020-00466-2.

Price, Morgan N., Paramvir S. Dehal, and Adam P. Arkin. 2009. FastTree: Computing large minimum evolution trees with profiles instead of a distance matrix. *Molecular Biology and Evolution* 26 (7): 1641–1650. https://doi.org/10.1093/molbev/msp077. https://pubmed.ncbi. nlm.nih.gov/19377059; https://www.ncbi.nlm.nih.gov/pmc/articles/PMC2693737/.

———. 2010. FastTree 2 – Approximately maximum-likelihood trees for large alignments. *PloS One* 5 (3): –e9490. https://doi.org/10.1371/journal.pone.0009490. https://www.ncbi.nlm.nih. gov/pubmed/20224823. https://www.ncbi.nlm.nih.gov/pmc/articles/PMC2835736/.

Pruesse, Elmar, Christian Quast, Katrin Knittel, Bernhard M. Fuchs, Wolfgang Ludwig, Jörg Peplies, and Frank Oliver Glöckner. 2007. SILVA: A comprehensive online resource for quality checked and aligned ribosomal RNA sequence data compatible with ARB. *Nucleic Acids Research* 35 (21): 7188–7196. https://doi.org/10.1093/nar/gkm864.

Pruesse, Elmar, Jörg Peplies, and Frank Oliver Glöckner. 2012. SINA: Accurate high-throughput multiple sequence alignment of ribosomal RNA genes. *Bioinformatics* 28 (14): 1823–1829. https://doi.org/10.1093/bioinformatics/bts252.

Quast, Christian, Elmar Pruesse, Pelin Yilmaz, Jan Gerken, Timmy Schweer, Pablo Yarza, Jörg Peplies, and Frank Oliver Glöckner. 2012. The SILVA ribosomal RNA gene database project: Improved data processing and web-based tools. *Nucleic Acids Research* 41 (D1): D590–D596. https://doi.org/10.1093/nar/gks1219.

———. 2013. The SILVA ribosomal RNA gene database project: Improved data processing and web-based tools. *Nucleic Acids Research* 41 (Database issue): D590–D596. https://doi.org/10. 1093/nar/gks1219. https://pubmed.ncbi.nlm.nih.gov/23193283. https://www.ncbi.nlm.nih.gov/ pmc/articles/PMC3531112/.

Ritari, Jarmo, Jarkko Salojärvi, Leo Lahti, and Willem M. de Vos. 2015. Improved taxonomic assignment of human intestinal 16S rRNA sequences by a dedicated reference database. *BMC Genomics* 16 (1): 1056. https://doi.org/10.1186/s12864-015-2265-y.

Rosen, Michael J., Benjamin J. Callahan, Daniel S. Fisher, and Susan P. Holmes. 2012. Denoising PCR-amplified metagenome data. *BMC Bioinformatics* 13: 283–283. https://doi.org/10.1186/ 1471-2105-13-283. https://www.ncbi.nlm.nih.gov/pubmed/23113967. https://www.ncbi.nlm. nih.gov/pmc/PMC3563472/.

Saitou, N., and M. Nei. 1987. The neighbor-joining method: A new method for reconstructing phylogenetic trees. *Molecular Biology and Evolution* 4 (4): 406–425. https://doi.org/10.1093/ oxfordjournals.molbev.a040454.

Sankaran, Kris, and Susan Holmes. 2014. structSSI: Simultaneous and selective inference for grouped or hierarchically structured data. *Journal of Statistical Software* 1 (13). https://doi. org/10.18637/jss.v059.i13. https://www.jstatsoft.org/v059/i13.

Schloss, Patrick D., Sarah L. Westcott, Thomas Ryabin, Justine R. Hall, Martin Hartmann, Emily B. Hollister, Ryan A. Lesniewski, Brian B. Oakley, Donovan H. Parks, Courtney J. Robinson, Jason W. Sahl, Blaz Stres, Gerhard G. Thallinger, David J. Van Horn, and Carolyn F. Weber. 2009. Introducing mothur: Open-source, platform-independent, community-supported software for describing and comparing microbial communities. *Applied and Environmental Microbiology* 75 (23): 7537–7541. https://doi.org/10.1128/AEM.01541-09. https://journals.asm.org/doi/ abs/10.1128/AEM.01541-09.

Sierra, Maria A., Qianhao Li, Smruti Pushalkar, Bidisha Paul, Tito A. Sandoval, Angela R. Kamer, Patricia Corby, Yuqi Guo, Ryan Richard Ruff, and Alexander V. Alekseyenko. 2020. The influences of bioinformatics tools and reference databases in analyzing the human oral microbial community. *Genes* 11 (8): 878.

Studier, J.A., and K.J. Keppler. 1988. A note on the neighbor-joining algorithm of Saitou and Nei. *Molecular Biology and Evolution* 5 (6): 729–731. https://doi.org/10.1093/oxfordjournals. molbev.a040527.

Tang, Zheng-Zheng, Guanhua Chen, Alexander V. Alekseyenko, and Hongzhe Li. 2017. A general framework for association analysis of microbial communities on a taxonomic tree. *Bioinformatics (Oxford, England)* 33 (9): 1278–1285. https://doi.org/10.1093/bioinformatics/btw804. https://www.ncbi.nlm.nih.gov/pubmed/28003264. https://www.ncbi.nlm.nih.gov/pmc/articles/ PMC5408811/.

Vernon, Keith. 1988. The founding of numerical taxonomy. *The British Journal for the History of Science* 21 (2): 143–159.

Wang, Qiong, George M. Garrity, James M. Tiedje, and James R. Cole. 2007. Naive Bayesian classifier for rapid assignment of rRNA sequences into the new bacterial taxonomy. *Applied and Environmental Microbiology* 73 (16): 5261–5267. https://doi.org/10.1128/AEM.00062-07. https://pubmed.ncbi.nlm.nih.gov/17586664. https://www.ncbi.nlm.nih.gov/pmc/articles/PMC1 950982/.

Woese, Carl R., and George E. Fox. 1977. Phylogenetic structure of the prokaryotic domain: The primary kingdoms. *Proceedings of the National Academy of Sciences* 74 (11): 5088–5090. https://doi.org/10.1073/pnas.74.11.5088.   https://www.pnas.org/content/pnas/74/11/5088. full.pdf.

Woese, C.R., O. Kandler, and M.L. Wheelis. 1990. Towards a natural system of organisms: Proposal for the domains Archaea, Bacteria, and Eucarya. *Proceedings of the National Academy of Sciences of the United States of America* 87 (12): 4576–4579. https://doi.org/10.1073/pnas. 87.12.4576. https://pubmed.ncbi.nlm.nih.gov/2112744. https://www.ncbi.nlm.nih.gov/pmc/arti cles/PMC54159/.

Xia, Y. 2020. Correlation and association analyses in microbiome study integrating multiomics in health and disease. *Progress in Molecular Biology and Translational Science* 171: 309–491. https://doi.org/10.1016/bs.pmbts.2020.04.003.

Xia, Yinglin, Jun Sun, and Ding-Geng Chen. 2018. Bioinformatic analysis of microbiome data. In *Statistical Analysis of Microbiome Data with R*, 1–27. Singapore: Springer.

Xiao, Jian, Hongyuan Cao, and Jun Chen. 2017. False discovery rate control incorporating phylogenetic tree increases detection power in microbiome-wide multiple testing. *Bioinformatics* 33 (18): 2873–2881. https://doi.org/10.1093/bioinformatics/btx311.

Xiao, Jian, Li Chen, Stephen Johnson, Yue Yu, Xianyang Zhang, and Jun Chen. 2018. Predictive modeling of microbiome data using a phylogeny-regularized generalized linear mixed model. *Frontiers in Microbiology* 9 (1391). https://doi.org/10.3389/fmicb.2018.01391. https://www. frontiersin.org/article/10.3389/fmicb.2018.01391.

Yilmaz, Pelin, Laura Wegener Parfrey, Pablo Yarza, Jan Gerken, Elmar Pruesse, Christian Quast, Timmy Schweer, Jörg Peplies, Wolfgang Ludwig, and Frank Oliver Glöckner. 2013. The SILVA and "all-species Living Tree Project (LTP)" taxonomic frameworks. *Nucleic Acids Research* 42 (D1): D643–D648. https://doi.org/10.1093/nar/gkt1209.

———. 2014. The SILVA and "all-species Living Tree Project (LTP)" taxonomic frameworks. *Nucleic Acids Research* 42 (Database issue): D643–D648. https://doi.org/10.1093/nar/gkt1209. https://www.ncbi.nlm.nih.gov/pubmed/24293649. https://www.ncbi.nlm.nih.gov/pmc/articles/ PMC3965112/.

Yoon, Seok-Hwan, Sung-Min Ha, Soonjae Kwon, Jeongmin Lim, Yeseul Kim, Hyungseok Seo, and Jongsik Chun. 2017. Introducing EzBioCloud: A taxonomically united database of 16S rRNA gene sequences and whole-genome assemblies. *International Journal of Systematic and Evolutionary Microbiology* 67 (5): 1613–1617. https://doi.org/10.1099/ijsem.0.001755. https:// www.ncbi.nlm.nih.gov/pubmed/28005526. https://www.ncbi.nlm.nih.gov/pmc/articles/ PMC5563544/.

Chapter 6
Clustering Sequences into OTUs

Abstract This chapter describes and illustrates taxonomic classification of the representative sequences and clustering of OTUs. First it introduces some preliminary procedures of clustering sequences into OTUs. Then it describes VSEARCH and q2-vsearch. Next three sections introduce and illustrate three approaches of clustering sequences into OTUs using q2-vsearch: closed-reference clustering, *de novo* clustering, and open-reference clustering.

Keywords Clustering · OTUs · VSEARCH · q2-vsearch · Closed-reference clustering · *De novo* clustering · Open-reference clustering · q2-cutadapt · Quality-filter · *Q-score* · *Q-score-joined* · USEARCH, UCLUST

Chapter 4 described and illustrated how to generate feature table and feature data (i.e., representative sequences). Chapter 5 described and illustrated how to assign taxonomy and build phylogenetic tree. In this chapter, we will describe and illustrate taxonomic classification of the representative sequences and clustering of OTUs. We first introduce some preliminary procedures of clustering sequences into OTUs (Sect. 6.1). Then we introduce VSEARCH and q2-vsearch (Sect. 6.2). In the next three sections, we introduce and illustrate cluster sequences into OTUs using q2-vsearch: closed-reference clustering (Sect. 6.3), *De novo* Clustering (Sect. 6.4), and open-reference clustering (Sect. 6.5), respectively. Finally, we provide a brief summary in Sect. 6.6.

6.1 Introduction to Clustering Sequences into OTUs

OTUs are used pragmatically as proxies for potential microbial species represented in a sample. The developers of QIIME 2 recommend working with Amplicon Sequence Variants (ASVs); thus QIIME 2 workflow by default does not include a

---

Supplementary Information The online version contains supplementary material available at https://doi.org/10.1007/978-3-031-21391-5_6.

147

typical OTU picking step. However, OTU picking step is the traditional approach to generate feature table or OTU table for downstream data analysis, and currently some bioinformatic centers still use this approach to generate data. Specifically, OTU picking step is an option in QIIME 2 and the **q2-vsearch** plugin is available for this analysis. Thus, here we still want to introduce the OTU picking technique.

To cluster sequences, several preliminary works need to be done: (1) merging paired-end reads, (2) removing non-biological sequences, (3) trimming all reads to the same length, (4) discarding low-quality reads, and (5) dereplicating the reads.

6.1.1   Merge Reads

Whether or not need to merge reads is depending on how the sequences will be denoised or clustered into ASVs or OTUs. The sequences need to be jointed when using **Deblur** or OTU clustering methods, which can be achieved via the QIIME 2 **q2-vsearch** plugin with the join-pairs method in **Deblur**. When the sequences were denoised using **DADA2**, then merging reads is not necessary because **DADA2** performs read merging automatically after denoising each sequence.

6.1.2   Remove Non-biological Sequences

Any non-biological sequences such as primers, sequencing adapters, and PCR spacers should be removed before clustering. There are comprehensive methods for removing non-biological sequences from paired-end or single-end data in the **q2-cutadapt** plugin. We refer the interested readers to QIIME 2 documentation files for details.

As recall in Chap. 4, DADA2 can remove biological sequences when the denoising function is called to denoise sequences. When calling the denoise functions, we can specify the values for **--p-trim** parameter to remove base pairs from the 5′ end of the reads. In this case, ASVs were obtained through denoising sequences using **DADA2** method, in which the non-biological sequences have been removed, so we do not need to perform this step again. If sub-OTUs/ASVs were obtained through **Deblur**, then we need to remove non-biological sequences because **Deblur** does not have this functionality yet.

6.1.3   Trim Reads Length

The raw reads need to be trimmed to the same length before OTU clustering. QIIME 2 recommends first denoising reads. Denoising reads involves a length trimming step, and then the length trimming step can optionally pass to the ASVs through a clustering algorithm. Thus, currently QIIME 2 does not have a function to trim reads to the same length directly.

6.1.4   Discard Low-Quality Reads

Low-quality reads will be discarded through quality filtering using the **quality-filter** plugin. Different types of quality filtering are available in QIIME 2, including the *q-score* method for single- or paired-end sequences (i.e., *SampleData [PairedEndSequencesWithQuality | SequencesWithQuality]*), **q-score-joined** for joined reads (i.e., *SampleData[JoinedSequencesWithQuality]*) after merging. We refer the readers to Chap. 4 (Sect. 4.4.3 Preliminary Works for Denoising with Deblur).

6.1.5   Dereplicate Sequences

All types of clustering first need to dereplicate the sequences. In QIIME 2, dereplicate-sequences can be performed via the **q2-vsearch** plugin.

6.2   Introduction to VSEARCH and q2-vsearch

Example 6.1: Miseq_SOP, Examples 5.1 and 5.2, Cont.
In this section, we'll cover these three OTU picking methods via QIIME 2 using the example dataset we used in Chaps. 4 and 5. In this case, because ASVs were obtained through denoising sequences using **DADA2** method, the reads were already merged, so a merging step can be omitted.

After quality filtering and denoising DNA sequences, to obtain datasets suitable for downstream statistical analyses, sequences are identified by assigning them to taxonomic groups or cluster them into OTUs. Typically, there are three ways to assign sequences to OTUs (Lawley and Tannock 2017; Whelan and Surette 2017; De Filippis et al. 2018): closed-reference clustering, *de novo* clustering, and open-reference clustering. QIIME 2 currently supports *de novo*, closed-reference, and open-reference clustering (Rideout et al. 2014).

VSEARCH (Rognes et al. 2016) is a versatile open source tool for processing and preparing metagenomics, genomics, and population genomics nucleotide sequence data. It was designed as an alternative to the **USEARCH** (Edgar 2010) based on a fast heuristic algorithm for searching nucleotide sequences. It performs optimal global sequence alignment of the query using full dynamic programming. Its functionalities include performing searching, clustering, chimera detection, and subsampling, paired-end reads merging and dereplication.

Currently two options are available in QIIME 2 for clustering of sequences or features into OTUs using **vsearch**: (1) using demultiplexed, quality-controlled sequence data (i.e., a SampleData[Sequences] artifact). Currently this option is performed in two steps. A single command is expected in the future release of

QIIME 2. (2) Using dereplicated, quality-controlled data in feature table and feature representative sequences (i.e., the FeatureTable[Frequency] and FeatureData [Sequence]artifacts). These artifacts could be generated using a variety of analysis pipelines, such as **qiime vsearch dereplicate-sequences**, and **qiime dada2 denoise** or **qiime deblur denoise** commands. The second option is performed in one step. The FeatureTable[Frequency] (in this case, FeatureTableMiSeq_SOP.qza) and FeatureData[Sequence] (RepSeqsMiSeq_SOP.qza) artifacts have already been generated in Chap. 4. We can directly use them to cluster sequences into OTUs.

Traditionally, the 97% threshold was used for approximating to species (Stackebrandt and Goebel 1994; Schloss and Handelsman 2005; Seguritan and Rohwer 2001; Westcott and Schloss 2017). Currently more stringent cut-offs was suggested to avoid over-classification of the representative sequences because it could result in spurious OTUs. Given much larger datasets currently available, around 99% for full-length sequences and around 100% for the V4 hypervariable region are considered as optimal identity thresholds (Edgar 2018).

6.3   Closed-Reference Clustering

Closed-reference clustering (Caporaso et al. 2012; Navas-Molina et al. 2013) is a phylotype-based method, also called as phylotyping (Schloss and Westcott 2011) or taxonomy-dependent method (Sun et al. 2012).

6.3.1   Introduction

Closed reference clustering is to group those sequences that match the same reference sequence in a database with a certain similarity together. That is, this method bins sequences into groups within a well curated database of known sequences, first comparing each query sequence to an annotated reference taxonomy database via the sequence classification or searching methods (Liu et al. 2017a, b; Rodrigues et al. 2017), then grouping the sequences that are matched to the same reference sequence into the same OTU. The algorithm behind closed-reference clustering is first to cluster the sequences in the FeatureData[Sequence] artifact against a reference database, and then to collapse the features in the FeatureTable into new features that are clusters of the input features.

6.3.2   Implement Cluster-Features-Closed-Reference

Below, we do closed-reference clustering with the **cluster-features-closed-reference** method via the **qiime vsearch** plugin. The **cluster-features-closed-reference** method is a wrap of the VSEARCH function **--usearch_global**. To perform closed-reference clustering of a feature table (in this case, FeatureTableMiSeq_SOP.qza), we download a reference database Greengenes (16S rRNA) with the 13_8 (most recent version) from QIIME 2 website (https://docs.qiime2.org/2022.2/data-resources/). The Greengene database (gg_13_8_otus.tar.gz) includes the 99_otus. fasta file. We first import this fasta file as a FeatureData[Sequence] artifact representing the Greengenes 13_8 99% OTUs.

```
source activate qiime2-2022.2
mkdir QIIME2R-Bioinformatics
cd QIIME2R-Bioinformatics
qiime tools import \
  --input-path gg_13_8_otus/rep_set/99_otus.fasta\
  --output-path 99_otus.qza\
  --type 'FeatureData[Sequence]'
```

Imported gg_13_8_otus/rep_set/99_otus.fasta as DNASequencesDirectoryFormat
to 99_otus.qza

In general, closed-reference OTU clustering prefers to be performed at a higher percent identity. Here, we perform clustering at 99% identity against the Greengenes 13_8 99% OTUs.

```
qiime vsearch cluster-features-closed-reference \
  --i-table FeatureTableMiSeq_SOP.qza \
  --i-sequences RepSeqsMiSeq_SOP.qza \
  --i-reference-sequences 99_otus.qza \
  --p-perc-identity 0.99 \
  --o-clustered-table TableCR99.qza \
  --o-clustered-sequences RepSeqsCR99.qza \
  --o-unmatched-sequences UnmatchedCR99.qza
```

Saved FeatureTable[Frequency] to: TableCR99.qza
Saved FeatureData[Sequence] to: RepSeqsCR99.qza
Saved FeatureData[Sequence] to: UnmatchedCR99.qza

In above commands, the **--i-reference-sequences** flag is used to include reference database to cluster against with. This reference input file should be a .qza file containing a fasta file with the sequences to use as references, with QIIME 2 data type FeatureData[Sequence]. SILVA or GreenGenes for 16S rRNA gene sequences are most often used as the references input file, while other standard references such as UNITE for ITS data are also used. Still others prefer to curate their own databases.

After implementing closed-reference clustering, we obtain a FeatureTable[Frequency] artifact (TableCR99.qza) and a FeatureData[Sequence] artifact (RepSeqsCR99.qza). Note that The FeatureData[Sequence] artifact (in this case RepSeqsCR99.qza or UnmatchedCR99.qza) is **not** the sequences defining the features in the FeatureTable, but rather the collection of feature ids and their sequences that didn't match the reference database at 99% identity.

6.3.3   Remarks on Closed-Reference Clustering

Closed-reference clustering as a phylotype-based method directly assigns sequences based on their distance (similarity) to phylotypes, i.e., reference sequences, whereas distance-based methods group sequences based on their distance (similarity) between sequences to OTUs.

The phylotype-based methods have several appealing features, including:

- Easily linking a sequence to previously identified microbes, computational efficiency, and stable classification.
- Have the strengths of speed, potential for trivial parallelization (Westcott and Schloss 2015).
- Closed-reference clustering methods cluster sequence reads against a reference dataset; thus the OTUs obtained from this method can be used to do alpha- and beta-diversity estimations and directly compare OTUs across studies (Westcott and Schloss 2015; He et al. 2015).
- Sequence reads from different marker gene regions can be clustered together if the reference dataset consists of full-length marker genes (He et al. 2015).
- OTU clustering can be parallelized for large datasets (He et al. 2015), which is suitable for meta-analysis.

However, the phylotype-based methods also have critical challenges:

- The success of assignment is highly contingent on sequencing platform and reference database (Tyler et al. 2014). Thus, when reference databases are incomplete because a large portion of taxa in a sample is unknown or has not yet been well defined, and hence not recorded in databases, then they cannot be assigned to an OTU. Thus, it is impossible to analyze novel sequences detected in an experiment via previously unidentified taxonomic lineages (Tyler et al. 2014; Schloss and Westcott 2011).
- Due to largely being dependent on the completeness of the reference database, these clustering methods do not perform well if many novel organisms exist in the sequencing data (Schloss and Westcott 2011; Chen et al. 2016).
- Especially, the fundamental problem of the closed-reference approach is that two query sequences matched to the same reference sequence at a higher same (e.g., 97%) similarity may only have a lower similarity (e.g., 94%) to each other (Westcott and Schloss 2015). This is the issue of adverse triplets, which is common in practice (Edgar 2018).

In summary, because closed-reference clustering methods are largely dependent on the completeness of the reference database, they are often employed to annotate sequences (Sun et al. 2012) rather than to detect novel sequences.

6.4   *De Novo* Clustering

De novo clustering is a distance-based method (Schloss and Westcott 2011), also called as taxonomy-independent (Sun et al. 2012), OTU-based (Zongzhi Liu et al. 2008; Chen et al. 2013), taxonomy-unsupervised (Sul et al. 2011), or *de novo* (Navas-Molina et al. 2013; Edgar 2010) clustering methods.

6.4.1   *Introduction*

De novo clustering clusters sequences into groups based on sequence identity or genetic distances alone. It first clusters all sequences into OTUs based on the pairwise sequence distances to compare each sequence against each other rather than to compare against a reference database (Forster et al. 2016), then group sequences into OTUs by implementing a clustering algorithm with a specified threshold.

The algorithm behind *de novo* clustering is first to cluster all sequences in the FeatureData[Sequence] artifact against one another (rather than against a reference database) based on the pairwise sequence distances, and then to collapse features in the FeatureTable into new features that are clusters of the input features, i.e., classify reads that have a similarity greater than a threshold (typically 97% or 99% identity) as the same OTU.

6.4.2   *Implement Cluster-Features-De-Novo*

De novo clustering of a feature table can be performed as follows. Here, we perform clustering at 99% identity by specifying 99% identity in **--p-perc-identity** parameter, which wraps the VSEARCH **--cluster_size** function, to create 99% OTUs. First, store the artifacts of FeatureTableMiSeq_SOP.qza and RepSeqsMiSeq_SOP.qza into the directory QIIME2R-Bioinformatics. Then type: cd QIIME2R-Bioinformatics in terminal after activating QIIME 2 to link the datasets to this folder. Finally, call the **cluster-features-de-novo method** via the **qiime vsearch** plugin to implement the *de novo* clustering.

```
qiime vsearch cluster-features-de-novo \
  --i-table FeatureTableMiSeq_SOP.qza \
  --i-sequences RepSeqsMiSeq_SOP.qza \
  --p-perc-identity 0.99 \
  --o-clustered-table TableDn99MiSeq_SOP.qza \
  --o-clustered-sequences RepSeqsDn99MiSeq_SOP.qza
```

Saved FeatureTable[Frequency] to: TableDn99MiSeq_SOP.qza
Saved FeatureData[Sequence] to: RepSeqsDn99MiSeq_SOP.qza

Above commands generate two artifacts: a FeatureTable[Frequency] (TableDn99MiSeq_SOP.qza) with the BIOMV210DirFmt format, and a FeatureData[Sequence](RepSeqsDn99MiSeq_SOP.qza)   with   the DNASequencesDirectoryFormat format. We review them through the QIIME2 viewer. The FeatureData[Sequence] artifact contains the centroid sequence defining each OTU cluster.

6.4.3   Remarks on De Novo Clustering

De novo clustering methods do overcome most limitations of phylotype-based methods and have several advantages. The *de novo* clustering approach:

- Carries out the clustering step independently without references for a phylotype-database. Thus, these methods outperform phylotype-based reference methods for assigning 16S rRNA gene sequences to OTUs and have been preferably used across the field (Westcott and Schloss 2015).
- Is optimal for samples that contain many bacteria that have no reference sequences in the public databases.
- Particularly, it was demonstrated that *de novo* clustering methods significantly outperform the approaches of closed-reference clustering and open-reference clustering for picking OTUs (Schloss 2016; Jackson et al. 2016).

However, *de novo* clustering methods also have several weaknesses, such as:

- It is computationally intensive (cost of hierarchical clustering), relatively slow, and larger memory required due to higher sequencing error rates in expanding sequencing throughput, the difficult choice of linkage method for clustering (Schloss and Westcott 2011; Westcott and Schloss 2015).
- It tends to produce a very large number of OTUs.
- The OTUs obtained from both *de novo* clustering and open-reference OTU clustering methods affect alpha-diversity analyses (e.g., rarefaction curves), beta-diversity analyses such as principal component analysis and distance-based ordination (e.g., principal coordinate analysis), and the identification of differentially represented OTUs by a hypothesis testing, such as ADONIS's R value (He et al. 2015).
- Especially, the *de novo* clustering methods have one fundamental problem: the clustering results (i.e., OTU assignments) are strongly influenced by or sensitive to the input order of the sequences (Mahé et al. 2014; He et al. 2015).

In summary, *de novo* clustering methods have been attracted more attention and have become the preferred option for researchers (Schloss 2010; Cai et al. 2017).

6.5 Open-Reference Clustering

Open-reference clustering is a hybrid of the closed-reference clustering and *de novo* clustering approaches (Navas-Molina et al. 2013; Rideout et al. 2014).

6.5.1 Introduction

Open-reference clustering combines the closed-reference and *de novo* methods and sequentially performs closed-reference clustering and *de novo* clustering, in which a closed-reference clustering method is first used to assign OTUs, and the unassigned sequences outputted by the closed-reference method are then grouped by a *de novo* clustering method (Westcott and Schloss 2017).

For example, in QIIME-uclust, the "pick_open_reference_otus.py" script implements the latest QIIME open reference OTU clustering (Rideout et al. 2014). The algorithm behind open-reference clustering is: it first performs closed-reference clustering against a reference database (e.g., Greengenes v.13.8, 97% OTU database), using clustering method UCLUST (Edgar 2010), which exploits USEARCH to search large sequence databases and to assign sequences to clusters. Then, it subsamples (default proportion of subsampling = 0.001) those reads that do not map in this first step and performs a *de novo* OTU clustering step. Next, remaining unmapped reads are subsequently closed-reference clustered against these *de novo* OTUs. Finally, it performs another step of *de novo* clustering on the remaining unmapped reads.

6.5.2 Implement Cluster-Features-Open-Reference

Similar to the closed-reference clustering, open-reference clustering (Rognes et al. 2016) can be performed using the **cluster-features-open-reference** method via the **qiime vsearch** plugin. Also similar to the closed-reference clustering, open-reference OTU clustering is generally performed at a higher percent identity.

```
qiime vsearch cluster-features-open-reference \.
  --i-table FeatureTableMiSeq_SOP.qza \
  --i-sequences RepSeqsMiSeq_SOP.qza \
  --i-reference-sequences 99_otus.qza \
  --p-perc-identity 0.99 \
  --o-clustered-table TableOR99.qza \
  --o-clustered-sequences RepSeqsOR99.qza \
  --o-new-reference-sequences NewRefSeqsOR99.qza
```

Saved FeatureTable[Frequency] to: TableOR99.qza
Saved FeatureData[Sequence] to: RepSeqsOR99.qza
Saved FeatureData[Sequence] to: NewRefSeqsOR99.qza

Open-reference clustering generated a FeatureTable[Frequency] artifact TableOR99.qza, and two FeatureData[Sequence] artifacts: RepSeqsOR99.qza and NewRefSeqsOR99.qza. The first FeatureData[Sequence] artifact represents the clustered sequences, while the second artifact represents the new reference sequences, composed of the reference sequences used for input, and the sequences clustered as part of the internal *de novo* clustering step.

6.5.3 Remarks on Open-Reference Clustering

Open-reference clustering combines the closed-reference and *de novo* methods and sequentially performs closed-reference clustering and *de novo* clustering; thus, theoretically this method should reserve the strengths of both closed-reference and *de novo* clustering. For example, it was evaluated that open-reference clustering method is a much more effective compared to using *de novo* methods alone, and was recommended for assigning OTUs along implementing using uclust in QIIME (He et al. 2015).

However, open-reference clustering methods have the weaknesses:

- It blends the strengths and weaknesses of the other methods and was reviewed to have potential problems when using these two methods together due to the different OTU definitions employed by commonly used closed-reference and *de novo* clustering implementations and associated with database quality and classification error (Westcott and Schloss 2015, 2017).
- OTU clustering tends to exaggerate the number of unique organisms found within a sample (Edgar 2017). Especially open-reference OTU clustering consistently picks up more number of OTUs (QIIME) than the number of ASVs (DADA2) (Sierra et al. 2020).
- Particularly a recent study (Prodan et al. 2020) showed that QIIME-uclust (used in QIIME 1) produced large number of spurious OTUs and inflated alpha-diversity measures, and suggested QIIME-uclust should be avoided in future studies.

In summary, the performance of open-reference clustering has not been consistently confirmed.

6.6 Summary

This chapter described clustering sequences into OTUs via QIIME 2. First, several preliminary steps of OTU clustering was described including merging reads, removing non-biological sequences, trimming reads length, discard low-quality reads, and dereplicating sequences. Then, the VSEARCH and q2-vsearch were introduced, which perform versatile bioinformatic functions including searching, clustering, chimera detection and subsampling, paired-end reads merging, and dereplication. Next, three approaches of OTU clustering methods including closed-reference clustering, *de novo* clustering, and open-reference clustering were described and illustrated. Their advantages and disadvantages were also discussed. In Chap. 7, we will introduce the original OTU methods in numerical taxonomy.

References

Cai, Yunpeng, Wei Zheng, Jin Yao, Yujie Yang, Volker Mai, Qi Mao, and Yijun Sun. 2017. ESPRIT-Forest: Parallel clustering of massive Amplicon Sequence data in subquadratic time. *PLoS Computational Biology* 13 (4): e1005518.

Caporaso, J. Gregory, Christian L. Lauber, William A. Walters, Donna Berg-Lyons, James Huntley, Noah Fierer, Sarah M. Owens, Jason Betley, Louise Fraser, Markus Bauer, Niall Gormley, Jack A. Gilbert, Geoff Smith, and Rob Knight. 2012. Ultra-high-throughput microbial community analysis on the Illumina HiSeq and MiSeq platforms. *The ISME Journal* 6 (8): 1621–1624. https://doi.org/10.1038/ismej.2012.8. https://www.ncbi.nlm.nih.gov/pubmed/22402401, https://www.ncbi.nlm.nih.gov/pmc/PMC3400413/.

Chen, Wei, Clarence K. Zhang, Yongmei Cheng, Shaowu Zhang, and Hongyu Zhao. 2013. A comparison of methods for clustering 16S rRNA sequences into OTUs. *PLoS One* 8 (8): e70837. https://doi.org/10.1371/journal.pone.0070837.

Chen, Shi-Yi, Feilong Deng, Ying Huang, Xianbo Jia, Yi-Ping Liu, and Song-Jia Lai. 2016. bioOTU: An improved method for simultaneous taxonomic assignments and operational taxonomic units clustering of 16s rRNA gene sequences. *Journal of Computational Biology* 23 (4): 229–238.

De Filippis, F., E. Parente, T. Zotta, and D. Ercolini. 2018. A comparison of bioinformatic approaches for 16S rRNA gene profiling of food bacterial microbiota. *International Journal of Food Microbiology* 265: 9–17. https://doi.org/10.1016/j.ijfoodmicro.2017.10.028.

Edgar, Robert C. 2010. Search and clustering orders of magnitude faster than BLAST. *Bioinformatics* 26 (19): 2460–2461. https://doi.org/10.1093/bioinformatics/btq461.

———. 2017. Accuracy of microbial community diversity estimated by closed- and open-reference OTUs. *PeerJ* 5: e3889. https://doi.org/10.7717/peerj.3889.

———. 2018. Updating the 97% identity threshold for 16S ribosomal RNA OTUs. *Bioinformatics* 34 (14): 2371–2375. https://doi.org/10.1093/bioinformatics/bty113.

Forster, Dominik, Micah Dunthorn, Thorsten Stoeck, and Frédéric Mahé. 2016. Comparison of three clustering approaches for detecting novel environmental microbial diversity. *PeerJ* 4: e1692.

He, Yan, J. Gregory Caporaso, Xiao-Tao Jiang, Hua-Fang Sheng, Susan M. Huse, Jai Ram Rideout, Robert C. Edgar, Evguenia Kopylova, William A. Walters, Rob Knight, and Hong-Wei Zhou. 2015. Stability of operational taxonomic units: An important but neglected property for analyzing microbial diversity. *Microbiome* 3: 20–20. https://doi.org/10.1186/s40168-015-0081-x.

https://www.ncbi.nlm.nih.gov/pubmed/25995836, https://www.ncbi.nlm.nih.gov/pmc/PMC4438525/.

Jackson, Matthew A., Jordana T. Bell, Tim D. Spector, and Claire J. Steves. 2016. A heritability-based comparison of methods used to cluster 16S rRNA gene sequences into operational taxonomic units. *PeerJ* 4: e2341.

Lawley, Blair, and Gerald W. Tannock. 2017. Analysis of 16S rRNA gene amplicon sequences using the QIIME software package. In *Oral Biology*, 153–163. Springer.

Liu, Zongzhi, Todd Z. DeSantis, Gary L. Andersen, and Rob Knight. 2008. Accurate taxonomy assignments from 16S rRNA sequences produced by highly parallel pyrosequencers. *Nucleic Acids Research* 36 (18): e120–e120. https://doi.org/10.1093/nar/gkn491. https://pubmed.ncbi.nlm.nih.gov/18723574, https://www.ncbi.nlm.nih.gov/pmc/articles/PMC2566877/.

Liu, Zhunga, Quan Pan, Jean Dezert, Jun-Wei Han, and You He. 2017a. Classifier fusion with contextual reliability evaluation. *IEEE Transactions on Cybernetics* 48 (5): 1605–1618.

Liu, Zhun-Ga, Quan Pan, Jean Dezert, and Arnaud Martin. 2017b. Combination of classifiers with optimal weight based on evidential reasoning. *IEEE Transactions on Fuzzy Systems* 26 (3): 1217–1230.

Mahé, Frédéric, Torbjørn Rognes, Christopher Quince, Colomban de Vargas, and Micah Dunthorn. 2014. Swarm: Robust and fast clustering method for amplicon-based studies. *PeerJ* 2: e593. https://doi.org/10.7717/peerj.593.

Navas-Molina, José A., Juan M. Peralta-Sánchez, Antonio González, Paul J. McMurdie, Yoshiki Vázquez-Baeza, Xu Zhenjiang, Luke K. Ursell, Christian Lauber, Hongwei Zhou, Se Jin Song, James Huntley, Gail L. Ackermann, Donna Berg-Lyons, J. Susan Holmes, Gregory Caporaso, and Rob Knight. 2013. Advancing our understanding of the human microbiome using QIIME. *Methods in Enzymology* 531: 371–444. https://doi.org/10.1016/b978-0-12-407863-5.00019-8. https://www.ncbi.nlm.nih.gov/pubmed/24060131, https://www.ncbi.nlm.nih.gov/pmc/PMC4517945/.

Prodan, Andrei, Valentina Tremaroli, Harald Brolin, Aeilko H. Zwinderman, Max Nieuwdorp, and Evgeni Levin. 2020. Comparing bioinformatic pipelines for microbial 16S rRNA Amplicon Sequencing. *PLoS One* 15 (1): e0227434. https://doi.org/10.1371/journal.pone.0227434.

Rideout, Jai Ram, Yan He, Jose A. Navas-Molina, William A. Walters, Luke K. Ursell, Sean M. Gibbons, John Chase, Daniel McDonald, Antonio Gonzalez, Adam Robbins-Pianka, Jose C. Clemente, Jack A. Gilbert, Susan M. Huse, Hong-Wei Zhou, Rob Knight, and J. Gregory Caporaso. 2014. Subsampled open-reference clustering creates consistent, comprehensive OTU definitions and scales to billions of sequences. *PeerJ* 2: e545. https://doi.org/10.7717/peerj.545.

Rodrigues, Matias, F. João, Thomas S.B. Schmidt, Janko Tackmann, and Christian von Mering. 2017. MAPseq: Highly efficient k-mer search with confidence estimates, for rRNA sequence analysis. *Bioinformatics* 33 (23): 3808–3810. https://doi.org/10.1093/bioinformatics/btx517.

Rognes, Torbjørn, Tomáš Flouri, Ben Nichols, Christopher Quince, and Frédéric Mahé. 2016. VSEARCH: A versatile open source tool for metagenomics. *PeerJ* 4: e2584. https://doi.org/10.7717/peerj.2584.

Schloss, Patrick D. 2010. The effects of alignment quality, distance calculation method, sequence filtering, and region on the analysis of 16S rRNA gene-based studies. *PLoS Computational Biology* 6 (7): e1000844–e1000844. https://doi.org/10.1371/journal.pcbi.1000844. https://www.ncbi.nlm.nih.gov/pubmed/20628621, https://www.ncbi.nlm.nih.gov/pmc/PMC2900292/.

———. 2016. Application of a database-independent approach to assess the quality of operational taxonomic unit picking methods. *Msystems* 1 (2): e00027–e00016.

Schloss, Patrick D., and Jo Handelsman. 2005. Introducing DOTUR, a computer program for defining operational taxonomic units and estimating species richness. *Applied and Environmental Microbiology* 71 (3): 1501–1506. https://doi.org/10.1128/aem.71.3.1501-1506.2005. https://aem.asm.org/content/aem/71/3/1501.full.pdf.

Schloss, P.D., and S.L. Westcott. 2011. Assessing and improving methods used in operational taxonomic unit-based approaches for 16S rRNA gene sequence analysis. *Applied and Environmental Microbiology* 77 (10): 3219–3226.

Seguritan, V., and F. Rohwer. 2001. FastGroup: A program to dereplicate libraries of 16S rDNA sequences. *BMC Bioinformatics* 2: 9–9. https://doi.org/10.1186/1471-2105-2-9. https://www.ncbi.nlm.nih.gov/pubmed/11707150, https://www.ncbi.nlm.nih.gov/pmc/articles/PMC59723/.

Sierra, Maria A., Qianhao Li, Smruti Pushalkar, Bidisha Paul, Tito A. Sandoval, Angela R. Kamer, Patricia Corby, Yuqi Guo, Ryan Richard Ruff, and Alexander V. Alekseyenko. 2020. The influences of bioinformatics tools and reference databases in analyzing the human oral microbial community. *Genes* 11 (8): 878.

Stackebrandt, E., and B.M. Goebel. 1994. Taxonomic note: A Place for DNA-DNA reassociation and 16S rRNA sequence analysis in the present species definition in bacteriology. *International Journal of Systematic and Evolutionary Microbiology* 44 (4): 846–849. https://doi.org/10.1099/00207713-44-4-846. https://www.microbiologyresearch.org/content/journal/ijsem/10.1099/00207713-44-4-846.

Sul, Woo Jun, James R. Cole, C. Ederson da, Qiong Wang Jesus, Ryan J. Farris, Jordan A. Fish, and James M. Tiedje. 2011. Bacterial community comparisons by taxonomy-supervised analysis independent of sequence alignment and clustering. *Proceedings of the National Academy of Sciences of the United States of America* 108 (35): 14637–14642. https://doi.org/10.1073/pnas.1111435108. https://pubmed.ncbi.nlm.nih.gov/21873204, https://www.ncbi.nlm.nih.gov/pmc/articles/PMC3167511/.

Sun, Yijun, Yunpeng Cai, Susan M. Huse, Rob Knight, William G. Farmerie, Xiaoyu Wang, and Volker Mai. 2012. A large-scale benchmark study of existing algorithms for taxonomy-independent microbial community analysis. *Briefings in Bioinformatics* 13 (1): 107–121. https://doi.org/10.1093/bib/bbr009. https://www.ncbi.nlm.nih.gov/pubmed/21525143, https://www.ncbi.nlm.nih.gov/pmc/PMC3251834/.

Tyler, Andrea D., Michelle I. Smith, and Mark S. Silverberg. 2014. Analyzing the human microbiome: A "how to" guide for physicians. *The American Journal of Gastroenterology* 109: 983. https://doi.org/10.1038/ajg.2014.73.

Westcott, Sarah L., and Patrick D. Schloss. 2015. *De Novo* clustering methods outperform reference-based methods for assigning 16S rRNA gene sequences to operational taxonomic units. *PeerJ* 3: e1487. https://doi.org/10.7717/peerj.1487.

———. 2017. OptiClust, an improved method for assigning amplicon-based sequence data to operational taxonomic units. *mSphere* 2 (2): e00073–e00017. https://doi.org/10.1128/mSphereDirect.00073-17. https://www.ncbi.nlm.nih.gov/pubmed/28289728, https://www.ncbi.nlm.nih.gov/pmc/articles/PMC5343174/.

Whelan, Fiona J., and Michael G. Surette. 2017. A comprehensive evaluation of the sl1p pipeline for 16S rRNA gene sequencing analysis. *Microbiome* 5 (1): 100. https://doi.org/10.1186/s40168-017-0314-2.

Chapter 7
OTU Methods in Numerical Taxonomy

Abstract This chapter investigates the original OTU methods in numerical taxonomy. First, it briefly introduces the history of numerical taxonomy. Second, it introduces the principles and aims of numerical taxonomy. Third, it briefly describes the philosophy of numerical taxonomy. Fourth, it introduces the formation and characteristics of commonly clustering-based OTU methods, which have large impact on microbiome study in both concept and methodology. Fifth, it introduces statistical hypothesis testing of OTUs. Sixth, it describes the characteristics of clustering-based OTU methods.

Keywords Numerical taxonomy · Characters · Taxa · Sample size calculation · Taxonomic rank · Phenetic taxonomy · Phenetics · Natural classification · Biological classifications · Taxonomic structure · Operational taxonomic units (OTUs) · Taxonomic resemblance · R-Technique · Q-Technique · Similarity coefficients · Commonly clustering-based OTU methods · Similarity/resemblance matrix · Single linkage clustering · Complete linkage clustering · Average linkage clustering · Factor analysis · Ordination methods · Discriminant analysis

In Chap. 6 we introduced how to cluster sequences into OTUs using closed-reference clustering, de novo clustering, and open-reference clustering. In this chapter we first briefly introduce the history of numerical taxonomy (Sect. 7.1). Then we introduce the principles and aims of numerical taxonomy (Sect. 7.2). Next, we briefly describe the philosophy of numerical taxonomy (Sect. 7.3). Followed that in Sect. 7.4, we focus on the formation and characteristics of commonly clustering-based OTU methods, which have a large impact on microbiome study in both concept and methodology. In Sect. 7.5, we discuss statistical hypothesis testing of OTUs. Section 7.6 describes the characteristics of clustering-based OTU methods. Finally, we conclude this chapter with a summary (Sect. 7.7).

In general, all the methods used in numerical taxonomy can be called numerical taxonomic methods. Here, we define the methodology of numerical taxonomy in the non-rigorous way and name the methods used in numerical taxonomy as "OUT-based methods" or "OTU methods," which include defining OTUs, measuring taxonomic resemblance (similarity/dissimilarity), clustering-based OTU methods, factor analysis, ordination, and discriminant function analysis because in numerical taxonomy all the used methods mainly regard construction of taxonomic structure by classifying and identifying OTUs. We specifically refer to all the clustering methods used in numerical taxonomy as clustering-based OTU methods.

Numerical taxonomy has directly affected several research fields, e.g., phylogenetics, genetics, and paleontology (Sokal and Sneath 1963), especially ecology, psychometrics (Sneath 1995), and microbiome. The clustering-based OTU methods in microbiome studies were originated in OTU methods in numerical taxonomy founded in 1963 by microbiologist Peter H. A. Sneath and population geneticist Robert R. Sokal. The lineage between the fields of numerical ecology and numerical taxonomy had been described in (Legendre 2019).

7.1 Brief History of Numerical Taxonomy

The concept of OTU links to the field of numerical taxonomy, founded mainly by the three independent groups (Vernon 1988): (1) Michener and Sokal, (2) Sneath, and (3) Cain and Harrison. These groups in 1957 and 1958 independently published several papers. Sokal and Sneath also considered of Rogers and Tanimoto as the fourth group in contribution of establishing numerical taxonomy (Sneath 1995; Sokal 1985).

- Michener and Sokal (Michener and Sokal 1957; Sokal 1958a) studied coding and scaling multistate characters, parallelism and convergence, and equal weighting of characters. Sokal's work (Sokal 1985) originated from how to efficiently classify group of organisms and contributed his biological statistics. He developed the average-linkage clustering methods, which were inspired by various types of primitive cluster and factor analyses in the 1940s by Holzinger and Harman (1941), Cattell (1944), Tryon's cluster analysis (Tryon 1934), and as well Thurstone on factor analysis (Vernon 1988).
- Sneath (Sneath 1957a, b) presented his principles and methodology, which were built on a similarity coefficient and Jaccard's coefficient and single-linkage clustering (Sokal 1985).
- Cain and Harrison (1958) were the first to differentiate clearly phenetic from cladistic relationships (Sokal 1985; Sneath 1995) and to apply average distance to numerical taxonomy (Sokal 1985). Cain's two other papers (Cain 1958, 1959) also contributed to the early establishment of numerical taxonomy (Sneath 1995) in which he discusses the logic and memory in Linnaeus's system of taxonomy and the deductive and inductive methods in post-Linnaean taxonomy, respectively.
- Rogers and Tanimoto (1960) developed the early methods of probabilistic distance coefficients and clustering numerical taxonomy (Sokal 1985; Sneath 1995). George Estabrook, a distinguished student from this group, also made contributions to numerical taxonomy and especially made outstanding contributions to numerical phylogenetic inference (Sokal 1985; Sneath 1995).

All the early papers from Michener and Sokal, Sneath, and Cain and Harrison contain remarkably similar ideas on the theory nature and practice of taxonomy (Vernon 1988) that laid the theoretical foundation of numerical taxonomy (Sneath and Sokal 1962; Rogers and Tanimoto 1960). Other early indirect contributors (Sneath 1995) to the establishment of numerical taxonomy included:

- Gilmour (1937, 1940, 1951) who provided some key concepts of information content and predictivity that they adapted from the philosophies of science of Mill (1886) and Whewell (1840).
- Simpson's clear thought on systematics and the problems confronting them had major influence on Sokal and Sneath and contributed to the founding of numerical taxonomy, although not from methodology (Simpson 1944, 1961).

Previously, the methods for quantifying the classificatory process in systematics had been called "quantitative systematics." In 1960, Sokal and Sneath renamed these methods collectively as "numerischen taxonomie" ("numerical taxonomy") (Sokal 1960, 1961; Rohlf and Sokal 1962) (On September 15, 1982, Sokal said in fact they coined the name "numerical taxonomy," This Week's Citation Classic, CC/Number 46, November 15, 1982). Sokal and Sneath in particular became acknowledged fathers of numerical taxonomy (Vernon 1988; Goodfellow et al. 1985). For the detail founding history of numerical taxonomy, we reference the interested readers to the review articles by Sokal (1985), Vernon (1988), and Sneath (1995) as well as references therein.

First methods and theory of numerical taxonomy were developed during 1957–1961 (Sneath and Sokal 1973), while in 1962, Rohlf (1962) first conducted hypothesis test of the nonspecificity; Sokal and Rohlf (1962) first used criterion of goodness of a classification to compare dendrograms in numerical taxonomic studies. In 1962, when the book *Principles of Numerical Taxonomy* (Sokal and Sneath 1963) was still in preparation, Sneath and Sokal published an article called "Numerical Taxonomy" in *Nature* (Sneath and Sokal 1962). In this article, the comprehensive theory and methods of numerical taxonomy were first published. In this *Nature* article, Sneath and Sokal defined their named field "Numerical Taxonomy" as "the numerical evaluation of the affinity or similarity between taxonomic units and the ordering of these units into taxa on the basis of their affinities" (Sneath and Sokal 1962, p. 48). "The term may include the drawing of phylogenetic inferences from the data by statistical or other mathematical methods to the extent to which this is possible. These methods will almost always involve the conversion of information about taxonomic entities into numerical quantities" (Sokal and Sneath 1963, p. 48). In their 1973 book, the term "numerical taxonomy" (Sneath and Sokal 1973; Sokal 1985) was clearly defined as "the grouping by numerical methods of taxonomic units into taxa on the basis of their character states. The term includes the drawing of phylogenetic inferences from the data by statistical or other mathematical methods to the extent to which this is possible. These methods require the conversion of information about taxonomic entities into numerical quantities" (Sneath and Sokal 1973, p. 4).

7.2   Principles of Numerical Taxonomy

Numerical taxonomy is an empirical science on a classification study and a multivariate technique focusing on problems of classification and based on multiple
characters. The basic ideas of numerical taxonomy were to use large numbers of
unweighted characters through statistical techniques to estimate similarity between
organisms. In other words, numerical taxonomy is based on shared characters and by
clustering their similarity values to create groups and to construct the classifications
without using any phylogenetic information (Sokal 1963). It aims to (1) obtain
repeatability and objectivity [see (Sokal 1963, pp. 49–50; Sneath and Sokal 1973,
p. 11)]; (2) quantitatively measure resemblance from numerous equally weighted
characters (Sokal 1963, pp. 49–50); (3) construct taxa from character correlations
leading to groups of high information content (Sokal 1963); and (4) separate phenetic and phylogenetic considerations (Sokal 1963).

   Numerical taxonomy has been built on seven fundamental principles (Sneath and
Sokal 1973, p. 5): "(1) The greater the content of information in the taxa of a
classification and the more characters on which it is based, the better a given
classification will be. (2) A priori, every character is of equal weight in creating
natural taxa. (3) Overall similarity between any two entities is a function of their
individual similarities in each of the many characters in which they are being
compared. (4) Distinct taxa can be recognized because correlations of characters
differ in the groups of organisms under study. (5) Phylogenetic inferences can be
made from the taxonomic structures of a group and from character correlations,
given certain assumptions about evolutionary pathways and mechanisms. (6) Taxonomy is viewed and practiced as an empirical science. And (7) Classifications are
based on phenetic similarity."

   Based on the 7 principles, Sokal and Sneath criticized that the practices of
previous taxonomy may yield arbitray and artificail taxa (Sokal 1963, p. 9) because
previous taxonomy recognized and defined taxa were based on three often not
clearly separated kinds of evidence (Sokal 1963, p. 7): (1) *Resemblances* (a taxon
is formed based on phonetically more resemble entities), (2) *Homologous characters*
(a taxon is formed by entities sharing characters of common origin), and (3) *Common
line of descent* (a taxon is formed with membership in a common line of descent).
The evidence of type 3 is rare and hence taxa are usually inferred from types 1 and
2, while the conclusions based on homologies (evidence of type 2) are often
deducted from phylogenetic speculations (evidence of type 3).

   Different from the traditional subjective approach, the approach of numerical
taxonomy is to classify organisms by statistical procedures (Vernon 1988). The
underlying reason to propose a new field of taxonomy was a dissatisfaction with the
evolutionary basis of taxonomy (Vernon 1988) or a major concern of producing a
classification in accordance with phylogeny. In other words, for numerical taxonomy, relationships must be determined in a non-historical sense and only then the
most likely lines of descent can be decided on it (Michener and Sokal 1957).

The principles or framework of numerical taxonomy consists of several core ideas and statistical procedures. The main point is how to construct a taxonomic system. Understanding these principles will facilitate better understanding of numerical taxonomy, why the methodology of numerical taxonomy has been adopted into microbiome study, as well as the current challenges using the methodology in microbiome study. Construction of a taxonomic system is to find taxonomic evidence, which is relevant to defining characters and taxa.

In this section, we review how numerical taxonomy defines characters and taxa (the fundamental taxonomic units) as well as obtains estimations of affinity between taxa from the data (characters).

7.2.1   Definitions of Characters and Taxa

Numerical taxonomy defines characters entirely based on the differences between individuals [see (Sokal and Sneath 1963, p. 62; Sneath and Sokal 1973, p. 72)]; thus a character is a property or "feature which varies from one kind of organism to another" (Michener and Sokal 1957) or anything that can be considered as a variable logically (rather than functionally or mathematically) independent of any other thing (Cain and Harrison 1958).

Unit characters are defined in terms of information theory [see (Sokal and Sneath 1963, p. 63; Sneath and Sokal 1973, p. 72)], which is to allow unit characters to be treated in a broad way, that is, to convey information. Thus, a unit character (or a "feature") is an attribute possessed by an organism for yielding a single piece of information.

During 1960s–early 1970s, in most cases it is difficult to make genetic inferences from phenetic studies of characters. The considerations of defining unit characters in terms of information theory were premature; thus, Sokal and Sneath proposed a narrow working definition of unit character (phenotypic characters). A unit character is defined as taxonomic character of two or more states, which within the study at hand cannot be subdivided logically, except for subdivision brought about by changes in the method of coding (Sokal and Sneath 1963, p. 65). Sokal and Sneath roughly grouped taxonomic characters into four categories [see Sokal and Sneath (Sokal and Sneath 1963, pp. 93–95; Sneath and Sokal 1973, pp. 90–91)]: (1) morphological characters (external, internal, microscopic, including cytological and developmental characters); (2) physiological and chemical characters; (3) behavioral characters; and (4) ecological and distributional characters (habitats, food, hosts, parasites, population dynamics, geographical distribution).

One important impact of defining unit character in terms of information lies on the factor that many bits of information contained in the simplest organisms can be interpreted in terms of molecular genetics.

Among the taxonomic characters the chemical characters are very important. They are the nucleotide bases of the nucleic acids (often the DNA of the genome) in their specific sequences (Sneath and Sokal 1973, p. 94). These characters are created by the chemical methods: the nucleic acid pairing technique which was pioneered by Doty et al. (1960), and advanced particularly by McCarthy and Bolton (1963). The characters of nucleotide bases of the nucleic acids are closer to the genotype, and the chemical data can be arranged in the order, and hence their superiority over to morphological characters have been discussed among biochemists; but at that time (1973) numerical taxonomy was not optimistic on these characters in their practical applications (Sneath and Sokal 1973, p. 94). The study of DNA sequences in numerical taxonomy has not largely been explored. But the problems or challenges of how to define these kinds of characters and more precisely and profoundly analyze the nature of phenetic similarity have been described (Sneath and Sokal 1973).

Actually, the microbiome study is just based on these characters. The later microbiome sequecing study originated in coding genetic information and defining unit character in terms of information.

Sokal and Sneath suggested all known characters should be included to avoid bias (or spurious resemblance between characters). They also suggested that each character should be given equal weight because no method can satisfactorily allocate weights on characters.

Sokal and Sneath adopted Gilmour naturalness as a criterion for natural classifications in phenetics and thought that most taxa in nature (not only biological taxa) are polythetic. The term "nature" is used in the sense implied that suggests a true reflection of the arrangement of the organisms in nature. For them, no single state is either essential or sufficient to make an organism a member of the group; in a polythetic taxon membership is based on the greatest number of shared character states. Sokal and Sneath defined "natural" taxa as the particular groups of organisms that are studying. In other words, natural taxa are based on the concept of "affinity," which is measured by taking all characters into consideration and the taxa are separated from each other by means of correlated features (Sokal and Sneath 1963, p. 16). Thus, naturalness is a matter of degree rather than an absolute criterion of a classification (Sokal 1986). Depending on the number of shared character states, different classifications will consequently be more or less natural.

7.2.2 Sample Size Calculation: How Many Characters?

First, we need to point out that numerical taxonomy uses the notations of sample size and variable in a different way from many multivariate statistics. In multivariate statistics, n is used for sample size, and p for variable number. However, in numerical taxonomy, sample size represents the number of OTUs (denoted as t); character represents the variable (the variable number is denoted as n). In microbiome study, we can consider n as sample size and p as variable number in an OTU table that is originated in numerical taxonomy. In later numerical ecology, similar usages of variable and object have been adapted. In numerical ecology, the terms descriptor and variable are used interchangeably to refer to the attributes, characters, or features for describing or comparing the objects of the study, while the term objects is used to refer to the sites, quadrats, observations, sampling units, individual organisms, or subjects (Legendre and Legendre 2012, p. 33). Legendre and Legendre (2012) have point out that ecologists also called observation units as "samples"; and they used the term sample in its statistical sense, which refers to a set of sampling observations. They called individuals or OTUs (operational taxonomic units) used in numerical taxonomy as objects (p. 33). This usage is consistent with that of numerical taxonomy. However, in microbiome study, the terms sample and variable are used in the different way from numerical taxonomy. In microbiome study, both terms of sample and variable are used in their statistical senses, in which samples are used to refer to observation units, whereas OTUs are used as variables. Thus, n is used for sample size, and p or m for variable number or how many OTUs or taxa. Therefore, when we talk about the statistical issues such as large p and small n problems, we have used them in this way (i.e., in most multivariate statistics) [See (Xia et al. 2018; Xia 2020; Xia and Sun 2022) and also Chap. 18 in this book]. Another different use in numerical taxonomy is R-and Q-Techniques (see Sect. 7.4.2.2 for details).

Power and sample size calculations are challenging in experimental studies and clinical trials, especially for microbiome related projects. It is very challenging to estimate the power and sample size in numerical taxonomy and microbiome studies. At the beginning when numerical taxonomy was established, Sokal and Sneath recognized that it was very complicated to obtain the required number of characters to estimate the similarity with given confidence interval and significance of probability level (Sokal and Sneath 1963; Sneath and Sokal 1973).

Generally, the number of characters chosen to estimate of an overall similarity is based on the matches asymptote hypothesis and an empirical observation: a given estimate of phenetic similarity is favored with much different phenetic information provided by large number of characters. Overall, numerical taxonomy is a large sample method; as a general rule, it emphasizes to use more rather than fewer characters. To achieve overall predictivity classification and overall phenetic similarity, taxa should be based on many characters.

In their 1963 and 1973 books, Sokal and Sneath separately considered at least four methods to estimate sample size in numerical taxonomy.

(1) The reliability of the estimates of 95% confidence limits of the correlation coefficients (Sokal and Sneath 1963, pp. 115–116). The algorithm is that as sample size (number of characters) increases, the confidence limits of the similarity coefficients will be narrowed. Thus, if the confidence limits will be getting sufficiently narrow, then sampling error is not likely to be mistaken for real differences in relationship. Based on this criterion, Michener and Sokal (1957) suggested at least sixty characters are needed to give sufficiently narrow confidence limits. This criterion is obviously arbitrary.

(2) A sampling strategy of phenotypic characters based on molecular considerations. This strategy considers each character as a sample of the genome and determines when a required percentage of the genome has been sampled at a

required confidence level. However, since any reasonable number of characters represents a relatively small percentage of the genome, this approach is premature and the links between the molecular basis of the genome and the phenetic characters are limited (Sneath 1995). Thus, this approach cannot be used to calculate sample size (Sokal 1963, p. 117).

(3) An empirical evidence-based numerical classification. Sokal and Rohlf (1970) found how many independent individuals coding can describe the same variation patterns and additional independently varying characters would not substantially change the classification or the underlying character-correlational structure. Later, Sneath and Sokal (1973, p. 108) suggested performing a factor analysis to check how many reasonable number of characters is employed to stabilize common factors (taxonomic structure).

(4) A simulation experiment was suggested to test the different correlational structure to see the effect of the addition of new characters, that is, to check how many reasonable numbers of characters is employed to stabilize the matrix correlation of successive similarity matrices. Two distinct models or distribution of characters (J-shaped curve similar to a Poisson distribution with a low mean and uniform distribution) were also used in this verification of the asymptotic approach to taxonomic stability (Sneath and Sokal 1973, p. 108).

This large sampling theory and experience-based sample size calculation and power analysis are continued to be used in microbiome study. However, these approaches do not often lead to a satisfactory result. Sokal (1985) acknowledged that the genetically based estimate of an overall similarity based on the matches asymptote hypothesis of Sokal and Sneath (1963) to choose the characters is not useful to solve the increasing complex relations underlying molecular genetics. Sample size calculation and power analysis are especially challenging in microbiome research, because in a microbiome study, we not only need to determine how many samples are required to obtain sufficient power in classification of OTUs, but also need to determine how many samples in statistical analysis could lead to a robust result.

The sample size estimation and power analysis for microbiome study have double challenges:

On one hand we really do not know or rarely considered how many samples that are needed to process sequencing to obtain OTUs or ASVs; on the other hand, we also do not know for sure how many samples are required for downstream statistical analysis to obtain sufficient power to detect the hypothesis testing of outcome. Currently, the most proposed methods in microbiome study are based on simulation and use the detected effect sizes in a given database or study as the input sample size.

7.2.3   Equal Weighting Characters

For Sokal and Sneath, "weighting" (Sneath and Sokal 1973, p. 110) or a priori weighting (Sneath and Sokal 1973, p. 109) is for general taxonomic analyses or general nature of natural classifications and is commenced before a classification. They advocated to give each character (feature) equal weight in the computation of a resemblance coefficient or in creating taxonomic groups (natural taxa). Equal weighting is based on seven reasons (Sneath and Sokal 1973, pp. 112–113): (1) we cannot decide how to weight the features, (2) we cannot prejudge the importance of characters, (3) the concept of taxonomic importance has no exact meaning, (4) we cannot give exact rules for differential weighting, (5) the nature of a taxonomy is for general purposes, and it is inappropriate to give greater weight to features for general use, (6) the property of "naturalness" lies on its high implied information by a natural group, which is irrelevant to its importance, and (7) the use of many characters makes the effective weight to be clear that each character contributes to the similarity coefficients.

We should point out, at the beginning, Sokal and Sneath emphasized the importance of equal weighting of characters and the impossibility of unequal weighting. However, to respond to the weighting approaches proposed in numerical taxonomy literature, Sokal (1986) recognized, in practice, strictly equal weighting of characters is impossible because the omitted characters have zero weight, and the similarity is mainly determined by the characters with many states than those with few states. Particularly, in their 1973 book, they stated that unequal weighting in phenetic taxonomy can be carried out by an explicit algorithm, such as the rarity of features or their commonness in the whole set of OTUs can be used to weight the characters (Sneath and Sokal 1973, p. 111). However, they thought that the results using unequal weighting likely will be little different from those using equal weighting and most work in phenetic numerical taxonomy has avoided explicit weighting because no algorithm is convincing for doing so.

For the detail discussions on weighting, the interested reader is referred to (Sokal 1963, pp. 50–51; pp. 118–120; Sneath 1957b; Michener and Sokal 1957; Sneath and Sokal 1973, pp. 109–113).

7.2.4   Taxonomic Rank

Generally three criteria for taxonomic rank in numerical taxonomy have been applied (Sneath and Sokal 1973, p. 60): (1) Both phenetic and genetic criteria are commonly used although for biological species and below level, these criteria may be in conflict. (2) In the absence of data on breeding and in all apomictic groups (which is the most major cases in systematics), the phenetic similarity between the individuals and on phenotypic gaps are used to define species. (3) The higher rank of categories is per forcedly defined using the phenetic criteria.

In numerical taxonomy, the species are based on the phenetic similarity between the individuals and on phenetic gaps. The taxonomic rank is in the sense of phenetic rank, which is based on affinity alone (Sokal and Sneath 1963, pp. 30–31). However, since the end of 1960s, numerical taxonomists has recognized that there are no criteria for any absolute measure of taxonomic rank, and criteria of phenetic rank is difficult to apply to all fields (Sneath and Sokal 1973, p. 61). For example, in protein and DNA, it was shown that widely different phenetic differences exist at the same taxonomic ranks. It was difficult to use the age of clades to measure the taxonomic rank. Overall, Sneath and Sokal noticed that it was difficult to define the taxa of low rank and to arrange them hierarchically (Sneath and Sokal 1973, p. 62).

In summary, Sneath and Sokal (1973, p. 290) believed that in practice taxonomists should usually use phenetic criteria to measure rank and that the most satisfactory criterion of a rank of a single taxon is some measure of its diversity. Of course, the criteria of taxonomic rank are challenging and debatable not only in numerical taxonomy but also in microbiome study.

7.3   Phenetic Taxonomy: Philosophy of Numerical Taxonomy

Here, to better understand phenetic taxonomy, and related methods that have been adopted into microbiome study, we describe the standing points of philosophy of numerical taxonomy. We should distinguish phenetic taxonomy (or phenetics) from phenetic techniques in numerical phenetics. The former is a theory or philosophy or of classification, and the latter are the methods developed to apply the theory.

Three common taxonomic philosophies exist in literature: phenetic philosophy, phylogenetic (cladistic) philosophy, and evolutionary philosophy. Overall, the philosophy of numerical taxonomy is the movement from the metaphysics of essentialism to the science of experience, from deductive methods to inclusive methods, from evolutional phylogenetics to numerical phenetics based on similarity.

7.3.1   Phenetics: Numerical-Based Empiricism Versus Aristotle's Essentialism

Some descriptions of Aristotle's metaphysics of essentialism (metaphysical essentialism) and Darwin' theory of evolution will provide a basic background to understand numerical phenetics.

7.3.1.1 Aristotle's Metaphysics of Essentialism

Essentialists believe true essences exist. Aristotle defined essence as "the what it was to be" for a thing (Aristotle 2016) or sometimes used the shorter term "the what it is" for approximating the same idea. Aristotle defined essence as analyzed entities. For example, in his logical works (Aristotle 2012), Aristotle links essence to definition—"a definition is an account (*logos*) that signifies an essence" or "there will be an essence only of those things whose *logos* is a definition," "the essence of each thing is what it is said to be intrinsically." Aristotle's essence is related to a thing. In his classic definition of essence, a thing's essence is that which it is said to be per se, or is the property without which the thing would not exist as itself. Therefore, essence is constitutive of a thing, which is eternal, universal, most irreducible, unchanging. The classical example of essence definition is that Aristotle defined the human essence: to be human is to be a "rational animal." The "rational" is the species or specific difference, which separates humans from everything else in the class of animals.

Another important essentialist concept is forms. For Plato, forms are abstract objects, existing completely outside space and time. Aristotle criticized and rejected Plato's theory of forms but not the notion of form itself. For Aristotle, forms inform the thing's particular matter. Thus, forms do not exist independently of things—every form is the form of some things. He called "the form" of a particular thing (Albritton 1957).

The naturalists in seventeenth and eighteenth centuries who attempted to classify living organisms were actually inspired by the terms and definitions from the Aristotelian system of logic. Darwin traced evolutionary ideas (the principle of natural selection) back to Aristotle in his later editions of the book of *The Origin of Species* (Darwin 1872), and commented on Aristotle on forms.

Aristotelian system attempted to discover and define the *essence* of a taxonomic group ("real nature" or "what makes the thing what it is"). Aristotelian logic is a system of analyzed entities rather than a system of unanalyzed entities.

7.3.1.2 Natural Classification: Numerical-Based Empiricism

Numerical taxonomy is a largely empirical science (Sokal 1985; Sneath 1995).

Natural classification was developed based on the philosophy of empiricism and was affected by the philosophy of the inductive sciences. Some key concepts of information content and predictivity provided by John Gilmour (Sneath 1995) who was one of the early contributors to the founding of numerical taxonomy were adapted from the Victorian philosophers of science John Stuart Mill (1886) and William Whewell (1840).

Sokal and Sneath criticized that Aristotelian logic cannot lead to biological taxonomy because the properties of taxonomy cannot be inferred from the definitions [see (Sokal and Sneath 1963, p. 12; Sokal 1962; Sneath and Sokal 1973,

p. 20)]. Thus, they emphasized that natural taxa do not necessarily possess any single specified feature, which is the Aristotelian concept of an *essence* of a taxon, because natural or "general" classification or natural groups are in logic *unanalyzed entities* (Sokal and Sneath 1963, pp. 19–20).

Sneath and Sneath (1962) differentiated the terms "polythetic" and "monothetic" in taxonomic groups. The monothetic groups define a unique set of features that is both sufficient and necessary for membership in the group thus defined. The monothetic groups have the risk to misclassify natural phenetic groups and hence do not yield "natural" taxa. In contrast, in a polythetic group, organisms that have the greatest number of shared character states are placed together, and hence no single state is either essential to group membership or is sufficient to make an organism a member of the group.

For Sneath and Sokal, natural taxa are usually not fully polythetic, and it is possible that they are never fully polythetic; however, in practice, they suggest considering the possibility of a taxon being fully polythetic due to limited knowledge of whether any characters are common to all members (Sneath and Sokal 1973, pp. 20–21).

Numerical taxonomists emphasize that biological taxonomy is a system of unanalyzed entities. Their properties cannot be inferred from the definitions, instead through classification. In his "Taxonomy and Philosophy" (Gilmour 1940), Gilmour stated that classification enables us to make generalizations and predictions. The more information obtained by classification enables us to make more predictions. The more predictions the classification is more "natural."

For Gilmour, the most "natural" system/classification contains the most information, which is highly predictive and serves the most general purpose (Vernon 1988), while less "natural" classification is very useful for specific purposes. Generally Sokal and Sneath held with Gilmour's review of "natural" classification. However, classification serves as special purpose. Such classification is a special classification and is often "arbitrary." Thus Gilmour's classification conveys less information than a general or "natural" one. For Sokal and Sneath, a natural taxonomic group or natural taxa of any rank can be classified. For example, the genetic natural classification can be supported (Vernon 1988). They defined a natural classification in phenetics as one in which the members of each taxon at each level are on the average more similar to each other than they are to members of other taxa at corresponding levels (Sokal 1986).

The philosophical attitude of numerical taxonomy in systematics is empiricism and consequently is not committed to the existence of biological species (Sokal and Crovello 1970). Put together, numerical taxonomy is both operational and empirical. Numerical-based natural classification is an empiricism. Empiricism in taxonomy is the only reasonable approach for arranging organized nature (Sokal 1963, p. 55). Sneath and Sokal (1973, p. 362) emphasized that numerical taxonomy is an empiricism and operationism and thought that the biological species definition was difficult to apply and in practice was nonoperational.

7.3.2 Classification: Inductive Theory Versus Darwin' Theory of Evolution

Sneath (1995) considered Sokal's and his numerical taxonomy as the greatest advance in systematics since Darwin or even since Linnaeus because Darwin had relatively little effect on taxonomic practice (Hull 1988). This statement emphasizes that the numerical taxonomy has the different taxonomic methodology from Darwin's taxonomic method.

Darwin in 1859 published his *On the Origin of Species* (or, more completely, *On the Origin of Species by Means of Natural Selection, or the Preservation of Favoured Races in the Struggle for Life*) (Charles Darwin 1859), which is considered to be the foundation of evolutionary biology. Darwin's scientific theory introduced in this book stated that populations evolve over the course of generations through a process of natural selection. In the 1930s and 1940s, the modern evolutionary synthesis had integrated Darwin's concept of evolutionary adaptation through natural selection with Gregor Mendel's theories of genetic inheritance (McBride et al. 2009) to become a central concept of modern evolutionary theory and even unifying concept of the life sciences.

The arguments of Darwin's theory of evolution are combined with facts/observations and the inferences/speculations drawn from them (Mayr 1982). His theory of classification stated in Chapter XIII of his book on the origin of species (Darwin 1859) starts by observing that classification is more than mere resemblance, and it includes propinquity of descent (the only known cause of the similarity of organic beings) which is bonded and hidden by various degrees of modification (pp. 413–414). For Darwin, the natural system is founded on descent with modification and the characters that have been shown having true affinity between any two or more species, are those which have been inherited from a common parent, and, in so far, "all true classification is genealogical" (Darwin 1859, p. 420).

Obviously above Darwin's theory of classification is deductive although he said later that organic beings can be classed in many ways, either artificially by single characters or more naturally by a number of characters. Thus classification is not related to genealogical succession (Darwin 1886, p. 364). Darwin's work contributed to the recognition of species as real entities (Sokal and Crovello 1970).

The influence of Darwin's theory of evolution lies on the fact that systematics was considered as phylogenetically, in which a taxon was interpreted as a monophyletic array of related forms. However, the change of philosophy was not accompanied with a change in method (Sokal 1963, p. 20). Thus, during the early 1950s, some taxonomists began to suspect the appropriateness of the evolutionary approach to classification in phylogenetic taxonomy and reexamined the process of classification. For example, Mayr and Simpson recommended highly weighting those characters which indicated evolutionary relationships. However, for those characters with the absence of fossils and knowledge of breeding patterns, this was very challenging and thus the phylogenetic importance of characters could be given due to much speculation.

The trend to reject phylogenetic speculation in classification has been reviewed as background that Sokal and Michener, Cain and Harrison, and Sneath developed their ideas of the numerical taxonomy (Vernon 1988). Under the influence of the Darwinian theory of evolution, the natural classifications of earlier taxonomists can be explained and understood in terms of descent from a common ancestor. That is, they now interpreted a taxon as a monophyletic array of related forms (Sneath and Sokal 1973, p. 40). However, the change of the Darwinian philosophy did not bring a methodology change in taxonomy. It only changed its terminology.

Essentially the methodology of the theory of classification in numerical taxonomy is inductive, and hence it was arising against Darwinism's deductive theory of classification. Sokal and Sneath (1963, p. 21) criticized that the phylogenetic approach cannot be used to establish natural taxa because it lacks the fossil record. They emphasized that classification cannot describe both affinity and descent in a single scheme [see (Michener and Sokal 1957; Sokal 1963, p. 27)]. For them, deductive reasoning in tracing phylogenies is dangerous (Sokal 1963, p. 29).

7.3.3 Biological Classifications: Phenetic Approach Versus Cladistic Approach

As a field, numerical taxonomy provides explicit principles of phenetics through numerical approaches to phylogeny and emphasizes the distinction between phenetic and cladistic relationships (Sneath 1995), which was first clearly stated by Cain and Harrison (1960). Numerical taxonomy has a broader scope than phenetics. It includes both phenetic and phylogenetic approaches. It includes the drawing of phylogenetic inferences from the data by statistical or other mathematic methods to the extent that this was possible [see (Sokal 1963, p. 48; Sneath 1995)].

A phenetic classification attempts to establish taxa either based on maximum similarity among the OTUs or on maximum predictive value (homogeneity of character states) (Sokal 1986). This approach assumes but never rigorously demonstrated (Farris 1979) that high similarity among individuals in a taxon would result in high predictive values for many of the characters.

For an approach of phenetic phylogenetic classifications, most phenetic methods do not attempt to maximize similarities in taxa but settle for proven suboptimal solutions such as average-linkage clustering (Sokal 1986). However, once a phenetic classification has been established, it is possible to attempt to arrive the phylogenetic evidence and phylogenetic deductions.

Sneath and Sokal (1973) divided and refined phylogenetic relationships into phenetic and cladistic relationships (p. 41). Sokal and Sneath clearly made distinction between phenetic and cladistic classifications (Sokal 1985) and considered it as an important advance in taxonomic thinking for separation of overall similarity (phenetics) from evolutionary branching sequences (cladistics) (Sneath and Sokal 1973, p. 10). Thus, the phylogeny that the numerical taxonomy aimed to establish is first to make the distinction between phenetic and cladistic relationships (Sneath 1995).

The theory of numerical classification provided a theoretical background of annotation of phylogenetics in 16S RNA sequencing analysis. Thus, we summarize several main arguments of the theory of numerical classification below. The interested reader is referred to the books by Sokal and Sneath (1963, p. 56, pp. 101–102, pp. 216–230, pp. 227–230) and Sneath and Sokal (1973, pp. 42–60, pp. 216–220):

- Natural taxonomic groups have the evolutionary basis, and the observed phenetic diversity of living creatures is a reflection of evolution (Sokal 1963). In other words, natural taxonomic groups just reflect the restriction of evolution and the accumulation with the time genetic difference.
- Phylogeny or cladistic relationships cannot be used to construct phyletic classification. Numerical classification principally is a natural classification, which is to be natural rather than artificial. It truly reflects the arrangement of the organisms in nature. Generally natural taxa are considered to be those monophyletic taxa that reflected phylogenetic history.
- Even if phylogeny could be used to create a valid phylogenetic classification, it is not necessarily desirable. Phyletic classifications could be chaotic in classification of viruses and especially in bacteriophages, and phenetic classification is the only way to classify them. And the validity of a phylogenetic classification may be in a serious doubt.
- Overall phenetic similarities between organisms can be used to make cladistic deductions. Phenetic clusters based on living organisms are more *likely* to be monophyletic; in the absence of direct evidence, phenetic resemblance is the best indication of cladistic relationships.
- A phenetic approach should be chosen to classify natural taxa. Numerical taxonomy uses phenetic evidence to establish a classification. Although both phenetic and cladistic approaches have problems to the study of taxonomic relationships, the phenetic approach is the better choice to be used in classifying nature. First, the problems of estimating phenetic relationships are largely technical challenges and can be overcome, whereas the problems of estimating cladistic sequences are difficult to be solved. Especially numerical cladistics cannot provide necessarily cladistic sequences. Second, since in the fossil record only very small portion of phylogenetic history is known, in the vast majority of cases phylogenies are unknown and possibly unknowable; it is difficult to use the phylogenetic approach in systematics/taxonomy to make use of phylogeny for classification. Thus, it is almost impossible to establish natural taxa based on phylogenies. In contrast, a phenetic classification is at least a self-sufficient, factual procedure and hence can provide the best classification in most cases although it cannot explain all biological phenomena.
- In general, a phenetic classification is superior to a cladistic classification from two viewpoints: domain and function.
- It is difficult to define or obtain clear decisions on monophyly. Taxa in orthdox taxonomy should be in phenetic groups rather than in general monophyletic

groups (clades). There are three reasons: (1) Phenetic and phyletic relations are two formally independent taxonomic dimensions. (2) It is expected that the phenetic groups in great majority of examples will indeed be clades based on the assumption that close phenetic similiarity is usually due to close relationship by ancestry. (3) In general classfication, phenetic taxa is perfered to cladal taxa, while employing clades has the advantages only in the study of evolution.

- Phenetics aims to give information-rich groups, while the goal of phylogenetics is to reconstruct evolutionary history. Phenetic groupings can be verified by phenetic criteria but cannot be proven to correspond to reality, whereas phylogenetic groupings must correspond to reality, but cannot be verified (Sneath 1995). The reason is that information-rich groups do not necessarily have a historical basis. Actually, the theories of phenetics involve three types of theory (Sneath 1995): (1) theory of homology (i.e., what should be compared with what), (2) theory of information-rich groupings (which implies high predictivity) (Mill 1886; Gilmour 1940), and (3) Mill's (1886) theory of general causes (i.e., phenetic groups are due to as yet undiscovered causes, which linked numerical taxonomy more closely to Darwin's the origin of species) (Sneath 1995).

We should point out, at the beginning, numerical taxonomy still attempted to establish phylogenetic classifications (Michener and Sokal 1957). However, subsequently, the numerical taxonomists realized that because different lineages may evolve at different rates, it is impossible to draw a phyletic tree in which all the lineages evolve at the same rate and or to derive evolutionary rates from similarity coefficients among recent forms. Thus, they claimed that phenetic classification is the only consistent and objective classification (Sokal and Sneath 1963, pp. 229–230).

In summary, Sokal and Sneath agreed with Bigelow (1961) that a horizontal (phenetic) classification is more satisfactory than a vertical (phyletic) classification (Sokal and Sneath 1963, p. 104). In terms of vertical classification and horizontal classification (Simpson 1945), numerical taxonomy groups taxa based on horizontal classification. However, in numerical taxonomy, a phenetic taxonomic group does not need to be necessarily identical with a phyletic group (Sokal and Sneath 1963, p. 32).

7.4   Construction of Taxonomic Structure

Phenetic taxonomy (classification) aims to arrange operational taxonomic units (OTUs) into stable and convenient classification (Sokal 1986), which is to group the most similar OTUs. Thus, various statistical methods that serve the goals of grouping OTUs used in numerical taxonomy can be called OTU-based methods. Among these methods, the concept of OTUs, clustering analysis methods, the coefficients of similarity, distance, association, correlation, and ordination techniques have the most impacts on microbiome study. Actually, ordination techniques have been most fully developed in ecology. The ordination methods used in later numerical taxonomy were adopted or influenced by ecology. We comprehensively investigate ordination methods in Chap. 10. In this section, we focus on the review of the concept of OTUs, clustering analysis methods, the coefficients of similarity, distance, association, and correlation.

The empirical inclusive methodology of numerical taxonomy has impacted many other research fields. Numerical taxonomy has expanded into ecology and psycho-metrics at the later 1970s and early 1980s. The numerous applications in ecology were summarized by Legendre and Legendre (1983; Sneath 1995). The methodol-ogy of numerical taxonomy and especially the clustering methods have been adopted in numerical ecology and later microbiome fields. For example, single-linkage, complete-linkage, and average-linkage clustering methods were adopted and espe-cially the average-linkage clustering method was advocated by the microbiome software **mothur.**

Various clustering analysis methods were developed and employed to measure homogeneity of OTUs regarding character states to maximize homogeneity (Sneath and Sokal 1973; Sokal 1985). We term these methods that were adopted from numerical taxonomy as the commonly clustering-based OTU methods. Here we characterize them as below.

7.4.1 Defining the Operational Taxonomic Units

The taxonomic units that numerical taxonomy classifies are not the fundamental taxonomic units, but are the hierarchical level of taxonomic unit. They are termed as *"operational taxonomic units (OTU's)"* [see (Sokal and Sneath 1963, p. 121) and (Sneath and Sokal 1973, p. 69)]. Here, OTUs or "the organizational level of a unit character" was used to define the primary taxonomic entities (taxa/species/individ-uals), which may be used as exemplars of taxonomic groups (Sneath and Sokal 1962). In the beginning, Sokal and Sneath used the term "operational taxonomic unit (OTU)" for the unit that is to be classified (Sokal 1963, p. 111), but deliberately used the term OTU's instead of OTUs for representing the "operational taxonomic units" (Sokal 1963; Sneath and Sokal 1973). The term OTUs was used for representing "operational taxonomic units (OTUs)" in later time." Such as the term "operational taxonomic units (OTUs)" was used in Sokal 1986's review paper (Sokal 1986).

By defining taxonomic unit as operational:

First it makes numerical taxonomy to get rid of relying on the validity of species defined by the conventional methods to define taxa. Thus, numerical taxonomy does not care how many species concepts and species definitions already have so far (Sokal and Sneath 1963, p. 121).

Second, it emphasizes that most taxa are phenetic, rather than to satisfy the genetic definition of species status (Sokal and Sneath 1963, p. 121). Thus, the concept of

OTU is different from Mayr's genetic definition of species (Mayr 1999) (first published in 1942).

Third, the term "taxon" (plural taxa) is retained for its proper function; that is, we can use the term taxon to indicate any sort of taxonomic group (Sneath and Sokal 1962), i.e., taxa are groups of OTUs or an abbreviation for taxonomic group of any nature or rank (Sneath and Sokal 1973). Thus, a taxon can be used to denote a variable for one or more characters, i.e., used as OTU's higher taxa (e.g., genera, families, and orders) (Sokal and Sneath 1963, pp. 121–122).

7.4.2   Estimation of Taxonomic Resemblance

In Sokal and Sneath (1963), "resemblance," "similarity," and "affinity" are used interchangeably to imply a solely phenetic relationship. However, since affinity means kinship, hence by descent. Thus in Sneath and Sokal (1973), to avoid confusion the term "affinity" was not used; instead the terms "resemblance" and "similarity" are used interchangeably. Similarly the term relationship was used (see Sect. 7.5.1).

7.4.2.1   Data Matrix/OTU Table

Since the classification must be based on all available information, similarity must be computed from data matrixes describing the characters of the OTUs being classified. A data matrix or an OTU table is formed with columns representing taxa (or OTUs) and rows representing characters (Sneath and Sokal 1962; Sokal 1963; Sokal and Sneath 1963). The columns of this matrix represent OTUs, which can stand for any taxonomic unit from individual through species up to higher categories (Camin and Sokal 1965). The rows of this matrix represent characters, which can be the length of an organ or the form of margin of a leaf or the presence or absence of an attribute (such as ability to fly) (Sneath and Sokal 1962). The microbiome data matrix structure mainly combines or adopts the data matrix structures from numerical taxonomy and numerical ecology. The term of OTU table in microbiome study was adopted from numerical data matrix (see Table 7.1). The main data matrix

Table 7.1 Numerical taxonomy data matrix (or OTU table)

Characters	OTUs	
(Attributes/descriptors/variables)	(Taxa/species/individuals/specimens/organisms/stands/analysis units/observations)	
1	A B C D E (Group A)	F G H I J (Group B)
2		
3		

Table 7.2 Ecological data matrix

Objects	Descriptors		
(Sampling sites/sampling units/observations/locations/subjects/ individuals/stands/quadrats/"taxa" such as woodland, prairie, or moorland in plant ecology)	(Species/attributes/variables)		
1 2 3	A B C D E (Group A)		F G H I J (Group B)

Table 7.3 Microbiome abundance data matrix

OTUs/ASVs	Samples		
(Taxa/species/variables)	(Subjects)		
1 2 3	A B C D E (Group A)		F G H I J (Group B)
Samples	OTUs/ASVs		
(Subjects)	(Taxa/species/variables)		
1 2 3	A B C D E (Group A)		F G H I J (Group B)

structures in numerical taxonomy, ecology, and microbiome are summarized in Tables 7.1, 7.2, and 7.3, respectively.

In numerical taxonomy data matrix (or OTU table) (Table 7.1), rows represent characters (attributes/descriptors/variables), while **columns** represent OTUs (taxa/analysis units/observations).

The setting of ecological data matrix is different from that in a numerical taxonomic matrix, in which **rows** represent *objects* (subjects/stands/quadrats), while **columns** represent *descriptors* (species). Although data matrices in ecology and numerical taxonomy have different structures (Legendre and Legendre 2012, p. 60), however, knowing the distinctions between attributes and individuals is very important because it is often confused between attributes and individuals in ecology. In ecology, attributes are species and individuals are objects (Table 7.2).

The structures of microbiome abundance data matrix (or OUT/ASV table) are relatively simple. It is a hybrid of numerical data matrix and numerical taxonomic matrix, in which **rows** represent OTUs (ASVs/sub-OTUs/variables), while **columns** represent *samples* (subjects) or reverse (Table 7.3). Here samples are analysis units; OTUs or species are variables. We do not need to distinct attributes and individuals. However, when we use microbiome software, we need to specify which data matrix structure is used. For example, the **phyloseq** package has been developed based on ecological software vegan package. When creating data object, it asks the user to specify whether taxa are_rows = TRUE or FALSE, which is confirming the structure of data matrix: whether rows represent taxa/OTUs/ASVs. In its initial version, the **phyloseq** package even used species to stand for taxa.

7.4.2.2   R-Technique and Q-Technique

Data matrices in Tables 7.1, 7.2 and 7.3 can be examined using two kinds of techniques (Cattell 1952): the R-technique and the Q-technique (originated by Stephenson under the name of inverted factor technique (Stephenson 1936, 1952, 1953)). In general, in an association matrix, a so-called Q matrix is formed to compare all pairs of individuals, while a so-called R matrix (dispersion or correlation matrix) is formed for comparisons between the variates. Numerical taxonomy starts logically with the choice of the t entities to be classified (OTUs) and n attributes or variables (generally characters in terms of biological work), which gives the n x t matrix. The R-technique examines the association of pairs of characters (rows) over all OTUs (columns), while the Q-technique, in contrast, examines the association of pairs of OTUs (columns) over all characters (rows). In numerical taxonomy, the R-technique refers to correlations among characters based on OTUs, considering the correlation of pairs of characters present in a population, while the Q-technique refers to the qualifications of relations between pairs of taxa, frequently species, based on a preferably large number of characters (Sokal 1963, pp. 125–126), considering the correlations of pairs of individuals present in a population in terms of their characters. In other words, Q-technique refers to the study of similarity between pairs of OTUs, and R-technique refers to the study of similarity between pairs of characters (Sneath and Sokal 1973, p. 256).

The adoption of using the R-technique and the Q-technique in numerical taxonomy has been criticized at the beginning after the publication of Sokal and Sneath's 1963 book. The duality of the R or Q problem is one of some fundamental (Williams and Dale 1966) and statistical (Sneath 1967) problems in numerical taxonomy. For example, factor analysis compares characters but finally results in a grouping of OTUs. To overcome the confusion over the use of Q and R technique, Williams and Dale (1966) suggested that (1) Q and R should refer to the purpose of the analysis, i.e., refer to the matrix, and (2) A-space and I-space rather than R-space or Q-space should be used that is operated upon when the relationships are represented in a hyperspace.

The organisms can be represented as points in a multivariate character space (A-space, or attribute space), and conversely the characters can be represented in an I-space (individual space) (Sneath 1967). A-space and I-space refer to the model: the former is used for a model in attribute-space and the latter is used for a model in the individual-space. A matrix of inter-individual distances implies relationships between points in an A-space, while an inter-individual correlation matrix implies angles in an I-space. They both are Q-technique. An attribute-correlation matrix is R-technique which concerns with angles, while an individual-distance matrix is Q-technique which concerns with point.

Sneath (1967; Sneath and Sokal 1973) accepted the two suggestions of Williams and Dale (1966). In numerical taxonomy, R-technique leads to a classification of characters, Q-technique to one of OTUs. But they emphasized formally both R-technique and Q-technique have the same main mathematical steps because by transposing the data matrix in an R study, the characters (rows) become the individuals comparable to the former OTUs, and the actual OTUs or taxa (columns) become the characteristics (attributes) over which association is computed (Sneath and Sokal 1973, p. 116). In either space the character values are the coordinates of the points in the hyperspace. Based on the distinction between A-space and I-space in Williams and Dale (1966), in numerical taxonomy, A-space (attribute space) has formally n dimensions (one for each attribute or character) with t points that represent the OTUs. I-space (individual space) has formally t dimensions (one for each OTU) with n points representing the attributes or characters (Sneath and Sokal 1973) although less than n (in A-space) or t (in I-space) dimensions respectively could be after attempting the dimensionality reduction (p. 116).

Cattell (1966) restricted Q- and R-techniques to factor analysis, and applied Q-technique as a customary practice to numerical taxonomy in terms of cluster analysis of OTUs. Both Q- and R-techniques are employed in cluster as well as factor analysis although numerical taxonomy mainly emphasized the Q-technique. By using the notations of A-space and I-space, most numerical taxonomic methods are operated by Q-technique on an A-space, but some cluster and factor analyses are Q-techniques operating on an I-space.

Overall, Sneath and Sokal emphasized that Q- and R-techniques have been adapted in the general sense in numerical taxonomy. Conveniently a Q-study is used for the quantification of relations between organisms according to their characters aiming to produce classifications of organisms; while an R-study refers to classifying characters into groups according to the organisms that possess them, in which the characters become in effect the OTUs, and the procedures are similar to Q-study, but operate on the transpose of the n x t matrix [See (Sneath 1967; Sneath and Sokal 1973, pp. 115–116)].

Basically, the uses of Q- and R-techniques (or analyses) in numerical taxonomy have been adopted in numerical ecology (Legendre and Legendre 2012, pp. 266–267), whereas Q measures (Q mode study) are used for the association coefficients (the relationships) among objects given the set of descriptors and R measures (R mode analysis) are used for the association coefficients (the relationships) among descriptors (variables) for the set of objects in the study.

In summary, Q analyses refer to the studies that start with computing an association matrix among objects, whereas R analyses refer to the studies that start with computing an association matrix among descriptors (variables). Numerical taxonomy defines a character as a variable (feature) and OTU as an analysis unit, which leads to the different understanding of R-technique and Q-technique in some other taxonomists and ecology. Later microbiome study adopted the methods of numerical taxonomy, but not the definition of character, where samples are observations and OTUs are variables (features/characteristics). The way data matrix is used in microbiome study is consistent with ecology. In microbiome study, Q- and R-techniques (or analyses) are not explicitly differentiated but in some software that are adopted from ecology the terms and according data matrices are often used. In terms of R-technique for relationship among variables and Q-technique for relationship among individuals, the microbiome study mainly uses R-technique to find the relationship among variables (OTUs/ASVs/taxa) over the samples.

7.4.2.3   Similarity Coefficients

In numerical taxonomy, the term "similarity coefficient" is most commonly used, although in the strict sense the term "coefficient of resemblance" would be more appropriate. The term "coefficients of similarity" or "coefficients of dissimilarity" is also used. Numerical taxonomy roughly groups four types of coefficients—distance coefficients, association coefficients, correlation coefficients, and probabilistic similarity coefficients. They collectively refer to as coefficients of resemblance or similarity and are used to compute resemblances between taxa.

The space that measures similarity does not need to be Euclidean in the strict sense, but it should be metric, in other words, determined by a metric function (Williams and Dale 1966; Johnson 1970).

Based on the requirements for similarity coefficients, Sneath and Sokal (1973) described the four axioms over the entire set of OTUs that the measures of dissimilarity or functions (converting similarity coefficients into measures of dissimilarity) should be satisfied (p. 120). This is very important when we apply these similarity coefficients or develop the new measures of similarity. However, in microbiome literature, this topic has been rarely mentioned. Thus, here we briefly describe these four axioms before we formally introduce similarity coefficients.

Let φ be a real nonnegative number and define the function for pairs of OTUs a, b, and c, and then, the measures of dissimilarity have the property of a metric (Williams and Dale 1966; Johnson 1970, pp. 120–121). These four axioms can be described below.

Axiom 1: $\varphi(a, b) \geq 0$, and $\varphi(a, a) = \varphi(b, b) = 0$. It states that identical OTUs are indistinguishable while nonidentical ones may or may not be distinguishable by the dissimilarity function.

Axiom 2: $\varphi(a, b) = \varphi(b, a)$. This is the symmetry axiom. It states that the value of function φ from a to b is the same as from b to a.

Axiom 3: $\varphi(a, c) \leq \varphi(a, b) + \varphi(b, c)$. This is the triangle inequality axiom. It states that the function between a and c cannot be greater than the sum of the functions between a and b and b and c. In other words, the sum of the lengths of two lines is greater than the length of third line.

Axiom 4: If $a \neq b$, then $\varphi(a, b) > 0$. This axiom states that if a and b differ in terms of their character states, then the function between them must be greater than zero.

All these four axioms are true for the Euclidean distances (which additionally obeys Pythagoras' theorem: $a^2 + b^2 = c^2$, i.e., the lengths of the square whose side is the hypotenuse (the side opposite the right angle) is equal to the sum of the lengths of the squares on the triangle's other two sides), and thus they are metrics. However, not all measures of dissimilarity are metrics in numerical taxonomy. For example, there exist the cases that the difference in character states between two OTUs a and

b is known; however, the function between a and b may still be zero. Thus, the axiom 4 is not fulfilled for all OTUs in the taxon (which are called as pseudometric or semimetric systems) (Sneath and Sokal 1973, p. 121). For ease of understanding of the relationships, numerical taxonomy mainly focuses on the coefficients whose properties lead to metric spaces.

The dendogram requires that the pair-function is monotonically increasing (for a dissimilarity function) or decreasing (for a similarity function). To insure the pair-function, in numerical taxonomy, the third axiom is relaxed as:

Axiom 3*: $\varphi(a, c) \leq \max [\varphi(a, b), \varphi(b, c)]$.

Here the pair-function is called an ultrametric pair-function for any pair of OTUs. A phenogram is one example of this function (Sneath and Sokal 1973, p. 121).

Distance Coefficients
They are used to measure the distance between OTUs in a space defined in various ways, including Euclidean distance in a character space of one or more dimensions and non-Euclidean metrices. In the numerical taxonomy, the two commonly used distances are Manhattan distance (mean character difference) (Cain and Harrison 1958) and Euclidean distance (taxonomic distance) (Sokal 1961). Distance coefficients are the converse (or complement) of similarity coefficients. The distance is a measure that is complement of similarity, which are, in fact, measures of dissimilarity. Actually, we can consider complementary functions of association or correlation coefficients as measures of dissimilarity in an analogous way. It was shown that the maximum distance possible between OTUs would be $\sqrt{2}$ in two-dimensional model, would be $\sqrt{3}$ in three-dimensional model, and $\sqrt{n}$ in an n-space, a hyperspace of n dimensions (Sokal 1963, pp. 143–144). Table 7.4 summarizes some measures of taxonomic distance compared by Sokal and Sneath.

Association Coefficients
They are used for categorical data (two-state characters or multi-state characters) to measure the agreement of the states in the two data columns representing the OTUs in numerical taxonomy.

The association coefficients are also called similarity coefficients, relationship or matching coefficients. In a Q-type 2×2 data matrix or table, a parametric resemblance value between the two OTUs is expressed as a percentage of matching characters. It measures a sample proportion (percentage) of matches from an infinitely large population of character matches which could be attempted (Sokal and Sneath 1963). Three kinds of association coefficients have been proposed for use in numerical taxonomy.

A conventional 2×2 frequency table can facilitate formulating the association coefficients (Table 7.5).

In the above schematic, the marginal totals are the sums of these frequencies; n represents the sum of the four frequencies, the number of non-missed characters in the study. The number of matches or agreements is defined as m ($m = a + d$), and the number of mismatches is defined as u ($u = b + c$), hence $n = m + u$ (Table 7.5).

Table 7.4 Selected measures of taxonomic distance in numerical taxonomy

Definition	Reference and comments
The mean character difference of (Cain and Harrison 1958): $$M.C.D. = \frac{1}{n} \sum_{i=1}^{n} \mid X_{ij} - X_{ik} \mid$$	It is defined as the absolute (positive) values of the differences between the OTUs for each character It is a special case of Manhattan or city-block metric It has the advantages of its simplicity and it is a metric It also has some disadvantages, such as always underestimating the Euclidean distance between the taxa in space, lacking of the desirable attributes of the taxonomic distance or its square (Sneath and Sokal 1973)
The taxonomic distance of (Sokal 1961): $$\Delta_{jk} = \left[\sum_{i=1}^{n} \left(X_{ij} - X_{ik} \right)^2 \right]^{1/2},$$ And the average distance is defined as: $$d_{jk} = \sqrt{ \frac{\sum_{i=1}^{n} \left(X_{ij} - X_{ik} \right)^2}{n} }.$$	The taxonomic distance between two OTUs j and k is defined in (Sokal 1961; Rohlf 1965; Sokal 1963) This is a generalized version of the Euclidean distance between two points in an n-dimensional space Like to $M. C. D.$, it is a special case of Manhattan or city-block metric It was first used by Heincke as early as 1898 (Sokal and Sneath 1963) d_{jk}^2 was used in (Sokal 1961) and suggested to standardize character state codes for each character The expected value of d very closely approaches $\sqrt{2}$ and the expected variance of $d \approx 1/n$ which approaches zero as n tends to infinity (Sneath and Sokal 1973)
The coefficient of diverge of (Clark 1952): $$CD_{jk} = \sqrt{ \frac{\sum_{i=1}^{n} \left(\frac{X_{ij} - X_{ik}}{X_{ij} + X_{ik}} \right)^2}{n} }$$	The ratio varies between zero and unity Originally each OTU represents a number of specimens, while single values are used in numerical taxonomy (Sokal 1963)
The coefficient of racial likeness (Sokal and Sneath 1963): $$C.R.L = \sqrt{ \frac{1}{2n} \sum_{i=1}^{n} \left(X_{ij} - X_{ik} \right)^2 } - \frac{2}{n}$$	This formula was developed based on Karl Pearson (1926)'s coefficient of racial likeness (Pearson 1926), in which the rows of the Q-type matrix are standardized The OTUs are represented by only single values (with a variance of one) rather than means of the ith character for entity j (Sokal 1963) The strengths of this coefficient (Sneath and Sokal 1973, p. 185) are (1) taking the variance of the estimates into consideration and (2) permitting estimates of the degree of misclassification of individuals in two populations

(continued)

Table 7.4   (continued)

Definition	Reference and comments
The generalized distance developed by Mahalanobis (1936) and Rao (1948): $$D^2_{JK} = \delta'_{JK} W^{-1} \delta_{JK},$$ Where W^{-1} is the inverse of the pooled variance-covariance (dispersion) matrix within samples with dimension $n \times n$ δ_{JK} is a vector of differences between means of samples J and K for all characters	The generalized distance is equivalent to distances between mean discriminant values in a generalized discriminant function, and therefore measuring the distance as a function of the overlap between pairs of populations and transforming the original distances so as to maximize the power of discrimination between individual specimens in the constructed groups (i.e., the OTUs) (Sneath and Sokal 1973, p. 128) It is computed by first obtaining the maximal variance between pairs of groups relative to the pooled variance within groups and then maximizing the difference between pairs of means of characters (Sneath and Sokal 1973, p. 127) The generalized distance has the same strengths as the coefficient of racial likeness (Sneath and Sokal 1973, p. 185): (1) accounting for the variance of the estimates into consideration and (2) permitting estimates of the degree of misclassification of individuals in two populations It is particularly useful in numerical taxonomy (Sneath and Sokal 1973, p. 127) However, it also has some weaknesses (Sneath and Sokal 1973, p. 128): (1) it may distort the original distances considerably; (2) small eigenvalues may make a large haphazard contribution to the distance when a large number of characters is employed; (3) its value is heavily dependent on assumptions of multivariate normal distributions within the OTUs, which will usually not hold above the population or species level

Table 7.5   The 2×2 table for computing association coefficients

		OTU(Taxon) j		
		1	0	
OTU(Taxon) k	1	a	b	a + b
	0	c	d	c + d
		a + c	b + d	$n = a + b + c + d$

In summary, generally the association coefficients include both positive and negative matches in their calculation of coefficients. Both similarity and dissimilarity coefficients are used for measuring resemblance. For binary variables, Jaccard (similarity) coefficient was used in Sneath (1957b) and single-matching coefficient was used in Sokal and Michener (1958). Sneath and Sokal (1973) advocated

Table 7.6 Selected coefficients of association for two-state characters in numerical taxonomy

Definition	Reference and comments
The coefficient of (Jaccard 1908): $S_J = \frac{a}{a+u} = \frac{a}{a+b+c}$	As used by (Sneath 1957a, b; Jaccard 1908) Excludes negative matches in its calculation of coefficient This is the simplest of the coefficients, which has been widely used in R-type and Q-type ecology studies It is appropriate if without considering inclusion of the negative matches in its calculation because it is simplest coefficient (Sokal 1963, p. 139) The values of the coefficient of Jaccard range between 0 and 1 It has three disadvantages (Sneath and Sokal 1973): (1) the complements of S_J coefficient are nonmetric (pseudometric), (2) do not necessarily obey the triangle inequality, and (3) do not consider matches in negative character states (actually the coefficient of Jaccard is appropriate when excluding negative matches is considered as appropriate)
The simple matching coefficient (Sokal and Michener 1958): $S_{SM} = \frac{m}{m+u} = \frac{m}{n} = \frac{a+d}{a+b+c+d}$	First suggested by Sokal and Michener (1958) Equals to the coefficient of Jaccard but includes the negative matches Restricts to dichotomous characters only If considering all matches, Sokal and Sneath preferred the simple matching coefficient over that of Rogers and Tanimoto because it is simpler and easier to interpret (Sokal 1963, p. 139)
The coefficient of (Rogers and Tanimoto 1960): $S_{RT} = \frac{m}{m+2u} = \frac{a+d}{a+2b+2c+d}$	This similarity ratio was originally developed to include characters with more than two states and account for missing information Sokal and Sneath limited this coefficient to the case of two states per character and with complete information (Sokal 1963, p. 134) It includes the negative matches
The coefficient of (Hamann 1961): $S_H = \frac{m-\mu}{n} = \frac{a+d-b-c}{a+b+c+d}$	Balances the number of matched and unmatched pairs in the numerator The values rang from −1 to +1 This coefficient has an undesirable property and independence probably does not have a clear meaning in 2×2 table in numerical taxonomy (Sokal 1963, p. 134) The concept of balancing matches against mismatches does not appear of special utility in the estimation of similarity (Sneath and Sokal 1973, p. 133)

(continued)

Table 7.6 (continued)

Definition	Reference and comments
The coefficient of (Yule and Kendall 1950): $S_Y = \frac{ad - bc}{ad + bc}$	Balances the number of matched and unmatched pairs in the numerator The values rang from -1 to $+1$ It has rarely been employed in numerical taxonomy (Sneath and Sokal 1973, p. 133)
The phi coefficient (Yule and Kendall 1950): $S_\phi = \frac{ad - bc}{\sqrt{(a+b)(c+d)(a+c)(b+d)}}$	Balances the number of matched and unmatched pairs in the numerator The values rang from -1 to $+1$ It is important because of its relation to χ^2 $(\chi^2 = \phi^2 n)$, which permits a test of significance (Sokal 1963, p. 135) However, the test of significance is unmeaningful due to the problem of heterogeneity of column vectors (Sokal and Sneath 1963, p. 135)
The general similarity coefficient of (Gower 1971): $$S_G = \frac{\left(\sum\limits_{i=1}^{n} w_{ijk} S_{ijk} \right)}{\sum\limits_{i=1}^{n} w_{ijk}}$$ This is one version of Gower's coefficient for two individuals j and k, which is obtained by assigning a score $0 \le S_{ijk} \le 1$ and a weight w_{ijk} for character i The weight w_{ijk} is set to 1 for a valid comparison for character i and to 0 when the value of the state for character i is unknown for one or both OTUs	It is applicable to all three types of characters: two-state, multistate (ordered and qualitative), and quantitative (Sneath and Sokal 1973, pp. 135–136) In a data matrix consisting of two-state characters only, Gower's general coefficient therefore becomes the coefficient of Jaccard (S_J). Typically S_{ijk} is 1 for matches and 0 for mismatches and sets $w_{ijk} = 0$ when two OTUs match for the negative state of a two-state character For multistate characters (ordered or qualitative) S_{ijk} is 1 for matches between states for that character and is 0 for a mismatch. Due to without consideration of the number of states in the multistate character, the coefficient resembles a simple matching coefficient applied to a data matrix involving multistate characters For quantitative characters Gower sets $S_{ijk} = 1 - (|X_{ij} - X_{ik}|/R_i)$, where X_{ij} and X_{ik} are scores of OTU j and k for character i, and R_i is the range of character i in the sample or over the entire known population. It results in $S_{ijk} = 1$ when character states are identical and $S_{ijk} = 0$ when the two character states in OTUs j and k span the extremes of the range of the character The general similarity coefficient of Gower has the advantages: (1) these coefficient matrices are positive semi-definite which make it possible to represent the t OTUs as a set of points in Euclidean space (Sneath and Sokal 1973, p. 136); (2) can obtain a convenient representation by taking the distance between the jth

(continued)

Table 7.6 (continued)

Definition	Reference and comments
	and kth OTUs proportional to $\sqrt{1 - S_{jk}}$ (Gower 1966) However, this similarity coefficient also has the disadvantage: the positive semidefinite property of the similarity matrix could be lost due to NCs (no comparisons) (this is also true of R-type or character- correlation matrices involving substantial numbers of NCs) (Sneath and Sokal 1973, p. 136)

negative matches and hence favored the modified Gower's coefficient which allows negative matches in two-state characters. They thought that the modified Gower's coefficient appeared to be a very attractive index for expressing phenetic similarity between two OTUs based on mixed types of characters (p. 136).

Correlation Coefficients

Correlation coefficients are special cases of angular coefficients. In numerical taxonomy, correlation coefficients are used to measure proportionality and independence between pairs of OTU vectors, in which both the product-moment correlation coefficient for continuous variables and other correlation coefficients for ranks or two-state characters have been employed. The product-moment correlation coefficient is most often used in early numerical taxonomy; other correlation coefficients such as the rank correlation have also been occasionally used (Daget and Hureau 1968; Sokal 1985).

For continuous variables, the Pearson product-moment correlation coefficient was most often used in both psychology and ecology Q-type studies previously (Stephenson 1936). At the end of 1950s, it was first proposed to use in numerical taxonomy by (Michener and Sokal 1957; Sokal and Michener 1958), and has been used since then and become the most frequently employed similarity coefficients in numerical taxonomy for calculating between pairs of OTUs (Sneath and Sokal 1973, p. 137). The coefficient, computed between taxa j and k, is given as:

$$r_{jk} = \frac{\sum_{i=1}^{n} \left(X_{ij} - \overline{X}_j\right)\left(X_{ik} - \overline{X}_k\right)}{\sqrt{\sum_{i=1}^{n} \left(X_{ij} - \overline{X}_j\right)^2 \left(X_{ik} - \overline{X}_k\right)^2}}, \qquad (7.1)$$

where X_{ij} denotes the character state value of character i in OTU j, $\overline{X}_j$ is the mean of all state values for OTU j, and n is the number of character sampled. The values of correlation coefficients range from -1 to $+1$.

Regarding the Pearson product-moment correlation coefficient in application of numerical taxonomy, we can summarize three points as below.

(1) Theoretically it is possible that high negative correlations between OTUs exist, but practically they are unlikely in actual data (Sneath and Sokal 1973, p. 138). The argument for this statement is that Sneath and Sokal thought that a study has a few OTUs, then "pairs of OTU's antithetical for a sizeable number of characters are improbable" (Sneath and Sokal 1973, p. 138). It actually stated that the components of OTUs are not compositional.

(2) Pearson correlation analysis is based on the bivariate normal frequency distribution. However, numerical taxonomic data are not normally distributed. Thus, it cannot be directly applied into numerical taxonomic data (Sneath and Sokal 1973, p. 138). Three reasons that make numerical taxonomic data unlikely to be normally distributed: (i) the heterogeneity of the column vectors; (ii) dependence of characters (rows of the data matrix); and (iii) different characters being measured in widely varying scales. Therefore it is invalid to conduct statistical hypothesis testing of correlation coefficients in numerical taxonomy based on the usual assumptions.

(3) For this reason, numerical taxonomists generally advocated standardization of the rows of the data matrix (i.e., standardization of characters over all OTUs) to make the mean of the column vectors frequently approaches 0 and to allow representing inter-OTU relations by using a relatively simple geometrical model (Rohlf and Sokal 1965; Sneath and Sokal 1973, p. 138). The goal of standardization of the characters is to make the vectors carrying the OTUs to be unit length and the column means to be zero (i.e., approximately normalize the column vectors of the OTUs), in such that correlations between OTUs are functions of the angle between the lines connecting them to the origin. Thus, the cosine of the angle between any two vectors is an exact measure of the correlation coefficient (Sneath and Sokal 1973, pp. 138–139).

Although the Pearson product-moment correlation coefficient was criticized since it was first proposed in Michener and Sokal (1957), it is the most commonly used similarity coefficients in numerical taxonomy and currently is still used in some microbiome studies. Sokal and Sneath advocated the Pearson product-moment correlation coefficient because they thought it has three strengths or merits: (1) This formula is based on moments around the mean, accounting for the magnitudes of mismatches between taxa for characters with more than two states. Thus, this coefficient is superior to some association coefficients. For example, it actually resembles the three coefficients of association with range of limits in $[-1, +1]$: Hamann, Yule, and phi coefficients. (2) This coefficient has the merit with it: negative correlation between taxa is at least theoretically possible (Sokal 1963, p. 141). But we should point out that correlation coefficients also have some weaknesses (Sneath and Sokal 1973, pp. 138–139):

(1) They are generally nonmetric functions although the semichord distances can be transformed to yield a Euclidean metric. When correlation coefficients are

converted to their complementary form (i.e., distances), generally the axiom of triangle inequality is violated (Sneath and Sokal 1973, p. 138). Perfect correlation could also occur between nonidenticals, such as in two column vectors when one is the other multiplied by a scalar. (2) The correlation coefficient in numerical taxonomy has an undesirable property because in the cases with few characters, the values of the correlation coefficient will be affected by the direction of the coding of the character states (Eades 1965). Thus, it requires that all characters have the same directional and dimensional properties (Minkoff 1965).

However, overall Sneath and Sokal (1973, p. 140) believe that correlation coefficients are usually the most suitable measure in numerical taxonomy to evaluate OTUs. They do not think above undesirable properties would not manifest themselves in large data matrices involving many OTUs with the large number of characters. Additionally, when the interpretation of taxonomic structure is based on phenograms, it has been observed that the correlation coefficients are the most appropriate measure.

In summary, due to heterogeneity of OTUs, the basic assumptions of the bivariate normal frequency distribution cannot be met, so in psychology and numerical taxonomy literature, the character scores are often standardized or a percentage scale or a ratio of the variable are used before applying this coefficient (Sokal and Sneath 1963, p. 142).

Probabilistic Similarity Coefficients

Probabilistic similarity coefficients were developed in 1960s; Sneath and Sokal have reviewed these similarity coefficients in their 1973 book (Sneath and Sokal 1973).

First we should point out that in numerical taxonomy, the characters are the variables; OTUs are observations. Thus, here, the distribution refers to the distribution of the frequencies of the character states over the set of OTUs. While as Sneath and Sokal (1973) pointed out, probabilistic similarity indices are more relevant to ecological classification, where the abundance of a species (species are the equivalent of characters in numerical taxonomy, and stands correspond to OUTs) in a sampling unit is a stochastic function (p. 141).

Probabilistic similarity coefficients were developed based on the assumption that it is less probable to agree among rare character states than to agree for frequent character states and therefore rarer character states should be given more heavy or enhanced weight (Sneath and Sokal 1973, p. 140). The technique of probabilistic similarity had been developed to avoid using similarity coefficients and instead to go straight to use the data matrix for a classification, in which the establishing partitions are guided by only a criterion of goodness of the classification (Sneath and Sokal 1973, p. 120).

Goodall's Probabilistic Similarity Index (Goodall 1964, 1966a) is a similarity index to measure the complement of the probability that two OTUs would have the observed (or a greater) degree of similarity if attribute values were assorted at random. It was shown by Sneath and Sokal (1973) that this index resulted in very good general agreement with the analyses using simple Jaccard similarity indices.

One important kind of similarity coefficients is the information statistics. It has the advantage of additive property. Thus information statistics permits the partitioning of indices for larger groups into information statistic for subgroups (Sneath and Sokal 1973).

In ecology, information statistics is used to measure the frequencies or probabilities of the occurrence of the various species over the collection of stands. Sneath and Sokal introduced one probabilistic similarity coefficient based on information-theoretical concepts into numerical taxonomy to measure the homogeneity of the system by partitioning or subpartitioning sets of OTUs (Sneath and Sokal 1973, p. 120).

The Probabilistic Similarity Index Described by Sneath and Sokal (1973, pp. 141–143) was developed based on information statistics. Technically "information" is a measure of disorder, variance, or confusion, or "surprisal" as Orloci (1969) put out, rather than a measure of the amount of knowledge about a group. Given a probability distribution for the occurrence of various character states over the t available OTUs (Goodall 1966b), the measure of disorder for character i based on a measure of information or entropy is defined as follows.

$$H(i) = -\sum_{g=1}^{m_i} p_{ig} \ln p_{ig}, \tag{7.2}$$

where m_i is the number of different states in character i, and p_{ig} is the observed proportion of the t OTUs exhibiting state g for character i. $\sum_{g=1}^{m_i} p_{ig} = 1$. When the n characters are not correlated, the total information of the group can be obtained by summing the separate values of $H(i)$: $I = t\sum_{i=1}^{n} H(i) = 1$.

The constructed information or entropy can also be applied to ecological data. In an ecological setting characters (rows) are species and OTUs (columns) are stands or quadrats, and hence for data matrices in ecology, this formula is to measure the frequencies of any one species that are recorded over a series of stands (Sneath and Sokal 1973, p. 142). To conduct hypothesis testing of "states" m_i, these frequencies must be grouped into classes to provide m_i "states" (e.g., "absent," "rare," "common," and "abundant") of a species (or character) i.

An approximately x^2 distribution with $2n$ degrees of freedom can be described by this formula in terms of frequencies for taxon J containing t_J OTUs. Where the quantity I measures the total information (entropy, surprisal) in the entire sample with the lower value of I indicating the more homogeneous a taxon will be. Thus, one goal of clustering OTUs using information statistics is to obtain taxa with low values of I (Sneath and Sokal 1973, p. 142).

Compared to most resemblance coefficients that only apply to pairs of single OTUs or pairs of taxa, another advantage of the information statistics is it's applicable to sets of OTUs. Sneath and Sokal (1973, pp. 143–145) introduced an index based on two concepts: **joint information and mutual information** discussed by (Estabrook 1967; Orloci 1969). Given the correlation of characters, joint information

$I(h, i)$ is the union of the information content of two characters over a set of OTUs; mutual information $I(h; i)$ is their intersection. The relation between these quantities is given below:

$$I(h) + I(i) - I(h, i) = \mathrm{I}(h; i). \tag{7.3}$$

The mutual information $I(h; i)$ is the ratio of the sums of the information content exclusive to characters h and i divided by the total information possessed jointly by h and i.

Overall, information statistics have important properties (Sneath and Sokal 1973, p. 120), such as are usually additive, are distributed as chi-square, and are probabilistic, and therefore a statistical hypothesis testing can be conducted through a standard statistical means. Additionally, the special cases of these coefficients can be represented as distance coefficients.

The denoising methods that are used to classify ASVs or sub-OTUs in current microbiome research actually take the same approaches as in probabilistic similarity coefficients (see Chap. 8 for details).

The choice of which resemblance coefficient not only is determined by the nature of the original data matrix, such as association coefficients for binary and categorical data matrices and distance and correlation for continuous variables, but also is based on the type of resemblance. Sokal (1985) reviewed that classifications have been based on correlation coefficients than based on similarity using distance-based measures. For binary data, correlations are not especially helpful and standardization does not help either.

In summary, regarding these three coefficients of association, distance measures, and correlation, Sokal and Sneath (1963, p. 167) thought that association and distance are easier to interpret conceptually and relatively simpler than correlation. Among these three coefficients, they recommended using the simple matching coefficient and Jaccard coefficient in case of excluding negative matches, the coefficient of taxonomic distance d_{jk} based on standardized characters, and Pearson's product-moment r_{jk}, respectively. In their 1973 book Sneath and Sokal favor more strongly than in their 1963 book on binary similarity coefficient because of the following: (1) its simplicity and (2) possible relationship to information theory; (3) the similarity between fundamental units of variation can be estimated; and (4) related to natural measures of similarity or distance between fundamental genetic units (amino acid or nucleotide sequences) (Sneath and Sokal 1973, p. 147).

7.4.3 Commonly Clustering-Based OTU Methods

Taxa (taxonomic structure/taxonomic system) are constructed through resembling similarity (the resemblance matrix) at the same or different rank levels hierarchically by using a measure of resemblance, i.e., grouping together OTUs to form taxa. The taxonomic system that is to be constructed should be "natural" in an empirical sense, and thus should be highly predicable and be formed with a nested hierarchy.

7.4.3.1 Similarity/Resemblance Matrix

A resemblance or similarity matrix is a square matrix consisting of the OTUs in both rows and columns and the entries in the matrix with the estimates of the resemblances/similarity coefficients for every OTU compared with every other OTU. This square matrix is $t \times t$ ordered matrix with t representing the number of OTUs (Table 7.7).

In above matrix S represents any similarity coefficient and S_{jk} is the resemblance/similarity coefficient in the jth row and the kth column. The entries of resemblance matrix can be used to include both "similarities" and "dissimilarities." Sine based on the property of symmetry, the resemblance of a to b is the same as that of b to a, and then the top right part of the matrix is a mirror image of the lower left part. Thus, in practice, only the lower left triangle is given (Table 7.8). The lower left triangle matrix is often provided by some computer software. For example, the function **vegdist()** in ecology package **vegan** provides the lower left triangle matrix if Jaccard or Sørensen dissimilarity measures are used. For details, the readers are referred to Sneath and Sokal (1973, pp. 190–192) and Chap. 10 of this book.

The main techniques used for constructing taxonomic structure are clustering techniques, which are applicable to all three types of similarity coefficients (distance, association, and correlation coefficients). The more complicated method factor analysis can be used to construct taxonomic structure.

Four sets of methods have been developed in numerical taxonomy: (1) clustering, (2) factor analysis, (3) ordination, and (4) discriminant analysis. We can collectively

Table 7.7 Resemblance matrix of OTUs

OUTs	OUTs					
		1	2	3	...	t
	1	S_{11}	S_{12}	S_{13}	...	S_{1t}
	2	S_{21}	S_{22}	S_{23}	...	S_{2t}
	3	S_{31}	S_{32}	S_{33}	...	S_{3t}
	⋮	⋮	⋮	⋮		⋮
	t	S_{t1}	S_{t2}	S_{t3}	...	S_{tt}

Table 7.8 Lower left triangle resemblance matrix of OTUs

OUTs	OUTs				
		1	2	3	...	t
	1					
	2	S_{21}				
	3	S_{31}	S_{32}			
	⋮					
	t	S_{t1}	S_{t2}	S_{t3}	...	

call them as "numerical taxonomic methods" or "OTU methods" since they are used to classify or identify the OTUs. The first two sets of methods were mainly proposed or developed in Sokal and Sneath's early work, such as in their 1963 book,whereas the last two sets of methods were mainly adopted into numerical taxonomy in their later works (e.g., 1973 book). The first three sets of methods are used for classification, and the discriminant analysis is used for both classification and identification. We describe them separately below.

One goal of clustering in numerical taxonomy is to recognize the patterns of distribution of OTUs and groups of OTUs (taxa) in a space (commonly a hyperdimensional, phenetic A-space) (Sneath and Sokal 1973, p. 192).

7.4.3.2   Definitions of Cluster in Numerical Taxonomy

When the patterns are described, they are typically conducted at least within a three-space (mathematically a higher-dimensional space): multidimensional Euclidean, non-Euclidean, and nonmetric spaces. Generally, regular and random distributions are the two observed dispersion patterns that can be compared. The clusters in numerical taxonomy are defined as the sets of OTUs in phenetic hyperspace that exhibit neither random nor regular distribution patterns and that meet one or more of various criteria.

In numerical taxonomy clusters are defined from different perspectives. We summarize these particular cluster definitions in Table 7.9. For details of these cluster definitions, the reader is referred to Sneath and Sokal's 1973 book (Sneath and Sokal 1973, pp. 194–200).

7.4.3.3   Commonly Clustering Methods in Defining Groups of OTUs

Numerical classification can be performed by arranging of organisms in one of three ways (Sneath and Sokal 1973, p. 200): (1) hierarchical clustering systems of partitions (nested, mutually exclusive systems), (2) nonhierarchic clustering systems of partitions, and (3) ordination-the positioning of OTUs in relation to multidimensional character axes in an A space. In this section, we introduce clustering including both hierarchical clustering and non-hierarchic clustering. In the next section, we will briefly review ordination techniques in numerical taxonomy, and the more detailed introduction will be presented in Chap. 10 when microbiome study is covered.

In numerical taxonomy, cluster analysis is referred to as a large class of numerical techniques for defining groups of related OTUs based on high similarity coefficients. Clustering methods typically represent the similarity matrix among taxonomic units by dendrograms (a treelike arrangement). The grouped OTUs by clustering may form a coarser partition from a finer one, or may separate an entire set of OTUs into increasingly finer partitions.

Table 7.9   Cluster definitions in numerical taxonomy

Definition and description
Cluster center
The OTUs are within the center of average organism or centroid
For the hypothetical OTU (the hypothetical modal organism), it should be at the corner that is nearest to the centroid
For an actual OTU, it is the centrotype. That is the OTU with the highest mean resemblance to all other OTUs of the cluster or the OTU nearest to the centroid for Euclidean distance models, or the OTU having the highest loading on a factor representing the cluster in a Q analysis
Density
A cluster should be denser than other areas of the space under consideration
It can be expressed as number of OTUs per unit hypervolume or as the density/volume ratio of a convex hyperdimensional set
Variance
A cluster for a satisfactory partition should be that variance within the subsets is minimized relative to variance among the subsets in analysis of variance (univariate and multivariate)
Dimension
A dimension of a cluster can be measured in various ways
In discriminatory methods the cluster is defined in terms of a radius of the cluster, i.e., a specified majority of OTUs lie within this radius
In ordination methods, the ellipticity of a cluster can be estimated from the standard deviations of the OTUs on the several factor axes (the square roots of the eigenvalues)
Number of members in a cluster
Some clustering methods either
Allocate weights according to the number of members in a cluster, or
Require that a group shall contain a given minimum or maximum number of OTUs (the cluster size) to recognize a cluster
Connectivity
To be members of a cluster, individuals have to be related (i.e., connected) to a certain minimal degree, such as within some kind of closeness by distance in the hyperspace
It requires that each OTU is connected at the minimal acceptable level to every other OTU either directly or via other OTUs
Straggliness
It implies much greater extension in a few of the dimensions than in the others
Ellipticity can be revealed by examining factor analysis plots, or calculating the standard deviations
The highly straggly clusters can also be shown by a histogram of distances of OTUs from the centroid to see whether a sharp mode is well away from the origin, or is flattened, falling off from the origin
Gaps and moats
The gap or moat by which the distance from each member of a cluster to the closest one in another subset should be sufficiently large to separate or isolate the subsets
A moat can be the absolute distance between the closest members of different subsets, or the average of such distances for all members of both subsets; or may be some ratio of the distance between the centroids of clusters and the radial vectors of the cluster envelopes to their centroids
The gap between clusters is a criterion of measuring a new type of cluster, for which the greatest distance between any pair of OTUs within the cluster is less than the smallest gap from a cluster member to any OTU outside the cluster

Clustering can be described from eight aspects: (1) agglomerative versus divisive; (2) hierarchic versus non-hierarchic; (3) non-overlapping versus overlapping; (4) sequential versus simultaneous; (5) local versus global criteria; (6) direct versus iterative solutions; (7) weighted versus unweighted; and (8) non-adaptive versus adaptive. Among these eight methods, the **sequential, agglomerative, hierarchic, non-overlapping** (the acronym **SAHN**) clustering methods are the most frequently employed strategies for finding clusters (Sneath and Sokal 1973, p. 214).

The grouping of OTUs in numerical taxonomy has been most often employed by **sequential** clustering and by **agglomerative** techniques and rarely by divisive techniques. **Agglomerative** techniques start with a set of t separate entities, group them into successively fewer than t sets, arriving eventually at a single set containing all t OTUs. In contrast, divisive techniques start with all t OTUs in one set, subdividing it into one or more subsets, until each of the subsets is subdivided. Numerical taxonomy defines hierarchies in terms of nestedness as well as mutual exclusiveness. Thus, **hierarchic** clustering is to classify organisms into nested, mutually exclusive taxa. The hierarchic classifications are referred to as stratified. In a **non-overlapping** method, taxa at any one rank are mutually exclusive; that is, if the OTUs have been contained within one taxon, they cannot also be members of a second taxon of the same rank. For the details, the interested readers are referred to Sneath and Sokal's 1973 book (Sneath and Sokal 1973, pp. 202–214).

Here, we summarize the commonly SAHN clustering methods used for defining groups of OTUs. Among them, four mainly described cluster methods are (1) single linkage clustering, (2) complete linkage clustering, (3) average linkage clustering, and (4) central or nodal clustering.

7.4.3.3.1   Single Linkage Clustering

Single linkage clustering (Sneath 1957a) establishes connections between OTUs and clusters and between two clusters by single links between pairs of OTUs (Sneath and Sokal 1973, p. 218). First, it finds and clusters together those mutually most similar pairs (least dissimilar pairs) with the highest possible similarity coefficient. Then, it successively lowers the level of admission by steps of equal magnitude and examines distances from new candidates for fusion with the established clusters. Thus, in single linkage clustering, a candidate OTU for an extant cluster has similarity to that cluster equal to its similarity to the closest member within the cluster (Sneath and Sokal 1973, p. 218).

Remarks on Single Linkage Clustering

Single linkage clustering is also called clustering by single linkage (Sneath's method) (Sokal and Sneath 1963) or the nearest neighbor clustering (Lance and Williams 1966) or minimum method (Johnson 1967). It was introduced to taxonomy by Florek et al. (1951a, b) and Sneath (1957a). Single linkage clustering has the disadvantages: (1) sensitive to noise in the data (Milligan 1996); (2) may form complicated serpentine clusters, which is not very desirable in biological taxonomy

(Sokal and Sneath 1963, pp. 192–193); and (3) leads to a peculiar feature: chaining (elongate growth), which especially occurs when a number of equidistant points or near equidistant points are frequently found in taxonomy and ecology. Such chaining causes phenograms not to be very informative because the phenograms do not well show the information on intermediate or connecting OTUs (Sneath and Sokal 1973, p. 223).

7.4.3.3.2 Complete Linkage Clustering

Complete linkage clustering (Sørensen 1948) joints two clusters by their similarity that exists between the farthest pair of members, one in each cluster (Sneath and Sokal 1973, p. 222). It requires a given OTU joining a cluster at a certain similarity coefficient must have relation at that level or above with every member of the cluster. Thus, in complete linkage clustering, a candidate OTU for admission to an extant cluster has similarity to that cluster equal to its similarity to the farthest member within the cluster (Sneath and Sokal 1973, p. 222).

Remarks on Complete Linkage Clustering

Complete linkage clustering is also called clustering by complete linkage (Sørensen's method) (Sokal 1963) or farthest/furthest neighbor clustering (Lance and Williams 1966) or maximum method (Johnson 1967). Complete linkage clustering has the disadvantages:

(1) Generally leads to tight, hyersphencal, discrete clusters that join others only with difficulty and at relative low overall similarity values (Sneath and Sokal 1973, p. 222).
(2) Likely forms very compact and well-defined clusters, which look like clouds of OTUs in a hyperspace and like hyperspheroids in that space (Sokal 1963, p. 192) or small tight compact clusters (Sneath and Sokal 1973, p. 223).
(3) Different initial levels of similarity coefficients could result in different number of clusters (Sokal 1963).

7.4.3.3.3 Average Linkage Clustering

Average linkage clustering (Sokal and Michener 1958) clusters any individual OTU into a cluster on some kind of average similarity or dissimilarity between a candidate OTU or cluster and an extant cluster (Sneath and Sokal 1973, p. 228).

Remarks on Average Linkage Clustering

Average linkage clustering is also called clustering by average linkage (the group method of Sokal and Michener) (Sokal 1963), and earlier it was called group methods in numerical taxonomy (Sneath and Sokal 1973). Average linkage clustering techniques were developed to avoid the extremes resulted in either single linkage or complete linkage clustering. This clustering can be based on all three types of similarity coefficient matrices although the correlation coefficient matrices are originally suggested. Sometimes Spearman's sums of variables method (Spearman 1913) was used for calculation of the correlation coefficients with average linkage clustering in numerical taxonomy. The average cluster method is recommended for use because single and complete linkage clustering methods have their disadvantages (Sokal and Sneath 1963, pp. 192–193).

7.4.3.3.4 Central or Nodal Clustering

Central or nodal clustering (Rogers and Tanimoto 1960) starts with an association coefficient (similarity coefficient). It calculates the value R_i based on a matrix of similarity coefficients to represent the number of nonzero similarity coefficients that a given OTU i has with other OUTs. R_i value is considered as the similarity coefficients greater than the expected value (or a given criterion) to give useful clusters (Sokal and Sneath 1963). In practice, the R_i values, which are indices of typicality, are considered only similarity coefficients larger than 0.65 (Silvestri et al. 1962). The larger the R_i value, the more "typical" a given OTU is to group under the study.

Remarks on Central or Nodal Clustering
Central or nodal clustering is also called the method of Rogers and Tanimoto (Sokal and Sneath 1963). It has the advantage to obtain a measure of the homogeneity of clusters because it computes a coefficient of inhomogeneity, which is desirable in biology (Sokal 1963, p. 193). However, it also has the disadvantages: (1) defines only a series of primary nodes without connecting them into a dendrogram, and (2) is much indeterminant to obtain the clustering solutions (Sokal and Sneath 1963, p. 193).

In the literature, sometimes clustering methods are categorized based on whether the clustering is arithmetic average or centroid clustering, and weighted or unweighted clustering (Sneath and Sokal 1973, p. 228).

7.4.3.3.5 Weighted Versus Unweighted Clustering

Weighted clustering was introduced to weight the individuals unequally, i.e., give merging branches in a dendrogram equal weight regardless of the number of OTUs carried on each branch. Thus, it attempts to give equal importance to phyletic lineages (Sokal and Michener 1958).

Unweighted clustering gives equal weight to each OTU in clusters, evaluates the similarity of OTU with another cluster (or OTU), and thus it less distorts the compared phenograms with original similarity matrices (Sneath 1969) (also see Sneath and Sokal 1973, p. 228).

Remarks on Weighted Clustering

Weighted method is preferred when very disparate sample sizes are used to represent several taxa. The benefit of using weighted method is when the sparsely represented sample is given more importance, it results in less biased estimates of intercluster distances (Sneath and Sokal 1973, p. 229).

7.4.3.3.6 Arithmetic Average Versus Centroid Clustering

Arithmetic average clustering (Sneath and Sokal 1973, p. 228) "computes the arithmetic average of the similarity (or dissimilarity) coefficients between an OTU candidate for admission and members of an extant cluster, or between the members of two clusters about to fuse." It does not take into account similarity or dissimilarity (taxonomic distance) coefficients between members within the cluster; thus, it does not consider the density of the points that constitute the extant cluster (or a candidate cluster about to fuse with it) when evaluating the resemblance between the two entities.

Centroid clustering (Sneath and Sokal 1973, p. 228) "finds the centroid of the OTUs forming an extant cluster in an A-space and measures the dissimilarity (usually Euclidean distance) of any candidate OTU or cluster from this point."

Arithmetic average clustering can be grouped into unweighted pair-group method using arithmetic averages (UPGMA) and weighted pair-group method using arithmetic averages (WPGMA) (Sneath and Sokal 1973, p. 218). UPGMA computes the average similarity or dissimilarity of a candidate OTU to an extant cluster, weighting each OTU in that cluster equally, regardless of its structural subdivision (Sneath and Sokal 1973, p. 230). WPGMA differs from UPGMA in that it weights the member most recently admitted to a cluster equal with all previous members (Sneath and Sokal 1973, p. 234).

Centroid clustering can be grouped into unweighted centroid and weighted centroid (Sneath and Sokal 1973), or just centroid and median (Lance and Williams 1967), respectively. Centroid clustering can also be grouped into unweighted pair-group centroid method (UPGMC) and weighted pair-group centroid method (WPGMC). WPGMC was also called the median method (Lance and Williams 1967). It was first developed by (Gower 1967a). WPGMC weights the most recently admitted OTU in a cluster equally to the previous members of the cluster (Sneath and Sokal 1973, p. 235).

Remarks on Arithmetic Average Clustering and Centroid Clustering

Arithmetic average clustering results in the similar general taxonomic structure as complete linkage analysis, although some fine distinctions exist between them. The disadvantage of arithmetic average clustering is that it does not have easily discernible geometric interpretation as in centroid clustering.

Centroid clustering recognizes the dissimilarities among the points constituting a cluster and hence it has a simple geometric interpretation.

UPGMA is also called group average (Lance and Williams 1967). It works for both dissimilarity and similarity coefficients. It is the most frequently used clustering method in numerical taxonomy and adopted in ecology and microbiome studies. WPGMA has the general taxonomic structure that is identical to UPGMA and shares the properties of UPGMA (Sneath and Sokal 1973, p. 234). But compared to UPGMA, three differences exist in WPGMA (Sneath and Sokal 1973, p. 234): (1) It distorts the overall taxonomic relationships in favor of the most recent arrival within a cluster. (2) In WPGMA the larger clusters are further apart from each other due to the greater dissimilarity in the later joined clusters. (3) The average distances between the clusters are increased because of this heavier weighting in WPGMA.

UPGMC computes the centroid of the OTUs that join to form clusters, and then computes distances or their squares between these centroids, and thus it is attractive conceptually (Sneath and Sokal 1973, pp. 234–235). However, the results of UPGMC procedures are not monotonic due to not meeting the ultrametric inequality (Sneath and Sokal 1973, p. 235). WPGMC shares some properties with UPGMC (i.e., a lack of monotonicity). The heavier weight used to late joiners of clusters causes the taxonomic structure of WPGMC to have some differences from that in UPGMC as implied by the phenogram (Sneath and Sokal 1973, p. 235).

It was shown (Sneath and Sokal 1973, p. 240) by comparing the phenograms from the different clustering methods that the levels of linkage between clusters are on the average highest (least dissimilar) for single linkage and lowest for complete linkage, and intermediate for the average linkage methods. To overcome the disadvantages of single and complete linkage analyses, several modified versions of single and complete linkage clustering have been developed. For example, to avoid the extremes of single linkage (chaining) and complete linkage (small tight compact clusters) clustering, Sneath (1966) proposed four modified linking strategies to link two clusters: (1) integer link linkage, (2) proportional link linkage, (3) absolute resemblance linkage, and (4) relative resemblance linkage. Lance and Williams (1967) proposed a flexible clustering strategy to constrain the linear formula to overcome chaining by single linkage and extremely compact clustering as observed in complete linkage analysis. Shepherd and Willmott (1968) proposed a modified single linkage clustering to prevent chaining. The interested readers are referred to Sneath and Sokal (1973) and references therein for further details of these modified single and complete linkage analyses.

In summary, the average linage clustering method developed by Sokal and Michener has been recommended as clustering method for numerical taxonomy (Sneath and Sokal 1973) because intuitively within-taxon similarity an average measure rather than the minimum or maximum thresholds is more acceptable (Sokal 1985). It can be used as a variable-group method or a pair-group method (Sokal 1963, p. 194). However, using resemblances of taxa to achieve an optimal classification is more complex than maximizing the cophenetic correlation coefficient. Overall criterion should not only consider the average similarity within a taxon and its variance, but also consider the rank of a given taxonomic subset of the entire study (Sokal 1985).

7.4.3.4   Factor Analysis

Taxonomic structure can also be constructed through factor analysis. Based on (Sokal and Sneath 1963), it is Sokal (1958b) who first used factor analysis to analyze taxonomic relationships from a similarity matrix of correlation coefficients, in which factor analysis is used for describing the complex interrelationships among taxa in terms of the smaller number of factors. Here, OTU is used as an observed variable, while taxon is used as a latent factor. The factor loading represents the degree of similarity between an OTU and the average aspect of the taxon. The higher the factor loading, the more typical the OTU belongs to the taxon. Later multiple factor analysis (Harman 1976) was initially used to phenetic classification by (Rohlf and Sokal 1962), in which centroid factor extraction with subsequent rotation is performed to obtain "simple structure" for revealing the interrelationships among OTUs.

Employing factor analysis to classification actually is to define taxa (e.g., species) in terms of similarity of OTUs by using factor analysis. The later bioinformatic work using factor analysis to define species originated in this work of numerical taxonomy. However, the challenge is how many factors should be extracted to define the taxa. In factor analysis the number of factors is usually arbitrarily chosen. Additionally, factor analysis cannot draw straight line across a dendrogram to yield the groups, suggesting that the factor analysis only can approximately, but not exactly form the groups at the same hierarchical level (Rohlf and Sokal 1962; Sokal 1963).

In their 1973 book (Sneath and Sokal 1973, pp. 98–103), Sneath and Sokal discussed the advantages and disadvantages of factor analysis. First, factor analysis has the following advantages:

(1) It provides important implications for choice of characters. For example, phenetic similarity may be estimated based on the extracted k factors rather than choosing using n characters. In this way such characters are used as necessary to evaluate the factor endowment of each OTU.
(2) It theoretically reduces the number of phenetic dimensions necessary to visualize the organisms.
(3) It is possible to use factor analysis quickly to find whether more characters should be measured.

However, factor analysis has both practical and theoretical problems. Practically, before a factor analysis is conducted, factor scores and correlations of the characters are needed; thus factor analysis is not really efficient. Additionally, similar as it is a challenge to find clusters of highly correlated characters in cluster analysis, it is even risk to find factors when the characters are highly correlated. Thus, Sneath and Sokal suggested that we might still use the whole set of characters to estimate phenetic similarity from factor scores (Sneath and Sokal 1973, p. 99). Theoretically, it is difficult to know how to weight the extracted factors in measuring the similarity between OTUs. Thus, we need more research to develop an independent criterion for weighting factors.

7.4.3.5   Ordination Methods

Ordinations also present classifications, in which the data are first required to be transformed before being turned into ordinations. The ordination methods in phenetic taxonomy are referred to the methods that describe the occupation of phenetic space by the OTUs being ordinated (Sokal 1986). The ordination methods in application to phenetic taxonomy differentiate from clustering methods, in which OTUs are arranged in a continuous hyperspace with its dimensions being defined by the characters of the OTUs, whereas clustering methods assign OTUs to groups (Sokal and Sneath 1963; Sneath and Sokal 1973; Clifford and Stephenson 1975; Dunn and Everitt 1982; de Queiroz and Good 1997). Both clustering and ordination methods are based on similarities. However, clustering methods aim to base nested hierarchical taxonomies on similarity, whereas ordination methods do not attempt to give hierarchical results, which was reviewed as more accurately representing similarities than do by clustering methods (de Queiroz and Good 1997).

Narrowly OTU-method in numerical taxonomy only includes commonly used clustering methods and factor analysis. Based on (de Queiroz and Good 1997), Sokal pioneered the biological application of ordination techniques, e.g., in (Rohlf and Sokal 1962). However, in their 1963 book, Sokal and Sneath did not include ordination techniques as an approach to phenetic taxonomy. Ordination techniques were described as a potential approach to phenetic taxonomy in their 1973 book (Sneath and Sokal 1973), including principal component analysis (PCA), principal coordinate analysis (PCoA), nonmetric multidimensional scaling (NMDS), and seriation. PCA, PCoA, and NMDS have been readily adopted in microbiome study. We will introduce and illustrate PCA, PCoA, and NMDS as well as other ordination methods with real microbiome data in Chap. 10 (Sect. 10.3). Here, we briefly describe seriation method. For details, the interested readers are referred to Sneath and Sokal (1973) and Legendre and Legendre (2012) and the references therein.

Seriation is a simple type of (one-dimensional) ordination previously developed in anthropology (Petrie 1899; Czekanowski 1909) and archaeology (Craytor and Johnson Jr 1968; Johnson Jr 1968; Kendall 1971, 1988). It was first applied to ecology by (Kulczynski 1927). Seriation in archaeology (Hodson 1970; Cowgill 1968) and ecology (Legendre and Legendre 2012) is used as a clustering analysis.

Seriation is to permute the rows and columns of a similarity matrix in such a way as to arrange the highest similarities (or the lowest distances) near the principal diagonal of the similarity (resemblance) matrix and with the similarity values decreasing orderly away from the diagonal. Thus, seriation method essentially is how to operationally define order among the coefficients in a similarity matrix (Sneath and Sokal 1973, pp. 250–251).

Remarks on Seriation Method
The seriation method has been reviewed having some shortcomings (Sneath and Sokal 1973, p. 251). For example, (1) this technique requires an iterative algorithm

to order all possible arrangements of OTUs and hence is computationally expensive (Craytor and Johnson Jr 1968); (2) it is prone to be local optima (Kendall 1963; Craytor and Johnson Jr 1968); (3) it is only applicable to very specialized data sets such as in cases of true linear trend development; (4) it has been reviewed that seriation is inferior to NMDS and principal axis methods (Cowgill 1968; Kendall 1969; Hodson 1970), as well as less specific and efficient than the clustering methods in analysis of a usual similarity matrix since seriation does not directly generate clusters of objects (Legendre and Legendre 1998, p. 372).

Seriation also has the attractive property because it can be applied to both non-symmetric and symmetric similarity or distance matrices (Legendre and Legendre 2012, p. 403). This interesting property may attract the microbiome researchers to apply this method in their studies since as in ecology, the similarity or distance matrices in microbiome may be not symmetric. The R package seriation has been developed for performing seriation analysis (Hahsler et al. 2008).

7.4.3.6   Discriminant Analysis in Identifying OTUs

Numerical taxonomy makes distinction between classification and identification. As we described in Sect. 7.2.3, at the beginning, unequal weighting was not promoted in numerical taxonomy. However, Sokal and Sneath recognized that strictly equal weighting is impossible in their later publications, such as in (Sneath and Sokal 1973). They distinguished identification from classification, in which identification or assignment is used to assign a new entity (Dagnelie 1966).

Classification is used in the sense of making classes, clusters, or taxa. By contrast, identification is used for "identifying" new individuals (OTUs/taxa). However, when we identify an individual as new one, we actually know that an existing class is unacceptable. The cluster analyses can effectively work on the classification. But to identify new individuals, it requires that some strategies can combine the two procedures of classification and identification. The discriminant functions (analyses) are the main methods used in identification.

Along with the distinction between classification and identification, Sneath and Sokal believed that weighting is needed. The goal of weighting characters in numerical taxonomy is for identification. Sneath and Sokal (1973, p. 400) thought that weighting can maximize the probability of correctly identifying unknown specimens from a few close or overlapping taxa through a set of multivariate statistical methods.

Sneath and Sokal have specifically introduced three discriminant methods in describing or discriminate OTUs (or taxa): discriminant function analysis (DFA), and multiple discriminant analysis in Mahalanobis' D-space and canonical variates analysis. We can view these multivariate statistical methods as extensions of taxon distance models in Sect. 7.4.2.3. The purpose of taxon distance models in Sect. 7.4.2.3 is to classify the OTUs, while the purpose of discriminant analyses is to perform both classification and identification, i.e., to discriminate the OTUs (or taxa).

7.4.3.6.1 Discriminant Function Analysis (DFA)

DFA or linear discriminant analysis (LDA) or linear discriminant function analysis (LDFA) is a generalization of Fisher's linear discriminant method (Fisher 1936). LDFA is a member of the family of classification methods known as supervised learning, in which group membership of individuals is used to maximize group differences during the process. In this method, a linear combination of features that characterizes or separates two or more classes of objects is to be identified and to be used as a linear classifier or for dimensionality reduction before later classification. LDA requires that groups (i.e., the class label) are known a priori.

In numerical taxonomy, the linear discriminant function is a linear function z of characters describing OTUs that weight and separate the characters in such a way to identify two taxa: maximizing number of the OTUs in one taxon has high values for z and maximizing number of the OTUs in another taxon has low values, so that z serves as a much better discriminant of the two taxa than does any one character taken singly (Sneath and Sokal 1973, p. 400). By performing the linear discriminant function, (1) the n characters are almost always reduced to a smaller set, m; and (2) the function also has maximal variance between groups relative to the pooled variance within groups.

The dichotomous discriminant function that was originally developed by (Fisher 1936) was applied to two taxa, and was later generalized to multiple taxa. In LDA the pooled variances and covariances between the m characters within each taxon are calculated. They are calculated by a weighted average of the variances and covariances of the characters in taxa J and K in its original form, while in its generalized form, the variances and covariances are calculated from pooling all the taxa being considered. Additionally, the means of each character for each taxon are calculated to represent the centroids of the taxa. Suppose J, K, and L represent three taxa for two character axes, X_1 and X_2. LDA can be briefly summarized as the following eleven steps. For the details, the readers are referred to Sneath and Sokal (1973, pp. 401–405).

Step 1: The linear discriminant function generates a set of taxa (i.e., three taxa J, K, and L) for all the character axes (i.e., X_1 and X_2). The generated taxa are represented by the clusters of individuals (OTUs).

Step 2: After clustering individuals (OTUs), we now only concern the descriptive parameters of the clusters (i.e., three taxa J, K, and L), i.e., the centroids and the dispersions of the clusters and do not concern with the values for the individuals.

Step 3: First calculate the variances and covariances between the characters, yielding a set of (here three) $m \times m$ matrices (here $m = 2$); then average these variances and covariances to give a pooled within-groups variance-covariance matrix W.

Step 4: Calculate the discriminant function between J and K. Multiply the inverted W matrix by the vector $\delta_{JK} = \left[\left(\overline{X}_{1J} - \overline{X}_{1K} \right), \left(\overline{X}_{2J} - \overline{X}_{2K} \right), \ldots, \left(\overline{X}_{mJ} - \overline{X}_{mK} \right) \right]$ to obtain the discriminant function as a vector z: $z_{JK} = W^{-1} \delta_{JK}$. This vector consists of a series of weights, w_i, for characters 1, 2, ..., m, it is symbolized as $z_1, z_2, \ldots, z_m$.

Step 5: Calculate a discriminant score DS (here the discriminant score scale for taxa J and K). Multiply these weights by the observed character values of the unknown individual, and sum them to give a discriminant score

$$DS: DS_u = z_1 X_{1u} + z_2 X_{2u} + \ldots + z_m X_{mu}.$$

Step 6: Use this discriminant score to calculate three reference scores for the centroid of J, the centroid of K, and the point midway between them: DS_J, DS_K, and $DS_{0.5}$ respectively for discriminating between members of J and K. They are given by:

$$DS_J = \overline{X}_J z', DS_K = \overline{X}_K z' \text{ and } DS_{0.5} = \frac{1}{2}\left(\overline{X}_J + \overline{X}_K\right) z' = \frac{1}{2}(DS_J + DS_K).$$

Step 7: Calculate the length of the line between the centroids of J and K and allocate the unknown to taxon. First, obtain the square root of Mahalanobis' D^2, which is the length of the line between the centroids of J and K measured in discriminant function units (the absolute difference between the scores DS_J and DS_K is also equal to D^2); then allocate the unknown to taxon J if the observed score for an unknown, DS_u, lies on the DS_J side, otherwise to taxon K if on the D DS_K side. A different discriminant function, and discriminant scores, is calculated for each pair of taxa. The square root of Mahalanobis' D^2 (Mahalanobis 1936) is used to measure the length of the line between the centroids of two taxa J and K in discriminant function units, which equals to the absolute difference between the discriminant scores of these two taxa DS_J and DS_K.

Step 8: Interpret discriminant function vectors. The vector angle between two discriminant function vectors can be interpreted as measuring contrasts of form of the taxa, where the small angle measures similar contrasts of form and large angles represent distinct contrasts.

Step 9: Perform statistical hypothesis testing of the centroids using the distance scores. The distance DS_{JK}^2 between the scores DS_J and DS_K (Sneath and Sokal 1973, pp. 403–404) has two important uses. One is that we can perform a null statistical hypothesis through a F-test to test whether the centroids are significantly different with a testing ratio:

$$\frac{D_{JK}^2 (t_J t_K)(t_J + t_K - m - 1)}{(t_J + t_K)(t_J + t_K - 2)m}, \text{ with } m \text{ and } (t_J + t_K - m - 1) \text{ degrees of freedom.}$$

The test ratio is related to Hotellings' T^2 (Rao 1952):

$$T^2 = \frac{t_J t_K}{t_J + t_K} D_{JK}^2. \tag{7.4}$$

Step 10: Perform variable (character) selection. We also can use the distance scores to determine each character contributing to D^2. If the characters with little discriminatory power can be identified, then drop them from the analysis. Alternatively, we can choose the best few characters from the set by the F test. If we do not consider correlations between characters, the percent contribution of character i is $100 \times (z_i \delta_i / D^2)$, where z_i and δ_i are the ith elements of vectors z_{JK} and δ_{JK}.

Step 11: Calculate the probability of misclassification. A primary purpose of a discriminant function is to minimize the probability of wrong assignment of unknown individuals. We can calculate the probability of misclassification of assignment of unknown individuals by assuming a multivariate normal distribution and also that the unknown does belong to J or K (and not to some distant cluster).

7.4.3.6.2   Multiple Discriminant Analysis (MDA)

The original dichotomous discriminant function has two limitations (Sneath and Sokal 1973, p. 404): (1) The algorithm used in the original dichotomous discriminant function implies that the unknown either belongs to one or the other of the two taxa being considered. Thus, when an unknown taxon is located far off in the space, belongs to a quite different cluster, and hence it really belongs to a third cluster. However, using the dichotomous discriminant function, this unknown taxon may have almost any discriminant score, and thus is assigned to one or the other of the two taxa under consideration. (2) It cannot test against a large number of discriminant functions when there are many taxa.

MDA in Mahalanobis' D-space (Mahalanobis 1936) was developed to largely overcome the above two problems of the original dichotomous DFA. MDA is a multivariate dimensionality reduction technique. It is not directly used to perform classification, instead merely supports classification by compressing the signal down to a lower-dimensional space amenable to classification. MDA typically projects the multivariate signal down to a dimensional space with less one dimension of the number of categories. Mahalanobis (1936) generalized calculation of D^2 between the centroids from the discriminant scores to between any pair of points so that the square root of D^2 become simply the Euclidean distance in the D-space. The Mahalanobis' D^2 between any pair of points f and g is given:

$$D_{fg}^2 = \delta_{fg}' W^{-1} \delta_{fg}, \tag{7.5}$$

MDA method transforms the original space into a new space, where the original axes are stretched and also skewed, and results in not any more right angles between axes.

The length of a unit in dimension equals to the original units times by the square root of the element of the matrix w^{-1} (Sneath and Sokal 1973, p. 405). The direction cosine between the dimensions has an interpretation of correlation coefficients (Gower 1967b).

7.4.3.6.3 Canonical Variate Analysis (CVA)

CVA is a widely used method for analyzing group structure in multivariate data. It is mathematically equivalent to a one-way MANOVA (multivariate analysis of variance) and often called canonical discriminant analysis. The reason that we can perform analyses on distances between taxa or individuals, in particular by principal coordinate analysis is that the square root of D (the Mahalanobis' D^2), the difference between discriminant scores of two taxa can be represented in an orthogonal system of axes (though the orientation is arbitrary) (Gower 1966, 1967b) (also see (Sneath and Sokal 1973, p. 406)).

CVA and MDA are equivalent except in minor particulars. We can obtain the coordinates of the points in an orthogonal system such as by principal coordinate analysis of D, and the orthogonal axes are the canonical variates, which can be used also for discrimination to readily identify unknowns by seeing whether they fall within critical distances of taxon centroids.

Remarks on DFA and Mahalanobis' D^2
DFA and the Mahalanobis' D^2 have the advantages: (1) they are less sensitive to general size factors than taxonomic distance (Sneath and Sokal 1973, p. 406), although an unknown of the same shape as a member of a taxon may appear outside that taxon if it differs much in size. (2) Discriminant functions have most value for very close clusters that partly overlap (Sneath and Sokal 1973, pp. 406–407).

However, DFA also has the disadvantages (Sneath and Sokal 1973, p. 406):

(1) The pooled within-groups variance-covariance matrix W cannot be inverted, unless "generalized inverses" are used if any character is invariant in each of the taxa.
(2) It is difficult to choose the limited set of the best characters from the large number, which should be employed in numerical taxonomy. Characters with means that are well separated in relation to the variances and that are not highly correlated with other characters are in general the best, but optimal methods of choosing them pose statistical and computational problems (Feldman et al. 1969).
(3) It has been shown by Dunn and Varady (1966) that rather large numbers of individuals are required in each taxon for reliable discriminant functions.
(4) The gain in discriminatory power over simpler methods may not be great (Sokal 1965) particularly with 0, 1 characters (Gilbert 1968; Kurczynski 1970) and simple discriminants based on equal weight for each character can be quite effective (e.g., Kim et al. 1966).

7.5   Statistical Hypothesis Testing of OTUs

Statistical hypothesis testing of OTUs is challenging in performing both statistical inference of similarity coefficients and clustering.

7.5.1   Hypothesis Testing on Similarity Coefficients of OTUs

Numerical taxonomy concerns more on general significance of the similarity matrix among all the OTUs or the joint significance of the entire matrix. In numerical taxonomy, it is not important to conduct statistical hypothesis testing significance of each individual similarity coefficient. There are two reasons behind this (Sokal 1963, pp. 153–154). We briefly describe below.

7.5.1.1   Heterogeneity of the Column Vectors of the Data Matrix

In the OTU Data Matrix, the similarity between OTUs j and k are measured over a set of n characters. The problem is that these OTUs are not taken as random samples from a common population, but instead are really taken from a heterogeneous sample wherein each variate estimates a different character. In other words, the character states for any one OTU are not samples that estimate a single mean and variance. Thus, because the column vectors of the data matrix are heterogeneous, it is difficult to assume the sampling parametric distributions for justifying statistical test; and hence it is inappropriate to make ordinary tests of significance (Sokal and Sneath 1963, p. 313). Since a parametric model cannot be easily specified, in ecology and microbiome study, a permutation test is often used to perform a hypothesis testing.

7.5.1.2   Hypothesis Testing of Individual Similarity Coefficients

An appropriate parametric statistical test in multivariate statistics requires the variables (in this case, characters) are independent so that there are p variables (characters) in the data matrix, and then the data matrix has p dimension with each of the p variables representing one dimension of the data. The problem is that in OTU data matrix, the characters (rows of the data matrix) are correlated with each other, and thus a similarity coefficient based on p characters would not reflect p independent dimensions of variation (Sneath and Sokal 1973, p. 163). Thus, standardization of characters is often recommended in numerical taxonomy to alleviate the heterogeneity of column vectors. However, standardization of every character state to have a population mean of zero and variance of unity only mitigate some degree of heterogeneity of column vectors. Actually, put another way, statistical hypothesis tests in numerical taxonomy only have partially succeeded for entire classifications (similarity matrices), but not for individual coefficients (Sneath and Sokal 1973, pp. 163–164). In numerical taxonomy, scaling continuous characters (variables) (lengths, percentages, ratios) is recommended by either standardization or ranging scaling methods. Because of scaling characters, correlation or taxonomic distance matrices between OTUs hence are altered. Two states of characters should be coded as binary, while ordered discrete multistate characters (qualitative traits and counts) can be analyzed as continuous characters (Sokal 1986).

These two problems actually reveal the challenges of directly conducting the hypothesis testing of individual similarity coefficients in numerical taxonomy, and ecology as well as in later microbiome study. Performing hypothesis testing pairwise correlation or partial correlation also poses a challenge. In a similar reason, beta diversity measures such as Bray-Curtis, Jaccard, and Sørensen dissimilarities are usually used as input data for ordination and association analysis rather than directly being tested. This links up the concepts and topics in multivariate statistics, with which we will discuss in Chapters 10 and 11.

7.5.2 Hypothesis Testing on Clustering OTUs

One critical step in numerical taxonomy is to assemble the OTUs into groups of higher rank using the similarity coefficients. In other words, the taxonomic groups are formatted by using cluster analysis (or multiple factor analysis, which is somewhat similar to cluster analysis) using the similarity coefficients as criteria of rank (Sneath and Sokal 1962; Sokal and Sneath 1963). As we described above, it is because the OTUs (or taxonomic groups) are formed using cluster analysis, we call the OTU methods as clustering-based OTU methods. However, it is also difficult to perform hypothesis testing on clustering OTUs. The cluster analytic procedures of numerical taxonomy have been frequently criticized due to lack of a sound basis in statistical inference (Cormack 1971; Sokal 1986).

Several different approaches of significance tests for clusters have been comprehensively reviewed and discussed in literature (Dubes and Jain 1979; Bock 1981, 1985, 1996, 2002, 2012; Perruchet 1983).

Generally it is not recommended to globally apply a cluster test to a large or high-dimensional data set; however, performing a "local" cluster test (e.g., the maximum F test with 2 classes) to a specified part of the data is often useful for providing evidence for or against a prospective clustering tendency (Bock 1985).

Most cluster analyses test the null hypothesis that there is no structure at all such as test for hierarchical structure versus random data (Rohlf and Fisher 1968). However, this test of clustering is less important in biological classification because it is most likely that the organisms representing a taxonomic group must have taxonomic structure (Sokal 1986). The advances of clustering methods from classical models to the approaches developed until 2002 was reviewed by Bock (2002).

   To address the problem of assessing the validity of clusters produced by classical clustering methods, some advances were made in testing for significant clusters, including permutation test for significance (Good 1982; Park et al. 2009) and Monte Carlo test for assessing the value of a U-statistic based on the sets of pairwise dissimilarities to identify genuine clusters in a classification (Gordon 1994).

   However, probabilistic clustering models for noisy and outlying data (Bock 2002, 2012) may remedy some degrees of limitation of classical clustering methods. Probabilistic clustering models were developed for testing for homogeneity versus clustering; that is, to test the discriminating (distinguishing) of homogeneity and heterogeneity in multidimensional observations. Typically statistical methods in probabilistic models discriminate whether the observations are sampled from "homogeneous" population (the null hypothesis) or from "heterogeneous" (or "clustering") population (the alternative hypothesis). Bock (2002) reviewed and discussed two model-based probabilistic clustering models which lead to a new "robust" clustering method: (1) modeling noisy data in clustering (Banfield and Raftery 1993); (2) modeling outliers and robust clustering (Gallegos 2001, 2002, 2003).

Remarks on Hypothesis Testing on Similarity Coefficients and Clustering of OTUs

Summary statistics are commonly used in various fields of research. Like in numerical taxonomy, currently in microbiome study, the techniques of summary statistics, such as alpha and beta diversity measures, ordination, correlation, and partial correlation analyses, and clustering are all used to summarize data and measure taxonomic structure.

   However, clustering in numerical taxonomy is based on phenetic resemblances (affinity) only without considering phyletic connotations (Sokal 1963, p. 52). The question still remains to be asked: (1) how it is appropriate to adopt similarity coefficients and clustering of OTUs into microbiome data? (2) how the statistical hypothesis testing of similarity coefficients and clustering of OTUs is clinically relevant? We feel that overall test of similarity or dissimilarity structure using clustering methods does not provide much meaningful clinical insights in current microbiome study. Further functional studies in experimental models and validation using different cohorts are needed.

7.6   Some Characteristics of Clustering-Based OTU Methods

Describing the main characteristics of numerical taxonomy and especially major characteristics of clustering-based OTU methods will not only facilitate understanding the field of numerical taxonomy per se, but also help understanding the advantages and limitations of the concept of OTUs and the clustering-based OTU methods that were adopted in microbiome study. In this section, we summarize some basic questions and the main characteristics of numerical taxonomy and especially of clustering-based OTU methods.

7.6.1 Some Basic Questions in Numerical Taxonomy

Numerical taxonomy has provided its unique resolutions to some basic questions in taxonomy or classification in systematics. We describe the main ones below.

Regarding the Rates of Evolution

Sokal and Sneath definitely stated that no absolute evolutionary rates exist, and all groups are ranked based on investigating and evaluating its characters and phenetic resemblances and thus evolutionary rates are related to the higher ranking groups and to neighboring taxa (Sokal 1963, p. 239). For them, the affinity values provided by the techniques of numerical taxonomy can be used to measure the evolutionary rate (Sokal 1963, p. 241).

Regarding the Formation of Species

Whether phenetic similarity is concordant with taxonomic species, and whether there exist the correlation (correspondence) between genetic relationships and phenetic groupings are debatable.

Some researchers thought that a very close correspondence between them cannot be found, while others thought that they could find reasonable agreement between phenetic relations and the ease of hybridization. Sokal and Sneath emphasized that there exists a sharp distinction between phenetic grouping and genetic groupings (Sokal and Sneath 1963, p. 243). They thought that phenetic, genetic, and geographical groupings cannot always be equated each other. Thus, two distinct genetic groups would be within a single phenetic group. However, they believed that phonetically distinct groups identified by numerical taxonomy will usually be found to be genetically distinct.

Regarding the Pattern of Branching of Taxonomic Dendrograms

If a pattern of taxonomic dendrograms occurs at random, then the number of side branches per stem in a given interval is distributed with a Poisson distribution with the number of OTUs being the total number of side branches plus one. Obviously, a Poisson distribution is not appropriate to describe the pattern of branching of taxonomic dendrograms and the phenetic distributions. In taxonomic dendrograms, given a Poisson distribution, the variance is much greater than the mean number of branches per phenon, and the observed number with no branches (monotypic phenons) is greater than the expected [see (Sokal and Sneath 1963, pp. 245–246; Sneath and Sokal 1973, p. 307)]. Sokal and Sneath (1963, p. 246) called this kind of over-dispersed and zero-inflated distribution as a clustered distribution and thought that it has affected on the construction of hierarchies. In their 1973 book (Sneath and Sokal 1973, p. 308), they further described three properties of phenetic distributions: (1) there exist some dense clusters of lower taxa; (2) there exist a great number of isolated monotypic taxa; and (3) some constancy exists in the organized nature or in the constructed rank levels which is indicated by the consistency of the slope of the logarithm of the number of included taxa.

7.6.2   Characteristics of Numerical Taxonomy

In this subsection, we describe some characteristics of numerical taxonomy.

7.6.2.1   Summary of Characteristics of Numerical Taxonomy

Numerical taxonomy has several characteristics; we can summarize and discuss below.

1. **Construction of classifications does not necessarily require phylogeny.**

   In numerical taxonomy classifications are called nature or "general" classifications, and the constructed taxa have "natural" properties or are "natural" taxa/ "natural taxa." For Sneath and Sokal (1962), "natural taxonomic groups" are "polytypic" (Beckner 1959) or better called as "polythetic" (meaning "many-arrangement"), which is against "monothetic" (meaning "one-arrangement").

   Sokal and Sneath tried to avoid confusing phylogeny with classfication (Huxley 1869) and emphasized that taxonomic classifications are distinctive from phylogenetic classifications. Basically numerical taxonomy does not attempt to force phylogenetic criteria on to taxonomic groupings. Sokal and Sneath separated the taxonomic process (which is to be based on affinity or resemblance) from phylogenetic speculation (Sneath and Sokal 1962). They emphasized that taxonomic relationships are "static" (Michener 1957) or preferably are called "phenetic" (Sneath and Sokal 1962; Cain and Harrison 1960), which are to be evaluated purely based on the resemblance among the materials. In numerical taxonomy it is logically fallacious to make phylogenetics deductions to arriving at taxonomies. Sokal and Sneath descibed a fallacy of circular reasoning (Sokal and Sneath 1963, p. 21):

   "Since we have only an infinitesimal portion of phylogenetic theory in the fossil record, it is almost impossible to establish nature taxa on a phylogenetic basis. Conversely, it is unsound to derive a definitive phylogeny from a tentative natural classification."

   To avoid this fallacy of circular reasoning, Sokal and Sneath used the operational homology of Adansonian taxonomy which involves fewer assumptions to classify characters. In summary, biological taxonomy (which is in logic unanalyzed or inductive) is not Aristotelian logic (which is in logic analyzed or deductive).

2. **Natural taxonomic classifications should be constructed based on similarity coefficients using many characters.**

In numerical taxonomy, taxa are based on correlations between features (Sneath and Sokal 1962), and taxonomic rank is defined by similarity (Sneath 1957a). The phenetic resemblance between the taxonomic entities is estimated using the similarity and dissimilarity coefficients (Sneath and Sokal 1962).

Three convenient types of coefficients (association, correlation, and distance) were suggested in numerical taxonomy as suitable measures of similarity between pairs of taxa (Sokal 1961). Specifically association coefficient is utilized to measure "present-absent" (binary) characters, which was first described by Sneath (Sneath 1957a). Binary data have attractive properties of simplicity and the potential to decompose variation among OTUs into fundamental units of information. The simple matching coefficient (including its one-complement coefficient, the average Manhattan distance coefficient) and the Jaccard coefficient were recommended for use in numerical taxonomy among the many binary association coefficients that most have been used in ecology (Sokal and Sneath 1963; Sokal 1986).

The ordinary Pearson product-moment correlation coefficient is still preferred to use as measures of similarity between OTUs because it is less sensitive to size differences, although it is non-metrics and non-invariant to the direction (Sokal 1986). In contrast, cosine coefficient is metrics and invariant to the direction of correlation coefficients. For continuous or ordered multistate characters (variables), both taxonomic distances and Manhattan distances can be used to calculate similarity and dissimilarity. Distance coefficients determine the similarities between two taxa as a function of their distance in a multiple dimensional space with the characters as coordinates. They are employed to measure continuous or more than two states (ordered multistate) of the characters (variables) (Sokal 1958a; Morishima and Oka 1960). Between taxonomic distances and Manhattan distances that can calculate similarity and dissimilarity, Sokal (1986) preferred taxonomic distances because of their geometric and heuristic properties. Based on Sokal (1961), the use of distance to measure taxonomic similarity can be located earliest by Heincke (1898) and others (Anderson and Abbe 1934; Pearson 1926; Rao 1952; Mahalanobis 1936). Both correlation and cosine coefficients are sensitive to data points close to the centroid. However, classifications in numerical taxonomy are preferred to be based on similarity using correlation (or cosine) coefficients than by using distance-based measures because the former usually provide better effects of classifications to exhibit size and shape differences in the data (Sokal and Rohlf 1980; Sokal 1986).

Additionally, Sokal and Snealth emphasized that natural taxonomic classifications need to be randomly chosen and should utilize large numbers of characters. Sokal (1986) provided an excellent summary of their thoughts on similarity coefficients. For the details, the interested reader is referred to this review (Sokal 1986).

3. **Characters should be equally weighted.**

In constructing general classifications, characters should have equal weight. Numerical taxonomy emphasizes equal weighting and treats all taxonomic characters as of equal importance through using large characters in computing the taxonomic similarity (Sneath and Sokal 1962). Equal weighting allows the ready mathematical treatment of characters, and the use of the estimates of resemblance (rather than key characters) to create taxa, i.e., to form the "natural" taxonomic groups.

4. Taxonomy approach is quantitative and numerical.

Taxonomic groups are numerically established by cluster analysis of values of similarity. To exploit the taxonomic structure, numerical taxonomy uses different clustering methods to group OTUs into clusters for achieving the taxonomic structure. Three commonly used clustering methods in numerical taxonomy are single linkage clustering, complete linkage clustering, and average linkage clustering methods. Among these three clustering methods, the average-linkage clustering method was promoted; and particularly the unweighted pair-group method using arithmetic averages (UPGMA) was also preferred for use because the average clustering method can present a compromise solution to form a cluster (Sokal 1986).

It is also because single-linkage and complete-linkage clustering methods represent the two extreme criteria and both could not give taxonomically satisfactory results (Hartigan 1985), although single linkage clustering has several elegant mathematical properties (Sokal 1986). Thus they both are used in special cases only (Sokal 1963; Sneath and Sokal 1973).

Graphically clustering methods are used for representing the similarity matrix among taxonomic units through dendrograms (a tree-like arrangement). In dendrograms taxonomic entities are arranged in a hierarchic non-overlapping manner by means of sequential, agglomerated algorithms (Sokal 1986). Sneath and Sokal (1962) called the established groups simply as "phenons" to differentiate it from the terms in the evolutionary, nomenclatural connotations such as genera or families.

5. The concept of operational taxonomic units (OTUs) is purely "operational," and thus it is not biologically motivated.

OTUs were introduced as a pragmatic definition to classify groups of organisms, allowing for taxonomy being classified quantitatively or numerically. Thus, the concept of OTUs and the field of numerical taxonomy were established against the subjective nature of classification in classical Linnaean taxonomy. Through creating the concept of OTUs, the term "taxon" can still be retained for its proper function role played in taxonomic groupings. Thus, this indicates that OTUs are not true taxa instead are operated as taxa. OTUs are used in constructing classifications without considering any phylogenetic information. Actually by using the OTUs, the classification was considered purely as a study of similarity, of "static" relationships in which phylogeny played no part (Vernon 1988). Thus, when constructing phylogenetic classifications, the static relationships and phylogeny should be separately assessed.

7.6.2.2   Remarks on Characteristics of Numerical Taxonomy

The approach of phonetic numerical taxonomy (or phenetics) and commonly clustering-based OTU methods have several challenges.

First, phonetic numerical taxonomy (Gilmour naturalness) assumes that high similarity (in fact largely clustered) will also lead to high predictivity. This has been identified as an unfinished problem (Farris 1979; Sokal 1985) because the linkage between similarity and predictivity has not been defined clearly. Actually, on one hand, it is challenging to show how to maximize similarity; on the other hand, the relationship between classification and predictivity has not yet been established although a close relation has obviously been observed between classifications in which objects within any given taxon resemble each other closely and predictivity of characters within that taxon (Sokal 1985). Predictivity is the most important criterion of a classification in terms of Gilmour naturalness. Thus, although it is difficult to achieve, generally high predictivity is the goal of a natural classification (Sokal 1985).

For example, phenetic clustering is commonly used in specifically and biologically estimating phylogeny or phylogeny reconstruction, and is also used to analyze relationships among potentially conspecific populations, as well as geographic variation and genetic continuity, and consequently, of species limits (Highton et al. 1989). Both above applications of phenetic clustering have been criticized having serious limitations because of their assumptions are questionable (de Queiroz and Good 1997). When phenetic clustering is used for estimating phylogeny, it assumes that similarity is correspondent to recency of common ancestry, while when it is used for analyzing patterns of geographic variation and genetic continuity among populations, it assumes that similarity is correspondent to degree of genetic continuity.

The fundamental weakness of numerical taxonomy and hence its clustering-based OTU methods is its questionable assumptions. Because similarity relationships do not strictly conform to the pattern of a nested hierarchy, and thus attempts to base nested hierarchical taxonomies on similarity will never be fully satisfactory (de Queiroz and Good 1997).

Second, the classifications in biological taxonomy have the property that the resulting classification should be hierarchical and nonoverlapping; and hence requires the resulting classification represents faithfully as possible the actual similarities among all pairs of OTUs resemblance matrix, and the measures (e.g., the cophenetic correlation coefficient) can express the goodness of a classification (Sokal and Rohlf 1962; Sokal 1985). However, it was shown that different clustering methods have different goodness of fits by cophenetic correlation coefficients to the same resemblance matrix (Sokal 1985). Both artificial data (Sneath 1966) and real data (Sokal 1985) have shown average-linkage clustering outperformed complete and single linkage clustering methods to measure the resemblances.

Overall, in one hand, importantly phenetic clustering OTUs expects to result in nested hierarchical structure. However, similarity that does not exhibit a strictly nested hierarchical structure has important consequences for its analysis using phenetic clustering (de Queiroz and Good 1997). On the other hand, even random distributions of OTUs will result in the formation of clusters [see(Sneath and Sokal 1973, p. 252; Dunn and Everitt 1982, p. 94; de Queiroz and Good 1997)].

Third, whether clustering methods or ordination analysis methods better represent taxonomic structure is still arguable. The phenetic criteria of goodness of classification are restricted to hierarchic Linnaean classifications. However, non-Linnaean modes of data analysis would provide better fits to the original resemblance matrix than dendrograms generated by hierarchic cluster analysis methods; thus, ordinations could summarize the taxonomic structure, and hence various ordination analysis methods have been employed in phenetic taxonomy; on the other hand, the hierarchic clustering analysis models have been used as standard methods for classification in the narrow sense (Sokal 1985). Given robust methods for similarity coefficients, Sokal (1985) considered that cluster analysis may ultimately result in obtaining stable and repeatable phenetic classification although it is not possible to achieve complete stability.

7.6.2.3   Remarks on Impact of Numerical Taxonomy on Microbiome Research

The philosophy, OTU concept, framework of statistical analysis of phonetic numerical taxonomy, and especially commonly clustering-based OTU methods have been adopted into microbiome research. All these have provided a solid foundation in the development of microbiome research, especially in the theoretical and conceptual frameworks. The philosophy and methodology of numerical taxonomic procedure have posed large impacts on early microbiome study. However, the limitations and challenges of phonetic numerical taxonomy both as a method for clustering OTUs, and as a statistical analysis method for assessing taxonomies and microbiomes still remain, and even more severe when using OTU methods to define species.

Numerical taxonomy (numerical classification) is a numerical-based empiricism. It has a direct impact on microbiome study in several perspectives. We summarize the main perspectives below.

First, OTU concept and its clustering-based OTU method. This may be one of the major impacts of numerical taxonomy on microbiome study. This has been especially reflected in the early development of bioinformatic software such as in the open source software package **mothur** for bioinformatics data processing.

Second, similarity and distance measurements. Various approaches of grouping OTUs into different levels of taxonomy in early bioinformatic analysis of microbiome sequencing data are just using the algorithms that are related to similarity or distance measurements developed in numerical taxonomy.

Third, ordination and discriminate analysis. Using ordination analysis to represent taxonomic structure and discriminant analysis to classify and identify OTUs (or taxa) provide meaningful insights into current microbiome study. For example, in Sneath and Sokal 1973 book, they made distinction between clustering and ordination, and discussed the advantages of ordination over clustering. They also made distinction between classification and identification, and described several discriminant analysis methods in the role of identifying new taxa. These contributions should be *acknowledged.*

Forth, correlation between clusters of OTUs and taxa. Overall correlated clusters of OTUs empirically observed by numerical taxonomic methodology may suggest a causal explanation or a causally associated with some mechanisms.

However, both philosophy and methodology of numerical taxonomy have their own fundamental weaknesses. When the philosophy and methodology of numerical taxonomy, and especially the clustering-based OTU methods, are adopted into microbiome study to define species, these weaknesses have not disappeared but somehow have been highlighted.

First, Currently the Use of Species in Microbiome Study Is Confusing
It confuses the phenetic species with the biological species (or used by microbiologists). Sokal and Crovello (1970) criticized the biological species concept because the concept of biological species is based on modern evolutionary theory using the population as its basic unit, and it is considerably difficult to define a species. They emphasized that the biological criteria for species should be dependent on phenetic considerations. In their 1973 book (Sneath and Sokal 1973, pp. 362–367), Sneath and Sokal discussed various species concepts and definitions and especially described three criteria that are used to define three major species concepts: genetic, phenetic, and nomenclatural. They are respectively referred to (1) genospecies (a group of organisms exchanging genes), (2) taxospecies (a phenetic cluster), and (3) nomenspecies (organisms bearing the same binomen) in Ravin (1963).

Second, OTU Methods Cannot Be Used to Define a Biological Species
Sneath and Sokal particularly discussed the biological species concept that was foremost advocated and defined by Ernst Mayr (1963), in which a biological species is defined as comprising "groups of actually or potentially interbreeding populations which are reproductively isolated from other such groups." However, they had no intention of using their OTU methods to define a biological species; instead they (Sneath and Sokal 1973, p. 364) explicitly stated that the OUT method defines the phenetic species but not the biological species and that the biological species concept is in practice nonoperational (Sokal and Crovello 1970) and is misleading because the species described by conventional phenetic criteria does not have the genetic properties ascribed to the biological species (Blackwelder 1962; Sokal 1962).

The methodology and especially the clustering based-OTU methods in numerical taxonomy have played an important role in defining OTUs in bioinformatic analysis of microbiome. In the meantime, we want to emphasize that both the philosophy and methodology of numerical taxonomic procedure have their limitations. When the clustering based-OTU methods are used in microbiome study to define species, tasks have been actually assigned: not only for construction of taxonomy but also implicitly requiring for a biological or functional role that the defined species should have. This is beyond the original means described in numerical taxonomy.

Numerical taxonomy along with its clustering based-OTU methods was a reform movement that arose in response to a dissatisfaction with the subjective and nonquantitative nature of taxonomy as it existed in the late 1950s (Vernon 1988). However, the applications of phenetic clustering methods within the context of taxonomy and in the biological sciences have never been entirely satisfactory (de Queiroz and Good 1997). The application of clustering based-OTU methods to bioinformatic classification of OTUs in the field of microbiome still neither is entirely satisfactory.

Actually, a reform movement in response to a dissatisfaction using clustering based-OTU methods to define taxonomy that has already started in the bioinformatic analysis of microbiome sequences. We will comprehensively review the bioinformatic analysis methods and the movement beyond OTUs in Chap. 8.

Numerical taxonomy in its original setting mainly is a purely descriptive science; its potential to be used for formulating hypotheses has not been fully developed in numerical taxonomy. The potential for formulating hypotheses is developed by microbiome researchers in their differential abundance analysis and association analysis. We will cover statistical analysis methods and models in Chaps. 9, 10, 11, 12, 13, 14, 15, 16, 17 and 18.

7.7 Summary

In this chapter we described the original OTU methods in numerical taxonomy.

First, the brief history of numerical taxonomy was introduced. Then, the principles of numerical taxonomy were described, including definitions of characters and taxa, sample size calculation, and equally weighting characters and taxonomic rank. Followed briefly introduced phenetic taxonomy, the philosophy of numerical taxonomy, and methodology, including the numerical-based empiricism, the numerical-based empiricist natural classification, and biological classifications. We focused on how to construct taxonomic structure, which includes defining the operational taxonomic units (OTUs), estimating taxonomic resemblance, commonly clustering-based OTU methods, factor analysis, ordination methods, discriminant analysis, and canonical variate analysis. Particularly, the advantages and disadvantages of single linkage, complete linkage, and average linkage clustering were described and discussed. Next, the problems of statistical hypothesis testing of OTUs using similarity coefficients and clustering methods were described and discussed. Finally, some characteristics of clustering-based OTU methods and their applications to microbiome research were investigated and commended. In Chap. 8, we will introduce and describe the movement of moving beyond the concept of OTUs and OTU methods in bioinformatic analysis of microbiome data and generally in microbiome study.

References

Albritton, Rogers. 1957. II. Forms of particular substances in Aristotle's metaphysics. *The Journal of Philosophy* 54 (22): 699–708.

Anderson, Edgar, and Ernst C. Abbe. 1934. A quantitative comparison of specific and generic differences in the Betulaceae. *Journal of the Arnold Arboretum* 15 (1): 43–49.

Aristotle. 2012. *The organon: The works of Aristotle on logic.* (Trans E.M. Edghill, A.J. Jenkinson, G.R.G. Mure and W.A. Pickard-Cambridge; Ed. Roger Bishop Jones). CreateSpace Independent Publishing Platform.

———. 2016. *Metaphysics* (Trans. Charles David Chanel Reeve). Indianapolis: Hackett Publishing Company, Inc.

Banfield, Jeffrey D., and Adrian E. Raftery. 1993. Model-based Gaussian and non-Gaussian clustering. *Biometrics* 49: 803–821.

Beckner, Morton. 1959. *The biological way of thought.* Columbia University Press.

Bigelow, R.S. 1961. Higher categories and phylogeny. *Systematic Zoology* 10 (2): 86–91.

Blackwelder, Richard E. 1962. Animal taxonomy and the new systematics. *Survey of Biological Progress* 4 (31–35): 53–57.

Bock, H.H. 1981. Statistical testing and evaluation methods in cluster analysis. In *Proceedings on the Golden Jubilee Conference in Statistics: Applications and New Directions, December 1981, Calcutta, Indian Statistical Institute*, 1984, 116–146

———. 1985. On some significance tests in cluster analysis. *Journal of Classification* 2 (1): 77–108. https://doi.org/10.1007/BF01908065.

Bock, Hans H. 1996. Probabilistic models in cluster analysis. *Computational Statistics & Data Analysis* 23 (1): 5–28. https://doi.org/10.1016/0167-9473(96)88919-5, https://www.sciencedirect.com/science/article/pii/0167947396889195.

Bock, Hans-Hermann. 2002. Clustering methods: From classical models to new approaches. *Statistics in Transition* 5 (5): 725–758.

Bock, H.H. 2012. Probabilistic aspects in cluster analysis. In *Conceptual and numerical analysis of data: Proceedings of the 13th conference of the gesellschaft für klassifikation eV, University of Augsburg*, April 10–12, 1989.

Cain, Arthur James. 1958. Logic and memory in Linnaeus's system of taxonomy. *Proceedings of the Linnean Society of London* 169: 144–163.

———. 1959. Deductive and inductive methods in post-Linnaean taxonomy. *Proceedings of the Linnean Society of London* 170: 185–217.

Cain, Arthur J., and G.A. Harrison. 1958. An analysis of the taxonomist's judgment of affinity. *Proceedings of the Zoological Society of London* 131: 85–98.

Cain, A.J., and G.A. Harrison. 1960. Phyletic weighting. *Proceedings of the Zoological Society of London* 135 (1): 1–31. https://doi.org/10.1111/j.1469-7998.1960.tb05828.x, https://zslpublications.onlinelibrary.wiley.com/doi/abs/10.1111/j.1469-7998.1960.tb05828.x.

Camin, Joseph H., and Robert R. Sokal. 1965. A method for deducing branching sequences in phylogeny. *Evolution* 19: 311–326.

Cattell, Raymond. 1944. A note on correlation clusters and cluster search methods. *Psychometrika* 9 (3): 169–184.

Cattell, Raymond B. 1952. *Factor analysis: An introduction and manual for the psychologist and social scientist.* Oxford: Harper.

Cattell, R.B. 1966. The data box: Its ordering of total resources in terms of possible relational systems. In *Handbook of multivariate experimental Psycholvgy*, ed. R.B. Cattell, 67–128. Chicago: Rand McNally. 959 pp.

Clark, Philip J. 1952. An extension of the coefficient of divergence for use with multiple characters. *Copeia* 1952 (2): 61–64.

Clifford, Harold Trevor, and William Stephenson. 1975. *Introduction to numerical classification.* Academic.

Cormack, R.M. 1971. A review of classification. *Journal of the Royal Statistical Society. Series A (General)* 134: 321–367.

Cowgill, George L. 1968. Archaeological applications of factor, cluster, and proximity analysis. *American Antiquity* 33 (3): 367–375.

Craytor, William Bert, and Le Roy Johnson Jr. 1968. *Refinements in computerized item seriation.* University of Oregon Museum of Natural History.

Czekanowski, J. 1909. Zur Differentialdiagnose der Neandertalgruppe. Korrespondenz-Blatt deutsch. *Gesellschaft für Anthropologie. Ethnologie und Urgeschichte* 40: 44–47.

Daget, Jacques, and J.C. Hureau. 1968. Utilisation des statistiques d'ordre en taxonomie numérique. *Bulletin du Muséum d'Histoire Naturelle, París* 40 (3): 465–473.

Dagnelie, P. 1966. A propos des différentes méthodes de classification numérique. *Revue de statistique appliquée* 14 (3): 55–75.

Darwin, Charles. 1859. *On the origin of species by means of natural selection, or the preservation of favoured races in the struggle for life.* A New Edition, Revised and Augmented by the Author, Ed. London: John Murray.

———. 1872. *The origin of species by means of natural selection, or the preservation of favoured races in the struggle for life.* 6th ed. London: John Murray.

———. 1886. *The origin of species.* 6th ed. London: John Murray.

de Queiroz, Kevin, and David A. Good. 1997. Phenetic clustering in biology: A critique. *The Quarterly Review of Biology* 72 (1): 3–30.

Doty, P., J. Marmur, J. Eigner, and C. Schildkraut. 1960. Strand separation and spcific recombination in deoxyribonucleic acids: Physical chemical studies. *Proceedings of the National Academy of Sciences of the United States of America* 46 (4): 461–476. https://doi.org/10.1073/pnas.46.4.461, https://pubmed.ncbi.nlm.nih.gov/16590628, https://www.ncbi.nlm.nih.gov/pmc/articles/PMC222859/.

Dubes, Richard, and Anil K. Jain. 1979. Validity studies in clustering methodologies. *Pattern Recognition* 11 (4): 235–254.

Dunn, G., and B.S. Everitt. 1982. *An introduction to mathematical taxonomy.* Cambridge: Cambridge University Press.

Dunn, Olive Jean, and Paul D. Varady. 1966. Probabilities of correct classification in discriminant analysis. *Biometrics* 22: 908–924.

Eades, David C. 1965. The inappropriateness of the correlation coefficient as a measure of taxonomic resemblance. *Systematic Zoology* 14 (2): 98–100.

Estabrook, George F. 1967. An information theory model for character analysis. *Taxon* 16: 86–97.

Farris, James S. 1979. The information content of the phylogenetic system. *Systematic Zoology* 28 (4): 483–519. https://doi.org/10.2307/2412562, http://www.jstor.org/stable/2412562.

Feldman, Sydney, Donald F. Klein, and Gilbert Honigfeld. 1969. A comparison of successive screening and discriminant function techniques in medical taxonomy. *Biometrics* 25: 725–734.

Fisher, R.A. 1936. The use of multiple measurements in taxonomic problems. *Annals of Eugenics* 7 (2): 179–188. https://onlinelibrary.wiley.com/doi/abs/10.1111/j.1469-1809.1936.tb02137.x.

Florek, Kazimierz, Jan Łukaszewicz, Julian Perkal, Hugo Steinhaus, and Stefan Zubrzycki. 1951a. Sur la liaison et la division des points d'un ensemble fini. *Colloquium Mathematicum* 2: 282–285.

———. 1951b. Taksonomia wrocławska. *Przegląd Antropologiczny* 17: 193–211.

Gallegos, M.T. 2001. *Robust clustering under general normal assumptions.* Technical report MIP-0103 September 2001.

Gallegos, María Teresa. 2002. Maximum likelihood clustering with outliers. In *Classification, clustering, and data analysis*, 247–255. Springer.

Gallegos, M.T. 2003. Clustering in the presence of outliers. In *Exploratory data analysis in empirical research*, 58–66. Springer.

Gilbert, Ethel S. 1968. On discrimination using qualitative variables. *Journal of the American Statistical Association* 63 (324): 1399–1412.

Gilmour, John S.L. 1937. A taxonomic problem. *Nature* 139 (3529): 1040–1042.

Gilmour, S.L. 1940. Taxonomy and philosophy. In *The new systematics*, 404–424. Oxford.

Gilmour, J.S.L. 1951. The development of taxonomic theory since 1851. *Nature* 168 (4271): 400–402. https://doi.org/10.1038/168400a0.

Good, I.J. 1982. C129. An index of separateness of clusters and a permutation test for its statistical significance. *Journal of Statistical Computation and Simulation* 15 (1): 81–84. https://doi.org/10.1080/00949658208810568.

Goodall, David W. 1964. A probabilistic similarity index. *Nature* 203 (4949): 1098–1098.

———. 1966a. A new similarity index based on probability. *Biometrics* 22: 882–907.

Goodall, D.W. 1966b. Numerical taxonomy of bacteria—Some published data re-examined. *Microbiology* 42 (1): 25–37.

Goodfellow, M., D. Jones, and F.G. Priest. 1985. Delineation and description of microbial populations using numerical methods. In *Computer-assisted bacterial systematics*, ed. M. Goodfellow, F.G. Priest, and D. Jones . London/Orlando: Published for the Society for General Microbiology by Academic Press.Microbiology Society for General. Accessed from https://nla.gov.au/nla.cat-vn18193. Special publications of the Society for General Microbiology; 15

Gordon, A.D. 1994. Identifying genuine clusters in a classification. *Computational Statistics & Data Analysis* 18 (5): 561–581. https://doi.org/10.1016/0167-9473(94)90085-X, https://www.sciencedirect.com/science/article/pii/016794739490085X.

Gower, John C. 1966. Some distance properties of latent root and vector methods used in multivariate analysis. *Biometrika* 53 (3–4): 325–338.

———. 1967a. A comparison of some methods of cluster analysis. *Biometrics* 23: 623–637.

———. 1967b. Multivariate analysis and multidimensional geometry. *Journal of the Royal Statistical Society: Series D (The Statistician)* 17 (1): 13–28.

———. 1971. A general coefficient of similarity and some of its properties. *Biometrics* 27: 857–871.

Hahsler, Michael, Kurt Hornik, and Christian Buchta. 2008. Getting things in order: An introduction to the R package seriation. *Journal of Statistical Software* 25 (3): 1–34.

Hamann, Ulrich. 1961. Merkmalsbestand und verwandtschaftsbeziehungen der farinosae: ein beitrag zum system der monokotyledonen. *Willdenowia* 2: 639–768.

Harman, Harry H. 1976. *Modern factor analysis*. University of Chicago Press.

Hartigan, John A. 1985. Statistical theory in clustering. *Journal of Classification* 2 (1): 63–76.

Heincke, F.R. 1898. *Naturgeschichte des Herings. Abhandl. d. Deutsch.* Berlin: Seefischerei-Vereins, II.

Highton, Richard, George C. Maha, and Linda R. Maxson. 1989. *Biochemical evolution in the slimy salamanders of the Plethodon glutinosus complex in the Eastern United States/57.* Urbana: University of Illinois Press.

Hodson, F.R. 1970. Cluster analysis and archaeology: Some new developments and applications. *World Archaeology* 1 (3): 299–320. https://doi.org/10.1080/00438243.1970.9979449.

Holzinger, K.J., and H.H. Harman. 1941. *Factor analysis*. Chicago: University of Chicago Press.

Hull, D.L. 1988. *Science as a process: An evolutionary account of the social and conceptual development of science*. Chicago: University Chicago Press.

Huxley, Thomas Henry. 1869. *An introduction to the classification of animals*. John. Churchill & Sons.

Jaccard, Paul. 1908. Nouvelles recherches sur la distribution florale. *Bulletin de la Societe Vaudoise des Sciences Naturelles* 44: 223–270.

Johnson, Stephen C. 1967. Hierarchical clustering schemes. *Psychometrika* 32 (3): 241–254.

Johnson, Lawrence A.S. 1970. Rainbow's end: The quest for an optimal taxonomy. *Systematic Zoology* 19 (3): 203–239.

Johnson, LeRoy, Jr. 1968. *Item seriation as an aid for elementary scale and cluster analysis*. Eugene: Museum of Natural History, University of Oregon.

Kendall, David G. 1963. A statistical approach to flinders petries sequence-dating. *Bulletin of the International Statistical Institute* 40 (2): 657–681.

Kendall, David. 1969. Incidence matrices, interval graphs and seriation in archeology. *Pacific Journal of Mathematics* 28 (3): 565–570.

Kendall, David G. 1971. Seriation from abundance matrices. In *Mathematics in the archaeological and historical sciences*, vol. 215, 52. Edinburgh: Edinburgh University Press.

Kendall, D.G. 1988. Seriation. In *Encyclopedia of statistical sciences*, ed. S. Kotz and N.L. Johnson. New York: Wiley.

Kim, Ke Chung, Byron W. Brown Jr, and Edwin F. Cook. 1966. A quantitative taxonomic study of the Hoplopleura hesperomydis complex (Anoplura, Hoplopleuridae), with notes on a. posteriori taxonomic characters. *Systematic Zoology* 15 (1): 24–45.

Kulczynski, S. 1927. Zespoly roslin w Pieninach. (Die Pflanzen-associationen der Pieninen). *Bulletin International Academy Polish Science Letter Cl Science Mathematis and National Series B Suppl II* 57: 203.

Kurczynski, T.W. 1970. Generalized distance and discrete variables. *Biometrics* 26: 525–534.

Lance, Godfrey N., and William T. Williams. 1966. Computer programs for hierarchical polythetic classification ("similarity analyses"). *The Computer Journal* 9 (1): 60–64.

Lance, Godfrey N., and William Thomas Williams. 1967. A general theory of classificatory sorting strategies: 1. Hierarchical systems. *The Computer Journal* 9 (4): 373–380.

Legendre, Pierre. 2019. Numerical Ecology. In *Encyclopedia of ecology*, ed. B.D. Fath, 487–493. Oxford: Elsevier.

Legendre, L., and P. Legendre. 1983. *Numerical ecology*. First English edition. Amsterdam: Elsevier.

Legendre, P., and L. Legendre. 1998. *1998. Numerical ecology."* Second English edition. Amsterdam: Elsevier.

Legendre, Pierre, and Louis Legendre. 2012. *Numerical ecology*. Elsevier.

Mahalanobis, Prasanta Chandra. 1936. On the generalized distance in statistics. *Proceedings of the National Institute of Science of India* 2: 49–55.

Mayr, Ernst. 1963. Animal species and evolution. In *Animal species and evolution*, 797. Cambridge, MA: Harvard University Press.

———. 1982. *The growth of biological thought: Diversity, evolution, and inheritance*. Harvard University Press.

———. 1999. *Systematics and the origin of species, from the viewpoint of a zoologist*. Harvard University Press.

McBride, Paul D., Len N. Gillman, and Shane D. Wright. 2009. Current debates on the origin of species. *Journal of Biological Education* 43 (3): 104–107.

McCarthy, Brian J., and Ellis T. Bolton. 1963. An approach to the measurement of genetic relatedness among organisms. *Proceedings of the National Academy of Sciences of the United States of America* 50 (1): 156.

Michener, Charles D. 1957. Some bases for higher categories in classification. *Systematic Zoology* 6 (4): 160–173.

Michener, Charles D., and Robert R. Sokal. 1957. A quantitative approach to a problem in classification. *Evolution* 11 (2): 130–162.

Mill, John Stuart. 1886. *A system of-logic, ratiocinative and inductive, being a connected view of the principles of evidence and the method of scientific investigation. 1843*. London: Longmans, Green. Reprint: London: Longmans, Green 19: 30.

Milligan, Glenn W. 1996. Clustering validation: Results and implications for applied analyses. In *Clustering and classification*, 341–375. World Scientific.

Minkoff, Eli C. 1965. The effects on classification of slight alterations in numerical technique. *Systematic Zoology* 14 (3): 196–213.

Morishima, Hiroko, and Hiko-Ichi Oka. 1960. The pattern of interspecific variation in the genus Oryza: Its quantitative representation by statistical methods. *Evolution* 14 (2): 153–165. https://doi.org/10.2307/2405822, http://www.jstor.org/stable/2405822.

Orloci, L. 1969. Information theory models for hierarchic and non-hierarchic classifications. In *Numerical taxonomy*, 148–164. London: Academic.

Park, P.J., J. Manjourides, M. Bonetti, and M. Pagano. 2009. A permutation test for determining significance of clusters with applications to spatial and gene expression data. *Computational Statistics & Data Analysis* 53 (12): 4290–4300. https://doi.org/10.1016/j.csda.2009.05.031.

Pearson, Karl. 1926. On the coefficient of racial likeness. *Biometrika* 18 (1/2): 105–117.

Perruchet, Christophe. 1983. Significance tests for clusters: Overview and comments. In *Numerical taxonomy*, 199–208. Berlin/Heidelberg: Springer.

Petrie, William Matthew Flinders. 1899. *Sequences in prehistoric remains*. Vol. 331. Bobbs-Merrill.

Rao, C. Radhakrishna. 1948. The utilization of multiple measurements in problems of biological classification. *Journal of the Royal Statistical Society. Series B (Methodological)* 10 (2): 159–203.

———. 1952. *Advanced statistical methods in biometric research*. New York: Wiley.

Ravin, Arnold W. 1963. Experimental approaches to the study of bacterial phylogeny. *The American Naturalist* 97 (896): 307–318.

Rogers, David J., and Taffee T. Tanimoto. 1960. A computer program for classifying plants. *Science* 132 (3434): 1115–1118.

Rohlf, F. James. 1962. *A numerical taxonomic study of the genus Aedes (Diptera: Culicidae) with emphasis on the congruence of larval and adult classifications*. University of Kansas, Entomology.

———. 1965. Coefficients of correlation and distance in numerical taxonomy. *University of Kansas Science Bulletin* 5: 109–126.

Rohlf, F. James, and David R. Fisher. 1968. Tests for hierarchical structure in random data sets. *Systematic Biology* 17 (4): 407–412.

Rohlf, F. James, and Robert R. Sokal. 1962. The description of taxonomic relationships by factor analysis. *Systematic Zoology* 11 (1): 1–16.

Rohlf, F.J., and R.R. Sokal. 1965. Coefficients of correlation and distance in numerical taxonomy. *University of Kansas Science Bulletin* 45: 3–27.

Shepherd, M.J., and A.J. Willmott. 1968. Cluster analysis on the Atlas computer. *The Computer Journal* 11 (1): 57–62.

Silvestri, L., M. Turri, L.R. Hill, and E. Gilardi. 1962. A quantitative approach to the systematics of actinomycetes based on overall similarity. In *Symposium of the Society for General Microbiology*. Cambridge: Cambridge University Press.

Simpson, George Gaylord. 1944. *Tempo and mode in evolution*. Columbia University Press.

———. 1945. *The principles of classification and a classification of mammals*. Vol. 85. New York: Bulletin of the American Museum of Natural History.

———. 1961. *Principles of animal taxonomy*. Columbia University Press.

Sneath, Peter H.A. 1957a. The application of computers to taxonomy. *Microbiology* 17 (1): 201–226.

———. 1957b. Some thoughts on bacterial classification. *Microbiology* 17 (1): 184–200.

Sneath, P.H.A. 1966. A comparison of different clustering methods as applied to randomly-spaced points. *Classification Society Bulletin* 1 (2): 2.

———. 1967. Some statistical problems in numerical taxonomy. *Journal of the Royal Statistical Society. Series D (The Statistician)* 17 (1): 1–12. https://doi.org/10.2307/2987198, http://www.jstor.org/stable/2987198.

———. 1969. Evaluation of clustering methods. In *Numerical taxonomy. Proceedings of the colloquium in numerical taxonomy held in the University of St. Andrews, September 1968*, ed. Alfred John Cole, 257–271. Academic: London, 324 pp.

———. 1995. Thirty years of numerical taxonomy. *Systematic Biology* 44 (3): 281–298. https://doi.org/10.2307/2413593, http://www.jstor.org/stable/2413593.

Sneath, Peter H.A., and P.H.S. Sneath. 1962. *The construction of taxonomic groups*. Cambridge University Press.

Sneath, Peter H.A., and Robert R. Sokal. 1962. Numerical taxonomy. *Nature* 193 (4818): 855–860.

————. 1973. *Numerical taxonomy. The principles and practice of numerical classification.* San Francisco: W. H. Freeman and Company.

Sokal, Robert R. 1958a. A statistical method for evaluating systematic relationships. *University of Kansas Scientific Bulletin* 38: 1409–1438.

Sokal, R.R. 1958b. Quantification of systematic relationships and of phylogenetic trends. *Proceedings of the 10th International Congress of Entomology* 1: 409–415.

————. 1960. Die Grundlagen der numerischen Taxonomie. *Proceedings of the 11th International Congress of Entomology* 1: 7–12.

Sokal, Robert R. 1961. Distance as a measure of taxonomic similarity. *Systematic Zoology* 10 (2): 70–79.

————. 1962. Typology and empiricism in taxonomy. *Journal of Theoretical Biology* 3 (2): 230–267.

————. 1963. The principles and practice of numerical taxonomy. *Taxon* 12 (5): 190–199. https:// doi.org/10.2307/1217562, http://www.jstor.org/stable/1217562.

————. 1965. Statistical methods in systematics. *Biological Reviews* 40 (3): 337–389.

————. 1985. The priciples of numerical taxonomy: Twnty-five years later. In *Computer-assisted bacterial systematics. Priest special publications of the Society for General Microbiology 15,* ed. M. Goodfellow and F.G. Priest , 1–20. London/Orlando: Published for the Society for General Microbiology by Academic Press.D. Jones and Microbiology Society for General

————. 1986. Phenetic taxonomy: Theory and methods. *Annual Review of Ecology and Systematics* 17 (1): 423–442.

Sokal, Robert R., and Theodore J. Crovello. 1970. The biological species concept: A critical evaluation. *The American Naturalist* 104 (936): 127–153.

Sokal, Robert R., and F. James Rohlf. 1962. The comparison of dendrograms by objective methods. *Taxon* 11 (2): 33–40. https://doi.org/10.2307/1217208, https://onlinelibrary.wiley.com/doi/abs/10.2307/1217208.

Sokal, R.R., and C.D. Michener. 1958. A statistical method for evaluating systematic relationships. *University of Kansas Scientific Bulletin* 38: 1409–1438. https://books.google.com/books?id=o1 BlHAAACAAJ.

Sokal, R.R., and P.H.A. Sneath. 1963. *Principles of numerical taxonomy.* San Francisco: WH Freeman.

Sokal, Robert R., and F. James Rohlf. 1970. The intelligent ignoramus, an experiment in numerical taxonomy. *Taxon* 19 (3): 305–319.

————. 1980. An experiment in taxonomic judgment. *Systematic Botany* 5: 341–365.

Sørensen, Th.A. 1948. A method of establishing groups of equal amplitude in plant sociology based on similarity of species content and its application to analyses of the vegetation on Danish commons. *Biologiske Skrifter* 5: 1–34.

Spearman, Charles. 1913. Correlations of sums or differences. *British Journal of Psychology* 5 (4): 417.

Stephenson, William. 1936. The inverted factor technique. *British Journal of Psychology* 26 (4): 344–361.

————. 1952. Some observations on Q technique. *Psychological Bulletin* 49 (5): 483.

————. 1953. *The study of behavior; Q-technique and its methodology.* University of Chicago Press.

Tryon, R.C. 1934. *Cluster analysis.* Ann Arbor: Edwards Brothers.

Vernon, Keith. 1988. The founding of numerical taxonomy. *The British Journal for the History of Science* 21 (2): 143–159.

Whewell, W. 1840. *The philosophy of the inductive sciences: Founded upon their history.* London: J.W. Parker.

Williams, William T., and Michael B. Dale. 1966. Fundamental problems in numerical taxonomy. In *Advances in botanical research,* 35–68. Elsevier.

Xia, Yinglin. 2020. Correlation and association analyses in microbiome study integrating multiomics in health and disease. *Progress in Molecular Biology and Translational Science* 171: 309–491.

Xia, Yinglin, and Jun Sun. 2022. *Statistical data analysis of microbiomes and metabolomics*. Vol. 13. American Chemical Society.

Xia, Yinglin, Jun Sun, and Ding-Geng Chen. 2018. *Statistical analysis of microbiome data with R*. Vol. 847. Springer.

Yule, G. Udny, and M. Kendall. 1950. *An introduction to the theory of statistics*. New York: Hafner.

Chapter 8
Moving Beyond OTU Methods

Abstract This chapter investigates the movement of moving beyond OTU methods and discusses the necessity and possibility of this movement. First, it describes clustering-based OTU methods and the purposes of using OTUs and definitions of species and species-level analysis in microbiome studies. Then, it introduces the OTU-based methods that move toward single-nucleotide resolution. Third, it describes moving beyond the OTU methods. Finally, it discusses the necessity and possibility of moving beyond OTU methods as well as the issues of sub-OTU methods, assumption of sequence similarity predicting the ecological similarity, and functional analysis and multi-omics integration.

Keywords Clustering-based OTU methods · Hierarchical clustering OTU methods · Heuristic clustering OTU methods · Taxonomy · OTUs · Sequencing error · Species and species-level analysis · Eukaryote species · Prokaryote or bacterial species · 16S rRNA method · Physiological characteristics · Single-nucleotide resolution-based OTU methods · Distribution-based clustering (DBC) · Swarm2 · Entropy-based methods · Oligotyping · Denoising-based methods · Pyrosequencing flowgrams · Cluster-free filtering (CFF) · DADA2 · UNOISE2 · UNOISE3 · Deblur · SeekDeep · Sub-OTU methods · Sequence similarity · Ecological similarity · Functional analysis · Multi-omics integration

In Chap. 7 we introduced clustering-based OTU methods in numerical taxonomy. In this chapter we will discuss the movement of moving beyond OTU methods and investigate the necessity and possibility of this movement. We divide the presentation into four sections: We first describe clustering-based OTU methods and the purposes of using OTUs and definitions of species and species-level analysis in microbiome studies (Sect. 8.1). We then introduce the OTU-based methods that move toward single-nucleotide resolution (Sect. 8.2). Section 8.3 describes moving beyond the OTU methods and discusses the limitations of OTU methods. Section 8.4 describes the movement of moving beyond OTU methods and discusses the necessity and possibility of moving beyond OTU methods. Finally, we summarize the contents of this chapter (Sect. 8.5).

8.1 Clustering-Based OTU Methods in Microbiome Study

In microbiome studies, there exist two widely used approaches of sequence similarity: clustering-based OTU methods and phylotype-based (phylotyping) methods. Both approaches use similarity measure and group sequences into small units, which are rooted in clustering methods in the original numerical taxonomy.

8.1.1 Common Clustering-Based OTU Methods

Clustering-based OTU methods typically first perform a hierarchical cluster analysis and then assign the sequence reads into OTUs based on a distance threshold, i.e., cluster their similarity to other sequences in the community (Schloss and Westcott 2011; Huse et al. 2010; Schloss and Handelsman 2005; Schloss et al. 2009; Sun et al. 2009).

In contrast, phylotype-based methods classify sequences into taxonomic bins (i.e., phylotypes) based on their similarity to reference sequences (Schloss and Westcott 2011; Huse et al. 2008; Liu et al. 2008; Stackebrandt and Goebel 1994). For example, taxonomy-supervised analysis (Sul et al. 2011) belongs to the phylotyping methods, in which sequences are allocated into taxonomy-supervised "taxonomy bins" to provide 16S rRNA gene classification based on the existing bacterial taxonomy from ribosomal RNA databases, such as Ribosomal Database Project (RDP) (Cole et al. 2007), Greengenes (DeSantis et al. 2006a), and SILVA (Pruesse et al. 2007). Because the processing of taxonomy bins is rooted in polyphasic taxonomy (Colwell 1970), the taxonomy-supervised analysis is believed reflecting physiological, morphological, and genetic information (Sul et al. 2011). SortMeRNA (Kopylova et al. 2012) suited for closed-reference OTU clustering also belongs to the phylotyping methods that query sequences are searched against a reference database and then alignments are evaluated based on optimal regions of similarity between two sequences.

For the benefits and challenges of using phylotype-based methods, the interested reader is referred to the paper by Schloss and Westcott (2011) and references cited therein for more detailed description. For how to process taxonomic binning and overview of existing web tools for taxonomic assignment and phylotyping of metagenome sequence samples, the interested reader is referred to Dröge and McHardy (2012). For a comprehensive evaluation of the performance of open-source clustering software (namely, OTUCLUST, Swarm, SUMACLUST, and SortMeRNA) against UCLUST and USEARCH in QIIME, hierarchical clustering methods in mothur, and USEARCH's most recent clustering algorithm, UPARSE, the interested reader is referred to Kopylova et al. (2016) and references cited therein. One example for comparison of various methods used to cluster 16S rRNA gene sequences into OTUs is from Jackson et al. (2016).

Within the framework of clustering-based OTU methods, various nonhierarchical clustering algorithms have been developed to improve the resolution of assigning 16S rRNA gene sequences to OTUs, including **heuristic clustering**, **network-based clustering**, and **model-based clustering**.

Although hierarchical clustering and heuristic clustering methods have been criticized due to either the computational complexity (hierarchical clustering) (Barriuso et al. 2011) or considerable loss in clustering quality (greedy clustering) (Sun et al. 2012; Chen et al. 2013a; Bonder et al. 2012), these two clustering methods are still used to assign OTUs in microbiome literature. In the following, we will introduce hierarchical clustering OTU methods (Sect. 8.1.2) and heuristic clustering OTU methods (Sect. 8.1.3), respectively.

8.1.2 Hierarchical Clustering OTU Methods

Single-linkage (also called nearest-neighbor) clustering, complete-linkage (also called furthest-neighbor) clustering, and average-linkage (also known as average-neighbor or UPGMA: unweighted pair-group method by using arithmetic averages) clustering are the three commonly used hierarchical clustering algorithms (Sokal and Rohlf 1995; Legendre and Legendre 1998). As we reviewed in Chap. 7, hierarchical clustering-based OTU methods are rooted in numerical taxonomy.

Single-linkage algorithm defines the distance between two clusters using the minimum distance between two sequences in each cluster. Complete-linkage algorithm defines the distance between two clusters using the maximum distance between two sequences in each cluster. Average-linkage algorithm defines the distance between two clusters using the average distance between each sequence in one cluster to every sequence in the other cluster. For more detailed description of these three clustering methods, the reader is referred to Chap. 7. Theses hierarchical clustering methods were widely adopted in the early development of microbiome software. For example, **DOTUR** (Distance-Based OTU and Richness) (2005) (Schloss and Handelsman 2005) is a clustering-based OTU method as the name suggested. It assigns sequences to OTUs by using either the furthest (i.e., complete-linkage), average (i.e., unweighted pair-group method by using arithmetic averages), or nearest (i.e., single-linkage) neighbor algorithm for each distance level (Schloss and Handelsman 2005). All these hierarchical clustering algorithms are commonly used in various disciplines (Sokal and Rohlf 1995; Legendre and Legendre 2012).

RDP (the Ribosomal Database Project) (2009) (Cole et al. 2009) is the taxonomy assignment alignment and analysis tools for quality-controlled bacterial and archaeal small-subunit rRNA research. RDP provides taxonomy-independent alignment, in which the trimmed reads are first aligned and then reads are clustered into OTUs at multiple pairwise distances using the complete-linkage clustering algorithm (Cole et al. 2009). In early release RDP has included a pairwise distance matrix suitable for the DOTUR molecular macroecology package (Cole et al. 2007). RDP pipeline was reviewed having better performance by using quality scores and performing a structural alignment but is only accurate at high sequence differences (Quince et al. 2009).

In 2009, in order to perform hypothesis tests and describe and compare communities using OTU-based methods, **mothur** (Schloss et al. 2009) integrated other software features into its implementation, including pyrosequencing pipeline (RDP) (Cole et al. 2009); NAST, SINA, and RDP aligners (Cole et al. 2009; DeSantis et al. 2006a, b; Pruesse et al. 2007); DNADIST (Felsenstein 1989); DOTUR (Schloss and Handelsman 2005); CD-HIT (Li and Godzik 2006); SONS (Schloss and Handelsman 2006b); LIBSHUFF (Schloss et al. 2004; Singleton et al. 2001); TreeClimber (Maddison and Slatkin 1991; Martin 2002; Schloss and Handelsman 2006a); and UniFrac (Lozupone and Knight 2005).

Although mothur includes many functions for performing hypothesis tests and describing and comparing communities, on the basis, mothur is the improved version of DOTUR: it uses the OTU-based methods of DOTUR (Schloss and Handelsman 2005) and CD-HIT (Li and Godzik 2006) for assigning sequences to OTUs, via implementing three de novo clustering algorithms, i.e., clustering sequences using the furthest, nearest, or UPGMA algorithms from DOTUR or using a nearest-neighbor-based approach from CD-HIT. Thus, mothur is one of representative hierarchical clustering methods for picking OTUs.

In 2009, **ESPRIT** (Sun et al. 2009) was proposed to use the k-mer (substrings of length k) distance to quickly compute pairwise distances of reads and classify sequence reads into OTUs at different dissimilarity levels via the complete-linkage algorithm. ESPRIT uses massively parallel pyrosequencing technology, i.e., via large collections of 16S rRNA pyrosequences for estimating species richness.

In 2010, **SLP** (a single-linkage pre-clustering) (Huse et al. 2010) was proposed to overcome the effect of sequencing errors and decrease the inflation of OTUs because the complete-linkage algorithm is sensitive to sequencing artifacts (Huse et al. 2010). SLP uses an average-linkage clustering to more accurately predict expected OTU richness in environmental samples.

In 2013, a modified version of **mcClust** (Cole et al. 2014; Fish et al. 2013) was proposed to allow the complete-linkage clustering to compute cluster and incorporate algorithmic changes for lowering the time complexity and speeding up clustering. **HPC-CLUST** (Matias Rodrigues and von Mering 2013) was proposed for large sets of nucleotide sequences via a distributed implementation of the complete- and average-linkage hierarchical clustering algorithms with high optimization.

In 2015, **oclust** (Franzén et al. 2015) was developed to improve OTU quality and decrease variance using long-read 16S rRNA gene amplicon sequencing by implementing the complete-linkage clustering.

8.1.3 Heuristic Clustering OTU Methods

In the early microbiome research, the commonly used hierarchical clustering methods have been adopted from numerical taxonomy (Sokal and Sneath 1963) and numerical ecology (Legendre and Legendre 1998). However, hierarchical clustering algorithms have the high computational complexity in time and space. Thus, to trade off between computational efficiency and accuracy, various heuristic methods have been proposed to reduce the computational complexity in sequence comparison by using greedy clustering strategy (e.g., **CD-HIT** (Huang et al. 2010; Li and Godzik 2006; Fu et al. 2012) and UCLUST (Edgar 2010)) instead of hierarchical clustering.

Heuristic clustering methods process input sequences one by one rather than compute pairwise distances of all sequences. Typically a greedy incremental clustering strategy and a seed are used to initiate its clustering (Chen et al. 2016). Because heuristic clustering methods just compare each sequence with the seed sequences, they are able to process massive sequence datasets. Thus, the heuristic clustering methods are considered as more attractive than clustering methods for picking OTUs in 16S rRNA datasets (Cai and Sun 2011; Wei et al. 2021) and have been applied by the human microbiome project (HMP) (Peterson et al. 2009).

The two best-known heuristic methods for improving the resolution of clustering OTUs are **CD-HIT** (Li and Godzik 2006; Huang et al. 2010; Fu et al. 2012) and **USEARCH** (Edgar 2010). In 2010, **QIIME** (pronounced "chime": "quantitative insights into microbial ecology"), the software tool for downstream analyses, included the functions for choosing OTUs (Caporaso et al. 2010). Especially, QIIME has used the heuristic recursive algorithm UCLUST (Edgar 2010) as the default clustering method since UCLUST's publication (corresponding to QIIME version 1.0.0) to cluster sequences into OTUs; i.e., it first identifies the centroid of the OTU based on its abundance and then includes all sequences that have sequence identity above a prefixed threshold with the centroid into this OTU. We summarize the heuristic clustering OUT methods in Table 8.1.

8.1.4 Limitations of Clustering-Based OTU Methods

The early methodological developments of numerical taxonomy provided a rich source for modern ecology and microbiome research. For example, the concepts of OTU, similarity, and using similarity to define taxonomic rank and using distance as a measure of taxonomic similarity have been adopted in current ecology and microbiome research.

However, OTU methods in microbiome and numerical taxonomy have different assumptions, purposes, and approaches. Numerical taxonomy assumes that characters can be grouped based on their similarities. It employs commonly used clustering to perform classification of characters. Bioinformatic analysis of microbiome assumes that sequence similarity can predict taxonomic similarity. It has dual purpose: grouping sequences and estimating diversity. Its final goal is to detect the association between microbiome health and the development of the disease. Thus, microbiome research is not restricted to commonly used clustering and heuristic clustering.

Table 8.1 Heuristic clustering methods for constructing OTUs

Heuristic clustering	References
CD-HIT	Huang et al. (2010) and Fu et al. (2012)
Aims to improve better accuracy, scalability, flexibility, and speed for clustering and comparing biological sequences	
Sorts the sequence before clustering by the length of sequences	
USEARCH	Edgar (2010)
Is an algorithm for sequence database search that seeks high-scoring global alignments	
Aims to implement greedy clustering and typically achieves good sensitivity for identity	
Sorts the sequence before clustering by sequence abundance	
UCLUST	
Is a heuristic clustering algorithm that employs USEARCH to assign sequences to clusters	
Aims to take the advantages of UBLAST and USEARCH algorithms enabling sensitive local and global search of large sequence databases at exceptionally high speeds to achieve high throughput	
Both QIIME and UPARSE are based on USEARCH	
GramClust	Russell et al. (2010)
Aims to use the inherent grammar distance metric of each pairwise sequence to cluster a set of sequences	
Determines partitioning for a set of biological sequences with higher statistical accuracy than both CD-HIT-EST and UCLUST	
ESPRIT-Tree	Cai and Sun (2011)
Is a new version of ESPRIT with focusing on the quasilinear time and space complexity	
Aims to avoid using the seed sequences to represent clusters like the most existing clustering methods	
Initially constructs a pseudometric-based partition tree for a coarse representation of the entire sequences	
Then iteratively finds the closest pairs of sequences or clusters and merges them into a new cluster	
Is able to analyze very large 16S rRNA pyrosequences in quasilinear computational time with the effectiveness and accuracy comparable to other greedy heuristic clustering algorithms	
DNACLUST	DNAclust, Ghodsi et al. (2011)
Aims specifically to cluster highly similar DNA sequences for phylogenetic marker genes (e.g., 16S rRNA)	
Uses a greedy clustering strategy but via a novel sequence alignment and k-mer-based filtering algorithms to accelerate the clustering	
DySC (Dynamic Seed-based Clustering)	Zheng et al. (2012)
Aims to use a dynamic seeding strategy for greedy clustering 16S rRNA reads	

First uses the traditional greedy incremental clustering strategy to form the pending clusters Then converts the pending cluster into a fixed cluster when it reaches a threshold size and reselects a new fixed seed The new fixed seed is defined as the sequence that maximizes the sum of k-mers shared between the fixed read and other reads in one cluster	Namiki et al. (2013)
LST-HIT Is a DNA sequence clustering method based on CD-HIT Aims to speed up the computation due to the intrinsic difficulty in parallelization Uses a novel filtering technique based on the longest common subsequence (LST) to remove dissimilar sequence pairs before performing pairwise sequence alignment	
DBC454 Is a taxonomy-independent (i.e., unsupervised clustering) method for fungal ITS1 (internal transcribed spacer 1) sequences of 454 reads using a density-based hierarchical clustering procedure Aims to focus on reproducibility and robustness	Pagni et al. (2013)
MSClust Is a multiseed-based heuristic clustering method Aims to use an adaptive strategy to generate multiseeds for one cluster Either assigns one query sequence to one cluster if the average distance between the sequence and seeds is smaller than the user-defined threshold Or marks the sequence as unassigned	Chen et al. (2013a)
UPARSE Is an improved version of USEARCH with adding the chimera detection for seed sequences Aims to highly accurately sequence microbial amplicon reads and to reduce amplification artifacts Uses a greedy algorithm to perform chimera filtering and OTU clustering simultaneously through quality filtering, trimming reads, optionally discarding singleton reads, and then clustering the remaining reads Uses the similar algorithms that used in AmpliconNoise (Quince et al. 2011), i.e., inferring errors in a sequence using parsimony The generated OTU's sequences were reported with ≤1% incorrect bases; such accurate OTUs are more closer to the species in a community	Edgar (2013)
LSH Is a greedy clustering algorithm that uses the locality-sensitive hashing (LSH-based similarity function) to cluster similar sequences and make individual groups (OTUs)	Rasheed et al. (2013)

(continued)

Table 8.1 (continued)

Heuristic clustering	References
Aims to accelerate the pairwise sequence comparisons by using LSH algorithm and to improve the quality of sequence comparisons via incorporating a matching criterion The assigned OTUs can be utilized for computation of different species diversity/richness metrics (e.g., species richness estimation)	Mercier et al. (2013)
SUMACLUST Is a de novo clustering method Aims to perform exact sequence alignment, comparison, and clustering rather than semiglobal alignments implemented in CD-HIT and USEARCH Incrementally constructs the clusters via comparing an abundance-ordered list of input sequences with the representative already-chosen sequences based on a greedy clustering strategy SUMACLUST has been integrated into QIIME 1.9.0	
Swarm Is a de novo clustering method based on an unsupervised single-linkage clustering method Aims to address two fundamental problems suffered by greedy clustering methods: arbitrary global clustering thresholds and input-order dependency induced by centroid selection First uses a local threshold clustering to generate an initial set of OTUs (nearly identical amplicons) by iteratively agglomerating similar sequences Then uses internal structure of clusters and amplicon abundances to refine the clustering results into sub-OTUs Swarm has been integrated into QIIME 1.9.0	Swarm (Mahé et al. 2014, 2015b)
OTUCLUST Is a greedy clustering method (like SUMACLUST) Is a de novo clustering algorithm specifically designed for the inference of OTUs Aims to perform exact sequence alignment Compares an abundance-ordered list of input sequences with the representative already-chosen sequences Performs sequence de-duplication and chimera removal via UCHIME (Edgar et al. 2011) Is one component of software pipeline MICCA for the processing of amplicon metagenomic datasets that combines quality filtering, clustering of OTUs, taxonomy assignment, and phylogenetic tree inference	Albanese et al. (2015)
VSEARCH Is an alternative to the USEARCH tool based on a heuristic algorithm for searching nucleotide sequences	VSEARCH (Rognes et al. 2016)

Description	Reference
Aims to perform optimal global sequence alignment of the query using full dynamic programming Performs searching, clustering, chimera detection (reference-based or de novo) and subsampling, paired-end reads merging, and dereplication (full length or prefix)	Westcott and Schloss (2017)
OptiClust Is an OTU assignment algorithm that iteratively reassigns sequences to new OTUs to maximize the Matthews correlation coefficient (MCC) Aims to improve the quality of the OTU assignments Requires a distance matrix for constructing OTUs because it is a distance-based algorithm	
ESPRIT-Forest Is an improved method of ESPRIT and ESPRIT-Tree Aims to cluster massive sequence data in a sub-quadratic time and space complexity while inheriting the same pipeline of ESPRIT and ESPRIT-Tree for preprocessing, hierarchical clustering, and statistical analysis First organizes sequences into a pseudometric-based partitioning tree for sublinear time searching of nearest neighbors Then uses a new multiple-pair merging criterion to construct clusters in parallel using multiple threads	Cai et al. (2017)
DBH Is a de Bruijn (DB) graph-based heuristic clustering method Aims to reduce the sensitivity of seeds to sequencing errors by just selecting one sequence as the seed for each cluster First forms temporary clusters using the traditional greedy clustering approach Then builds a DB graph for this cluster and generates a new seed to represent this cluster when the size of a temporary cluster reaches the predefined minimum sequence number	Wei and Zhang (2017)
Fuzzy Aims to improve the clustering quality via the clustering based on fuzzy sets Takes into account uncertainty when producing OTUs using the fuzzy OTU-picking algorithm	Bazin et al. (2019)
DMSC Is a dynamic multiseed clustering (DMSC) method for generating OTUs First uses greedy incremental strategy to generate a series of clusters based on the distance threshold When the sequence number in a cluster reaches the predefined minimum size, then the multicore sequence (MCS) is selected as the seeds of the cluster The average distance to MCS and the distance standard deviation in MCS will determine whether a new sequence is added to the cluster The MCS will be dynamically updated until no sequence is merged into the cluster	Wei and Zhang (2019)

Theoretically any methods that can be used to achieve the goals are appropriate. The newly developed methods such as denoising, refining OTU clustering, model-based methods, and filtering-free methods are all targeting the final goal of bioinformatics in microbiome research.

However, the main motivation of moving beyond the clustering-based OTU methods in microbiome research is that current application of OTU methods to microbiome data has several limitations. Here we describe and discuss the clustering-based OTU methods and especially their limitations that have motivated microbiome research to move beyond them.

8.1.4.1 Assumptions of OTU Methods in Numerical Taxonomy and Microbiome

OTU methods used in numerical taxonomy and microbiome have different aims. Adopting OTU methods into microbiome studies is somewhat inappropriate and thus is controversial.

Although at the beginning the concept of OTU is convenient for bioinformatic analysis of microbiome sequencing data, the use of the OTU methods in microbiome literature deviates from the considerations of the original authors. As reviewed in Chap. 7, the original authors of numerical taxonomy strictly separated taxonomic classifications from phylogenetic classifications and treated the latter as "phylogenetic speculation" and the former as "taxonomic procedure." The original authors considered OTUs as operational not functional; the functionality remained to taxa; cluster analysis of similarity is merely on datasets at hand and requires that the characters are from larger and randomly chosen samples; taxonomic similarity does not represent the phylogenetic similarity. In contrast, by assigning sequences to a particular OTU, actually it assumes that the similarities among sequences are closely related to the phylogeny, and therefore the similarities of sequences can most likely be used to derive phylogenetic divergence or ecologically rank the similarity of taxa. However, the assumption that sequence similarity of 16S hypervariable regions is a good proxy for phylogenetic and therefore ecological similarity is problematic (Prosser et al. 2007; Tikhonov et al. 2015; Preheim et al. 2013) because sequence similarity does not imply ecological similarity. It was demonstrated that sequence similarity very poorly predicted ecological similarity (Tikhonov et al. 2015). The application of OTU methods in microbiome studies actually forces OTUs to play the functional role of taxa.

8.1.4.2 Challenges of Defining Species Using OTU Methods

OTU methods have been used in microbiome studies to classify taxa in classical Linnaean taxonomy at different taxonomic levels with similar DNA sequences.

Because 16S rRNA genes evolve at different rates in different organisms, determining taxonomic rank by sequence identity thresholds is approximate at the best

(Woese 1987). Hierarchical clustering methods generally use a predefined clustering threshold to the hierarchical tree and then group sequences within the threshold into one OTU. For example, because a natural entity "species" cannot be identified as a group of strains that is genetically well separated from its phylogenetic neighbors, sequences are often clustered into OTUs as proxies for species, although sequence divergence is not evenly distributed in the 16S rRNA region. Thus, defining a species by a polyphasic approach is only a pragmatic approach.

Previously, the threshold value of 70% was used in the phylogenetic definition of a species ("approximately 70% or greater DNA-DNA relatedness and with 5°C or less") (Wayne et al. 1987). This approach of applying DNA similarity to study bacterial species was well received and well proven and has been acknowledged since the 1970s (Stackebrandt and Goebel 1994; Johnson 1985; Steigerwalt et al. 1976). In 1994 Stackebrandt and Goebel (1994) proposed to use the "similarities of 97% and higher" sequence identity as the canonical clustering threshold when few 16S rRNA sequences were available. In practice the 3% dissimilarity is often chosen as the cutoff value to define bacteria species (Sogin et al. 2006; Sun et al. 2009; Schloss 2010; Huse et al. 2007) although different cutoff values of similarity have been studied such as 99.6–95.6% similarity (equating 1–13 bp) (Hathaway et al. 2017).

However, the cutoff value of similarities of 97% was based on the observation that species having 70% or greater DNA similarity usually have more than 97% sequence identity with these 3% or 45-nucleotide differences being concentrated mainly in certain hypervariable regions (Stackebrandt and Goebel 1994) given the primary structure of the 16S rRNA is highly conserved. Thus, the criterion of 97% similarities or 3% dissimilarities (Schloss and Handelsman 2005) is observed and experienced. It is not really able to define a natural entity "species." Especially, it was shown that 97% threshold is not optimal for an approximation to species because it is too low. To define a species, at least 99% of OTUs should be used (Edgar 2018). In summary, this practical approach of using OTUs to define a species is some kinds of arbitrary, subjective, and therefore inaccurate.

8.1.4.3   Criteria of Defining the Levels of Taxonomy

Like the 3% dissimilarities of sequences that are used to define a **species** (Borneman and Triplett 1997; Stackebrandt and Goebel 1994; Hugenholtz et al. 1998; Konstantinidis et al. 2006; Goris et al. 2007), accordingly in practice the 95% sequence similarity (95% homology) or the dissimilarities of 5% by species are used to define a **genus** (Wolfgang Ludwig et al. 1998; Everett et al. 1999; Sait et al. 2002; Schloss and Handelsman 2005); 10% dissimilarities (90% homology) are used to define a **family** (Schloss et al. 2009); 20% dissimilarities (80% homology) are used to define a **phylum** (Schloss and Handelsman 2005); an interkingdom identity range of 70–85% is used to define a **kingdom** (Woese 1987; Borneman and Triplett 1997); and 45% dissimilarities (55% homology) are used to define a **domain** (Stackebrandt and Goebel 1994).

This practice of using an arbitrary fixed global clustering threshold to group amplicons into molecular OTUs has been criticized (Mahé et al. 2015b) because global clustering thresholds have rarely been justified and are not applicable to all taxa and marker lengths (Caron et al. 2009; Nebel et al. 2011; Dunthorn et al. 2012; Brown et al. 2015). Actually it is impossible to use an accurate distance-based threshold to define taxonomic levels (Schloss and Westcott 2011). Therefore, the criteria of defining the levels of taxonomy are arbitrary, subjective, and inaccurate.

8.1.4.4 Clustering Algorithms and the Accuracy of OTU Clustering Results

As reviewed in Chap. 7, the original OTU methods are based on cluster analysis. However, when the clustering-based OTU methods were adopted in microbiome studies, several limitations have been detected that are associated with the clustering algorithms.

Hierarchical Clustering Methods Versus Heuristic Clustering Methods

In practice it is difficult to choose which clustering algorithms to use even if we could use a constant threshold to define OTUs. Because, in general, different clustering algorithms have assessed having different clustering quality, such as by comparing hierarchical and heuristic clustering algorithms in OTU construction, hierarchical clustering algorithms are assessed as more accurate (Sun et al. 2010). However, heuristic clustering methods are able to process massive sequence datasets. Thus, heuristic clustering methods are more attractive than hierarchical clustering methods for picking OTUs in 16S rRNA datasets (Cai and Sun 2011; Wei et al. 2021) and have been applied by the human microbiome project (HMP) (Peterson et al. 2009). However, most heuristic clustering methods just randomly select one sequence as the seed to represent the cluster and use this seed fixedly which results in the assigned OTUs sensitive to/depending on the selected seeds. Actually no real consensus has *achieved* on how the random seed is chosen, suggesting that OTU assignments are not definitive (Schloss and Westcott 2011).

Complete-Linkage, Single-Linkage, or Average-Linkage Clustering

Among the commonly used hierarchical clustering algorithms, a unanimous conclusion on which hierarchical clustering algorithm is optimal has not been reached.

The complete-linkage was originally recommended in numerical ecology because it is often desirable in ecology when one wishes to delineate clusters with clear discontinuities (Legendre and Legendre 1998), while the single-linkage clustering is implemented (Wei et al. 2012) in BlastClust (BLASTLab 2004) and GeneRage (Enright and Ouzounis 2000). Other researchers demonstrated that the average-linkage algorithm produces more robust OTUs than other hierarchical and heuristic clustering algorithms (i.e., CD-HIT, UCLUST, ESPRIT, and BlastClust) (Schloss and Westcott 2011; Sun et al. 2012; Quince et al. 2009). The average-linkage algorithm was further evaluated as having higher clustering quality compared to the other common clustering algorithms and even compared to greedy heuristic clustering algorithms, including distance-based greedy clustering (DGC) and abundance-based greedy clustering (AGC) (Westcott and Schloss 2015; Ye 2011). In such analysis, an OTU is generally considered presenting a bacterial species (Kuczynski et al. 2012). Therefore, using the average-linkage algorithm for clustering 16S rRNA gene sequences into OTUs has been suggested (Huse et al. 2010; Schloss and Westcott 2011). The suggestion of using the average-linkage clustering in microbiome data is in line with the suggestion of the original authors of numerical taxonomy (see Chap. 7 for details).

Read Preprocessing, Input-Order Dependency, and Accuracy of Clustered OTUs

The accuracy of clustered OTUs is also determined by removal of low-quality reads and of chimera sequences. It was shown that strict quality filtering may lead to a loss of information, which particularly affects the least abundant species, whereas if reads are clustered with low quality, then spurious OTUs may be produced which may result in overestimating the complexity of the analyzed samples (Albanese et al. 2015). It was also shown that amplicon input order can result in controversial results of OTU assignments (Koeppel and Wu 2013; Mahé et al. 2014, 2015b).

Challenges of Evaluating Clustering Algorithms

Different clustering algorithms and associated clustering thresholds have significantly impacted on the performance of clustering OTUs (Franzén et al. 2015). Since actually optimal thresholds are all higher than 97% (Edgar 2018). Thus, algorithms cannot be meaningfully ranked by OTU quality at 97% of similarity. For example, when comparing clustering algorithms, single-linkage, complete-linkage, average-linkage, AGC (Ye 2011), and OptiClust (OC) (Westcott and Schloss 2017), it was shown that no algorithm is consistently better than any other (Edgar 2018). In summary, no real consensus has emerged on the clustering algorithm.

The controversial and challenges of clustering approaches are due to using different criteria to assess clustering quality or test accuracy of clustering, such as simulation, information theoretic-based normalized distance measurement (Chen et al. 2013b; Baldi et al. 2000; van Rijsbergen 1979), stability of OTU assignments (He et al. 2015), and Matthew's correlation coefficient (MCC) (Schloss and Westcott 2011; Westcott and Schloss 2015), or combining the MCC and the F-score (Preheim et al. 2013; Baldi et al. 2000).

Different criteria often reach totally different conclusions. For example, many hierarchical and greedy clustering methods are unable to stably assign or produce OTUs. Thus, some researchers, such as He et al. (2015), thought that clustering all sequences into one OTU is completely useless for downstream analysis. By using MCC, Westcott and Schloss (2015) further demonstrated that the stability of OTUs actually suffered a lack of quality because the stability concept focused on the precision of the assignments and did not reflect the quality of the OTU assignments. Actually, this point was also raised in the numerical taxonomy.

In summary, the commonly used hierarchical and heuristic-based OTU methods suffered several fundamental problems, such as (1) the assumption of sequence similarity implying ecological similarity, (2) using arbitrary criteria to define the level of taxonomy (e.g., the dissimilarity of 3% OTUs is used to define species), (3) seed sensitivity, (4) input-order dependency, and (5) read preprocessing.

8.1.5   Purposes of Using OTUs in Microbiome Study

In microbiome study, the concept of OTUs was borrowed from numerical taxonomy. OTUs are used to serve the triple purpose: First is to define different taxonomic levels of taxonomy. Second is expected to reduce the impact of amplicon sequencing error on ecological diversity estimations and measures of community composition via grouping errors together with the error-free sequence (Eren et al. 2016). Third, the refined OTUs were repurposed to play the functionality of species.

8.1.5.1   Defining Taxonomy

Initially when the OTU concept was proposed to be used in microbiome sequencing, the purpose is for the taxonomic grouping, i.e., identifying bacterial population boundaries (Callahan et al. 2017). However, as we reviewed above and others such as Murat Eren et al. (2016) and Benjamin Callahan et al. (2017), the original purpose of using OTUs has not been achieved by the clustering OTU methods. For example, it is problematic to define OTUs using 97% sequence similarity threshold in the closed-reference approach because two sequences might be 97% similar to the same reference sequence, maybe less similarity (e.g., only 94%) shared by each other (de Queiroz and Good 1997; Westcott and Schloss 2015). Thus, the use of 97% OTUs to construct "species" has not been largely approved. In practice a 16S rRNA gene sequence similarity threshold range of 98.7–99% was recommended for testing the genomic uniqueness of a novel isolate(s) (Erko Stackebrandt and Ebers 2006). Similarly, it has not been largely approved to use the 95%, 90% of sequence similarity to define a genus and a family, respectively, and so on.

8.1.5.2   Reducing the Impact of Sequencing Error

The second purpose of using OTU concept and methods is to differentiate true diversity from sequencing errors (Callahan et al. 2017). This purpose is different from and orthogonal to the first one. When the OTU concept was initially proposed and used in microbiome study, OTUs were not expected to reduce the impact of amplicon sequencing error. The task of correcting errors has been repurposed to OTU (Rosen et al. 2012). In other words, this task is an expectation or was weighted as having the function of reducing sequence errors because the use of 97% sequence similarity threshold has successfully reduced the impact of erroneous OTUs on diversity estimations (Eren et al. 2016). After the second role was repurposed to OTUs, clustering reads into OTUs by sequence similarity became a standard approach to filter the noise (Tikhonov et al. 2015).

However, the second purpose has not be achieved either. The use of OTUs for differentiating true diversity from sequencing errors is only appropriate for higher levels of taxonomy (e.g., phylum). When OTU methods were used for probing finer-scale diversity in lower levels of taxonomy, they have intrinsically high false-positive and false-negative rates because using arbitrary cutoff values not only overestimate diversity when there exist errors larger than the OTU-defining cutoff but also cannot resolve real diversity at a scale finer (Rosen et al. 2012). It was also shown that clustering reads into OTUs greatly underestimates ecological richness (Tikhonov et al. 2015). In other words, the use of OTUs has not reduced the controversy on diversity estimations and measures of community composition.

Taken together, both purposes of using OTU-based methods have problems to be achieved and the two major challenges remained: one is to identify bacterial population boundaries and another is to differentiate true diversity from sequencing errors (Preheim et al. 2013).

The third purpose of using OTUs to play the functionality of species in microbiome research is more complicated. To better understand whether this purpose is achieved, we write Sect. 8.1.6 to provide a theoretical background to discuss definitions of species and species-level analysis.

8.1.6 Defining Species and Species-Level Analysis

Defining species is a core component of defining taxonomy, which is an important purpose that the 16S rRNA approach assumed to achieve. In Chap. 7, we reviewed three major species concepts that have been suggested to distinguish species in taxonomic studies: (1) genospecies (a group of organisms exchanging genes), (2) taxospecies (a phenetic cluster), and (3) nomenspecies (organisms bearing the same binomen). We also discussed the difference between a phenetic species and a biological species.

Historically the prokaryote taxonomy was reviewed (Rosselló-Mora and Amann 2001) having largely focused on the nomenclature of taxa (Heise and Starr 1968; Buchanan 1955; Cowan 1965; Sneath 2015; Trüper 1999) rather than the practical circumscription of the species concept applied to prokaryotes. Today's prokaryotic species concept results from empirical improvements of what has been thought to be a unit (Rosselló-Mora and Amann 2001). However, the underlying idea of defining a species is not only to formulate a unique unit for analysis but also to determine the functionality/genomic (i.e., a genotypic/genomic cluster definition) and phenotypic properties (i.e., not only biochemical or physiological properties but also chemotaxonomical markers, such as fatty acid profiles that are related to the species and its host).

The topics on eukaryotic and prokaryotic species concepts are very complicated and still debatable. Many books and review articles have contributed to these topics. We here have no intention of fully reviewing either eukaryote or prokaryotic species concept. The interested reader is referred to these books (Claridge et al. 1997b; Wheeler and Meier 2000) and articles (Rosselló-Mora and Amann 2001; Sokal and Crovello 1970; Mishler 2000; Hey 2006; Zink and McKitrick 1995; Balakrishnan 2005; Platnick 2000; Mishler 1999). Below we just summarize various definitions of species and provide a history background of species concepts in microbiome study.

8.1.6.1   Eukaryote Species Concepts

The species concept has been discussed among microbiologists and eukaryote taxonomists (Claridge et al. 1997a). The species concepts for prokaryotes and eukaryotes are different (May 1986, 1988); briefly review of species concept for eukaryotes will definitely help to delineate the species concept for prokaryotes.

Until the end of the 1990s, at least 22 species concepts have already been used (Mayden 1997), which reflects that it is unable to define a species or is difficult to find a unified species or lacks of consensus criterion to define a species. Hull (1997) stated that a concept is defined requiring to meet at least one of the three most common criteria or goals: (1) universality (or generality), (2) applicability, and (3) theoretical significance. Scientists define their species concepts to be as general as possible, while biologists have much difficulty to formulate a species concept that encompasses all organisms (Hull 1997). These 22 species concepts can be categorized into 1 of these 3 most common criteria. They are either more general or practical or theoretically significant species concepts. Rosselló-Mora and Amann (2001) reviewed the prokaryotic species concept and thought that the important requirements for a concept are based on whether the resulting classification scheme is stable, operational, and predictive. We describe some representative eukaryote species concepts in Table 8.2.

8.1.6.2   Prokaryote or Bacterial Species Concepts

The concepts of prokaryote (or bacterial) species were adopted from the concepts of eukaryote species. A brief history may be helpful to understand prokaryote (or bacterial) species concepts. Although in the seventeenth and eighteenth centuries the "infusion animalcules" were already observed due to invention of microscopes (Xia and Sun 2022), however, initially no classification was attempted because it lacked of a useful fossil record and was difficult to identify diagnostic characteristics from these small organisms. The first attempt to systematically arrange microorganisms was made at the end of the eighteenth century by Otto Müller (Nekhaev et al. 2015). However, the most important steps in the development of microbiology and especially classification of microorganisms were taken with the advances of new

Table 8.2 Selective eukaryote species concepts

Biological species concept (BSC)
BSC defines a species as
"groups of actually or potentially interbreeding natural populations which are reproductively isolated from other such groups" (Mayr 1940, 1942) and (Mayr 1963) (p. 19)
In his later publications (Mayr 1969, 2004, 2015), Mayr dropped the much criticized phrase "potentially interbreeding" from the definition and defined a species as:
"groups of interbreeding natural populations that are reproductively (genetically) isolated from other such groups" (Mayr 2004)
Evolutionary species concept (ESC)
ESC defines a species as
"a lineage (an ancestral-descendant sequence of populations) evolving separately from others and with its own unitary evolutionary role and tendencies" (Simpson 1961)
"a single lineage of ancestor-descendant populations which maintains its identity from other such lineages and which has its own evolutionary tendencies and historical fate" (Wiley 1978)
"an entity composed of organisms which maintains its identity from other such entities through time and over space, and which has its own independent evolutionary fate and historical tendencies" (Mayden 1997; Wiley and Mayden 2000)
Phenetic or polythetic species concept (PhSC)
PhSC defines a species as
"the species level is that at which distinct phenetic clusters can be observed" (Sneath 1976) (p. 437)
Phylogenetic species concept (PSC)
PSC defines a species as
"the smallest biological entities that are diagnosable and/or monophyletic" (Mayden 1997)
"the smallest diagnosable monophyletic unit with a parenteral pattern of ancestry and descent" (Rosselló-Mora and Amann 2001)

techniques. We categorize the development of prokaryote or bacterial species concepts into the following four stages.

1. The 1870s to Early 1950s: Phenotypic Classification (Phenotypic Definition of Species)

The ability to isolate organisms in pure cultures started in 1872 cultivating pure colonies of chromogenic bacteria by Joseph Schroeter (Logan 2009). Cultivating microorganisms facilitates bacteria classification based on the phenotypic description of these organisms. However, early bacterial species was often defined based on monothetic groups (a unique set of features). This is a phenotypic definition of species, which lies on a database with an accurate morphologic and phenotypic (e.g., biochemical) description of type strains or typical strains and comparing the isolate to be identified to the database using standard methods to determine these characteristics for the isolate (Clarridge 2004). The limitation is that these sets of phenotypic properties were subjectively selected (Goodfellow et al. 1997).

Among the more than 20 species concepts, **Mayr's BSC** is the most widely known and most controversial concept (Rosselló-Mora and Amann 2001; Sokal 1973). BSC attempts to unify genetics, systematics, and evolutionary biology (Claridge et al. 1997a). However, it lacks of practicability and hence generally it was agreed that Mayr's BSC should be abandoned (Rosselló-Mora and Amann 2001; Sokal 1973). Therefore, BSC cannot presently be applied to prokaryotes (Stackebrandt and Goebel 1994).

ESC defines a species as a lineage that is explicitly temporal, treating the units as lineages extended in time (Hull 1997). It is the most theoretically committed species concept and the only one being a primary concept due to its accommodating all types of species known (Mayden 1997; Hull 1997). The disadvantages of ESC (Rosselló-Mora and Amann 2001) are as follows: (1) It has no pragmatic significance for the prokaryotes in analyzing the current state of knowledge about this group of organisms. (2) It cannot recognize an evolutionary fate nor historical tendencies of the prokaryotes due to lacking a useful fossil record. Although ESC has been recommended to be used for animals (Cracraft 1997), however, it is not yet possible to be adopted for prokaryotes due to (1) rather incomplete knowledge of evolutionary tempo and mode of evolutionary changes in prokaryotes and (2) difficult predictions in accounting for the possibilities of horizontal gene transfers between distant groups (Rosselló-Mora and Amann 2001).

2. The Late 1950s to Early 1960s: Phenetic Classification

The development of numerical taxonomy in the late 1950s in parallel to application of computer for multivariate analyses makes bacteria classification toward an objective approach (Sneath 1989). The numerical taxonomy began to establish when modern biochemical analytical techniques advanced to study the distributions of specific chemical constituents (i.e., amino acids, proteins, sugars, and lipids in bacteria) (Rosselló-Mora and Amann 2001; Logan 2009).

Phenetic or polythetic (Van Regenmortel 1997) species concept (**PhSC**) describes "the species level is that at which distinct phenetic clusters can be observed" (Sneath 1976) (p. 437). This is a pragmatic and similarity species concept. It is based on statistically covarying characteristics which are not necessarily universal among the members of the taxa (Sokal and Crovello 1970; Hull 1997). PhSC has some advantages, including the following: (1) It is theory neutral or theory-free and thus can avoid the controversial issue and difficulty (Hull 1997). (2) It is operational and very stable and a valuable pragmatic concept (Rosselló-Mora and Amann 2001). (3) It is universal, *monistic,* and applicable, which mostly meets the primary requirements for being a concept, and hence has the most valuable characteristics for scientists to consider it as one of the most persuasive concepts (Hull 1997). PhSC has been recommended for higher organisms as well as most similarly for applying to prokaryotic species (Sokal and Crovello 1970).

3. The Late 1960s: Genomic Classification

The development of molecular biological techniques in the early 1960s, including the techniques of initially overall base compositions of DNAs, DNA-DNA hybridization (Brenner et al. 1969), and defining clusters of strains (Krieg 1988), leads to classify bacterial species by comparing their genomes. A DNA similarity concept was also developed, and DNA-DNA hybridization became a standard technique for classifying bacterial species (genotypic definition of species) (Wayne et al. 1996).

However, DNA-DNA hybridization is a difficult technique and does not always correlate with other definitions of species (Fournier et al. 2003). Moreover, a DNA similarity group often depended on phenotypic characters such as use of nucleic acid homologies (Johnson 1973). Thus, classification of bacterial species using only DNA similarity (DNA-DNA similarities) is in practice useless. A bacterial species classification was recommended that must be provided with diagnostic phenotypic properties (Wayne et al. 1987).

4. The Late 1970s to the Mid-1980s: 16S rRNA Classification

In the late 1970s, with the advent of cataloging ribosomal ribonucleic acids (rRNAs) (Stackebrandt et al. 1985), the rRNA and especially 16S rRNA sequences were shown to be a very useful molecular marker for phylogenetic analyses (Ludwig and Schleifer 1994).

Clustering-based OTU methods have been combined with the 16S rRNA sequencing. The 16S rRNA sequencing method has been widely used for the prokaryotic classification (Olsen et al. 1994), and the datasets generated from 16S rRNA sequencing studies have also been increasingly used to propose new bacterial species (Stackebrandt and Goebel 1994).

The determination of relationships between distantly related bacteria has been evidenced a remarkable breakthrough in the late 1970s by cataloging 16S ribosomal ribonucleic acids (rRNAs) (Stackebrandt et al. 1985) and DNA-RNA hybridization (De Ley and De Smedt 1975) and in the mid-1980s by the full sequence analysis of rRNA (Rosselló-Mora and Amann 2001). In the 1980s, determining phylogenetic relationships of bacteria and all life forms by comparing a stable part of the genetic code (commonly the 16S rRNA gene) became a new standard for identifying bacteria (Woese et al. 1985; Woese 1987).

The reason that the 16S rRNA gene is chosen as the gene to sequence is mainly because it has two important properties (Woese 1987; Clarridge 2004; Xia et al. 2018):

1. It is highly conserved (Dubnau et al. 1965; Woese 1987). Given its conservation the 16S rRNA gene marks evolutionary distance and relatedness of organisms; thus the evolutionary rates in the 16S rRNA gene sequence can be estimated through comparing studies of nucleotide sequences (Kimura 1980; Pace 1997; Thorne et al. 1998; Harmsen 2004). Currently it has accepted that 16S rRNA sequence analysis can be used for prokaryotic classification to identify bacteria or to assign close relationships at the genus and species levels (Clarridge 2004).
2. It is universal in bacteria; thus the relationships among all bacteria can be measured (Woese et al. 1985; Woese 1987). This is also an important property. It was reviewed that through the comparison of the 16S rRNA gene sequences, it not only can classify strains at multiple levels, including the species and subspecies level, but also differentiate organisms at the genus level across all major phyla of bacteria (Clarridge 2004).

Due to above two important properties, bacterial phylogeny (the genealogical trees among the prokaryotes) can be established by comparing the 16S rRNA gene sequences (Ludwig et al. 1998). PSC includes monophyletic species concept (MSC) and diagnostic species concept (DSC) (Hull 1997). Both are defined as phylogenetic (or genealogical) concepts with a minimal time dimension (Rosselló-Mora and Amann 2001).

When DNA was discovered, prokaryote classification was based solely on phenotypic characteristics. With the development of numerical taxonomy (Sneath and Sokal 1973), the individuals are treated as operational taxonomic units that are polythetic (they can be defined only in terms of statistically covarying characteristics), resulting in a more objective circumscription of prokaryotic units. The discovery of genetic information gave a new dimension to the species concept for microorganisms, which enable at least a first rough insight into phylogenetic relationships. Thus, the species concept for prokaryotes evolved into a mostly phenetic or polythetic. This means that species are defined by a combination of independent, covarying characters, each of which may occur also outside the given class, thus not being exclusive of the class (Van Regenmortel 1997).

Compared to DNA-RNA hybridization approach, the 16S rRNA gene sequence approach is much simpler and thus has replaced DNA-RNA hybridization and become the new gold standard to define a species (Fournier et al. 2003; Harmsen 2004). However, as we will describe in Sect. 8.1.6.3, using the 16S rRNA method to define a species still remains challenging. Before we move forward to definition of species in 16S rRNA method, we quote some microbiological definitions of species.

Microbiologist unofficially defines a species for pure culture (essential for the classification of new prokaryotic species) as:

• "a microbial species is a concept represented by a group of strains, that contains freshly isolated strains, stock strains maintained in vitro for varying periods of time, and their variants (strains not identical with their parents in all characteristics), which have in common a set or pattern of correlating stable properties that separates the group from other groups of strains (Gordon 1978)."

Microbiologist unofficially also generally describes a prokaryote species to be included uncultured organisms which constitute the largest proportion of living prokaryotes as:

• "a group of strains that show a high degree of overall similarity and differ considerably from related strain groups with respect to many independent characteristics (Colwell et al. 1995)" or
• "a collection of strains showing a high degree of overall similarity, compared to other, related groups of strains (Colwell et al. 1995)."

8.1.6.3 16S rRNA Method and Definition of Species

Using the 16S rRNA similarities to define bacteria species remains challenging. Much early in the 1870s, Ferdinand Cohn investigated whether the similarity exits between bacteria and whether bacteria, like animals and plants, can be arranged in distinct taxa (Cohn 1972; Schlegel and Köhler 1999). Cohn considered the form genera as natural entities but considered species as largely artificial. Cohn emphasized that character of the classification system is artificial. The original authors of OTU method Sokal and Sneath (1963) also criticized their then current taxonomy made little increase in understanding the nature and evolution of the higher categories (Sokal and Sneath 1963) (p. 5).

When clustering-based OTU methods emerged into the 16S rRNA sequencing analysis, an important concept, the phylogenetic species concept (PSC), was developed. PSC compares the 16S rRNA similarities with DNA similarity (DNA-DNA similarities), assuming that nucleic acid (DNA) reassociation can determine whether taxa are phylogenetically homogeneous. PSC defines a species that would generally include strains with "approximately 70% or greater DNA-DNA relatedness and with 5°C or less ΔT_m" (Wayne et al. 1987). Here, T_m is the melting temperature or the thermal denaturation midpoint, and ΔT_m is the difference between the homoduplex DNA T_m and the heteroduplex DNA T_m. Thus, ΔT_m is a reflection of the thermal stability of the DNA duplexes.

However, as reviewed in Sect. 8.1.4.2, the rationale for using DNA reassociation as the gold standard to delineate species originates from the observations that a high degree of correlation existed between DNA similarity and chemotaxonomic, genomic, serological, and numerical phenetic similarity (Stackebrandt and Goebel 1994). The 16S rRNA sequencing method does not have sufficient power to correctly define bacterial species, which was reviewed in several articles including Ash et al. (1991); Amann et al. (1992); Fox et al. (1992); Martinez-Murcia et al. (1992); Stackebrandt and Goebel (1994); and Johnson et al. (2019).

Thus, it suggests that the 16S rRNA method cannot be used for defining bacteria species. We can summarize four arguments as follows:

1. The 97% sequence similarity used by 16S rRNA method is based on the assumption that 70% DNA similarity defines a species. Organisms that have 70% or greater DNA similarity will not necessarily have at least 96% DNA sequence identity. The threshold value of 70% does not consider the possibility that the tempo and mode of changes differ in different prokaryotic strains (Stackebrandt and Goebel 1994). DNA hybridization is significantly higher resolution of power than that of sequence analysis (Amann et al. 1992). Thus, the 70% DNA similarity cannot be used to define a species. Moreover, species having 70% DNA similarity usually have more than 97% sequence identity (Stackebrandt and Goebel 1994). Thus, even if the 70% DNA similarity can be used to define a species, the 97% sequence identity cannot be used to define a species.

2. The 16S rRNA has a highly conserved primary structure; there does not exist a linear correlation between the two phylogenetic parameters DNA-DNA similarity percent and 16S rRNA similarity for closely related organisms (Stackebrandt and Goebel 1994). Sequence similarities and DNA reassociation values obtained for the same strain pairs do not have a linear relationship (Erko Stackebrandt and Ebers 2006; Stackebrandt and Goebel 1994), and actually 16S rRNA sequence

identity is not sufficient to guarantee species identity (Fox et al. 1992; Rosselló-Mora and Amann 2001; Martinez-Murcia et al. 1992).

3. The 3% or 45-nucleotide differences are not evenly distributed over the primary structure of the molecule; instead they are highly likely to concentrate in the hypervariable regions. Thus, it is impossible to exactly describe the genealogy of the molecule because the false identities are simulated and the actual number of evolutionary events is masked (Grimont 1988; Stackebrandt and Goebel 1994).

4. Technically, 16S variable regions (i.e., ranging from single variable regions, such as V4 or V6, to three variable regions, such as V1–V3 or V3–V5) cannot be targeted with short-read sequencing platforms (e.g., using Illumina sequencing platform to produce ≤300 bases of short sequences) to achieve the taxonomic resolution afforded by sequencing the entire (~1500 bp) gene (Johnson et al. 2019).

5. Particularly, the examples described below (Rosselló-Mora and Amann 2001) suggest it is difficult to use the 16S rRNA similarities to define bacteria species, because (1) different species could have identical (Probst et al. 1998) or nearly identical 16S rRNA sequences (Fox et al. 1992; Martinez-Murcia et al. 1992); (2) within a single species, there exists a microheterogeneity of the 16S rRNA genes (Bennasar et al. 1996; Ibrahim et al. 1997); and (3) in exceptional cases, single organisms with two or more 16S rRNA genes could have relatively high sequence divergence (Mylvaganam and Dennis 1992; Nübel et al. 1996).

In summary, clustering sequence similarity of 16S rRNA can never be solely used to define bacterial species. On the one hand, it was recognized that the DNA hybridization (DNA reassociation) remains the optimal method for measuring the degree of relatedness between highly related organisms and hence cannot be replaced by 16S rRNA method for defining species (Stackebrandt and Goebel 1994; Rosselló-Mora and Amann 2001). On the other hand, it has been accepted that a prokaryotic species classification should be defined by integrating both phenotypic and genomic parameters, as well as should analyze and compare as many phenotypic and genomic parameters as possible (Wayne et al. 1987). This approach is known as "polyphasic taxonomy" (Vandamme et al. 1996). Based on this polyphasic approach, in 2001 a more pragmatic prokaryote species concept (called a phylo-phenetic species concept) has been described as "a monophyletic and genomically coherent cluster of individual organisms that show a high degree of overall similarity in many independent characteristics, and is diagnosable by a discriminative phenotypic property" (Rosselló-Mora and Amann 2001). Obviously, this is far beyond the clustering-based OTU method can achieve.

8.1.6.4 16S rRNA Method and Physiological Characteristics

Whether the 16s rRNA method can inference the physiological characteristics is debatable. On the one hand, the 16S rRNA approach has the practical advantage for species identification and has been used frequently in bacterial phylogenetics, species delineation, and microbiome studies; for example, it is recommended to include the ribosomal sequence for describing new prokaryotic species (Ludwig 1999; Chakravorty et al. 2007). Some researchers considered that comparative analysis of 16S rRNA is a very good method to find a first phylogenetic affiliation in both potentially novel and poorly classified organisms (Goodfellow et al. 1997). Thus, although the 16S rRNA sequence analysis is not sufficient for numerically delineating borders of the prokaryotic species, it can indicate the ancestry pattern of the taxon studied, as well as confirm the monophyletic nature in grouping the members (Rosselló-Mora and Amann 2001).

On the other hand, other researchers have questioned the validity of 16S rRNA as a marker for phylogenetic inferences and thought that it is very difficult to infer the physiological characteristics of an organism by clustering the rRNA sequence method (Rosselló-Mora and Amann 2001; Cohan 1994; Gupta 1998; Gribaldo et al. 1999). Thus, using arbitrary threshold of 97% similarity (or 3% dissimilarity) to define OTUs that could be interpreted as a *proxy* for bacterial *species* is computationally convenient, but at the expense of accurate ecological inference (Eren et al. 2016) because 3% OTUs are often phylogenetically mixed and inconsistent (Koeppel and Wu 2013; Nguyen et al. 2016; Eren et al. 2014).

Recently, Hassler et al. (2022) found that phylogenies of the 16S rRNA gene and its hypervariable regions lack concordance with core genome phylogenies, and they concluded that the 16S rRNA gene has a poor phylogenetic performance and has far-reaching consequences.

By comparing core gene phylogenies to phylogenies constructed using core gene concatenations to estimate the strength of signal for the 16S rRNA gene, its hypervariable regions, and all core genes at the intra- and inter-genus levels, Hassler et al. (2022) showed that at both intra- and inter-genus taxonomic levels, the 16S rRNA gene was recombinant and subject to horizontal gene transfer, suffering from intragenomic heterogeneity, accumulating recombination and an unreliable phylogenetic signal. The poor phylogenetic performance using the 16S rRNA gene not only results in incorrect species/strain delineation and phylogenetic inference but also has the potential to confound community diversity metrics if incorporating phylogenetic information in the analysis such as using Faith's phylogenetic diversity and UniFrac (thus, their use to measure the diversity is not recommended) (Hassler et al. 2022). Additionally, taxonomic abundance and hence measures of microbiome diversity may also be inflated and confounded by the multiple copies within a genome and wide range in 16S rRNA gene among genomes (De la Cuesta-Zuluaga and Escobar 2016; Louca et al. 2018).

8.2   Moving Toward Single-Nucleotide Resolution-Based OTU Methods

In summary, using OTUs as the atomic unit of analysis has both benefits and controversies. This consequence is often due to unacknowledged factor that OTUs fail to serve the dual tasks well; and especially the connection between OTUs and species is largely unfounded (Stackebrandt and Ebers 2006). In order to overcome the limitations of clustering-based OTU methods in bioinformatic analysis of microbiome sequencing data, the reform movements of moving beyond OTUs and clustering-based OTU methods have begun in microbiome research. Two strategies for these reform movements that represent overlapping goals are moving toward single-nucleotide resolution and moving beyond the OTU-based methods with denoising-based methods. In this section, we describe the first strategy, moving toward single-nucleotide resolution-based OTU methods, and in Sect. 8.3, we investigate the second strategy: moving beyond the OTU-based methods with denoising-based methods.

8.2.1   Concept Shifting in Bioinformatic Analysis

In bioinformatic analysis of microbiome data, a concept shifting regarding analysis units began with a dissatisfactory of using similarity/dissimilarity threshold to define OTUs.

As we reviewed in Sects. 8.1.5 and 8.1.6, OTUs used in microbiome study have three purposes: (1) to define taxonomic levels, (2) to improve the accuracy of ecological diversity estimations and measures of community composition, and (3) to play the functionality of species. All these are done through defining certain particular taxonomic levels by borrowing ecological taxonomic concepts into the context of high-throughput marker-gene sequencing of microbial communities, such as to classify microbial species (Rosen et al. 2012); the 97% sequence identity (or 3% ribosomal) OTUs are used to define the level of "like species." When we do this way, we actually assume that the greater similar sequences more likely represent phylogenetically similar organisms. Defining OTUs in this way is expected to simplify the complexity of the large datasets (Westcott and Schloss 2015), to facilitate taxonomy-independent analyses, such as alpha diversity, beta diversity, and taxonomic composition, and thus to effectively reduce the computational resources (Human Microbiome Project 2012; He et al. 2015).

However, it was assessed (Chen et al. 2013b) that using a constant threshold to define OTUs at a specific taxonomic level, e.g., using 3% dissimilarity to define species, is not ideal because clustering reads into OTUs vastly underestimates ecological richness (Tikhonov et al. 2015) and hence may end up with an incorrectly estimated number of OTUs/taxa.

Thus, some researchers began reconsidering the concept of OTUs and clustering-based OTU methods.

Some heuristic-based OTU methods have already tried to address the problem of how to choose an optimal distance threshold to define OTUs to represent different taxonomic levels. AGC (abundance-sorted greedy clustering) (Ye 2011) was developed for identifying and quantifying abundant species from pyrosequences of 16S rRNA by consensus alignment, aiming to avoid inflation of species diversity estimated from error-prone 16S rRNA pyrosequences. UPARSE (Edgar 2013) reported that the generated OTU's sequences with ≤1% incorrect bases are more closer to the species in a community, and Swarm's (Mahé et al. 2014) sub-OTUs solution is also an effort to refine the OTU clustering.

Both model-based and network-based clustering OTU methods have also been developed to address the problem of selecting optimal distance threshold to define OTUs (Wei et al. 2021).

Such model-based clustering OTU methods include CROP (Clustering 16S rRNA for OTU Prediction) (Hao et al. 2011), BEBaC (Cheng et al. 2012), and BC (Jääskinen et al. 2014). Such network-based clustering OTU methods include M-pick (Wang et al. 2013), (Wei and Zhang 2015), and DMclust (Wei et al. 2017).

However, within the framework of OTU methods, all these clustering methods including hierarchical, heuristic-based OTU methods and model-based and network-based OTU methods remain using the concept of OTUs; the difference between them is how to cluster sequences to OTUs and what cutoff values are used to refine OTUs as closer to species.

Until 2011, most researchers still believed that the solutions of OTU-based methods can be improved to quickly and accurately assign sequences to OTUs and hence the obtained taxonomic information from those OTUs can greatly improve OTU-based analyses. The OTU-based methods are considered being robust and can overcome many of the challenges encountered with phylotype-based methods (Schloss and Westcott 2011).

There are three motivations behind the development of new methods:

- The first is to overcome the weakness of classical OTU approach based on an arbitrary sequence identity threshold because 97% similarity is not accurate and cannot approximate biology species.
- The second is to overcome the amplicon sequencing errors. The classical OTU approach assumed that it could reduce problems caused by erroneous sequences. But it cannot solve the problem of erroneously clustering OTUs. It also reduces phylogenetic resolution because sequences below the identity threshold cannot be differentiated (Amir et al. 2017).
- The third is to facilitate merging datasets generated by bioinformatic tools. Because the *de novo* OTUs were assessed having problem of merging OTUs and although this problem could be reduced in closed-reference and open-reference OTU picking (Rideout et al. 2014), it remains a challenge to integrate large datasets into a single OTU space (Amir et al. 2017).

The last two motivations are associated with downstream statistical analysis. To avoid these weaknesses, the new approaches focus on sub-OTU methods or algorithms to detect ecological differences by a single base pair and try to approve that the subunits are ecological and functional meaningfulness, while expecting sub-OTUs are feasible for analysis. The new direction of bioinformatic analysis of microbial community has been challenging the importance of the solution of OTUs and OTU-based methods. It aims to define atomic analysis units of microbiome data and to search for algorithms that can provide single-nucleotide resolution without relying on arbitrary percent similarity thresholds of OTUs (Eren et al. 2016) or refining the OTUs.

8.2.2   Single-Nucleotide Resolution Clustering-Based OTU Methods

Several computational methods or algorithms have been developed to improve the resolution of 16S data analysis beyond the threshold of 3% dissimilarity OTUs. We introduce the two most important algorithms below: distribution-based clustering (Sect. 8.2.2.1) and Swarm2 (Sect. 8.2.2.2).

8.2.2.1   Distribution-Based Clustering (DBC)

In 2013, Preheim et al. (2013) used DBC to refine the OTU for single-nucleotide resolution but without relying on sequence data alone to serve the dual purpose for the use of OTUs: identifying taxonomic groups and eliminating sequencing errors. Alternatively, the DBC method compares the distribution of sequences across samples. In other words, it uses ecological information (distribution of abundance across multiple biological samples) to supplement sequence information.

DBC algorithm is implemented under the framework of clustering OTU methods. It first uses Jukes-Cantor-corrected genetic distances or Jensen-Shannon divergence (JSD) as appropriate to measure the genetic distances among sequences and then implements either complete algorithm (all sequences are analyzed together in the analysis) or parallel algorithm (sequences are pre-clustered with a heuristic approach) or different algorithms (e.g., nearest-neighbor single-linkage clustering). DBC first uses the chi-squared test to determine whether two sequences have similar distributions across libraries. Then, based on ecological distribution to form OTUs, so 16S rRNA sequences that even differ by only 1 base but that are found in different samples are still put into different OTUs. Conversely, the sequences drawn from the same underlying distribution across samples are grouped together into the same OTU regardless of interoperon variation or sequence variation or random sequencing errors. The underlying assumption is that bacteria in different populations are often highly correlated in their abundance across different samples (Preheim et al. 2013).

Thus, 16S rRNA sequences derived from the same population regardless of their variations or random sequencing error will have the same underlying distribution across sampled environments.

It was shown (Preheim et al. 2013) that DBC has the advantages. We summarize them below:

1. DBC algorithm uses both genetic distance and the distribution of relative abundances of sequences across samples to identify the appropriate grouping for each taxonomic lineage and to detect many methodological errors. Thus, it may avoid the limitation that will necessarily either over-cluster or under-cluster sequences using genetic information alone.

2. DBC method is more accurate and sensitive in identifying true input sequences, clustering sequencing, and methodological errors, in which the accuracy is in terms of the F-score and the Matthew's correlation coefficient (MCC) (Baldi et al. 2000). The DBC method is more accurate (more correct OTUs, fewer spurious/incorrect OTUs) than other OTU methods including closed-reference (i.e., phylotyping), de novo clustering, and open-reference (i.e., a hybrid of phylotyping and de novo clustering) and UCLUST at grouping reads into OTUs. It is also sensitive enough to differentiate between OTUs that differ by a single base pair for showing evidence of differing ecological roles. In other words, DBC predicts the most accurate OTUs when sequences are distributed in an ecologically meaningful way across samples. It more accurately represents the input sequences based on the total number of OTUs and predicts fewer overall OTUs than the de novo and open-reference methods. Because detecting the differences between closely related organisms is crucial (Shapiro et al. 2012), thus, DBC can be chosen to distinguish the signal from the noise of sequencing errors and to form accurate OTU for identifying evolutionary and ecological mechanisms (Preheim et al. 2013).

However, DBC as an OTU clustering algorithm has two important challenges:

1. DBC is unreliable to conduct cross-sample comparisons for low-count sequences (Tikhonov et al. 2015).
2. DBC currently requires very long run time on very large datasets and prohibitively long even for moderately sized datasets. This is a severe limitation of DBC (Preheim et al. 2013) and has been criticized (Tikhonov et al. 2015).

In summary, DBC is a clustering OTU method or a refined OTU method. The difference from other OTU methods is that it uses ecology to model multiple samples to refine the OTU. As a refined OTU method, DBC is still within traditional framework of OTU methods.

8.2.2.2 Swarm2

The de novo amplicon clustering methods share two fundamental problems (reviewed in Sect. 6.4). One uses arbitrary fixed global clustering thresholds; another is input-order dependency induced by centroid selection, i.e., the clustering results are strongly influenced by the input order of amplicons. To solve these two fundamental problems, in 2014, Mahé et al. (2014) proposed an amplicon clustering algorithm Swarm to fine-scale OTUs without relying on arbitrary global clustering thresholds and input-order dependency.

Swarm is a highly scalable and high-resolution, input-order independent amplicon clustering. Swarm first uses a local threshold to cluster nearly identical amplicons iteratively and then uses clusters' internal structure and amplicon abundances to refine its results. Swarm was defined as a fast and exact, agglomerative, unsupervised (de novo) single-linkage clustering implementing with a *growth* and a *breaking* two phases. Swarm has the advantages:

1. Swarm was expected to reduce the influence of clustering parameters and to produce robust OTUs (Mahé et al. 2014).
2. Swarm2 (Mahé et al. 2015b) directly integrated the clustering and breaking phases and hence improved the performances in terms of computation time and reducing under-grouping by grafting low abundant OTUs (e.g., singletons and doubletons) onto larger ones. The improvement is done through improving Swarm's scalability to linear complexity and adding the fastidious option (Mahé et al. 2015b).

However, Swarm has the disadvantage: it cannot resolve at the single-base level (Hathaway et al. 2017).

Like distribution-based clustering (Sect. 8.2.2.1), Swarm and Swarm2 aim to provide single-nucleotide resolution (Mahé et al. 2014, 2015b). Other bioinformatic software that aim to single-nucleotide resolution include oligotyping (Section 8.3.1), denoising-based methods (Sect. 8.3.2) such as cluster-free filtering (Sect. 8.3.2.3), DADA2 (Sect. 8.3.2.4), UNOISE2 (Sect. 8.3.2.5), Deblur (Sect. 8.3.2.6), and SeekDeep (Sect. 8.3.2.7).

8.3 Moving Beyond the OTU Methods

We can use the following four characteristics to describe the beyond OTU methods: (1) to avoid the concept of OTUs, especially without relying on using 97% sequence identity to approximate species, (2) to avoid using clustering method, (3) to focus on single-nucleotide (a single base pair) resolution to provide more accurate estimates of diversity, and (4) to expect more clinical or functional relevancy. We use these four criteria especially the first three criteria to assess on whether or not the newly developed methods are moving beyond the traditional OTU methods.

8.3.1 Entropy-Based Methods: Oligotyping

In molecular biology, the term oligotyping mainly consists of dual DNA sequence analysis methods or functions of taxonomy and sequencing; that is, the approach of oligotyping is to use primary DNA sequencing to identify organism taxonomy (taxonomy), while improving the accuracy of DNA sequencing (sequencing).

In 2013, Eren et al. (2013) described a supervised computational method called oligotyping in analysis of 16S rRNA gene data and demonstrated that oligotyping is able to differentiate closely related microbial taxa. Oligotyping decomposes marker-gene amplicons with a different way from clustering OTUs. The basic theory underlying the oligotyping workflow is minimum entropy decomposition (MED) (Eren et al. 2014), the unsupervised oligotyping, which relies on Shannon entropy (Shannon 1948), a measure of information uncertainty (Jost 2006). MED is an algorithm for fine-scale resolution for Illumina amplicon data.

The oligotyping workflow is implemented via the following three procedures: First, MED identifies variable nucleotide positions among the entire sequencing data based on subtle nucleotide variation (Ramette and Buttigieg 2014). Second, MED only chooses those significantly variate positions to partition reads into oligotypes ("MED nodes": representing homogeneous OTUs), i.e., sensitively partitioning the high-throughput marker-gene sequences (Eren et al. 2014). Third, MED produces a sample-by-OTU table for downstream analyses (Buttigieg and Ramette 2014).

It was shown (Eren et al. 2014, 2016) that the MED strategy and algorithm along with the underlying information theory-based decomposition process have the benefits:

1. Could resolve closely related but distinct taxa that differ by as little as one nucleotide at the sequenced region.
2. Could not only allow identifying nucleotide positions that likely carry phylogenetically important signal, and enable finer representing the microbial diversity in a wide range of ecosystems without relying on extensive computational heuristics and user supervision, but also improve the ecological signal for downstream analyses.
3. Particularly, it was shown (Hathaway et al. 2017) that like SeekDeep (Sect. 8.3.2.7), MED was able to achieve 100% haplotype recovery using 454 and Ion Torrent pyrosequencing reads, as well as Illumina MiSeq reads on the Illumina platform.

However, it was also shown (Hathaway et al. 2017) that the MED algorithm appears to get trouble as a haplotype's abundance increases.

8.3.2   Denoising-Based Methods

Typically many sequences generated by any sequencing platform contain at least one error (Amir et al. 2017). When microbiome researchers noticed that the sequences contain sequencing errors, the use of OTU methods has been changed (Gaspar 2018) to deal with sequencing errors.

Sequencing errors are extremely important confounding factors for detecting low-frequency genetic variants (Ma et al. 2019) and thus impact the underlying biology due to inaccurate taxon identification and inflated diversity statistics (Amir et al. 2017). Although these errors seldom affect statistical hypothesis testing of differences between two communities, clinically higher precision could be deviated by these errors (Amir et al. 2017). Removing or reducing the sequencing errors will reduce false-negative results and improve the accuracy of describing the microbial community. Thus, one direction of methodology development in bioinformatic analysis of microbiome data is to handle sequencing errors (Ma et al. 2019).

One way to deal with sequencing errors is to denoise them. Denoising-based or error-correction methods exploit the predictable structure of certain error types to attempt to reassign or eliminate noisy reads (Huse et al. 2010; Quince et al. 2011; Rosen et al. 2012). Most denoising algorithms aim to assign erroneous reads to their most likely source, to make the abundance estimates of true sequences more accurate (Tikhonov et al. 2015).

Several types of error are introduced by PCR amplification followed by sequencing (Edgar 2016b). Of which sequencing error, PCR single-base substitutions, and PCR chimeras are the three important sources of error (Quince et al. 2011): (1) point errors caused by substitution and gap errors due to incorrect base pairing and polymerase slippage, respectively (Turnbaugh et al. 2010), (2) PCR chimeras caused by extending an incomplete amplicon prime into a different biological template (Haas et al. 2011), and (3) spurious species due to contaminants from reagents and other sources (Edgar, 2013) as well as introduced when reads are assigned to incorrect samples due to *cross talk* (Carlsen et al. 2012).

Thus, to infer accurate biological template sequences from noisy reads, the denoising-based or error-correction methods typically take two steps (Edgar 2016b): first, denoising or correcting point errors to obtain an accurate set of amplicon sequences and then filtering the chimeric amplicons. Among them, DADA2, UNOISE3, and Deblur are the three most widely used denoising packages. We introduce them, respectively, as below.

8.3.2.1   Denoising-Based Methods Versus Clustering-Based Methods

In 2018, Almeida et al. (2018) compared the default classifiers of MAPseq (Matias Rodrigues et al. 2017), mothur, QIIME, and QIIME 2 using synthetic simulated datasets to evaluate their accuracy when paired with both different reference databases (i.e., Greengenes (McDonald et al. 2012), NCBI (Federhen 2012), RDP (Cole et al. 2014) and SILVA (Yilmaz et al. 2014)) and variable subregions (V1-V2, V3-V4, V4, and V4-V5) of the 16S rRNA gene. This study showed that (Almeida et al. 2018):

1. QIIME 2 outperformed MAPseq, mothur, and QIIME in terms of overall recall and F-scores at both genus and family levels as well as with the lowest distance estimates between the observed and simulated (predicted) samples, but at the expense of CPU time and memory usage. In contrast, MAPseq showed the highest precision, with miscall rates consistently <2%.
2. Compared to using Greengenes, generally a higher recall was yielded using the SILVA database.
3. The performance of taxonomic assignment for each tool can be considerably influenced varied up to 40% depending on the 16S rRNA subregion targeted. Overall, the V1–V2 and V3–V4 subregions performed the best across most of the software tools. However, the V1–V2 subregions were not recommended to be used for classification of complex community samples because the V1–V2 primers did not match almost 70% of the sequences across the four reference databases.
4. This study concluded that use of either QIIME 2 or MAPseq is optimal for 16S rRNA gene profiling, where the recall rate (sensitivity) was estimated as the percentage of sequences assigned to the expected taxa for each biome, while precision (specificity) was calculated as the fraction of sequences from these predicted taxa out of all those from the taxa observed. The distances were estimated with either the Bray-Curtis or Jaccard dissimilarity indices at the genus level. The F-score was calculated as $F - \text{score} = 2 \times \frac{\text{precision} \times \text{recall}}{\text{precision} + \text{recall}}$, where precision is defined as $\frac{\text{True positives (TP)}}{\text{True positives (TP)} + \text{False positives (FP)}}$ and recall is defined as $\frac{\text{True positives (TP)}}{\text{True positives (TP)} + \text{False negatives (FN)}}$.

In the same year, Nearing et al. (2018) conducted a thorough comparison of three of the most widely used denoising packages (DADA2, UNOISE3, and Deblur) which use high-resolution ASVs and an open-reference 97% OTU clustering pipeline. In literature, USEARCH's most recent clustering algorithm has been referenced as UNOISE or UNOISE in USEARCH (Hathaway et al. 2017), UNOISE3 or USEARCH's unoise3 (Nearing et al. 2018), and USEARCH-UNOISE3 (Prodan et al. 2020). UNOISE3 is used to denote the denoising method that uses the "unoise3" command in USEARCH. The different studies used different names, but all referenced the method paper UNOISE2 (Edgar 2016b). Thus, in this book, we use these terms interchangeably to denote improved error correction for Illumina 16S and ITS amplicon sequencing from USEARCH. In this study (Nearing et al. 2018):

1. The mock community analyses showed that these three denoising pipelines generate very different numbers of ASVs that significantly impact alpha diversity metrics although they produced similar microbial compositions based on relative abundance.

2. The real analyses showed that the three packages were consistent in their per-sample compositions, resulting in only minor differences of the intra-sample distances based on weighted UniFrac and Bray-Curtis dissimilarity.
3. Overall, open-reference OTU clustering approach consistently identified considerably more OTUs than the number of ASVs generated by the denoising pipelines in all datasets tested.
4. It was shown that DADA2 tended to find more ASVs than UNOISE3 and Deblur (Deblur called the least amount of ASVs), suggesting that it could be better at finding rare organisms, but at the expense of possible false positives.
5. It also showed that UNOISE3 is very faster than DADA2 and Deblur, with DADA2 being the lowest with 1200 times lower than UNOISE3.
6. This study concluded that all pipelines result in similar general community structure; however, the number of ASVs/OTUs and resulting alpha diversity metrics varies considerably. Thus when attempting to identify rare organisms from possible background noise, we should consider that determining species richness within low-diverse samples could be problematic for the denoising pipelines (Nearing et al. 2018).

In 2020, Prodan (Prodan et al. 2020) compared six bioinformatic software for the analysis of amplicon sequence data: three for assigning OTUs (QIIME-uclust, mothur, and USEARCH-UPARSE) and three for generating ASVs (Callahan et al. 2017) or "sub-OTUs" (Amir et al. 2017) or "zero noise OTUs" (Edgar 2016b) (DADA2, Qiime2-Deblur, and USEARCH-UNOISE3). The overview of the strengths and weaknesses of different outputs from these six software was assessed in terms of the sensitivity, specificity, and degree of consensus. It was demonstrated that (Prodan et al. 2020):

1. DADA2 offered the best sensitivity but at the expense of decreased specificity compared to USEARCH-UNOISE3 and Qiime2-Deblur.
2. USEARCH-UNOISE3 had the best balance between resolution and specificity (Prodan et al. 2020).
3. Compared to ASV-level software, OTU-level software USEARCH-UPARSE and mothur performed well, but with lower specificity, whereas QIIME-uclust produced large number of spurious OTUs as well as inflated alpha diversity measures.

8.3.2.2 Pyrosequencing Flowgrams

Some important denoising-based methods for pyrosequencing flowgrams (patterns of intensities in each read) include:

- **PyroNoise** (Quince et al. 2009, 2011). The PyroNoise algorithm is the first amplicon sequencing error-correction method that was designed for clustering the flowgrams of 454 pyrosequencing reads using a distance measure to model sequencing noise. PyroNoise is used to accurately construct OTUs from 16S rRNA sequence data and assign pyrosequenced reads to known taxa (Quince et al. 2009). It may be more robust to noise. However, PyroNoise is computationally expensive and is difficult to implement for most users and thus has received limited application (Schloss et al. 2011).

- **DeNoiser** (Reeder and Knight 2010) is another denoising pyrosequencing algorithm that was proposed for denoising pyrosequencing amplicon reads by exploiting rank-abundance distributions. Like PyroNoise, DeNoiser was reviewed as computationally expensive and difficult for most users and has limited application (Schloss et al. 2011).
- **SLP** (single-linkage pre-clustering) (Huse et al. 2010) was proposed to address the problem that the complete-linkage clustering significantly increases the number of predicted OTUs and inflates richness estimates. SLP implements its algorithm via four steps: First it orders the unique sequences by frequency. Second, from the most abundant to the least sequence, it assigns them to clusters through testing each subsequent sequence against the growing list of clusters using the single-linkage algorithm. Third, it adds a new sequence to the cluster and stops testing against subsequent clusters if the sequence has a pairwise distance less than 0.02 (2%) to any of the sequences already in the cluster. It was shown that SLP can more accurately predict expected OTUs if followed by an average-linkage clustering based on pairwise alignments (Huse et al. 2010) and is not computationally expensive (Schloss et al. 2011).
- **AmpliconNoise** (Quince et al. 2011), a development of the PyroNoise algorithm, was proposed to remove noise from pyrosequenced amplicons to avoid inflation of diversity estimates of OTUs. It uses the error-model-based denoising algorithm to distinguish noise from true sequence diversity in data and uses expectation-maximization (EM) to inference the true sequences. It was shown that AmpliconNoise can separately remove 454 sequencing errors and PCR single-base errors to obtain accurate estimates of OTU number (Quince et al. 2011).
- **APDP** (Amplicon Pyrosequencing Denoising Program) (Morgan et al. 2013) was proposed to process raw sequence datasets into a set of validated sequences with compatible formats for facilitating downstream analyses. APDP uses both between-samples abundance distribution of similar sequences and within-samples frequency and diversity of sequences to distinguish real sequences from errors generated by PCR and sequencing. It was shown that APDP can effectively remove errors from both deeply sequenced datasets comprising biological and technical replicates and efficiently denoise single-sample datasets and thus provides more conservative or accurate biological diversity estimates (Morgan et al. 2013).
- Other 454 sequence programs for correcting pyrosequencing errors include **Acacia** (Bragg et al. 2012) and **HECTOR** (Wirawan et al. 2014).

Compared to denoising pyrosequencing flowgrams, Illumina denoisers have been developed more recently. Some Illumina methods can also be used with the 454 sequencing platform, such as cluster-free filtering (CFF) and SeekDeep. Below we will introduce CFF in Sect. 8.3.2.3, DADA2 in Sect. 8.3.2.4, UNOISE2 in Sect. 8.3.2.5, Deblur in Sect. 8.3.2.6, and SeekDeep in Sect. 8.3.2.7.

8.3.2.3 Cluster-Free Filtering (CFF)

In 2015, Tikhonov et al. (2015) proposed a cluster-free filtering (CFF) denoiser approach to achieve sub-OTU resolution with ecologically differing by as little as one nucleotide (nt) (99.2% similarity). The CFF denoiser was proposed in the contexts: on the one hand, to improve estimates of ecological diversity by reducing the impact of amplicon sequencing error, researchers have tried various denoising algorithms to reassign or eliminate noisy reads; however, the challenging still remains because no approach can fully address the denoiser issues. On the other hand, the alternative DBC approach is necessarily unreliable due to computational cost and severe limitation for low-count sequences in the cross-sample comparisons. Under this background, the focuses of the CFF denoiser are not to further try to improve the existing OTU clustering approaches and also not to attempt to identify rare species. Instead, the CFF denoiser combines error-model-based denoising and systematic cross-sample comparisons to resolve the fine (sub-OTU) structure of moderate-to-high-abundant taxonomy data (Tikhonov et al. 2015).

CFF employs cross-sample comparisons to achieve sub-OTU resolution. CFF was developed based on the observation that sequence similarity need not imply dynamical similarity and vice versa in time-series data. In contrast, the maximum correlation between the time traces (normalized counts versus observation day) of two sequences depends on their abundance measured by Poisson model. Thus CFF denoiser (Tikhonov et al. 2015) first defines the "dynamical similarity" of two traces as the Pearson correlation of their abundance, normalized by their maximum possible correlation. Then it models the abundance traces of these two sequences by an additive Poisson counting noise, setting an upper bound on the correlation coefficient for low-abundance sequences because Poisson sampling noise becomes non-negligible for low-abundance sequences. Finally it uses the Hamming distance metric to measure the pairwise sequence distances based on normalized counts.

Three important characteristics that characterize CFF as a method of moving beyond the traditional OTU methods are as follows: (1) aims to differ 16S sequences by one nucleotide or a single base pair, which is similar to oligotyping and distribution-based clustering approaches; (2) does not rely on clustering similar sequences together, which is similar to oligotyping; and (3) does not assume that sequence similarity implies ecological similarity. Instead CFF actually considers sequence similarity is a very poor predictor of ecological similarity (Tikhonov et al. 2015). Thus CFF uses highly conservative filtering criteria, assuming that sequence similarity contains only true biological sequences, that is, there are no false positives. Regarding this point, CFF is similar to Sokal and Sneath's numerical taxonomy (Sokal and Sneath 1963).

As a hybrid of error-model-based denoising and distribution-based clustering, CFF does not require manual supervision, which makes it different from oligotyping. In the meanwhile, CFF models an entire community instead of an isolated OTU, which is similar to DBC method. However, CFF denoiser and DBC method have different goals: DBC is an OTU clustering algorithm, whereas CFF denoiser is to identify sub-OTU structure of moderate-to-high-abundance community members. As a denoiser, CFF focuses on the substitution errors (Tikhonov et al. 2015), rather than all the three main sources of errors that often occur in Illumina platform to generate 16S data: PCR substitutions, PCR chimeras, and substitution errors due to Illumina base call errors. The CFF denoiser method has some advantages including:

1. CFF demonstrated "sequence similarity need not imply ecological similarity, and vice versa," and the ecological relatedness can be independently assessed. This is different from the tag-sequencing data analysis and clustering-based OTU methods in microbiome study because typically these methods assume that sequence similarity of 16S hypervariable regions can be used as a proxy for phylogenetic, and therefore ecological, relatedness (Tikhonov et al. 2015). This idea of CFF is similar to the important idea of separating taxonomic procedure from phylogenetic speculation in the original OTU methods (Sokal and Sneath 1963).
2. CFF can achieve a single-nucleotide (nt) sub-OTU resolution (Tikhonov et al. 2015).
3. Particularly, CFF can achieve the same accuracy as DADA, which was considered as the best denoiser in 2015 (Tikhonov et al. 2015), but has less execution time than DADA. We can ignore the advantage of execution time since the longer time used in DADA largely due to its exact treatment of probabilities for processing low abundant sequences while CFF denoiser discards sequences of low abundance prior to denoising. In other words, the speed of CFF is at the cost of its limitation.

However, CFF also has limitations: (1) It is specifically designed to be run on large multi-sample datasets. (2) It is not a replacement for OTU clustering; it focuses on moderate-to-high-abundance sequences and hence discards low-abundance sequences. Thus, CFF is unsuitable for studying population-level alpha or beta diversity (Tikhonov et al. 2015), which probably is an important limitation of this method.

8.3.2.4 DADA2

DADA2 and q2-dada2 plugin were introduced in Sect. 4.2.1. In line with providing single-nucleotide resolution without using arbitrary taxonomic similarity thresholds to define molecular OTUs, Callahan et al. (2017) proposed to use exact "sequence variants" to replace OTUs in marker-gene data analysis and developed software package DADA2 (DADA: Divisive Amplicon Denoising Algorithm) for modeling and correcting Illumina-sequenced amplicon errors. The proposal of using ASVs instead of the traditional OTUs was due to the limitations of OTU methods, such as clustering sequence reads into OTUs often eliminates biological information present in the data, and OTUs are not species, and their construction is not necessitated by amplicon errors (Callahan et al. 2016a), and especially was motivated by a series of new computational methods (Eren et al. 2013, 2016; Tikhonov et al. 2015; Edgar 2016b; Amir et al. 2017).

DADA2 is a parametric model that relies on input read abundances and distances. It assumes that true reads are likely to be more abundant and less abundant reads are likely error-derived which may be only a few base differences away from a more abundant sequence (Prodan et al. 2020; Callahan et al. 2016a). DADA2 was developed based on the DADA (Rosen et al. 2012). DADA was built on AmpliconNoise's error-modeling approach (Quince et al. 2011). The distinctive feature of DADA is its *divisive* hierarchical clustering algorithm, while previous methods, including AmpliconNoise and simple OTU clustering, use *agglomerative* approach.

The core denoising algorithm (Divisive Amplicon Denoising Algorithm) in the DADA2 is based on a Poisson model to estimate the errors in Illumina-sequenced amplicon reads. This error model estimates the probability for each amplicon read with each sequence is produced from sample sequence as a function of sequence composition and quality. The Poisson model assumes that the number of amplicon reads (the abundance) of each sequence is consistent with the error model and the p-value of the null hypothesis is calculated. The calculated p-values or quality scores (base Q-scores) are used as the division criteria for an iterative partitioning algorithm; sequencing reads will continue to be divided until all partitions are consistent with being produced from their central sequence.

Like other new methods (e.g., Deblur) for denoising sequences, the general goal of DADA2 is to provide single-nucleotide resolution and without using arbitrary taxonomic similarity thresholds to define molecular OTUs, which supposedly represent the true biological sequences present in the data. Thus, an exact "sequence variants" was proposed to replace OTUs in marker-gene data analysis. Both DADA and DADA2 use a model-based approach to correct amplicon errors without constructing OTUs. However, the applications of DADA and DADA2 are different: DADA was used to identify fine-scale variation in 454-sequenced amplicon data (Rosen et al. 2012), while DADA2 was used for modeling and correcting Illumina-sequenced amplicon errors (Rosen et al. 2012; Callahan et al. 2016a). As an extended and improved version of DADA, DADA2 is reference-free and applicable to any genetic locus. **DADA2** generates a parametric error model that is trained on the entire sequencing run and then applies that model to correct and collapse the sequence errors into amplicon sequence variants (ASVs). This approach has its advantages because:

1. DADA2 builds unique error models for each sequencing run (Nearing et al. 2018; Callahan et al. 2016b).
2. DADA2 exactly infers sample sequences and resolves differences of as little as one nucleotide (Callahan et al. 2016b).
3. DADA method was shown to outperform AmpliconNoise's and other previous methods in both speed and accuracy (Rosen et al. 2012).
4. DADA2 R package can implement the full amplicon workflow from filtering, dereplication, sample inference, and chimera identification to merging the paired-

end reads (Callahan et al. 2016a). Inferring ASVs (also called RSVs: ribosomal sequence variants) was shown in amplicon bioinformatic workflow (Callahan et al. 2016a, b, 2017).

5. Particularly, it was demonstrated that the DADA2 methods were more accurate than the four OTU methods: UPARSE (Edgar 2013), MED (minimum entropy decomposition) (Eren et al. 2015), mothur (average linkage) (Schloss et al. 2009), and QIIME (uclust) (Caporaso et al. 2010).

However, DADA2 approach also has its limitations:

1. As reviewed above, it was shown (Nearing et al. 2018; Prodan et al. 2020) that DADA2 had the best sensitivity and resolution, but at the expense of producing more number of spurious ASVs compared to USEARCH-UNOISE3 and Qiime2-Deblur.
2. It was also shown (Hathaway et al. 2017) that like MED and UNOISE, DADA2 could create larger false haplotypes especially for 454 technique. In DADA2, when the number of false haplotypes is minimized, the sensitivity is also lost, especially for the lower read depth input, in which the low-abundance one-off haplotypes are missed (Hathaway et al. 2017). For Ion Torrent and 454 pyrosequencing data, both UNOISE and DADA2 cannot achieve 100% haplotype recovery compared to 100% haplotype recovery in SeekDeep and MED (Hathaway et al. 2017).
3. It was also shown (Tikhonov et al. 2015) that DADA has a long execution time largely due to its exact treatment of probabilities. This is critically important to process sequences with an abundance of just a few counts.

Overall, DADA2 has been recognized as so far the best denoising-based method for focusing on the highest possible biological resolution (e.g., differentiating closely related strains) (Prodan et al. 2020).

8.3.2.5 UNOISE2 and UNOISE3

To improve error correction in Illumina 16S and ITS amplicon sequencing, distinguishing the sequence errors caused by PCR and sequencing from true biological variation, Edgar (2016b) proposed UNOISE2, an updated version of the UNOISE algorithm (Edgar and Flyvbjerg 2015) for denoising (error-correcting) Illumina amplicon reads.

UNOISE2 uses a one-pass clustering strategy without using quality (Q) scores. The clustering strategy is different from DADA2's model-based approach, which uses quality scores in an iterative divisive partitioning clustering strategy. In UNOISE2, both algorithms of UCHIME2 (Edgar 2016a) and DADA2 are incorporated to reduce the number of incorrect sequences. However, both UNOISE2 and DADA2 were reviewed as still generating some amount of false positives with DADA2 likely to increase the number of false-positive chimera predictions (Edgar 2016b). In UNOISE2, the obtained predicted biological sequences are called as

ZOTUs (zero-radius OTUs) (Edgar 2016b, 2018). The ZOTU is similar to Callahan et al.'s ASV. In DADA2, the 97% similarity of OTUs is replaced by ASVs as the standard unit of marker-gene analysis and reporting; in UNOISE2, similarly the [97%] OTUs are replaced by ZOTUs (Edgar 2018).

The UNOISE3 denoising method first uses the "unoise3" command to rank sequences in decreasing order of abundance and then discards those sequences with counts less than the specified minimum abundance threshold (the default value is 8). By using 100% identity threshold, each distinct sequence defines a separate OTU. Both OTU-level clustering with UPARSE and ASV-level denoising with UNOISE3 are implemented via USEARCH, and actually the author of USEARCH (Edgar 2010) recommended that UPARSE and UNOISE3 should be performed together.

UNOISE2 and UNOISE3 have the advantages:

1. UNOISE2 has comparable or better accuracy than DADA2 demonstrated by the author of UNOISE2 (Edgar 2016b).
2. USEARCH-UNOISE3 (UNOISE2) had arguably the best overall performance compared to DADA2 and Qiime2-Deblur if combinedly considering its high sensitivity with excellent specificity (Prodan et al. 2020).
3. Especially USEARCH-UNOISE3 and Qiime2-Deblur had the perfect specificity, producing no spurious OTUs/ASVs based on the mock sample sequencing data analysis (Prodan et al. 2020).

However, UNOISE also has limitations:

1. Like DADA2, UNOISE was designed and tested on Illumina reads. It does not work well on Ion Torrent and 454 data and hence cannot achieve 100% haplotype recovery (Edgar 2021; Hathaway et al. 2017).
2. UNOISE collapses one-off errors if the abundance ratio of two sequences achieves a certain threshold, recommends not using singlet sequences, and hence decreases haplotype recovery at lower read depths. This in part contributes to UNOISE's speed and makes it to be the fastest algorithm compared to MED, DADA2, and SeekDeep (Hathaway et al. 2017). However, this also costs UNOISE to be unable to detect new haplotypes that differ by only one nucleotide (Hathaway et al. 2017).

8.3.2.6 Deblur

We illustrated Qiime2-Deblur in Chap. 4 (Sect. 4.4). Amir et al. (2017) developed a novel sub-OTU (sOTU) method called Deblur to fast and accurately identify exact amplicon sequences and to integrate large datasets. Like DADA2 and UNOISE2, Deblur aims to identify real ecological differences between taxa with a single base pair of amplicons differentiation. Deblur in concept is similar as DADA2 and UNOISE2, but uses a different algorithm to obtain single-nucleotide resolution. Deblur performs each sample independently and compares sequence-to-sequence

Hamming distances within a sample to an upper-bound error profile by combining with a greedy algorithm (Amir et al. 2017).

The Deblur algorithm is implemented via three steps (Amir et al. 2017): First, Deblur sorts sequences by abundance. Second, Deblur subtracts the number of predicted error-derived reads from the counts of neighboring reads based on their read-to-read Hamming distance using an upper bound on the error probability based on the most to least abundant sequence. Finally, Deblur removes any sequence whose abundance drops to 0 after a subtraction. Thus, after applying Deblur, the invalid (i.e., noise) sequences are removed, and only reads likely to have been presented to the sequencer are retained.

Like DADA2 and UNOISE2, in Deblur the stable sOTUs were proposed to be used for replacing [97%] OTUs. Deblur has been wrapped into QIIME 2. Deblur has the advantages:

1. Deblur is comparable or better than DADA2 and UNOISE2 in terms of performance characteristics and stability (i.e., obtaining the same sOTU across different samples) (Amir et al. 2017). For example, although all three methods can identify sOTUs with single-nucleotide differences which are close to the ground truth, Deblur has the greater stability than DADA2 and UNOISE2.
2. Unlike DADA2 and UNOISE2, Deblur performs each sample independently without requiring operation on the full study and therefore can be easily parallelized to very large projects.
3. Particularly, in one study that compared six bioinformatic software for the analysis of amplicon sequence data including QIIME-uclust, mothur, and USEARCH-UPARSE for generating OTUs and DADA2, Qiime2-Deblur, and USEARCH-UNOISE3 for generating ASVs, Qiime2-Deblur is one of the only two pipelines (another is USEARCH-UNOISE3) that showed perfect specificity on the mock sample sequencing data, producing no spurious OTUs/ASVs (Prodan et al. 2020).

Deblur also has the disadvantages, such as Deblur has poor performance compared to DADA2 and open-reference OTU clustering when it is used to detect low abundant taxa in the extreme dataset at 97% identity (Nearing et al. 2018). For comparing Deblur and DADA2, the reader is also referred to Sect. 4.4.6.

8.3.2.7 SeekDeep

SeekDeep (Hathaway et al. 2017) is an error-correction method for de novo (i.e., reference-free) analysis of amplicons. SeekDeep was developed in the contexts of comparing to three pipelines that aim for single-base resolution to determine the local PCR amplicon haplotypes (simply as haplotypes for brevity), MED (Eren et al. 2014), DADA2 (Callahan et al. 2016a), and UNOISE in USEARCH (Edgar 2016b), as well as two OTU-based clustering pipelines which cannot resolve at the single-base level: USEARCH (aka UCLUST/UPARSE) (Edgar 2016b) and Swarm (Edgar 2013). SeekDeep aims for single-base resolution to provide improved resolution toward species and strains. The central algorithm of SeekDeep is the qluster (for quality clustering) that improves the correction of PCR and sequencing errors through base quality values and k-mer frequencies and other multiple key ways. SeekDeep can be utilized with Ion Torrent, 454, and Illumina sequencing technologies. SeekDeep consists of four components (Mahé et al. 2014): (1) *extractor* for de-multiplexing and read filtering, (2) *qluster* for rapid and accurate clustering based on quality, (3) *processClusters* for replicate and population comparisons, and (4) *popClusteringViewer* for viewing and manipulating final results. The first three main components are central to generating clustering results, and the fourth is an additional component to aid in viewing and sharing the results.

It was shown (Hathaway et al. 2017) that SeekDeep has the advantages:

1. SeekDeep is able to resolve sequences differing by only a single base.
2. SeekDeep has greater consistency even at low frequencies.
3. Particularly, SeekDeep has matched or outperformed the five pipelines including MED, DADA2, UNOISE in USEARCH, USEARCH, and Swarm in terms of haplotype recovery (especially one-off haplotypes), accuracy of predicting the abundance (accurate abundance estimates), and the number and abundances of false haplotypes.

However, Hathaway et al. (2017) also acknowledged that SeekDeep creates more false haplotypes than DADA2 and UNOISE although the false haplotypes have much lower abundance.

8.3.2.8 Remarks on Denoising-Based Methods

Denoising-based methods are widely used for identifying low-abundance or rare species against a noisy background, often with the aim of improving estimates of ecological diversity (Hathaway et al. 2017), such as Chao1 richness.

Using 97% identity may merge phenotypically different strains with distinct sequences into a single cluster (Tikhonov et al. 2015; Callahan et al. 2016a, b, c). In contrast, the denoising usually results in a set of predicted biological sequences or valid OTUs, which often have single-base resolution. Thus, compared to OTU methods, in most cases denoising algorithms provide the maximum possible biological resolution and hence superior to conventional OTU methods.

However, the denoising-based methods cannot fully address all the issues of bioinformatic analysis of microbiome data although these methods were proposed for identifying rare species and improving the accuracy of ecological diversity estimates, because (1) all error models are necessarily approximate, and no denoising algorithm can deal with errors that are beyond their model capabilities, and (2) the denoising-based methods are particularly problematic when they are used for calling low-abundance species. As reviewed above, DADA2 had the best sensitivity and resolution, but also produced more number of spurious ASVs, while UNOISE2 was fast, but was not recommended for using singlet sequences due to decreased haplotype recovery at lower read depths.

8.4 Discussion on Moving Beyond OTU Methods

The concept of OTU and OTU methods were originated from numerical taxonomy. OTU is an operational taxonomic unit or the organizational level of a unit character. It was created in numerical taxonomy to facilitate taxonomic classification against previous subjective approaches of taxonomy. The OTU methods in numerical taxonomy can be characterized with three concepts and one assumption. The three concepts are OTU, similarity, and clustering. The assumption is that taxonomic classification can be done through clustering the similarities of characters with large random sampling. It separates taxonomic classification from phylogenetic speculation. In numerical taxonomy, the OTUs are used for classification of characters via commonly used clustering with the assumption that characters can be grouped based on their similarities.

In early microbiome studies, the concept of OTU and OTU methods were adopted for convenience, and using the 97% sequence identity for approximating species is arbitrary and was based on researchers' experiences but not substantial experiments. In microbiome, the use of OTUs serves the dual purpose: grouping sequences (the third purpose of playing species functionality is the extension of the first purpose) and estimating diversity. The final purpose of using OTUs is to detect the association between microbiome and health and the development of the disease. When the concept of OTU and OTU methods were adopted into microbiome study, not only the commonly used clustering methods but also various methods including heuristic clustering, model-based methods were developed to improve and refine OTUs, all with the assumption that sequence similarity can predict taxonomic similarity.

Under the framework of OTU methods, the dual purpose of using OTUs has been demonstrated failing to achieve. In the meanwhile, various new methods have been proposed in recent years with the goal to achieve single-nucleotide resolution of sequences and not necessarily relying on arbitrary sequence identity and clustering methods. Among different methods that were newly proposed, the denoising-based methods have been demonstrated their promising. However, either the failure of dual purpose of using OTU methods or the promising of new bioinformatic sequence analysis raises a number of interesting questions which deserve further discussion.

8.4.1 Necessity of Targeting Single-Base Resolution

First, whether it is necessary to target single-base resolution?
On the one hand, clinically and for downstream analysis of microbiome data, a consistent single-base resolution of analysis units is necessary. Clinically, consistently differentiating single-base differences of analysis units is particularly a necessity for studying eukaryotic intraspecies populations and for mutation detection. It is crucial for seeking to detect and quantify minority haplotypes that may be

represented by *a single-nucleotide polymorphism* (SNP), a nucleotide difference in a single DNA building block. SNP is the most common type of genetic variation among people. SNPs have been used to detect strains (also refers to subspecies) of clinical relevance or to predict phenotypic characteristics when they are stably associated with other parts of the bacterial haplotype (Fitz-Gibbon et al. 2013). In the oncology and infectious disease fields, such studies are becoming increasingly common. In these studies, these sequences often only differ from the wild type by a single base when using marker regions to bacterial strains or monitoring for pathogen drug resistance mutations. Thus, it is crucial to accurately quantify these low-abundance and genetically similar strains (Miotto et al. 2015; Hathaway et al. 2017). Another example provided by Tikhonov et al. (2015) is malaria research, where the strains within an infected individual are often defined by the sequence of a single amplicon, and these sequences usually differ by only a single base, representative of a SNP within the larger parasite population.

Recently, Johnson et al. (2019) suggested that intragenomic variation between 16S gene copies must be necessarily accounted for in modern bioinformatic analysis. They also demonstrated that full-length 16S intragenomic copy variants can potentially provide taxonomic resolution of bacterial communities at species and strain level when they are appropriately treated, where the full 16S gene refers to the ~1500 bp 16S rRNA gene comprising nine variable regions (V1–V9) interspersed throughout the highly conserved 16S sequence. Hassler et al. (2022) found that alignment SNP count strongly predicted concordance for any given gene and the strongest concordance was displayed by the SNPs from non-ribosomal protein coding genes and the weakest concordance was shown by the SNPs from rRNA genes. Statistically, single-base resolution of 16S amplicon clustering improves accuracy and sensitivity of traditional OTUs and thus extracts maximal information for downstream analysis (Tikhonov et al. 2015).

On the other hand, the techniques for obtaining a consistent single-base resolution of atomic analysis units are still challenging. For example, as we reviewed, DADA2 could have the best sensitivity and resolution, but also is prone to inflate spurious ASVs, while UNOISE2 could be fast, but without recommending using singlet sequences because of decreased haplotype recovery at lower read depths. SeekDeep may create more false haplotypes than DADA2 and UNOISE2.

8.4.2 Possibility of Moving Beyond Traditional OTU Methods

Second, whether it is possible to move beyond the traditional OTU-based methods?

Regardless of the names either called "oligotypes," "ASVs," "ZOTUs," "sOTU," or others, the authors of all the newly developed methods explicitly or inexplicitly want to replace the traditional OTUs by their proposed sub-OTUs.

New Methods Outperform the Traditional OTU Methods We review the possibilities of using sub-OTUs to replace OTUs from four perspectives below:

1. The new available methods provide better resolution than OTU methods. The sub-OTU methods can obtain single-nucleotide resolution, which makes it possible to identify real ecological differences between taxa or organisms in communities whose amplicons differ by a very small number of nucleotides (i.e., a single base pair, a single nucleotide) (Hathaway et al. 2017; Eren et al. 2013; Callahan et al. 2016a; Amir et al. 2017). For example, the identified oligotypes by oligotyping offer a comparable or higher level of resolution than OTUs at 97%, while the traditional OTU clustering at 97% identity level does not have so high resolution.

2. The new available methods are more accurate than OTU methods with demonstrating as good or better sensitivity and specificity than OTU methods. For example, oligotyping and ASV methods can separate different environments more efficiently and have increased power to discriminate better ecological distribution patterns that taxonomical classification and OTU clustering at 97% identity level cannot detect (Eren et al. 2013, 2014, 2015; Callahan et al. 2016a). ASVs even can capture all biological variation present in the data (Callahan et al. 2017). Swarm2 can distinguish higher-resolution clusters or different taxa, while traditional UCLUST cannot detect (Callahan et al. 2017). Deblur also has similar or better sensitivity and specificity while it substantially reduces computational demands relative to similar sOTU methods (Mahé et al. 2015b).

3. Like PiCRUST (Amir et al. 2017), sub-OTUs are functional and biological or clinically meaningful. For example, the identified subpopulations of a single species by a sequence or the sub-OTU can provide functional information: insight into functional relatedness of community members (Langille et al. 2013). Sequences differing by as little as one nucleotide (99.2% similarity) within the 16S rRNA gene can be ecologically distinct and reveal clinical or epidemiological associations that would be missed by genus-level or species-level categorization of 16S rRNA data (Tikhonov et al. 2015).

4. The sub-OTUs facilitate downstream analysis due to their precision, reusability across studies, and reproducibility in future datasets. The sub-OTU methods, such as oligotyping, cluster-free filtering, and DADA2, were shown not only comparable or better than traditional clustering analysis at explaining the structure of the bacteria dataset but also can avoid vast underestimates of ecological richness, improve accurate measurement of diversity and dissimilarity, provide new insight into factors shaping community assembly and the prevalence of strain exchange between communities and invasion/extinction dynamics of OTU subpopulations, and resolve distinct subpopulations with high dynamical similarity. Thus, due to its precision, the sub-OTU methods could potentially allow amplicon methods to probe strain-level variation (Eren et al. 2011, 2013; Callahan et al. 2016a). For traditional OTU methods, after filtering the sequences and removing the chimer, a sample-by-OTU feature table is generated to serve as the basis for further analysis. Most sub-OTU methods also can obtain a sample-by-sub-OTU feature

table for downstream analysis. Probably, more importantly, sub-OTUs are reusable across studies and comprehensive and reproducible in future datasets because they are more precise and combine the benefits for subsequent analysis of closed-reference and de novo OTUs (Tikhonov et al. 2015). We can continue our discovery by exploring the present and future large datasets and through reuse of existing rich datasets (Callahan et al. 2017). Some sub-OTUs, such as DADA2, can easily process Illumina samples on a laptop, and a workflow from raw data to downstream analysis was shown in recent publications (Callahan et al. 2016a, 2017; Amir et al. 2017). In summary, all these advantages of sub-OTUs, especially more accurate abundance estimates of sub-OTUs, facilitate downstream statistical analyses using sophisticated parametric or nonparametric models (Callahan et al. 2016b).

Incorporating Sub-OTU Methods in Bioinformatic Tools Based on above review, the sub-OTU methods are better than OTU methods. Most of them are stable and able to integrate and have good performance and open-source license. Using sub-OTU methods in marker-gene data analysis will be a trend in bioinformatic analysis of microbiome data or at least is an alternative method of OTU approach. The sub-OTUs are the good alternative to OTUs as the analysis units of microbiome data.

The usage of these new sub-OTU methods is facilitated by using them in datasets and bioinformatics pipelines. For example, since its introduction, Swarm v1 has been used in a variety of datasets (Callahan et al. 2016a; de Vargas et al. 2015; Filker et al. 2015; Lima-Mendez et al. 2015; Mahé et al. 2015a). QIIME already offers Swarmv1.2, to facilitate its usage, and Swarm v2 can be included in QIIME2 and in Galaxy (Oikonomou et al. 2015; Mahé et al. 2015b). QIIME 2 has already incorporated DADA2 by the use of exact "ASVs" rather than "OTUs." Because of its stability and open-source license, Deblur is easy for commercial adoption and peer scrutiny and to be integrated (Goecks et al. 2010). The R package DADA2 was developed to show its usage in analysis of microbiome data (Amir et al. 2017).

8.4.3   Issues of Sub-OTU Methods

Third, whether there still exist issues in sub-OTU methods after moving beyond the use of OTUs and OTU-based methods?
Sub-OTU methods open potential new directions for a more accurate depiction of microbiome communities through marker-gene amplicons. However, the new methods also have the issues and come with new questions.

First, although the new methods can distinguish highly similar sequence variants, there is still no guarantee that the resulting sub-OTUs are necessarily phylogenetically and ecologically meaningful due to limitations of the selected 16S rRNA gene (Callahan et al. 2016a; Berry et al. 2017). Thus, the inference based on sub-OTUs (i.e., ASVs) does not solve all problems (Callahan et al. 2017). Second, not all new methods are suitable for studies of microbiome communities. For example, in the cluster-free filtering, the low-abundance sequences are discarded; thus, we cannot use it to study population-level alpha or beta diversity (Callahan et al. 2017). Third, to be more accessible, computational and ecological issues still need to be addressed for new methods. To move beyond the use of OTUs, the microbial ecologists and the developers of widely used software platforms still need further works (Tikhonov et al. 2015).

8.4.4 Prediction of Sequence Similarity to Ecological Similarity

Fourth, whether we need to avoid the assumption that the sequence similarity predicts the ecological similarity or phylogeny?
In literature, on the one hand, it was exhibited that 16S-based methods have limitations, and on the other hand, it was demonstrated that even a purely 16S-based study can exhibit ecologically significant distinctions to provide insight into functional relatedness of community members (Eren et al. 2016; Langille et al. 2013; Tikhonov et al. 2015). Thus, the assumption of the sequence similarity predicting the ecological similarity or phylogeny remains unanswered.

8.4.5 Functional Analysis and Multi-omics Integration

Fifth, whether functional analysis and multi-omics integration are necessary and important in microbiome study?
Discussion on the advantages and disadvantages of the OTU methods and the movement of moving beyond OTU methods highlights the importance of two topics in bioinformatic and statistical analysis of microbiome data: (1) What are the optimal analysis units of microbiome data? (2) What kind of science microbiome should be?

Regarding the first topic, compared to OTU methods, in general the methods that provide ASVs or single-nucleotide base resolution are more close to microbiome sequence data and serve the purposes of microbiome study. We have shown that targeting single-base resolution is necessary and the movement of moving beyond traditional OTU-methods is possible in bioinformatic analysis of microbiome data. However, as we reviewed above, the ASVs and single-nucleotide base resolution methods also have limitations, and not all OTU methods are inferior to all the ASV methods. Actually, all statistical models are used to approximate the real data but are not expected to get insight into all the complexities of reality; i.e., every single model will never represent the exact real data and rare achieve all the purposes of the analysis and the study supposed to be; the goal of model comparisons is to find the model that is sufficiently close to the reality because it is useful even if this model cannot describe exactly the reality. The challenge is how to choose the parsimonious bioinformatic models that provide remarkably useful approximations to the real sequencing data and extract maximal information for statistical analysis to obtain insight into clinical relevance. Here, Ockham's razor [William of Ockham (c. 1287–1347)] or the principle of parsimony or law of parsimony is applicable. The principle of parsimony states that we should choose the simplest model while approximating to reality because "entities should not be multiplied beyond necessity" (Schaffer 2015).

Regarding the second topic: what kind of science is microbiome? These are the questions: How to define microbiome? What research theme in microbiome study? What are the unique data structure and characteristics? The interested reader is referred to Xia and Sun (2022).

Briefly, we define microbiome as an experimental science. As an experimental science, microbiome does not solely lie on direct observation and measurement of objects/samples and phenomena to identify association or correlation, but more importantly conducts experiments to find the functions and mechanism/causality that play in the human health and the development of disease. Like other experimental sciences, "inference from the particular to the general must be attended with some degree of uncertainty, but this is not the same as to admit that such inference cannot be absolutely rigorous, for the nature and degree of uncertainty may itself be capable of rigorous expression" [see Fisher (1935)'s Design of Experiments (p. 4)].

Actually, for microbiome study, the accurate coding of sequence data by a bioinformatic tool is just the first step. We need multiple approaches, especially including experiments and clinical trials to conform and validate the findings of bioinformatic analyses.

Microbiome study as a research field is not merely descriptive and observational. Describing diversity and visualizing the differences between the experimental groups are necessary. But the final goal of microbiome study is to find the causality between microbiome and the human health and the development of disease and to improve human health and prevent disease using the findings from microbiome research. The diversity concept was adopted from macroecology. The alpha diversity including richness and evenness are important in macroecology. But are they equally important in microbial ecology if considering their differences and complicities of microbial ecology? The answer is no.

Currently two approaches in microbiome research tend to be more important although they are more complicated: One is functional study, which is to find out the functionality of microbiota, identify the top microbes, and use them in experiment study to test their functions. Another is to integrate microbiome with other omics, such as metabolomics, metaproteomics, transcriptomics, genomics, and epigenomics to find the biomarkers that are associated with favorable and adverse outcomes. Finally the findings from microbiome research are expected to apply to human health and treat the human diseases.

Overall, we want to highlight the importance of functional, experimental studies with multiple omics' biological design. Like other experimental sciences, the final purpose of microbiome study is to find the mechanism of microbiome and the links to biological and environmental factors and to establish causality of them. Thus, an experimental and functional study is more important than an operational method. However, any experimental and functional study needs statistical methods to confirm its effect.

8.5 Summary

In this chapter we first described how the commonly clustering-based OTU methods were adopted in microbiome study and how the heuristic clustering OTU methods have been developed to advance the assignment of OTUs in microbiome study. We also described the limitations of clustering-based OTU methods and the purposes for the use of OTUs in microbiome study and discussed defining species and species-level analysis. Next, we introduced the concept shifting in bioinformatic analysis and the two single-nucleotide resolution-based OTU methods: distribution-based clustering (DBC) and Swarm2, which provided the contexts of moving beyond the OTU methods. Then, we focused on two kinds of moving beyond the OTU methods: oligotyping, the entropy-based single-nucleotide resolution non-clustering OTU method, and denoising-based methods including pyrosequencing flowgrams and Illumina denoisers. In bioinformatic analysis of microbiome data, pyrosequencing techniques were early developed, while Illumina techniques represented the current development. Thus we focused on five newly developed Illumina denoisers: (1) cluster-free filtering (CFF), (2) DADA2, (3) UNOISE2 and UNOISE3, (4) Deblur, and (5) SeekDeep. We also introduced overall comparisons of denoising-based and clustering-based methods and commented on the denoising-based methods. In the final parts of this chapter, we discussed on the movement of moving beyond OTU methods, highlighting the necessity of targeting single-base resolution, discussing the possibility of moving beyond traditional OTU methods, the issues of sub-OTU methods, the assumption of sequence similarity predicting to the ecological similarity, as well as the functional analysis and multi-omics integration. Starting with Chap. 9 through Chap. 18, we will focus on investigating statistical analysis of microbiome data. In Chap. 9, we will introduce alpha diversity.

References

Albanese, Davide, Paolo Fontana, Carlotta De Filippo, Duccio Cavalieri, and Claudio Donati. 2015. MICCA: A complete and accurate software for taxonomic profiling of metagenomic data. *Scientific Reports* 5 (1): 9743. https://doi.org/10.1038/srep09743.

Almeida, Alexandre, Alex L. Mitchell, Aleksandra Tarkowska, and Robert D. Finn. 2018. Benchmarking taxonomic assignments based on 16S rRNA gene profiling of the microbiota from commonly sampled environments. *GigaScience* 7 (5): giy054.

Amann, Rudolf I., Chuzhao Lin, Rebekah Key, Larry Montgomery, and David A. Stahl. 1992. Diversity among fibrobacter isolates: Towards a phylogenetic classification. *Systematic and Applied Microbiology* 15 (1): 23–31.

Amir, Amnon, Daniel McDonald, Jose A. Navas-Molina, Evguenia Kopylova, James T. Morton, Zhenjiang Zech Xu, Eric P. Kightley, Luke R. Thompson, Embriette R. Hyde, Antonio Gonzalez, and Rob Knight. 2017. Deblur rapidly resolves single-nucleotide community sequence patterns. *mSystems* 2 (2). https://doi.org/10.1128/mSystems.00191-16. https:// msystems.asm.org/content/msys/2/2/e00191-16.full.pdf.

Ash, Carol, John A.E. Farrow, Matthias Dorsch, Erko Stackebrandt, and Matthew D. Collins. 1991. Comparative analysis of Bacillus anthracis, Bacillus cereus, and related species on the basis of reverse transcriptase sequencing of 16S rRNA. *International Journal of Systematic and Evolutionary Microbiology* 41 (3): 343–346.

Balakrishnan, Rohini. 2005. Species concepts, species boundaries and species identification: A view from the tropics. *Systematic Biology* 54 (4): 689–693.

Baldi, Pierre, Søren Brunak, Yves Chauvin, Claus A.F. Andersen, and Henrik Nielsen. 2000. Assessing the accuracy of prediction algorithms for classification: An overview. *Bioinformatics* 16 (5): 412–424. https://doi.org/10.1093/bioinformatics/16.5.412.

Barriuso, Jorge, Jose R. Valverde, and Rafael P. Mellado. 2011. Estimation of bacterial diversity using next generation sequencing of 16S rDNA: A comparison of different workflows. *BMC Bioinformatics* 12 (1): 473. https://doi.org/10.1186/1471-2105-12-473.

Bazin, Alexandre, Didier Debroas, and Engelbert Mephu Nguifo. 2019. A de novo robust clustering approach for amplicon-based sequence data. *Journal of Computational Biology* 26 (6): 618–624.

Bennasar, Antonio, Ramon Rossello-Mora, Jorge Lalucat, and Edward R.B. Moore. 1996. 16S rRNA gene sequence analysis relative to genomovars of Pseudomonas stutzeri and proposal of Pseudomonas balearica sp. nov. *International Journal of Systematic and Evolutionary Microbiology* 46 (1): 200–205.

Berry, Michelle A., Jeffrey D. White, Timothy W. Davis, Sunit Jain, Thomas H. Johengen, Gregory J. Dick, Orlando Sarnelle, and Vincent J. Denef. 2017. Are oligotypes meaningful ecological and phylogenetic units? A case study of microcystis in freshwater lakes. *Frontiers in Microbiology* 8 (365). https://doi.org/10.3389/fmicb.2017.00365. https://www.frontiersin.org/article/ 10.3389/fmicb.2017.00365.

BLASTLab. 2004. *Using BLASTClust to make non-redundant sequence sets*. Last Modified Spring, 2004.   https://www.ncbi.nlm.nih.gov/Web/Newsltr/Spring04/blastlab.html.   Accessed 1 Dec 2004.

Bonder, Marc J., Sanne Abeln, Egija Zaura, and Bernd W. Brandt. 2012. Comparing clustering and pre-processing in taxonomy analysis. *Bioinformatics* 28 (22): 2891–2897. https://doi.org/10. 1093/bioinformatics/bts552.

Borneman, J., and E.W. Triplett. 1997. Molecular microbial diversity in soils from eastern Amazonia: Evidence for unusual microorganisms and microbial population shifts associated with deforestation. *Applied and Environmental Microbiology* 63 (7): 2647–2653. https://doi.org/10. 1128/aem.63.7.2647-2653.1997. https://pubmed.ncbi.nlm.nih.gov/9212415, https://www.ncbi. nlm.nih.gov/pmc/articles/PMC168563/.

Bragg, Lauren, Glenn Stone, Michael Imelfort, Philip Hugenholtz, and Gene W. Tyson. 2012. Fast, accurate error-correction of amplicon pyrosequences using Acacia. *Nature Methods* 9 (5): 425–426. https://doi.org/10.1038/nmeth.1990.

Brenner, Don J., George R. Fanning, Adrian V. Rake, and Karl E. Johnson. 1969. Batch procedure for thermal elution of DNA from hydroxyapatite. *Analytical Biochemistry* 28: 447–459.

Brown, Emily A., Frédéric J.J. Chain, Teresa J. Crease, Hugh J. MacIsaac, and Melania E. Cristescu. 2015. Divergence thresholds and divergent biodiversity estimates: Can metabarcoding reliably describe zooplankton communities? *Ecology and Evolution* 5 (11): 2234–2251. https://doi.org/10.1002/ece3.1485. https://www.ncbi.nlm.nih.gov/pubmed/260 78859, https://www.ncbi.nlm.nih.gov/pmc/PMC4461424/.

Buchanan, R.E. 1955. Taxonomy. *Annual Reviews in Microbiology* 9 (1): 1–20.

Buttigieg, Pier Luigi, and Alban Ramette. 2014. A guide to statistical analysis in microbial ecology: A community-focused, living review of multivariate data analyses. *FEMS Microbiology Ecology* 90 (3): 543–550. https://doi.org/10.1111/1574-6941.12437.

Cai, Yunpeng, and Yijun Sun. 2011. ESPRIT-Tree: hierarchical clustering analysis of millions of 16S rRNA pyrosequences in quasilinear computational time. *Nucleic Acids Research* 39 (14): e95. https://doi.org/10.1093/nar/gkr349. https://www.ncbi.nlm.nih.gov/pubmed/21596775, https://www.ncbi.nlm.nih.gov/pmc/PMC3152367/.

Cai, Yunpeng, Wei Zheng, Jin Yao, Yujie Yang, Volker Mai, Qi Mao, and Yijun Sun. 2017. ESPRIT-Forest: Parallel clustering of massive amplicon sequence data in subquadratic time. *PLoS Computational Biology* 13 (4): e1005518.

Callahan, Benjamin J., Paul J. McMurdie, Michael J. Rosen, Andrew W. Han, Amy Jo A. Johnson, and Susan P. Holmes. 2016a. DADA2: High-resolution sample inference from Illumina amplicon data. *Nature Methods* 13 (7): 581–583. https://doi.org/10.1038/nmeth.3869. https://www.ncbi.nlm.nih.gov/pubmed/27214047, https://www.ncbi.nlm.nih.gov/pmc/articles/PMC4927377/.

Callahan, Ben J., Kris Sankaran, Julia A. Fukuyama, Paul J. McMurdie, and Susan P. Holmes. 2016b. Bioconductor workflow for microbiome data analysis: From raw reads to community analyses. *F1000Research* 5: 1492. https://doi.org/10.12688/f1000research.8986.2. https://www.ncbi.nlm.nih.gov/pubmed/27508062, https://www.ncbi.nlm.nih.gov/pmc/PMC4955027/.

Callahan, B.J., K. Sankaran, J.A. Fukuyama, P.J. McMurdie, and S.P. Holmes. 2016c. Bioconductor workflow for microbiome data analysis: From raw reads to community analyses [version 1; referees: 3 approved]. *F1000Research* 5 (1492). https://doi.org/10.12688/f1000research.8986.1. http://openr.es/7gi.

Callahan, Benjamin, Paul McMurdie, and Susan Holmes. 2017. Exact sequence variants should replace operational taxonomic units in marker-gene data analysis. *The ISME Journal* 11: 2639.

Caporaso, J. Gregory, Justin Kuczynski, Jesse Stombaugh, Kyle Bittinger, Frederic D. Bushman, Elizabeth K. Costello, Noah Fierer, Antonio Gonzalez Peña, Julia K. Goodrich, Jeffrey I. Gordon, Gavin A. Huttley, Scott T. Kelley, Dan Knights, Jeremy E. Koenig, Ruth E. Ley, Catherine A. Lozupone, Daniel McDonald, Brian D. Muegge, Meg Pirrung, Jens Reeder, Joel R. Sevinsky, Peter J. Turnbaugh, William A. Walters, Jeremy Widmann, Tanya Yatsunenko, Jesse Zaneveld, and Rob Knight. 2010. QIIME allows analysis of high-throughput community sequencing data. *Nature Methods* 7 (5): 335–336. https://doi.org/10.1038/nmeth.f.303.

Carlsen, Tor, Anders Bjørnsgaard Aas, Daniel Lindner, Trude Vrålstad, Trond Schumacher, and Håvard Kauserud. 2012. Don't make a mista(g)ke: Is tag switching an overlooked source of error in amplicon pyrosequencing studies? *Fungal Ecology* 5 (6): 747–749. https://doi.org/10.1016/j.funeco.2012.06.003. https://www.sciencedirect.com/science/article/pii/S1754504812000918.

Caron, David A., Peter D. Countway, Pratik Savai, Rebecca J. Gast, Astrid Schnetzer, Stefanie D. Moorthi, Mark R. Dennett, Dawn M. Moran, and Adriane C. Jones. 2009. Defining DNA-based operational taxonomic units for microbial-eukaryote ecology. *Applied and Environmental Microbiology* 75 (18): 5797–5808. https://doi.org/10.1128/aem.00298-09. https://www.ncbi.nlm.nih.gov/pubmed/19592529, https://www.ncbi.nlm.nih.gov/pmc/PMC2747860/.

Chakravorty, S., D. Helb, M. Burday, N. Connell, and D. Alland. 2007. A detailed analysis of 16S ribosomal RNA gene segments for the diagnosis of pathogenic bacteria. *Journal of Microbiological Methods* 69 (2): 330–339. https://doi.org/10.1016/j.mimet.2007.02.005.

Chen, Wei, Yongmei Cheng, Clarence Zhang, Shaowu Zhang, and Hongyu Zhao. 2013a. MSClust: A multi-seeds based clustering algorithm for microbiome profiling using 16S rRNA sequence. *Journal of Microbiological Methods* 94 (3): 347–355.

Chen, Wei, Clarence K. Zhang, Yongmei Cheng, Shaowu Zhang, and Hongyu Zhao. 2013b. A comparison of methods for clustering 16S rRNA sequences into OTUs. *PLoS One* 8 (8): e70837. https://doi.org/10.1371/journal.pone.0070837.

Chen, Shi-Yi, Feilong Deng, Ying Huang, Xianbo Jia, Yi-Ping Liu, and Song-Jia Lai. 2016. bioOTU: An improved method for simultaneous taxonomic assignments and operational taxonomic units clustering of 16s rRNA gene sequences. *Journal of Computational Biology* 23 (4): 229–238.

Cheng, Lu, Alan W. Walker, and Jukka Corander. 2012. Bayesian estimation of bacterial community composition from 454 sequencing data. *Nucleic Acids Research* 40 (12): 5240–5249. https://doi.org/10.1093/nar/gks227.

Claridge, M.F., H.A. Dawah, and M.R. Wilson. 1997a. Practical approaches to species concepts for living organisms. In *Species: The units of biodiversity*, ed. M.F. Claridge, H.A. Dawah, and M.R. Wilson, 1–15. London: Chapman and Hall.

Claridge, Michael F., Hassan A. Dawah, and Michael R. Wilson. 1997b. *Species: The units of biodiversity*. London: Chapman and Hall.

Clarridge, Jill E., 3rd. 2004. Impact of 16S rRNA gene sequence analysis for identification of bacteria on clinical microbiology and infectious diseases. *Clinical Microbiology Reviews* 17 (4): 840–862. https://doi.org/10.1128/CMR.17.4.840-862.2004. https://pubmed.ncbi.nlm.nih.gov/1 5489351, https://www.ncbi.nlm.nih.gov/pmc/articles/PMC523561/.

Cohan, Frederick M. 1994. Genetic exchange and evolutionary divergence in prokaryotes. *Trends in Ecology & Evolution* 9 (5): 175–180.

Cohn, Ferdinand. 1972. *Untersuchungen Ueber Bacterien*. Breslau: J.U. Kern.

Cole, J.R., B. Chai, R.J. Farris, Q. Wang, A.S. Kulam-Syed-Mohideen, D.M. McGarrell, A.M. Bandela, E. Cardenas, G.M. Garrity, and J.M. Tiedje. 2007. The ribosomal database project (RDP-II): Introducing myRDP space and quality controlled public data. *Nucleic Acids Research* 35 (Database issue): D169–D172. https://doi.org/10.1093/nar/gkl889.

Cole, James R., E. Qiong Wang, J. Fish Cardenas, Benli Chai, Ryan J. Farris, A.S. Kulam-Syed-Mohideen, Donna M. McGarrell, T. Marsh, and George M. Garrity. 2009. The Ribosomal Database Project: Improved alignments and new tools for rRNA analysis. *Nucleic Acids Research* 37 (suppl_1): D141–D145.

Cole, James R., Qiong Wang, Jordan A. Fish, Benli Chai, Donna M. McGarrell, C. Yanni Sun, Titus Brown, Andrea Porras-Alfaro, Cheryl R. Kuske, and James M. Tiedje. 2014. Ribosomal Database Project: Data and tools for high throughput rRNA analysis. *Nucleic Acids Research* 42 (D1): D633–D642. https://doi.org/10.1093/nar/gkt1244.

Colwell, R.R. 1970. Polyphasic taxonomy of the genus vibrio: numerical taxonomy of Vibrio cholerae, Vibrio parahaemolyticus, and related Vibrio species. *Journal of Bacteriology* 104 (1): 410–433. https://doi.org/10.1128/jb.104.1.410-433.1970. https://pubmed.ncbi.nlm.nih.gov/54 73901, https://www.ncbi.nlm.nih.gov/pmc/articles/PMC248227/.

Colwell, R.R., R.A. Clayton, B.A. Ortiz-Conde, D. Jacobs, and E. Russek-Cohen. 1995. *The microbial species concept and biodiversity*. Microbial diversity and ecosystem function: Proceedings of the IUBS/IUMS Workshop held at Egham, UK, 10–13 August 1993.

Cowan, S.T. 1965. Principles and practice of bacterial taxonomy – A forward look. *Microbiology* 39 (1): 143–153.

Cracraft, Joel. 1997. Species concepts in systematics and conservation biology-An ornithological viewpoint. In *Species: The units of biodiversity*, ed. M.F. Claridge, H.A. Dawah, and M.R. Wilson. London: Chapman and Hall.

De la Cuesta-Zuluaga, Jacobo, and Juan S. Escobar. 2016. Considerations for optimizing microbiome analysis using a marker gene. *Frontiers in Nutrition* 3: 26.

De Ley, J., and J. De Smedt. 1975. Improvements of the membrane filter method for DNA: rRNA hybridization. *Antonie Van Leeuwenhoek* 41 (3): 287–307.

de Queiroz, Kevin, and David A. Good. 1997. Phenetic clustering in biology: A critique. *The Quarterly Review of Biology* 72 (1): 3–30.

de Vargas, Colomban, Stéphane Audic, Nicolas Henry, Johan Decelle, Frédéric Mahé, Ramiro Logares, Enrique Lara, Cédric Berney, Noan Le Bescot, Ian Probert, Margaux Carmichael, Julie Poulain, Sarah Romac, Sébastien Colin, Jean-Marc Aury, Lucie Bittner, Samuel Chaffron, Micah Dunthorn, Stefan Engelen, Olga Flegontova, Lionel Guidi, Aleš Horák, Olivier Jaillon,

Gipsi Lima-Mendez, Julius Lukeš, Shruti Malviya, Raphael Morard, Matthieu Mulot, Eleonora Scalco, Raffaele Siano, Flora Vincent, Adriana Zingone, Céline Dimier, Marc Picheral, Sarah Searson, Stefanie Kandels-Lewis, Silvia G. Acinas, Peer Bork, Chris Bowler, Gabriel Gorsky, Nigel Grimsley, Pascal Hingamp, Daniele Iudicone, Fabrice Not, Hiroyuki Ogata, Stephane Pesant, Jeroen Raes, Michael E. Sieracki, Sabrina Speich, Lars Stemmann, Shinichi Sunagawa, Jean Weissenbach, Patrick Wincker, and Eric Karsenti. 2015. Eukaryotic plankton diversity in the sunlit ocean. *Science* 348 (6237). https://doi.org/10.1126/science.1261605. http://science.sciencemag.org/content/sci/348/6237/1261605.full.pdf.

DeSantis, Todd Z., Philip Hugenholtz, Neils Larsen, Mark Rojas, Eoin L. Brodie, Keith Keller, Thomas Huber, Daniel Dalevi, Hu Ping, and Gary L. Andersen. 2006a. Greengenes, a chimera-checked 16S rRNA gene database and workbench compatible with ARB. *Applied and Environmental Microbiology* 72 (7): 5069–5072.

DeSantis, T.Z., K. Philip Hugenholtz, Eoin L. Keller, N. Larsen Brodie, Y.M. Piceno, R. Phan, and Gary L. Andersen. 2006b. NAST: A multiple sequence alignment server for comparative analysis of 16S rRNA genes. *Nucleic Acids Research* 34 (suppl_2): W394–W399.

Dröge, Johannes, and Alice C. McHardy. 2012. Taxonomic binning of metagenome samples generated by next-generation sequencing technologies. *Briefings in Bioinformatics* 13 (6): 646–655. https://doi.org/10.1093/bib/bbs031.

Dubnau, David, Issar Smith, Pierre Morell, and Julius Marmur. 1965. Gene conservation in Bacillus species. I. Conserved genetic and nucleic acid base sequence homologies. *Proceedings of the National Academy of Sciences of the United States of America* 54 (2): 491.

Dunthorn, Micah, Julia Klier, John Bunge, and Thorsten Stoeck. 2012. Comparing the hyper-variable V4 and V9 regions of the small subunit rDNA for assessment of ciliate environmental diversity. *Journal of Eukaryotic Microbiology* 59 (2): 185–187. https://doi.org/10.1111/j.1550-7408.2011.00602.x. https://onlinelibrary.wiley.com/doi/abs/10.1111/j.1550-7408.2011.00602.x.

Edgar, Robert C. 2010. Search and clustering orders of magnitude faster than BLAST. *Bioinformatics* 26 (19): 2460–2461. https://doi.org/10.1093/bioinformatics/btq461.

———. 2013. UPARSE: Highly accurate OTU sequences from microbial amplicon reads. *Nature Methods* 10 (10): 996–998. https://doi.org/10.1038/nmeth.2604.

———.. 2016a. UCHIME2: Improved chimera prediction for amplicon sequencing. *bioRxiv*. https://doi.org/10.1101/074252. https://www.biorxiv.org/content/biorxiv/early/2016/09/09/074252.full.pdf.

———. 2016b. UNOISE2: Improved error-correction for Illumina 16S and ITS amplicon sequencing. *bioRxiv*. https://doi.org/10.1101/081257. https://www.biorxiv.org/content/biorxiv/early/2016/10/15/081257.full.pdf.

———. 2018. Updating the 97% identity threshold for 16S ribosomal RNA OTUs. *Bioinformatics* 34 (14): 2371–2375. https://doi.org/10.1093/bioinformatics/bty113.

Edgar, Robert. 2021. *usearch v11*. http://www.drive5.com/usearch/manual/faq_unoise_not_illumina.html. Accessed 2 Sept 2021.

Edgar, Robert C., and Henrik Flyvbjerg. 2015. Error filtering, pair assembly and error correction for next-generation sequencing reads. *Bioinformatics* 31 (21): 3476–3482. https://doi.org/10.1093/bioinformatics/btv401.

Edgar, Robert C., Brian J. Haas, Jose C. Clemente, Christopher Quince, and Rob Knight. 2011. UCHIME improves sensitivity and speed of chimera detection. *Bioinformatics (Oxford, England)* 27 (16): 2194–2200. https://doi.org/10.1093/bioinformatics/btr381. https://www.ncbi.nlm.nih.gov/pubmed/21700674, https://www.ncbi.nlm.nih.gov/pmc/PMC3150044/.

Enright, Anton J., and Christos A. Ouzounis. 2000. GeneRAGE: A robust algorithm for sequence clustering and domain detection. *Bioinformatics* 16 (5): 451–457. https://doi.org/10.1093/bioinformatics/16.5.451.

Eren, A. Murat, Marcela Zozaya, Christopher M. Taylor, Scot E. Dowd, David H. Martin, and Michael J. Ferris. 2011. Exploring the diversity of Gardnerella vaginalis in the genitourinary tract microbiota of monogamous couples through subtle nucleotide variation. *PLoS One* 6 (10):

e26732. https://doi.org/10.1371/journal.pone.0026732. https://www.ncbi.nlm.nih.gov/pubmed/22046340, https://www.ncbi.nlm.nih.gov/pmc/PMC3201972/.

Eren, A. Murat, Loïs Maignien, Woo Jun Sul, Leslie G. Murphy, Sharon L. Grim, Hilary G. Morrison, and Mitchell L. Sogin. 2013. Oligotyping: Differentiating between closely related microbial taxa using 16S rRNA gene data. *Methods in Ecology and Evolution* 4 (12): 1111–1119. https://doi.org/10.1111/2041-210x.12114. https://www.ncbi.nlm.nih.gov/pubmed/24358444, https://www.ncbi.nlm.nih.gov/pmc/PMC3864673/.

Eren, A. Murat, Hilary G. Morrison, Pamela J. Lescault, Julie Reveillaud, Joseph H. Vineis, and Mitchell L. Sogin. 2014. Minimum entropy decomposition: Unsupervised oligotyping for sensitive partitioning of high-throughput marker gene sequences. *The ISME Journal* 9 (4): 968–979. https://doi.org/10.1038/ismej.2014.195. https://www.ncbi.nlm.nih.gov/pubmed/2532 5381, https://www.ncbi.nlm.nih.gov/pmc/PMC4817710/.

Eren, A. Murat, Mitchell L. Sogin, Hilary G. Morrison, Joseph H. Vineis, Jenny C. Fisher, Ryan J. Newton, and Sandra L. McLellan. 2015. A single genus in the gut microbiome reflects host preference and specificity. *The ISME Journal* 9 (1): 90–100. https://doi.org/10.1038/ismej.2014.97. https://www.ncbi.nlm.nih.gov/pubmed/24936765, https://www.ncbi.nlm.nih.gov/pmc/PMC4274434/.

Eren, A. Murat, Mitchell L. Sogin, and Loïs Maignien. 2016. Editorial: New insights into microbial ecology through subtle nucleotide variation. *Frontiers in Microbiology* 7 (1318). https://doi.org/10.3389/fmicb.2016.01318. https://www.frontiersin.org/article/10.3389/fmicb.2016.01318.

Everett, Karin D.E., Robin M. Bush, and Arthur A. Andersen. 1999. Emended description of the order Chlamydiales, proposal of Parachlamydiaceae fam. nov. and Simkaniaceae fam. nov., each containing one monotypic genus, revised taxonomy of the family Chlamydiaceae, including a new genus and five new species, and standards for the identification of organisms. *International Journal of Systematic and Evolutionary Microbiology* 49 (2): 415–440.

Federhen, Scott. 2012. The NCBI taxonomy database. *Nucleic Acids Research* 40 (D1): D136–D143.

Felsenstein, Joseph. 1989. PHYLIP—Phylogeny inference package. *Cladistics* 5: 164–166.

Filker, Sabine, Anna Gimmler, Micah Dunthorn, Frédéric Mahé, and Thorsten Stoeck. 2015. Deep sequencing uncovers protistan plankton diversity in the Portuguese Ria Formosa solar saltern ponds. *Extremophiles* 19 (2): 283–295. https://doi.org/10.1007/s00792-014-0713-2.

Fish, Jordan, Benli Chai, Qiong Wang, Sun Yanni, C. Titus Brown, James Tiedje, and James Cole. 2013. FunGene: The functional gene pipeline and repository. *Frontiers in Microbiology* 4 (291). https://doi.org/10.3389/fmicb.2013.00291. https://www.frontiersin.org/article/10.3389/fmicb.2013.00291.

Fisher, R.A. 1935. *Design of experiments*. Edinburgh: Oliver & Boyd.

Fitz-Gibbon, Sorel, Shuta Tomida, Bor-Han Chiu, Lin Nguyen, Christine Du, Minghsun Liu, David Elashoff, Marie C. Erfe, Anya Loncaric, Jenny Kim, Robert L. Modlin, Jeff F. Miller, Erica Sodergren, Noah Craft, George M. Weinstock, and Huiying Li. 2013. Propionibacterium acnes strain populations in the human skin microbiome associated with acne. *The Journal of Investigative Dermatology* 133: 2152–2160.

Fournier, Pierre-Edouard, J. Stephen Dumler, Gilbert Greub, Jianzhi Zhang, Wu Yimin, and Didier Raoult. 2003. Gene sequence-based criteria for identification of new rickettsia isolates and description of Rickettsia heilongjiangensis sp. nov. *Journal of Clinical Microbiology* 41 (12): 5456–5465. https://doi.org/10.1128/JCM.41.12.5456-5465.2003. https://pubmed.ncbi.nlm.nih.gov/14662925, https://www.ncbi.nlm.nih.gov/pmc/articles/PMC308961/.

Fox, George E., Jeffrey D. Wisotzkey, and Peter Jurtshuk Jr. 1992. How close is close: 16S rRNA sequence identity may not be sufficient to guarantee species identity. *International Journal of Systematic and Evolutionary Microbiology* 42 (1): 166–170.

Franzén, O., J. Hu, X. Bao, S.H. Itzkowitz, I. Peter, and A. Bashir. 2015. Improved OTU-picking using long-read 16S rRNA gene amplicon sequencing and generic hierarchical clustering. *Microbiome* 3: 43. https://doi.org/10.1186/s40168-015-0105-6.

Fu, Limin, Beifang Niu, Zhengwei Zhu, Wu Sitao, and Weizhong Li. 2012. CD-HIT: Accelerated for clustering the next-generation sequencing data. *Bioinformatics* 28 (23): 3150–3152. https://doi.org/10.1093/bioinformatics/bts565.

Gaspar, John M. 2018. NGmerge: Merging paired-end reads via novel empirically-derived models of sequencing errors. *BMC Bioinformatics* 19 (1): 536. https://doi.org/10.1186/s12859-018-2579-2.

Ghodsi, Mohammadreza, Bo Liu, and Mihai Pop. 2011. DNACLUST: Accurate and efficient clustering of phylogenetic marker genes. *BMC Bioinformatics* 12 (1): 271. https://doi.org/10.1186/1471-2105-12-271.

Goecks, Jeremy, Anton Nekrutenko, James Taylor, and Team Galaxy. 2010. Galaxy: A comprehensive approach for supporting accessible, reproducible, and transparent computational research in the life sciences. *Genome Biology* 11 (8): R86–R86. https://doi.org/10.1186/gb-2010-11-8-r86. https://www.ncbi.nlm.nih.gov/pubmed/20738864, https://www.ncbi.nlm.nih.gov/pmc/PMC2945788/.

Goodfellow, M., G.P. Manfio, and J. Chun. 1997. Towards a practical species concept for cultivable bacteria. In *Species: The units of biodiversity*, ed. M.F. Claridge, H.A. Dawah, and M.R. Wilson, 25–59. London: Chapman & Hall.

Gordon, Ruth E. 1978. A species definition. *International Journal of Systematic and Evolutionary Microbiology* 28 (4): 605–607.

Goris, Johan, Konstantinos T. Konstantinidis, Joel A. Klappenbach, Tom Coenye, Peter Vandamme, and James M. Tiedje. 2007. DNA–DNA hybridization values and their relationship to whole-genome sequence similarities. *International Journal of Systematic and Evolutionary Microbiology* 57 (1): 81–91.

Gribaldo, S., V. Lumia, R. Creti, E. Conway de Macario, A. Sanangelantoni, and P. Cammarano. 1999. Discontinuous occurrence of the hsp70 (dnaK) gene among Archaea and sequence features of HSP70 suggest a novel outlook on phylogenies inferred from this protein. *Journal of Bacteriology* 181 (2): 434–443. https://doi.org/10.1128/JB.181.2.434-443.1999. https://pubmed.ncbi.nlm.nih.gov/9882656, https://www.ncbi.nlm.nih.gov/pmc/articles/PMC93396/.

Grimont, P.A. 1988. Use of DNA reassociation in bacterial classification. *Canadian Journal of Microbiology* 34 (4): 541–546. https://doi.org/10.1139/m88-092.

Gupta, Radhey S. 1998. What are archaebacteria: Life's third domain or monoderm prokaryotes related to Gram-positive bacteria? A new proposal for the classification of prokaryotic organisms. *Molecular Microbiology* 29 (3): 695–707.

Haas, Brian J., Dirk Gevers, Ashlee M. Earl, Mike Feldgarden, Doyle V. Ward, Georgia Giannoukos, Dawn Ciulla, Diana Tabbaa, Sarah K. Highlander, and Erica Sodergren. 2011. Chimeric 16S rRNA sequence formation and detection in Sanger and 454-pyrosequenced PCR amplicons. *Genome Research* 21 (3): 494–504.

Hao, Xiaolin, Rui Jiang, and Ting Chen. 2011. Clustering 16S rRNA for OTU prediction: A method of unsupervised Bayesian clustering. *Bioinformatics (Oxford, England)* 27 (5): 611–618. https://doi.org/10.1093/bioinformatics/btq725. https://www.ncbi.nlm.nih.gov/pubmed/21233169, https://www.ncbi.nlm.nih.gov/pmc/PMC3042185/.

Harmsen, Dag. 2004. 16S rDNA for diagnosing pathogens: A living tree. *Asm News* 70: 19–24.

Hassler, Hayley B., Brett Probert, Carson Moore, Elizabeth Lawson, Richard W. Jackson, Brook T. Russell, and Vincent P. Richards. 2022. Phylogenies of the 16S rRNA gene and its hypervariable regions lack concordance with core genome phylogenies. *Microbiome* 10 (1): 104. https://doi.org/10.1186/s40168-022-01295-y.

Hathaway, Nicholas J., Christian M. Parobek, Jonathan J. Juliano, and Jeffrey A. Bailey. 2017. SeekDeep: single-base resolution de novo clustering for amplicon deep sequencing. *Nucleic Acids Research* 46 (4): e21. https://doi.org/10.1093/nar/gkx1201.

He, Yan, J. Gregory Caporaso, Xiao-Tao Jiang, Hua-Fang Sheng, Susan M. Huse, Jai Ram Rideout, Robert C. Edgar, Evguenia Kopylova, William A. Walters, Rob Knight, and Hong-Wei Zhou. 2015. Stability of operational taxonomic units: An important but neglected property for analyzing microbial diversity. *Microbiome* 3: 20–20. https://doi.org/10.1186/s40168-015-0081-x.

https://www.ncbi.nlm.nih.gov/pubmed/25995836, https://www.ncbi.nlm.nih.gov/pmc/PMC4438525/.

Heise, Helen, and Mortimer P. Starr. 1968. Nomenifers: Are they christened or classified? *Systematic Biology* 17 (4): 458–467.

Hey, Jody. 2006. On the failure of modern species concepts. *Trends in Ecology & Evolution* 21 (8): 447–450. https://doi.org/10.1016/j.tree.2006.05.011. https://www.sciencedirect.com/science/article/pii/S0169534706001649.

Huang, Ying, Beifang Niu, Ying Gao, Fu Limin, and Weizhong Li. 2010. CD-HIT Suite: A web server for clustering and comparing biological sequences. *Bioinformatics* 26 (5): 680–682. https://doi.org/10.1093/bioinformatics/btq003.

Hugenholtz, P., B.M. Goebel, and N.R. Pace. 1998. Impact of culture-independent studies on the emerging phylogenetic view of bacterial diversity. *Journal of Bacteriology* 180 (18): 4765–4774. https://doi.org/10.1128/JB.180.18.4765-4774.1998. https://pubmed.ncbi.nlm.nih.gov/9733676, https://www.ncbi.nlm.nih.gov/pmc/articles/PMC107498/.

Hull, D.L. 1997. The ideal species concept-and why we can't get it. In *Species: The units of biodiversity*, ed. M.F. Claridge, H.A. Dawah, and M.R. Wilson, 357–380. London: Chapman and Hall.

Human Microbiome Project, Consortium. 2012. A framework for human microbiome research. *Nature* 486 (7402): 215–221. https://doi.org/10.1038/nature11209. https://www.ncbi.nlm.nih.gov/pubmed/22699610, https://www.ncbi.nlm.nih.gov/pmc/PMC3377744/.

Huse, Susan M., Julie A. Huber, Hilary G. Morrison, Mitchell L. Sogin, and David Mark Welch. 2007. Accuracy and quality of massively parallel DNA pyrosequencing. *Genome Biology* 8 (7): R143–R143. https://doi.org/10.1186/gb-2007-8-7-r143. https://www.ncbi.nlm.nih.gov/pubmed/17659080, https://www.ncbi.nlm.nih.gov/pmc/PMC2323236/.

Huse, Susan M., Les Dethlefsen, Julie A. Huber, David Mark Welch, David A. Relman, and Mitchell L. Sogin. 2008. Exploring microbial diversity and taxonomy using SSU rRNA hypervariable tag sequencing. *PLoS Genetics* 4 (11): e1000255. https://doi.org/10.1371/journal.pgen.1000255.

Huse, S.M., D.M. Welch, H.G. Morrison, and M.L. Sogin. 2010. Ironing out the wrinkles in the rare biosphere through improved OTU clustering. *Environmental Microbiology* 12 (7): 1889–1898.

Ibrahim, Ashraf, Peter Gerner-Smidt, and Werner Liesack. 1997. Phylogenetic relationship of the twenty-one DNA groups of the genus Acinetobacter as revealed by 16S ribosomal DNA sequence analysis. *International Journal of Systematic and Evolutionary Microbiology* 47 (3): 837–841.

Jääskinen, V., V. Parkkinen, L. Cheng, and J. Corander. 2014. Bayesian clustering of DNA sequences using Markov chains and a stochastic partition model. *Statistical Applications in Genetics and Molecular Biology* 13 (1): 105–121. https://doi.org/10.1515/sagmb-2013-0031.

Jackson, Matthew A., Jordana T. Bell, Tim D. Spector, and Claire J. Steves. 2016. A heritability-based comparison of methods used to cluster 16S rRNA gene sequences into operational taxonomic units. *PeerJ* 4: e2341. https://doi.org/10.7717/peerj.2341.

Johnson, John L. 1973. Use of nucleic-acid homologies in the taxonomy of anaerobic bacteria. *International Journal of Systematic and Evolutionary Microbiology* 23 (4): 308–315.

———. 1985. 2 DNA reassociation and RNA hybridisation of bacterial nucleic acids. In *Methods in microbiology*, 33–74. San Diego: Elsevier.

Johnson, Jethro S., Daniel J. Spakowicz, Bo-Young Hong, Lauren M. Petersen, Patrick Demkowicz, Lei Chen, Shana R. Leopold, Blake M. Hanson, Hanako O. Agresta, Mark Gerstein, Erica Sodergren, and George M. Weinstock. 2019. Evaluation of 16S rRNA gene sequencing for species and strain-level microbiome analysis. *Nature Communications* 10 (1): 5029. https://doi.org/10.1038/s41467-019-13036-1.

Jost, Lou. 2006. Entropy and diversity. *Oikos* 113 (2): 363–375. https://doi.org/10.1111/j.2006.0030-1299.14714.x. https://onlinelibrary.wiley.com/doi/abs/10.1111/j.2006.0030-1299.14714.x.

Kimura, Motoo. 1980. A simple method for estimating evolutionary rates of base substitutions through comparative studies of nucleotide sequences. *Journal of Molecular Evolution* 16 (2): 111–120.

Koeppel, Alexander F., and Martin Wu. 2013. Surprisingly extensive mixed phylogenetic and ecological signals among bacterial Operational Taxonomic Units. *Nucleic Acids Research* 41 (10): 5175–5188. https://doi.org/10.1093/nar/gkt241. https://www.ncbi.nlm.nih.gov/pubmed/23571758, https://www.ncbi.nlm.nih.gov/pmc/PMC3664822/.

Konstantinidis, Konstantinos T., Alban Ramette, and James M. Tiedje. 2006. The bacterial species definition in the genomic era. *Philosophical Transactions of the Royal Society of London. Series B, Biological Sciences* 361 (1475): 1929–1940. https://doi.org/10.1098/rstb.2006.1920. https://pubmed.ncbi.nlm.nih.gov/17062412, https://www.ncbi.nlm.nih.gov/pmc/articles/PMC1764935/.

Kopylova, Evguenia, Laurent Noé, and Hélène Touzet. 2012. SortMeRNA: Fast and accurate filtering of ribosomal RNAs in metatranscriptomic data. *Bioinformatics* 28 (24): 3211–3217. https://doi.org/10.1093/bioinformatics/bts611.

Kopylova, Evguenia, Jose A. Navas-Molina, Céline Mercier, Xu Zhenjiang Zech, Frédéric Mahé, Yan He, Hong-Wei Zhou, J. Torbjørn Rognes, Gregory Caporaso, and Rob Knight. 2016. Open-source sequence clustering methods improve the state of the art. *mSystems* 1 (1): e00003–e00015.

Krieg, Noel R. 1988. Bacterial classification: An overview. *Canadian Journal of Microbiology* 34 (4): 536–540.

Kuczynski, Justin, Jesse Stombaugh, William Anton Walters, J. Antonio González, Gregory Caporaso, and Rob Knight. 2012. Using QIIME to analyze 16S rRNA gene sequences from microbial communities. *Current Protocols in Microbiology* 1: Unit-1E.5. https://doi.org/10.1002/9780471729259.mc01e05s27. http://www.ncbi.nlm.nih.gov/pmc/articles/PMC4477843/.

Langille, Morgan G.I., J. Jesse Zaneveld, Gregory Caporaso, Daniel McDonald, Dan Knights, Joshua A. Reyes, Jose C. Clemente, Deron E. Burkepile, Rebecca L. Vega, Rob Knight Thurber, Robert G. Beiko, and Curtis Huttenhower. 2013. Predictive functional profiling of microbial communities using 16S rRNA marker gene sequences. *Nature Biotechnology* 31 (9): 814–821. https://doi.org/10.1038/nbt.2676. http://www.ncbi.nlm.nih.gov/pmc/articles/PMC3819121/.

Legendre, P., and L. Legendre. 1998. *Numerical ecology, amsterdam, second English edition: Developments in environmental modeling*. Vol. 20. New York: Elsevier Science. 853pp.

Legendre, Pierre, and Louis Legendre. 2012. *Numerical ecology*. Amsterdam: Elsevier.

Li, Weizhong, and Adam Godzik. 2006. Cd-hit: A fast program for clustering and comparing large sets of protein or nucleotide sequences. *Bioinformatics* 22 (13): 1658–1659. https://doi.org/10.1093/bioinformatics/btl158.

Lima-Mendez, Gipsi, Karoline Faust, Nicolas Henry, Johan Decelle, Sébastien Colin, Fabrizio Carcillo, Chaffron Samuel, J. Cesar Ignacio-Espinosa, Simon Roux, Flora Vincent, Lucie Bittner, Youssef Darzi, Jun Wang, Stéphane Audic, Léo Berline, Gianluca Bontempi, Ana M. Cabello, Laurent Coppola, Francisco M. Cornejo-Castillo, Francesco d'Ovidio, Luc De Meester, Isabel Ferrera, Marie-José Garet-Delmas, Lionel Guidi, Elena Lara, Stéphane Pesant, Marta Royo-Llonch, Guillem Salazar, Pablo Sánchez, Marta Sebastian, Caroline Souffreau, Céline Dimier, Marc Picheral, Sarah Searson, Stefanie Kandels-Lewis, Gabriel Gorsky, Fabrice Not, Hiroyuki Ogata, Sabrina Speich, Lars Stemmann, Jean Weissenbach, Patrick Wincker, Silvia G. Acinas, Shinichi Sunagawa, Peer Bork, Matthew B. Sullivan, Eric Karsenti, Chris Bowler, Colomban de Vargas, and Jeroen Raes. 2015. Determinants of community structure in the global plankton interactome. *Science* 348 (6237). https://doi.org/10.1126/science.1262073. http://science.sciencemag.org/content/sci/348/6237/1262073.full.pdf.

Liu, Zongzhi, Todd Z. DeSantis, Gary L. Andersen, and Rob Knight. 2008. Accurate taxonomy assignments from 16S rRNA sequences produced by highly parallel pyrosequencers. *Nucleic Acids Research* 36 (18): e120–e120. https://doi.org/10.1093/nar/gkn491. https://pubmed.ncbi.nlm.nih.gov/18723574, https://www.ncbi.nlm.nih.gov/pmc/articles/PMC2566877/.

Logan, Niall A. 2009. *Bacterial systematics*. Hoboken: Wiley.

Louca, Stilianos, Michael Doebeli, and Laura Wegener Parfrey. 2018. Correcting for 16S rRNA gene copy numbers in microbiome surveys remains an unsolved problem. *Microbiome* 6 (1): 41. https://doi.org/10.1186/s40168-018-0420-9.

Lozupone, Catherine, and Rob Knight. 2005. UniFrac: A new phylogenetic method for comparing microbial communities. *Applied and Environmental Microbiology* 71 (12): 8228–8235.

Ludwig, W. 1999. The role of rRNA as a phylogenetic marker in the context of genomics. *USFCC Newsletter* 29: 2–6.

Ludwig, W., and K.H. Schleifer. 1994. Bacterial phylogeny based on 16S and 23S rRNA sequence analysis. *FEMS Microbiology Reviews* 15 (2-3): 155–173.

Ludwig, Wolfgang, Oliver Strunk, Sabine Klugbauer, Norbert Klugbauer, Michael Weizenegger, Judith Neumaier, Marianne Bachleitner, and Karl Heinz Schleifer. 1998. Bacterial phylogeny based on comparative sequence analysis. *Electrophoresis* 19 (4): 554–568.

Ma, Xiaotu, Ying Shao, Liqing Tian, Diane A. Flasch, Heather L. Mulder, Michael N. Edmonson, Yu Liu, Xiang Chen, Scott Newman, Joy Nakitandwe, Yongjin Li, Benshang Li, Shuhong Shen, Zhaoming Wang, Sheila Shurtleff, Leslie L. Robison, Shawn Levy, John Easton, and Jinghui Zhang. 2019. Analysis of error profiles in deep next-generation sequencing data. *Genome Biology* 20 (1): 50–50. https://doi.org/10.1186/s13059-019-1659-6. https://pubmed.ncbi.nlm.nih.gov/30867008, https://www.ncbi.nlm.nih.gov/pmc/articles/PMC6417284/.

Maddison, Wayne P., and Montgomery Slatkin. 1991. Null models for the number of evolutionary steps in a character on a phylogenetic tree. *Evolution* 45 (5): 1184–1197.

Mahé, Frédéric, Torbjørn Rognes, Christopher Quince, Colomban de Vargas, and Micah Dunthorn. 2014. Swarm: Robust and fast clustering method for amplicon-based studies. *PeerJ* 2: e593. https://doi.org/10.7717/peerj.593.

Mahé, Frédéric, Jordan Mayor, John Bunge, Jingyun Chi, Tobias Siemensmeyer, Thorsten Stoeck, Benjamin Wahl, Tobias Paprotka, Sabine Filker, and Micah Dunthorn. 2015a. Comparing high-throughput platforms for sequencing the V4 region of SSU-rDNA in environmental microbial eukaryotic diversity surveys. *Journal of Eukaryotic Microbiology* 62 (3): 338–345. https://doi.org/10.1111/jeu.12187. https://onlinelibrary.wiley.com/doi/abs/10.1111/jeu.12187.

Mahé, Frédéric, Torbjørn Rognes, Christopher Quince, Colomban de Vargas, and Micah Dunthorn. 2015b. Swarm v2: Highly-scalable and high-resolution amplicon clustering. *PeerJ* 3: e1420. https://doi.org/10.7717/peerj.1420. https://www.ncbi.nlm.nih.gov/pubmed/26713226, https://www.ncbi.nlm.nih.gov/pmc/PMC4690345/.

Martin, Andrew P. 2002. Phylogenetic approaches for describing and comparing the diversity of microbial communities. *Applied and Environmental Microbiology* 68 (8): 3673–3682.

Martinez-Murcia, A.J., S. Benlloch, and M.D. Collins. 1992. Phylogenetic interrelationships of members of the genera Aeromonas and Plesiomonas as determined by 16S ribosomal DNA sequencing: Lack of congruence with results of DNA-DNA hybridizations. *International Journal of Systematic and Evolutionary Microbiology* 42 (3): 412–421.

Matias Rodrigues, João F., and Christian von Mering. 2013. HPC-CLUST: Distributed hierarchical clustering for large sets of nucleotide sequences. *Bioinformatics* 30 (2): 287–288. https://doi.org/10.1093/bioinformatics/btt657.

Matias Rodrigues, João F., Thomas S.B. Schmidt, Janko Tackmann, and Christian von Mering. 2017. MAPseq: Highly efficient k-mer search with confidence estimates, for rRNA sequence analysis. *Bioinformatics* 33 (23): 3808–3810. https://doi.org/10.1093/bioinformatics/btx517.

May, Robert M. 1986. Biological diversity: How many species are there? *Nature* 324 (6097): 514–515.

———. 1988. How many species are there on earth? *Science* 241 (4872): 1441–1449.

Mayden, R.L. 1997. A hierarchy of species concepts: The denouement in the saga of the species problem. In *Species: The units of biodiversity*, ed. M.F. Claridge, H.A. Dawah, and M.R. Wilson, 381–424. London: Chapman and Hall.

Mayr, Ernst. 1940. Speciation phenomena in birds. *The American Naturalist* 74 (752): 249–278. http://www.jstor.org.proxy.cc.uic.edu/stable/2457576.

———. 1942. *Systematics and the origin of species*. New York: Columbia University Press.

————. 1963. *Animal species and evolution*. Cambridge, MA: Harvard University Press.

————. 1969. The biological meaning of species. *Biological Journal of the Linnean Society* 1 (3): 311–320.

————. 2004. *What makes biology unique? Considerations on the autonomy of a scientific discipline*. Cambridge: Cambridge University Press.

————. 2015. *Principles of systematic zoology*. New Delhi: Scientific Publishers.

McDonald, Daniel, Morgan N. Price, Julia Goodrich, Eric P. Nawrocki, Todd Z. DeSantis, Alexander Probst, Gary L. Andersen, Rob Knight, and Philip Hugenholtz. 2012. An improved Greengenes taxonomy with explicit ranks for ecological and evolutionary analyses of bacteria and archaea. *The ISME Journal* 6 (3): 610–618. https://doi.org/10.1038/ismej.2011.139. https://www.ncbi.nlm.nih.gov/pubmed/22134646, https://www.ncbi.nlm.nih.gov/pmc/PMC3280142/.

Mercier, C., F. Boyer, A. Bonin, and E. Coissac. 2013. *SUMATRA and SUMACLUST: Fast and exact comparison and clustering of sequences*. Programs and Abstracts of the SeqBio 2013 Workshop. Abstract, (Citeseer), 27–29.

Miotto, Olivo, Roberto Amato, Elizabeth A Ashley, Bronwyn MacInnis, Jacob Almagro-Garcia, Chanaki Amaratunga, Pharath Lim, Daniel Mead, Samuel O Oyola, and Mehul Dhorda. 2015. "Genetic architecture of artemisinin-resistant Plasmodium falciparum." *Nature Genetics 47* (3): 226–234.

Mishler, Brent D. 1999. Getting rid of species. In *Species: New interdisciplinary essays*, 307. Cambridge, MA: MIT Press.

————. 2000. The phylogenetic species concept (sensu Mishler and Theriot): monophyly, apomorphy, and phylogenetic species concepts. In *Species concepts and phylogenetic theory, a debate*, vol. 44, 54. New York: Columbia University Press.

Morgan, Matthew J., Anthony A. Chariton, Diana M. Hartley, Leon N. Court, and Christopher M. Hardy. 2013. Improved inference of taxonomic richness from environmental DNA. *PLoS One* 8 (8): e71974. https://doi.org/10.1371/journal.pone.0071974. https://www.ncbi.nlm.nih.gov/pubmed/23991013, https://www.ncbi.nlm.nih.gov/pmc/PMC3753314/.

Mylvaganam, Shanthini, and Patrick P. Dennis. 1992. Sequence heterogeneity between the two genes encoding 16S rRNA from the halophilic archaebacterium Haloarcula marismortui. *Genetics* 130 (3): 399–410.

Namiki, Youhei, Takashi Ishida, and Yutaka Akiyama. 2013. Acceleration of sequence clustering using longest common subsequence filtering. *BMC Bioinformatics* 14 (Suppl 8): S7. https://doi.org/10.1186/1471-2105-14-S8-S7. https://pubmed.ncbi.nlm.nih.gov/23815271, https://www.ncbi.nlm.nih.gov/pmc/articles/PMC3654901/.

Nearing, Jacob T., Gavin M. Douglas, André M. Comeau, and Morgan G.I. Langille. 2018. Denoising the Denoisers: An independent evaluation of microbiome sequence error-correction approaches. *PeerJ* 6: e5364. https://doi.org/10.7717/peerj.5364. https://www.ncbi.nlm.nih.gov/pubmed/30123705, https://www.ncbi.nlm.nih.gov/pmc/articles/PMC6087418/.

Nebel, Markus, Cornelia Pfabel, Alexandra Stock, Micah Dunthorn, and Thorsten Stoeck. 2011. Delimiting operational taxonomic units for assessing ciliate environmental diversity of small subunit rRNA gene sequences. *Environmental Microbiology Reports* 3: 154.

Nekhaev, Ivan O., T. Shiøtte, and Maxim V. Vinarski. 2015. Type materials of European freshwater molluscs described by Otto Friedrich Müller. *Archiv für Molluskenkunde* 144 (1): 51–64.

Nguyen, Nam-Phuong, Tandy Warnow, Mihai Pop, and Bryan White. 2016. A perspective on 16S rRNA operational taxonomic unit clustering using sequence similarity. *npj Biofilms and Microbiomes* 2 (1): 16004. https://doi.org/10.1038/npjbiofilms.2016.4.

Nübel, Uea, Bert Engelen, Andreas Felske, Jiri Snaidr, Alois Wieshuber, Rudolf I. Amann, Wolfgang Ludwig, and Horst Backhaus. 1996. Sequence heterogeneities of genes encoding 16S rRNAs in Paenibacillus polymyxa detected by temperature gradient gel electrophoresis. *Journal of Bacteriology* 178 (19): 5636–5643.

Oikonomou, Andreas, Sabine Filker, Hans-Werner Breiner, and Thorsten Stoeck. 2015. Protistan diversity in a permanently stratified meromictic lake (Lake Alatsee, SW Germany).

Environmental Microbiology 17 (6): 2144–2157. https://doi.org/10.1111/1462-2920.12666. https://onlinelibrary.wiley.com/doi/abs/10.1111/1462-2920.12666.

Olsen, Gary J., Carl R. Woese, and Ross Overbeek. 1994. The winds of (evolutionary) change: Breathing new life into microbiology. *Journal of Bacteriology* 176 (1): 1–6.

Pace, Norman R. 1997. A molecular view of microbial diversity and the biosphere. *Science* 276 (5313): 734–740.

Pagni, Marco, Hélène Niculita-Hirzel, Loïc Pellissier, Anne Dubuis, Ioannis Xenarios, Antoine Guisan, Ian R. Sanders, Jérôme Goudet, and Nicolas Guex. 2013. Density-based hierarchical clustering of pyro-sequences on a large scale – The case of fungal ITS1. *Bioinformatics (Oxford, England)* 29 (10): 1268–1274. https://doi.org/10.1093/bioinformatics/btt149. https://pubmed. ncbi.nlm.nih.gov/23539304, https://www.ncbi.nlm.nih.gov/pmc/articles/PMC3654712/.

Peterson, Jane, Susan Garges, Maria Giovanni, Lu Pamela McInnes, Jeffery A. Wang, Vivien Bonazzi Schloss, Jean E. McEwen, Kris A. Wetterstrand, and Carolyn Deal. 2009. The NIH human microbiome project. *Genome Research* 19 (12): 2317–2323.

Platnick, Norman I. 2000. A defense of the phylogenetic species concept (sensu Wheeler and Platnick). In *Species concepts and phylogenetic theory: A debate*, 185–197. New York: Columbia University Press.

Preheim, Sarah P., Allison R. Perrotta, Antonio M. Martin-Platero, Anika Gupta, and Eric J. Alm. 2013. Distribution-based clustering: Using ecology to refine the operational taxonomic unit. *Applied and Environmental Microbiology* 79 (21): 6593–6603. https://doi.org/10.1128/aem. 00342-13. https://www.ncbi.nlm.nih.gov/pubmed/23974136, https://www.ncbi.nlm.nih.gov/ pmc/PMC3811501/.

Probst, Andreas J., Christian Hertel, Lothar Richter, Lars Wassill, Wolfgang Ludwig, and Walter P. Hammes. 1998. Staphylococcus condimenti sp. nov., from soy sauce mash, and Staphylococcus carnosus (Schleifer and Fischer 1982) subsp. utilis subsp. nov. *International Journal of Systematic and Evolutionary Microbiology* 48 (3): 651–658.

Prodan, Andrei, Valentina Tremaroli, Harald Brolin, Aeilko H. Zwinderman, Max Nieuwdorp, and Evgeni Levin. 2020. Comparing bioinformatic pipelines for microbial 16S rRNA amplicon sequencing. *PLoS One* 15 (1): e0227434. https://doi.org/10.1371/journal.pone.0227434.

Prosser, James I., Brendan Bohannan, Tom Curtis, Richard Ellis, Mary K. Firestone, Rob P. Freckleton, Jessica L. Green, Laura Green, Ken Killham, Jack Lennon, Andrew Osborn, Martin Solan, Christopher van der Gast, and J. Peter Young. 2007. Essay – The role of ecological theory in microbial ecology. *Nature Reviews Microbiology* 5: 384.

Pruesse, Elmar, Christian Quast, Katrin Knittel, Bernhard M. Fuchs, Wolfgang Ludwig, Jörg Peplies, and Frank Oliver Glöckner. 2007. SILVA: A comprehensive online resource for quality checked and aligned ribosomal RNA sequence data compatible with ARB. *Nucleic Acids Research* 35 (21): 7188–7196. https://doi.org/10.1093/nar/gkm864.

Quince, Christopher, Anders Lanzén, Thomas P. Curtis, Russell J. Davenport, Neil Hall, Ian M. Head, L. Fiona Read, and William T. Sloan. 2009. Accurate determination of microbial diversity from 454 pyrosequencing data. *Nature Methods* 6: 639. https://doi.org/10.1038/nmeth. 1361. https://www.nature.com/articles/nmeth.1361#supplementary-information.

Quince, Christopher, Anders Lanzen, Russell J. Davenport, and Peter J. Turnbaugh. 2011. Removing noise from pyrosequenced amplicons. *BMC Bioinformatics* 12 (1): 38. https://doi.org/10. 1186/1471-2105-12-38.

Ramette, Alban, and Pier Luigi Buttigieg. 2014. The R package otu2ot for implementing the entropy decomposition of nucleotide variation in sequence data. *Frontiers in Microbiology* 5 (601). https://doi.org/10.3389/fmicb.2014.00601. https://www.frontiersin.org/article/10.33 89/fmicb.2014.00601.

Rasheed, Zeehasham, Huzefa Rangwala, and Daniel Barbará. 2013. 16S rRNA metagenome clustering and diversity estimation using locality sensitive hashing. *BMC Systems Biology* 7 (Suppl 4): S11. https://doi.org/10.1186/1752-0509-7-S4-S11. https://pubmed.ncbi.nlm.nih. gov/24565031, https://www.ncbi.nlm.nih.gov/pmc/articles/PMC3854655/.

Reeder, Jens, and Rob Knight. 2010. Rapidly denoising pyrosequencing amplicon reads by exploiting rank-abundance distributions. *Nature Methods* 7 (9): 668–669. https://doi.org/10.1038/nmeth0910-668b.

Rideout, Jai Ram, Yan He, Jose A. Navas-Molina, William A. Walters, Luke K. Ursell, Sean M. Gibbons, John Chase, Daniel McDonald, Antonio Gonzalez, Adam Robbins-Pianka, Jose C. Clemente, Jack A. Gilbert, Susan M. Huse, Hong-Wei Zhou, Rob Knight, and J. Gregory Caporaso. 2014. Subsampled open-reference clustering creates consistent, comprehensive OTU definitions and scales to billions of sequences. *PeerJ* 2: e545. https://doi.org/10.7717/peerj.545.

Rognes, T., T. Flouri, B. Nichols, C. Quince, and F. Mahé. 2016. VSEARCH: A versatile open source tool for metagenomics. *PeerJ* 4: e2584. https://doi.org/10.7717/peerj.2584.

Rosen, Michael J., Benjamin J. Callahan, Daniel S. Fisher, and Susan P. Holmes. 2012. Denoising PCR-amplified metagenome data. *BMC Bioinformatics* 13: 283–283. https://doi.org/10.1186/1471-2105-13-283. https://www.ncbi.nlm.nih.gov/pubmed/23113967, https://www.ncbi.nlm.nih.gov/pmc/PMC3563472/.

Rosselló-Mora, Ramon, and Rudolf Amann. 2001. The species concept for prokaryotes. *FEMS Microbiology Reviews* 25 (1): 39–67.

Russell, David J., Samuel F. Way, Andrew K. Benson, and Khalid Sayood. 2010. A grammar-based distance metric enables fast and accurate clustering of large sets of 16S sequences. *BMC Bioinformatics* 11 (1): 601. https://doi.org/10.1186/1471-2105-11-601.

Sait, Michelle, Philip Hugenholtz, and Peter H. Janssen. 2002. Cultivation of globally distributed soil bacteria from phylogenetic lineages previously only detected in cultivation-independent surveys. *Environmental Microbiology* 4 (11): 654–666. https://doi.org/10.1046/j.1462-2920.2002.00352.x. https://sfamjournals.onlinelibrary.wiley.com/doi/abs/10.1046/j.1462-2920.2002.00352.x.

Schaffer, Jonathan. 2015. What not to multiply without necessity. *Australasian Journal of Philosophy* 93 (4): 644–664. https://doi.org/10.1080/00048402.2014.992447. S2CID 16923735.

Schlegel, H.G., and W. Köhler. 1999. Bacteriology paved the way to cell biology: A historical account. In *Biology of the prokaryotes*, ed. J.W. Lengeler, G. Drews, and H.G. Schlegel. Stuttgart: Thieme.

Schloss, Patrick D. 2010. The effects of alignment quality, distance calculation method, sequence filtering, and region on the analysis of 16S rRNA gene-based studies. *PLoS Computational Biology* 6 (7): e1000844. https://doi.org/10.1371/journal.pcbi.1000844. https://www.ncbi.nlm.nih.gov/pubmed/20628621, https://www.ncbi.nlm.nih.gov/pmc/PMC2900292/.

Schloss, Patrick D., and Jo Handelsman. 2005. Introducing DOTUR, a computer program for defining operational taxonomic units and estimating species richness. *Applied and Environmental Microbiology* 71 (3): 1501–1506. https://doi.org/10.1128/aem.71.3.1501-1506.2005. https://aem.asm.org/content/aem/71/3/1501.full.pdf.

———. 2006a. Introducing TreeClimber, a test to compare microbial community structures. *Applied and Environmental Microbiology* 72 (4): 2379–2384.

———. 2006b. Introducing SONS, a tool for operational taxonomic unit-based comparisons of microbial community memberships and structures. *Applied and Environmental Microbiology* 72 (10): 6773–6779. https://doi.org/10.1128/AEM.00474-06. https://pubmed.ncbi.nlm.nih.gov/17021230, https://www.ncbi.nlm.nih.gov/pmc/articles/PMC1610290/.

Schloss, Patrick D., and Sarah L. Westcott. 2011. Assessing and improving methods used in operational taxonomic unit-based approaches for 16S rRNA gene sequence analysis. *Applied and Environmental Microbiology* 77 (10): 3219–3226. https://doi.org/10.1128/aem.02810-10. https://www.ncbi.nlm.nih.gov/pubmed/21421784, https://www.ncbi.nlm.nih.gov/pmc/PMC3126452/.

Schloss, Patrick D., Bret R. Larget, and Jo Handelsman. 2004. Integration of microbial ecology and statistics: A test to compare gene libraries. *Applied and Environmental Microbiology* 70 (9): 5485–5492.

Schloss, Patrick D., Sarah L. Westcott, Thomas Ryabin, Justine R. Hall, Martin Hartmann, Emily B. Hollister, Ryan A. Lesniewski, Brian B. Oakley, Donovan H. Parks, Courtney J. Robinson,

Jason W. Sahl, Blaz Stres, Gerhard G. Thallinger, David J. Van Horn, and Carolyn F. Weber. 2009. Introducing mothur: Open-source, platform-independent, community-supported software for describing and comparing microbial communities. *Applied and Environmental Microbiology* 75 (23): 7537–7541. https://doi.org/10.1128/aem.01541-09. https://aem.asm.org/content/aem/75/23/7537.full.pdf.

Schloss, Patrick D., Dirk Gevers, and Sarah L. Westcott. 2011. Reducing the effects of PCR amplification and sequencing artifacts on 16S rRNA-based studies. *PLoS One* 6 (12): e27310. https://doi.org/10.1371/journal.pone.0027310.

Shannon, C.E. 1948. A mathematical theory of communication. *The Bell System Technical Journal* 27 (3): 379–423. https://doi.org/10.1002/j.1538-7305.1948.tb01338.x.

Shapiro, B. Jesse, Jonathan Friedman, Otto X. Cordero, Sarah P. Preheim, Sonia C. Timberlake, Gitta Szabó, Martin F. Polz, and Eric J. Alm. 2012. Population genomics of early events in the ecological differentiation of bacteria. *Science (New York, N.Y.)* 336 (6077): 48–51. https://doi.org/10.1126/science.1218198. https://pubmed.ncbi.nlm.nih.gov/22491847, https://www.ncbi.nlm.nih.gov/pmc/articles/PMC3337212/.

Simpson, George Gaylord. 1961. *Principles of animal taxonomy*. New York: Columbia University Press.

Singleton, David R., Michelle A. Furlong, Stephen L. Rathbun, and William B. Whitman. 2001. Quantitative comparisons of 16S rRNA gene sequence libraries from environmental samples. *Applied and Environmental Microbiology* 67 (9): 4374–4376.

Sneath, P.H.A. 1976. Phenetic taxonomy at the species level and above. *Taxon* 25 (4): 437–450.

———. 1989. Numerical taxonomy. In *Bergey's manual of systematic bacteriology*, ed. S.T. Williams, M.E. Sharpe, and J.G. Holt, 2303–2305. Baltimore: Williams & Wilkins.

Sneath, Peter H.A. 2015. Bacterial nomenclature. In *Bergey's manual of systematics of archaea and bacteria*, 1–8. Hoboken: Wiley.

Sneath, P.H.A., and R.R. Sokal. 1973. *Numerical taxonomy: The principles and practice of numerical classification*. San Francisco: W. H. Freeman and Company.

Sogin, Mitchell L., Hilary G. Morrison, Julie A. Huber, David Mark Welch, Susan M. Huse, Phillip R. Neal, Jesus M. Arrieta, and Gerhard J. Herndl. 2006. Microbial diversity in the deep sea and the underexplored "rare biosphere". *Proceedings of the National Academy of Sciences of the United States of America* 103 (32): 12115–12120. https://doi.org/10.1073/pnas.0605127103. https://www.ncbi.nlm.nih.gov/pubmed/16880384, https://www.ncbi.nlm.nih.gov/pmc/PMC1524930/.

Sokal, Robert R. 1973. The species problem reconsidered. *Systematic Zoology* 22 (4): 360–374. https://doi.org/10.2307/2412944. http://www.jstor.org.proxy.cc.uic.edu/stable/2412944.

Sokal, Robert R., and Theodore J. Crovello. 1970. The biological species concept: A critical evaluation. *The American Naturalist* 104 (936): 127–153.

Sokal, Robert R., and F. James Rohlf. 1995. *Biometry*. New York: W. H. Freeman and Company.

Sokal, R.R., and P.H.A. Sneath. 1963. *Principles of numerical taxonomy*. San Francisco: WH Freeman.

Stackebrandt, Erko, and J. Ebers. 2006. Taxonomic parameters revisited: Tarnished gold standards. *Microbiology Today* 8: 6–9.

Stackebrandt, E., and B.M. Goebel. 1994. Taxonomic note: A place for DNA-DNA reassociation and 16S rRNA sequence analysis in the present species definition in bacteriology. *International Journal of Systematic and Evolutionary Microbiology* 44 (4): 846–849. https://doi.org/10.1099/00207713-44-4-846. https://www.microbiologyresearch.org/content/journal/ijsem/10.1099/00207713-44-4-846.

Stackebrandt, E., W. Ludwig, and G.E. Fox. 1985. 16 S ribosomal RNA oligonucleotide cataloguing. In *Methods in microbiology*, 75–107. San Diego: Elsevier.

Steigerwalt, Arnold G., G. Richard Fanning, Mary Alyce Fife-Asbury, and Don J. Brenner. 1976. DNA relatedness among species of Enterobacter and Serratia. *Canadian Journal of Microbiology* 22 (2): 121–137.

Sul, Woo Jun, James R. Cole, C. Ederson da, Qiong Wang Jesus, Ryan J. Farris, Jordan A. Fish, and James M. Tiedje. 2011. Bacterial community comparisons by taxonomy-supervised analysis independent of sequence alignment and clustering. *Proceedings of the National Academy of Sciences of the United States of America* 108 (35): 14637–14642. https://doi.org/10.1073/pnas. 1111435108. https://pubmed.ncbi.nlm.nih.gov/21873204, https://www.ncbi.nlm.nih.gov/pmc/articles/PMC3167511/.

Sun, Yijun, Yunpeng Cai, Li Liu, Yu Fahong, Michael L. Farrell, William McKendree, and William Farmerie. 2009. ESPRIT: Estimating species richness using large collections of 16S rRNA pyrosequences. *Nucleic Acids Research* 37 (10): e76. https://doi.org/10.1093/nar/gkp285. https://pubmed.ncbi.nlm.nih.gov/19417062, https://www.ncbi.nlm.nih.gov/pmc/articles/PMC2 691849/.

Sun, Yijun, Yunpeng Cai, Volker Mai, William Farmerie, Yu Fahong, Jian Li, and Steve Goodison. 2010. Advanced computational algorithms for microbial community analysis using massive 16S rRNA sequence data. *Nucleic Acids Research* 38 (22): e205. https://doi.org/10.1093/nar/gkq872. https://www.ncbi.nlm.nih.gov/pubmed/20929878, https://www.ncbi.nlm.nih.gov/pmc/PMC3001099/.

Sun, Yijun, Yunpeng Cai, Susan M. Huse, Rob Knight, William G. Farmerie, Xiaoyu Wang, and Volker Mai. 2012. A large-scale benchmark study of existing algorithms for taxonomy-independent microbial community analysis. *Briefings in Bioinformatics* 13 (1): 107–121. https://doi.org/10.1093/bib/bbr009. https://www.ncbi.nlm.nih.gov/pubmed/21525143, https://www.ncbi.nlm.nih.gov/pmc/PMC3251834/.

Thorne, Jeffrey L., Hirohisa Kishino, and Ian S. Painter. 1998. Estimating the rate of evolution of the rate of molecular evolution. *Molecular Biology and Evolution* 15 (12): 1647–1657.

Tikhonov, M., R.W. Leach, and N.S. Wingreen. 2015. Interpreting 16S metagenomic data without clustering to achieve sub-OTU resolution. *The ISME Journal* 9 (1): 68–80.

Trüper, Hans G. 1999. How to name a prokaryote? Etymological considerations, proposals and practical advice in prokaryote nomenclature. *FEMS Microbiology Reviews* 23 (2): 231–249.

Turnbaugh, Peter J., Christopher Quince, Jeremiah J. Faith, Alice C. McHardy, Tanya Yatsunenko, Faheem Niazi, Jason Affourtit, Michael Egholm, Bernard Henrissat, Rob Knight, and Jeffrey I. Gordon. 2010. Organismal, genetic, and transcriptional variation in the deeply sequenced gut microbiomes of identical twins. *Proceedings of the National Academy of Sciences* 107 (16): 7503–7508. https://doi.org/10.1073/pnas.1002355107. https://www.pnas.org/content/pnas/10 7/16/7503.full.pdf.

Van Regenmortel, M.H.V. 1997. Viral species. In *Species: The units of biodiversity*, ed. M.F. Claridge, H.A. Dawah, and M.R. Wilson, 17–24. London: Chapman & Hall.

van Rijsbergen, C.V. 1979. *Information retrieval*. 2nd ed. Boston: Butterworth.

Vandamme, Peter, Bruno Pot, P. De Monique Gillis, Karel Kersters Vos, and Jean Swings. 1996. Polyphasic taxonomy, a consensus approach to bacterial systematics. *Microbiological Reviews* 60 (2): 407–438.

Wang, Xiaoyu, Jin Yao, Yijun Sun, and Volker Mai. 2013. M-pick, a modularity-based method for OTU picking of 16S rRNA sequences. *BMC Bioinformatics* 14 (1): 43. https://doi.org/10.1186/1471-2105-14-43.

Wayne, L.G., D.J. Brenner, R.R. Colwell, P.A.D. Grimont, O. Kandler, M.I. Krichevsky, L.H. Moore, W.E.C. Moore, RGEea Murray, and E.S.M.P. Stackebrandt. 1987. Report of the ad hoc committee on reconciliation of approaches to bacterial systematics. *International Journal of Systematic and Evolutionary Microbiology* 37 (4): 463–464.

Wayne, L.G., R.C. Good, E.C. Bottger, R. Butler, M. Dorsch, T. Ezaki, W. Gross, V. Jonas, J. Kilburn, and P. Kirschner. 1996. Semantide-and chemotaxonomy-based analyses of some problematic phenotypic clusters of slowly growing mycobacteria, a cooperative study of the International Working Group on Mycobacterial Taxonomy. *International Journal of Systematic and Evolutionary Microbiology* 46 (1): 280–297.

Wei, Ze-Gang, and Shao-Wu Zhang. 2015. MtHc: A motif-based hierarchical method for clustering massive 16S rRNA sequences into OTUs. *Molecular BioSystems* 11 (7): 1907–1913. https://doi.org/10.1039/C5MB00089K.

————. 2017. DBH: A de Bruijn graph-based heuristic method for clustering large-scale 16S rRNA sequences into OTUs. *Journal of Theoretical Biology* 425: 80–87. https://doi.org/10.1016/j.jtbi.2017.04.019. https://www.sciencedirect.com/science/article/pii/S0022519317301832.

————. 2019. DMSC: A dynamic multi-seeds method for clustering 16S rRNA sequences into OTUs. *Frontiers in Microbiology* 10 (428). https://doi.org/10.3389/fmicb.2019.00428. https://www.frontiersin.org/article/10.3389/fmicb.2019.00428.

Wei, Dan, Qingshan Jiang, Yanjie Wei, and Shengrui Wang. 2012. A novel hierarchical clustering algorithm for gene sequences. *BMC Bioinformatics* 13: 174–174. https://doi.org/10.1186/1471-2105-13-174. https://pubmed.ncbi.nlm.nih.gov/22823405, https://www.ncbi.nlm.nih.gov/pmc/articles/PMC3443659/.

Wei, Z.G., S.W. Zhang, and Y.Z. Zhang. 2017. DMclust, a density-based modularity method for accurate OTU picking of 16S rRNA sequences. *Molecular Informatics* 36 (12). https://doi.org/10.1002/minf.201600059.

Wei, Ze-Gang, Xiao-Dan Zhang, Ming Cao, Fei Liu, Yu Qian, and Shao-Wu Zhang. 2021. Comparison of methods for picking the operational taxonomic units from amplicon sequences. *Frontiers in Microbiology* 12 (474). https://doi.org/10.3389/fmicb.2021.644012. https://www.frontiersin.org/article/10.3389/fmicb.2021.644012.

Westcott, Sarah L., and Patrick D. Schloss. 2015. De novo clustering methods outperform reference-based methods for assigning 16S rRNA gene sequences to operational taxonomic units. *PeerJ* 3: e1487. https://doi.org/10.7717/peerj.1487.

————. 2017. OptiClust, an improved method for assigning amplicon-based sequence data to operational taxonomic units. *mSphere* 2 (2). https://doi.org/10.1128/mSphereDirect.00073-17. https://msphere.asm.org/content/msph/2/2/e00073-17.full.pdf.

Wheeler, Quentin D., and Rudolf Meier. 2000. *Species concepts and phylogenetic theory: A debate*. New York: Columbia University Press.

Wiley, Edward O. 1978. The evolutionary species concept reconsidered. *Systematic Zoology* 27 (1): 17–26.

Wiley, E.O., and R.L. Mayden. 2000. The evolutionary species concept. In *Species concepts and phylogenetic theory: A debate*, ed. Q.D. Wheeler and R. Meier, 70–89. New York: Columbia University Press.

Wirawan, Adrianto, Robert S. Harris, Yongchao Liu, Bertil Schmidt, and Jan Schröder. 2014. HECTOR: A parallel multistage homopolymer spectrum based error corrector for 454 sequencing data. *BMC Bioinformatics* 15 (1): 131. https://doi.org/10.1186/1471-2105-15-131.

Woese, Carl R. 1987. Bacterial evolution. *Microbiological Reviews* 51 (2): 221–271.

Woese, C.R., E. Stackebrandt, T.J. Macke, and G.E. Fox. 1985. A phylogenetic definition of the major eubacterial taxa. *Systematic and Applied Microbiology* 6 (2): 143–151.

Xia, Y., and J. Sun. 2022. *An integrated analysis of microbiomes and metabolomics*. Washington, DC: American Chemical Society. https://doi.org/10.1021/acsinfocus.7e5003. ISBN 9780841299542.

Xia, Yinglin, Jun Sun, and Ding-Geng Chen. 2018. Bioinformatic analysis of microbiome data. In *Statistical analysis of microbiome data with R*, 1–27. Singapore: Springer.

Ye, Yuzhen. 2011. Identification and quantification of abundant species from pyrosequences of 16S rRNA by consensus alignment. In *Proceedings IEEE international conference on bioinformatics and biomedicine*, 2010, 153–157. https://doi.org/10.1109/bibm.2010.5706555. https://www.ncbi.nlm.nih.gov/pubmed/22102981, https://www.ncbi.nlm.nih.gov/pmc/PMC3217275/.

Yilmaz, Pelin, Laura Wegener Parfrey, Pablo Yarza, Jan Gerken, Elmar Pruesse, Christian Quast, Timmy Schweer, Jörg Peplies, Wolfgang Ludwig, and Frank Oliver Glöckner. 2014. The SILVA and "all-species living tree project (LTP)" taxonomic frameworks. *Nucleic Acids Research* 42 (D1): D643–D648.

Zheng, Zejun, Stefan Kramer, and Bertil Schmidt. 2012. DySC: Software for greedy clustering of 16S rRNA reads. *Bioinformatics* 28 (16): 2182–2183. https://doi.org/10.1093/bioinformatics/bts355.

Zink, Robert M., and Mary C. McKitrick. 1995. The debate over species concepts and its implications for ornithology. *The Auk* 112 (3): 701–719.

Chapter 9
Alpha Diversity

Abstract Population diversity is one important characteristic of a microbiome community, which is highly related to its environment. The change of diversity may indicate that an environment of human body has undergone a change (e.g., from a healthy to a disease condition), or the environment is disrupted by some factors (e.g., antibiotic use or a change of immune response). Thus, one task of microbiome community analysis is to analyze diversity of the gut microbiota and find host and/or environment factors that can modulate the taxonomic composition. Generally alpha, beta, and gamma are three kinds of diversity in microbiome community. Among them, alpha diversity and beta diversity are two most commonly used diversity measures in microbiome research. This chapter focuses on alpha diversity analysis. First, it introduces abundance-based alpha diversity metrics and phylogenetic metrics. Then, it explores alpha diversity and abundance by some common plots. Next, it describes statistical hypothesis testing of alpha diversity using R and QIIME 2.

Keywords Alpha diversity · Chao 1 · Abundance-based coverage estimator (ACE) · Shannon diversity · Simpson diversity · Inverse Simpson diversity · Simpson evenness · Pielou's evenness · Phylogenetic diversity · Phylogenetic entropy · Phylogenetic quadratic entropy · Kruskal-Wallis test · Alpha-phylogenetic method

One important characteristic of a microbiome community is its population diversity. The community diversity is highly related to its environment, so if a change of diversity has been detected, then it may either indicate that an environment in the human body has undergone a change (e.g., from a healthy to a disease condition) or indicate that the environment is disrupted by some factors (e.g., antibiotic use or a change of immune response). Thus, one task of microbiome community analysis is to find host and/or environment factors that can modulate the taxonomic composition (classified groups of closely related microbiota) and diversity (distribution of microbiota within or between communities) of the gut microbiota.

Supplementary Information The online version contains supplementary material available at https://doi.org/10.1007/978-3-031-21391-5_9.

In microbiome community, generally there are three kinds of diversity including alpha, beta, and gamma diversities (Whittaker 1972). The alpha diversity and beta diversity are two common concepts that are used to measure the diversity of microbiome. In Chap. 9, we focus on alpha diversity analysis. Chapters 10 and 11 will introduce beta diversity analysis. This chapter is organized as follows. Section 9.1 introduces abundance-based alpha diversity metrics. Section 9.2 introduces phylogenetic metrics. Section 9.3 explores alpha diversity and abundance by some common plots. Section 9.4 describes statistical hypothesis testing of alpha diversity. Section 9.5 introduces alpha diversity analysis in QIIME 2. Finally, we briefly summarize this chapter in Sect. 9.6.

9.1 Abundance-Based Alpha Diversity Metrics

Although previously in the literature species diversity and species richness are sometimes used interchangeably, now it tends to use "species diversity" in the form of a species diversity index, that is, as an expression or index of some relation between number of species and number of individuals, and use "species richness" to refer to the number of species (in a given area or in a given sample) (Spellerberg and Fedor 2003).

Alpha diversity estimates the diversity within a single community (within-sample), which reflects both the number of species present (species richness) and the distribution of the number of organisms per species (species evenness), i.e., their relative abundance (evenness or equitability) and their taxonomic distribution. Several non-phylogenetic and phylogenetic metrics are available to measure alpha diversity. The non-phylogenetic metrics are based solely on abundance information, weight relatively rare, and mid-abundant and abundant species, while the phylogenetic metrics further utilize phylogenetic tree information.

9.1.1 Chao 1 Richness and Abundance-Based Coverage Estimator (ACE)

Species richness estimators estimate the total number of species present in a sample or community. Chao1 (Chao 1984) and ACE (abundance-based coverage estimator) (Chao and Lee 1992) were developed specifically for estimating richness of species from samples and for the general class-estimation problem, respectively (Colwell and Coddington 1994). In this section, we focus on Chao1 richness, and in Sect. 9.1.2, we introduce ACE.

9.1.1.1 The Measures of Chao 1 Richness

Two nonparametric methods of species richness for presence/absence data were developed by Anne Chao that were called "Chao 1" and "Chao 2" in the literature (Colwell and Coddington 1994). Chao1 index is a simple estimator to estimate the true number of species in the sample (or classes in a population) (Chao 1984); thus it is an abundance-based estimator of species richness.

Chao 1 index is based upon the number of rare species (or rare classes, i.e., taxa or OTUs) found in a sample (Chao 1984). The formula is given as below:

$$S_{Chao1} = S_{obs} + \frac{n_1^2}{2n_2},\qquad(9.1)$$

where S_{Chao1} is the estimated number of species, S_{obs} is the observed number of species in a sample in total, and n_1 and n_2 are the number of singletons and doubletons, respectively. In other words, n_1 is the number of observed species that are represented by only a single individual in that sample, and n_2 is the number of observed species represented by exactly two individuals in that sample.

Chao 1 estimator is based on the distribution of individuals among species and the data on singletons and doubletons. Colwell and Coddington (1994) thought that the same approach can be applied to the distribution of species among samples based on only presence-absence data. They call this estimator "Chao 2":

$$S_{Chao2} = S_{obs} + \frac{L^2}{2M},\qquad(9.2)$$

where L is the number of species that occur in only one sample ("unique" species) and M is the number of species that occur in exactly two samples. In concept, Chao 1 and Chao 2 are very similar. However, Colwell and Coddington's approach, i.e., interpretation of Chao estimator from the distribution of individuals among species to the distribution of species among samples and from the data on singletons and doubletons to the data on only presence-absence, provides an alternative interpretation of Chao estimator. For example, n_1 could be the number of species (taxa, OTUs) represented only once in the samples (unique species, taxa, OTUs) or a single read in that community, while n_2 could be the number of species (taxa, OTUs) represented only twice in the samples. The confidence intervals are constructed by slightly modifying the percentile method based on bootstrap distributions (Efron 1981, 1982).

Note that Chao index only uses the singletons and doubletons to estimate the number of missing species (Chao 1984) (see formula 9.1), so this index gives more weight to the low-abundance species. Therefore, Chao index is particularly useful for datasets skewed toward the low-abundance species (Hughes et al. 2001), as is likely to be the case with microbiome data.

However, from above formula, notice that if the number of singleton taxa n_1 is getting larger, that is, a sample contains many singletons; in such case, it is likely that the sample exists more undetected species; then the *Chao 1* index will estimate greater species richness than it would for a sample without rare species.

In 1987 (Chao 1987) Chao derived a closed-form solution for the variance of S_{Chao1} as below:

$$Var(S_{Chao1}) = n_2 \left(\frac{m^4}{4} + m^2 + \frac{m^2}{2} \right), \tag{9.3}$$

where $m = \frac{n_1}{n_2}$. This formula estimates the precision of Chao1 from multiple samples, replacing the more complex variance estimation technique presented in Chao (1984) (Colwell and Coddington 1994). The chao1 confidence interval for richness was developed in 2004 (Colwell et al. 2004).

9.1.1.2 The Measures of ACE

ACE (Chao and Lee 1992; Chao and Yang 1993) is a nonparametric method for estimating the number of species (classes) using sample coverage, which is defined as the sum of the probabilities of the observed species (classes):

$$S_{ACE} = S_{abund} + \frac{S_{rare}}{C_{ACE}} + \frac{n_1}{C_{ACE}} \gamma^2_{ACE}, \tag{9.4}$$

where S_{abund} and S_{rare} are the number of abundant and rare species (taxa, OTUs), respectively. C_{ACE} is the sample abundance coverage estimator, n_1 is the number (frequency) of singletons, and γ^2_{ACE} is the estimated coefficient of variation for rare species (taxa, OTUs), which is defined as

$$\gamma^2_{ACE} = \max \left[\frac{S_{rare}}{C_{ACE}} \frac{\sum_{i=1}^{10} i(i-1)n_1}{N_{rare}(N_{rare} - 1)} - 1, 0 \right]. \tag{9.5}$$

Chao and Yang (1993) discussed a theoretical optimal stopping rule based on the minimization of the testing cost and expected penalty due to the unremoved bugs. The optimal detection value is chosen as 10. The number of bugs detected more than 10 times is then added to the resulting estimate.

ACE method (see Eq. 9.4) divides the number of observed species into the frequencies abundant species (those with more than ten individuals in the sample) and the frequencies rare species (those with fewer than ten individuals) (Kim et al. 2017). The ACE method has some properties (Chao and Lee 1992; Chao and Yang 1993; Hughes et al. 2001; Sornplang and Piyadeatsoontorn 2016; Kim et al. 2017): (1) it only considers the presence or absence information of abundant species because abundant species would be discovered anyway; (2) it does not require the exact frequencies for the abundant species; (3) it does require the exact frequencies for the rare species because the estimation of the number of missing species is based entirely on these rare species.

9.1.1.3   Calculating Chao 1 Richness and ACE Using Ampvis2 Package

In this section, we use both ampvis2 and microbiome packages to illustrate the alpha diversity calculations based on two datasets. The first example is breast cancer QTRT1 mouse gut data (Zhang et al. 2020). The second example is mouse gut microbiome we generated from raw sequencing using QIIME 2 in Chap. 4 and has been used for bioinformatic analysis in Chaps. 5 and 6.

Example 9.1: Mouse Gut Microbiome Regulated by QTRT1 Gene in the Development of Breast Cancer
Sun and her team in 2020 conducted a murine microbiome experiment to explore whether and how the enzyme queuosine (Q)-tRNA ribosyltransferase catalytic subunit 1 (QTRT1) affects tumorigenesis and the microbiome. Queuosine is a micronutrient from diet and microbiome. Transfer RNA (tRNA) Q-modifications occur specifically in four cellular tRNAs at the wobble anticodon position. Because the microbiome product queuine is required for its installation, so tRNA Q-modification in human cells depends on the gut microbiome. This study aims to (1) examine microbiome-dependent Q-tRNA modification and its impact on breast tumor growth and gene expression and (2) identify the mechanisms of tRNA Q-modification-dependent cellular phenotypes in vitro and in vivo. Single clones of complete QTRT1-knockout (KO) in human breast cancer MCF7 cells were generated. The Q-modification levels of the MCF7 wild-type (WT) and QTRT1-KO cells were measured using the standard APB gel method. In this study, 20 KO mice ($n = 10$ for each of before and post treatments) and 20 WT mice ($n = 10$ for each of before and post treatments) were investigated. DNA was extracted from fecal samples, and 16S rRNA gene V3–V4 variable regions were barcoded and sequenced using MiSeq according to the Illumina protocol. After DNA sequencing, all possible raw paired-end reads were evaluated and merged using the PEAR (Zhang et al. 2014). Quality-filtered sequences were performed using QIIME pipeline (Caporaso et al. 2010) at a 97% similarity threshold utilizing the USEARCH algorithm (Edgar 2010) and the reference database silva_132_16S.97 (Glöckner et al. 2017). The microbiome data for downstream analysis include a total of 40 samples across 20 mice, resulting in 635 unique OTUs/taxa. The OTUs ranged from 39,813 to 65,762 for the individual sample, with a mean of 55,502.7 and median of 56,093 and total reads of 2,220,108.

Throughout this book, we will call this dataset as "Breast cancer QTRT1 mouse gut microbiome" for simplification.

First, we install the package **ampvis2** (current version 2.7.17, March 2022) and load the data into R. To install the ampvis2 package, enter the following commands in R or RStudio.

```
install.packages("remotes")
remotes::install_github("MadsAlbertsen/ampvis2")
```

To fast install the ampvis2 package, we can use multicore processors by setting the Ncpus argument as below:

```
install.packages("remotes")
remotes::install_github("MadsAlbertsen/ampvis2", Ncpus = 6)
```

Because today most computers' CPU can run eight processes simultaneously, thus unless you know your computer has a CPU with more (logical) cores than 8, setting it to 6 (Ncpus = 6) is a good starting point.

Windows users should also install **Rtools4** on Windows. Starting with R 4.0.0 (released April 2020), R for Windows uses a toolchain bundle called **rtools4** (https://mirrors.dotsrc.org/cran/bin/windows/Rtools/rtools40.html). Rtools4 is based on *msys2*, which makes easier to build and maintain R itself as well as the system libraries needed by R packages on Windows.

Here, we use breast cancer QTRT1 mouse gut data. The **ampvis2** package can import data directly from the commonly used amplicon pipelines: QIIME 2, mothur, and UPARSE. The **amp_import_biom()** function is used to import the data in BIOM format generated from QIIME 2 and mothur, while the **amp_import_uparse()** function is used to import the data from the UPARSE pipeline. Then the data is loaded into a single ampvis2 object by the **amp_load()** function, which results in a list of data frames. The data that we use here are already in the format of data frames. Below we load otu_table and metadata into R:

```
> setwd("~/Documents/QIIME2R/Ch9_AlphaDiversity")
> otu_tab <- read.csv("otu_table_L7_MCF7_amp.csv", check.names =
FALSE)
> meta_tab <- read.csv("metadata.csv", check.names = FALSE)
```

The function **amp_load()** is very useful; it checks and combines the data into a single ampvis2 object, making it easier to manipulate, filter, and subset all elements of the data at once for analysis. Please note that ampvis2 expects OTU ID's to be the first column named OTU/ASV in "otutable." Now we combine the otu_table and meta table into a single ampvis2 object "ds":

```
> library(ampvis2)
> ds <- amp_load(otutable = otu_tab,
+            metadata = meta_tab)
> ds
ampvis2 object with 3 elements.
Summary of OTU table:
   Samples    OTUs Total#Reads  Min#Reads   Max#Reads Median#Reads
      40      635   2220108      39813      65762       56093
  Avg#Reads
   55502.7
```

Assigned taxonomy:

```
Kingdom  Phylum   Class    Order   Family  Genus   Species
635(100%) 635(100%) 635(100%) 635(100%) 635(100%) 635(100%) 635(100%)
```

Metadata variables: 4.
SampleID, Group, Time, Group4.
We can explore the individual data frames in the list with **View()** function.

```
> View(ds$metadata)
> View(otu_tab)
> View(meta_tab)
```

The **ampvis2** package uses the **amp_alphadiv()** function to perform alpha diversity analysis, which calculates alpha diversity indices for each sample using the **vegan** function **diversity()**, where the read abundances are first rarefied using **rrarefy()** function by the size of the rarefy argument and combined with the metadata. The syntax is as below:

```
amp_alphadiv(data, measure = NULL, richness = FALSE, rarefy = NULL)
```

The function has four arguments: **data**, **measure**, **richness**, and **rarefy** and return a dataframe, where the argument:

- **data** is the required argument, listing as loaded with the amp_load() function.
- **measure** is the listing vector that listed one or more alpha diversities to be calculated including "observed" for Observed taxa (OTUs), "shannon" for Shannon diversity, "simpson" for Simpson diversity, and "invsimpson" for inverse Simpson diversity.
- **richness** is used to calculate sample richness estimates (Chao 1 and ACE) which is calculated by the estimate() function. The default is FALSE, without calculating richness.
- **rarefy** rarefies species richness to the specified value before calculating alpha diversity and/or richness. By default(NULL), samples will be not rarefied before calculation.

It has been shown that using *rarefying* of counts is an inappropriate approach for detection of differentially abundant species (McMurdie and Holmes 2014). Here we do not rarefy the sample counts. By default, if no measure(s) are chosen, all diversity indices will be returned. We can also provide a list of vector to calculate more than one diversity, such as measure = c("shannon", "simpson") to calculate Shannon and Simpson alpha diversity indices simultaneously. However, here, we illustrate the function **amp_alphadiv()** to calculate each diversity one by one. The following R commands will calculate all the alpha diversity indices including Chao 1 and ACE:

```
> alpha_amp <- amp_alphadiv(ds,
+               measure =,
+               richness = TRUE,
+               rarefy = NULL)
> alpha_amp
          SampleID Group  Time  Group4 Reads ObservedOTUs Shannon Simpson
invSimpson Chao1  ACE
Sun040.BH5 Sun040.BH5   KO Before KO_BEFORE 39813        110  2.385 0.8270
5.779 141.0 156.1
Sun039.BH4 Sun039.BH4   KO Before KO_BEFORE 42056        112  2.378 0.8336
6.009 160.0 170.5
Sun027.BF2 Sun027.BF2   WT Before WT_BEFORE 42729        101  2.276 0.7784
4.512 153.8 162.9
......
```

We can explore the results in the data frame using **View()** and **head()** functions.

```
> View(alpha_amp)
> head(alpha_amp)
```

9.1.1.4   Calculating Chao 1 Richness and ACE Using Microbiome Package

In Chap. 2 (Sect. 2.3.2), we introduced the **microbiome** package (current version 1.14.0, March 2022). If this package has not been installed in your computer, open R and enter the following R commands to install:

```
library(BiocManager)
#source("https://bioconductor.org/install")
#useDevel()
BiocManager::install(version='devel')
BiocManager::install("microbiome")
```

Alternatively, we can install the development version (potentially unstable) by entering the following R commands in R:

```
# Alternative install the development version
# Load the devtools package
library(devtools)
# Install the package
install_github("microbiome/microbiome")
```

Example 9.2: Mouse Gut Microbiome Data
In order to show the workflow from bioinformatic data generation to downstream statistical analysis, here we use the same mouse gut microbiome data generated by DADA2 within QIIME 2. Thus we use **qiime2R** package to export the .qza and .tsv

files into R. Entering the following R commands to install this package, then call the function **qza_to_phyloseq()** to create a phyloseq object. To do so, we also need the **phyloseq** package available:

```
> install.packages("remotes")
> remotes::install_github("jbisanz/qiime2R")
> library(qiime2R)
> library(phyloseq)
> physeq<-qza_to_phyloseq(features="FeatureTableMiSeq_SOP.qza",
taxonomy = "TaxonomyMiSeq_SOP.qza", tree = "RootedTreeMiSeq_SOP.
qza", metadata="SampleMetadataMiSeq_SOP.tsv")
> physeq
phyloseq-class experiment-level object
otu_table() OTU Table:         [ 392 taxa and 360 samples ]
sample_data() Sample Data:     [ 360 samples by 11 sample variables ]
tax_table() Taxonomy Table:    [ 392 taxa by 7 taxonomic ranks ]
phy_tree() Phylogenetic Tree:  [ 392 tips and 389 internal nodes ]
```

Two ways can be used to calculate Chao 1 richness using the **microbiome** package: one is to call the function **alpha()** and specify the index = "chao1", and another way is to directly call the function **richness(),** which also calculates the observed taxa:

```
> library(microbiome)
> library(knitr)
```

The following R commands call the function **alpha()** and specify the index = "chao1" to calculate Chao 1 richness:

```
> chao1 <- alpha(physeq, index = "chao1")
> head(chao1,3)
        chao1
F3D0    111
F3D1    102
F3D11    89
```

The following R commands directly call the function **richness() to** calculate Chao 1 richness and the observed taxa:

```
> rich <- richness(physeq)
> head(rich,3)
       observed chao1
F3D0       111  111
F3D1       102  102
F3D11       89   89
```

However, current version (version 1.14.0) of microbiome package does not provide the capability to estimate the measures of ACE, which is available via **estimate_richness()** function in the **phyloseq** package:

```
> library(phyloseq)
> richness_physeq <- estimate_richness(physeq, split = TRUE, measures
= c("Observed", "Chao1", "ACE"))
> head(richness_physeq,3)
      Observed Chao1 se.chao1 ACE se.ACE
F3D0      111  111        0 111  4.727
F3D1      102  102        0 102  4.284
F3D11      89   89        0  89  4.290
```

9.1.2   Shannon Diversity

Shannon and Simpson diversity indices are most often used to measure the diversity of communities. Both indices take both abundance and relative abundances (evenness) of the species present into account. Thus they provide more information about the community composition than simple species richness (i.e., the number of species present).

9.1.2.1   The Measures of Shannon Diversity

The measure of Shannon diversity index (Shannon 1948), an index of both species richness and evenness, is the most commonly used and specifically the most popular in the ecological and microbiome literatures. Shannon's index (labeled as H) was originally developed by Claude Shannon in 1948 to quantify the entropy (uncertainty or information content) in strings of text. The uncertainty is measured by the Shannon function H, which is based on communication theory and measures how to predict the next letter in a message or communication (a common question in communication) (Shannon 1948). Shannon's index (H) is given as below:

$$H = - \sum_{i=1}^{S} p_i \ln p_i, \tag{9.6}$$

where p_i is the proportion (or relative abundance) of individuals of species i in the community and S is the total number of species in the community (richness), so $\sum_{i=1}^{S} p_i = 1$. Thus, Shannon's diversity index is to first calculate the proportion of species i relative to the total number of species (p_i) and then multiplied by the natural logarithm of this proportion ($\ln p_i$). The resulting product is summed across species and multiplied by -1. In summary, Shannon's index measures the uncertainty: How difficult would it be to predict correctly the species of the next individual collected? Put another way, it measures the amount of uncertainty, so that the larger the value of H, the greater the uncertainty.

It is worth mentioning that Shannon's index has some properties:

- Strictly speaking, the Shannon measure of information content should be used only on random samples drawn from a large community in which the total number of species is known (Pielou 1966).
- Theoretically the value H could be very large because it increases as the number of species in the community increases and as the distribution of individuals among the species becomes even (Lemos et al. 2011; Magurran 2013). But in practice for biological communities, H does not seem to be larger than 5.0 (Washington 1984).
- Shannon's index "gives more importance" to less common categories (e.g., rare species in the case of microbiome) and places a greater weight on species richness than species evenness (Schloss and Handelsman 2006; Schloss et al. 2009).
- Because Shannon's index values increase as the number of sample size increases, thus normalization is crucial and suggested to avoid biases when it is compared between samples (Lemos et al. 2011).

Note that Shannon index is also known as the Shannon-Wiener index, the Shannon-Weaver index, and the Shannon entropy (Spellerberg and Fedor 2003). Based on Spellerberg and Fedor (2003), the uses of the "Shannon-Wiener" index and the "Shannon-Weaver" index are due to confusion. Actually, Shannon's mathematical theory of communication including the first form of the present Shannon expression (index) was published in the 1948s *Bell System Technical Journal* (Shannon 1948). In 1949, Shannon published this information jointly with Warren Weaver, a mathematician, in the book *The Mathematical Theory of Communication* (Shannon et al. 1949), which is very similar to the original published papers, and the entropy "H" in 1948 was taken account of by Shannon in the first part of this 1949 book. In his paper Shannon appreciated the impact of the mathematician Norbert Wiener's basic philosophy and theory on his communication theory and cited several of Wiener's publications. Thus, Spellerberg and Fedor (2003) suggested that the "mislabeling" and the confusion of the Shannon index "H" (as referred to by Krebs 1999) as the Shannon and Weaver index are partly due to the joint authorship of Shannon and Weaver's book, while given the fact that in the late 1940s, Shannon had built on the work of Wiener, it seems preferable to refer to "H" (the species diversity index) as the "Shannon index" or the "Shannon and Wiener index" (Spellerberg and Fedor 2003).

9.1.2.2   Calculating Shannon Diversity Using ampvis2 Package

Example 9.3: Breast Cancer QTRT1 Mouse Gut Microbiome, Example 9.1 Cont.
The following R commands calculate Shannon index for QTRT1 dataset using the **amp_alphadiv()** in ampvis2 package:

```
> shannon_amp <- amp_alphadiv(ds,
+               measure = "shannon",
+               richness = FALSE,
+               rarefy = NULL)


> View(shannon_amp)
> head(shannon_amp,3)
    SampleID Group Time    Group4 Reads Shannon
1 Sun040.BH5    KO Before KO_BEFORE 39813    2.385
2 Sun039.BH4    KO Before KO_BEFORE 42056    2.378
3 Sun027.BF2    WT Before WT_BEFORE 42729    2.276
```

9.1.2.3   Calculating Shannon Diversity Using Microbiome Package

Example 9.4: Mouse Gut Microbiome Data, Example 9.2 Cont.
To calculate Shannon diversity, we call the function **alpha()** in microbiome package and specify the index = "shannon":

```
> shannon <- alpha(physeq, index = "shannon")
> head(shannon,3)
      diversity_shannon
F3D0        3.885
F3D1        3.992
F3D11       2.893
```

9.1.3   Simpson Diversity

Simpson's diversity indices refer to four closely related indices: (1) Simpson's index (D), (2) Simpson's index of diversity 1 – D, (3) Simpson's reciprocal index 1 / D, and (4) Simpson's index of evenness.

9.1.3.1 The Measures of Simpson Diversity

Simpson in 1949 (Simpson 1949) proposed a new nonparametric measure which combines species richness and evenness. The new concept of diversity states that diversity is inversely related to "the probability that two individuals chosen at random and independently from the population will be found to belong to the same group" (Simpson 1949), i.e., will belong to the same species. Although in ecological literature the new concept of diversity is synonymous with diversity (Hurlbert 1971), it actually is about species heterogeneity (Good 1953).

Simpson's Index (D_0)
Simpson's index (D_0) measures the probability that two individuals randomly selected from a sample will belong to the same species. This is Simpson's original index, which is given as

$$D_0 = \sum (n/N)^2, \tag{9.7}$$

where n is the total number of individuals (organisms) of a particular species and N is the total number of individuals (organisms) of all species. Another version of the formula for calculating Simpson's original index (D_0) is given as

$$D_0 = \frac{\sum n(n-1)}{N(N-1)}. \tag{9.8}$$

This version of the formula will give consistent and acceptable results. In literature, Simpson's original index in (9.7) is also written as $D_O = \sum_{i=1}^{S} p_i^2$, where S is the total number of species in the community and p_i is the proportion of individuals (or relative abundance) of species i in the community. Simpson's index gives more weight to the more abundant species in a sample. Additional rare species to a sample does not change the value of D_0 too much.

The value of D_0 ranges between 0 and 1, with 0 representing infinite diversity (i.e., the lower value of D_0 the more diversity) and 1 suggesting no diversity (i.e., the bigger value of D_0, the lower diversity). Obviously Simpson's original index cannot provide an intuitive and logical interpretation of diversity.

Simpson's Diversity Index ($1-D_0$)
To provide a straightforward interpretation, D_0 is often subtracted from 1 which defines Simpson's diversity index as below:

$$D = 1 - \frac{\sum n(n-1)}{N(N-1)}. \tag{9.9}$$

In the literature, Simpson's diversity index is also written as $D = 1 - \sum_{i=1}^{S} p_i^2$. The values of Simpson's diversity index also range from 0 to 1, but now with lower values (close to 0) indicating lower sample diversity, while greater values (close to 1) indicating greater sample diversity. Thus, Simpson's diversity index measures the probability that two individuals randomly selected from a sample will belong to different species. In contrary to Shannon's index, Simpson's diversity index "gives more importance" to more common species.

Simpson's Reciprocal Index ($1/D_0$)
Another way to avoid nonintuitive nature of Simpson's index is to take the reciprocal of the index, which defines inverse Simpson's index ($1/D_0$). The inverse Simpson index is given as below:

$$D_I = \frac{1}{\sum_{i=1}^{S} p_i^2},\tag{9.10}$$

which is the reciprocal of Simpson's original index ($1/D_O$).

Simpson's Index of Evenness
As we stated above, heterogeneity contains both species richness and evenness; thus, researchers naturally try to separately measure the evenness component from richness. The null hypothesis of evenness is all species in a hypothetical community are equally common. However, most communities contain a few dominant species and many species that are relatively uncommon. Evenness measures attempt to quantify this unequal representation against the null hypothesis. As the independent measures of species richness, many different measures of evenness (or equitability) have been proposed in the ecology and microbiome literatures. Simpson's index of evenness is one of them. Simpson's index of evenness is obtained from reciprocal of Simpson's original index $D_O = \sum_{i=1}^{S} p_i^2$, which is defined as

$$E = \frac{1}{S \sum_{i=1}^{S} p_i^2},\tag{9.11}$$

where S is the number of species in the sample. This index also ranges from 0 to 1 and is relatively unaffected by the rare species in the sample.

Simpson index also has some properties:

- Simpson's diversity index measures the species dominance and reflects the probability of two individuals that belong to the same species being randomly chosen. It varies from 0 to 1 and the index increases as the diversity decreases (Simpson 1949).

- Simpson diversity puts a higher weight on species evenness more than species richness in its measurement (Schloss and Handelsman 2006; Schloss et al. 2009).
- Same to Shannon diversity, normalization is considered as crucial and important to avoid biases when comparing Simpson diversity between different samples (Lemos et al. 2011).

9.1.3.2 Calculating Simpson Diversity and Inverse Simpson Diversity Using Ampvis2 Package

Example 9.5: Breast Cancer QTRT1 Mouse Gut Microbiome, Example 9.1 Cont.

The following R commands calculate Simpson diversity for QTRT1 dataset using the **amp_alphadiv()** in ampvis2 package:

```
> simpson_amp <- amp_alphadiv(ds,
+             measure = "simpson",
+             richness = FALSE,
+             rarefy = NULL)
> View(simpson_amp)
> head(simpson_amp,3)
  SampleID Group  Time   Group4 Reads Simpson
1 Sun040.BH5  KO Before KO_BEFORE 39813 0.8270
2 Sun039.BH4  KO Before KO_BEFORE 42056 0.8336
3 Sun027.BF2  WT Before WT_BEFORE 42729 0.7784
```

The following R commands calculate inverse Simpson index for QTRT1 dataset using the **amp_alphadiv()** in ampvis2 package:

```
> invsimpson_amp <- amp_alphadiv(ds,
+             measure = "invsimpson",
+             richness = FALSE,
+             rarefy = NULL)
> View(invsimpson_amp)
> head(invsimpson_amp,3)
  SampleID Group  Time   Group4 Reads invSimpson
1 Sun040.BH5  KO Before KO_BEFORE 39813    5.779
2 Sun039.BH4  KO Before KO_BEFORE 42056    6.009
3 Sun027.BF2  WT Before WT_BEFORE 42729    4.512
```

9.1.3.3 Calculating Simpson Diversity, Inverse Simpson Diversity, and Simpson Evenness Using Microbiome Package

Example 9.6: Mouse Gut Microbiome Data, Example 9.2 Cont.

To calculate Simpson diversity, we call the function **alpha()** and specify the index = "simpson" in the microbiome package:

```
> simpson <- alpha(physeq, index = "simpson")
> head(simpson,3)
     evenness_simpson dominance_simpson
F3D0       0.2548          0.03535
F3D1       0.3343          0.02933
F3D11      0.1113          0.10098
```

To calculate inverse Simpson diversity, we call the function **alpha()** and specify the index = "inverse_simpson" in the microbiome package:

```
> inv_simpson <- alpha(physeq, index = "inverse_simpson")
> head(inv_simpson,3)
     diversity_inverse_simpson
F3D0           28.285
F3D1           34.098
F3D11           9.903
```

To calculate Simpson evenness, we call the function **evenness()** and specify the index = "simpson" in the microbiome package:

```
> even_simpson <- evenness(physeq, "simpson")
> head(even_simpson,3)
 simpson
F3D0 0.2548
F3D1 0.3343
F3D11 0.1113
```

9.1.4   Pielou's Evenness

Pielou's evenness (Pielou 1966) is the most common measure of evenness. Here, we describe and illustrate it below.

9.1.4.1   The Measures of Pielou's Evenness

Pielou's evenness measures diversity along with species richness. The formula of Pielou's evenness index is given as below:

$$J = \frac{H}{\log(S)}, \tag{9.12}$$

where H is Shannon's diversity index and S is the total number of species observed in a sample. The value of Pielou's evenness ranges from 0 (no evenness) to 1 (complete evenness). Many diversity indices, such as Simpson's diversity and Shannon's diversity, incorporate evenness. However, it has been shown that the diversity indices which concentrate totally on evenness are fraught with problems, including dependence on species counts (McCune and Grace 2002). Pielou's index has a particular problem: it is a ratio of a relatively stable index, H, and one that is strongly dependent on sample size, S.

9.1.4.2 Calculating Pielou Evenness Using Microbiome Package

Example 9.7: Mouse Gut Microbiome Data, Example 9.2, Cont.
To calculate Pielou evenness, we call the function **alpha()** and specify the index = "pielou" in microbiome package:

```
> pielou <- alpha(physeq, index = "pielou")
> head(pielou,3)
      evenness_pielou
F3D0       0.8249
F3D1       0.8631
F3D11      0.6444
```

The following R commands calculate all available evenness measures in the microbiome package:

```
> even <- evenness(physeq, "all")
> head(even,3)
      camargo pielou simpson  evar bulla
F3D0   0.9118 0.8249 0.2548 0.3774 0.5021
F3D1   0.8991 0.8631 0.3343 0.3939 0.5597
F3D11  0.8549 0.6444 0.1113 0.2462 0.2859
```

The following R commands calculate Pielou evenness using **evenness()** in the microbiome package:

```
> even_pielou <- evenness(physeq, "pielou")
> head(even_pielou,3)
   pielou
F3D0 0.8249
F3D1 0.8631
F3D11 0.6444
```

9.2 Phylogenetic Alpha Diversity Metrics

Phylogenetic metrics not only use abundance information but also further utilize phylogenetic tree information. Several phylogenetic metrics are available to measure alpha diversity. In this subsection, we introduce three commonly used phylogenetic metrics: (1) phylogenetic diversity, (2) phylogenetic entropy, and (3) phylogenetic quadratic entropy.

9.2.1   Phylogenetic Diversity

Faith's phylogenetic diversity (PD) (Faith 1992) is a measure of biodiversity, based on phylogeny (the tree of life), that is, incorporating phylogenetic difference between species. It can be considered as "a phylogenetic generalization of species richness" (Chao et al. 2016). Faith thought that the branch lengths on the tree are informative because they count the relative number of new features arising along that part of the tree. Thus, he defined and calculated the PD of a set of species as "equal to the sum of the lengths of all those branches that are members of the corresponding minimum spanning path" (Faith 1992), where "branch" is a segment of a cladogram, and the minimum spanning path is the minimum distance between the two nodes. The formula of PD is given as below:

$$PD = \sum_{branches\,b} \ell(b), \tag{9.13}$$

where $\ell(b)$ is the length of a branch b. As discussed in Faith (1992), PD is a "feature diversity" and so PD counts features. Faith's PD has been reviewed as a good measure of phylogenetic diversity. However, PD is only based on phylogeny. It is important to incorporate abundance information into phylogenetic diversity measures for conservation (Chao et al. 2016). We will illustrate how to calculate Faith's PD using QIIME 2 in Sect. 9.5.3.

9.2.2   Phylogenetic Entropy

Phylogenetic entropy (PE) (Allen et al. 2009) generalizes or extends the Shannon index (Shannon entropy) to incorporate phylogenetic distances among species. Thus, it is a diversity index (measure) that accounts for both phylogeny and species abundances, including species richness, evenness, and distinctness. The formula of PE is given as below:

$$H_p = - \sum_{branches\,b\ of\ T} \ell(b)p(b)\ln p(b), \tag{9.14}$$

where T is a rooted phylogenetic tree for the community, $\ell(b)$ is the length of a branch b of T, and $p(b)$ is the proportion of individuals in the (present-day) community who are represented by leaves descending from b.

Given defining PE this way, the Shannon index $H = - \sum_{i=1}^{S} p_i \ln p_i$ becomes a special case of the PD if all species are equally distinct or, equivalently, if T has uniform branch lengths (Allen et al. 2009).

Shannon entropy does not obey the replication principle (or "doubling property"), which states that the total diversity of pooling two equally diverse and equally large groups with no shared species should be two times the diversity of a single group (Chao et al. 2010, 2016). PE (as a phylogenetic generalization of Shannon entropy) was reviewed having the same interpretational problem as Shannon index (Chao et al. 2010, 2016).

9.2.3 *Phylogenetic Quadratic Entropy (PQE)*

Phylogenetic quadratic entropy (PQE) (Rao 1982; Warwick and Clarke 1995), the most established quadratic diversity, measures the average taxonomic or phylogenetic distance between individual organisms. PQE that was introduced by Rao (1982) is a generalization of the Gini-Simpson index (Chao et al. 2016). It was independently rediscovered by Warwick and Clarke (1995) under the name "taxonomic diversity" (Allen et al. 2009; Warwick and Clarke 1995). The formula is given as below:

$$Q = \sum_{i<j} d_{ij} p_i p_j, \qquad (9.15)$$

where d_{ij} is the taxonomic or phylogenetic distance between species s_i and s_j; d_{ij} can be equated with the branch length of the shortest path between the corresponding leaves of a phylogenetic tree. p_i and p_j are the proportion of individuals in species s_i and s_j, respectively.

Although Rao's quadratic entropy Q was the first diversity measure that accounts for both phylogeny and species abundances, however, like the Gini-Simpson index, its parent measure, it does not obey the replication principle. Thus it has the same interpretational problem as Gini-Simpson measure (Chao et al. 2010, 2016).

9.3 Exploring Alpha Diversity and Abundance

Alpha diversities and microbial abundances can be plotted using ampvis2 and microbiome packages. Comparing and visualizing groups based on the differences or similarities is also important. Various plots can be used to achieve different goals. We choose some useful plots to illustrate the capabilities of these two packages.

9.3.1   Heatmap

In this subsection, we illustrate how to generate heatmap using ampvis2 package.

Example 9.8: Breast Cancer QTRT1 Mouse Gut Microbiome, Example 9.1 Cont.

The ampvis2 package generates plots using the ggplot2 package. The heatmap is generated by the **amp_heatmap()** function. This function by default aggregates data to phylum level and shows the top 10 phyla, ordered by mean read abundance across all samples. Before generating heatmap, let's review a short summary of the breast cancer QTRT1 mouse gut data:

```
> ds
ampvis2 object with 3 elements.
Summary of OTU table:
   Samples     OTUs Total#Reads  Min#Reads  Max#Reads
      40       635   2220108      39813      65762
Median#Reads  Avg#Reads
   56093       55502.7
```

Assigned taxonomy:

```
 Kingdom  Phylum   Class    Order   Family   Genus  Species
635(100%) 635(100%) 635(100%) 635(100%) 635(100%) 635(100%) 635(100%)
Metadata variables: 4
SampleID, Group, Time, Group4
```

First we use the **amp_heatmap()** function to generate the heatmap for all four groups which by default aggregates to phylum level and shows the top 10 phyla, ordered by mean read abundance across all samples (Fig. 9.1):

```
> # Figure 9.1
> amp_heatmap(ds, group_by = "Group4")
```

In below arguments, we select genus as the aggregated level, showing top 10 genera, adding family level to get additional taxonomic information, with the abundance values in plots. The samples are grouped by lockout before and post, as wild type before and post 4 groups (Fig. 9.2):

```
> # Figure 9.2
> amp_heatmap(ds,
+        group_by = "Group4",
+        tax_aggregate = "Genus",
+        tax_add = "Family",
+        tax_show = 10,
+        color_vector = c("white", "red"),
+        plot_colorscale = "sqrt",
```

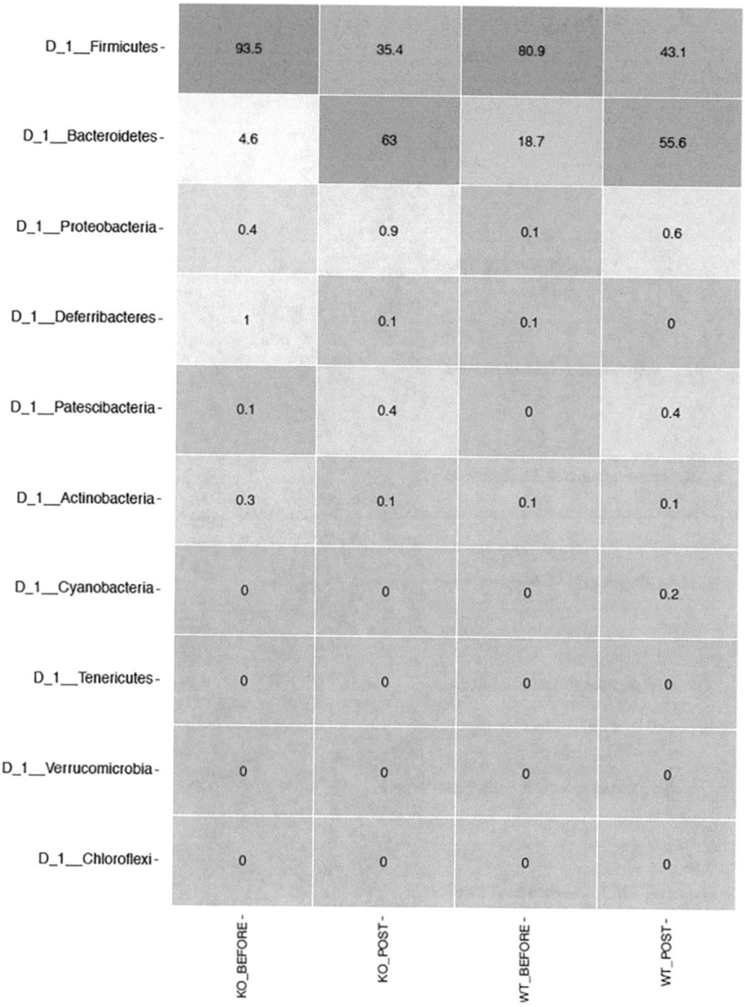

	KO_BEFORE	KO_POST	WT_BEFORE	WT_POST
D_1__Firmicutes	93.5	35.4	80.9	43.1
D_1__Bacteroidetes	4.6	63	18.7	55.6
D_1__Proteobacteria	0.4	0.9	0.1	0.6
D_1__Deferribacteres	1	0.1	0.1	0
D_1__Patescibacteria	0.1	0.4	0	0.4
D_1__Actinobacteria	0.3	0.1	0.1	0.1
D_1__Cyanobacteria	0	0	0	0.2
D_1__Tenericutes	0	0	0	0
D_1__Verrucomicrobia	0	0	0	0
D_1__Chloroflexi	0	0	0	0

Fig. 9.1 Heatmap of QTRT1 data aggregating to phylum level with the top 10 phyla, ordered by mean read abundance across all samples

```
+        plot_values = TRUE) +
+  theme(axis.text.x = element_text(angle = 45, size=10, vjust = 1),
+     axis.text.y = element_text(size=8),
+     legend.position="right")
```

Before generating the heatmap, we can filter and subset the data and then perform the function to obtain a specific heatmap to meet our study design. The following R commands subset the post treatment data for the analysis using the **amp_subset_samples()** function:

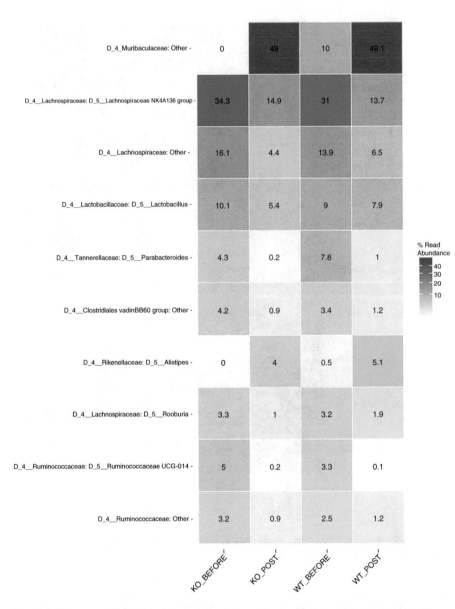

Fig. 9.2 Heatmap of QTRT1 data aggregating to genus level with the top 10 genera, ordered by mean read abundance across all samples

```
> ds_sub <- amp_subset_samples(ds, Time%in% c("Post"), minreads =
1000)
20 samples and 249 OTUs have been filtered
Before: 40 samples and 635 OTUs
After: 20 samples and 386 OTUs
```

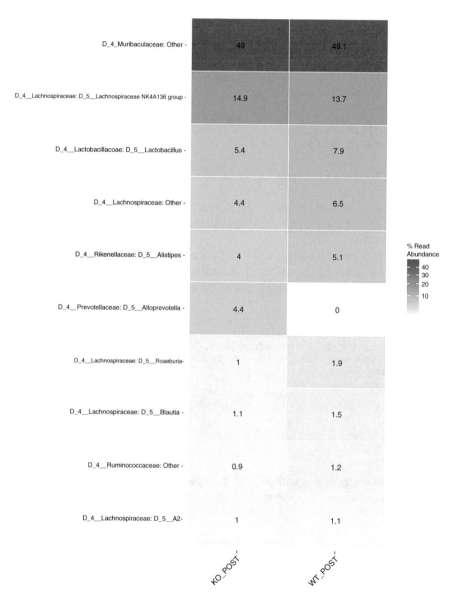

Fig. 9.3 Heatmap of QTRT1 data for post treatments aggregating to genus level with the top 10 genus, ordered by mean read abundance across all samples

Figure 9.3 is generated using the post samples only and minimum reads equal to 1000:

```
> # Figure 9.3
> amp_heatmap(ds_sub,
```

```
+          group_by = "Group4",
+          tax_aggregate = "Genus",
+          tax_add = "Family",
+          tax_show = 10,
+          color_vector = c("white", "red"),
+          plot_colorscale = "sqrt",
+          plot_values = TRUE) +
+  theme(axis.text.x = element_text(angle = 45, size=10, vjust = 1),
+       axis.text.y = element_text(size=8),
+       legend.position="right")
```

We can subset the data based on two or more variables using "&" to separate the conditions or simply use the function more than once. The "!" (logical NOT operator) indicates "except" and is useful to remove outliers. The minreads = cut-off values argument removes any sample(s) with total amount of reads below the chosen threshold.

We can also subset the data based on the taxonomy using the **amp_subset_taxa()** function. For example, the following R commands subset the data to the taxa of D_3__Aminicenantales and D_3__Blastocatellales with a vector specifying their names separated by a comma:

```
> ds_Aminicenantales_Blastocatellales <- amp_subset_taxa(ds,
tax_vector=c("D_3__Aminicenantales", " D_3__Blastocatellales"))
634 OTUs have been filtered
Before: 635 OTUs
After: 1 OTUs
```

9.3.2 Boxplot

In this subsection, we illustrate how to generate boxplot using ampvis2 and microbiome packages, respectively.

9.3.2.1   Generating Boxplot Using Ampvis2 Package

We use the **amp_boxplot()** function to generate boxplots for the four groups, showing top 5 genera and adding family information. The plot again is ordered by mean read abundance across all samples (Fig. 9.4):

```
> # Figure 9.4
> amp_boxplot(ds,
+          group_by = "Group4",
+          tax_show = 5,
+          tax_add = "Family")
```

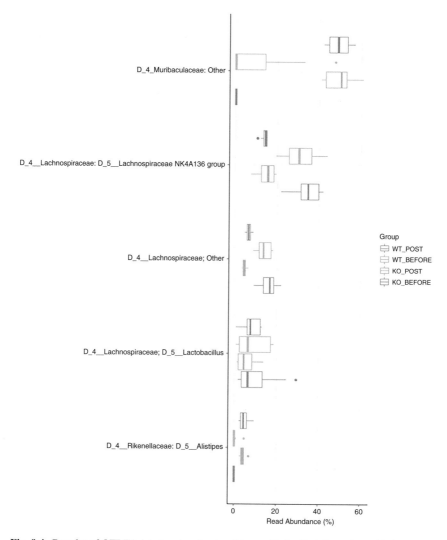

Fig. 9.4 Boxplot of QTRT1 data for showing top 5 taxa with family information added

9.3.2.2 Generating Boxplot Using Microbiome Package

Example 9.9: Mouse Gut Microbiome Data, Example 9.2 Cont.

Next, we show steps to visualize the differences and/or similarities between groups using the microbiome package. By doing the plots, we need the useful R package ggpubr (current version 0.4.0, March 2022). The following R commands call the microbiome, ggpubr, and other two R packages:

```
> library(microbiome)
> library(ggpubr)
> library(knitr)
> library(dplyr)
```

Let's first prune the taxa and then calculate the Shannon diversity:

```
> physeq1 <- prune_taxa(taxa_sums(physeq) > 0, physeq)
```

The Shannon diversity is re-calculated as below.

```
> shannon1 <- alpha(physeq1, index = "shannon")
> head(shannon1,3)
      diversity_shannon
F3D0          3.885
F3D1          3.992
F3D11         2.893
```

We pool out the metadata from the physeq1 object:

```
> physeq1_meta <- meta(physeq1)
> head(physeq1_meta)
```

And then add the diversity table to metadata:

```
> physeq1_meta$Shannon <- shannon1$diversity_shannon
> head(physeq1_meta,3)
      BarcodeSequence ForwardPrimerSequence ReversePrimerSequence
F3D0        <NA>            <NA>                 <NA>
F3D1        <NA>            <NA>                 <NA>
F3D11       <NA>            <NA>                 <NA>
                ForwardRead                    ReverseRead
F3D0   F3D0_S188_L001_R1_001.fastq.gz  F3D0_S188_L001_R2_001.fastq.gz
F3D1   F3D1_S189_L001_R1_001.fastq.gz  F3D1_S189_L001_R2_001.fastq.gz
F3D11  F3D11_S198_L001_R1_001.fastq.gz  F3D11_S198_L001_R2_001.
fastq.gz
         Group   Sex Time DayID DPW    Description Shannon
F3D0    F3D0  Female Early D000   0 QIIME2RAnalysisSet  3.885
F3D1    F3D1  Female Early D001   1 QIIME2RAnalysisSet  3.992
F3D11   F3D11 Female Early D011  11 QIIME2RAnalysisSet  2.893
```

Now, we can generate the boxplots using the **ggboxplot()** function from the **ggpubr** package. The following call function generates boxplots with jittered points. The outline colors are specified for changing by groups (in this case, Time). We customize color palette for two groups as blue and red and specify adding jitter points and changing the shape by groups (Fig. 9.5):

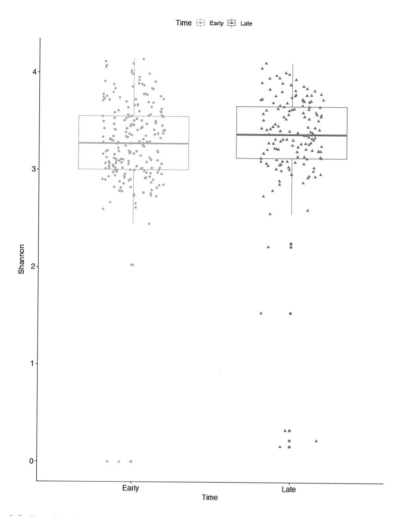

Fig. 9.5 Boxplot of gut microbiome data for Shannon diversity by early and late time points in mice

```
> # Figure 9.5
> p <- ggboxplot(physeq1_meta, x = "Time", y = "Shannon",
+          color = "Time", palette =c("#00AFBB", "#FC4E07"),
+          add = "jitter", shape = "Time")
> p
```

The following R commands specify the comparisons of groups and add *P*-values for comparing groups (Fig. 9.6). Please note that here comparisons of Shannon diversity that are conducted between early and late time points may not be appropriate and interest of the reader's study. We just use this dataset for illustrating the commands:

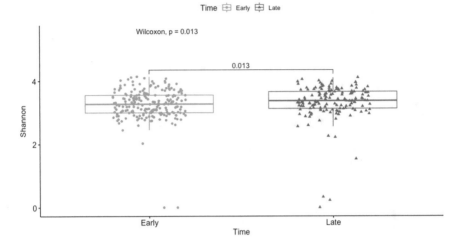

Fig. 9.6 Boxplot of mouse gut microbiome data for Shannon diversity by early and late time points with *p*-value generated by Wilcoxon rank sum test

```
> # Figure 9.6
> my_comparisons <- list(c("Early", "Late"))
> p + stat_compare_means(comparisons = my_comparisons) + # Add pairwise
comparisons p-value
+    stat_compare_means(label.y = 5.5)  # Add global p-value
```

Here the pairwise comparison *P*-value is the same as the global *P*-value because there are only two groups (early and late) in this comparison.

9.3.3   Violin Plot

Example 9.10: Mouse Gut Microbiome Data, Example 9.2 Cont.
For violin plot, we continue to use the same phyloseq object "physeq1" as in boxplot and compare differences in Shannon diversity between early and late time points.

Now we can use the **ggviolin()** function in the **ggpubr** package to create a violin plot (Fig. 9.7):

```
> # Figure 9.7
> p1 <- ggviolin(physeq1_meta, x = "Time", y = "Shannon",
+        add = "boxplot", fill = "Time", palette = c("#a6cee3",
"#b2df8a"))
> print(p1)
```

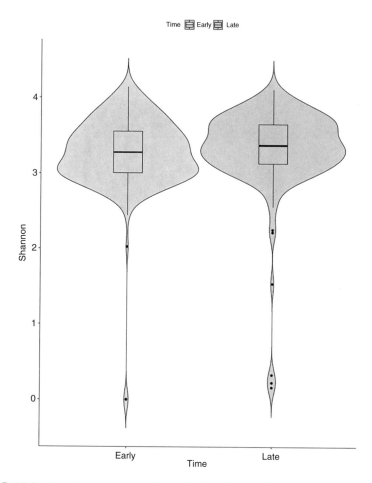

Fig. 9.7 Violin plot of mouse gut microbiome data for Shannon diversity by early and late time points

Here we only have two levels of groups; in the case of more than two groups, a pairwise comparison can be conducted. For the purpose of illustration, we show how to get the variable and create a list of pairwise comparisons:

```
> time <- levels(physeq1_meta$Time)
> comp_time<- combn(seq_along(time), 2, simplify = FALSE, FUN =
function(i)time[i])
> print(comp_time)
[[1]]
[1] "Early" "Late"
```

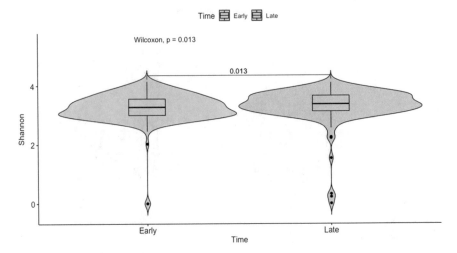

Fig. 9.8 Violin plot of mouse gut microbiome data for Shannon diversity by early and late time points with *p*-value generated by Wilcoxon rank sum test

The following R commands add the statistics (*P*-values) to the violin plot generated by pairwise comparisons using a nonparametric test (in this case, Wilcoxon rank sum test) (Fig. 9.8):

```
> # Figure 9.8
> p1 <- p1 + stat_compare_means(comparisons = comp_time) + # Add
pairwise comparisons p-value
+    stat_compare_means(label.y = 5.5)          # Add global p-value
> print(p1)
```

9.4   Statistical Hypothesis Testing of Alpha Diversity

Example 9.11: Breast Cancer QTRT1 Mouse Gut Microbiome, Example 9.1 Cont.
Depending on the number of groups and the distribution of the alpha diversity measures, we can conduct a statistical hypothesis testing of alpha diversity using a two-sample Welch's *t*-test, a Wilcoxon rank sum test, a one-way ANOVA, or a Kruskal-Wallis test (Xia et al. 2018). In above boxplots and violin plots, the *P*-values generated are from comparisons of Shannon diversity between early and late time points using Wilcoxon rank sum test. Here, we use the nonparametric Kruskal-Wallis test to compare groups (in the case of two groups, it equals to Wilcoxon rank sum test).

9.4.1   Summarize the Diversity Measures

The following commands calculate Shannon diversity using the **amp_alphadiv()** function in ampvis2 package based on the "ds" object created by this package previously:

```
> # Shannon index
> shannon_amp <- amp_alphadiv(ds,
+                  measure = "shannon",
+                  richness = FALSE,
+                  rarefy = NULL
+ )
```

We use the **Summarize()** in the **FSA** package (current version 0.9.3, March 2022) to summarize the Shannon diversity measures to obtain the basic statistics per group including the number of samples, mean, median, SD, minimum and maximum values, and Q1 and Q3:

```
> library(FSA)
> Summarize(Shannon ~ Group4, data = shannon_amp)
  Group4 n mean sd min Q1 median Q3 max
1 KO_BEFORE 10 2.480 0.0982 2.369 2.390 2.462 2.563 2.620
2 KO_POST 10 2.076 0.2680 1.570 1.995 2.090 2.155 2.462
3 WT_BEFORE 10 2.373 0.1747 1.917 2.401 2.412 2.446 2.561
4 WT_POST 10 2.080 0.1922 1.809 1.891 2.142 2.235 2.327
```

9.4.2   Plot Histogram of the Diversity Distributions

We use the **histogram()** in the **lattice** package (current version 0.20.45, March 2022) to obtain a distribution plot for each group. We specify the layout of individual plots in panel of two columns and two rows (Fig. 9.9):

```
> # Figure 9.9
> library(lattice)
> histogram(~ Shannon|Group4, data=shannon_amp, layout=c(2,2))
```

9.4.3   Kruskal-Wallis Test

We perform a Kruskal-Wallis test of Shannon diversity by calling the **kruskal.test()** function:

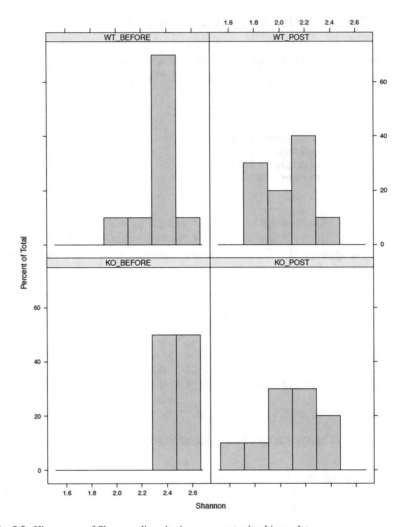

Fig. 9.9 Histograms of Shannon diversity in mouse gut microbiome data

```
> kruskal.test(Shannon ~ Group4, data = shannon_amp)
  Kruskal-Wallis rank sum test

data: Shannon by Group4
Kruskal-Wallis chi-squared = 20, df = 3, p-value = 0.0002
```

As the *P*-value is less than the significance level 0.05, we can conclude that there are significant differences between the groups. The Kruskal-Wallis chi-squared value of 20 is greater than the chi-square statistic of 7.815 that is generated by the following **qchisq()** function with 3 degrees of freedom. So, the significance between groups were further confirmed:

```
> qchisq(0.950, 3)
[1] 7.815
```

9.4.4   Perform Multiple Comparisons

From the output of the Kruskal-Wallis test, we know that there is a significant difference between groups. To find out which pairs of groups are different, we use the following R commands to conduct multiple comparisons using the Tukey method for adjusting *P*-values:

```
> shannon_amp$Group4 <- as.factor(shannon_amp$Group4)
> library(DescTools)
> adj_tukey = NemenyiTest(x = shannon_amp$Shannon,
+               g = shannon_amp$Group4,
+               dist="tukey")
> adj_tukey
```

Nemenyi's test of multiple comparisons for independent samples (tukey).

```
                      mean.rank.diff  pval
KO_POST-KO_BEFORE           -17.4    0.0048 **
WT_BEFORE-KO_BEFORE          -4.6    0.8153
WT_POST-KO_BEFORE           -19.2    0.0014 **
WT_BEFORE-KO_POST            12.8    0.0683 .
WT_POST-KO_POST              -1.8    0.9860
WT_POST-WT_BEFORE           -14.6    0.0269 *
---
Signif. codes:  0 '***' 0.001 '**' 0.01 '*' 0.05 '.' 0.1 ' ' 1
```

The following R commands conduct multiple comparisons using the Benjamini and Hochberg method for adjusting *P*-values:

```
> library(FSA)
> adj_bh = dunnTest(shannon_amp$Shannon ~ shannon_amp$Group4,
data=shannon_amp, method="bh")
> adj_bh
Dunn (1964) Kruskal-Wallis multiple comparison
 P-values adjusted with the Benjamini-Hochberg method.
```

```
                  Comparison      Z P.unadj   P.adj
1 KO_BEFORE - KO_POST     3.3282 0.0008742 0.002623
2 KO_BEFORE - WT_BEFORE 0.8799 0.3789374 0.454725
3 KO_POST - WT_BEFORE    -2.4483 0.0143534 0.021530
4 KO_BEFORE - WT_POST    3.6724 0.0002402 0.001441
5 KO_POST - WT_POST      0.3443 0.7306271 0.730627
6 WT_BEFORE - WT_POST    2.7926 0.0052289 0.010458
```

It's also possible to use the function **pairwise.wilcox.test**() to calculate pairwise comparisons between group levels and correct for multiple testing as below:

```
> pairwise.wilcox.test(shannon_amp$Shannon, shannon_amp$Group4,
+            p.adjust.method = "BH")
  Pairwise comparisons using Wilcoxon rank sum exact test
data: shannon_amp$Shannon and shannon_amp$Group4
      KO_BEFORE KO_POST WT_BEFORE
KO_POST  0.003    -     -
WT_BEFORE 0.423    0.053  -
WT_POST  0.00006  0.853  0.002
P value adjustment method: BH
```

9.5   Alpha Diversity Analysis in QIIME 2

QIIME 2's diversity analyses are available through the q2-diversity plugin, which supports calculating alpha and beta diversity metrics, performing related statistical tests, and generating interactive visualizations. Both alpha and beta diversity measures, as well as phylogenetic and non-phylogenetic diversity measures, can be generated with a single command qiime diversity **core-metrics-phylogenetic** command. However, to show QIIME 2's capabilities, in this section we will first use the **core-metrics-phylogenetic** command to calculate alpha and beta diversity metrics together, and then we will illustrate how to calculate each alpha diversity using individual alpha command. In Chap. 10, we will illustrate how to calculate each beta diversity using beta command. For diversity analysis, a sequence/feature/OTU table and a phylogenetic tree are required. And a sampling depth for random subsampling needs to be specified.

9.5.1   Calculate Alpha Diversity Using
Core-Metrics-Phylogenetic Method

Here, we use a single command to generate all the phylogenetic and non-phylogenetic diversity measures available in QIIME 2. The alpha diversity measures calculated by the qiime diversity **core-metrics-phylogenetic** method include Shannon's diversity, Pielou's evenness, Faith's phylogenetic diversity, and observed OTUs. The beta diversity measures calculated by this command include weighted/unweighted UniFrac (Lozupone and Knight 2005; Lozupone et al. 2007), Bray-Curtis, and Jaccard indices.

Because most diversity metrics are sensitive to different sampling depths across different samples, in order to compute samples with uneven sequencing depth, the core-metrics-phylogenetic method will randomly subsample or rarefy the counts from each sample to a user-specified value (i.e., sampling depth) when QIIME 2 computes these alpha and beta diversity metrics. Thus, it is crucial to provide the even sampling (i.e., rarefaction) depth to the parameter --p-sampling-depth. The algorithm works this way: The value provided to --p-sampling-depth will set up a total count for each sample in the resulting table when subsampling the counts in each sample without replacement. For example, suppose we specify 1000 for this parameter; then if the total count for any sample(s) are smaller than 1000, those samples will be dropped from the diversity analysis. The recommendation of choosing this value is to review the information presented in the Feature table.qzv QIIME 2 artifact and in particular the *Interactive Sample Detail* tab in that visualization. Typically, choose a value that is as high as possible (so that we can retain more sequences per sample) while excluding as few samples as possible.

Example 9.12: Mouse Gut Microbiome Data, Example 9.2 Cont.
Qiime tools view FeatureTableMiSeq_SOP.qzv.

Above command activates qiime 2 view. In the *Interactive Sample Detail* tab, we can see that sample M1D141 has 1153 sequences; then there are 10 samples from M3D8 to M3D149; the sequences decrease from 815, 694, . . . to the fewest sequence 0. Here we will choose 1153 based on the number of sequences in the sample M1D141. The value of 1153 is considerably or relatively higher than the number of sequences in the samples that have fewer sequences (from 694 to 0). Thus, choosing this value we can retain more sequences per sample while excluding few samples:

```
source activate qiime2-2022.2
cd QIIME2R-Bioinformatics
qiime diversity core-metrics-phylogenetic \
  --i-phylogeny RootedTreeMiSeq_SOP.qza \
  --i-table FeatureTableMiSeq_SOP.qza \
  --p-sampling-depth 1153 \
  --m-metadata-file SampleMetadataMiSeq_SOP.tsv \
  --output-dir CoreMetricsResults
```

```
Saved FeatureTable[Frequency] to: CoreMetricsResults/rarefied_table.qza
Saved SampleData[AlphaDiversity] to: CoreMetricsResults/faith_pd_vector.qza
Saved SampleData[AlphaDiversity] to: CoreMetricsResults/observed_features_vector.qza
Saved SampleData[AlphaDiversity] to: CoreMetricsResults/shannon_vector.qza
Saved SampleData[AlphaDiversity] to: CoreMetricsResults/evenness_vector.qza
Saved DistanceMatrix to: CoreMetricsResults/unweighted_unifrac_distance_matrix.qza
Saved DistanceMatrix to: CoreMetricsResults/weighted_unifrac_distance_matrix.qza
Saved DistanceMatrix to: CoreMetricsResults/jaccard_distance_matrix.qza
Saved DistanceMatrix to: CoreMetricsResults/bray_curtis_distance_matrix.qza
Saved PCoAResults to: CoreMetricsResults/unweighted_unifrac_pcoa_results.qza
Saved PCoAResults to: CoreMetricsResults/weighted_unifrac_pcoa_results.qza
Saved PCoAResults to: CoreMetricsResults/jaccard_pcoa_results.qza
Saved PCoAResults to: CoreMetricsResults/bray_curtis_pcoa_results.qza
Saved Visualization to: CoreMetricsResults/unweighted_unifrac_emperor.qzv
Saved Visualization to: CoreMetricsResults/weighted_unifrac_emperor.qzv
Saved Visualization to: CoreMetricsResults/jaccard_emperor.qzv
Saved Visualization to: CoreMetricsResults/bray_curtis_emperor.qzv
```

Above command generates alpha and beta diversity measures at 1153 sequences per sample. It also generates PCoA plots automatically. QIIME artifact files generated by this command can be opened by changing .qza to .gz and then double clicking to unzip. The measures are .tsv files, for example, the Shannon diversity for the first few samples like below:

```
    shannon
F3D0    5.597038025432271
F3D1    5.7812812011984835
F3D11   4.175566318851661
```

9.5.2   Calculate Alpha Diversity Using Alpha Method

In QIIME 2, a user-specified alpha diversity metric for all samples in a feature table can be calculated using the alpha and alpha-phylogenetic methods for abundance-based (i.e., non-phylogenetic) alpha diversity metrics and phylogenetic alpha diversity metrics, respectively. Here, we continue to illustrate most frequently used abundance-based alpha diversity metrics using the mouse gut microbiome data in Example 9.12.

9.5.2.1   Shannon Index

The following commands calculate Shannon's index which calculates richness and diversity using a natural logarithm and accounts for both abundance and evenness of the taxa present in samples:

```
source activate qiime2-2022.2
QIIME2R-Bioinformatics
qiime diversity alpha \
  --i-table FeatureTableMiSeq_SOP.qza \
  --p-metric shannon\
  --o-alpha-diversity ShannonVector.qza
```

```
Saved SampleData[AlphaDiversity] to: ShannonVector.qza
```

The input parameter **--i-table** is used to specify the feature table containing the samples for which the alpha diversity metric will be calculated. The parameter **--p-metric** is used to specify the alpha diversity metric to be calculated. The output parameter **--o-alpha-diversity** is used to specify the output file, which you can give a meaningful name.

9.5.2.2   Chao1 Index and Chao1 Confidence Interval

The following commands calculate Chao 1 index which estimates diversity from abundant data and estimates number of rare taxa missed from under-sampling:

```
qiime diversity alpha \
  --i-table FeatureTableMiSeq_SOP.qza \
  --p-metric chao1\
  --o-alpha-diversity Chao1Vector.qza
```

```
Saved SampleData[AlphaDiversity] to: Chao1Vector.qza
```

The following commands calculate chao1 confidence interval for richness estimator:

```
qiime diversity alpha \
  --i-table FeatureTableMiSeq_SOP.qza \
  --p-metric chao1_ci \
  --o-alpha-diversity Chao1CIVector.qza
```

```
Saved SampleData[AlphaDiversity] to: Chao1CIVector.qza
```

9.5.2.3   Observed Features

The following commands calculate observed features (here, OTUs):

```
qiime diversity alpha \
  --i-table FeatureTableMiSeq_SOP.qza \
  --p-metric observed_features\
  --o-alpha-diversity ObservedFeaturesVector.qza
```

```
Saved SampleData[AlphaDiversity] to: ObservedFeaturesVector.qza
```

9.5.2.4   Simpson Index and Simpson's Evenness

The following commands calculate Simpson's index to measure the relative abundance of the different species making up the sample richness:

```
qiime diversity alpha \
  --i-table FeatureTableMiSeq_SOP.qza \
  --p-metric simpson \
  --o-alpha-diversity SimpsonVector.qza
```

Saved SampleData[AlphaDiversity] to: SimpsonVector.qza

The following commands calculate Simpson's evenness, the diversity that account for the number of organisms and number of species:

```
qiime diversity alpha \.
  --i-table FeatureTableMiSeq_SOP.qza \
  --p-metric simpson_e \
  --o-alpha-diversity SimpsonEVector.qza
```

Saved SampleData[AlphaDiversity] to: SimpsonEVector.qza

9.5.2.5   Pielou's Evenness

The following commands calculate **Pielou's** evenness to measure the relative evenness of species richness:

```
qiime diversity alpha \
  --i-table FeatureTableMiSeq_SOP.qza \
  --p-metric pielou_e \
  --o-alpha-diversity PielouEVector.qza
```

Saved SampleData[AlphaDiversity] to: PielouEVector.qza

9.5.3   Calculate Alpha Diversity Using Alpha-Phylogenetic Method

QIIME 2 implements the phylogenetic alpha diversity metrics using the **alpha-phylogenetic** method.

The following commands calculate Faith's phylogenetic diversity:

```
qiime diversity alpha-phylogenetic \
  --i-table FeatureTableMiSeq_SOP.qza \
  --i-phylogeny RootedTreeMiSeq_SOP.qza \
  --p-metric faith_pd \
  --o-alpha-diversity FaithPDVector.qza
```

Saved SampleData[AlphaDiversity] to: FaithPDVector.qza

The input parameter **--i-phylogeny** is used to specify the phylogenetic tree containing the tip identifiers that correspond to the feature identifiers in the table and is only used for the alpha-phylogenetic command for calculating phylogenetic diversity metrics.

9.5.4   Test for Differences of Alpha Diversity Between Groups

The following commands conduct significant testing of Shannon's diversity for all categorical variables in sample metadata:

```
qiime diversity alpha-group-significance \
  --i-alpha-diversity CoreMetricsResults/shannon_vector.qza \
  --m-metadata-file SampleMetadataMiSeq_SOP.tsv \
  --o-visualization ShannonGroupSignificance.qzv
```

Saved Visualization to: ShannonGroupSignificance.qzv

The **alpha-group-significance** command generates boxplots of the alpha diversity values and runs all-group and pairwise Kruskal-Wallis tests, a nonparametric ANOVA to compare significant differences among all groups. We can activate qiime2.view to check the significant testing results. Three things are automatically done for each categorical data in metadata columns including alpha diversity boxplots, Kruskal-Wallis (all groups), and Kruskal-Wallis (pairwise). For example, for time variable, the results show that there are total 349 samples (208 for early and 141 for late times). Both all-group and pairwise Kruskal-Wallis show that later time has more Shannon diversity than early time with P-value of 0.003227 and Q-value of 0.003227. However, there is no difference of Shannon diversity between male and female mice with P-value of 0.826126 and Q-value of 0.826126. In this case, the results from all-group and pairwise Kruskal-Wallis are the same, and P-value is also equal to Q-value because the time variable only has two points. We can download the boxplots as SVG format or download raw data as TSV format. The Kruskal-Wallis testing results can be downloaded as CSV files.

The following commands conduct significant testing of Pielou's evenness for all categorical variables in sample metadata:

```
qiime diversity alpha-group-significance \
  --i-alpha-diversity CoreMetricsResults/evenness_vector.qza \
  --m-metadata-file SampleMetadataMiSeq_SOP.tsv \
  --o-visualization EvennessGroupSignificance.qzv
```

Saved Visualization to: EvennessGroupSignificance.qzv

The following commands conduct significant testing of Faith's phylogenetic diversity for all categorical variables in sample metadata:

```
qiime diversity alpha-group-significance \
  --i-alpha-diversity CoreMetricsResults/faith_pd_vector.qza \
  --m-metadata-file SampleMetadataMiSeq_SOP.tsv \
  --o-visualization FaithPDGroupSignificance.qzv
```

Saved Visualization to: FaithPDGroupSignificance.qzv

The following commands conduct significant testing of observed OTUs for all categorical variables in sample metadata:

```
qiime diversity alpha-group-significance \
  --i-alpha-diversity
CoreMetricsResults/observed_features_vector.qza \
  --m-metadata-file SampleMetadataMiSeq_SOP.tsv \
  --o-visualization ObservedFeaturesGroupSignificance.qzv
```

Saved Visualization to: ObservedFeaturesGroupSignificance.qzv

9.5.5   Alpha Rarefaction in QIIME 2

An alpha rarefaction analysis explores alpha diversity as a function of sampling depth, which is used to determine if an environment has been sequenced to a sufficient depth.

Alpha rarefaction is conducted by randomly subsampling the data at a series of sequence depths and plotting the alpha diversity metrics calculated from the random subsamples as a function of the sequencing depth. If a given sample has a plateau on the rarefaction curve, then it provides evidence that the sample has been sequenced to a sufficient depth to capture the majority of taxa.

In QIIME 2, we can use the qiime diversity alpha-rarefaction visualizer to conduct alpha rarefaction analysis. This visualizer will generate rarefaction curves using one or more alpha diversity metrics (by default based on Shannon diversity and observed features, i.e., OTUs measures) at multiple sampling depths, in steps between 1 (optionally controlled with --p-min-depth) and the value provided as --p-max-depth. In the case, if the phylogenetic tree is provided using the **--i-phylogeny** parameter, then this visualizer will also generate phylogenetic diversity-based curves.

Below we run the qiime diversity alpha-rarefaction visualizer to generate alpha rarefaction curves showing taxon accumulation as a function of sequence depth. Figure 9.10 can be reproduced using the following QIIME 2 commands:

```
# Figure 9.10
qiime diversity alpha-rarefaction \
  --i-table FeatureTableMiSeq_SOP.qza \
```

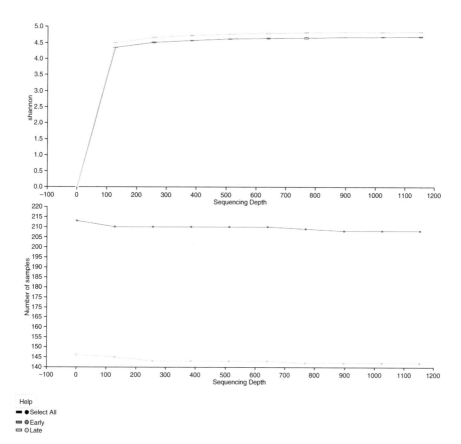

Fig. 9.10 The alpha rarefaction curves based on Shannon diversity with the results averaged by times

```
--i-phylogeny RootedTreeMiSeq_SOP.qza \
--p-max-depth 1153 \
--m-metadata-file SampleMetadataMiSeq_SOP.tsv \
--o-visualization AlphaRarefaction.qzv
```

Saved Visualization to: AlphaRarefaction.qzv

The above command generates two plots. The alpha rarefaction plot at top is primarily used to determine if the richness of the samples has been fully observed or sequenced. If the slope of lines in the plot approaches zero or "level out" at some sampling depth along the x-axis, then it suggests collecting additional sequences (taxa) beyond that sampling depth likely to be either rare microorganism or the result of the sequencing error. In other words, the slope of lines does not approach zero; then it indicates either too few sequences were collected and thus the richness of the samples hasn't been fully observed yet or the data remains having a lot of sequencing error; thus it is mistaken to use this data for diversity analysis.

The bottom plot is generated when sample metadata has grouping groups. It illustrates the number of samples that remain in each group when the feature table is rarefied to each sampling depth. The lines show the number of samples remains in each group when the feature table is rarefied to each sampling depth.

The essential function of the bottom plot is used to check whether the data presented in the top plot is reliable. Please note that we cannot specify a sampling depth that is larger than the total frequency that was obtained for a sample because it is impossible to calculate the diversity measure for this sample at the sampling depth.

Two patterns were noticed from this alpha rarefaction curve: One is that both early and late time points appear to plateau. The Shannon diversity measure does generally continue to increase as a function of the sequencing depth; however, the accumulation increases slowly, suggesting that we have sufficient sample sequence depth to have captured the majority of taxa present in the sample. Another pattern we can see from this plotting is that the late samples have a significantly higher Shannon diversity than the early time. This result is consistent with the Kruskal-Wallis test in Sect. 9.5.4.

9.6   Summary

In this chapter, we first introduced some most common used abundance-based alpha diversity measures and their calculations, including Chao 1 richness and abundance-based coverage estimator (ACE), Shannon diversity, Simpson diversity indices, Pielou's evenness, and also three phylogenetic-based diversity measures: phylogenetic diversity (PD), phylogenetic entropy (PE), and phylogenetic quadratic entropy (PQE). We then illustrated three exploration tools for alpha diversity analysis: heatmap, boxplot, and violin plot. Next, we illustrated how to summarize and visualize alpha diversity measures and conduct statistical hypothesis testing of alpha diversity using Kruskal-Wallis test as well as perform multiple comparisons to adjust for P-values. Finally, we focused on alpha diversity analysis in QIIME 2 including how to (1) calculate alpha diversity using core-metrics-phylogenetic method, (2) calculate abundance-based alpha diversity using alpha method, (3) calculate Faith's PD using alpha-phylogenetic method, (4) conduct statistical hypothesis testing of alpha diversity using Kruskal-Wallis test, and (5) perform multiple comparisons to adjust for P-values. We also illustrate how to generate alpha rarefaction curve in QIIME 2. In Chaps. 10 and 11, we will focus on investigating beta diversity analysis. In Chap. 10, we will introduce beta diversity metrics and their calculation using R and QIIME 2 as well as explore beta diversity using ordination methods. In Chap. 11, we will introduce statistical hypothesis testing of beta diversity.

References

Allen, B., M. Kon, and Y. Bar-Yam. 2009. A new phylogenetic diversity measure generalizing the shannon index and its application to phyllostomid bats. *The American Naturalist* 174 (2): 236–243. https://doi.org/10.1086/600101.

Caporaso, J. Gregory, Justin Kuczynski, Jesse Stombaugh, Kyle Bittinger, Frederic D. Bushman, Elizabeth K. Costello, Noah Fierer, Antonio Gonzalez Pena, Julia K. Goodrich, and Jeffrey I. Gordon. 2010. QIIME allows analysis of high-throughput community sequencing data. *Nature Methods* 7 (5): 335–336.

Chao, Anne. 1984. Nonparametric estimation of the number of classes in a population. *Scandinavian Journal of Statistics* 11 (4): 265–270. www.jstor.org/stable/4615964.

Chao, A. 1987. Estimating the population size for capture-recapture data with unequal catchability. *Biometrics* 43: 783–791.

Chao, Anne, and Shen-Ming Lee. 1992. Estimating the number of classes via sample coverage. *Journal of the American Statistical Association* 87 (417): 210–217. https://doi.org/10.2307/2290471. www.jstor.org/stable/2290471.

Chao, Anne, and C.K. Mark Yang. 1993. Stopping rules and estimation for recapture debugging with unequal failure rates. *Biometrika* 80 (1): 193–201. https://doi.org/10.2307/2336768. http://www.jstor.org.proxy.cc.uic.edu/stable/2336768.

Chao, Anne, Chun-Huo Chiu, and Lou Jost. 2010. Phylogenetic diversity measures based on hill numbers. *Philosophical Transactions of the Royal Society B: Biological Sciences* 365 (1558): 3599–3609.

———. 2016. Phylogenetic diversity measures and their decomposition: A framework based on hill numbers. In *Biodiversity conservation and phylogenetic systematics: Preserving our evolutionary heritage in an extinction crisis*, ed. Roseli Pellens and Philippe Grandcolas, 141–172. Cham: Springer International Publishing.

Colwell, R.K., and J.A. Coddington. 1994. Estimating terrestrial biodiversity through extrapolation. *Philosophical Transactions of the Royal Society of London. Series B, Biological Sciences* 345 (1311): 101–118. https://doi.org/10.1098/rstb.1994.0091. http://www.ncbi.nlm.nih.gov/pubmed/7972351.

Colwell, Robert K., Chang Xuan Mao, and Jing Chang. 2004. Interpolating, extrapolating, and comparing incidence-based species accumulation curves. *Ecology* 85 (10): 2717–2727.

Edgar, Robert C. 2010. Search and clustering orders of magnitude faster than BLAST. *Bioinformatics* 26 (19): 2460–2461.

Efron, Bradley. 1981. Nonparametric standard errors and confidence intervals. *The Canadian Journal of Statistics / La Revue Canadienne de Statistique* 9 (2): 139–158. https://doi.org/10.2307/3314608. http://www.jstor.org.proxy.cc.uic.edu/stable/3314608.

———. 1982. *The jackknife, the bootstrap and other resampling plans*. SIAM.

Faith, Daniel P. 1992. Conservation evaluation and phylogenetic diversity. *Biological Conservation* 61 (1): 1–10. https://doi.org/10.1016/0006-3207(92)91201-3. http://www.sciencedirect.com/science/article/pii/0006320792912013.

Glöckner, Frank Oliver, Pelin Yilmaz, Christian Quast, Jan Gerken, Alan Beccati, Andreea Ciuprina, Gerrit Bruns, Pablo Yarza, Jörg Peplies, and Ralf Westram. 2017. 25 years of serving the community with ribosomal RNA gene reference databases and tools. *Journal of Biotechnology* 261: 169–176.

Good, I.J. 1953. The population frequencies of species and the estimation of population parameters. *Biometrika* 40: 237–264.

Hughes, J.B., J.J. Hellmann, T.H. Ricketts, and B.J. Bohannan. 2001. Counting the uncountable: Statistical approaches to estimating microbial diversity. *Applied and Environmental Microbiology* 67 (10): 4399–4406. https://doi.org/10.1128/aem.67.10.4399-4406.2001.

Hurlbert, S.H. 1971. The non-concept of species diversity: A critique and alternative parameters. *Ecology* 52: 577–589.

Kim, B.R., J. Shin, R. Guevarra, J.H. Lee, D.W. Kim, K.H. Seol, J.H. Lee, H.B. Kim, and R. Isaacson. 2017. Deciphering diversity indices for a better understanding of microbial communities. *Journal of Microbiology and Biotechnology* 27 (12): 2089–2093. https://doi.org/10.4014/jmb.1709.09027.

Krebs, C.J. 1999. *Ecological methodology*. 2nd ed. Menlo Park: Addison-Wesley Longman.

Lemos, Leandro N., Roberta R. Fulthorpe, Eric W. Triplett, and Luiz F.W. Roesch. 2011. Rethinking microbial diversity analysis in the high throughput sequencing era. *Journal of Microbiological Methods* 86 (1): 42–51. https://doi.org/10.1016/j.mimet.2011.03.014. https://www.sciencedirect.com/science/article/pii/S0167701211001138.

Lozupone, Catherine, and Rob Knight. 2005. UniFrac: A new phylogenetic method for comparing microbial communities. *Applied and Environmental Microbiology* 71 (12): 8228–8235. https://doi.org/10.1128/aem.71.12.8228-8235.2005. https://aem.asm.org/content/aem/71/12/8228.full.pdf.

Lozupone, Catherine A., Micah Hamady, Scott T. Kelley, and Rob Knight. 2007. Quantitative and qualitative β diversity measures lead to different insights into factors that structure microbial communities. *Applied and Environmental Microbiology* 73 (5): 1576–1585.

Magurran, Anne E. 2013. *Measuring biological diversity*. Wiley.

McCune, B., and J.B. Grace. 2002. *Analysis of ecological communities*. MJM Press.

McMurdie, Paul J., and Susan Holmes. 2014. Waste not, want not: Why rarefying microbiome data is inadmissible. *PLoS Computational Biology* 10 (4): e1003531. https://doi.org/10.1371/journal.pcbi.1003531.

Pielou, E.C. 1966. The measurement of diversity in different types of biological collections. *Journal of Theoretical Biology* 13: 131–144.

Rao, C. Radhakrishna. 1982. Diversity and dissimilarity coefficients: a unified approach. *Theoretical Population Biology* 21 (1): 24–43. https://doi.org/10.1016/0040-5809(82)90004-1. http://www.sciencedirect.com/science/article/pii/0040580982900041.

Schloss, Patrick D., and Jo Handelsman. 2006. Introducing SONS, a tool for operational taxonomic unit-based comparisons of microbial community memberships and structures. *Applied and Environmental Microbiology* 72 (10): 6773–6779. https://doi.org/10.1128/AEM.00474-06. https://pubmed.ncbi.nlm.nih.gov/17021230; https://www.ncbi.nlm.nih.gov/pmc/articles/PMC1610290/.

Schloss, P.D., S.L. Westcott, T. Ryabin, J.R. Hall, M. Hartmann, E.B. Hollister, R.A. Lesniewski, B.B. Oakley, D.H. Parks, C.J. Robinson, J.W. Sahl, B. Stres, G.G. Thallinger, D.J. Van Horn, and C.F. Weber. 2009. Introducing mothur: open-source, platform-independent, community-supported software for describing and comparing microbial communities. *Applied and Environmental Microbiology* 75 (23): 7537–7541. https://doi.org/10.1128/aem.01541-09.

Shannon, Claude E. 1948. A mathematical theory of communication. *Bell System Technical Journal* 27 (3): 379–423.

Shannon, Claude Elwood, and Warren Weaver. 1949. *The mathematical theory of communication, by CE Shannon* (and recent contributions to the mathematical theory of communication), W. Weaver. Urbana, USA: University of illinois Press.

Simpson, E.H. 1949. Measurement of diversity. *Nature* 163 (4148): 688–688. https://doi.org/10.1038/163688a0.

Sornplang, Pairat, and Sudthidol Piyadeatsoontorn. 2016. Probiotic isolates from unconventional sources: a review. *Journal of Animal Science and Technology* 58 (1): 26. https://doi.org/10.1186/s40781-016-0108-2.

Spellerberg, Ian F., and Peter J. Fedor. 2003. A tribute to Claude Shannon (1916–2001) and a plea for more rigorous use of species richness, species diversity and the 'Shannon-Wiener' Index. *Global Ecology and Biogeography* 12 (3): 177–179. https://doi.org/10.1046/j.1466-822x.2003.00015.x.

Warwick, R.M., and K.R. Clarke. 1995. New 'biodiversity' measures reveal a decrease in taxonomic distinctness with increasing stress. *Marine Ecology Progress Series* 129 (1/3): 301–305. http://www.jstor.org.proxy.cc.uic.edu/stable/24855596.

Washington, H.G. 1984. Diversity, biotic and similarity indices. A review with special relevance to aquatic ecosystems. *Water Research* 18: 653–694.

Whittaker, Robert H. 1972. Evolution and measurement of species diversity. *Taxon* 21 (2–3): 213–251.

Xia, Yinglin, Jun Sun, and Ding-Geng Chen. 2018. Univariate community analysis. In *Statistical analysis of microbiome data with R*, ed. Yinglin Xia, Jun Sun, and Ding-Geng Chen, 251–283. Singapore: Springer Singapore.

Zhang, Jiajie, Kassian Kobert, Tomáš Flouri, and Alexandros Stamatakis. 2014. PEAR: A fast and accurate Illumina Paired-End reAd mergeR. *Bioinformatics* 30 (5): 614–620.

Zhang, Jilei, Rong Lu, Yongguo Zhang, Żaneta Matuszek, Wen Zhang, Yinglin Xia, Tao Pan, and Jun Sun. 2020. tRNA queuosine modification enzyme modulates the growth and microbiome recruitment to breast tumors. *Cancers* 12 (3): 628.

Chapter 10
Beta Diversity Metrics and Ordination

Abstract Chapter 9 investigated within-sample (alpha-) diversity. This chapter focuses on between-sample (beta-) diversity metrics and their ordination. Beta diversity measures the difference between two samples or communities. Beta diversity analysis requires a distance or dissimilarity measure matrix as input. This chapter first introduces abundance-based and phylogenetic beta diversity metrics, respectively; then introduces ordination methods and ordination plots. Next, it illustrates the beta diversity metrics and ordination in QIIME 2. Finally, it conducts some general remarks on ordination and clustering.

Keywords Beta diversity · Bray-Curtis index · Jaccard index · Sørensen index · vegan package · Unweighted UniFrac · Weighted UniFrac · Generalized UniFrac · pldist · Rarefaction · Ordination · Principal component analysis (PCA) · Principal coordinate analysis (PCoA) · Non-metric multidimensional scaling (NMDS) · Correspondence analysis (CA) · Detrended correspondence analysis (DCA) · Redundancy analysis (RDA) · Canonical correspondence analysis (CCA) · Clustering · QIIME 2 · Emperor plots

In Chap. 9, we focused on within-sample (alpha) diversity. In this chapter, we focus between-sample (beta) diversity metrics and their ordination. Beta diversity measures the difference between two samples or communities. When we conduct beta diversity analysis, a distance or dissimilarity measure matrix is required. This chapter is organized this way. Section 10.1 introduces abundance-based beta diversity metrics. Section 10.2 introduces phylogenetic beta diversity metrics. Section 10.3 introduces ordination methods and ordination plots. Section 10.4 introduces beta diversity metrics and ordination in QIIME 2. In Sect. 10.5, we conduct some general remarks on ordination and clustering. Finally we briefly summarize this chapter in Sect. 10.6.

Supplementary Information The online version contains supplementary material available at https://doi.org/10.1007/978-3-031-21391-5_10.

10.1   Abundance-Based Beta Diversity Metrics

Beta diversity measures the difference between two communities or samples. Like alpha diversity, we can group beta diversity metrics into two categories: abundance-based beta diversity dissimilarity and phylogenetic beta diversity dissimilarity. So far there are more than two dozens of beta diversity measures available in the literature of ecology (Koleff et al. 2003; Jari Oksanen and Tonteri 1995), of which the abundance-based dissimilarities Bray-Curtis index (Bray and Curtis 1957), Jaccard index, and Sørensen index are most commonly used and adopted for microbiome studies. Several phylogenetic beta diversity metrics have been specifically proposed for microbiome data including unweighted/weighted UniFrac distances (Lozupone and Knight 2005; Lozupone et al. 2007) and generalized UniFrac distances (Chen et al. 2012).

We can also categorize the beta diversity indices into two broad classes of similarity measures: binary similarity coefficients and quantitative similarity coefficients. Binary similarity coefficients only measure the presence/absence data that are available for the species in a community, whereas quantitative similarity coefficients measure the relative abundance that are available for each species.

The methods for estimating alpha diversity are fairly straightforward. In contrast, measurement of beta diversity is controversial (Ellison 2010), because some beta diversity measures are designed solely to determine whether communities are significantly different and others are to measure the distance between pairs of communities that satisfy the requirements of a distance metric. For example, Jaccard and Bray-Curtis coefficients measure the distance between communities based on the species that they contain (Lozupone and Knight 2005). The key point to selecting a proper measure of beta diversity is based on microbiome hypothesis testing and the methods that must be tailored to the hypothesis, rather than vice versa.

The measures of beta diversity typically are not reported alone. In contrast, the matrices are used as inputs in the functions of ordination plots and hypothesis testing. But they can be calculated independently in some software. For example, Koleff et al. (2003) reviewed 24 indices of beta diversity including Bray-Curtis index, Jaccard index, and Sørensen index (Koleff et al. 2003). All commonly used indices can be found using **betadiver()** function in the **BiodiversityR** package. The function **betadiver()** for indices of beta diversity in the community ecology package **vegan** (vegan function **vegdist()**) can directly calculate any of the 24 indices of beta diversity. Xia et al. illustrated the calculations of Bray-Curtis index, Jaccard index, and Sørensen index via the vegan package in their book (Xia et al. 2018a). QIIME 2 can calculate unweighted UniFrac and weighted UniFrac distances. GUniFrac can calculate generalized UniFrac distances as well as unweighted UniFrac and weighted UniFrac distances. While we reported each measure of alpha diversity in Chap. 9, for beta diversity in this chapter, we focus on introductions of the most commonly used beta diversity measures and their calculation as inputs of matrices for the functions of ordination. These beta diversity measures can also be used for statistical hypothesis testing in Chap. 11.

10.1.1   Bray-Curtis Dissimilarity

Bray-Curtis index of dissimilarity (Bray and Curtis 1957), a distance measure of matrix, was developed and named after J. Roger Bray and John T. Curtis. It is the most widely used beta diversity in ecology and microbiome research fields. The Bray-Curtis coefficient has been evaluated to be one of the most reliable performers in terms of robustness in a simulation study (Faith et al. 1987).

10.1.1.1   The Measures of Bray-Curtis Index

Bray-Curtis dissimilarity was developed based on counts at each sample to quantify the compositional dissimilarity between two different samples; that is to measure the relative abundances of species. For microbiome abundance data, the measures of distance coefficients are not really distances. They actually measure "dissimilarity." So we call Bray-Curtis index as distance (dissimilarity) coefficients.

For the simplest case, there are two species in two community samples. The *smaller* the distance, the *more similar* the two communities are. When a distance coefficient is zero, the two communities are identical. The intuitively appealing feature of distance coefficients (although they are not really distances in microbiome case) for the microbiome researchers is that they can be visualized. Euclidian, Manhattan, and Bray-Curtis coefficients all measure the distance (dissimilarity). The formula of Euclidian distance is given as below:

$$d_{jk} = \sqrt{\sum_{i=1}^{n} \left(X_{ij} - X_{ik}\right)^2} \tag{10.1}$$

The formula of Manhattan distance is given as below:

$$d_M(j,k) = \sum_{i=1}^{n} \left|X_{ij} - X_{ik}\right| \tag{10.2}$$

where d_{jk} and $d_M(j, k)$ are Euclidean distance between samples j and k and Manhattan distance between samples j and k, respectively. X_{ij} and X_{ik} are the number of individuals of species i in sample j and sample k, respectively. n is the total number of species in samples.

Bray-Curtis dissimilarity was defined based on Euclidean distance. The formula of Bray and Curtis' dissimilarity index is given as follows:

$$BC = \frac{\sum\limits_{i=1}^{n} \left| X_{ij} - X_{ik} \right|}{\sum\limits_{i=1}^{n} \left(X_{ij} + X_{ik} \right)} \tag{10.3}$$

where BC is the Bray-Curtis measure of dissimilarity; X_{ij} and X_{ik} are the number of individuals in species i in samples j and k, respectively; and n is the total number of species in samples.

Bray-Curtis measure is the standardized Manhattan metric (Bray and Curtis 1957), so its values range from 0 (similar) to 1 (dissimilar). Bray-Curtis measure has the properties:

- It ignores cases in which the species is absent in both community samples. Therefore, Bray-Curtis measure is not affected by joint absences and is sufficiently robust for marine ecology data (Field and McFarlane 1968).
- It gives more weight to abundant species than to rare ones and hence is dominated by the abundant species so that rare species add very little to the value of the coefficient. This property has been reviewed intuitively by most ecologists (Field et al. 1982).

10.1.1.2 Calculating Bray-Curtis Index Using the vegan Package

Example 10.1: Breast Cancer QTRT1 Mouse Gut Microbiome
In Chap. 9, we used this dataset to illustrate calculation of alpha diversity (Zhang et al. 2020). Here, we continue to use this dataset to illustrate calculation of beta diversity.

The **vegdist()** function is used to compute dissimilarity indices that are most commonly used by community ecologists. All indices use quantitative data, and the binary index can be calculated using an appropriate argument. One syntax of the **vegdist()** function is given as below:

```
vegdist(x, method="bray", binary=FALSE, diag=FALSE, upper=FALSE, na.
rm = FALSE)
```

where the argument:

- **x** is the community data matrix.
- **method** is used to specify the dissimilarity index, such as "bray" and "jaccard."
- **binary** is used to specify performing presence/absence standardization before analysis using decostand()function.
- **diag** is used for computing diagonals.
- **upper** is used for returning only the upper diagonal.
- **na.rm** is used to specify pairwise deletion of missing observations when computing dissimilarities.

The R package **vegan** (current version 2.5.7, March 2022) is a community ecology package. This package provides tools for descriptive community ecology; it has most basic functions of diversity analysis, community ordination and dissimilarity analysis for community ecology, and other data types as well (Jari Oksanen et al. 2020), such as microbiome data. The following R commands calculate Bray-Curtis dissimilarity via the **vegdist()** function in vegan package:

```
> setwd("~/Documents/QIIME2R/Ch10_BetaDiversity")
> abund_tab=read.csv("otu_table_L7_MCF7_vegan.csv",row.names=1,
check.names=FALSE)
> abund_tab_t<-t(abund_tab)
> head(abund_tab_t,3)
> library(vegan)
> BC<-vegdist(abund_tab_t, "bray")
> head(BC)
[1] 0.6141 0.1895 0.6118 0.2544 0.6261 0.2412
```

10.1.2 Jaccard Dissimilarity

Jaccard index, developed by the vegetation scientist Paul Jaccard in 1900 (Jaccard 1900), is the first similarity coefficient used to analyze vegetation survey data, and nowadays this coefficient is still in wide use in all fields including ecology and microbiome to analyze multivariate presence/absence observational data.

10.1.2.1 The Measures of Jaccard Index

Both Jaccard and Sørensen indices are the most often used binary similarity coefficients among the more than 20 binary similarity measures in the literature. A 2 × 2 contingency table can facilitate the calculation of the coefficients (or association) of presence-absence binary data (Table 10.1).

where a is the number of species in sample A and sample B (joint occurrences), b is the number of species in sample B but not in sample A, c is the number of species in sample A but not in sample B, and d is the number of species absent in both samples (zero-zero matches). The Jaccard index is given as below:

Table 10.1 A 2 × 2 contingency table for defining beta diversity

| | | Sample A | |
		No. of species present	No. of species absent
Sample B	**No. of species present**	a	b
	No. of species absent	c	d

$$S_j = \frac{a}{a+b+c},\tag{10.4}$$

where S_j is the Jaccard similarity coefficient as defined in above presence-absence matrix and a, b, and c are as defined in Table 10.1.

Jaccard's dissimilarity coefficient is defined as $1 - S_j$ via this similarity.

10.1.2.2 Calculate Jaccard Index Using the vegan Package

Jaccard's index can be calculated using the **vegdist()** function in vegan package as below:

```
> jaccard <-vegdist(abund_tab_t, "jaccard",binary=TRUE)
> head(jaccard)
[1] 0.5563 0.4821 0.5906 0.4593 0.5494 0.4458
```

Please note that specifying "jaccard" in the function **vegdist()** returns Jaccard similarity instead of Jaccard dissimilarity. The Jaccard dissimilarity is obtained by $1 - S_j$ (i.e., $1 -$ jaccard).

Bray-Curtis and Jaccard indices are rank-order similar. The **vegdist()** function by default uses Bray-Curtis which is semimetric. Based on the vegan manual, it probably should be preferred using the metric Jaccard index instead of the default semimetric Bray-Curtis index. Jaccard index is computed as 2BC / (1 + BC), where BC is Bray-Curtis dissimilarity. However, the quantitative version of Jaccard should probably be called Ružička index (Jari Oksanen 2020, pp. 273–277). Thus, we specify binary=TRUE in the **vegdist()** function to obtain the binary version of Jaccard index.

10.1.3 Sørensen Dissimilarity

10.1.3.1 The Measures of Sørensen Index

Sørensen's index (1948) is very similar to the Jaccard index, which is given as below:

$$S_S = \frac{2a}{2a+b+c},\tag{10.5}$$

where S_S is Sørensen's similarity coefficient. This index can also be modified to a coefficient of *dissimilarity*: $1 - S_S$.

The Sorensen and Jaccard coefficients are thought as very closely correlated (Baselga and Orme 2012). The range of all similarity coefficients for binary data is supposed to be 0 (no similarity) to 1 (complete similarity). In fact, this is not true for all coefficients.

The best known index of beta diversity is based on the ratio of total number of species in a collection of samples S and the average richness per sample $\bar{\alpha}$ (Tuomisto 2010), which is given as below:

$$\beta = S/\bar{\alpha} - 1 \tag{10.6}$$

Subtraction of one means that $\beta = 0$ when there are no excess species or no heterogeneity between samples. A drawback of above formula is that S increases with sample size, but α is expected to be constant, so the beta diversity increases with sample size. This really caucuses problem in ecology and also microbiome studies. Thus, to overcome this drawback, Whittaker suggested using pairwise comparison of samples to find the index (Whittaker 1960), i.e., to study the beta diversity of pairs of sites or samples (Marion et al. 2017). The new index is called the Sørensen index of dissimilarity, which is given as below:

$$\beta = \frac{a+b+c}{(2a+b+c)/2} - 1 = \frac{b+c}{2a+b+c}, \tag{10.7}$$

where a is the number of shared species in two samples and b and c are the numbers of species unique to each sample, respectively. $\bar{\alpha} = (2a+b+c)/2$ is the average richness per one sample $S = a + b + c$.

10.1.3.2 Calculate Sørensen Index Using the vegan Package

The Sørensen index of dissimilarity can be calculated for all samples using vegan function **vegdist()** with argument binary = TRUE for suggesting binary data:

```
> Sorensen <-vegdist(abund_tab_t,binary=TRUE)
> head(Sorensen)
[1] 0.3853 0.3176 0.4191 0.2981 0.3787 0.2868
```

Please note that specifying binary=TRUE in the function **vegdist** () returns Sørensen similarity instead of Sørensen dissimilarity. The Sørensen dissimilarity is obtained by

$$1 - S_S \text{ (i.e., 1-Sorensen).}$$

After we obtain Bray, Jaccard, and Sørensen diversity indices, we can conduct hypothesis testing and statistical analysis on them. Typically, these dissimilarity matrices can be analyzed by a multivariate technique such as analysis of similarities (ANOSIM) or permutational MANOVA (PERMANOVA) (see Chap. 11 for details).

10.1.3.3 Calculate Matrices of Bray-Curtis, Jaccard, and Sørensen
Indices Using the vegan Package

As illustration, the following R commands calculate Bray-Curtis, Jaccard, and Sørensen indices together:

```
> library(matrixStats)
> library(vegan)
> library(permute) #Loading vegan required package
> library(lattice) #Loading vegan required package

> sim_matrix <- list(BC=vegdist(abund_tab_t, method="bray"),# Bray-
Curtis index
+        JAC=as.matrix(vegdist(abund_tab_t, 'jaccard', binary=TRUE)),
# Jaccard index
+           SOR=as.matrix(vegdist(abund_tab_t,binary=TRUE))# Sørensen
index
+           )
> head(sim_matrix)
```

The calculation of distance/dissimilarity matrices mainly involves two packages: **matrixStats** and **vegan**. The **matrixStats** (current version 0.61.0, March 2022) package was developed to provide high-performing functions operating on rows and columns of matrices (and to vectors) (Bengtsson 2020).

10.2 Phylogenetic Beta Diversity Metrics

Multivariate analyses of microbial communities typically first need one distance metric to measure distances or dissimilarities between microbial communities and then to conduct comparisons based on the measurements.

The unique fraction (UniFrac distance) metrics, the phylogenetic distance measures, which account for the phylogenetic relationship among the taxa, are very powerful methods because they exploit the degree of divergence between different sequences. UniFrac distance metrics are often used to summarize the overall microbiota variability. In this subsection, we introduce three phylogenetic beta diversity metrics (unweighted UniFrac, weighted UniFrac, and GUniFrac) that were specifically developed for microbiome data.

10.2.1 Unweighted UniFrac

In 2005, UniFrac distance metric (Lozupone and Knight 2005), a phylogenetic-based method, was proposed to measure the phylogenetic distance between sets of taxa in a phylogenetic tree, taking the natural hierarchical structure of the data into account. UniFrac distance metric aims to enable objective comparison between microbiome samples from different conditions. UniFrac does not rely on statistical testing for differences in each individual taxon; instead it directly conducts statistical hypothesis testing of two samples by comparing the taxonomic distance between the sets of taxa from each sample (Lozupone and Knight 2005).

UniFrac is based on phylogenetic information considering the presence/absence of species without weighting the relative abundances (Lozupone and Knight 2005). The formula of UniFrac is given as below:

$$d_U = \frac{\sum_{i=1}^{m_b} b_i \, | \, I(p_i^A > 0) - I(p_i^B > 0) \, |}{\sum_{i=1}^{m_b} b_i},$$

(10.8)

where b_i, for $i = 1, \ldots, m_b$, denotes the length of ith branch of the phylogenetic tree and p_i^A and p_i^B denote the cumulative proportions of all taxa descending from the ith branch for communities A and B, respectively. $I(\cdot)$ is the binary indicator function. Because of the probability that the rare taxa sequenced are directly related to the presence/absence of species, thus the unweighted UniFrac could most efficiently detect the variability in community membership or the abundance of rare lineages (Chen et al. 2012). However the drawback of the unweighted UniFrac distance is that it may have lower power in detecting change in moderately abundant lineages (Chen et al. 2012).

10.2.2   Weighted UniFrac

In 2007, weighted UniFrac distance metric (Lozupone et al. 2007) was proposed to incorporate phylogenetic information with abundance information. The 2005 original UniFrac distance was developed using only the presence/absence data (Lozupone and Knight 2005), while the weighted UniFrac was developed based on the relative abundance of each taxon adding a proportional weighting to the original UniFrac. Because the version of weighted UniFrac was proposed, the original UniFrac distance was called as the unweighted UniFrac to differentiate from the weighted UniFrac. The formula of weighted UniFrac is given as below:

$$d_w = \frac{\sum_{i=1}^{m_b} b_i \, | \, p_i^A - p_i^B \, |}{\sum_{i=1}^{m_b} b_i (p_i^A + p_i^B)},$$

(10.9)

Unweighted UniFrac distance considers only species presence and absence information and counts the fraction of branch length unique to either community. In contrast, weighted UniFrac distance uses species abundance information and weights the branch length with abundance difference. Thus, the weighted UniFrac is most sensitive to detect change in abundant lineages. However, like unweighted

UniFrac, the weighted UniFrac distance may be underpowered in detecting change in moderately abundant lineages (Chen et al. 2012).

10.2.3   GUniFrac

Both unweighted and weighted UniFrac distance metrics have become the most widely used phylogenetic distance measures since their developments. However, they were evaluated having limitations (Chen et al. 2012), assign too much weight either to rare lineages (unweighted UniFrac distance) or to most abundant lineages (weighted UniFrac distances); thus, they may not be very powerful in detecting change in moderately abundant lineages. To overcome this limitation, a generalized version of UniFrac distance (GUniFrac) was proposed in 2012 (Chen et al. 2012) to capture the variability of taxa that have the middle abundances. The generalized UniFrac distance is defined as below:

$$d_G^{(\alpha)} = \frac{\sum_{i=1}^{m_b} b_i \left(p_i^A + p_i^B \right)^{\alpha} \left| \frac{p_i^A - p_i^B}{p_i^A + p_i^B} \right|}{\sum_{i=1}^{m_b} b_i (p_i^A + p_i^B)^{\alpha}}, \tag{10.10}$$

where the extra parameter α is used to control the weight on abundant lineages so that the distance is not dominated by highly abundant lineages. $d_G^{(0.5)}$ denotes the distance with $\alpha = 0.5$ which overall has been shown to be very robust and can efficiently capture the microbiota variability in the moderately abundant lineages.

10.2.4   pldist

The **pldist** (paired and longitudinal UniFrac ecological dissimilarity) method (Plantinga et al. 2019) is another approach of distance-based analysis. The goal of pldist method is to reduce intersubject variation by modifying the distance metric to accommodate related samples. It first summarizes within-individual (or within-pair) shifts in microbiome composition and then compares these compositional shifts across individuals (or pairs). The pldist consists of two paired dissimilarities (unweighted PUniFrac, generalized PUniFrac) and two longitudinal UniFrac dissimilarities (unweighted LUniFrac and generalized LUniFrac), in which the LUniFrac dissimilarities are the extensions of the PUniFrac dissimilarities, respectively. The pldist method uses the centered log-ratio transformation (CLR) to account for data compositionality. It can incorporate phylogenetic and abundance-based dissimilarities, such as Gower's distance (Gower 1971), Bray-Curtis dissimilarity (Bray and Curtis 1957), and Jaccard distance (Jaccard 1912). UniFrac-based metrics are based in part on GUniFrac (Chen et al. 2012).

The PUniFrac and LUniFrac dissimilarities can be used in any analysis where a beta diversity matrix is required, that is, ordination, clustering, and global hypothesis testing, such as permutation-based methods (e.g., PERMANOVA) and kernel machine regression-based association tests (e.g., MiRKAT) (Zhao et al. 2015), and MiRKAT-S (Plantinga et al. 2017; Wu et al. 2016; Zhan et al. 2017; Zhao et al. 2015). The pldist method explicitly considers changes in microbiome over time (Plantinga et al. 2019). However, as a compositional method, pldist replaces zeros with a small pseudocount. This practice is arguable.

10.2.5 Calculate (Un)Weighted UniFrac and GUniFrac Distances Using the GUniFrac Package

The syntax of the GUniFrac in the GUniFrac package is given as

```
GUniFrac(otu.tab, tree, alpha = c(0, 0.5, 1))
```

where the argument:

- **otu.tab** is the OTU count table with row presenting the sample and column presenting the OTU.
- **tree** is the rooted phylogenetic tree of R class "phylo."
- **alpha** is used to control weight on abundant lineages.

After implementing the **GUniFrac()** function, it returns a LIST containing unifracs, which is a three-dimensional array containing all the UniFrac distance matrices.

Excepting for calculating the GUniFrac distance, the **GUniFrac** package can also calculate the unweighted and weighted UniFrac and variance-adjusted weighted UniFrac distances.

Example 10.2: Breast Cancer QTRT1 Mouse Gut Microbiome, Example 10.1 Cont.

Here, we continue to use Example 10.1 to illustrate the calculations of GUniFrac distance family measures. Since the function **GUniFrac()** only accepts rooted tree, while Example 10.1 dataset does not have a tree data, so we first generate a tree data using the phyloseq package.

Create a phyloseq Object Using the phyloseq Package

The following R commands create a phyloseq object. For the detail descriptions, please see Chap. 2 (Sect. 2.3.1.1):

```
> otu<-read.csv("otu_table_L7_MCF7_phyloseq.csv",row.names = 1)
> tax<-read.csv("tax_table_L7_MCF7_phyloseq.csv",row.names = 1)
> sam<-read.csv("metadata.csv",row.names = 1)
```

```
> otumat<-as.matrix(otu)
> taxmat<-as.matrix(tax)
> meta_tab = sample_data(sam)

> library("phyloseq")
> otu_tab = otu_table(otumat, taxa_are_rows = TRUE)
> tax_tab = tax_table(taxmat)

> physeq = phyloseq(otu_tab, tax_tab, meta_tab)
> physeq
phyloseq-class experiment-level object
otu_table() OTU Table: [ 635 taxa and 40 samples ]
sample_data() Sample Data: [ 40 samples by 3 sample variables ]
tax_table() Taxonomy Table: [ 635 taxa by 7 taxonomic ranks ]
```

Generate a Rooted Tree Using the Ape Package

After creating a phyloseq object, we now can use the ape package to create a rooted tree dataset as below:

```
> library("ape")
> random_tree = rtree(ntaxa(physeq), rooted=TRUE, tip.
label=taxa_names(physeq))
```

Now we rebuild phyloseq data object by combining all the four data components, which creates a dataset for downstream statistical analysis:

```
> physeq2 = phyloseq(otu_tab, tax_tab, meta_tab, random_tree)
> physeq2
phyloseq-class experiment-level object
otu_table() OTU Table: [ 635 taxa and 40 samples ]
sample_data() Sample Data: [ 40 samples by 3 sample variables ]
tax_table() Taxonomy Table: [ 635 taxa by 7 taxonomic ranks ]
phy_tree() Phylogenetic Tree: [ 635 tips and 634 internal nodes ]
```

But here, our purpose is to generate a rooted tree dataset for illustrating the calculations of GUniFrac distance family measures. So we output the tree dataset:

```
> write.tree(random_tree, "Rooted_tree_QTRT1.tre")
```

Now we illustrate how to calculate the UniFrac distance family measures step by step via the GUniFrac package as follows.

Step 1: Load abundance table (otu-table) and rooted tree and transform into appropriate formats.

```
> abund_tab=read.csv("otu_table_L7_MCF7_vegan.csv",row.names=1,
check.names=FALSE)
> Rooted_tree=read.tree("Rooted_tree_QTRT1.tre")
```

```
> head(abund_tab,3) # otu table with an otu by sample matrix.
     Sun071.PG1 Sun027.BF2 Sun066.PF1 Sun029.BF4 Sun068.PF3 Sun026.BF1
OTU_1        0          0          0          0          0          0
OTU_2        0          0          0          0          1          0
OTU_3        0          0          0          0          1          0
```

```
> abund_tab_t<-t(abund_tab) # otu table with a sample by otu matrix.
> head(abund_tab_t,3)
            OTU_1 OTU_2 OTU_3 OTU_4 OTU_5 OTU_6 OTU_7 OTU_8 OTU_9 OTU_10 OTU_11
Sun071.PG1      0     0     0     0     0     0     0     0     0      0      0
            OTU_12 OTU_13 OTU_14 OTU_15 OTU_16 OTU_17 OTU_18 OTU_19 OTU_20
Sun071.PG1      0      0      0      0      0      0      0      0      0
            OTU_21 OTU_22 OTU_23 OTU_24 OTU_25 OTU_26 OTU_27 OTU_28 OTU_29
Sun071.PG1      0      0      0      0      0      1      0      0      0
```

Step 2: Calculate the UniFracs.

```
> library(matrixStats)
> library(ape) #Loading GUniFrac required package
> library(GUniFrac)

> unifracs <- GUniFrac(abund_tab_t, Rooted_tree)$unifracs
> unifracs_family <- list(UniFrac=unifracs[, , c('d_UW')],#
Unweighted UniFrac
+          WUniFrac=unifracs[, , c('d_1')],# Weighted UniFrac
+          GUniFrac=unifracs[, , c('d_0.5')]# GUniFrac with alpha 0.5
+          )
```

The calculation of distance/dissimilarity matrices mainly involves three packages: **matrixStats**, **vegan**, and **GUniFrac**. The GUniFrac (current version 1.4, March 2022) package was developed to calculate generalized UniFrac distance for comparing microbial communities (Chen et al. 2022). It contains an extra parameter controlling the weight on abundant lineages to ensure that the distance is not dominated by highly abundant lineages. The GUniFrac package is also able to calculate the unweighted and weighted UniFrac and variance-adjusted weighted UniFrac distances as well as to implement a permutation-based multivariate analysis of variance using multiple distance matrices. Here, we used this package to calculate the unweighted and weighted UniFrac and generalized UniFrac distances.

The arguments "**d_UW**", "**d_1**", and "**d_0.5**" are specified to calculate the distances for unweighted UniFrac, weighted UniFrac, and GUniFrac with alpha 0.5, respectively:

```
> head(unifracs_family,3)
$UniFrac
           Sun071.PG1 Sun027.BF2 Sun066.PF1 Sun029.BF4 Sun068.PF3 Sun026.BF1
Sun071.PG1     0.0000     0.3160     0.2724     0.3501     0.2739     0.3171
Sun027.BF2     0.3160     0.0000     0.2999     0.2562     0.3181     0.2314
```

```
Sun066.PF1   0.2724   0.2999   0.0000   0.3308   0.2721   0.3138
------


$WUniFrac
       Sun071.PG1 Sun027.BF2 Sun066.PF1 Sun029.BF4 Sun068.PF3 Sun026.BF1
Sun071.PG1   0.00000   0.60041   0.11045   0.48766   0.14562   0.5235
Sun027.BF2   0.60041   0.00000   0.60027   0.31149   0.68021   0.2024
Sun066.PF1   0.11045   0.60027   0.00000   0.51515   0.10918   0.5441
------


$GUniFrac
       Sun071.PG1 Sun027.BF2 Sun066.PF1 Sun029.BF4 Sun068.PF3 Sun026.BF1
Sun071.PG1   0.0000   0.5978   0.2415   0.5025   0.2714   0.5476
Sun027.BF2   0.5978   0.0000   0.5423   0.3506   0.6222   0.2496
Sun066.PF1   0.2415   0.5423   0.0000   0.4967   0.2172   0.5393
------
```

10.2.6   Remarks on Rarefaction for Alpha and Beta Diversity Analysis

The appropriateness of using rarefaction in microbiome data is a controversial topic. On the one hand, rarefaction is still recommended for alpha and beta diversity analysis and especially for unweighted UniFrac distance measure and confounded scenarios to address issue of the different sequencing depths in different samples (Weiss et al. 2017; de Cárcer et al. 2011). On the other hand, rarefaction has been criticized as suffering from a great power loss due to discarding a lot of reads (McMurdie and Holmes 2014). In the literature, some researchers treated different UniFrac distances differently, while some others treated them equally. For example, there is the case that the unweighted UniFrac distance matrix was calculated based on rarefied data, while other distance matrices including weighted, information, and ratio UniFrac using the non-rarefied data (Wong et al. 2016). In contrast, other researchers recommended and used the rarefied OTU table for calculating all UniFrac and abundance-based distances to reduce potential sequence depth-dependent bias (Zhang et al. 2018) (also see the GUniFrac package manual in 2012).

The rarefaction can be done using either function **rrarefy()** or **rarefy()** from the **vegan** package although their uses are different. Rarefaction also can be done in the phyloseq and microbiome packages. The function **rarefy_even_depth()** in the microbiome R package can be used to rarefy microbiome data.

10.3 Ordination Methods

Visualization is an effective explorative procedure for providing insight into the data patterns in the dataset and help to formulate statistical hypothesis testing of data. It is usually the first step in any statistical analysis. Visualization is especially important in the analysis of high-dimensional ecological and omics datasets. Because microbiome sequencing datasets are high-dimensional, they typically have larger number of microbial taxa but with smaller number of samples. Thus, it could cause the large P small N problem (see Sect. 18.3 of Chap. 18 for details). One important step before visualization is to reduce the high dimensionality of microbial taxa. Ordination is an effective dimension reduction technique to optimally represent (dis)similarities between samples in an ordination. It typically arranges samples that are similar in high-dimensional space to be represented close together in two or three dimensions.

10.3.1 Introduction to Ordination

In microbiome study, the ordination methods were mainly adopted from ecology and numerical taxonomy.

10.3.1.1 Brief History of Ordination

The term "ordination" originates from the Latin *ordinatio* (meaning the action of setting in order), so ordination is essentially a method to arrange the analysis units in some order. In vegetation ecology, the process of arranging samples (or species) in relation to one or more gradients or axes of variation, that is, arranging vegetational units in some uni- or multidimensional order (Anderson 1965), placing vegetation samples in a coordinate space rather than dividing them into groups is called ordination (Goodall 1954a). For ease of representation and inspection, two- or three-dimensional ordination is usually employed.

The approach of ordination has its roots deeply embedded in ecological literature (Gleason 1926; Ramensky 1930; Anderson 1965). However, the ordination techniques that were applied for vegetation in Ramensky's 1930 article are informal, and in the early 1950s, such informal and largely subjective methods became widespread (Whittaker 1967). In 1954, the vegetation ecologist David Goodall (1954b) first used factor analysis as objective methods for classification of vegetation in community ecology (Legendre 2019).

Because Goodall first applied the term "ordination" to designate principal components analysis (factor analysis) to ordinate vegetation data, it was thought that he coined the term "ordination" (Minchin and Oksanen 2015). Perhaps it is more accurate to say that the term of ordination is a translation of Ramensky's Ordnung, in his German versions 1924 and 1930 articles (Ramensky 1930; Whittaker 1967). Bray and Curtis (1957) developed the polar ordination, which became the first widely used ordination technique in ecology. This term ordination is now widely used in community ecology and many other fields (Legendre 2019) including microbiome.

In 1982 Gauch (1982b) in his book *Multivariate Analysis in Community Ecology* used nontechnical terms to describe ordination and allowing ordination techniques to the general practitioner. ter Braak and Prentice (1988) developed a theoretical unification of ordination techniques, hence providing gradient analysis with a firm theoretical foundation (ter Braak and Prentice 1988).

10.3.1.2 Ordination Plots

Visualizing multidimensional data is not only important but also especially challenging. For a dataset with n variables (descriptors/OTUs), we need to draw n (n − 1)/2 of scatterplots to explore the data structure (Xia et al. 2018b, p. 208). Such large number of scatterplots is not only tedious to work on but also not informative.

Ordination is based on extracting the eigenvectors or factors of the data matrix, which is style of factor analysis technique. So it can reduce the data space. Ordination methods are such techniques that project multidimensional data into a reduced space (usually two or eventually no more than three dimensions). Ordination projects the multidimensional scatter diagram onto bivariate graphs whose axes are known to be of particular interest. In ordination, the data structure of two variables is typically visualized by a scatterplot of the objects (samples), called an ordination plot, in which objects (samples) are presented as points in an x/y-plane, with the ecological differences between objects (samples) being interpreted simply by the distance/dissimilarity between the points. Thus, ordination is essentially a method to order objects (samples) based on similarity and evaluate the differences between objects (samples) in microbial community composition. The goal of ordination is to choose the axes of these graphs to maximally represent the variability of the multidimensional data matrix, in a reduced dimensionality space (Legendre and Legendre 2012, p. 425).

Microbiome data are multidimensional and generally have many variables (i.e., species, taxa, or OTUs). The microbiome dataset is a collection of samples (subjects) positioned in a space where each variable or species (or taxa/OTUs) defines one dimension. Thus, there are as many dimensions as variables or species (or taxa/OTUs). We can summarize and visualize microbiome samples using ordination, in which the samples are visualized in two- or three-dimensional space allowing the densities of OTUs (or taxa) and relationships of important points of concentration to be observed. Thus, ordination primarily endeavors to represent sample and OTUs relationships as faithfully as possible in a low-dimensional space (Gauch, Jr. 1982a, b), i.e., achieving dimensionality reduction.

The application of ordination in microbiome study is mainly adopted from numerical taxonomy and ecology. In phenetic (numerical) taxonomy, the ordination method is to describe the occupation of phenetic space by the OTUs being ordinated. It assumes that positions of the OTUs in phenetic space faithfully reflect their true relationships. Thus we can make inferences from the patterns of OTUs in the space to the processes responsible for the patterns (Sokal 1986). Given a $n \times t$ matrix in numerical taxonomy, ordination is the placement of t OTUs in an A-space of dimensionality varying from 1 to n or $t - 1$, whichever is less (Sneath and Sokal 1973, p. 245). Here, the A-space refers to the low-dimension space obtained by the ordination methods, in which the OTUs are represented.

The motivation of using ordination is that it is not possible to use conventional methods to represent the set of OTUs with respect to more than three characteristics (variables). Thus, ordination is used as a method to summarize the information about relationships implied by the entire suite of characters (variables). In ecology, ecologists usually are interested in characterizing the main trends of variation of the objects with respect to all descriptors, not only a few of them (Legendre and Legendre 2012, p. 425). Microbial ecologists utilize a common multivariate statistical method to compare the microbial communities of different objects. However, as described above, it is challenging and not informative to use scatter plots of the objects with respect to all possible pairs of descriptors to inspect the data structure and pattern. Ecologists therefore are motivated to work on a reduced (i.e., lower) dimensional space. In a given matrix (or a data frame) with objects in rows and descriptors (including species or taxa) in columns, ecologists typically employ ordination to represent objects (stations, etc.) as points along one or several axes of reference (Gower 1984; Legendre and Legendre 2012).

In ordination, the differences and **distances** between samples with respect to their microbial community composition are same concepts. The distances between samples are essentially measured using a mathematical formula. The distance or (dis)-similarity coefficient represents the differences between the samples. The calculated distance or (dis)similarity coefficient consists of a symmetrical distance matrix containing a coefficient for each pair of samples.

Ordination analysis is usually applied to the normalized data in multivariate analysis including numerical taxonomy (Sneath and Sokal 1973, pp. 245–249) such as using PCA and PCoA, numerical ecology (Gower 1966; Legendre and Legendre 2012) such as using PCA and PCoA, and microbiome study (Weiss et al. 2017).

In summary, ordination is a popular approach for exploring microbial community composition in the context of sample metadata. In microbial ecology, there are seven most common ordination methods (Xia et al. 2018a, b, c), including five unconstrained ordinations, principal component analysis (PCA), principal coordinate analysis (PCoA), nonmetric multidimensional scaling (NMDS), correspondence analysis (CA), and detrended correspondence analysis (DCA), and two constrained ordinations: redundancy analysis (RDA) and constrained correspondence analysis (CCA). Of them, PCoA and NMDS are explicit distance measures, while PCA, CA, RDA, and CCA are implicit distance measures. For explicit distance measures, before performing the ordination, a distance matrix based on a suited distance measure needs to be calculated, while implicit distance measures do not need to calculate a distance matrix.

In this section, we introduce some commonly used ordination methods and illustrate ordination plots using real microbiome data via the ampvis2 package.

10.3.2   Ordination Plots in the ampvis2 Package

In this section, we will use the **amp_ordinate()** function in the **ampvis2** package to perform the ordination plots. The **amp_ordinate()** function is primarily based on vegan and ggplot2 (Wickham 2016) packages. It basically wraps numerous functions from the vegan (Jari Oksanen et al. 2018), ggplots (Wickham and Wickham 2007), and ape (Paradis and Schliep 2019) packages to generate ggplot2 ordination plots suited for analysis and comparison of microbial communities. The syntax of the function is simple and easy to use. Ordination in a multistep workflow can often be performed with just one function call. The function **amp_ordinate()** by default performs the following seven steps in order: (1) filters low abundant OTUs, (2) performs data transformation, (3) calculates a distance matrix when performing PCoA or NMDS that needs an explicit distance measure, (4) calculates both site (sample) and species (taxa/OTUs) scores by the chosen ordination method, (5) generates the ordination plot with numerous visual layers defined by the user, (6) fits the correlation of environmental variables to the ordination plot, and (7) returns a nice-looking plot or an interactive plotly plot.

To perform ordination in ampvis2 package, two input datasets are needed: (1) OTU table, which simply contains the read counts of each OTU in each sample, and (2) corresponding metadata table, in which the samples are used for all aesthetic options. For example, we can color or shape the sample points based on a group variable; or for constrained ordination methods, these samples can be used for environmental interpretation or fitting the correlation of environmental variables onto the ordination plot.

Currently the ampvis2 package supports seven different ordination methods including unconstrained ordinations, principal component analysis (PCA), principal coordinate analysis (PCoA) or metric multidimensional scaling (MMDS), correspondence analysis (CA), detrended correspondence analysis (DCA), and nonmetric multidimensional scaling (NMDS), and constrained ordinations: redundancy analysis (RDA) and canonical correspondence analysis (CCA).

Various data transformations are available for this function including "total", "max", "freq", "normalize", "range", "standardize", "pa" (presence/absense), "chi. square", "hellinger", "log", or "sqrt", which are performed from the **decostand()** function (standardization methods for community ecology) in vegan package. The package also uses Plotly graphs for interactive plots. The plots generated are interactive and have higher quality for publication.

One sample syntax of the **amp_ordinate()** function is as below:

```
ordination <- amp_ordinate(ds,
        type = "CCA",
        transform = "Hellinger",
        distmeasure = "bray",
        constrain = "Group",
        sample_color_by = "Group",
        sample_shape_by = "Time",
        sample_colorframe = TRUE,
        sample_colorframe_label = "Group",
        detailed_output = TRUE)
```

In above syntax, only four main arguments are involved in the actual calculations: (1) **type** = " ", (2) **transform** = " ", (3) **distmeasure** = " ", and (4) **constrain** = " " for constrained ordination (only used in RDA or CCA). The rests are just various plotting features. This function by default removes any OTU with an abundance no higher than 0.1% in any sample to improve the calculation time. This threshold can be manually adjusted by changing the filter_species = 0.1 argument. The interested reader can check the website for details (Andersen and Albertsen 2021).

10.3.3 Principal Component Analysis (PCA)

PCA is a well-established linear unconstrained ordination technique for dimensionality reduction and visualization. It is implicitly based on Euclidean distances among samples using quantitative data. PCA assumes that sample dissimilarities are well represented by Euclidean distances for both environmental and species/taxa data and also assumes that abundances are linearly related to environmental gradients. Thus, PCA was developed in the context of the classical statistical model of multivariate normality (Anderson 2001).

10.3.3.1 Introduction to PCA

PCA was invented in 1901 by Karl Pearson (1901) and was later independently developed and named by Harold Hotelling in the 1930s (Hotelling 1933, 1936). PCA orthogonally transforms a set of samples of possibly correlated variables into linearly uncorrelated and orthogonal principal components (PCs), which reduces the data dimension, visualizes the similarities between the biological samples, and filters noise (Jolliffe 2002). In a $n \times p$ data matrix where n is objects (samples) and p represents the variables (dimensions), then PCA has the properties: a large proportion of the dispersion engendered by a large number of variables over the objects (samples) may be accounted for by a smaller dimension. And under a framework of low dimensionality, PCA may thus explain a large portion of the variation of the original data (Sneath and Sokal 1973, p. 245) (also see Legendre and

Legendre 2012, p. 430). In PCA, the eigenvalue of a dispersion matrix is equal to the variance corresponding to the successive principal axes. Under a multinormal distribution, the principal axis corresponding to the largest eigenvalue is the line that goes through the dimension that accounts for the greatest amount of variance from the sample. In the same way, the second principal axis (second shorter and orthogonal to the first axis) accounts for the second largest amount of variance from the sample and so forth.

PCA has been reviewed having several limitations (Xia 2020). We summarize them here:

1. **PCA** suffers from double-zero problem. The problem of "double zeros" or "negative matches" is a situation when certain species is missing in both compared community samples for which similarity/distance is calculated (Sneath and Sokal 1973; Sokal and Sneath 1963; Legendre and Legendre 2012). How to deal with them is a major dichotomy among similarity coefficients. In ecology, the double-zero problem has unimodal distributions and Hutchinson's niche theory behind it, in which species are known to have unimodal distributions along environmental gradients (Whittaker 1967) and Hutchinson's (1957) niche theory (Hutchinson 1957) states that species are more likely to be found at sites where they encounter appropriate living conditions (i.e., species have ecological preferences) and hence the distribution of a species has its mode at this optimum value. For more details, the reader is referred to Legendre and Legendre (2012). Thus, in ecology, based on both statistical and biological considerations, ecologists are warned to beware of double zeros when using PCA. In ecological datasets many species may be missed in many samples, and PCA may be not suitable for heterogeneous compositional datasets with many zeros. In microbiome study, the biological rationale of double-zero problem is not yet confirmed. However, statistically, the problem of double zeros is a really challenging issue.
2. PCA has its limitation in application of the high-throughput data due to the identified Gaussian components and inconsistency when the number of variables is larger than the number of samples.
3. PCA has its difficulty in interpretation and visualization of the composite measures because the PCs are the linear combinations of the entire set of variables (OTUs) under consideration.
4. PCA is characterized by faithfully representing distances between the major groups or clusters but is notorious for falsifying distances between close neighbors (Rohlf 1968).
5. PCA results in economy of description, but the extracted factors do not necessarily have to be interpretable (Sneath and Sokal 1973, p. 247).
6. PCA has its potential artifacts because PCA uses Euclidean distance (calculated by the Pythagorean theorem) to measure sample dissimilarity when microbiome data have many zeros.

To avoid the problems caused by double zeros, non-multinormal distribution, and Euclidean distance, PCA is often applied on normalized and/or pre-transformed data using a transformation-based principal component analysis (tb-PCA) such as using

Hellinger, chord, or other transformation. The transformation ensures that PCA is implicitly based on non-Euclidean distance (i.e., Hellinger, chord, or other). One reason that a chord or Hellinger transformation and correspondence analysis (CA) is usually advocated for such cases.

However, the appealing of PCA (and its constrained version RDA) may be due to its simplicity. Also, with appropriate data transformation (e.g., the Hellinger transformation), PCA is quite useful to explore the most abundant OTUs/taxa and their numerical differences between samples.

10.3.3.2   Implement PCA

Example 10.3: Breast Cancer QTRT1 Mouse Gut Microbiome, Example 10.1 Cont.

In this section, we will use the same QTRT1 mouse gut dataset to illustrate how to generate seven ordination plots whose ordination methods are available in the ampvis2 package, including PCA, PCoA, NMDS, CA, RDA, and CCA. The data importation, loading, and creation of ampvis2 object are the same as in Chap. 9. We repeat the R commands here for convenience:

```
> setwd("~/Documents/QIIME2R/Ch10_BetaDiversity")
> otu_tab <- read.csv("otu_table_L7_MCF7_amp.csv", check.names =
FALSE)
> meta_tab <- read.csv("metadata.csv", check.names = FALSE)

> library(ampvis2)
> ds <- amp_load(otutable = otu_tab,
+         metadata = meta_tab)
```

We conduct PCA using the **amp_ordinate()** function. As an unconstrained ordination, PCA only requires three main arguments: **ds**, **type**, and **distmeasure**. We intentionally provide almost all available arguments for the function to explain plotting features:

```
> amp_ordinate(ds,
+         filter_species = 0.1,
+         type = "PCA",
+         distmeasure = "bray",
+         transform = "hellinger",
+         constrain = NULL,
+         x_axis = 1, y_axis = 2,
+         print_caption = FALSE,
+         sample_color_by = "Group4",
+         sample_color_order = NULL,
+         sample_shape_by = NULL,
+         sample_colorframe = FALSE,
+         sample_colorframe_label = NULL,
+         sample_colorframe_label_size = 3,
```

```
+        sample_label_by = NULL,
+        sample_label_size = 4,
+        sample_label_segment_color = "black",
+        sample_point_size = 2,
+        sample_trajectory = NULL,
+        sample_trajectory_group = NULL,
+        sample_plotly = NULL,
+        species_plot = FALSE,
+        species_nlabels = 0,
+        species_label_taxonomy = "Genus",
+        species_label_size = 3,
+        species_label_color = "grey10",
+        species_rescale = FALSE,
+        species_point_size = 2,
+        species_shape = 20,
+        species_plotly = FALSE,
+        envfit_factor = NULL,
+        envfit_numeric = NULL,
+        envfit_signif_level = 0.005,
+        envfit_textsize = 3,
+        envfit_textcolor = "darkred",
+        envfit_numeric_arrows_scale = 1,
+        envfit_arrowcolor = "darkred",
+        envfit_show = TRUE,
+        repel_labels = TRUE,
+        opacity = 0.8,
+        tax_empty = "best",
+        detailed_output = FALSE,
+        num_threads = 1L)
```

In the above arguments:

- **ds** is required, which is loaded with the **amp_load()** function.
- **filter_species** is used to remove low abundant OTUs across all samples below this threshold in percent. The default is 0.1; setting this value to 0 will overwrite the default setting and may drastically increase computation time.
- **type** is also required, presenting the one type of ordination methods: PCA, CA, DCA, NMDS, PCoA, RDA, and CCA. *Except the* PCoA() *function from the ape package, all other functions are from the vegan package.*
- **distmeasure** is required for distance-based ordination methods NMDS and PCoA. We can choose one distance measure from below: (1) "wunifrac" (weighted UniFrac distances, only for PCoA), which requires a rooted phylogenetic tree as input data; (2) "unifrac" (unweighted UniFrac distances, only for PCoA), which also requires a phylogenetic tree as input data; (3) "jsd" (Jensen-Shannon divergence, only for PCoA), which is based on Arumugam et al. (2011) and Bork (2021); and (4) any of the distance measures supported by the vegdist() function, including "manhattan," "Euclidean," "Canberra," "bray," "kulczynski," "jaccard," "gower," "altGower," "morisita," "horn," "mountford," "raup," "binomial," "chao," "cao," and "mahalanobis." The default distance measure in vegdist

() is "bray," representing the Bray-Curtis method, which is most often used in microbiome studies.

- **transform** is *recommended* in the ampvis2 package. Typically before ordination, we need to transform the abundances using one method that can be chosen from "total", "max", "freq", "normalize", "range", "standardize", "pa" (presence/absense), "chi.square", "hellinger", "log", or "sqrt". Based on literature (Buttigieg and Ramette 2014; Legendre and Legendre 2012; Legendre and Gallagher 2001) *and the authors of this function*, when performing PCA/RDA, it is a good choice using the Hellinger transformation because it will produce a more ecologically meaningful result. When performing CA/CCA, the Hellinger transformation can help reduce the impact of low abundant species. However, when performing distance-based ordination NMDS or PCoA, it is not recommended to use both transformation *and* a distance measure because this will obscure the chosen distance measure (*default:* "hellinger"). If this is not deliberate, consider transform = "none".
- **constrain** is *required for RDA and CCA. We can choose the* variable(s) in the metadata for constrained analyses with a vector for multiple variables.
- **x_axis** = 1(*default:* 1), **y_axis** = 2(*default:* 2), specify which axis from the ordination results to plot as the first and second axis, respectively.
- **sample_color_by** = "Group4", specify color sample points by the variable Group4 in the metadata.
- **sample_color_order** = NULL, specify not to order the colors in sample_color_by the order in a vector.
- **sample_shape_by** = NULL, specify not using shape sample points by a variable in the metadata.

The value generated by the **amp_ordinate()** function is a ggplot2 object. If detailed_output = TRUE is specified, then a list with a ggplot2 object and additional data will be generated.

We can use the **View()** to view the otu table and meta table:

```
> View(otu_tab)
> View(meta_tab)
```

The following R commands are used to generate PCA plot using the Bray-Curtis dissimilarity method (Fig. 10.1):

```
> #Figure 10.1
> amp_ordinate(ds,
+        type = "PCA",
+        distmeasure = "bray",
+        transform = "hellinger",
+        x_axis = 1, y_axis = 2,
+        sample_color_by = "Group4",
+        sample_color_order = NULL,
+        sample_shape_by = NULL,
+        sample_colorframe = FALSE,
```

Fig. 10.1 Ordination plot for QTRT1 data generated by PCA with sample points colored by the groups

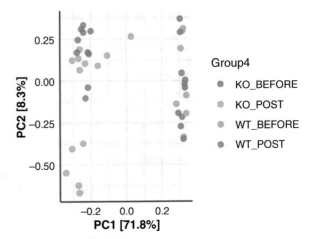

Fig. 10.2 Ordination plot for QTRT1 data generated by PCA with the sample points colored and framed by the groups

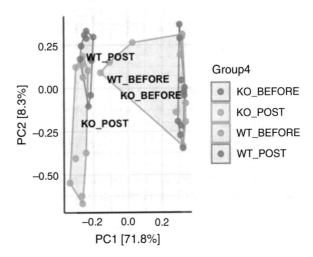

```
+          sample_colorframe_label = NULL,
+          sample_label_by = NULL,
+          sample_point_size = 2,
+          print_caption = FALSE,
+          sample_plotly = NULL)
```

In the following plot (Fig. 10.2), we change the specifications to sample_colorframe = "Group4", sample_colorframe_label = "Group4", and print_caption = TRUE to add colors and labels to differentiate groups and print caption in output. With specifying print_caption = TRUE, the caption is printed in output:

```
> #Figure 10.2
> amp_ordinate(ds,
```

```
+            type = "PCA",
+            distmeasure = "bray",
+            transform = "hellinger",
+            x_axis = 1, y_axis = 2,
+            sample_color_by = "Group4",
+            sample_color_order = NULL,
+            sample_shape_by = NULL,
+            sample_colorframe = "Group4",
+            sample_colorframe_label = "Group4",
+            sample_label_by = NULL,
+            sample_point_size = 2,
+            print_caption = TRUE,
+            sample_plotly = NULL)
```

10.3.4 Principal Coordinate Analysis (PCoA)

PCoA is a conceptual extension of PCA and advance in ordination technique. It is a metric (multidimensional) scaling method. Any Euclidean distance/similarity and non-Euclidean distance/similarity measures and their association coefficients can be used in PCoA. All types of variables (quantitative, semiquantitative, qualitative variables or even datasets with variables with mixed levels of precision) can be used. Thus, it is the most often used ordination technique in ecology and microbiome study.

10.3.4.1 Introduction to PCoA

PCoA or also known as metric multidimensional scaling (MMDS) (Richardson 1938) was developed by John Gower (1966). It uses spectral decomposition to approximate a matrix of distances/dissimilarities to reduce dimensions of data points (Gower 2005). In contrast to PCA, which typically tells us about the major relationships among a set of samples *and* how that relationship is determined by a set of variables, PCoA primarily tells us the similarity among the samples, and the individual data variables are less important. PCoA works by two steps: First, calculate the dissimilarity for every pair of samples in the high-dimensional space using a particular distance (dissimilarity) measure (e.g., Bray-Curtis dissimilarity) chosen by the user. Then, represent the samples in the reduced two dimensions such that their pairwise Euclidean distances approximate their corresponding distances in high-dimensional space as closely as possible.

Similar to PCA, RDA, CA, and CCA, **PCoA** is also based on **eigen analysis**, which *suggests each resulting axis is an eigenvector associated with an eigenvalue, and all axes are orthogonal to each other.* The unique information revealed by the axes are the inertia (in terms of ecology, variance) in the data and exactly how much inertia is indicated by the eigenvalue. The ordination result is plotted in an x/y scatterplot in this way: the largest eigenvalue is plotted on the first axis, and the second largest eigenvalue on the second axis. PCoA has these properties: The result of **PCoA** on distances obtained from standardized variables is identical with that obtained from PCA of product-moment correlation coefficients (J. C. Gower 1967). PCoA only approximates the solution of principal components when it is used to non-Euclidean distance.

PCoA has several advantages summarized as below:

1. PCoA is flexible to be used any distances (need not be Euclidean), and hence any distance/dissimilarity measure can be explicitly chosen (Gower 1966), including any common measures: Manhattan distances, Minkowski metric distances (Sneath and Sokal 1973), **Bray-Curtis** dissimilarity, **Pearson chi-squares, Jaccard, Chord**, and phylogenetic distance (e.g., UniFrac distance).
2. When PCoA is used in any Euclidean distance matrix, principal components can be computed without having either the original data matrix or a variance-covariance matrix of the variables or objects (samples) (Sneath and Sokal 1973) such as the matrices of characters by OTUs in numerical taxonomy, species by objects in ecology, or OTUs/species by samples in microbiome.
3. In PCoA ordination plot can be performed on sets of OTUs for which either original data matrices are not available or when such matrices cannot be obtained because of the nature of the data (Sneath and Sokal 1973).
4. PCoA appears to be less disturbed by missing values than PCA (Rohlf 1972).

10.3.4.2   Implement PCoA

Example 10.4: Breast Cancer QTRT1 Mouse Gut Microbiome, Example 10.1 Cont.
The following R commands are used to perform principal coordinate analysis without Hellinger data transformation and sample points being colored, framed, and labeled by groups (Fig. 10.3):

Fig. 10.3 Ordination plot for QTRT1 data generated by PCoA without Hellinger data transformation

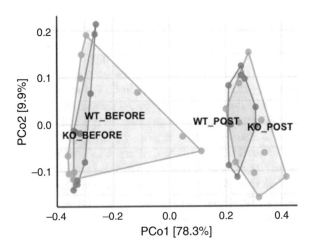

Fig. 10.4 Ordination plot
for QTRT1 data generated
by PCoA with Hellinger
data transformation

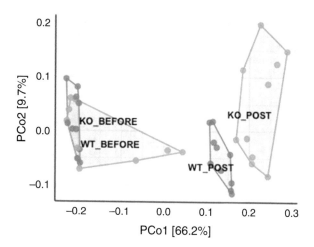

```
> #Figure 10.3
> amp_ordinate(ds,
+        type = "pcoa",
+        distmeasure = "bray",
+        transform = "none",
+        sample_color_by = "Group4",
+        sample_colorframe = TRUE,
+        sample_colorframe_label = "Group4") + theme(legend.position =
"blank")
```

```
> #Figure 10.4
> amp_ordinate(ds,
+        type = "pcoa",
+        distmeasure = "bray",
+        transform = "hellinger",
+        sample_color_by = "Group4",
+        sample_colorframe = TRUE,
+        sample_colorframe_label = "Group4") + theme(legend.position =
"blank")
```

By using Hellinger data transformation (Fig. 10.4), we obtain nice ordination plot
with two post groups being more differentiating each other. However, we get the
following warning message:

```
Warning message:
Using both transformation AND a distance measure is not recommended for
distance-based ordination (nMDS/PCoA). If this is not deliberate,
consider transform = "none".
```

By setting the sample_trajectory = "Time" and sample trajectory_group =
"Group4" arguments, the following R commands generate the plots to track changes

Fig. 10.5 Ordination plot for QTRT1 data generated by PCoA with Bray-Curtis measure and sample trajectory

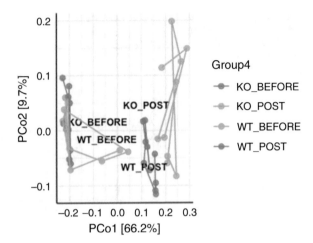

over time (in this case, before and post treatment) and reveal temporal patterns per groups (Fig. 10.5):

```
> #Figure 10.5
> amp_ordinate(ds,
+         type = "pcoa",
+         distmeasure = "bray",
+         sample_color_by = "Group4",
+         sample_colorframe_label = "Group4",
+         sample_trajectory = "Time",
+         sample_trajectory_group = "Group4")
```

10.3.5   Nonmetric Multidimensional Scaling (NMDS)

NMDS is the most general ordination technique. Like PCoA, any Euclidean distance/similarity and non-Euclidean distance/similarity measures and their association coefficients can be used in NMDS. Also similar to PCoA, all types of variables (quantitative, semiquantitative, qualitative variables, or even datasets with variables of mixed levels of precision) can be used. Thus, it is also most often used in ecology and microbiome study.

10.3.5.1   Introduction to NMDS

The intuitive ideas and general procedure of NMDS were provided by psychometrician Shepard (1962, 1966); however, the formal "goodness of fit" hypothesis testing procedure of NMDS was developed by mathematician, statistician, and psychometrician Kruskal (1964a, b; Mead 1992). NMDS was originally intended as a psychometric technique. In 1977, the utility of Kruskal's (1964a, b) NMDS for community ecology was independently discovered and demonstrated by Prentice (1977) and Fasham (1977). NMDS represents n objects/samples geometrically by n points to obtain the interpoint distances corresponding to the experimental dissimilarities between objects samples. The fundamental hypothesis of NMDS is that dissimilarities and distances are monotonically related (Kruskal 1964a). The goal of NMDS is to find n points whose interpoint distances closely agree with the given dissimilarities between n objects/samples. An ordination (a reduced-space scaling) would be perfect if the rank order of the computed (fitted) distances were monotonically related to and were identical to the observed (original) ordinal distances/ dissimilarity function (Sneath and Sokal 1973, p. 249). In other words, a reduced-space scaling would be perfect if all points in the Shepard diagram fell exactly on the regression line (straight line, smooth curve, or step function) and the value of the objective function would be zero (Legendre and Legendre 2012, p. 515).

In 1964, Kruskal (1964a) developed a measure of stress. In the same year, Kruskal (1964b) used an iterative technique to compute coordinates for the k-space, which aims to minimize the stress for any given Minkowski distance coefficient and for any dimensionality k.

Similar to PCoA, for the use of NMDS, we first need to calculate a matrix of sample dissimilarities using a chosen distance metric. However, in contrast to PCoA and also differencing from PCA, RDA, CA, and CCA (all these are eigen analysis-based methods using a direct eigen analysis algorithm), (1) NMDS instead is a rank-based method. It calculates the ranks of these distances among all samples and finds both a nonparametric and monotonic relationship between the dissimilarities and the Euclidean distances. Thus, NMDS focuses mainly on the ranking of dissimilarities rather than their numerical values (Xia 2020). (2) Different from the eigen analysis-based analyses, NMDS takes an *iterative* procedure to perform ordination and hence results in more than one single solution. (3) NMDS also uses a stress value rather than eigenvalues to determine the "success" of the ordination. The rule of thumb for stress value (goodness of fit) is as follows: greater than 0.20 is considered **unacceptable** (poor), 0.1 is acceptable (fair), 0.05 good, 0.025 excellent, and 0 "perfect." Here "perfect" means only that there is a perfect monotone relationship between dissimilarities and the distances (Kruskal 1964a). Or generally stress value above 0.20 is unlikely to be of interest; above 0.15 must still be cautious; from 0.1 to 0.15 is better; from 0.05 to 0.1 is satisfactory; below 0.05 is impressive (Kruskal 1964b), although it has been reviewed that these guidelines are oversimplistic because such a stress tends to increase with increasing numbers of samples (Clarke 1993).

NMDS has been reviewed having several advantages. We summarize them as follows:

1. NMDS has the great advantage that it can consider asymmetrical dissimilarity matrices (nonmetric distances or nonsymmetric matrices) as well as those with missing or tied dissimilarity values (Sneath and Sokal 1973, p. 250). Actually

NMDS has the feature that the more distances being missed, the easier to be computed as long as there are enough (dissimilarities) measures observed left to position each object with respect to a few of the others. This feature makes NMDS to be an appealing method for the analysis of matrices where missing pairwise distances often occur (Legendre and Legendre 2012, p. 513).

2. Similar to **PCoA**, NMDS is flexible to explicitly choose the distance measure from different measures including common measures: **Bray-Curtis** dissimilarity, **Pearson chi-squares, Jaccard, Chord,** as well as currently **UniFrac distance** which incorporates phylogeny.

3. Specially compared to PCA, NMDS has advantage in giving balance between the large inter-cluster distances and the fine differences between members of a given cluster (Rohlf 1970) and can summarize distances in fewer dimensions (i.e., lower stress in two dimensions) (Gower 1966) although a solution of PCoA remains easier to compute in most cases.

In summary, the advantages of NMDS include its robustness, independency of any underlying environmental gradient, and ability of handling missing values and different types of data at the same time. However, same as PCoA, NMDS only can investigate differences between samples but cannot plot both species (taxa/OTUs) scores and sample scores through a **biplot**.

10.3.5.2   Implement NMDS

Example 10.5: Breast Cancer QTRT1 Mouse Gut Microbiome, Example 10.1 Cont.

The following R commands perform nonmetric multidimensional scaling using Bray-Curtis dissimilarity without data transformation and generate sample points that are colored, framed, and labeled by groups (Fig. 10.6):

Fig. 10.6 Ordination plot for QTRT1 data generated by NMDS with Bray-Curtis measure

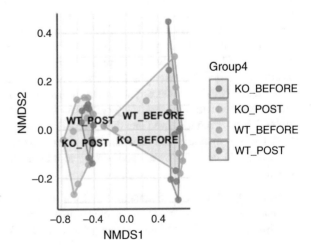

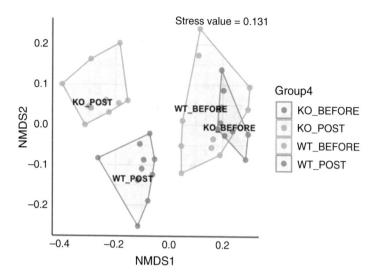

Fig. 10.7 Ordination plot for QTRT1 data generated by NMDS with Jaccard measure

```
> #Figure 10.6
> amp_ordinate(ds,
+       type = "nmds",
+       distmeasure = "bray",
+       transform = "none",
+       sample_color_by = "Group4",
+       sample_colorframe = TRUE,
+       sample_colorframe_label = "Group4")
```

The following R commands are used to perform NMDS using Jaccard dissimilarity without data transformation and sample points being colored, framed, and labeled by groups (Fig. 10.7):

```
> #Figure 10.7
> amp_ordinate(ds,
+       type = "nmds",
+       distmeasure = "jaccard", binary=TRUE,
+       transform = "none",
+       sample_color_by = "Group4",
+       sample_colorframe = TRUE,
+       sample_colorframe_label = "Group4")
```

The following R commands are used to perform nonmetric multidimensional scaling using Jaccard dissimilarity with Hellinger data transformation and sample points being colored, framed, and labeled by groups (Fig. 10.8):

```
> #Figure 10.8
> amp_ordinate(ds,
```

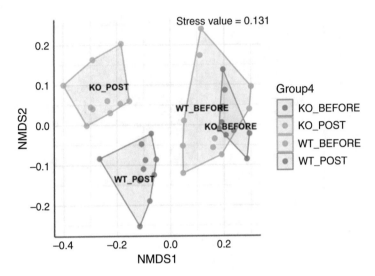

Fig. 10.8 Ordination plot for QTRT1 data generated by NMDS with Jaccard measure and Hellinger data transformation

```
+        type = "nmds",
+        distmeasure = "jaccard",binary=TRUE,
+        transform = "hellinger",
+        sample_color_by = "Group4",
+        sample_colorframe = TRUE,
+        sample_colorframe_label = "Group4")
```

10.3.6 Correspondence Analysis (CA)

CA uses χ^2 distance measure. It requires nonnegative, dimensionally homogeneous quantitative or binary data; specifically species (taxa or OTUs) abundance or presence/absence data can be used in ecology and microbiome study. CA has been used in ecology and microbiome study.

10.3.6.1 Introduction to CA

CA was developed or defined independently over time by a number of ways and several authors in different areas. Thus, CA has been known under different names, such as contingency table analysis (Fisher 1940), RQ technique (Hatheway 1971), reciprocal averaging (Hill 1973), correspondence analysis (Hill 1974), reciprocal ordering (1975) (Orlóci 2013), dual scaling (Nishisato 1980), and homogeneity analysis (Meulman 1982).

At least three approaches have been categorized in the literature (Greenacre 1988): (1) the dual scaling approach (Nishisato 1980; Healy and Goldstein 1976), (2) the geometric approach (Benzécri 1973; Greenacre 1984; Lebart et al. 1984), and (3) the approach that is considered as a method of weighted least-squares lower-rank approximation of a data matrix. The dual scaling approach aims to assign scale values to the categories of a set of discrete variables in order to maximize the variance of the resultant case scores (Greenacre 1988). This is like the analogy to PCA in the style of Hotelling (1933). The geometric (graphical) approach in CA was largely due to Benzécri (1973). It takes three steps (Greenacre 1988; Borg and Groenen 2005): First it normalizes row and column profiles by dividing the rows and columns with their respective totals so that it sums to one in each row and column, respectively. Second, it maps the row and column profiles, to points in high-dimensional Euclidean spaces. Third, it finds the low-dimensional subspaces closest to these points. The final step is achieved by rotating to principal axes such that the first dimension accounts for the maximum variance, the second dimension maximizes the remaining variance, and so on (Borg and Groenen 2005). This is analogous to the geometric approach to PCA pioneered by Pearson (1901). In the context of a two-way contingency table, the approach of weighted least-squares lower-rank approximation of a data matrix provides a decomposition of the usual Pearson chi-squared statistic for testing independence of rows and columns (Greenacre 1988).

The above three approaches coincide in the correspondence analysis of a two-way contingency table. Although CA in principle can be used on any rectangular table with nonnegative similarity values, it is a technique particularly suited for analyzing a contingency table of two categorical variables. The role of the rows and columns can be reversed by simply transposing the correspondence table.

This technique was reviewed (1980) (Nishisato 1980, 1984) that can trace its origin back to 1933 (Richardson and Kuder 1933).

CA was proposed for the analysis of contingency tables by Herman Otto Hartley (Hirschfeld 1935) and Fisher (1940) and later developed by Jean-Paul Benzécri (1973) to connect correlation and contingency. Since the 1960s and 1970s, CA have been used to analyze the sites by species tables in ecology, including Hatheway (1971) and Hill (1973, 1974) among others. Correspondence analysis introduced by Guttman (1941), Torgerson (1958, p. 338), and Hill (1973) is a method of scaling rather than of contingency table analysis (Hill 1974). The technique described in 1974 by Hill (1974) under the name "correspondence analysis" (a name translated from French "l'analyse factorielle des correspondances") (Benzécri 1973) is an analogue of PCA, which is appropriate to discrete rather than to continuous variates. Hill (1974) traced its first publication from Hirschfeld (1935). Since the name of correspondence analysis (CA) was introduced in 1973 by Hill to ecologists. CA quickly gained in popularity because of its better recovery of dimensional gradients than PCA and gradually replaced polar ordination and is used so far. Additionally, different from PCA, which is based on the linear relationship of taxa/species, CA assumes modal relationships of taxa relative to ecological gradients. Thus, this method is more attractive than PCA on the theoretical grounds.

Several important standard textbooks on correspondence analysis have been published including Nishisato (1980); Greenacre (1984); Lebart et al. (1984); ter Braak (1988a); van Rijckevorsel and de Leeuw (1988); and Gifi (1990). CA has been applied to microbiome data (Xia et al. 2018b).

CA is an **eigen analysis**-based method. The goal of CA is to find correspondence (interaction) between rows and columns of a contingency table (or to show the interaction of two variables in this table graphically) and to represent the correspondence in an ordination space. CA is a residual-based analysis. It analyzes residuals to capture the discrepancy between observed counts and the counts expected in the identical taxa composition over all samples. Thus, a certain mean-variance relationship for normalization of these residuals is implicitly assumed in CA.

CA conceptually is similar to **PCA**. CA can be considered as an equivalent of PCA on a contingency table of two categorical variables, in which every entry value provides the frequency of each combination of categories of the two variables. However, CA is based on the *Pearson chi-squared* measure and hence is used to analyze nominal or categorical data instead of continuous data. The three important differences between CA and PCA methods are as follows: (1) PCA maximizes the explained variance of measured variables, whereas CA maximizes the correspondence (similarity of frequencies) between rows (measured variables) and columns (samples) of a table (Yelland 2010); (2) PCA assumes a linear relationship among variables, whereas CA expects a unimodal model (Paliy and Shankar 2016); (3) CA uses weighted Euclidean distance (*or reciprocal averaging*, a variant of Euclidean distance) or chi-square distance to estimate the distances among samples in full CA ordination space.

CA and **MDS** (multidimensional scaling) also share several common properties and differ on other aspects. Borg and Groenen (2005) have provided an excellent comparison of CA and MDS techniques. We summarize the main points here. CA and MDS graphically display the objects (samples) as points in a low-dimensional space. However, several differences exist between them: (1) Basically, MDS is a one-mode technique and thus only analyzes one set of objects (samples), whereas CA is a two-mode technique and thus it displays both row and column objects (samples) as in unfolding (Borg and Groenen 2005). (2) MDS can use all types of data variables including quantitative/frequencies, qualitative/rankings, correlations, ratings, or nonnegative or negative, even mixed datasets, whereas CA restricts the data to be nonnegative. (3) MDS can accept any dissimilarity or similarity measures, whereas CA uses the χ^2 distance as a dissimilarity measure. (4) In MDS, the distances between all points can be directly interpreted, whereas in CA the relation between row and column points can only be assessed by projection and their points only can be interpreted, respectively. Therefore, we should interpret the results from CA with some care, similar to non-Euclidean MDS solutions (Borg and Groenen 2005).

In summary, unlike PCA, CA (and its constrained version CCA) is very sensitive to the less abundant and often unique species (or taxa/OTUs) in the samples. Thus, CA is a powerful method to investigate which species (or taxa/OTUs) characterizes each sample or group of samples. However, CA suffers from two major problems:

(1) It is vulnerable to produce a noticeable mathematical artifact called "arch effect" (ter Braak and Šmilauer 2015; Gauch 1977) sometimes also known as the "horse-shoe effect" (Kendall 1971), which is caused by the unimodal species response curves. (2) It suffers from compression of the ends of the gradient. Unimodal distribution refers to a distribution with one mode. In the unimodal species response curves, the expected abundance of species/taxon has a bell-shaped functional rela-tionship with a score with one optimal environmental condition (see Section 6.3.4 of Xia and Sun 2022a). Thus, the species will have a lesser abundance and hence have poor performance if any environment condition is greater or lesser than this opti-mum. (3) As a residual-based approach, CA is not well suitable to skewed data, and its mean-variance assumption is too rigid to account for the overdispersion (Warton et al. 2012).

10.3.6.2   Implement CA

Example 10.6: Breast Cancer QTRT1 Mouse Gut Microbiome, Example 10.1 Cont.
The following R commands perform CA using the Hellinger transformation and generate sample points that are colored, framed, and labeled by groups (Fig. 10.9):

```
> #Figure 10.9
> amp_ordinate(ds,
+        type = "CA",
+        transform = "Hellinger",
+        sample_color_by = "Group4",
+        sample_shape_by = "Group4",
+        sample_colorframe = TRUE,
+        sample_colorframe_label = "Group4",
+        species_plot = TRUE)
```

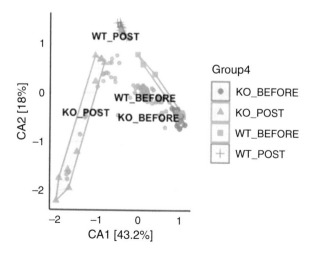

Fig. 10.9 Ordination plot for QTRT1 data generated by CA

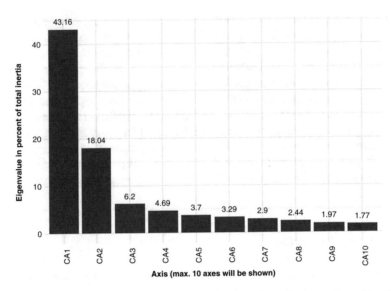

Fig. 10.10 Scree plot is used to obtain an overview of each axis' contribution to the total inertia and confirm those most significant axes based on QTRT1 data

The ordination plot generated with CA (Fig. 10.9) is different from that obtained with PCA (Fig. 10.1). Their species scores are also shown having big differences as seen in the biplot below.

If the detailed_output = TRUE is specified, then except ordination plot and scree plot, we will obtain a detail information on model, samples (sites), species (taxa/ OTUs), etc. (see Fig. 10.10):

```
> #Figure 10.10
> amp_ordinate(ds,
+          type = "CA",
+          transform = "Hellinger",
+          sample_color_by = "Group4",
+          sample_shape_by = "Group4",
+          sample_colorframe = TRUE,
+          sample_colorframe_label = "Group4",
+          species_plot = TRUE,
+          detailed_output = TRUE)
```

We print partial outputs as below:

```
$model
Call: cca(X = data$abund)

        Inertia Rank
Total 0.561
```

```
Unconstrained 0.561 39
Inertia is scaled Chi-square

Eigenvalues for unconstrained axes:
 CA1 CA2 CA3 CA4 CA5 CA6 CA7 CA8
 0.2420 0.1012 0.0348 0.0263 0.0208 0.0184 0.0163 0.0137
 (Showing 8 of 39 unconstrained eigenvalues)
```

Scree plot is a simple bar plot that plots all the axes and their eigenvalues. We can use it to get an overview of each axis' contribution to the total inertia and confirm those most significant axes. In the following bar plot, we can see CA1 explains 43.16% of total variability, which is equal to 0.2420/0.561 = 0.4314 in above model. Similarly, CA2 explains 18.04% of total variability, which is equal to 0.1012/0.561 = 0.1804.

10.3.7 Detrended Correspondence Analysis (DCA)

In Sect. 10.3.6.1, we described that CA suffers from two major problems: the arch effect and the edge effect (compression of the ends of the gradient). The first problem results in consequence that the second CA axis cannot easily be interpreted due to an artifact, while the second problem causes the position of objects/samples (and taxa/ species) along the first axis to be not necessarily related to the amount of change (or beta diversity) along the primary gradient. That is, CA particularly has two inherent distortions: (1) one-dimensional gradients tend to be distorted into an arch on the second ordination axis, and (2) samples tend to be unevenly spaced along the axis 1.

10.3.7.1 Introduction to DCA

Fortunately in 1979, Hill (1979) corrected some of the drawbacks/distortions of CA and thereby created improved ordination technique called detrended correspondence analysis (Hill and Gauch 1980). It remedies both the edge effect and the arch effect, which has been the most widely used indirect gradient analysis today. To overcome these problems, DCA flattens this arch and rescales the positions of samples along an axis. DCA aims to find the main factors or gradients in large, species-rich but usually sparse data matrices. Thus, DCA is frequently used to suppress these two artifacts inherent in most other multivariate analyses when applied to gradient data (Hill and Gauch 1980).

The software DECORANA (Detrended Correspondence Analysis) was developed to implement detrended correspondence analysis, which has become the backbone of many later software packages including vegan.

DCA is an iterative algorithm with three procedures behind it:

Step 1: Start by running a standard ordination to calculate correspondence analysis/ reciprocal averaging with either eigen analysis approach or reciprocal averaging (RA) approach (more intuitive) on the data.

Given a matrix of n rows of samples and p columns of taxa, RA initially assigns an arbitrarily chosen score to each sample. Then, it calculates scores for each taxon as a weighted average by multiplying the abundance of a taxon by the sample score and sums them across all samples and divides by the total abundance for that taxon. Next, it uses these taxon scores to calculate a new set of sample scores via the same procedure for each sample. Finally, it centers and standardizes the calculated sample scores. Alternately repeat the procedure of calculating sample and taxon scores until the scores are stabilized. For details on the algorithm of the "reciprocal averaging" technique, the readers are referred to Hill (1973) and ter Braak (1987c).

As described above, the standard ordination produces the initial horseshoe curve in which the first ordination axis distorts into the second axis. To improve the standard ordination technique, DCA uses the following detrending and rescaling algorithms.

Step 2: Detrend axes to effectively squash the curve flat. Several methods are available for detrending an axis. The simplest way is to divide the first axis into an arbitrary number of equal-length segments (default = 26, which has empirically produced acceptable results); within each segment, re-center the scores on the second axis to have mean value of zero. By implementing detrending, arch will be flattened onto the lower-order axis if there is arch present.

Step 3: Remove the edge effect by nonlinear rescaling of the axis. Before rescaling the species, curves are narrower near the ends of the axis than in the middle because of the edge effect. Thus, to remove the edge effect, it needs to rescale the axis to ensure that the ends are no longer compressed relative to the middle. The methods proposed to rescale include using polynomial regressions and using a sliding moving average window, which is the algorithm R used. The rescaling of an axis results in equalizing the weighted variance of taxon scores along the axis segments, such that 1 DCA unit approximates to the same rate of turnover all the way through the data.

Ter Braak (ter Braak and Prentice 1988), a minorly modified version that was printed in advance in ter Braak (1987c), has provided an excellent review and comparison of **DCA** and **CA** in the ecological literature. We summarize the main points as below. DCA has several advantages, including:

1. DCA often works remarkably well in practice (Hill and Gauch 1980; Gauch et al. 1981).

2. DCA was evaluated giving a much closer approximation to ML Gaussian ordination than CA did based on a two-dimensional species packing model using simulated data (ter Braak 1985b), in which this improvement was shown mainly due to the detrending, not to the nonlinear rescaling of axes.
3. DCA performed substantially better than CA when the two major gradients differed in length (Kenkel and Orlóci 1986). However, DCA also has several disadvantages, including:

   (a) DCA sometimes "collapsed and distorted" CA results when (a) there were few species per site and (b) the gradients were long (Kenkel and Orlóci 1986), in which few species per site were believed to be the real cause of the collapse (ter Braak 1987c).

   (b) DCA may flatten out some of the variation associated with one of the underlying gradients because an instability in the detrending-by-segments method causes this loss of information (Minchin 1987).

   (c) DCA has been thought being "overzealous" in correcting the "defects" in CA and "may sometimes lead to the unwitting destruction of ecologically meaningful information" (Pielou 1984, p. 197), mainly for the somewhat arbitrary and "overzealous" nature of its detrending process, but also because of the sometimes inappropriate imposition and non-robust behavior of an underlying χ^2 distance measure (Pielou 1984; Faith et al. 1987; Gower 1992; Clarke 1993).

DCA overall is a popular and reasonably robust approximation to Gaussian ordination method. DCA eliminates the arch effect that CA is suffered and is much more practical. However, to increase its robustness, two modifications are needed: First, since the edge effect is not too serious, the routine use of nonlinear rescaling was not advised (ter Braak and Prentice 1988). Second, the arch effect needs to be removed (Heiser 1987; ter Braak 1987c), but a more stable, less "zealous" method of detrending, namely, detrending by polynomials, was recommended (Hill and Gauch 1980; ter Braak and Prentice 1988).

Because some forms of nonmetric multidimensional scaling may be more robust (Kenkel and Orlóci 1986; Minchin 1987), some ecologists have suggested using MDS method although whether DCA or MDS can produce stronger distortions is not consistently confirmed. Here, we illustrate DCA using the same data used in Example 10.1.

10.3.7.2 Implement DCA

Example 10.7: Breast Cancer QTRT1 Mouse Gut Microbiome, Example 10.1 Cont.

```
> # Figure 10.11
> amp_ordinate(ds,
+    type = "DCA",
+    transform = "none",
```

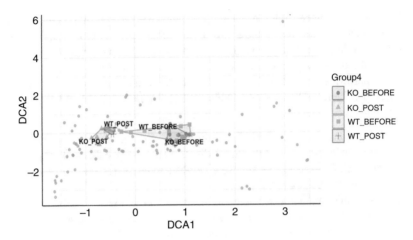

Fig. 10.11 Ordination plot for QTRT1 data generated by DCA

```
+     sample_color_by = "Group4",
+     sample_shape_by = "Group4",
+     sample_colorframe = TRUE,
+     sample_colorframe_label = "Group4",
+     species_plot = TRUE)
```

We perform DCA constraining on group variable and generate sample points that are colored, framed, and labeled by groups (Fig. 10.11):

10.3.8   Redundancy Analysis (RDA)

Canonical analysis simultaneously analyzes two or eventually more data tables. Canonical analysis allows us to perform either direct comparison (direct gradient analysis) or indirect comparison (indirect gradient analysis) of two data matrices such as species/taxa composition table and environmental/metadata variables table. In direct comparison analysis, the matrix of explanatory variables X intervenes in the calculation producing the ordination of response matrix Y, forcing the ordination vectors to be maximally related to combinations of the variables in X. In contrast, in indirect comparison, matrix X does not intervene in the calculation. Correlation or regression of the ordination vectors on X is computed a posteriori (Legendre and Legendre 2012).

10.3.8.1   Introduction to RDA

RDA method was first proposed by Rao (1964) and was later rediscovered by Wollenberg (Van Den Wollenberg 1977). RDA belongs to canonical analysis. In

RDA, each canonical ordination axis corresponds to a direction, in the multivariate scatter of objects/samples (matrix Y), which is maximally related to a linear combination of the explanatory variables X (Legendre and Legendre 2012). RDA assumes that variables from two datasets (e.g., an environmental dataset and a taxa abundance dataset) are asymmetrical and play different roles: ones are the "independent variables," and the others are the "dependent variables."

RDA can be understood in terms of two extensions: (1) It directly extends multiple regression to model multivariate response data. RDA is an asymmetric analysis of the response variables (Y matrix/table) and the explanatory variables (X table). (2) It also extends PCA because the canonical ordination vectors are linear combinations of the response variables Y. However, RDA differs from PCA on the factor that these ordination vectors in RDA not only could be computed on the matrix Y but also are constrained to be linear combinations of the variables in X. Thus, **RDA** is a constrained version of **PCA.**

Same as its unconstrained version **PCA, RDA** (PCA with instrumental variables) uses the **Euclidean distance** measure based on the Pythagorean theorem (Rao 1964). As constrained ordination, RDA was developed to assess how much variation in one set of variables can be explained by the variation in another set of variables. As a multivariate extension of simple linear regression into sets of variables, RDA summarizes the linear relations between multiple dependent and independent variables in a matrix, which is then incorporated into PCA (Rao 1964).

PCA and RDA ordination techniques have a better performance when species have monotonic distributions along the gradients. However, they both assume that the principal components (PCs) are constrained to be linear combinations of the explanatory variables. Thus, like PCA, RDA is inappropriate when relationship between response and environmental variables is unimodal rather than linear (Xia 2020, p. 383). Additionally, RDA usually does not completely explain the variation in the response variables (matrix Y) (Legendre and Legendre 2012, p. 641).

10.3.8.2   Implement RDA

Example 10.8: Breast Cancer QTRT1 Mouse Gut Microbiome, Example 10.1 Cont.

We perform RDA constraining on group variable using the Hellinger transformation and generate sample points that are colored, framed, and labeled by groups (Fig. 10.12):

```
> #Figure 10.12
> amp_ordinate(ds,
+      type = "RDA",
+      constrain = "Group4",
+      transform = "Hellinger",
+      sample_color_by = "Group4",
+      sample_shape_by = "Group4",
```

Fig. 10.12 Ordination plot
for QTRT1 data generated
by RDA

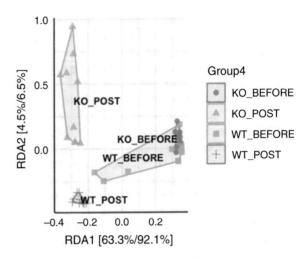

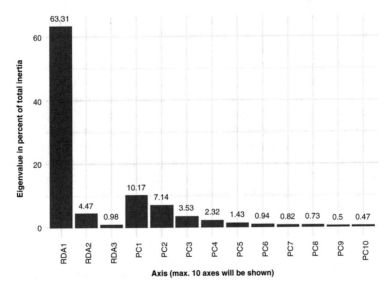

Fig. 10.13 Scree plot generated by RDA based on QTRT1 data

```
+        sample_colorframe = TRUE,
+        sample_colorframe_label = "Group4")
```

By specifying the detailed_output = TRUE argument, we obtain detailed output
and scree plot (Fig. 10.13):

```
> #Figure 10.13
> amp_ordinate(ds,
+        type = "RDA",
+        constrain = "Group4",
```

```
+        transform = "Hellinger",
+        sample_color_by = "Group4",
+        sample_shape_by = "Group4",
+        sample_colorframe = TRUE,
+        sample_colorframe_label = "Group4",
+        detailed_output = TRUE)
```

We print partial output as below. We can see the constrained axes explain 68.76% of total variances, while the unconstrained axes explain 31.24% of total variances. RDA1 can explain 63.31% (0.1527/0.2412) of total variances. Thus, the most variation of the data can be explained by the groups:

```
$model
Call: rda(formula = data$abund ~ Group4, data = data$metadata)

        Inertia Proportion Rank
Total          0.2412   1.0000
Constrained  0.1658   0.6876   3
Unconstrained 0.0753   0.3124   36
Inertia is variance

Eigenvalues for constrained axes:
 RDA1   RDA2   RDA3
0.1527 0.0108 0.0024

Eigenvalues for unconstrained axes:
  PC1    PC2    PC3    PC4    PC5    PC6    PC7    PC8
0.02452 0.01722 0.00851 0.00559 0.00346 0.00227 0.00197 0.00176
(Showing 8 of 36 unconstrained eigenvalues)
```

10.3.9   Canonical Correspondence Analysis (CCA)

CCA is the canonical form of CA. In CCA, any data table can be used for correspondence analysis to form a suitable response matrix Y for CCA: particularly, species/taxa presence-absence or abundance tables.

10.3.9.1   Introduction to CCA

CCA canonical asymmetric ordination method was developed by ter Braak (1986, 1987a, b). It was first implemented in the program CANOCO (ter Braak 1988a, b, 1990; ter Braak and Šmilauer 1998, 2011) and now is available in several R packages and functions including the ecological R package **vegan** and microbiome package **ampvis2**, which we use to illustrate CCA below.

CCA (ter Braak 1986) ushered in the biggest modern revolution in ordination methods. Previously ordination was mainly an "exploratory" method (Gauch,

Jr. 1982a, b), CCA technique coupled CA with regression methodologies, and hence it provides for hypothesis testing. Thus, ordination has been considered not mere "exploratory" analysis but also a hypothesis testing when canonical correspondence analysis (CCA) was introduced (ter Braak 1985a). CCA has been getting popular in community ecology since its introduction in 1986 (Wilmes and Bond 2004) and adopted to analyze microbiome data by microbiome researchers.

As a classical exploratory analysis method of contingency tables, CA (Benzécri 1973) is an unconstrained ordination technique, allowing for quantification of taxon contributions to the sample ordination, while CCA is a constrained analysis, in which the response variable set is constrained by the set of explanatory variables (ter Braak 1986), even allowing restricting the sample ordination to be explained by sample-specific variables.

Like **CA, CCA** is based on the *Pearson chi-squared* measure. Similar to RDA, CCA aims to find the relationship between two sets of variables. However, different from RDA which constructs a linear relationship among variables, CCA assumes a unimodal relationship and measures the separation based on the eigen values produced by CCA (Ram et al. 2005). In microbiome studies, we can use CCA to investigate taxa-environment relations, answering the specific questions about the response of taxa to environmental variables.

To distinguish CCA and CA from RDA and PCA, we can informally consider CCA and CA as more qualitative methods, while RDA and PCA as more quantitative methods. Thus, they are generally suitable to analyze different types of data: CCA, CA, and DCA are more suitable for analyzing the data with a unimodal distribution of the species (taxa/OTUs) abundances along the environmental gradient, while RDA and PCA are appropriate to the data showing a linear distribution.

CCA has several advantages, including:

1. Like DCA as a nonlinear ordination method, CCA is appropriate when the community variation is over a wider range (ter Braak and Prentice 1988), which is contrast to the linear ordination methods (PCA and RDA), which are appropriate when the community variation is within a narrow range. CCA and CA are chi-squared measure-based methods; theoretically they are more able to be used for ecological and microbiome data because these data are often sparse with many zeros with the fact that some species (taxa/OTUs) are present in some samples while absent in others.
2. CCA is the most powerful and is a much more practical technique in detecting relationships between species composition and environment and like other constrained methods particularly given the number of environmental variables is smaller than the number of sites (objects/samples) (ter Braak and Prentice 1988). Like its unconstrained version CA, CCA is very sensitive to the lower abundant and unique species (or taxa/OTUs) in the samples and is a powerful method to investigate which species (or taxa/OTUs) characterize each sample or group of samples.
3. CCA was evaluated as appropriate when it is used to describe how species/taxa respond to particular sets of observed environmental variables (McCune 1997).

However, like CA, CCA has also some disadvantages, including:

1. Like CA, CCA strongly assumes that species/taxa have unimodal response functions with respect to linear combinations of the environmental variables (ter Braak and Prentice 1988); that is, for each species/taxon, CCA expects its abundance shows a bell-shaped functional relationship with a particular sample along a gradient of environmental conditions (ter Braak 1986; Zhu et al. 2005).
2. As a residual-based method, CCA implicitly assumes a certain mean-variance relationship for normalized residuals. Such a stringent mean-variance assumption is too rigid to account for skewed data and the overdispersion of sequencing data such as microbiome data (Warton et al. 2012).
3. CCA can distort the representation of gradients in community structure when including noisy or irrelevant explanatory variables in the analysis and may mislead to misleading interpretations (McCune 1997).

10.3.9.2   Implement CCA

Example 10.9: Breast Cancer QTRT1 Mouse Gut Microbiome, Example 10.1 Cont.
We perform an interactive CCA with data transformation using Hellinger method and with Group4 as constraint variable. The plot generates sample points that are colored, framed, and labeled with groups and with different shapes for four groups (Fig. 10.14):

```
> #Figure 10.14
> amp_ordinate(ds,
+        type = "CCA",
+        constrain = "Group4",
+        transform = "Hellinger",
```

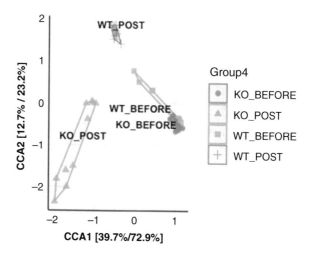

Fig. 10.14 Ordination plot for QTRT1 data generated by canonical correspondence analysis (CCA)

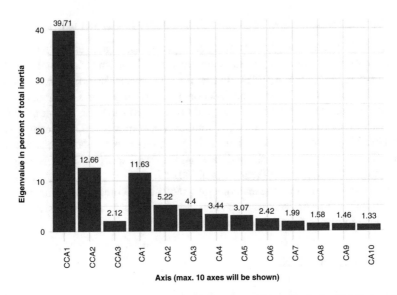

Fig. 10.15 Scree plot generated by canonical correspondence analysis (CCA) based on QTRT1 data

```
+           sample_color_by = "Group4",
+           sample_shape_by = "Group4",
+           sample_colorframe = TRUE,
+           sample_colorframe_label = "Group4")
```

Now we perform same CCA as Fig. 10.14. Additionally, we specify the detailed_output = TRUE to obtain detailed output information and scree plot (Fig. 10.15):

```
> #Figure 10.15
> amp_ordinate(ds,
+           type = "CCA",
+           constrain = "Group4",
+           transform = "Hellinger",
+           sample_color_by = "Group4",
+           sample_shape_by = "Group4",
+           sample_colorframe = TRUE,
+           sample_colorframe_label = "Group4",
+           detailed_output = TRUE)
```

The partial output is printed as below. We can see the constrained axes explain 54.5% of total variances, while the unconstrained axes explain 45.5% of total variances. CCA1 and CCA2 can explain 39.71% (0.2227/0.561) and 12.66%

(0.0710/0.561) of total variances, respectively. Thus, the most variation of the data can be explained by the groups:

```
$model
Call: cca(formula = data$abund ~ Group4, data = data$metadata)

          Inertia Proportion Rank
Total         0.561    1.000
Constrained   0.306    0.545  3
Unconstrained 0.255    0.455 36
Inertia is scaled Chi-square

Eigenvalues for constrained axes:
 CCA1   CCA2   CCA3
0.2227 0.0710 0.0119

Eigenvalues for unconstrained axes:
  CA1    CA2    CA3    CA4    CA5    CA6    CA7    CA8
0.0652 0.0292 0.0247 0.0193 0.0172 0.0136 0.0111 0.0088
(Showing 8 of 36 unconstrained eigenvalues)
```

10.4 Beta Diversity Metrics and Ordination in QIIME 2

As we described in Chap. 9, in QIIME 2, diversity analyses are available through the q2-diversity plugin. The analyses include calculating alpha and beta diversity, statistical testing group differences, and generating interactive visualizations. In this section, we illustrate calculations of beta diversity measures and ordination techniques in QIIME 2. In Chap. 11, we will illustrate how to perform statistical testing of beta diversity in QIIME 2 (Sect. 11.7).

10.4.1 Calculate Beta Diversity Measures

Example 10.10: Mouse Gut Microbiome Data, Example 9.12, Cont.
In Chap. 9 (Sect. 9.5.1), we generated all the alpha diversity and beta diversity measures available in QIIME 2 using a single-command qiime diversity **core-metrics-phylogenetic**. The calculated beta diversity measures include weighted/unweighted UniFrac, Bray-Curtis, and Jaccard. Here, we repeat this single command as below for recall:

```
source activate qiime2-2022.2
cd QIIME2R-Bioinformatics
qiime diversity core-metrics-phylogenetic \
  --i-phylogeny RootedTreeMiSeq_SOP.qza \
  --i-table FeatureTableMiSeq_SOP.qza \
```

```
--p-sampling-depth 1153 \
--m-metadata-file SampleMetadataMiSeq_SOP.tsv \
--output-dir CoreMetricsResults
```

The command generated several beta diversity artifacts and visualizations including:

- bray_curtis_distance_matrix.qza
- jaccard_distance_matrix.qza
- unweighted_unifrac_distance_matrix.qza,
- weighted_unifrac_distance_matrix.qza
- bray_curtis_pcoa_results.qza
- jaccard_pcoa_results.qza
- unweighted_unifrac_pcoa_results.qza
- weighted_unifrac_pcoa_results.qza
- bray_curtis_emperor.qzv
- jaccard_emperor.qzv
- unweighted_unifrac_emperor.qzv
- weighted_unifrac_emperor.qzv

10.4.2 *Explore Principal Coordinates (PCoA) Using Emperor Plots*

The PCoA plots are automatically generated along with beta diversity measures. These plots display a three-dimensional ordination and can be viewed. Figure 10.16 shows a static example of the interactive principal coordinates visualization of Bray-Curtis distance.

As recall, the **core-metrics-phylogenetic** command has already generated some emperor plots. When performing the qiime **emperor plot** command, we can specify an optional parameter --p-custom-axes to explore time-series data (in this case, DPW: days post weaning).

The qiime **emperor plot** command will use the PCoA results generated in **core-metrics-phylogeny** command. The PCoA results generated by the qiime **emperor plot** command will contain Axe 1 for principal coordinate 1 and Axe 2 for principal coordinate 2, respectively, and a third Axe (DPW: overlapped with Axe 2) for days post weaning. The third axis is used to explore how these samples changed over time.

The following qiime **emperor plot** command generates emperor plots for Bray-Curtis distance (Fig. 10.17):

```
# Figure 10.17
qiime emperor plot \
  --i-pcoa CoreMetricsResults/bray_curtis_pcoa_results.qza \
```

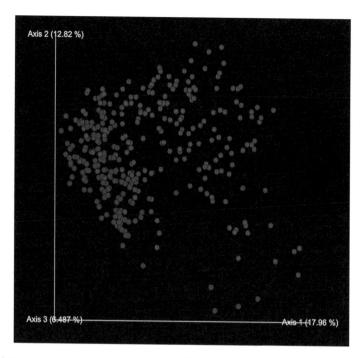

Fig. 10.16 A two-dimensional principal coordinate analysis ordination of the mouse gut microbiome samples based on Bray-Curtis distance. Red color labels early time and blue labels late time

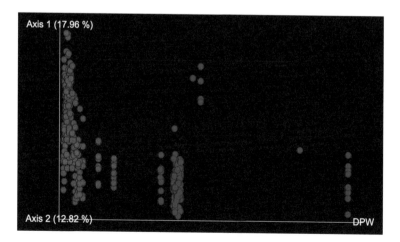

Fig. 10.17 Emperor plots for Bray-Curtis distance in mouse gut microbiome samples

```
--m-metadata-file SampleMetadataMiSeq_SOP.tsv \
--p-custom-axes DPW \
--o-visualization CoreMetricsResults/BrayCurtisEmperorDPW.qzv
```

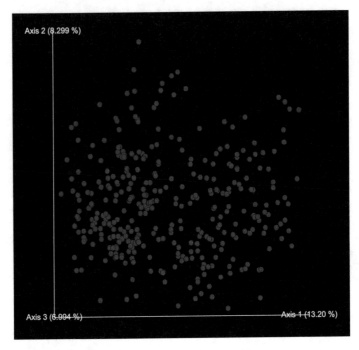

Fig. 10.18 Emperor plots for unweighted UniFrac distance in mouse gut microbiome samples

Saved Visualization to: CoreMetricsResults/BrayCurtisEmperorDPW.qzv

The following qiime **emperor plot** command generates emperor plots for unweighted UniFrac distance (Fig. 10.18):

```
# Figure 10.18
qiime emperor plot \
   --i-pcoa CoreMetricsResults/unweighted_unifrac_pcoa_results.qza \
   --m-metadata-file SampleMetadataMiSeq_SOP.tsv \
   --p-custom-axes DPW \
   --o-visualization CoreMetricsResults/UnweightedUniFracEmperorDPW.
qzv
```

Saved Visualization to:
CoreMetricsResults/UnweightedUniFracEmperorDPW.qzv

10.5   Remarks on Ordination and Clustering

Both clustering and ordination are often used in multivariate analysis. They are important components of a strategy for analyzing community data on community structure (multispecies distribution patterns) (Field et al. 1982; Clarke 1993).

Clustering most commonly represents the results by a dendrogram/phenogram, which is useful for summarizing the taxonomic relationships, particularly when the OTUs are well clustered and fit readily into a hierarchy (Sneath and Sokal 1973, p. 303), while ordination produces summaries of the variation in A-space that differ greatly in appearance and may differ in the taxonomic results to which they lead. Thus ordinations are especially valuable for understanding the taxonomic structure in more detail, which are preferably in three dimensions or as several two-dimensional diagrams (Sneath and Sokal 1973, p. 303).

Sneath and Sokal (1973) and Legendre and Legendre (1998, 2012) among others have provided an excellent review on the general performances of clustering and ordination methods in the numerical taxonomy and numerical ecology literatures. Since the materials have not been largely available in current microbiome literature, we summarize their main points as follows. Overall, both clustering and ordination methods have their own advantages and disadvantages, and no methods have satisfactorily used the similarity matrix itself to tell us whether clustering or ordination is most appropriate.

On the One Hand, Cluster Methods Have Their Advantages and Disadvantages
The merits of many clustering methods have been discussed at length in Section 5.4 "A Taxonomy of Clustering Methods" and Section 5.5 "Sequential Agglomerative, Hierarchic, Nonoverlapping Clustering Methods" (Sneath and Sokal 1973, pp. 201–245). For example, a cophenetic correlation may suggest that a phenogram is a reasonable representation of well-clustered phenetic distributions (Sneath and Sokal 1973), and dendrograms have the advantage of simplicity (Field et al. 1982). However, cluster methods have the disadvantages, such as the following: (1) They will yield clusters of some kind, whatever the structure of the data, even if the OTU distributions are random (Sneath and Sokal 1973). (2) It was shown that phenograms give poor representations of relationships between major clusters (Rohlf 1967, 1970). (3) The flexible sorting method does not very well cluster much material and express a preference for accentuating clustering (Williams and Lance 1968). In particular, dendrograms have been reviewed suffering four main disadvantages (Field et al. 1982): (1) The hierarchy is irreversible – once a sample has been placed in a group, its identity is lost. (2) Dendrograms only show inter-group relationships, in which the level of similarity indicated is the average inter-group value. (3) The sequence of samples (objects/individuals) in a dendrogram is arbitrary, and two adjacent samples are not necessarily the most similar. And (4) dendrograms tend to overemphasize discontinuities and may force a graded series into discrete classes.

On the Other Hand, Ordination Methods Also Have Their Advantages and Disadvantages
The advantages and disadvantages of individual ordination methods have been discussed at length in Sects. 10.3.3 for PCA, 10.3.4 for PCoA, 10.3.5 for NMDS, 10.3.6 for CA, 10.3.7 for DCA, 10.3.8 for RDA, and 10.3.9 for CCA. Here, we summarize some general discussions and main points about the disadvantages of ordination methods.

First, regarding what axes in an ordination are taxonomically meaningful is debatable in the numerical taxonomy literature. The first vector of a principal coordinate analysis was either omitted (Sheals 1965) or was thought as taxonomically meaningful (Sims 1966). Others thought that the first vector (Schnell 1970) or the fourth vector might represent a general size factor or reflected size (Seki 1969). Thus, there is no reason why the major axes of taxonomic variation should necessarily lie parallel to the direction that represents size (Sneath and Sokal 1973).

Second, the general disadvantages of ordination methods have been discussed, including the following: (1) Ordination may give no simple low-dimensional result with many OTUs and clusters (Williams and Lance 1968). (2) Clusters might have overlapped in two or three spaces although they are quite distinct in hyperspace. (3) Ordination may also be impracticable for very large numbers of characters (variables) and OTUs. (4) In particular, it was shown that like most clustering methods, Gower's principal coordinates ordination method revealed the major structure of community data, but the finer divisions (known to be ecologically significant) were less evident in ordinations than in clustering (Webb et al. 1967). This finding is in line with the observation that distances between close neighbors are not well represented by PCA ordination (Rohlf 1968).

Because both clustering and ordination methods have their own advantages and disadvantages, to leverage and to take advantage of the characteristics of clustering and ordination methods, combining the results from these two types of analyses on the same diagram using the same similarity/dissimilarity (or distance) matrix has been independently proposed in the numerical taxonomy and numerical ecology (Gower and Ross 1969; Rohlf 1970; Schnell 1970; Jackson and Crovello 1971; Sneath and Sokal 1973; Legendre 1976; Field et al. 1982; Clarke 1993; Legendre and Legendre 2012).

Whether sequentially performing ordination and clustering is also debatable without determinate conclusions (Sneath and Sokal 1973). On the one hand, first performing an ordination and then clustering on the reduced space is generally desirable, but information of largely unknown nature and quality has been lost during the ordination, and very low-order dimensionality cannot express much of the multidimensional phenetic relationships. Actually, it was reviewed (Sneath and Sokal 1973) that clustering on the reduced space by an ordination does not necessarily improve clustering: the analysis results are either very similar or identical to the full-dimensional space and even appear less satisfactory. On the other hand, it was shown (Hill et al. 1965) that clustering on the reduced dimensions is satisfactory, which provides good reasons for first clustering taxonomic groupings from the full A-space and then using ordinations for investigating the general pattern of variation (Sneath and Sokal 1973).

In numerical taxonomy literature, both clustering and ordination have been recommended either preferring to cluster first and then make an ordination (Williams and Lance 1968) or preferring ordination methods first or no specific sequencing order. In general, PCA and PCoA are recommended, while NMDS method may be preferred particularly when there are scale problems because it often gives results very comparable to PCA (Sneath and Sokal 1973, p. 305).

In ecological literature, classification (clustering) and ordination have been used and recommended as two sequential stages for analyzing community data on community structure or multispecies distribution patterns (Field et al. 1982). Clustering and ordination are commentary to each other. Thus, it can unambiguously confirm the inadequacy of a two-dimensional description if they display inconsistent results (Clarke 1993). In general, principal component analysis (PCA), principal coordinate analysis (PCoA), nonmetric multidimensional scaling (NMDS), and correspondence analysis (CA) are most often used for ordination of species in ecology. For example, under the title "Ordination in reduced space," PCA, PCoA, NMDS, and CA are introduced in various versions of Legendre and Legendre's numerical ecology (Legendre and Legendre 2012). The metric multidimensional scaling (MDS) ordination techniques (essentially PCoA) and in particular nonmetric MDS, which is much more flexible, have been recommended for implementing ordination (Field et al. 1982).

A cutoff value for ordination to be a satisfactory representation of taxonomic structure was evaluated as follows (Sneath and Sokal 1973, p. 304): (1) if the cophenetic correlation coefficient (or other suitable criterion) between the ordination and the original similarity matrix is ≥ 0.8, then it is high and desirable; (2) if the reduced dimensions displayed accounted for $\geq 40\%$ proportion of the variation, then the ordination may or may not be appreciable; and (3) if an ordination accounts for less than 40% proportion of the variation, then this ordination is unsatisfactory.

The cutoff value of 0.8 is also applied to clustering. It was reviewed as a cophenetic correlation is high (e.g., over 0.8), then it indicates that a phenogram adequately represents taxonomic structure and thus the phenogram is likely to be fairly satisfactory in this respect (Sneath and Sokal 1973, p. 304).

For the details on comparisons of overall performances of clustering and ordination methods, the readers are referred to Sneath and Sokal (1973, pp. 251–253), Legendre and Legendre (1998, pp. 482–486), and Legendre and Legendre (2012, pp. 522–526).

Here, we emphasize that microbiome data share some common characteristics with numerical taxonomy and numerical ecology, but microbiome study is essentially an experimental science; its definitions (Xia and Sun 2022a, b), research themes (Xia and Sun 2017; Xia 2020), and data characteristics (Xia et al. 2018c; Xia and Sun 2022a, b) are more complicated. Microbiome is dynamically integrated with environment and host as well as other omics such as metabolome and changes over time and the conditions of host and treatment. Thus, the goals of microbiome research are not only to identify the pattern of community data structure or multiple taxonomic pattern but also to find functions and mechanism of microbiome and/or core species/taxa, which we can use to prevent and treat diseases. Therefore, we think the statistical hypothesis testing methods and particularly the statistical methods that are suitable to address the unique characteristics of microbiome data (e.g., multivariate, overdispersed, and zero-inflated) are more important in microbiome study, while considering both clustering and ordination as exploratory techniques, which can visualize data pattern and facilitate hypothesis testing.

10.6   Summary

In this chapter, we first introduced three most common used abundance-based beta diversity metrics and their calculations, Bray-Curtis, Jaccard, and Sørensen dissimilarities, and three phylogenetic-based beta diversity metrics: unweighted, weighted, and generalized UniFrac. We then introduced ordination methods, including brief history of ordination methods, ordination plots in the ampvis2 package, and introduction and illustration of five unconstrained ordinations (PCA, PCoA, NMDS, CA, DCA) and two constrained ordinations (RDA, CCA) through the ampvis2 package. We next illustrated the calculations of beta diversity metrics using core-metrics-phylogenetic method and visualized principal coordinates using emperor plots in QIIME 2. We finally conducted some general remarks on ordination and clustering. In Chap. 11, we will introduce statistical hypothesis testing of beta diversity with applications in R and QIIME 2.

References

Andersen, Kasper Skytte, and Mads Albertsen. 2021. *Ordination plot*. https://madsalbertsen.github.io/ampvis2/reference/amp_ordinate.html.

Anderson, D.J. 1965. Classification and ordination in vegetation science: Controversy over a non-existent problem? *Journal of Ecology* 53 (2): 521–526. https://doi.org/10.2307/2257992. http://www.jstor.org/stable/2257992.

Anderson, M.J. 2001. A new method for non-parametric multivariate analysis of variance. *Austral Ecology* 26: 32–46.

Arumugam, Manimozhiyan, Jeroen Raes, Eric Pelletier, Denis Le Paslier, Takuji Yamada, Daniel R. Mende, Gabriel R. Fernandes, Julien Tap, Thomas Bruls, Jean-Michel Batto, Marcelo Bertalan, Natalia Borruel, Francesc Casellas, Leyden Fernandez, Laurent Gautier, Torben Hansen, Masahira Hattori, Tetsuya Hayashi, Michiel Kleerebezem, Ken Kurokawa, Marion Leclerc, Florence Levenez, Chaysavanh Manichanh, H. Bjørn Nielsen, Trine Nielsen, Nicolas Pons, Julie Poulain, Junjie Qin, Thomas Sicheritz-Ponten, Sebastian Tims, David Torrents, Edgardo Ugarte, Erwin G. Zoetendal, Jun Wang, Francisco Guarner, Oluf Pedersen, Willem M. de Vos, Søren Brunak, Joel Doré, María Antolín, François Artiguenave, Hervé M. Blottiere, Mathieu Almeida, Christian Brechot, Carlos Cara, Christian Chervaux, Antonella Cultrone, Christine Delorme, Gérard Denariaz, Rozenn Dervyn, Konrad U. Foerstner, Carsten Friss, Maarten van de Guchte, Eric Guedon, Florence Haimet, Wolfgang Huber, Johan van Hylckama-Vlieg, Alexandre Jamet, Catherine Juste, Ghalia Kaci, Jan Knol, Karsten Kristiansen, Omar Lakhdari, Severine Layec, Karine Le Roux, Emmanuelle Maguin, Alexandre Mérieux, Raquel Melo Minardi, Christine M'Rini, Jean Muller, Raish Oozeer, Julian Parkhill, Pierre Renault, Maria Rescigno, Nicolas Sanchez, Shinichi Sunagawa, Antonio Torrejon, Keith Turner, Gaetana Vandemeulebrouck, Encarna Varela, Yohanan Winogradsky, Georg Zeller, Jean Weissenbach, S. Dusko Ehrlich, Peer Bork, and H. I. T. Consortium Meta. 2011. Enterotypes of the human gut microbiome. *Nature* 473 (7346): 174–180. https://doi.org/10.1038/nature09944.

Baselga, Andres, and C. David L. Orme. 2012. betapart: An R package for the study of beta diversity. *Methods in Ecology and Evolution* 3: 808–812. https://doi.org/10.1111/j.2041-210X.2012.00224.x.

Bengtsson, Henrik. 2020. matrixStats: Functions that apply to rows and columns of matrices (and to vectors). *R package version 0.57.0*. https://github.com/HenrikBengtsson/matrixStats.

Benzécri, J.-P. 1973. *L'Analyse des Données. Volume II. L'Analyse des Correspondances*. Paris: Dunod.

Borg, Ingwer, and Patrick J.F. Groenen. 2005. Methods related to MDS. In *Modern multidimensional scaling: Theory and applications*, ed. Ingwer Borg and Patrick J.F. Groenen, 519–540. New York: Springer.

Bork, Peer. 2021. *Enterotyping: The original publication*. https://enterotype.embl.de/enterotypes. html#. Accessed 17 Mar 2021.

Bray, J. Roger, and John T. Curtis. 1957. An ordination of the upland forest communities of southern Wisconsin. *Ecological Monographs* 27 (4): 325–349. https://doi.org/10.2307/1942268. https://esajournals.onlinelibrary.wiley.com/doi/abs/10.2307/1942268.

Buttigieg, Pier Luigi, and Alban Ramette. 2014. A guide to statistical analysis in microbial ecology: A community-focused, living review of multivariate data analyses. *FEMS Microbiology Ecology* 90 (3): 543–550. https://doi.org/10.1111/1574-6941.12437.

Chen, Jun, Kyle Bittinger, Emily S. Charlson, Christian Hoffmann, James Lewis, Gary D. Wu, Ronald G. Collman, Frederic D. Bushman, and Hongzhe Li. 2012. Associating microbiome composition with environmental covariates using generalized UniFrac distances. *Bioinformatics (Oxford, England)* 28 (16): 2106–2113. https://doi.org/10.1093/bioinformatics/bts342. https://pubmed.ncbi.nlm.nih.gov/22711789, https://www.ncbi.nlm.nih.gov/pmc/articles/PMC3413390/.

Chen, Jun, Xianyang Zhang, and Lu Yang. 2022. GUniFrac: Generalized UniFrac distances. *R package version 1.1*. https://CRAN.R-project.org/package=GUniFrac. Last Modified 2022-04-05.

Clarke, K. Robert. 1993. Non-parametric multivariate analyses of changes in community structure. *Australian Journal of Ecology* 18 (1): 117–143.

de Cárcer, Daniel, Stuart E. Aguirre, Chris McSweeney Denman, and Mark Morrison. 2011. Evaluation of subsampling-based normalization strategies for tagged high-throughput sequencing data sets from gut microbiomes. *Applied and Environmental Microbiology* 77 (24): 8795–8798.

Ellison, A.M. 2010. Partitioning diversity. *Ecology* 91 (7): 1962–1963. http://www.ncbi.nlm.nih.gov/pubmed/20715615.

Faith, Daniel P., Peter R. Minchin, and Lee Belbin. 1987. Compositional dissimilarity as a robust measure of ecological distance. *Vegetatio* 69 (1): 57–68.

Fasham, M.J.R. 1977. A comparison of nonmetric multidimensional scaling, principal components and reciprocal averaging for the ordination of simulated coenoclines, and coenoplanes. *Ecology* 58 (3): 551–561.

Field, J.G., and G. McFarlane. 1968. Numerical methods in marine ecology. 1. A quantitative 'similarity' analysis of rocky shore samples in False Bay, South Africa. *African Zoology* 3 (2): 119–137.

Field, J.G., K.R. Clarke, and R.M. Warwick. 1982. A practical strategy for analysing multispecies distribution patterns. *Marine Ecology Progress Series* 8: 37–52.

Fisher, Ronald A. 1940. The precision of discriminant functions. *Annals of Eugenics* 10 (1): 422–429.

Gauch, Hugh G. 1977. *ORDIFLEX: A flexible computer program for four ordination techniques: Weighted averages, polar ordination, principal components analysis, and reciprocal averaging, Release B*. Ithaca: Ecology and Systematics, Cornell University.

Gauch, Hugh G., Jr., Robert H. Whittaker, and Steven B. Singer. 1981. A comparative study of nonmetric ordinations. *The Journal of Ecology*: 135–152.

Gauch, H.G., Jr. 1982a. Noise reduction by eigenvalue ordinations. *Ecology* 63: 1643–1649.

———. 1982b. *Multivariate analysis and community structure*. Cambridge: Cambridge University Press.

Gifi, Albert. 1990. *Nonlinear multivariate analysis*. Chichester: Wiley-Blackwell.

Gleason, H.A. 1926. The individualistic concept of the plant association. *Bulletin of the Torrey Botanical Club* 53 (1): 7–26. https://doi.org/10.2307/2479933. http://www.jstor.org/stable/24 79933.

Goodall, David William. 1954a. Vegetational classification and vegetational continua. *(Germ. summ.)*. Vol. Angew. PflSoziol., Wein, Festschr. Aichinger, I, 168–182.

Goodall, D.W. 1954b. Objective methods for the classification of vegetation. III. An essay in the use of factor analysis. *Australian Journal of Botany* 2 (3): 304–324.

Gower, J.C. 1966. Some distance properties of latent root and vector methods used in multivariate analysis. *Biometrika* 53 (3/4): 325–338. https://doi.org/10.2307/2333639. http://www.jstor.org. proxy.cc.uic.edu/stable/2333639.

———. 1967. Multivariate analysis and multidimensional geometry. *Journal of the Royal Statistical Society. Series D (The Statistician)* 17 (1): 13–28. https://doi.org/10.2307/2987199. http:// www.jstor.org/stable/2987199.

———. 1971. A general coefficient of similarity and some of its properties. *Biometrics* 27 (4): 857–871. https://doi.org/10.2307/2528823. www.jstor.org/stable/2528823.

———. 1984. Multivariate analysis: Ordination, multidimensional scaling and allied topics. In *Handbook of applicable mathematics*, ed. W. Lederman, 727–781. Chichester: Wiley.

Gower, John C. 1992. Generalized biplots. *Biometrika* 79 (3): 475–493.

———. 2005. Principal coordinates analysis. In *Encyclopedia of biostatistics*. Hoboken: Wiley.

Gower, John C., and Gavin J.S. Ross. 1969. Minimum spanning trees and single linkage cluster analysis. *Journal of the Royal Statistical Society: Series C: Applied Statistics* 18 (1): 54–64.

Greenacre, Michael J. 1984. *Theory and applications of correspondence analysis*. London: Academic Press.

———. 1988. Correspondence analysis of multivariate categorical data by weighted least-squares. *Biometrika* 75 (3): 457–467.

Guttman, Louis. 1941. The quantification of a class of attributes: A theory and method of scale construction. In *The prediction of personal adjustment*. New York: Social Science Research Council.

Hatheway, W.H. 1971. *Contingency-table analysis of rain forest vegetation*. International symposium on statistical ecology, New Haven, 1969.

Healy, M.J.R., and Harvey Goldstein. 1976. An approach to the scaling of categorized attributes. *Biometrika* 63 (2): 219–229.

Heiser, Willem J. 1987. Joint ordination of species and sites: the unfolding technique. In *Developments in numerical ecology*, 189–221. Berlin: Springer.

Hill, Mark O. 1973. Reciprocal averaging: An eigenvector method of ordination. *The Journal of Ecology*: 237–249.

———. 1974. Correspondence analysis: A neglected multivariate method. *Journal of the Royal Statistical Society: Series C: Applied Statistics* 23 (3): 340–354.

———. 1979. *DECORANA. A Fortran program for detrended correspondence analysis and reciprocal averaging*, Section of ecology and systematics. Ithaca: Cornell University.

Hill, Mark O., and Hugh G. Gauch. 1980. Detrended correspondence analysis: An improved ordination technique. In *Classification and ordination*, 47–58. New York: Springer.

Hill, L.R., L.G. Silvestri, P. Ihm, G. Farchi, and P. Lanciani. 1965. Automatic classification of staphylococci by principal-component analysis and a gradient method. *Journal of Bacteriology* 89 (5): 1393–1401.

Hirschfeld, H.O. 1935. A connection between correlation and contingency. *Mathematical Proceedings of the Cambridge Philosophical Society* 31 (4): 520–524. https://doi.org/10.1017/ s0305004100013517. https://www.cambridge.org/core/article/connection-between-correlation-and-contingency/D3A75249B56AF5DDC436938F1B6EABD1.

Hotelling, Harold. 1933. Analysis of a complex of statistical variables into principal components. *Journal of Educational Psychology* 24 (6): 417.

———. 1936. Relations between two sets of variates. *Biometrika* 28 (3/4): 321–377. https://doi. org/10.2307/2333955. http://www.jstor.org.proxy.cc.uic.edu/stable/2333955.

Hutchinson, G.E. 1957. Concluding remarks. *Cold Spring Harbor Symposia on Quantitative Biology* 22. https://doi.org/10.1101/SQB.1957.022.01.039.

Jaccard, P. 1900. Contribution au problème de l'immigration post-glaciare de la flore alpine. *Bulletin de la Société vaudoise des sciences naturelles* 36: 87–130. https://www.scopus.com/inward/record.uri?eid=2-s2.0-0013080309&partnerID=40&md5=6424a0a1835a5322485933 58d57a0824.

Jaccard, Paul. 1912. The distribution of the flora in the alpine zone.1. *New Phytologist* 11 (2): 37–50. https://doi.org/10.1111/j.1469-8137.1912.tb05611.x. https://nph.onlinelibrary.wiley.com/doi/abs/10.1111/j.1469-8137.1912.tb05611.x.

Jackson, R.C., and Theodore J. Crovello. 1971. A comparison of numerical and biosystematic studies in Haplopappus. *Brittonia* 23 (1): 54–70.

Jari Oksanen, F.G.B., Michael Friendly, Roeland Kindt, Pierre Legendre, Dan McGlinn, Peter R. Minchin, R.B. O'Hara, Gavin L. Simpson, Peter Solymos, and M. Henry H Stevens. 2018. Vegan: Community ecology package. *R Package Version 2 (6)*.

Jari Oksanen, F. Guillaume Blanchet, Michael Friendly, Roeland Kindt, Pierre Legendre, Peter R. Minchin, McGlinn Dan, R.B. O'Hara, Gavin L. Simpson, Solymos Peter, M. Henry, Eduard Szoecs, and Helene Wagner H. Stevens. 2020. vegan: Community ecology package. *R package version 2.5-7*. https://CRAN.R-project.org/package=vegan.

Jolliffe, I. 2002. *Principal component analysis*. New York: Springer.

Kendall, D.G. 1971. Seriation from abundance matrices. In *Mathematics in archaeologival and historical sciences*, ed. C.R. Hodson, D.G. Kendall, and P. Tautu. Edimburgh: Edimburgh University Press.

Kenkel, Norm C., and Laszlo Orlóci. 1986. Applying metric and nonmetric multidimensional scaling to ecological studies: Some new results. *Ecology* 67 (4): 919–928.

Koleff, Patricia, Kevin J. Gaston, and Jack J. Lennon. 2003. Measuring beta diversity for presence–absence data. *Journal of Animal Ecology* 72 (3): 367–382. https://doi.org/10.1046/j.1365-2656.2003.00710.x. https://besjournals.onlinelibrary.wiley.com/doi/abs/10.1046/j.1365-2656.2003.00710.x.

Kruskal, J.B. 1964a. Multidimensional scaling by optimizing goodness of fit to a nonmetric hypothesis. *Psychometrika* 29 (1): 1–27. https://doi.org/10.1007/bf02289565.

———. 1964b. Nonmetric multidimensional scaling: A numerical method. *Psychometrika* 29 (2): 115–129. https://doi.org/10.1007/bf02289694.

Lebart, Ludovic, Alain Morineau, and Kenneth M. Warwick. 1984. *Multivariate descriptive statistical analysis; Correspondence analysis and related techniques for large matrices*. New York: Wiley.

Legendre, Pierre. 1976. An appropriate space for clustering selected groups of Western North American Salmo. *Systematic Zoology* 25 (2): 193–195.

———. 2019. Numerical ecology. In *Encyclopedia of ecology*, ed. Brian Fath, 2nd ed., 487–493. Oxford: Elsevier.

Legendre, Pierre, and Eugene D. Gallagher. 2001. Ecologically meaningful transformations for ordination of species data. *Oecologia* 129 (2): 271–280. https://doi.org/10.1007/s004420100716.

Legendre, P., and L. Legendre. 1998. Numerical ecology. 2nd English ed. Amsterdam: Elsevier.

———. 2012. *Numerical ecology*. 3rd English Edition ed., Vol. 24. Amsterdam: Elsevier.

Lozupone, Catherine, and Rob Knight. 2005. UniFrac: A new phylogenetic method for comparing microbial communities. *Applied and Environmental Microbiology* 71 (12): 8228–8235. https://doi.org/10.1128/AEM.71.12.8228-8235.2005. https://pubmed.ncbi.nlm.nih.gov/16332807, https://www.ncbi.nlm.nih.gov/pmc/articles/PMC1317376/.

Lozupone, Catherine A., Micah Hamady, Scott T. Kelley, and Rob Knight. 2007. Quantitative and qualitative beta diversity measures lead to different insights into factors that structure microbial communities. *Applied and Environmental Microbiology* 73 (5): 1576–1585. https://doi.org/10.1128/AEM.01996-06. https://pubmed.ncbi.nlm.nih.gov/17220268, . https://www.ncbi.nlm.nih.gov/pmc/articles/PMC1828774/.

Marion, Zachary, James Fordyce, and Benjamin Fitzpatrick. 2017. Pairwise beta diversity resolves an underappreciated source of confusion in calculating species turnover. *Ecology* 98. https://doi.org/10.1002/ecy.1753.

McCune, Bruce. 1997. Influence of noisy environmental data on canonical correspondence analysis. *Ecology* 78 (8): 2617–2623.

McMurdie, Paul J., and Susan Holmes. 2014. Waste not, want not: why rarefying microbiome data is inadmissible. *PLoS Computational Biology* 10 (4): e1003531.

Mead, A. 1992. Review of the development of multidimensional scaling methods. *Journal of the Royal Statistical Society. Series D (The Statistician)* 41 (1): 27–39. https://doi.org/10.2307/2348634. http://www.jstor.org/stable/2348634.

Meulman, Jacqueline. 1982. *Homogeneity analysis of incomplete data*. Vol. 1. Leiden: Dswo Press.

Minchin, Peter R. 1987. An evaluation of the relative robustness of techniques for ecological ordination. In *Theory and models in vegetation science*, 89–107. Dordrecht: Springer.

Minchin, Peter R., and Jari Oksanen. 2015. Statistical analysis of ecological communities: Progress, status, and future directions. *Plant Ecology* 216 (5): 641–644.

Nishisato, Shizuhiko. 1980. *Analysis of categorical data – Dual scaling and its applications*, Mathematical expositions No. 24. Toronto: University of Toronto Press.

———. 1984. Forced classification: A simple application of a quantification method. *Psychometrika* 49 (1): 25–36.

Oksanen, Jari. 2020. *vegdist: Dissimilarity indices for community ecologists in package 'vegan': Community ecology package*. Last Modified November 28, 2020, https://cran.r-project.org, https://github.com/vegandevs/vegan. Accessed 18 Mar 2021.

Oksanen, Jari, and Tiina Tonteri. 1995. Rate of compositional turnover along gradients and total gradient length. *Journal of Vegetation Science* 6 (6): 815–824. https://doi.org/10.2307/3236395.

Orlóci, László. 2013. *Multivariate analysis in vegetation research*. Cham: Springer.

Paliy, O., and V. Shankar. 2016. Application of multivariate statistical techniques in microbial ecology. *Molecular Ecology* 25 (5): 1032–1057. https://doi.org/10.1111/mec.13536.

Paradis, Emmanuel, and Klaus Schliep. 2019. ape 5.0: An environment for modern phylogenetics and evolutionary analyses in R. *Bioinformatics* 35 (3): 526–528.

Pearson, Karl. 1901. LIII. On lines and planes of closest fit to systems of points in space. *The London, Edinburgh, and Dublin Philosophical Magazine and Journal of Science* 2 (11): 559–572. https://doi.org/10.1080/14786440109462720.

Pielou, Evelyn Chris. 1984. *The interpretation of ecological data: A primer on classification and ordination*. New York: Wiley.

Plantinga, Anna, Xiang Zhan, Ni Zhao, Jun Chen, Robert R. Jenq, and C.Wu. Michael. 2017. MiRKAT-S: A community-level test of association between the microbiota and survival times. *Microbiome* 5 (1): 17–17. https://doi.org/10.1186/s40168-017-0239-9. https://www.ncbi.nlm.nih.gov/pubmed/28179014, https://www.ncbi.nlm.nih.gov/pmc/PMC5299808/.

Plantinga, Anna M., Jun Chen, Robert R. Jenq, and Michael C. Wu. 2019. pldist: Ecological dissimilarities for paired and longitudinal microbiome association analysis. *Bioinformatics* 35 (19): 3567–3575. https://doi.org/10.1093/bioinformatics/btz120.

Prentice, Iain Colin. 1977. Non-metric ordination methods in ecology. *The Journal of Ecology*: 85–94.

Ram, R.J., N.C. Verberkmoes, M.P. Thelen, G.W. Tyson, B.J. Baker, R.C. Blake, M. Shah, R.L. Hettich, and J.F. Banfield. 2005. Community proteomics of a natural microbial biofilm. *Science* 308 (5730): 1915.

Ramensky, L.G. 1930. Zur Methodik der vergleichenden Bearbaitung und Ordnung von Pflanzen und anderen Objekten, die druch mehrere, verschiedenartig wirkende Factoren bestimmt werden. *Beiträge zur Biologie der Pflanzen* 18: 269–304.

Rao, C. Radhakrishna. 1964. The use and interpretation of principal component analysis in applied research. *Sankhyā: The Indian Journal of Statistics, Series A (1961–2002)* 26 (4): 329–358. http://www.jstor.org/stable/25049339.

Richardson, Marion Webster. 1938. Multidimensional psychophysics. *Psychological Bulletin* 35: 659–660.

Richardson, M.W., and G.F. Kuder. 1933. Making a rating scale that measures. *The Personnel Journal* 12: 36–40.

Rohlf, F. James. 1967. Correlated characters in numerical taxonomy. *Systematic Zoology* 16 (2): 109–126.

———. 1968. Stereograms in numerical taxonomy. *Systematic Biology* 17 (3): 246–255.

———. 1970. Adaptive hierarchical clustering schemes. *Systematic Biology* 19 (1): 58–82.

———. 1972. An empirical comparison of three ordination techniques in numerical taxonomy. *Systematic Biology* 21 (3): 271–280. https://doi.org/10.1093/sysbio/21.3.271.

Schnell, Gary D. 1970. A phenetic study of the suborder Lari (Aves) I. Methods and results of principal components analyses. *Systematic Biology* 19 (1): 35–57.

Seki, Tarow. 1969. A revision of the family Sematophyllaceae of Japan with special reference to a statistical demarcation of the family. *Journal of Science of the Hiroshima University* 2: 1–80.

Sheals, J.G. 1965. The application of computer techniques to Acarine taxonomy: A preliminary examination with species of the Hypoaspis-Androlaelaps complex (Acarina). *Proceedings of the Linnean Society of London* 176: 11.

Shepard, Roger N. 1962. The analysis of proximities: Multidimensional scaling with an unknown distance function. I. *Psychometrika* 27 (2): 125–140. https://doi.org/10.1007/BF02289630.

———. 1966. Metric structures in ordinal data. *Journal of Mathematical Psychology* 3 (2): 287–315.

Sims, R.W. 1966. The classification of the Megascolecoid earthworms: An investigation of Oligochaete systematics by computer techniques. *Proceedings of the Linnean Society of London* 177: 125.

Sneath, Peter H.A., and Robert R. Sokal. 1973. *Numerical taxonomy. The principles and practice of numerical classification*. San Francisco: W. H. Freeman and Company.

Sokal, Robert R. 1986. Phenetic taxonomy: Theory and methods. *Annual Review of Ecology and Systematics* 17 (1): 423–442.

Sokal, R.R., and P.H.A. Sneath. 1963. *Principles of numerical taxonomy*, xvi. San Francisco: W. H. Freeman and Company.

Sørensen, Thorvald Julius. 1948. *A method of establishing groups of equal amplitude in plant sociology based on similarity of species content and its application to analyses of the vegetation on Danish commons*. Copenhagen: I kommission hos E. Munksgaard.

ter Braak, C.J.F. 1985a. *CANOCO – A FORTRAN program for canonical correspondence analysis and detrended correspondence analysis*. Wageningen: IWIS-TNO.

ter Braak, Cajo J.F. 1985b. Correspondence analysis of incidence and abundance data: Properties in terms of a unimodal response model. *Biometrics*: 859–873.

———. 1986. Canonical correspondence analysis: A new eigenvector technique for multivariate direct gradient analysis. *Ecology* 67 (5): 1167–1179. https://doi.org/10.2307/1938672. https://esajournals.onlinelibrary.wiley.com/doi/abs/10.2307/1938672.

ter Braak, C.J.F. 1987a. Ordination. In *Data analysis in community and landscape ecology*, ed. C.J.F. ter Braak, O.F.R. van Tongeren, and R.H.G. Jongman, 91–274. Cambridge: Reissued in 1995 by Cambridge University Press.

ter Braak, Cajo J.F. 1987b. The analysis of vegetation-environment relationships by canonical correspondence analysis. *Vegetatio* 69 (1): 69–77.

———. 1987c. *Unimodal models to relate species to environment*. Wageningen: Wageningen University and Research.

———. 1988a. CANOCO – An extension of DECORANA to analyze species-environment relationships. *Vegetatio* 75 (3): 159–160. https://doi.org/10.1007/BF00045629.

ter Braak, Cajo J.F. 1988b. *CANOCO-a FORTRAN program for canonical community ordination by [partial][etrended][canonical] correspondence analysis, principal components analysis and redundancy analysis (version 2.1)*. MLV.

———. 1990. Update notes: Canoco, version 3.10.

ter Braak, Cajo J.F., and I. Colin Prentice. 1988. A theory of gradient analysis. In *Advances in ecological research*, 271–317. San Diego: Elsevier.

ter Braak, C.J.F., and P. Smilauer. 1998. *CANOCO release 4 reference manual and user's guide to Canoco for Windows – Software for canonical community ordination*. Ithaca: Microcomputer Power.

ter Braak, C., and P. Šmilauer. 2011. CANOCO reference manual and user's guide to CANOCO for Windows: software for Canonical Community Ordination (version 4) New York: Microcomputer Power Ithaca; 1998. *Eurasian Soil Science* 44 (2): 173–179.

ter Braak, Cajo, and Petr Šmilauer. 2015. Topics in constrained and unconstrained ordination. *Plant Ecology* 216: 683–696. https://doi.org/10.1007/s11258-014-0356-5.

Torgerson, Warren S. 1958. *Theory and methods of scaling*. New York: Wiley.

Tuomisto, H. 2010. A diversity of beta diversities: Straightening up a concept gone awry. 1. Defining beta diversity as a function of alpha and gamma diversity. *Ecography* 33: 2–22.

Van Den Wollenberg, Arnold L. 1977. Redundancy analysis an alternative for canonical correlation analysis. *Psychometrika* 42 (2): 207–219.

Van Rijckevorsel, Jan L.A., and J. de Leeuw. 1988. *Component and correspondence analysis; Dimension reduction by functional approximation*. Hoboken: Wiley.

Warton, David I., Stephen T. Wright, and Yi Wang. 2012. Distance-based multivariate analyses confound location and dispersion effects. *Methods in Ecology and Evolution* 3 (1): 89–101. https://doi.org/10.1111/j.2041-210X.2011.00127.x. https://besjournals.onlinelibrary.wiley.com/doi/abs/10.1111/j.2041-210X.2011.00127.x.

Webb, Leonard J., J. Geoffrey Tracey, William T. Williams, and Godfrey N. Lance. 1967. Studies in the numerical analysis of complex rain-forest communities: I. A comparison of methods applicable to site/species data. *The Journal of Ecology*: 171–191.

Weiss, Sophie, Xu Zhenjiang Zech, Shyamal Peddada, Amnon Amir, Kyle Bittinger, Antonio Gonzalez, Catherine Lozupone, Jesse R. Zaneveld, Yoshiki Vázquez-Baeza, Amanda Birmingham, Embriette R. Hyde, and Rob Knight. 2017. Normalization and microbial differential abundance strategies depend upon data characteristics. *Microbiome* 5 (1): 27. https://doi.org/10.1186/s40168-017-0237-y.

Whittaker, R.H. 1960. Vegetation of the Siskiyou Mountains, Oregon and California. *Ecological Monographs* 30: 279–338. https://doi.org/10.2307/1943563.

Whittaker, Robert Harding. 1967. Gradient analysis of vegetation. *Biological Reviews* 42 (2): 207–264.

Wickham, Hadley. 2016. *ggplot2: Elegant graphics for data analysis*. Cham: Springer.

Wickham, Hadley, and Maintainer Hadley Wickham. 2007. *The ggplot package*. https://cran.r-project.org/web/packages/ggplot2/index.html

Williams, W.T., and G.N. Lance. 1968. Choice of strategy in the analysis of complex data. *Journal of the Royal Statistical Society. Series D (The Statistician)* 18 (1): 31–43.

Wilmes, P., and P.L. Bond. 2004. The application of two-dimensional polyacrylamide gel electrophoresis and downstream analyses to a mixed community of prokaryotic microorganisms. *Environmental Microbiology* 6 (9): 911.

Wong, Ruth G., Jia R. Wu, and Gregory B. Gloor. 2016. Expanding the UniFrac toolbox. *PLoS ONE* 11 (9): e0161196. https://doi.org/10.1371/journal.pone.0161196.

Wu, Chong, Jun Chen, Junghi Kim, and Wei Pan. 2016. An adaptive association test for microbiome data. *Genome Medicine* 8 (1): 56–56. https://doi.org/10.1186/s13073-016-0302-3. https://www.ncbi.nlm.nih.gov/pubmed/27198579, https://www.ncbi.nlm.nih.gov/pmc/articles/PMC4872356/.

Xia, Y. 2020. Correlation and association analyses in microbiome study integrating multiomics in health and disease. *Progress in Molecular Biology and Translational Science* 171: 309–491. https://doi.org/10.1016/bs.pmbts.2020.04.003.

Xia, Yinglin, and Jun Sun. 2017. Hypothesis testing and statistical analysis of microbiome. *Genes & Diseases* 4 (3): 138–148. https://doi.org/10.1016/j.gendis.2017.06.001. http://www.sciencedirect.com/science/article/pii/S2352304217300351.

Xia, Y., and J. Sun. 2022a. *An integrated analysis of microbiomes and metabolomics.* Washington, DC: American Chemical Society.

―――. 2022b. *Statistical data analysis of microbiomes and metabolomics.* Washington, DC: American Chemical Society.

Xia, Yinglin, Jun Sun, and Ding-Geng Chen. 2018a. Exploratory analysis of microbiome data and beyond. In *Statistical analysis of microbiome data with R*, CSA book series in statistics, 191–249. Singapore: Springer.

―――. 2018b. Community diversity measures and calculations. In *Statistical analysis of microbiome data with R*, ed. Yinglin Xia, Jun Sun, and Ding-Geng Chen, 167–190. Singapore: Springer.

―――. 2018c. What are microbiome data? In *Statistical analysis of microbiome data with R*, 29–41. Singapore: Springer.

Yelland, Phillip M. 2010. An introduction to correspondence analysis. *The Mathematica Journal* 12 (1): 86–109.

Zhan, X., X. Tong, N. Zhao, A. Maity, M.C. Wu, and J. Chen. 2017. A small-sample multivariate kernel machine test for microbiome association studies. *Genetic Epidemiology* 41 (3): 210–220.

Zhang, J., Z. Wei, and J. Chen. 2018. A distance-based approach for testing the mediation effect of the human microbiome. *Bioinformatics* 34 (11): 1875–1883. https://doi.org/10.1093/bioinformatics/bty014.

Zhang, Jilei, Lu Rong, Yongguo Zhang, Żaneta Matuszek, Wen Zhang, Yinglin Xia, Tao Pan, and Jun Sun. 2020. tRNA queuosine modification enzyme modulates the growth and microbiome recruitment to breast tumors. *Cancers* 12 (3): 628. https://doi.org/10.3390/cancers12030628. https://pubmed.ncbi.nlm.nih.gov/32182756, https://www.ncbi.nlm.nih.gov/pmc/articles/PMC7139606/.

Zhao, Ni, Jun Chen, Ian M. Carroll, Tamar Ringel-Kulka, Michael P. Epstein, Hua Zhou, Jin J. Zhou, Yehuda Ringel, Hongzhe Li, and C.Wu. Michael. 2015. Testing in microbiome-profiling studies with MiRKAT, the microbiome regression-based Kernel Association Test. *American Journal of Human Genetics* 96 (5): 797–807. https://doi.org/10.1016/j.ajhg.2015.04.003. https://pubmed.ncbi.nlm.nih.gov/25957468, https://www.ncbi.nlm.nih.gov/pmc/articles/PMC4570290/.

Zhu, Mu, Trevor J. Hastie, and Guenther Walther. 2005. Constrained ordination analysis with flexible response functions. *Ecological Modelling* 187 (4): 524–536.

Chapter 11
Statistical Testing of Beta Diversity

Abstract Chapter 10 investigated various beta diversity metrics and ordination techniques. This chapter continues to contribute to beta diversity. First, it describes the general nonparametric methods for multivariate analysis of variance in ecological and microbiome data. Then, it mainly introduces two statistical hypothesis tests of beta diversity: analysis of similarity (ANOSIM) and permutational MANOVA (PERMANOVA), respectively. Next, it introduces analysis of multivariate homogeneity of group dispersions and pairwise PERMANOVA as well as how to identify core microbial taxa using the microbiome package. Following that, it introduces how to conduct statistical testing of beta diversity in QIIME 2.

Keywords Nonparametric MANOVA · Permutation tests · Analysis of similarity (ANOSIM) · Permutational MANOVA (PERMANOVA) · Homogeneity · Betadisper() · RVAideMemoire package · Emperor plots · dist() · vegdist() · adonis () · **adonis2()** · betadisper() · TukeyHSD() · pairwise.perm.manova() · p.adjust() · Bray-Curtis distance · Jaccard distance · Unweighted UniFrac distance · Weighted UniFrac distance

In Chap. 10, we investigated various beta diversity metrics and ordination techniques. This chapter will continue to contribute to beta diversity. First, we describe the general nonparametric methods for multivariate analysis of variance in ecological and microbiome data (Sect. 11.1). Then, we mainly introduce two statistical hypothesis tests of beta diversity: analysis of similarity (ANOSIM) and permutational MANOVA (PERMANOVA) in Sects. 11.2 and 11.3, respectively. Next, we introduce analysis of multivariate homogeneity of group dispersions (Sect. 11.4). Section 11.5 introduces pairwise PERMANOVA. Section 11.6 introduces how to identify core microbial taxa. Section 11.7 describes statistical testing of beta diversity in QIIME 2. Finally, we briefly summarize this chapter in Sect. 11.8.

Supplementary Information The online version contains supplementary material available at https://doi.org/10.1007/978-3-031-21391-5_11.

11.1   Introduction to Nonparametric MANOVA Using Permutation Tests

In Chap. 10 (Sect. 10.3.9.1), we reviewed that CCA (ter Braak 1986) ushered in the biggest modern revolution in ordination methods because it makes ordination to be not mere an "exploratory" method, but also a hypothesis testing method. Still an important advance in the analysis of multivariate data in ecology was the development of nonparametric methods for testing hypotheses concerning whole communities, particularly by using permutation techniques, such as works from Clarke (1988, 1993), Smith et al. (1990), and Biondini et al. (1991). The importance of testing hypotheses in multivariate analyses in ecology has been reviewed by Anderson (2001). As Anderson stated, before the development of nonparametric methods for testing hypotheses, particularly ANOSIM (Clarke 1993), the most widely available multivariate analyses in ecology focused on the reduction of dimensionality: either to produce and interpret patterns (ordination methods) or to place observations into natural groups by using numerical strategies (clustering). However, although "any inference from the particular to the general must be attended with some degree of uncertainty," "the mathematical attitude toward induction" is still rigorous in the sense that the degree of uncertainty can be expressed in terms of mathematical probability (Fisher 1971; Anderson 2001). In terms of mathematical probability, ordination and clustering methods do not rigorously express the nature and degree of uncertainty concerning a priori hypotheses; it is methods like Mantel's test (Mantel 1967), ANOSIM (Clarke 1993), and multi-response permutation procedures (Mielke Jr. et al. 1976) that allow such rigorous probabilistic statements to be made for multivariate ecological data (Anderson 2001).

The starting point of nonparametric methods for testing hypotheses is to use an analysis-of-variance approach and nonparametric techniques (permutation tests), but minimize the statistical assumptions about the data or be free of the assumption of multivariate normality required by parametric MANOVA. The minimized or less stringent primary statistical assumption is that the data are independently and identically distributed (exchangeability of replicates) (Legendre and Anderson 1999; Anderson 2001).

In the remaining of this chapter, we introduce statistical hypothesis testing of beta diversities through nonparametric MANOVA tests and other related nonparametric methods.

11.2   Analysis of Similarity (ANOSIM)

ANOSIM was developed under a unified framework (Clarke 1993) to (1) display the community patterns through clustering and ordination of samples, (2) identify the species principally responsible for determining sample groupings, (3) perform statistical tests for differences in space and time (analogues of parametric MANOVA, based on rank similarities), and (4) link the community differences to patterns in the environment (also dictated by rank similarities between samples). ANOSIM focuses on hypothesis testing, which is the third component of this framework.

ANOSIM is a nonparametric multivariate statistical method for analysis of changes in community structure based on a standardized rank correlation between two distance matrices. ANOSIM was developed under above described unified framework (Clarke 1993). As a distribution-free method of multivariate data analysis, ANOSIM is frequently used by community ecologists and microbiome researchers.

11.2.1 Introduction of ANOSIM

Clarke (1993) proposed a nonparametric method called "analysis of similarities" (ANOSIM) test to perform the null hypothesis testing of "no differences between sites (groups)" by permutations of the rank similarity matrix. Clarke (1993) thought that rank similarities among samples play a fundamental role in defining and visualizing the community pattern because of the following: (1) Generally relative levels and in particular the ranks of the similarity matrix have a natural interpretation, in which the ranks of similarity matrix summarize the data through statements such as "sample A is more similar to sample B than it is to sample C." (2) The rank similarity is intuitively appealing and very generally applicable to build a graphical representation of the sample patterns. (3) Particularly in effect, this is the only information that a successful NMDS ordination used.

ANOSIM is a nonparametric one-way ANOVA procedure for testing the hypothesis of no difference based on permutation test of the corresponding (rank) similarities among- and within-group samples in the underlying triangular similarity matrix (Clarke 1993). Like ANOVA, ANOSIM treats group membership or treatment levels as factors and models them as the explanatory variable. Analysis of similarity is based on the simple idea: if the tested groups are meaningful, then samples within the groups should be more similar in composition than samples from different groups. The test statistic is given below:

$$R = \frac{(\bar{r}_B - \bar{r}_W)}{(M/2)} \tag{11.1}$$

where R is test statistic, an index of relative within-group dissimilarity.

$\bar{r}_B$ is the mean of rank similarities from all pairs of samples between different groups.
$\bar{r}_W$ is the mean of all rank similarities among samples within groups.
$M = n(n - 1)/2$ and n is the total number of samples under consideration.

There are five main steps to conduct ANOSIM:

Step 1: calculate dissimilarity matrix.

Step 2: calculate rank dissimilarities and assign a rank of 1 to the smallest dissimilarity.

Step 3: calculate the mean among- and within-group rank dissimilarities.

Step 4: calculate test statistic R using the above formula $R = \frac{\bar{r}_B - \bar{r}_W}{M/2}$.

That is, the value of R taken is calculated based on the similarities only for the individual groups (sites), extracted from the matrix for all groups (sites) and re-ranked.

Step 5: test for significance. As in PERMANOVA test (see Sect. 11.3.1), the P-value of test statistic R is obtained by permutation: randomly assigning sample observations to groups. Then, the ranked similarity within and between groups is compared with the similarity that would be generated by random chance. The significance test is simply the fraction of permuted R's that are greater than the observed value of R. That is, it is the probability of an R given this large or larger. Under the null hypothesis H_0, "no differences between groups (sites)," there will be little effect on average to the value of R if the labels identifying which samples (replicates) belong to which groups (sites) are arbitrarily rearranged (Clarke 1993); that is, the total samples from all groups (sites) are just replicates from a single group (site) if H_0 is true. The rationale for a permutation test of H_0 is to examine all possible allocations of labels per group (site) to the total samples and recalculate the R statistic for each. The algorithm behind the hypothesis testing is same as PERMANOVA: if two groups of sampling units are really different in their species (or other taxa) composition, then compositional dissimilarities between the groups ought to be greater than those within the groups (see details in Sect. 11.3.1).

R is interpreted like a correlation coefficient which is a measure of "effect size" as other types of correlation coefficient such as Pearson's coefficient. The test statistic is to test there is no difference among groups under the null hypothesis:

- In this test statistic, the highest similarity is assigned to a rank of 1 (the lowest value).
- R lies in the range $(-1, 1)$.
- $R = 1$ only if all samples within groups are more similar to each other than any samples from different groups.
- $R = 0$ if the null hypothesis is true, which suggests that similarities between and within groups are the same on average.
- R will usually fall between 0 and 1, indicating some degree of discrimination between the groups.

If the null hypothesis is rejected, then $R \neq 0$, which suggests that all pairs of samples within groups are more similar than to any pair of samples from different groups. For example, in the case all the most similar samples are within the sample groups, then $R = 1$. Theoretically, it is also possible that $R < 0$, but practically such case is unlikely in ecological and microbiome studies; such an occurrence is more likely to indicate an incorrect labeling of samples. The extreme case is $R = -1$, which indicates that the most similar samples are all not in the groups.

11.2.2 Perform ANOSIM Using the Vegan Package

ANOSIM provides a way to statistically test whether there is a significant difference between two or more groups of sampling units. ANOSIM is implemented via the function **anosim()** in the vegan package. The function assumes that all ranked dissimilarities within groups have about equal median and range. The **anosim()** function requires a dissimilarity matrix as input data, which can be produced by the function dist() or vegdist(). A summary and plot methods are also available in ANOSIM to perform the post modeling analysis.

One syntax is given below:

```
anosim(data, grouping, permutations = 999, distance = "bray", strata = NULL)
```

where the argument:

- **data** is data matrix or data frame in which rows are samples and columns are response variable(s), or a dissimilarity object or a symmetric square matrix of dissimilarities.
- **grouping** is the grouping variable (a factor).
- **permutations** is the number of permutations to assess the significance of the ANOSIM statistic.
- **distance** is used to specify distance metric that measures the dissimilarity between two observations. When the input data is not a dissimilarity structure or a symmetric square matrix, then this argument needs to be specified.
- **strata** is an integer vector or factor which is used for specifying the strata for permutation. If supplied, observations are permuted only within the specified strata.

Example 11.1: Breast cancer QTRT1 Mouse Gut Microbiome, Example 10.1 Cont.
In this section, we illustrate ANOSIM test using the same breast cancer QTRT1 mouse gut microbiome dataset (Jilei Zhang et al. 2020). We perform ANOSIM test using the following steps.

Step 1: Load otu table into R or RStudio.

```
> setwd("~/Documents/QIIME2R/Ch11_BetaDiversity2")
> abund_tab=read.csv("otu_table_L7_MCF7_vegan.csv",row.names=1,
check.names=FALSE)
```

```
> meta_tab <- read.csv("metadata.csv", check.names = FALSE)
> head(abund_tab,3)
```

ANOSIM needs the input data to be a dissimilarity object or a matrix of dissimilarities, which is a numerical data matrix or data frame.

Step 2: Make the dataset to have rows being samples and columns being OTUs.

The anosim() function requires the input data matrix or data frame in which rows are samples and columns are OTUs (taxa). Here, the dataset with rows is OTUs and columns being samples:

```
> nrow(abund_tab)#OTUs
[1] 635
> ncol(abund_tab)#samples
[1] 40
```

Thus, we transform the "abund_tab" to the "abund_tab_t" in which rows are samples and columns are OTUs:

```
> abund_tab_t<-t(abund_tab)
> head(abund_tab_t)
> nrow(abund_tab_t)#samples
[1] 40
> ncol(abund_tab_t)#OTUs
[1] 635
```

Step 3: Perform ANOSIM tests of measures of dissimilarity.

Fit ANOSIM to test the Bray-Curtis dissimilarity.

```
> set.seed(123)
> library(vegan)
> bray<-vegdist(abund_tab_t, "bray")
> anosim(bray, meta_tab$Group4,permutations = 999)
> # Or
> anosim(abund_tab_t, meta_tab$Group4, permutations = 999, distance =
"bray")
> # Or
> fit_b <- anosim(bray, meta_tab$Group4,permutations = 999)
> summary(fit_b)
```

Partial output of Bray-Curtis dissimilarity is given below:

```
ANOSIM statistic R: 0.674
      Significance: 0.001
Dissimilarity ranks between and within classes:
      0% 25% 50% 75% 100% N
Between 8 282.8 480.5 630.2 780 600
```

```
KO_BEFORE 17 124.0 205.0 309.0 401 45
KO_POST 2 99.0 193.0 254.0 352 45
WT_BEFORE 20 204.0 326.0 407.0 436 45
WT_POST 1 25.0 73.0 128.0 239 45
```

Fit ANOSIM to test the Jaccard dissimilarity.

```
> jaccard<-vegdist(abund_tab_t, "jaccard",binary=TRUE)
> fit_j <- anosim(jaccard, meta_tab$Group4,permutations = 999)
> summary(fit_j)
```

Partial output of binary Jaccard dissimilarity is given below:

```
ANOSIM statistic R: 0.711
   Significance: 0.001
```

Dissimilarity ranks between and within classes:

```
  0%   25%   50%   75%   100%  N
Between 8 298.8 470.0 627.5 780.0 600
KO_BEFORE 3 39.0 89.0 191.5 481.0 45
KO_POST 1 46.0 145.0 275.0 575.5 45
WT_BEFORE 5 107.0 197.5 336.5 710.5 45
WT_POST 10 87.5 147.5 229.0 523.0 45
```

Fit ANOSIM to test the Sørensen dissimilarity.

```
> sørensen<-vegdist(abund_tab_t,method="bray",binary=TRUE)
> fit_s <- anosim(sørensen, meta_tab$Group4,permutations = 999)
> summary(fit_s)
```

Partial output of Sørensen (binary bray) dissimilarity is given below:

```
ANOSIM statistic R: 0.711
 Significance: 0.001
Dissimilarity ranks between and within classes:
 0% 25% 50% 75% 100% N
Between 8 298.8 470.0 627.5 780.0 600
KO_BEFORE 3 39.0 89.0 191.5 481.0 45
KO_POST 1 46.0 145.0 275.0 575.5 45
WT_BEFORE 5 107.0 197.5 336.5 710.5 45
WT_POST 10 87.5 147.5 229.0 523.0 45
```

Above outputs showed that performing ANOSIM tests of Bray-Curtis dissimilarity, Jaccard dissimilarity, and Sørensen dissimilarity give ANOSIM statistic R of 0.674, 0.711, and 0.711, respectively, at the statistical significance from permutation of 0.001. The P-value of 0.001 is less than 0.05, which indicates that within-group

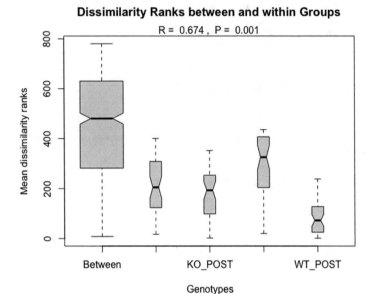

Fig. 11.1 Boxplot of Bray-Curtis dissimilarity based on QTRT1 mouse dataset

similarity is greater than between-group similarity at 0.05 significant level. Therefore, we can conclude that there is evidence that the within-group samples are more
similar than would be expected by random chance. In other words, there is evidence
that the between-group samples are more dissimilar.

Step 4: Plot the dissimilarity ranks between and within groups.

Finally, we plot the Bray-Curtis dissimilarity (Fig. 11.1):

```
> # Figure 11.1
> plot(fit_b,xlab="Genotypes",ylab="Mean dissimilarity ranks",
+   main="Dissimilarity Ranks between and within Groups", pch=17,
cex=1.2)
```

11.2.3   Remarks on ANOSIM

As described above, before the development of nonparametric methods for testing
hypotheses, particularly ANOSIM (Clarke 1993), the most widely available multivariate analyses in ecology focused on ordination and clustering (the methods for
reduction of dimensionality). The importance of ANOSIM lies on these three factors
(Clarke 1993): (1) It emphasizes it is relevant ecological assumptions that dictate the
choice of dissimilarity measure but not the mechanics of the ordination method. (2) It

not only avoids the difficulty of choosing measures and even the ordination technique itself but also does not need to consider whether species-environment relationships may be linear, monotonic, unimodal, or even multimodal forms (Clarke 1993). In practice all four combinations may be present. Canonical analysis has been proposed to avoid constraining species-environment relationships to linear, monotonic, unimodal, or multimodal forms. However, it remains unknown how well it stands up to further detailed scrutiny. The direct hypothesis testing approach of ANOSIM is more appealing. (3) The third is simplicity and validity of advocated nonparametric techniques.

As we remarked in Chap. 10 (Sect. 10.1.3.1), because Sørensen dissimilarity and Jaccard dissimilarity are similarly defined and due to ANOSIM rank-based testing property, the testing results of Jaccard dissimilarity and the Sørensen dissimilarity are almost identical in this example. The result of Bray-Curtis dissimilarity has been reported in previous publication (Jilei Zhang et al. 2020). We also compared the effects of normalization with fitting ANOSIM using relative abundance and absolute count data. The results showed that normalization has no large effect for Jaccard dissimilarity and the Sørensen dissimilarity, which may be due to their binary property. Their results including distance measures and statistical tests are almost identical (only upper quantiles of permutations are different) in this case. However, for Bray-Curtis dissimilarity, the relative abundance data returned much larger distance values than using absolute count data. The relative abundance data also give a smaller R statistic compared to using absolute count data (0.613 vs. 0.674).

11.3 Permutational MANOVA (PERMANOVA)

In this section, we first describe how PERMANOVA was developed, and next illustrate how to conduct PERMANOVA in microbiome data, and then briefly remark on this method.

11.3.1 Introduction to PERMANOVA

In 2001, Anderson (2001) proposed a nonparametric method for multivariate analysis of variance based on permutation tests called "nonparametric MANOVA," a permutational multivariate analysis of variance (PERMANOVA). PERMANOVA is used to perform the general multivariate null hypothesis of no differences in the composition and/or relative abundances of different species (taxa/variables) between two or more groups of samples.

PERMANOVA improves the previous nonparametric methods because it can be based on any measure of dissimilarity and by allowing a direct additive partitioning of variation for a multifactorial ANOVA model, while maintaining the flexibility of other nonparametric methods but lacking their formal assumptions. The test statistic is a multivariate analogue to Fisher's *F*-ratio and is calculated directly from any symmetric distance or dissimilarity matrix. *P*-values are obtained using permutations.

Before PERMANOVA, many nonparametric MANOVA-like methods for tests of differences among a priori groups of observations have been developed, including Mantel (1967); Mantel and Valand (1970); Mielke Jr. et al. (1976); Clarke (1988, 1993); and Legendre and Anderson (1999).

All these methods generally share two things (Anderson 2001): (1) They are based on measures of distance or dissimilarity between pairs of individual multivariate observations (refer to generally as distances) or their ranks and construct a statistic to compare these distances among observations in the same group versus those in different groups, following the conceptual framework of ANOVA. (2) They use permutations of the observations to obtain a probability associated with the null hypothesis of no differences among groups. However, previous nonparametric MANOVA-like methods have been reviewed suffering three weaknesses (Anderson 2001): The main drawback to all these methods is that they are not able to cope with multifactorial ANOVA to partition variation across the many factors that form part of the experimental design (Anderson 2001). Thus, they had to take multiple one-way analyses and qualitative interpretations of ordination plots to infer interactions or variability at different spatial scales. Some of these methods do allow partitioning for a complex design, but are restricted for use with metric distance measures (i.e., Euclidean distance), instead of using the semimetric Bray-Curtis measure of ecological distance which is ideal for ecological applications. Furthermore, these methods lack of appropriate permutational strategies for complex ANOVA, particularly for tests of interactions and directly statistical analysis of Bray-Curtis distances for any multifactorial ANOVA design.

PERMANOVA partitions variation based on any distance measure in any ANOVA design and is robust, interpretable by reference to the experimental design that either continuous or categorical variables and interaction terms can be specified in the model and do not need to formally assume the distributions of variables. We have recommended PERMANOVA for use in microbiome data (Xia et al. 2018). PERMANOVA is developed by the following two steps.

Step 1: Develop an *F*-ratio test statistic.

PERMANOVA is formulated based on the analysis and partitioning sums of square distance or dissimilarity matrix.

Let's consider a matrix of distances between every pair of observations and let $N = an$, the total number of observations (points), and d_{ij} be the distance between observation $i = 1, \ldots, N$ and observation $j = 1, \ldots, N$; then the total sum of squares SS_T is given below:

$$SS_T = \frac{1}{N} \sum_{i=1}^{N-1} \sum_{j=i+1}^{N} d_{ij}^2 \tag{11.2}$$

That is, add up the squares of all of the distances in the sub-diagonal (or upper-diagonal) half of the distance matrix (but not including the diagonal) and divide by N. SS_T is used to calculate average distance among all samples. Similarly, the within-group or residual sum of squares is given below:

$$SS_W = \frac{1}{n} \sum_{i=1}^{N-1} \sum_{j=i+1}^{N} d_{ij}^2 \varepsilon_{ij} \tag{11.3}$$

where ε_{ij} is an indicator and takes the value 1 if observation i and observation j are in the same group; otherwise it takes the value of zero. That is, add up the squares of all of the distances between observations that occur in the same group and divide by n, the number of observations per group. We can use SS_W to calculate average distance among samples. The sum of squares used to calculate average distance among groups can be obtained through subtracting SS_T by SS_W: $SS_A = SS_T$ - SS_W. Then, a pseudo-F-ratio to test the multivariate hypothesis is given:

$$F = \frac{SS_A/(a-1)}{SS_W/(N-a)} \tag{11.4}$$

The rationale for this test statistics lies in the fact if the points from different groups have different central locations (centroids in the case of Euclidean distances) in multivariate space, then the among-group distances will be relatively large compared to the within-group distances, and the resulting pseudo-F-ratio will be relatively large.

The development of this F-ratio test statistic is very flexible because the sums of squares in Eqs. (11.2) and (11.3) and the statistic in Eq. (11.4) from a distance matrix can been calculated using any distance measure. Actually, McArdle and Anderson (McArdle and Anderson 2001) have shown more generally how partitioning for any linear model can be done directly from the distance matrix to obtain the statistic in Eq. (11.4), regardless of the distance measure used. The statistic described in (11.4) holds another important property/characteristic: it gives the same value as the traditional parametric univariate F-statistic in the case of a Euclidean distance matrix calculated from only one variable (Anderson 2001).

Step 2: Obtain a P-value using permutations.

From Eq. (11.4), we can see that the bigger the ratio between $SS_A/(a-1)$ and $SS_W/(N-a)$ (called the signal-to-noise ratio), the larger the F-value, and thus it results in a smaller P-value. The question is: how can we obtain a P-value? The individual variables in ecology and microbiome are typically not normally distributed, and we do not expect that the Euclidean distance will necessarily be used for the analysis. As Anderson (2001) observed, even if each of the variables were normally distributed and the Euclidean distance can be used, the mean squares calculated for the multivariate data would not each consist of sums of independent χ^2 variables, because, although individual observations are expected to be independent, individual species (OTUs or taxa) variables are not independent of one another. Thus, traditional tabled P-values cannot be used.

Anderson (2001) proposed to create a distribution of the statistic under the null hypothesis using permutations of the observations (Edgington 1995; Manly 1997). The algorithm underlying this is that under the null hypothesis, the groups are not really different in terms of their composition and/or their relative abundances of species, as measured (e.g., by the Bray-Curtis distances); then the multivariate observations (rows) would be exchangeable among the different groups. Thus, we can randomly shuffle (permute) the rows that are identified to a particular group to obtain a new F-value (say, F^*). The random shuffling is repeated and the F^* is recalculated for all possible re-orderings of the rows. In such way, the entire empirical distribution of the pseudo-F-statistic under a true null hypothesis for the particular data is generated.

A P-value, which indicates the significance between groups, is calculated by comparing the F-value obtained with the original ordering of the rows to the empirical distribution created for a true null by permuting the labeled rows:

$$P = \frac{(\text{No. of } F^* \geq F)}{(\text{Total no. of } F^*)}, \tag{11.5}$$

where the original observed F-value is considered to be a member of the distribution of F^* under permutation (i.e., it is one of the possible orderings of the labels on the rows). Usually, a statistical significance level is set as $\alpha = 0.05$ for statistical tests.

In general there are $\frac{(kn)!}{(n!)^k k!}$ distinct ways of permuting the labels for n replicates at each of k sites (Clarke 1993). Although examining such a number of re-labeling is computationally possible and the precision of the P-value will increase with increasing numbers of permutations, however, the scale of calculation can quickly get out of hand with modest increases in replication. Thus, in practice it is usually not to calculate all possible permutations due to the time; P-value can be calculated using a large random subset of all possible permutations (Hope 1968), that is, usually is randomly sampled (with replacement). Generally, at least 1000 permutations are required for tests at a significance level of $\alpha = 0.05$ and at least 5000 permutations for tests with a significance level of $\alpha = 0.01$ (Manly 1997).

As presented above, permutations mean randomly assigning sample observations to groups. The P-value is calculated by comparing the permuted F-ratios to the observed F-ratio. The significance test is simply the fraction of permuted F-ratios that are greater than the observed F-ratio. Briefly, we can take one-way test as an example to explain what permutations actually do. In a one-way test, our interest is to see whether a statistic is either less than or greater than what can be expected by chance. The P-value is calculated from the proportion of permuted pseudo-F-

statistics which are greater than or equal to the observed F-statistic. In another words, we want to know whether the permuted datasets following the PERMANOVA yield a better resolution of groups relative to the actual dataset. If more than 5% of the permuted F-statistics has values greater than that of the observed F-statistic, the P-value is greater than 0.05. Then, we can conclude that any difference among groups is not statistically significant.

11.3.2 Perform PERMANOVA Using the Vegan Package

We can perform PERMANOVA through the functions **adonis()** and **adonis2()** in the vegan package. These two functions analyze and partition the sums of squares using distance/dissimilarity matrices and fit linear models (e.g., factors, polynomial regression) to distance matrices using a permutation test with pseudo-F-ratios.

The **adonis()** and **adonis2()** functions are directly based on the algorithms of Anderson (2001) to perform a sequential test of terms and McArdle and Anderson (2001) to perform sequential, marginal, and overall tests, respectively. Both functions can handle semimetric indices (e.g., Bray-Curtis dissimilarity) that produce negative eigenvalues. The **adonis2()** also allows using additive constants or square root of dissimilarities to avoid negative eigenvalues (Stevens and Oksanen 2020). If the random permutation is the same, the **adonis()** and **adonis2()** will provide the identical results of tests. However, **adonis2()** could be much slower compared to **adonis()**, in particular when several terms are in the model (Stevens and Oksanen 2020).

ADONIS (permutational multivariate analysis of variance using distance matrices) method is less sensitive to dispersion effects (different within-group variation) and flexible than both ANOSIM (Clarke 1993) and multi-response permutation procedure (MRPP) (Mielke Jr. et al. 1976), which only handle categorical predictors (Oksanen 2016). ADONIS allows ANOVA-like tests of the variance in beta diversity explained by continuous and/or categorical predictors, which is a recommended method in the vegan package.

In general, ADONIS can be used to analyze ecological and microbiome community data as well as genetic data. The formers typically are samples by taxa matrices, while genetic data often have thousands or millions of columns of gene expression with a limited number of samples of individuals. Since ADONIS method is to perform PERMANOVA (permutational multivariate analysis of variance) using distance matrices, thus a distance/dissimilarity measure is required when running PERMANOVA (ADONIS). In microbiome literature, four beta diversity measures, Bray-Curtis distance, Jaccard distance, and weighted and unweighted UniFrac distances, have often been reported (Linnenbrink et al. 2013), and "bray" (the Bray-Curtis distance) has been most widely used.

One example syntax of **adonis()** is given as below:

```
adonis(formula, data, permutations = 999, method = "bray", contr.
unordered = "contr.sum", contr.ordered = "contr.poly")
```

where the argument:

- **formula** = model formula, e.g., Y = A + B + C*D; Y must be either a community data matrix (data frame) or a dissimilarity matrix (data frame), e.g., inheriting from class "vegdist" or "dist". If Y is a community data matrix, the function **vegdist()** will be used to calculate pairwise distances/dissimilarities. A, B, C, and D are the independent variables, which may be factors or continuous variables. They can be transformed within the formula.
- **data** = the data frame for the independent variables.
- **permutations** = 999 is the specified number of replicate permutations used for the hypothesis tests (*F* tests).
- **method** = ""can be any method name used in the function **vegdist ()** to calculate pairwise distances if the left-hand side of the formula is a data frame or a matrix.
- **contr.unordered** = ""is used for contrasts of the design matrix; in general, R default uses dummy or treatment contrasts for unordered factors. However, the default contrasts in vegan package are different; they use "sum" or "ANOVA" contrasts.
- **contr.ordered** = ""is used for contrasts of the design matrix for ordered factors.

One example syntax of **adonis2()** is given as below:

```
adonis2(formula, data, permutations = 999, method = "bray", sqrt.dist =
FALSE, add = FALSE, by = "terms", parallel = getOption("mc.cores"))
```

where the argument:

- **sqrt.dist** = FALSE defines not to take square root of dissimilarities.
- **by** = "terms" is set to assess significance for each term (sequentially from first to last), setting **by** = "margin" will assess the marginal effects of the terms (each marginal term analyzed in a model with all other variables), and **by** = NULL will assess the overall significance of all terms together. The arguments will be passed on to anova.cca() function.
- **parallel** = is used to specify the number of parallel processes or a predefined socket cluster. With parallel = 1 uses ordinary, nonparallel processing.

Example 11.2: Breast Cancer QTRT1 Mouse Gut Microbiome, Example 11.1 Cont.

In this section, we use the same breast cancer QTRT1 mouse gut microbiome dataset to illustrate performing PERMANOVA through *R* packages vegan (version 2.5.7), phyloseq (version 1.36.0), and microbiome (version 1.14.0). We conduct hypothesis testing of whether the composition of microbiome communities varies over time and differentiates between groups (MCF7 and WT mice). Given the testing result of

adonis () is identical to **adonis2**(), below we will illustrate running **adonis2**() step by step.

Step 1: Read data into *R* and check data types.

```
> setwd("~/Documents/QIIME2R/Ch11_BetaDiversity2")
> otu<-read.csv("otu_table_L7_MCF7_phyloseq.csv",row.names = 1)
> tax<-read.csv("tax_table_L7_MCF7_phyloseq.csv",row.names = 1)
> sam<-read.csv("metadata.csv",row.names = 1)

> otumat<-as.matrix(otu)
> taxmat<-as.matrix(tax)

> class(otumat)
[1] "matrix"
> class(taxmat)
[1] "matrix"
> class(sam)
[1] "data.frame"
```

Step 2: Create a phyloseq object using otu, taxa, and meta tables.

```
> library("phyloseq")
> otu_tab = otu_table(otumat, taxa_are_rows = TRUE)
> tax_tab = tax_table(taxmat)
> meta_tab = sample_data(sam)

> physeq = phyloseq(otu_tab, tax_tab, meta_tab)
> physeq
phyloseq-class experiment-level object
otu_table() OTU Table: [ 635 taxa and 40 samples ]
sample_data() Sample Data: [ 40 samples by 3 sample variables ]
tax_table() Taxonomy Table: [ 635 taxa by 7 taxonomic ranks ]
```

Step 3: Transform to relative abundances.

The following *R* commands use the transform() function in the microbiome package to transform the count reads into relative or compositional abundances:

```
> library(microbiome)
> pseq_rel <- microbiome::transform(physeq, "compositional")
```

Step 4: Pick relative abundances and sample metadata.

```
> otu_tab_rel <- abundances(pseq_rel)
> meta_tab_rel <-meta(pseq_rel)
> head(meta_tab_rel)
```

Step 5: Perform PERMANOVA to test significance for group-level differences.

First, perform a sequential test of terms using adonis2().

Either **adonis()** or **adonis2()** can perform a sequential test of each term that is specified in their functions. In QTRT1 mouse dataset, three variables define three kinds of groups. Group has two levels, MCF7 knockout and wildtype, Time is coded as BEFORE and POST treatment, and Group4 presents Group and Time interaction. Here, we perform a sequential test of Group by Time for the Bray-Curtis dissimilarity. We also provide the *R* commands for running the tests using Jaccard and Sørensen dissimilarities.

Below we conduct a sequential test of the Bray-Curtis dissimilarity from Group to Time.

Both **adonis()** and **adonis2()** by default conduct a sequential test of terms to assess significance for each variable (sequentially from first to last) that is specified in formula: here, from Group, Time to Group by Time interaction. In **adonis2()**, we can also do the sequential test via specifying by = "terms" in the arguments:

```
> set.seed(123)
> library(vegan)
> adonis2(t(otu_tab_rel) ~ Group*Time, data = meta_tab_rel,
permutations=999, method = "bray")
```

Partial output is given below:

```
       Df SumOfSqs   R2    F Pr(>F)
Group    1   0.06 0.014 1.94 0.140
Time     1   3.16 0.700 96.65 0.001 ***
Group:Time 1   0.12 0.026 3.54 0.054 .
Residual 36   1.18 0.261
Total    39   4.51 1.000
---
Signif. codes: 0 '***' 0.001 '**' 0.01 '*' 0.05 '.' 0.1 ' ' 1
```

For **adonis2()**, specifying by = "term" will obtain the same results:

```
> set.seed(123)
> adonis2(t(otu_tab_rel) ~ Group*Time, by = "term", data =
meta_tab_rel, permutations=999, method = "bray")
```

Below we conduct a sequential test of the Jaccard dissimilarity from Group to Time.

```
> set.seed(123)
> adonis2(t(otu_tab_rel) ~ Group*Time,by = "term", data =
meta_tab_rel, permutations=999, method = "jaccard",binary=TRUE)
```

Partial output is given below:

```
        Df SumOfSqs   R2   F Pr(>F)
Group    1   0.21 0.043 2.07 0.010 **
Time     1   0.81 0.168 8.19 0.001 ***
Group:Time 1   0.24 0.051 2.47 0.007 **
Residual 36   3.57 0.739
Total   39   4.83 1.000
---
Signif. codes:  0 '***' 0.001 '**' 0.01 '*' 0.05 '.' 0.1 ' ' 1
```

Below we conduct a sequential test of the Sørensen dissimilarity from Group to Time.

```
> set.seed(123)
> adonis2(t(otu_tab_rel) ~ Group*Time,by = "term", data =
meta_tab_rel, permutations=999, method = "bray", binary=TRUE)
```

Partial output is given below:

```
        Df SumOfSqs   R2    F Pr(>F)
Group    1   0.097 0.045 2.34  0.013 *
Time     1   0.460 0.212 11.10 0.001 ***
Group:Time 1   0.123 0.057 2.96  0.002 **
Residual 36   1.493 0.687
Total   39   2.173 1.000
---
Signif. codes:  0 '***' 0.001 '**' 0.01 '*' 0.05 '.' 0.1 ' ' 1
```

Second, perform a marginal test using adonis2().

The marginal test will be performed by setting by = "margin" in **adonis2()**, which will assess the marginal effects of the terms (each marginal term analyzed in a model with all other variables).

Below we conduct a marginal test of the Bray-Curtis dissimilarity.

```
> set.seed(123)
> adonis2(t(otu_tab_rel) ~ Group*Time, by = "margin", data =
meta_tab_rel, permutations=999, method = "bray")
```

Partial output is given below:

```
        Df SumOfSqs   R2   F Pr(>F)
Group:Time 1    0.12 0.026 3.54  0.054 .
Residual  36   1.18 0.261
Total     39   4.51 1.000
---
Signif. codes:  0 '***' 0.001 '**' 0.01 '*' 0.05 '.' 0.1 ' ' 1
```

Below we conduct a marginal test of the Jaccard dissimilarity.

```
> set.seed(123)
> adonis2(t(otu_tab_rel) ~ Group*Time,by = "margin", data =
meta_tab_rel, permutations=999, method = "jaccard",binary=TRUE)
```

Partial output is given below:

```
        Df SumOfSqs   R2   F Pr(>F)
Group:Time 1    0.24 0.051 2.47  0.007 **
Residual  36   3.57 0.739
Total     39   4.83 1.000
---
Signif. codes:  0 '***' 0.001 '**' 0.01 '*' 0.05 '.' 0.1 ' ' 1
```

Below we conduct a marginal test of the Sørensen dissimilarity.

```
> adonis2(t(otu_tab_rel) ~ Group*Time, by = "margin", data =
meta_tab_rel, permutations=999, method = "bray",binary=TRUE)
```

Partial output is given below:

```
        Df SumOfSqs   R2   F Pr(>F)
Group:Time 1   0.123 0.057 2.96  0.006 **
Residual  36   1.493 0.687
Total     39   2.173 1.000
---
Signif. codes:  0 '***' 0.001 '**' 0.01 '*' 0.05 '.' 0.1 ' ' 1
```

Third, perform an overall test using adonis2().

Below we will use **adonis2()** to assess the overall significance of all terms together via specifying by = NULL in the arguments.

Below we conduct an overall test of the Bray-Curtis dissimilarity.

```
> set.seed(123)
> adonis2(t(otu_tab_rel) ~ Group*Time, by = NULL, data = meta_tab_rel,
permutations=999, method = "bray")
```

Partial output is given below:

```
    Df SumOfSqs  R2 F Pr(>F)
Model   3   3.34 0.739 34 0.001 ***
Residual 36   1.18 0.261
Total   39   4.51 1.000
---
Signif. codes:  0 '***' 0.001 '**' 0.01 '*' 0.05 '.' 0.1 ' ' 1
```

Below we conduct an overall test of the Jaccard dissimilarity.

```
> set.seed(123)
> adonis2(t(otu_tab_rel) ~ Group*Time, by = NULL, data = meta_tab_rel,
permutations=999, method = "jaccard",binary=TRUE)
```

Partial output is given below:

```
    Df SumOfSqs  R2   F Pr(>F)
Model   3   1.26 0.261 4.24  0.001 ***
Residual 36   3.57 0.739
Total   39   4.83 1.000
---
Signif. codes:  0 '***' 0.001 '**' 0.01 '*' 0.05 '.' 0.1 ' ' 1
```

Below we conduct an overall test of the Sørensen dissimilarity.

```
> adonis2(t(otu_tab_rel) ~ Group*Time, by = NULL, data = meta_tab_rel,
permutations=999, method = "bray",binary=TRUE)
```

Partial output is given below:

```
    Df SumOfSqs  R2   F Pr(>F)
Model   3   0.68 0.313 5.47  0.001 ***
Residual 36   1.49 0.687
Total   39   2.17 1.000
---
Signif. codes:  0 '***' 0.001 '**' 0.01 '*' 0.05 '.' 0.1 ' ' 1
```

Fourth, perform an overall test of Group4 using adonis2().

Next we will conduct the overall significance testing of the Group4 variable.

Below we conduct an overall test of the Group4 variable for the Bray-Curtis dissimilarity.

```
> set.seed(123)
> bray <- adonis2(t(otu_tab_rel) ~ Group4,
+               data = meta_tab_rel, permutations=999, method = "bray")
```

The partial output is given below:

```
    Df SumOfSqs   R2  F Pr(>F)
Group4   3    3.34 0.739 34  0.001 ***
Residual 36    1.18 0.261
Total    39    4.51 1.000
---
Signif. codes:  0 '***' 0.001 '**' 0.01 '*' 0.05 '.' 0.1 ' ' 1
```

Below we conduct an overall test of the Group4 variable for the Jaccard dissimilarity.

```
> set.seed(123)
> Jaccard <- adonis2(t(otu_tab_rel) ~ Group4,
+         data = meta_tab_rel, permutations=999, method = "jaccard",
binary=TRUE)
```

Partial output is given below:

```
    Df SumOfSqs R2 F Pr(>F)
Group4 3 1.26 0.261 4.24 0.001 ***
Residual 36 3.57 0.739
Total 39 4.83 1.000
---
Signif. codes:  0 '***' 0.001 '**' 0.01 '*' 0.05 '.' 0.1 ' ' 1
```

Below we conduct an overall test of the Group4 variable for the Sørensen dissimilarity.

```
> set.seed(123)
> Sørensen <- adonis2(t(otu_tab_rel) ~ Group4,
+ data = meta_tab_rel, permutations=999,method="bray", binary=TRUE)
```

Partial output is given below:

```
        Df SumsOfSqs MeanSqs F.Model   R2 Pr(>F)
Group4   3    0.68 0.2268  5.47 0.313 0.001 ***
Residuals 36   1.49 0.0415      0.687
Total    39   2.17          1.000
---
Signif. codes:  0 '***' 0.001 '**' 0.01 '*' 0.05 '.' 0.1 ' ' 1
```

11.3.3 Remarks on PERMANOVA

PERMANOVA method was developed within a nonparametrical framework and under the setting of multifactorial analysis of variance (ANOVA), and hence it has flexibility in the multivariate analysis of ecological and microbiome data. PERMANOVA has been reviewed (Xia et al. 2018) having at least two advantages over the traditional MANOVA and particularly in multivariate analysis of ecological and microbiome data (Anderson 2001).

First, it is a nonparametric method without assumption of multivariate normality required by parametric MANOVA. Because of using permutations, the PERMANOVA test requires no specific assumption regarding the number of variables or the nature of their individual distributions or correlations (Anderson 2001). The primary minimized statistical assumption is that the data are independently and identically distributed. Second, it has the flexibility to choose any distance measure (or on ranks of distances) appropriate for the data and hypothesis being tested. The semimetric Bray-Curtis dissimilarity has become most often used in ecological and microbiome studies because of its intuitive interpretation as "percentage difference" among other conveniences. However, it is not easy to calculate the central location directly from the sample data in multivariate Bray-Curtis space. The PERMANOVA method allows it to happen for comparing the performance of various measures of similarity (or dissimilarity) with different kinds of ecological and microbiome data. Actually, the PERMANOVA method combines the two benefits (Anderson 2001): (1) partitioning variation according to any ANOVA design from the traditional test statistics and (2) using any symmetric dissimilarity or distance measure (or their ranks) and providing a P-value using appropriate permutation methods like the most flexible nonparametric methods. Additionally, PERMANOVA has two important features (Anderson 2001): (1) Its statistic is analogous to Fisher's F-ratio and is constructed from sums of squared distances (or dissimilarities) within and between groups. (2) This statistic is equal to Fisher's original F-ratio in the case of one variable and when Euclidean distances are used. In summary, PERMANOVA method not only moves beyond the traditional MANOVA (e.g., Fisher et al.'s methods) but also generalizes the nonparametric multivariate methods using permutation tests such as Clarke's ANOSIM to construct test statistics to compare among-and-within groups based on ranks of dissimilarity.

As we stated in Chap. 10 (Sect. 10.1.3), the Sørensen dissimilarity is the binary (presence-absence) version of the Bray-Curtis dissimilarity, which can be obtained by specifying method = "bray", binary = TRUE or just specifying binary = TRUE in the vegan package (by default method = "bray"). In the literature, the presence-absence version of Bray-Curtis distance has been considered as equivalent to the Jaccard distance because their expressions are almost identical and the difference is usually ignorable (Tang et al. 2016; Zhang et al. 2018). Actually, these two distance measures return different values using the **vegdist()** in the vegan package (see Sect. 10.1.3.3 of Chap. 10). Their similarities are from the fact that binary version of the Bray-Curtis and Jaccard dissimilarities gives almost identical results when ANOSIM is used for statistical hypothesis testing, which tests dissimilarity ranks between and within classes (groups). For example, R statistic, P-values, and between- and within-group dissimilarity ranks are identical except for the upper quantiles of permutations (null model) (see Sect. 11.2.2). However, the results are different when PERMANOVA is used for the tests of these two distance measures. As seen from above outputs, the results are different including F-value and R^2.

11.4   Analysis of Multivariate Homogeneity of Group Dispersions

After conducting PERMANOVA to test the differences of beta diversity measures among groups, we next will perform multivariate analysis of homogeneity of group dispersions.

11.4.1   Introduction to the Function betadisper()

The multivariate analysis of homogeneity of group dispersions can be performed using the function **betadisper()**. The function betadisper() is a sister function of **adonis()** to implement the PERMDISP2 procedure for the analysis of homogeneity of group dispersions (variances) proposed by Anderson et al. (2006). It is a multi-variate analogue of Levene's test for homogeneity of variances. The procedure reduces the original distances to principal coordinates to handle the non-Euclidean distances between samples (objects) and group centroids. Later, this procedure has been adopted to analyze beta diversity. Several methods or functions are available in the procedure including anova(), scores(), plot(), boxplot(), eigenvals(), and TukeyHSD(). One sample syntax is given below:

```
betadisper (ds, group, type = c ("median","centroid"), bias.adjust =
TRUE, sqrt.dist = FALSE, add = FALSE)
```

where the argument:

- **ds** is a distance structure that returned by calling the functions **dist()**, **betadiver()**, or **vegdist()**.
- **group** is a vector describing the group structure, usually a factor or an object that can be coerced to a factor using the function as.factor() or a factor with a single level.
- **type** is used for specifying the type of analysis to perform. The default type is the spatial median as the group centroid. The permutation test for type = "centroid" had been noticed to give incorrect type I error and is anti-conservative (see the document of this function).
- **bias.adjust = TRUE** adjusts for small sample bias in beta diversity estimates.
- **sqrt.dist = FALSE** does not take square root of (Euclidean) dissimilarities.
- **add = TRUE** or **FALSE** is used to specify whether add a constant to the non-diagonal dissimilarities to ensure all nonnegative eigenvalues in the underlying principal coordinates.

11.4.2   Implement the Function betadisper()

Example 11.3: Breast Cancer QTRT1 Mouse Gut Microbiome, Example 11.1 Cont.

In this section, to ensure the PERMANOVA results reliable, let's assess whether or not variance homogeneity assumptions hold for the four levels of groups. The null hypothesis is that there is similar multivariate variance among these four levels of groups (analogous to variance homogeneity). We will use the same breast cancer QTRT1 mouse gut microbiome dataset to illustrate performing multivariate analysis of homogeneity of group dispersions using the function **betadisper()** in the vegan package.

Step 1: Perform variance homogeneity test using anova().

```
> # Calculate distance matrix
> library(vegan)
> dist <- vegdist(t(otu_tab_rel))

> # Calculate multivariate dispersions
> disp <- betadisper(dist, meta_tab_rel$Group4)
> anova(disp)
Analysis of Variance Table

Response: Distances
Df Sum Sq Mean Sq F value Pr(>F)
Groups 3 0.0704 0.02346 5.65 0.0028 **
Residuals 36 0.1495 0.00415
---
Signif. codes: 0 '***' 0.001 '**' 0.01 '*' 0.05 '.' 0.1 ' ' 1
```

Step 2: Perform variance homogeneity test using permutation test.

```
> # Permutation test for F-statistic
> permutest (disp, pairwise = TRUE, permutations = 999)
```

Response: Distances

```
       Df Sum Sq Mean Sq   F N.Perm Pr (>F)
Groups    3 0.0704 0.02346 5.65   999 0.006 **
Residuals 36 0.1495 0.00415
---
Signif. codes: 0 '***' 0.001 '**' 0.01 '*' 0.05 '.' 0.1 ' ' 1
```

The results show that Groups have significant different variances and hence the PERMANOVA results may be potentially explained by that.

The above permutation test for homogeneity of multivariate dispersions also provides both observed *P*-value and permuted *P*-value for pairwise comparisons. We extract the results as below:

Pairwise comparisons:

```
(Observed p-value below diagonal, permuted p-value above diagonal)
            KO_BEFORE  KO_POST WT_BEFORE WT_POST
KO_BEFORE              0.54300   0.10300    0.01
KO_POST     0.51982              0.03300    0.05
WT_BEFORE   0.10091   0.03762               0.00
WT_POST     0.01722   0.05324   0.00221
```

Step 3: Conduct pairwise comparisons using Tukey's method.

```
> # Tukey's honest significant differences
> (disp.HSD <- TukeyHSD (disp))
 Tukey multiple comparisons of means
  95% family-wise confidence level

Fit: aov (formula = distances ~ group, data = df)

$group
                        diff     lwr      upr     p adj
KO_POST-KO_BEFORE    -0.01613 -0.09374 0.06148 0.9433
WT_BEFORE-KO_BEFORE  0.06368 -0.01393 0.14129 0.1399
WT_POST-KO_BEFORE    -0.05223 -0.12984 0.02538 0.2843
WT_BEFORE-KO_POST    0.07981 0.00220 0.15742 0.0419
WT_POST-KO_POST      -0.03610 -0.11371 0.04151 0.5983
WT_POST-WT_BEFORE    -0.11591 -0.19352 -0.03830 0.0015
```

The set of confidence intervals (CIs) is created TukeyHSD.betadisper() function based on the studentized range statistic, Tukey's "Honest Significant Difference"

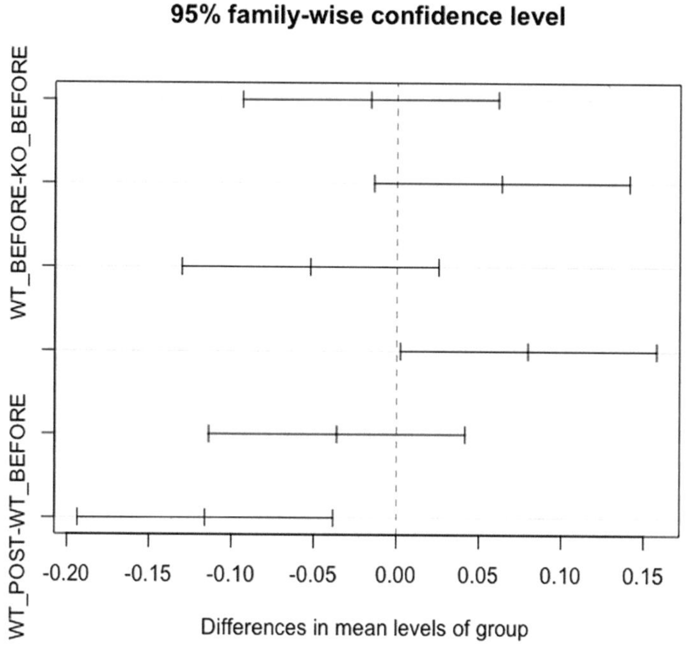

Fig. 11.2 The 95% family-wise confidence on differences in the mean levels of group

method. The CIs are on the differences between the mean distance to centroid of the levels of the grouping factor with the specified family-wise probability of coverage.

Step 4: Visualize homogeneity of multivariate dispersions by plot() and boxplot ().

The 95% family-wise confidence on differences in the mean levels of group can be plotted (Fig. 11.2):

```
> # Figure 11.2
> plot (disp.HSD)
```

The following *R* commands plot the groups and distances to centroids on the first two PCoA axes with data ellipses instead of hulls (Fig. 11.3):

```
> # Figure 11.3
> plot (disp, ellipse = TRUE, hull = FALSE) # 1 sd data ellipse
> plot (disp, ellipse = TRUE, hull = FALSE, conf = 0.90) # 90% data ellipse
```

The following *R* commands draw a boxplot of the distances to centroid for each group (Fig. 11.4):

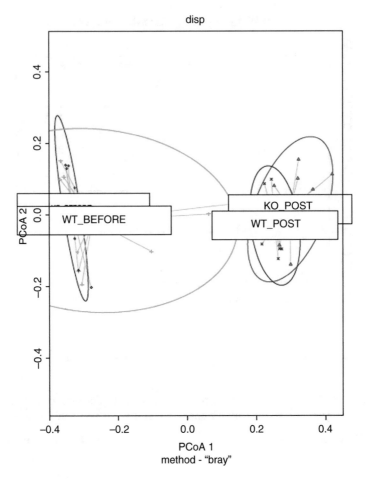

Fig. 11.3 Groups and distances to centroids on the first two PCoA axes with data ellipses

```
> # Figure 11.4
> boxplot(disp, ylab = "Distance to centroid", xlab = "Groups",
+ lty = "solid", lwd = 1, label.cex = 1)
```

11.5   Pairwise PERMANOVA

11.5.1   Introduction to Pairwise PERMMANOVA

In ANOVA and also PERMANOVA, if a null hypothesis of no difference among groups is rejected, then it suggests that there is a significant difference among the

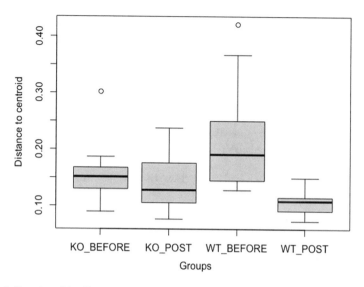

Fig. 11.4 Boxplot of the distances to centroid for each group

defined groups; however, there is no way to know which groups are significantly separated. One way to investigate which groups are different from each other is to conduct pairwise comparisons using an appropriate test and a statistical method to adjust the *P*-values for multiple comparisons, for example, as we did in Sect. 11.4.2 using permutation test and Tukey's method. However, the function **pairwise.perm. manova()** from the RVAideMemoire package (Testing and Plotting Procedures for Biostatistics, version 0.9.81.2 on February 21, 2022, by Maxime Hervé (2022)) is one of specifically designed statistical tools for pairwise comparisons of each group level with corrections for multiple testing after implementing permutational MANOVA. We refer this method as to pairwise PERMMANOVA to represent Pairwise Permutation MANOVA. Several *P*-value adjustment methods are available for the users to choose, including "bonferroni" (Bonferroni 1936), "holm" (Holm 1979), "hochberg" (Hochberg 1988) and "hommel" (Hommel 1988), "BH" or its alias "FDR" (Benjamini and Hochberg 1995), and "BY" (Benjamini and Yekutieli 2001). Tukey method is not available in this package. Type **p.adjust()** to check the options and references in R. If you do not want to adjust the *P*-value, use the pass-through option ("none").

11.5.2 Implement Pairwise PERMMANOVA Using the RVAideMemoire Package

Example 11.4: Breast Cancer QTRT1 Mouse Gut Microbiome, Example 11.1 Cont.

In this section, we will illustrate how to conduct pairwise comparisons of group using the function pairwise.perm.manova() and how to adjust the *P*-values for

multiple comparisons using various available statistical methods. We will still use
the same breast cancer QTRT1 mouse gut microbiome dataset for this illustration.
We first need to install and load the RVAideMemoire package:

```
install.packages("RVAideMemoire")
library(RVAideMemoire)
```

We then can use either one of the following three sample calls to conduct a
pairwise permutational MANOVA:

```
> pairwise.perm.manova(vegdist(t(otu_tab_rel),"bray"),
meta_tab_rel$Group4, nperm=999)
> # or
> bray<-vegdist(t(otu_tab_rel),"bray")
> pairwise.perm.manova(bray,meta_tab_rel$Group4,nperm=999)
> # or
> pairwise.perm.manova(bray, meta_tab_rel$Group4, test = c("Pillai",
+            "Wilks","Hotelling-Lawley", "Roy", "Spherical"),
+            nperm = 1000,progress = TRUE, p.method = "fdr")
```

where the argument:

- **vegdist**(t(otu_tab_rel) is specified to calculate distance matrix; this argument
  could also be either a typical matrix (one column per variable) or a data frame.
- **"bray"** is specified to calculate Bray-Curtis dissimilarity.
- **meta_tab_rel$Group4** is a grouping factor variable in meta matrix.
- **nperm** = 999 specifies the number of permutations to obtain the P-value.
- **test** = "Pillai" is used to choose the test statistic when the first argument (e.g.,
  bray) is a matrix.
- **progress** = TRUE or FALSE indicates whether or not display the progress bar.
- **p.method** = "fdr" specifies FDR method for P-value correction. We can choose
  p.method = "none" to pass through when we do not want to adjust the P-values.

Without adjusting P-values when p.method = "none" is chosen.

```
> pairwise.perm.manova(vegdist(t(otu_tab_rel),"bray"),
meta_tab_rel$Group4,
+            nperm=999, p.method = "none")
```

The partial output of pairwise comparisons using permutation MANOVAs on a
distance matrix is given below:

```
         KO_BEFORE KO_POST WT_BEFORE
KO_POST      0.001      -        -
WT_BEFORE    0.155   0.001       -
WT_POST      0.001   0.001    0.001

P value adjustment method: none
```

Adjust P-values using "fdr" method.

```
> pairwise.perm.manova(vegdist(t(otu_tab_rel),"bray"),
meta_tab_rel$Group4,
+                nperm=999, p.method = "fdr")
```

The partial output is given below:

```
              KO_BEFORE KO_POST WT_BEFORE
KO_POST      0.002        -        -
WT_BEFORE    0.161      0.002      -
WT_POST      0.002      0.002    0.002

P value adjustment method: fdr
```

For other methods for adjusting *P*-values, we will only provide the *R* commands but omit the outputs to save the text space.

Adjust P-values using "bonferroni" method.

```
> pairwise.perm.manova(vegdist(t(otu_tab_rel),"bray"),
meta_tab_rel$Group4,
+                nperm=999, p.method = "bonferroni")
```

Adjust P-values using "holm" method.

```
> pairwise.perm.manova(vegdist(t(otu_tab_rel),"bray"),
meta_tab_rel$Group4,
+                nperm=999, p.method = "holm")
```

Adjust P-values using "hochberg" method.

```
> pairwise.perm.manova(vegdist(t(otu_tab_rel),"bray"),
meta_tab_rel$Group4,
+                nperm=999, p.method = "hochberg")
```

Adjust P-values using "hommel" method.

```
> pairwise.perm.manova(vegdist(t(otu_tab_rel),"bray"),
meta_tab_rel$Group4,
+                nperm=999, p.method = "hommel")
```

Adjust P-values using "BH" method.

```
> pairwise.perm.manova(vegdist(t(otu_tab_rel),"bray"),meta_tab_rel
$Group4,
+ nperm=999, p.method = "BH")
```

Adjust P-values using "BY" method.

```
> pairwise.perm.manova(vegdist(t(otu_tab_rel),"bray"),meta_tab_rel
$Group4,
+ nperm=999, p.method = "BY")
```

11.6   Identify Core Microbial Taxa Using the Microbiome Package

When a significant association between microbiome abundance and an experiment or biological condition has been detected, microbiome researchers often want to investigate which core or top microbiota (microbial taxa) have contributed to the differences of abundance for further intervention design.

Example 11.5: Breast Cancer QTRT1 Mouse Gut Microbiome, Example 11.1 Cont.

For example, in our study of QTRT1 and gut microbiome in breast cancer (Jilei Zhang et al. 2020), we want to know which top 10 bacteria have contributed to the differences between QTRT1 knockout and wild type as well as before and post treatment.

Step 1: Perform PERMANOVA using the function adonis() and save the modeling results into the object "permanova".

```
> set.seed(123)
> permanova <- adonis(t(otu_tab_rel) ~ Group4,
+ data = meta_tab_rel, permutations=999, method = "bray")
```

Partial output is given below:

```
          Df SumsOfSqs MeanSqs F.Model  R2 Pr(>F)
Group4     3     3.34   1.112    34 0.739  0.001 ***
Residuals 36     1.18   0.033       0.261
Total     39     4.51            1.000
---
Signif. codes:  0 '***' 0.001 '**' 0.01 '*' 0.05 '.' 0.1 ' ' 1
```

A significant difference among these four levels of groups has been found. Now we investigate which top 10 taxa contribute to the separation of the groups.

Step 2: Use the function coefficients() to extract the coefficients of Group4 variable.

By checking the name of Group4 variable of the object "permanova" using head() or tail() function, we know the second column of $model.matrix is named as "Group41":

```
head(permanova)
# or
tail(permanova)
```

```
> # Show coefficients for the top taxa separating the groups
> coef <- coefficients(permanova)["Group41",]
> # Order and extract top 10 coefficients for the group variable
> top_coef <- coef[rev(order(abs(coef)))[1:10]]
```

Step 3: Plot the top taxa using the barplot() function (Fig. 11.5).

```
> # Figure 11.5
> par(mar = c(3, 14, 2, 1))
> barplot(sort(top_coef), horiz = T, las = 1, main = "Top 10 taxa")
```

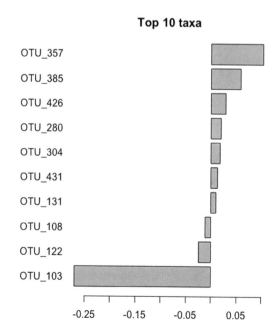

Fig. 11.5 Top 10 taxa in breast cancer QTRT1 mouse dataset

Table 11.1 Top 10 core species identified by PERMANOVA that result in knockout QTRT1 in the MCF7 cells

OTUID	Taxa	Description
OTU_357	D_5__Lachnospiraceae NK4A136 group	Positive
OTU_385	D_4__Lachnospiraceae	Positive
OTU_426	D_5_Ruminococcaceae UCG-014	Positive
OTU_280	D_5__Lactobacillus	Positive
OTU_304	D_4__Clostridiales vadinBB60 group	Positive
OTU_431	D_4__Ruminococcaceae	Positive
OTU_131	D_6__Parabacteroides goldsteinii CL02T12C30	Positive
OTU_108	D_5__Alloprevotella	Negative
OTU_122	D_5_Alistipes	Negative
OTU_103	D_4__Muribaculaceae	Negative

We identified ten most abundant taxa from taxa table, of which seven taxa (*Lachnospiraceae* NK4A136 group, *Lachnospiraceae*, *Ruminococcaceae* UCG-014, *Lactobacillus*, *Clostridiales* vadinBB60 group, *Ruminococcaceae*, and *Parabacteroides goldsteinii* CL02T12C30) are positively associated with the differences of group and three taxa (*Alloprevotella*, *Alistipes*, and *Muribaculaceae*) are negatively associated with the differences of group (see Table 11.1).

11.7   Statistical Testing of Beta Diversity in QIIME 2

As we described in Chap. 9, in QIIME 2, diversity analyses are available through the q2-diversity plugin. The analyses include calculating alpha and beta diversity, statistical testing group differences, and generating interactive visualizations. In Chap. 10 (Sect. 10.4), we introduced the calculations of beta diversity measures and the exploration of principal coordinates (PCoA) using emperor plots in QIIME 2. In this section, we introduce statistical testing of beta diversity between group differences in QIIME 2.

Example 11.6: Mouse Gut Microbiome Data, Example 9.12, Cont.
In Chap. 9 (Sect. 9.5.1) and Chap. 10 (Sect. 10.4.1), we generated all the alpha diversity and beta diversity measures available in QIIME 2 using a single-command qiime diversity **core-metrics-phylogenetic**. Here, we repeat this single command as below for recall:

```
source activate qiime2-2022.2
cd QIIME2R-Bioinformatics
qiime diversity core-metrics-phylogenetic \
 --i-phylogeny RootedTreeMiSeq_SOP.qza \
 --i-table FeatureTableMiSeq_SOP.qza \
 --p-sampling-depth 1153 \
```

```
--m-metadata-file SampleMetadataMiSeq_SOP.tsv \
--output-dir CoreMetricsResults
```

The command generated several beta diversity artifacts including (1) Bray-Curtis distance matrix (bray_curtis_distance_matrix.qza), (2) Jaccard distance matrix (jaccard_distance_matrix.qza), (3) unweighted UniFrac distance matrix (unweighted_unifrac_distance_matrix.qza), and (4) weighted UniFrac distance matrix (weighted_unifrac_distance_matrix.qza). We can use the qiime diversity **beta-group-significance** command to test the significant differences of beta diversity measures between sample groups.

The beta diversity group significance command generates boxplots of the beta diversity values to visualize the distance between samples aggregated by categorical variables or groups defined in the sample metadata table. It also performs statistical testing of significant differences among groups using the default method PERMANOVA (Anderson 2001) or optional method ANOSIM (Chapman and Underwood 1999). We will perform a statistical testing significant difference in the microbial community beta diversity measures of the time variable in the mouse gut microbiome example.

Different from diversity **alpha-group-significance** command, which tests for all categorical variables in sample metadata, the diversity **beta-group-significance** command compares only one sample metadata grouping at a time, so we need to specify the appropriate column name from the sample metadata file to conduct statistical testing.

11.7.1   Significant Testing of Bray-Curtis Distance

The following qiime diversity **beta-group-significance** command conducts significant testing of Bray-Curtis distance:

```
qiime diversity beta-group-significance \
--i-distance-matrix CoreMetricsResults/
bray_curtis_distance_matrix.qza\
--m-metadata-file SampleMetadataMiSeq_SOP.tsv \
--m-metadata-column Time \
--o-visualization CoreMetricsResults/BrayCurtisTimeSignificance.qzv
\
--p-pairwise
```

Saved Visualization to: CoreMetricsResults/BrayCurtisTimeSignificance.qzv

The above command outputs three results: overview, group significance plots, and pairwise permanova results. The overview results show an all-group PERMANOVA analysis on the Bray-Curtis dissimilarity measures for 2 time groups with sample size 349. The PERMANOVA method performs 999 permutations, resulting pseudo-*F* test statistic of 38.5453 with *P*-value of 0.001. Thus the test result reveals that the two time groups have significant differences in their community compositions. The pairwise permanova results show that the pseudo-*F* test statistic and *P*-value are the same as the results from overall test since in this case there are only two-group comparisons. The *Q*-value equals to the *P*-value of 0.001 since there is no adjustment here.

The raw distance can be downloaded as .tsv file. Two group significance plots are generated including distance to early and distance to late groups and can be saved as PDF format. The pairwise permanova results can be downloaded as CSV format.

11.7.2   Significant Testing of Jaccard Distance

The following qiime diversity **beta-group-significance** command conducts significant testing of Jaccard distance:

```
qiime diversity beta-group-significance \
 --i-distance-matrix CoreMetricsResults/jaccard_distance_matrix.qza \
 --m-metadata-file SampleMetadataMiSeq_SOP.tsv \
 --m-metadata-column Time \
 --o-visualization CoreMetricsResults/JaccardTimeSignificance.qzv \
 --p-pairwise
```

Saved Visualization to: CoreMetricsResults/JaccardTimeSignificance.qzv

11.7.3   Significant Testing of Unweighted UniFrac Distance

The following qiime diversity **beta-group-significance** command conducts significant testing of unweighted UniFrac distance:

```
qiime diversity beta-group-significance \
 --i-distance-matrix CoreMetricsResults/
unweighted_unifrac_distance_matrix.qza \
 --m-metadata-file SampleMetadataMiSeq_SOP.tsv \
 --m-metadata-column Time \
 --o-visualization CoreMetricsResults/
UnweightedUniFracTimeSignificance.qzv \
 --p-pairwise
```

Saved Visualization to: CoreMetricsResults/UnweightedUniFracTimeSignificance.qzv

11.7.4 Significant Testing of Weighted UniFrac Distance

The following qiime diversity beta-group-significance command conducts significant testing of weighted UniFrac distance:

```
qiime diversity beta-group-significance \
 --i-distance-matrix CoreMetricsResults/
weighted_unifrac_distance_matrix.qza \
 --m-metadata-file SampleMetadataMiSeq_SOP.tsv \
 --m-metadata-column Time \
 --o-visualization CoreMetricsResults/
WeightedUniFracTimeSignificance.qzv \
 --p-pairwise
```

Saved Visualization to: CoreMetricsResults/WeightedUniFracTimeSignificance.qzv

11.8 Summary

Chapter 10 focused on various beta diversity metrics and their calculation as well as exploring them using various ordination techniques. In this chapter, we continued to investigate beta diversity, but focused on statistical hypothesis testing. We first described nonparametric methods for multivariate analysis of variance using permutation tests in ecological and microbiome studies. We then focused on two multivariate nonparametric hypothesis testing methods of beta diversity: ANOSIM and PERMANOVA. We comprehensively described the development of these methods and illustrated their use in microbiome studies through the vegan, phyloseq, and microbiome packages. Followed that, we also illustrated how to analyze multivariate homogeneity of group dispersions, how to conduct pairwise PERMANOVA using the RVAideMemoire package, as well as how to detect core microbial taxa using the microbiome package. Finally, we introduced statistical hypothesis testing for differences of beta diversity between groups in QIIME 2: Bray-Curtis, Jaccard, unweighted UniFrac, and weighted UniFrac distances, respectively.

In Chap. 12, we will introduce the differential abundance analysis of microbiome data using metagenomeSeq.

References

Anderson, M.J. 2001. A new method for non-parametric multivariate analysis of variance. *Austral Ecology* 26: 32–46.

Anderson, Marti J., Kari E. Ellingsen, and Brian H. McArdle. 2006. Multivariate dispersion as a measure of beta diversity. *Ecology Letters* 9 (6): 683–693. https://doi.org/10.1111/j.1461-0248.2006.00926.x. https://onlinelibrary.wiley.com/doi/abs/10.1111/j.1461-0248.2006.00926.x.

Benjamini, Y., and Y. Hochberg. 1995. Controlling the false discovery rate: A practical and powerful approach to multiple testing. *Journal of the Royal Statistical Society Series B* 57: 289–300.

Benjamini, Y., and D. Yekutieli. 2001. The control of the false discovery rate in multiple testing under dependency. *Annals of Statistics* 29: 1165–1188.

Biondini, Mario E., Paul W. Mielke, and Edward F. Redente. 1991. Permutation techniques based on euclidean analysis spaces: A new and powerful statistical method for ecological research. In *Computer assisted vegetation analysis*, 221–240. Springer.

Bonferroni, C. E. 1936. Teoria statistica delle classi e calcolo delle probabilità. *Pubblicazioni del R Istituto Superiore di Scienze Economiche e Commerciali di Firenze*: Bonferroni, C. E., Teoria statistica delle classi e calcolo delle probabilità, Pubblicazioni del R Istituto Superiore di Scienze Economiche e Commerciali di Firenze 1936.

Chapman, M.G., and A.J. Underwood. 1999. Ecological patterns in multivariate assemblages: Information and interpretation of negative values in ANOSIM tests. *Marine Ecology Progress Series* 180: 257–265. http://www.jstor.org/stable/24852107.

Clarke, K. Robert. 1993. Non-parametric multivariate analyses of changes in community structure. *Australian Journal of Ecology* 18 (1): 117–143.

Clarke, K.R. 1988. Detecting change in benthic community structure. In *Proceedings of invited papers, 14th international biometric conference*. Belgium: Namour.

Edgington, Eugene S. 1995. *Randomization tests*. New York: M. Dekker.

Fisher, Ronald Aylmer. 1971. *The design of experiments*. 9th ed. New York: Hafner Press.

Hervé, Maxime. 2022. *RVAideMemoire: Testing and plotting procedures for biostatistics. R package version 0.9-81-2*. https://CRAN.R-project.org/package=RVAideMemoire.

Hochberg, Y. 1988. A Sharper Bonferroni procedure for multiple tests of significance. *Biometrika* 75: 800–803.

Holm, S. 1979. A simple sequentially rejective multiple test procedure. *Scandinavian Journal of Statistics* 6: 65–70.

Hommel, G. 1988. A stagewise rejective multiple test procedure based on a modified Bonferroni test. *Biometrika* 75: 383–386.

Hope, Adery C.A. 1968. A simplified Monte Carlo significance test procedure. *Journal of the Royal Statistical Society: Series B (Methodological)* 30 (3): 582–598.

Legendre, Pierre, and Marti J. Anderson. 1999. Distance-based redundancy analysis: Testing multispecies responses in multifactorial ecological experiments. *Ecological Monographs* 69 (1): 1–24. https://doi.org/10.2307/2657192. http://www.jstor.org/stable/2657192.

Linnenbrink, M., J. Wang, E.A. Hardouin, S. Kunzel, D. Metzler, and J.F. Baines. 2013. The role of biogeography in shaping diversity of the intestinal microbiota in house mice. *Molecular Ecology* 22 (7): 1904–1916.

Manly, B. 1997. *Randomization, bootstrap and Monte Carlo methods in biology*. 3rd ed. Boca Raton, FL: Chapman and Hall.

Mantel, Nathan. 1967. The detection of disease clustering and a generalized regression approach. *Cancer Research* 27 (2 Part 1): 209–220.

Mantel, Nathan, and Ranchhodbhai S. Valand. 1970. A technique of nonparametric multivariate analysis. *Biometrics*: 547–558.

McArdle, Brian, and M. Anderson. 2001. Fitting multivariate models to community data: A comment on distance-based redundancy analysis. *Ecology* 82: 290–297. https://doi.org/10.2307/2680104.

Jr Mielke, Paul W., Kenneth J. Berry, and Earl S. Johnson. 1976. Multi-response permutation procedures for a priori classifications. *Communications in Statistics-Theory and Methods* 5 (14): 1409–1424.

Oksanen, J., F. Guillaume Blanchet, et al. 2016. *Vegan: Community ecology package. R package*. http://CRAN.R-project.org/package=vegan.

Smith, Eric P., Kurt W. Pontasch, and Jr John Cairns. 1990. Community similarity and the analysis of multispecies environmental data: A unified statistical approach. *Water Research* 24 (4): 507–514.

Stevens, Martin Henry H., and Jari Oksanen. 2020. *Adonis: Permutational multivariate analysis of variance using distance matrices*. Package 'vegan': Community Ecology Package. Accessed March 18, 2021. https://cran.r-project.org/web/packages/vegan/vegan.pdf.

Tang, Zheng-Zheng, Guanhua Chen, and Alexander V. Alekseyenko. 2016. PERMANOVA-S: Association test for microbial community composition that accommodates confounders and multiple distances. *Bioinformatics* 32 (17): 2618–2625. https://doi.org/10.1093/bioinformatics/btw311.

ter Braak, Cajo J.F. 1986. Canonical correspondence analysis: A new eigenvector technique for multivariate direct gradient analysis. *Ecology* 67 (5): 1167–1179. https://doi.org/10.2307/1938672. https://esajournals.onlinelibrary.wiley.com/doi/abs/10.2307/1938672.

Xia, Yinglin, Jun Sun, and Ding-Geng Chen. 2018. Multivariate community analysis. In *Statistical analysis of microbiome data with R*, ed. Yinglin Xia, Jun Sun, and Ding-Geng Chen, 285–330. Singapore: Springer Singapore.

Zhang, J., Z. Wei, and J. Chen. 2018. A distance-based approach for testing the mediation effect of the human microbiome. *Bioinformatics* 34 (11): 1875–1883. https://doi.org/10.1093/bioinformatics/bty014.

Zhang, Jilei, Lu Rong, Yongguo Zhang, Żaneta Matuszek, Wen Zhang, Yinglin Xia, Tao Pan, and Jun Sun. 2020. tRNA queuosine modification enzyme modulates the growth and microbiome recruitment to breast tumors. *Cancers* 12 (3): 628. https://doi.org/10.3390/cancers12030628. https://pubmed.ncbi.nlm.nih.gov/32182756, https://www.ncbi.nlm.nih.gov/pmc/articles/PMC7139606/.

Chapter 12
Differential Abundance Analysis of Microbiome Data

Abstract Differential abundance analysis (DDA) is an active area in microbiome research. The main goal of DDA is to identify features (i.e., OTUs, ASVs, taxa, species) that are differentially abundant across sample groups. Microbiome data are sparse and have many zeros; thus to appropriately identify features, the models should be able to address the issues of over-dispersion, zero-inflation, and sparsity. This chapter first introduces the Zero-Inflated Gaussian (ZIG) and Zero-Inflated Log-normal (ZILN) mixture models. Then it illustrates the ZILN via the metagenomeSeq package. Next, it describes some additional statistical tests and some useful functions in metagenomeSeq. Next, some remarks on CSS, ZIG, and ZILN are also provided.

Keywords Zero-inflated Gaussian (ZIG) · Zero-inflated log-normal (ZILN) · Total sum scaling (TSS) · Cumulative sum scaling (CSS) · metagenomeSeq · Log normal permutation test · Presence-absence testing · Discovery odds ratio testing · Feature correlations · MRexperiment Object · cumNormMat() · MRcounts() · Normalization factors · calcNormFactors() · Library sizes · libSize() · Normalized counts · exportMat() · exportStats() · Taxa

In microbiome study, differential abundance analysis (DAA) is an active area of research. Over 12 years ago, the term "differential abundance" was coined as a direct analogy to differential expression from RNA-seq and has been adopted to use in microbiome literature (McMurdie and Holmes 2014; White et al. 2009; Paulson et al. 2013). The main goal of DAA is to identify features (i.e., OTUs, ASVs, taxa, species, etc.) that are differentially abundant across sample groups (i.e., features present in different abundances across sample groups).

Microbiome data are sparse and have many zeros; to appropriately identify features, the models that are used to analyze microbiome data should have the capability to address the issues of overdispersion, zero inflation, and sparsity. To

---

Supplementary Information The online version contains supplementary material available at https://doi.org/10.1007/978-3-031-21391-5_12.

address the overdispersion and zero inflation while identifying the microbial taxa that are associated with covariates, several specifically designed statistical models for microbiome data have been proposed. In this chapter, we first introduce the zero-inflated Gaussian (ZIG) and zero-inflated log-normal (ZILN) mixture models (Sect. 12.1). Then we illustrate the ZILN via the metagenomeSeq package (Sect. 12.2). Sections 12.3 and 12.4 introduce and illustrate some additional statistical tests and some useful functions in metagenomeSeq, respectively. Next, we provide some remarks on CSS, ZIG, and ZILN in Sect. 12.5. Finally, we complete this chapter with a brief summary in Sect. 12.6.

12.1 Zero-Inflated Gaussian (ZIG) and Zero-Inflated Log-Normal (ZILN) Mixture Models

Both ZIG mixture and ZILN models are implemented in the **metagenomeSeq** package (Paulson 2020) via the functions **fitZig()** and **fitFeatureModel()**, respectively. The function **fitZig()** relies heavily on the **limma** package (Smyth 2005). The ZIG mixture was proposed first in 2013 (Paulson et al. 2013), which estimates the probability whether a zero for a particular feature in a sample is a technical zero or not. The **fitFeatureModel()** was later reparametrized to fit a zero-inflated model for each specific OTU separately. Currently the ZILN model along with the **fitFeatureModel()** is recommended (Paulson 2020).

The proposal of ZIG mixture model was motivated by the observation that the depth of coverage in a sample is directly related to how many features are detected in a sample. It was shown that there exist a relationship between depth of coverage and OTU identification ubiquitous in marker-gene survey datasets. Thus, ZIG mixture model was proposed to use a probability mass to accommodate the excess zero counts (both sampling zeros and structural zeros) and a Gaussian distribution to model the mean group abundance (the nonzero observed counts) for each taxonomic feature to explicitly account for biases in DAA resulting from under-sampling of the microbial community.

The model directly estimates the probability that an observed zero is generated from the detection distribution owing to under-sampling (sampling zeros) or due to absence of the taxonomic feature in the microbial community (structural zeros), i.e., from the count distribution. An expectation-maximization algorithm is used to estimate the expected value of latent component indicators based on sample sequencing depth of coverage.

The model is implemented via the **metagenomeSeq** package, mainly by implementing two methods: the cumulative sum scaling (CSS) normalization and zero-inflated Gaussian (ZIG) distribution mixture model.

The ZIG mixture model was designed to use the CSS normalization technique to correct the bias in the assessment of differential abundance (DA) introduced by total sum (TSS) normalization.

12.1.1 Total Sum Scaling (TSS)

In RNA-seq and microbiome literature, TSS (Paulson et al. 2013; Chen et al. 2018) and proportion (McMurdie and Holmes 2014; Weiss et al. 2017) are referred as the same method. TSS uses the total read counts for each sample as the size factors to estimate the library size or scale the matrix counts. TSS normalizes count data by dividing taxon or OTU read counts by the total number of reads in each sample to convert the counts to proportion.

Assume raw data is given as count matrix $M(m, n)$, where m and n are the number of features (i.e., taxa, OTUs) and samples, respectively. Let counts c_{ij} represent the number of times taxonomic feature i was observed in sample j, the sum of counts for sample i is denoted as $s_j = \sum_i c_{ij}$, and then the normalized counts produced by the TSS normalization method is

$$\widetilde{c}_{ij} = \frac{c_{ij}}{s_j}. \tag{12.1}$$

At the beginning, microbiome researchers in practice simply used proportion or rarefying to normalize microbiome read counts. The proportion approach was directly adopted from RNA-seq field, while rarefying was inspired by the idea of rarefaction in ecology and hypergeometric model in RNA-seq data. In RNA-seq literature, TSS normalization has been evaluated to incorrectly bias differential abundance estimates (Bullard et al. 2010; Dillies et al. 2012) because a few genes could be sampled preferentially as sequencing yield increases in RNA-seq data derived through high-throughput technologies. For 16S rRNA-seq data, TSS or proportion also has been reviewed having limitations, including:

1. TSS is not robust to outliers (Chen et al. 2018), is inefficient to address heteroscedasticity, is unable to address overdispersion, and results in a high rate of false positives in tests for species. This systematic bias increases the type I error rate even after correcting for multiple testing (McMurdie and Holmes 2014).
2. TSS lies on the constant-sum constraint; hence it cannot remove compositionality; instead it is prone to create compositional effects, making nondifferential taxa or OTUs appear to be differential (Chen et al. 2018; Mandal et al. 2015; Tsilimigras and Fodor 2016; Morton et al. 2017).
3. TSS has a poor performance in terms of both FDR control and statistical power at the same false-positive rate, compared to other scaling or size factor-based methods such as GMPR and RLE because of strong compositional effects (Chen et al. 2018).
4. Actually, like to rarefying, TSS or proportion is built on the assumption that the individual gene or taxon count in each sample was randomly sampled from the reads in multiple samples. Thus, counts can be divided by total library size to convert to proportion, and gene expression analysis or taxon abundance analysis can be fitted via a Poisson distribution. Because in both RNA-seq data and 16S rRNA-seq data, the assumption cannot meet, the sample proportion approach is

inappropriate for detection of differentially abundant species (McMurdie and Holmes 2014) and should not be used for most statistical analyses (Weiss et al. 2017).

12.1.2   Cumulative Sum Scaling (CSS)

CSS normalization (Paulson et al. 2013) is a normalization method specifically designed for microbiome data. CSS aims to correct the bias in the assessment of differential abundance introduced by total sum normalization. CSS method is to divide raw counts into the cumulative sum of counts, up to a percentile determined using a data-driven approach in order to capture the relatively invariant count distribution for a dataset. In other words, the choices of percentiles are driven by empirical rather than theoretical considerations.

Let q_j^l denote the lth quantile of sample j (i.e., in sample j there are l taxonomic features with counts smaller than q_j^l). For $l = \lfloor .95\, m \rfloor$, q_j^l corresponds to the 95th percentile of the count distribution for sample j. Also denote $s_j^l = \sum\limits_{i|c_{ij} \le q_j^l} c_{ij}$ as the sum

of counts for sample j up to the lth quantile. Using this notation, the total sum $s_j = s_j^m$, CSS normalization method chooses a value $l \le m$ as a normalization scaling factor

for each sample to produce normalized counts $\widetilde{c}_{ij} = \left(\dfrac{c_{ij}}{s_j^l} \right)(N)$, where N is an appro-

priately chosen normalization constant.

In practice, the same constant N (the median is recommended using as scaling factor s_j^i across samples) is used to scale all samples so that normalized counts have interpretable units. Actually, CSS defines the sample-specific count distributions $l \le m$ as reference distribution (samples) and interprets counts for other samples as relative to the reference.

CSS requires to set the threshold as at least the 50th percentile. Then CSS calculate the normalization factors for samples as the sum over the taxa or OTUs counts up to the threshold. In such way, CSS method optimizes the threshold from the data to minimize the influence of variable high-abundant taxa or OTUs. For example, to mitigate the influence of larger count values in the same matrix column, CSS only scales the portion of each sample's count distribution that is relatively invariant across samples. CSS normalization can be implemented using metagenomeSeq package. By default, 50th percentile is set as the threshold.

12.1.3   ZIG and ZILN Models

ZIG mixture model assumes that CSS-normalized sample abundance measurements are approximately log-normally distributed given large numbers of samples. Thus, in

ZIG the normalized count data is first performed a logarithmic transform to control the variability of taxonomic feature measurements across samples.

Let n_A and n_B denote the microbiome count data for samples A and B from two populations, respectively, each with m features (OTUs); and let c_{ij} denote the raw count for sample j and feature i and $k(j) = I\{j \in groupA\}$ be the class indicator function. Then, the continuity-corrected $\log_2$ of the raw count data $y_{ij} = \log_2(c_{ij} + 1)$ is modeled by the zero-inflated model as a mixture of a point mass at zero $I_{\{0\}}(y)$ and a count distribution $f_{count}(y; \mu, \sigma^2) \sim N(\mu, \sigma^2)$. Given mixture parameters π_j, the density of the zero-inflated Gaussian distribution for feature i in sample j with s_j total counts is given as

$$f_{zig}\left(y_{ij}; s_j, \beta, \mu_i, \sigma_i^2\right) = \pi_j\left(s_j\right) \cdot I_{\{0\}}\left(y_{ij}\right) + \left(1 - \pi_j\left(s_j\right)\right) \cdot f_{count}\left(y_{ij}; \mu_i, \sigma_i^2\right), \quad (12.2)$$

while the mean model is specified as

$$E\left(y_{ij}|k(j)\right) = \pi_j \cdot 0 + \left(1 - \pi_j\right) \cdot \left(b_{i0} + \eta_i \log_2\left(\frac{s_j^l + 1}{N}\right) + b_{i1}k(j)\right), \quad (12.3)$$

where the parameter b_{i1} is an estimate of fold change in mean normalized counts between the two populations. The logged normalization factor $\log_2\left(\frac{s_j^l}{N}\right)$ term is used to capture OTU-specific normalization factors (e.g., feature-specific biases in PCR amplification) through parameter η_i. ZIG model can also be specified without OTU-specific normalization via including an offset term to serve as the normalization factor like in the linear model.

To model the dependency of the number of zero-valued features of a sample on its total number of counts, the mixture parameter $\pi_j(s_j)$ is modeled as a binomial process:

$$\log\left(\frac{\pi_j}{1 - \pi_j}\right) = \beta_0 + \beta_1 \cdot \log\left(s_j\right). \quad (12.4)$$

ZIG model was developed under the framework of linear modeling and standard conventions in DAA methods in gene expression; thus it has the advantage of controlling for confounding factors. Except the group membership variable, other clinical covariates can also be incorporated into the ZIG model to detect the confounding effect. Appropriate covariates can also be included to capture variability in the sampling process. The full set of estimates are denoted as $\theta_{ij} = \{\beta_0, \beta_1, b_{i0}, \eta_i, b_{i1}\}$. The mixture membership is treated as $\Delta_{ij} = 1$ if y_{ij} is generated from the zero point mass as latent indicator variables. Then the log likelihood in this extended model is given as

$$l\left(\theta_{ij}; y_{ij}, s_j\right) = \left(1 - \Delta_{ij}\right) \log f_{\text{count}}\left(y; \mu_i, \sigma_i^2\right) + \Delta_{ij} \log \pi_i\left(s_j\right)$$
$$+ \left(1 - \Delta_{ij}\right) \log\left\{1 - \pi_i\left(s_j\right)\right\}. \tag{12.5}$$

By this formation, this extended model is modeled as a mixture of a point mass at zero and a normal distribution, in which the part of the point mass at zero is used to model features (OTUs/taxa) that are usually sparse and have many zero counts and the normal distribution part is used to model the log transformed counts of the mean group abundance (the nonzero observed counts).The log maximum likelihood estimates are approximated by using the expectation-maximization algorithm. The function **fitZig()** estimates the probability whether a zero for a particular feature in a sample is a technical zero or not.

A moderated t-statistic fitFeatureModel is constructed to estimate fold change (b_{1i}) (only for the count component of the ZIG mixture model) and its standard error by empirical Bayes shrinkage of parameter estimates (Smyth 2005). The P-values are obtained by a hypothesis testing the null of $b_{1i} = 0$ using a parametric t-distribution. The zero-inflated log-normal model has been wrapped into the **fitFeatureModel()** to fit a zero-inflated model for each specific feature (OTU/taxon) separately. Currently, fitFeatureModel() is only implementable to binary covariate case. The moderated t-statistic is defined as below:

$$t_i = \frac{b_{1i}}{\left(\widetilde{s}_i^2 / \sum_j\left(1 - z_{ij}\right)\right)^{1/2}}. \tag{12.6}$$

where $\widetilde{s}_i^2 = \frac{d_0 s_0^2 + d_i s_i^2}{d_0 + d_i}$ is obtained by pooling all features' variances as described in Smyth (2005); s_i^2 and d_i are the observed feature variance and degrees of freedom, respectively; and d_0 and s_0^2 are estimated using the method of moments incorporating all feature variances and degrees of freedom, respectively. The multiple testing of features is corrected using the Q-value method (Storey and Tibshirani 2003). The choice of using a log-normal distribution in the count component of the mixture rather than a generalized linear model (e.g., negative binomial) in forming the **fitFeatureModel()** function is for computational efficiency and numerical stability as well as appropriateness of the log-normal distribution for the marker-gene survey study with moderate to large sample sizes (Soneson and Delorenzi 2013; Paulson et al. 2013).

12.2   Implement ZILN via metagenomeSeq

The **metagenomeSeq** package was designed to detect features (i.e., OTU, genus, species, etc.) that are differentially abundant between two or more groups of multiple samples. The software was also designed to implement the proposed CSS normalization technique and ZIG mixture model to account for sparsity due to under-sampling of microbial communities on disease association detection and testing of feature correlations (Paulson et al. 2013). The **metagenomeSeq** package implements two functions to (1) control for biases in measurements across taxonomic features via CSS and (2) account for varying depths of coverage via ZIG.

To facilitate modeling covariate effects or confounding sources of variability and interpreting results, the **metagenomeSeq** package uses linear model framework and provides a few visualization functions to examine discoveries. Generally, to implement **metagenomeSeq**, we need to take six steps: (1) load OTU, taxonomy data, and metadata; (2) create MRexperiment object; (3) normalize data via MRexperiment objects; (4) perform statistical testing of abundance or presence-absence; (5) visualize and save analysis results; and (6) visualize features. We illustrate how to implement ZIG using **metagenomeSeq** step by step as below.

Example 12.1: Breast Cancer QTRT1 Mouse Gut Microbiome

We introduced this dataset in Example 9.1 (Zhang et al. 2020) and used it to illustrate calculations and statistical hypothesis testing of alpha and beta diversities in Chaps. 9, 10, and 11, respectively. Here we continue to use this dataset to illustrate the metagenomeSeq package.

Step 1: Load OTU table, annotated taxonomy data, and metadata.

The **metagenomeSeq** package requires a feature/OTU table to store the count data in a delimited (tab by default) format with sample names along the first row and feature names along the first column. As recall in Chap. 2, a typical feature (OTU) table is given below:

	$sample_1$	$sample_2$	$\ldots$	$sample_n$
$feature_1$	c_{11}	c_{12}	$\ldots$	c_{1n}
$feature_2$	c_{21}	c_{22}	$\ldots$	c_{2n}
.				
.				
.				
$feature_m$	c_{m1}	c_{m2}	$\ldots$	c_{mn}

The above table presents m features in n samples, where the elements in a count matrix C (m, n), c_{ij}, are the number of reads annotated for a particular feature i (i.e., OTU, genus, species, etc.) in sample j.

To create a MRexperiment object, we need to load count data, taxonomy data, and metadata. First we set up working directory and install the **metagenomeSeq** package (version 1.36.0, April 10, 2022):

```
> setwd("~/Documents/QIIME2R/Ch12_Differential")
```

To install the **metagenomeSeq** package, type the following command in R (starting version 3.6):

```
> if (!requireNamespace("BiocManager", quietly = TRUE))
+ install.packages("BiocManager")
> BiocManager::install("metagenomeSeq")
```

The **metagenomeSeq** package requires a feature table, which is a count matrix with features along rows and samples along the columns. We can use the **loadMeta()** function to load count data, which is tab-delimited OTU matrix. The **loadMeta()** function loads the taxa (OTUs) and counts into a list:

```
> library(metagenomeSeq)
> # Load otu table
> otu_tab = loadMeta("otu_table_L7_MCF7_ZIG.txt")
> head(otu_tab$counts,3)
    Sun071.PG1 Sun027.BF2 Sun066.PF1 Sun029.BF4 Sun068.PF3 Sun026.BF1
OTU_1      0          0          0          0          0          0
OTU_2      0          0          0          0          1          0
OTU_3      0          0          0          0          1          0
------
> head(otu_tab$taxa,3)
  taxa
1 OTU_1
2 OTU_2
3 OTU_3
> dim(otu_tab$counts)
[1] 635 40
> dim(otu_tab$taxa)
[1] 635 1
```

Now we load the tax table. It requires that the row order of OTUs in taxonomy table (matrix with tab-delimited format) must match the row order of OTUs in OTU table (matrix with tab-delimited format):

```
> # Load taxa table
> tax_tab = read.delim( "tax_table_L7_MCF7_ZIG.txt", sep = "\t",
stringsAsFactors = FALSE, row.names = 1)
> head(tax_tab,3)
      Kingdom        Phylum          Class          Order
OTU_1 D_0__Archaea D_1__Thaumarchaeota D_2__Nitrososphaeria
D_3__Nitrososphaerales
OTU_2 D_0__Bacteria D_1__Acidobacteria D_2__Acidobacteriia
D_3__Solibacterales
OTU_3 D_0__Bacteria D_1__Acidobacteria D_2__Acidobacteriia
D_3__Solibacterales
                    Family        Genus Species
OTU_1        D_4__Nitrososphaeraceae        Other Other
OTU_2 D_4__Solibacteraceae (Subgroup 3) D_5__Bryobacter Other
OTU_3 D_4__Solibacteraceae (Subgroup 3)        Other Other
```

In above **read.delim** () function, specifying the argument "stringsAsFactors = FALSE" is important. By default, "stringsAsFactors" is set to TRUE in the read.

delim() and other functions including read.table(), read.csv(), and read.csv2() when importing the columns containing character strings into R as factors. The reason for setting strings as factors is to tell R to treat categorical variables into individual dummy variables for modeling functions like **lm()** and **glm()**. However, in microbiome or genomics study, it doesn't make sense to encode the names of the taxa or genes in one column of data as factors because they are essentially just labels and not to be used in any modeling function. So, we need to specify the "stringsAsFactors" argument into FALSE to change the default setting.

Phenotype data can be optionally loaded into R with **loadPhenoData()**. This function loads the data as a list. Now we use loadPhenoData () function to load the metadata into R:

```
> # Load metadata
> meta_tab = loadPhenoData("metadata_QtRNA.txt", tran = TRUE, sep =
"\t")
> head(meta_tab,3)
       MouseID Genotype Group  Time   Group4 Total.Read
Sun071.PG1   PG1     KO   1  Post  KO_POST    61851
Sun027.BF2   BF2     WT   0 Before WT_BEFORE   42738
Sun066.PF1   PF1     WT   0  Post  WT_POST    54043
```

```
> match_ord = match(colnames(otu_tab$counts), rownames(meta_tab))
> meta_tab = meta_tab[match_ord, ]
> head(meta_tab[1:2], 3)
       MouseID Genotype
Sun071.PG1   PG1     KO
Sun027.BF2   BF2     WT
Sun066.PF1   PF1     WT
```

The argument tran = (Boolean) is used to specify whether the covariates are along the columns and samples along the rows. If it is true, then tran should equal TRUE.

Step 2: Create MRexperiment object.

In R the S4 class system allows for object-oriented definitions. The **MRexperiment** object of **metagenomeSeq** was built using the **Biobase** package in Bioconductor and its virtual class, eSet (Gentleman et al. 2020). The **eSet** class (container) is used to contain high-throughput assays and experimental metadata. Classes derived from eSet contain one or more identical-sized matrices as assayData elements. To derive classes (e.g., ExpressionSet-class, SnpSet-class), we need to specify which elements must be present in the assayData slot. As a virtual class, eSet object cannot be instantiated directly. To assess eSet object, we need to find its slots. The slots of eSet object include:

- **assayData**: contains matrices with equal dimensions and with column number equal to nrow(phenoData), which is an AssayData-class.

- **phenoData**: contains experimenter-supplied variables describing sample (i.e., columns in assayData) phenotypes, which is an AnnotatedDataFrame-class.
- **featureData**: contains variables describing features (i.e., rows in assayData) unique to this experiment, which is an AnnotatedDataFrame-class.
- **experimentData**: contains details of experimental methods, which is a MIAME-class.
- **annotation**: label associated with the annotation package used in the experiment, which is character.
- **protocolData**: contains microarray equipment-generated variables describing sample (i.e., columns in assayData) phenotypes, which is an AnnotatedDataFrame-class.

The MRexperiment object uses a simple extension of eSet called **expSummary** (the S4 class with adding a single experiment summary slot, a data frame) to include the depth of coverage and the normalization factors for each sample. In MRexperiment object, the three main slots of **assayData**, **phenoData**, and **featureData** are most useful. All matrices are organized in the **assayData** slot, all phenotype data is stored in **phenoData**, and feature data (OTUs, taxonomic assignment to varying levels, etc.) is stored in **featureData**. Additional slots are available for reproducibility and annotation.

A MRexperiment object will be created by the function **newMRexperiment()**. This function takes a count matrix, phenoData (annotated data frame), and featureData (annotated data frame) as input. To create a MRexperiment object, call the **newMRexperiment()** and pass the counts, phenotype, and feature data as parameters. After the datasets are formatted as MRexperiment objects, the analysis is relatively easy using the metagenomeSeq package.

Below we create a MRexperiment object named "MRobj". First, we convert both phenoData and featureData into data frames using the **AnnotatedDataFrame()** function:

```
> phenoData = AnnotatedDataFrame(meta_tab)
> phenoData
An object of class 'AnnotatedDataFrame'
 rowNames: Sun071.PG1 Sun027.BF2 ... Sun062.PE2 (40 total)
 varLabels: MouseID Genotype ... Total.Read (6 total)
 varMetadata: labelDescription
```

```
> taxaData = AnnotatedDataFrame(tax_tab)
> taxaData
An object of class 'AnnotatedDataFrame'
 rowNames: OTU_1 OTU_2 ... OTU_635 (635 total)
 varLabels: Kingdom Phylum ... Species (7 total)
 varMetadata: labelDescription
```

Then we call the **newMRexperiment()** function to create the object "MRobj":

```
> MRobj = newMRexperiment(otu_tab$counts,phenoData=phenoData,
featureData=taxaData)
> # Links to a paper providing further details can be included
optionally.
> # experimentData(obj) = annotate::pmid2MIAME("21680950")
> MRobj
MRexperiment (storageMode: environment)
assayData: 635 features, 40 samples
 element names: counts
protocolData: none
phenoData
 sampleNames: Sun071.PG1 Sun027.BF2 ... Sun062.PE2 (40 total)
 varLabels: MouseID Genotype ... Total.Read (6 total)
 varMetadata: labelDescription
featureData
 featureNames: OTU_1 OTU_2 ... OTU_635 (635 total)
 fvarLabels: Kingdom Phylum ... Species (7 total)
 fvarMetadata: labelDescription
experimentData: use 'experimentData(object)'
Annotation:
```

Step 3: Normalize data via MRexperiment objects.

After the MRexperiment object was created, we can use the MRexperiment object to normalize the data, perform statistical tests (abundance or presence-absence), and visualize or save the analysis results.

The proposed method implements two data-driven methods for the percentile calculation:

(i) Calculate normalization factors using the function cumNorm().

The normalization starts with the first critical step: calculate the proper percentile by which to normalize counts. The normalization factors are stored in the experiment summary slot.

First, check if the samples were normalized.

We can use the **normFactors ()** function to access the normalization factors (aka the scaling factors) of samples in a MRexperiment object:

```
> # Access the normalization factors (aka the scaling factors)
> # Check if the samples were normalized
> head(normFactors(MRobj),3)
          [,1]
Sun071.PG1  NA
Sun027.BF2  NA
Sun066.PF1  NA
```

The above outputs showed that the samples were not normalized.

Second, calculate the proper percentile using the cumNormStat( )and cumNormStatFast( ).

To calculate the proper percentile for which to sum counts up to and scale by, we can use either **cumNormStat ()** or **cumNormStatFast()**.

The syntax of **cumNormStat ()** is given as:

```
cumNormStat(obj, qFlag = TRUE, pFlag = FALSE, rel = 0.1, ...),
```

where the argument **obj** is a matrix or MRexperiment object; **qFlag** is a flag that is used to either calculate the proper percentile using R's stepwise quantile function or approximate function; **pFlag** is a flag that is used to plot the relative difference of the median deviance from the reference; **rel** is used to specify a cutoff for the relative difference from one median difference from the reference to the next; and the argument **...** is used to specify additional plotting parameters, which is applicable if **pFlag** ==TRUE.

The **cumNormStatFast()** function is faster than the **cumNormStat()** function.

The syntax of **cumNormStatFast()** is given as:

```
cumNormStatFast(obj, pFlag = FALSE, rel = 0.1, ...)
```

Below we use the **cumNormStatFast()** and the **cumNormStat ()** functions to calculate the percentile, respectively. By using default, these two functions generate same percentile values:

```
> ptile = round(cumNormStatFast(MRobj,pFlag=FALSE),digits=2)
Default value being used.
> ptile
[1] 0.5
> ptile1 = round(cumNormStat(MR_obj,pFlag=FALSE),digits=2)
Default value being used.
> ptile1
[1] 0.5
```

Third, calculate the scaling factors using the cumNorm ().

The **cumNorm ()** function calculates each column's quantile and the sum up to quantile including that quantile. We can either first calculate the percentile using the **cumNormStatFast()/cumNormStat ()** and then use this percentile as input to the **cumNorm ()** function, like this p = cumNormStatFast(obj), obj = cumNorm(obj, p = p), or directly call the **cumNorm()** function like this: cumNorm(obj, p = cumNormStatFast(obj)):

```
> # Calculate the scaling factors
> QtRNAData = cumNorm(MRobj, p = ptile)
> QtRNAData
MRexperiment (storageMode: environment)
assayData: 635 features, 40 samples
 element names: counts
protocolData: none
phenoData
 sampleNames: Sun071.PG1 Sun027.BF2 ... Sun062.PE2 (40 total)
 varLabels: MouseID Genotype ... Total.Read (6 total)
 varMetadata: labelDescription
featureData
 featureNames: OTU_1 OTU_2 ... OTU_635 (635 total)
 fvarLabels: Kingdom Phylum ... Species (7 total)
 fvarMetadata: labelDescription
experimentData: use 'experimentData(object)'
Annotation:
```

We can also combine the cumNorm() and cumNormStatFast() functions in one step to calculate the scaling factors:

```
> QtRNAData1 = cumNorm(MRobj, p = cumNormStatFast(MRobj))
```

(ii) Calculate normalization factors using the function wrenchNorm ().

An alternative to normalizing counts (calculating normalization factors) using **cumNorm** is to use the **wrenchNorm** method. These two functions behave similarly; however, the difference is that the **cumNorm** method uses the percentile, while the **wrenchNorm** method takes the condition as argument. The condition is a factor with values that separate samples into phenotypic groups of interest (in this case, Group/Genotype). The authors of this package stated that when it is used appropriately, wrench normalization is preferable over cumulative normalization:

```
> conds = MRobj$Group
> QtRNAData2 = wrenchNorm(MRobj, condition = conds)
```

Step 4: Perform statistical testing of abundance.

The differential abundance analysis of features (e.g., OTUs) can be fitted using the functions **fitFeatureModel()** and **fitZig()**. The **fitFeatureModel()** was a latest development by reparametrizing the authors' original zero-inflation model to fit zero-inflated log-normal mixture model for each feature (OTU) separately. The **fitFeatureModel()** was recommended over **fitZig()** in the **metagenomeSeq** package.

The syntax is given as below:

```
fitFeatureModel(obj, mod, coef = 2, B = 1, szero = FALSE, spos = TRUE).
```

where the argument:

- **obj** is a MRexperiment object with count data.
- **mod** is used to specify the model for the count distribution.
- **coef** is the coefficient of interest to grab log fold changes.
- **B** is used to specify the number of bootstraps to perform. If B is specified greater than 1, then it performs permutation test.
- **szero** (TRUE/FALSE) is used to shrink zero component parameters.
- **spos** (TRUE/FALSE) is used to shrink positive component parameters.

After implementing the **fitFeatureModel()**, it returns a list of objects including **call** (the call made to fitFeatureModel), **fitZeroLogNormal** (a list of parameter estimates for the zero-inflated log-normal model), **design** (model matrix), **taxa** (taxa names), **counts** (count matrix), **pvalues** (calculated *P*-values), and **permuttedfits** (permutted z-score estimates under the null). The model outputs provide the useful summary tables: MRcoefs, MRtable, and MRfulltable.

Here, we use **fitFeatureModel()** to conduct hypothesis testing of mice microbiome differential abundance to compare the differences between genotype/group factors. This study was a longitudinal design with two measurements at before and post treatment. To illustrate this model, here we use the data at post treatment.

We can easily subset the samples based on the criteria of features (OTUs) and sample variables in MRexperiment-class object:

```
> # Subsetting MRexperiment-class object
> OTUsToKeep = which(rowSums(MRobj) >= 100)
> samplesToKeep = which(pData(MRobj)$Time =="Post")
> obj_Post = MRobj[OTUsToKeep, samplesToKeep]
```

After subsetting the data, we can process other steps such as remove NA in group and filter the data for effectively modeling:

```
> # Combine the cumNorm() and cumNormStatFast() functions to calculate
the scaling factors
> QtRNAData_Post = cumNorm(obj_Post, p = cumNormStatFast(MRobj))


> pd <- pData(QtRNAData_Post)
> mod <- model.matrix(~1 + Group, data = pd)


> QtRNA_Rest = fitFeatureModel(QtRNAData_Post, mod)
```

Step 5: Save and explore analysis results.

We can use the functions **MRcoefs()**, **MRtable()**, and **MRfulltable()** to view coefficient fits and related statistics and export the data with optional output values (Table 12.1):

```
> # Review coefficients
> head(MRcoefs(QtRNA_Rest))


> fulltable <- MRfulltable(QtRNA_Rest)
> # Write results table
> # Make the table
> library(xtable)
> table <- xtable(fulltable,caption = "Full Table of Modeling Results",
lable="tax_table_fitFeatureModel ")
> print.xtable(table,type="html",file =
"Ch12_1_MetagenomeSeq_Table_fitFeatureModel_QtRNA.html")
> write.csv(fulltable,file = paste
("Ch12_1_Results_MetagenomeSeq_Table_fitFeatureModel_QtRNA.csv",
sep = ""))
> # Review full table
> MRfulltable(QtRNA_Rest)
```

Step 6: Visualize Features.

MetagenomeSeq has developed several plotting functions to help with visualization and analysis of datasets.

We can use these plotting functions to gain insight into the overall structure and particular individual features of the dataset. In the following we illustrate some plots to review the data structure.

Step 6a: plotMRheatmap – heatmap of abundance estimates (Fig. 12.1).

The **plotMRheatmap()** plots heatmap of the "n" features with greatest variance across rows for normalized counts:

```
> Figure 12.1
> QtRNA_exp = pData(obj_Post)$Group
> heatmapColColors = brewer.pal(12, "Set3")[as.integer(factor
(QtRNA_exp))]
> heatmapCols = colorRampPalette(brewer.pal(9, "RdBu"))(50)
> # plotMRheatmap
> plotMRheatmap(obj = obj_Post, n = 100, norm = TRUE, log = TRUE, fun =
sd,
+          cexRow = 0.4, cexCol = 0.4, trace = "none", col = heatmapCols,
ColSideColors = heatmapColColors)
```

The above commands create a heatmap and hierarchical clustering of $\log_2$ transformed counts for the 100 OTUs with the largest overall variance. Red values indicate counts close to zero. Row color labels indicate OTU taxonomic class;

Table 12.1 Modeling results using fitFeatureModel based on post-QtRNA mouse microbiome data

	+Samples in group 0	+samples in group 1	Counts in group 0	counts in group 1	oddsRatio	Lower	Upper	FisherP	fisherAdjP	logFC	se	p-values	adjPvalues
OTU_126	2	10	3	10052	0	0	0.261985148691248	0.000714455822814956	0.0147177899499881	7.7552762392443	0.782297610849027	0	0
OTU_116	3	10	10	8130	0	0	0.388279386040879	0.00309597523219815	0.0631425748194016	6.63063439517916	0.735958182613116	0	0
OTU_101	3	6	3	3058	0.305415044366772	0.030053635388153	2.46429182813556	0.369849964277209	1	5.390903521682831	1.24753104634402	2.173509996872270-06	2.79839412097305o-05
OTU_125	1	10	2	3241	0	0	0.166690753277639	0.000119075970469159	0.00306620623958085	5.39137800592425	0.188856285124351	0	0
OTU_558	2	6	4	1863	0.184118110070328	0.0125264745480946	1.65925396045797	0.169802333889021	1	4.66860566955941	1.71741594144821	0.00656005395386594	0.0337842778624096
OTU_108	1	6	2	27889	0.0859350429538673	0.00146108353677648	1.05397598911988	0.0572755417956657	0.453798523457967	4.61980547754296	1.35624280574321	0.000658354420500995	0.00565087544263354
OTU_408	10	1	1044	1	Inf	5.9991330073025	Inf	0.000119075970469159	0.00306620623958085	-4.32710318121278	0.33177493173258	0	0
OTU_323	0	10	0	527	0	0	0.0897995510045625	1.0825088224690e-05	0.001111498408712031	3.39310932393747	0.388662351098798	0	0
OTU_97	1	6	3	4689	0.0859350429538673	0.00146108353677648	1.05397599891988	0.0572755417956657	0.453798523457967	3.18079174547734	1.10745461114788	0.00407670782992202	0.02210000477095674
OTU_367	1	6	1	850	0.0859350429538673	0.00146108353677648	1.05397599891988	0.0572755417956657	0.453798523457967	2.67579145567032	0.783777736855131	0.000640239150733191	0.00566508754 44263354

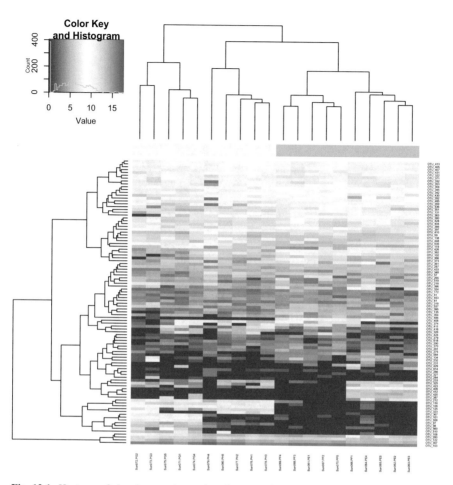

Fig. 12.1 Heatmap of abundance estimates based on post-QtRNA mouse microbiome data

column color labels indicate genotype (yellow = KO, green = WT). Notice that the samples are obviously clustered by genotypes.

Step 6b: plotCorr – heatmap of pairwise correlations (Fig. 12.2).

The plotCorr() plots a heatmap of pairwise correlations with the "n" features with greatest variance across rows for normalized or unnormalized counts. The following commands plot a correlation matrix for the same features:

```
> Figure 12.2
> # plotCorr
> plotCorr(obj = obj_Post, n = 50, norm = TRUE, log = TRUE, fun = cor,
+     cexRow = 0.25, cexCol = 0.25, trace = "none", dendrogram = "none",
col = heatmapCols)
Default value being used.
```

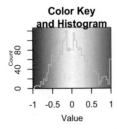

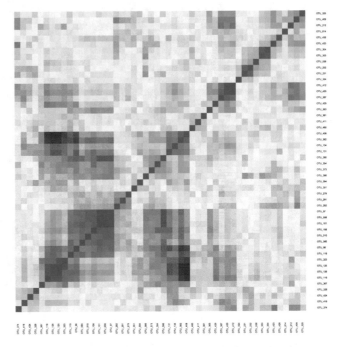

Fig. 12.2 Heatmap of pairwise correlations based on post-QtRNA mouse microbiome data

Step 6c: plotOrd – PCA/MDS components (Fig. 12.3).

The plotOrd () function plots either PCA or MDS coordinates for the distances of normalized or unnormalized counts and helps to visualize or uncover the batch effects or feature relationships. The following commands plot the MDS components of the data. Typically it is recommended to remove outliers before plotting. Here we do not find outliers and none of the data is removed:

```
> Figure 12.3
> class = factor(pData(obj_Post)$Group)
> # plotOrd
```

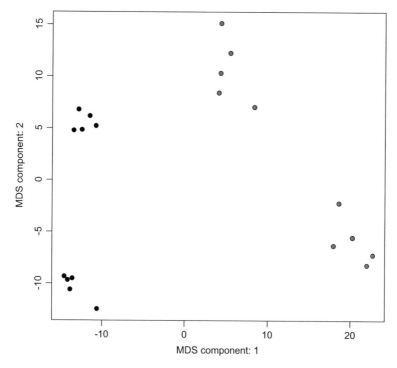

Fig. 12.3 Ordination plot with MDS components based on post-QtRNA mouse microbiome data

```
> plotOrd(obj_Post, tran = TRUE, usePCA = FALSE, useDist = TRUE,
+ bg = class, pch = 21)
```

We can also load the vegan package and set distfun = vegdist and use dist. method = 'bray' to generate the PCA using the Bray-Curtis dissimilarity method.

12.3   Some Additional Statistical Tests in metagenomeSeq

Below we will illustrate some useful tests in the **metagenomeSeq** package.

12.3.1   Log-Normal Permutation Test of Abundance

We can fit the same model as above using a standard log-normal linear model to provide a permutation-based P-values. This method was originally used by metastats (White et al. 2009; Paulson et al. 2011) for non-sparse large samples. The

fitLogNormal() function is a wrapper to perform the permutation test on the *t*-statistic. There is an option to choose the CSS normalization and use $\log_2$ transformation to transform the data. Here, we use 1000 permutations to provide *P*-value resolution to the thousandth. We specify the coefficient parameter equal to 2 to test the coefficient of interest (here, Group). A list of significant features is generated to hold the significant features:

```
> coeff = 2 # Here it is specified to test the coefficients of Group
> rest1 = fitLogNormal(obj = obj_Post, mod = mod, useCSSoffset = FALSE,
+           B = 1000, coef = coeff)

> # Extract p-values and adjust for multiple testing the p-values
(rest1$p)
> # that are calculated through permutation
> adjustedPvalues = p.adjust(rest1$p, method = "fdr")
> # Extract the absolute fold-change estimates
> foldChange = abs(rest1$fit$coef[, coeff])
> # Retain the significant features after adjusting
> sigList = which(adjustedPvalues <= 0.05)
> # Order the significant features
> sigList = sigList[order(foldChange[sigList])]
> # View the top taxa associated with the coefficient of interest
> taxa[sigList]
 [1] NA                      "Ruminococcus callidus et rel."
 [3] NA                      NA
 [5] "Clostridium orbiscindens et rel."    NA
 [7] NA                      NA
 [9] NA                      NA
[11] NA                      "Papillibacter cinnamivorans et rel."
[13] "Outgrouping clostridium cluster XIVa" NA
[15] NA                      "Clostridium leptum et rel."
[17] "Dorea formicigenerans et rel."    "Oscillospira guillermondii et
rel."
[19] NA                      NA
[21] NA                      "Clostridium symbiosum et rel."
[23] "Faecalibacterium prausnitzii et rel." "Coprococcus eutactus et
rel."
[25] "Lachnobacillus bovis et rel."
```

12.3.2 Presence-Absence Testing of the Proportion/Odds

The presence-absence testing is implemented to test the hypothesis that the proportion/odds of a given feature presented is higher/lower among one group of individuals compared to another. The test uses the **fitPA()** to perform Fisher's exact test to create a 2×2 contingency table and calculate *P*-values, odd's ratios, and confidence intervals. The function accepts either a MRexperiment object or matrix as input data:

```
> class = pData(obj_Post)$Group
> rest2 = fitPA(obj_Post, cl = class,adjust.method = "fdr")
> head(rest2,3)
       oddsRatio lower upper pvalues adjPvalues
OTU_22         0     0   Inf       1          1
OTU_57         0     0   Inf       1          1
OTU_59         0     0   Inf       1          1
> tail(rest2,3)
        oddsRatio   lower  upper pvalues adjPvalues
OTU_584    8.0182 0.638635 470.18  0.1409     0.9068
OTU_614    3.6108 0.230118 224.19  0.5820     1.0000
OTU_618    0.4625 0.006799  10.51  1.0000     1.0000
```

12.3.3 Discovery Odds Ratio Testing of the Proportion of Observed Counts

The discovery test is implemented to test the hypothesis that the proportion of observed counts for a feature of all counts are comparable between groups. The test uses the **fitDO()** to perform Fisher's exact test to create a 2 × 2 contingency table and calculate *P*-values, odd's ratios, and confidence intervals. The function accepts either a MRexperiment object or matrix as input data:

```
> class = pData(obj_Post)$Group
> rest3 = fitDO(obj_Post[1:50, ], cl = class, norm = FALSE, log = FALSE,
adjust.method = "fdr")
> head(rest3,3)
       oddsRatio lower upper pvalues adjPvalues
OTU_22         0     0   Inf       1          1
OTU_57         0     0   Inf       1          1
OTU_59         0     0   Inf       1          1
> tail(rest3,3)
        oddsRatio lower upper pvalues adjPvalues
OTU_345         0     0   Inf       1          1
OTU_348         0     0   Inf       1          1
OTU_351         0     0   Inf       1          1
```

12.3.4 Perform Feature Correlations

We can implement **correlationTest ()** and **correctIndices()** to conduct a pairwise test of the correlations of abundance features or samples for each row of a matrix or MRexperiment object. The function correlationTest () returns the basic Pearson, Spearman, Kendall correlation statistics for the rows of the input and the associated *P*-values. For a vector of length ncol(obj), this function also calculates the correlation of each row with the associated vector. However, we should be cautious when we infer correlation networks from genomic survey data (Friedman and Alm 2012):

```
> corr = correlationTest(obj_Post[1:5, ], method = "pearson",
alternative = "two.sided", norm = FALSE, log = FALSE)
> head(corr)
              correlation      p
OTU_22-OTU_57       NA      NA
OTU_22-OTU_59       NA      NA
OTU_22-OTU_61       NA      NA
OTU_22-OTU_81       NA      NA
OTU_57-OTU_59  0.664267 0.001401
OTU_57-OTU_61  0.009267 0.969070

> corr_s = correlationTest(obj_Post[1:5, ], method = "spearman",
alternative = "two.sided", norm = FALSE, log = FALSE)
> corr_k = correlationTest(obj_Post[1:5, ], method = "kendall",
alternative = "two.sided", norm = FALSE, log = FALSE)
```

12.4   Illustrate Some Useful Functions in metagenomeSeq

To facilitate analysis using the **metagenomeSeq** package, we illustrate some useful functions.

12.4.1 Access the MRexperiment Object

Three pairs of functions are used to access the raw or normalized count matrix, feature, and phenotype information, respectively. The **MRcounts ()** function is used for accessing the raw or normalized count matrix, the **featureData()** and **fData()** functions are used for accessing feature information, and the **phenoData()** and **pData()** functions are used for accessing phenotype information.

We use the **MRcounts()** function to access the raw count matrix:

```
> # Access raw or normalized counts matrix
> head(MRcounts(MRobj[, 1:6]),3)
      Sun071.PG1 Sun027.BF2 Sun066.PF1 Sun029.BF4 Sun068.PF3 Sun026.BF1
OTU_1      0        0        0        0        0        0
OTU_2      0        0        0        0        1        0
OTU_3      0        0        0        0        1        0
```

The feature information can be accessed with the **featureData()** and **fData()** functions. The differences are that the method of **featureData()** provides the S4 class information, while the method of **fData()** provides the information on data frame of taxonomy table:

```
> # Access feature information with the featureData method
> featureData(MRobj)
An object of class 'AnnotatedDataFrame'
 featureNames: OTU_1 OTU_2 ... OTU_635 (635 total)
 varLabels: Kingdom Phylum ... Species (7 total)
 varMetadata: labelDescription
```

```
> # Access feature information with the fData method
> head(fData(MRobj)[, -c(2, 10)], 3)
 Kingdom Class Order Family Genus Species
OTU_1 D_0__Archaea D_2__Nitrososphaeria D_3__Nitrososphaerales
D_4__Nitrososphaeraceae Other Other
OTU_2 D_0__Bacteria D_2__Acidobacteriia D_3__Solibacterales
D_4__Solibacteraceae (Subgroup 3) D_5__Bryobacter Other
OTU_3 D_0__Bacteria D_2__Acidobacteriia D_3__Solibacterales
D_4__Solibacteraceae (Subgroup 3) Other Other
```

The phenotype information can be accessed with the **phenoData ()** and **pData ()** functions. The differences are that the method of **phenoData ()** provides the S4 class information, while the method of **pData ()** provides the information on data frame of metadata table:

```
> # Access phenotype information with the phenoData method
> phenoData(MRobj)
An object of class 'AnnotatedDataFrame'
 sampleNames: Sun071.PG1 Sun027.BF2 ... Sun062.PE2 (40 total)
 varLabels: MouseID Genotype ... Total.Read (6 total)
 varMetadata: labelDescription
```

```
> # Access phenotype information with the pData method
> head(pData(MRobj), 3)
          MouseID Genotype Group  Time   Group4 Total.Read
Sun071.PG1   PG1     KO    1  Post  KO_POST    61851
Sun027.BF2   BF2     WT    0 Before WT_BEFORE   42738
Sun066.PF1   PF1     WT    0  Post  WT_POST    54043
```

12.4.2   Subset the MRexperiment Object

We can subset the MRexperiment-class object. The following commands subset the object "MRobj" with row sum greater than 100 features for before-treatment samples:

```
> # Subset MRexperiment-class object
> sub_feat = which(rowSums(MRobj) >= 100)
> sub_samp = which(pData(MRobj)$Time == "Before")
```

```
> obj_f = MRobj[sub_feat, sub_samp]
> obj_f
MRexperiment (storageMode: environment)
assayData: 103 features, 20 samples
 element names: counts
protocolData: none
phenoData
 sampleNames: Sun027.BF2 Sun029.BF4 ... Sun035.BG5 (20 total)
 varLabels: MouseID Genotype ... Total.Read (6 total)
 varMetadata: labelDescription
featureData
 featureNames: OTU_22 OTU_57 ... OTU_618 (103 total)
 fvarLabels: Kingdom Phylum ... Species (7 total)
 fvarMetadata: labelDescription
experimentData: use 'experimentData(object)'
Annotation:

> head(pData(obj_f), 3)
       MouseID Genotype Group  Time   Group4 Total.Read
Sun027.BF2   BF2      WT   0 Before WT_BEFORE     42738
Sun029.BF4   BF4      WT   0 Before WT_BEFORE     53760
Sun026.BF1   BF1      WT   0 Before WT_BEFORE     56523
```

Above outputs show that the sub-object has 20 samples in before-treatment.

12.4.3 Filter the MRexperiment Object or Count Matrix

We can use the **filterData** () function to filter the data to maintain a threshold of minimum depth or OTU presence. The following commands filter the data based on presenting 1 number of features after filtering samples by at least depth of coverage 100. This function can be conducted on a MRexperiment object or count matrix:

```
> filter_obj<- filterData(MRobj, present = 1, depth = 100)
```

12.4.4 Merge the MRexperiment Object

We can merge two MRexperiment-class objects using the **mergeMRexperiments()** function as below:

```
> MRobj1<-MRobj
> merge_obj = mergeMRexperiments(MRobj, MRobj1)
```

12.4.5   Call the Normalized Counts Using the cumNormMat() and MRcounts()

Below we use the function **cumNormMat()** to return the normalized matrix using the percentile generated by the **cumNormStatFast** () function and scale by 1000:

```
> cumNormMat(MRobj, p = cumNormStatFast(MRobj), sl = 1000)
> head(cumNormMat(MRobj, p = cumNormStatFast(MRobj), sl = 1000),3)
Default value being used.
 Sun071.PG1 Sun027.BF2 Sun066.PF1 Sun029.BF4 Sun068.PF3 Sun026.BF1
Sun072.PG2 Sun076.PH1 Sun063.PE3 Sun036.BH1 Sun024.BE4 Sun065.PE5
Sun033.BG3 Sun040.BH5 Sun030.BF5
OTU_1 0 0 0 0 0 0 0 0 0 0 0 0 0 0 0
OTU_2 0  0 0 0 800  0  00 0 0 0 0 0
OTU_3 0 0 0 0 800 0 0 0 0 0 0 0 0 0
------
```

where **p** is used to specify the pth quantile and **sl** is used to specify the value to scale by (default = 1000).

Below we use the function **MRcounts** () to access the counts slot of a MRexperiment object. The counts slot holds the raw count data representing (along the rows) the number of reads annotated for a particular feature and (along the columns) the sample.

The syntax is given as.

```
MRcounts(obj, norm = TRUE, log = FALSE, sl = 1000)
```

where **obj** presents a MRexperiment object, **norm** is the logical indicator for whether or not to return normalized counts; **log** is the logical indicator for whether or not to $\log_2$ transform scale; and **sl** is used to specify the value to scale by (default = 1000).

It returns raw counts when specifying norm = FALSE:

```
> head(MRcounts(MRobj, norm = FALSE, log = FALSE, sl = 1000),3)
 Sun071.PG1 Sun027.BF2 Sun066.PF1 Sun029.BF4 Sun068.PF3 Sun026.BF1
Sun072.PG2 Sun076.PH1 Sun063.PE3 Sun036.BH1 Sun024.BE4 Sun065.PE5
Sun033.BG3 Sun040.BH5 Sun030.BF5
OTU_1 0 0 0 0 0 0 0 0 0 0 0 0 0 0 0
OTU_2 0 0 0 0 1 0 0 0 0 0 0 0 0 0 0
OTU_3 0 0 0 0 1 0 0 0 0 0 0 0 0 0 0
------
```

It returns normalized counts when specifying norm = TRUE:

```
> head(MRcounts(MRobj, norm = TRUE, log = FALSE, sl = 1000),3)
Default value being used.
 Sun071.PG1 Sun027.BF2 Sun066.PF1 Sun029.BF4 Sun068.PF3 Sun026.BF1
```

```
Sun072.PG2 Sun076.PH1 Sun063.PE3 Sun036.BH1 Sun024.BE4 Sun065.PE5
Sun033.BG3 Sun040.BH5 Sun030.BF5
OTU_1 0 0 0 0 0 0 0 0 0 0 0 0 0 0 0
OTU_2 0 0 0 0 8 0 0 0 0 0 0 0 0 0 0
OTU_3 0 0 0 0 8 0 0 0 0 0 0 0 0 0 0
------
```

12.4.6 Calculate the Normalization Factors Using the calcNormFactors()

```
> nfactor = calcNormFactors(MRobj, p = cumNormStatFast(MRobj))
Default value being used.
> head(nfactor,3)
 normFactors
Sun071.PG1 200
Sun027.BF2 83
Sun066.PF1 162
```

```
> nfactor1 = calcNormFactors(MRobj, p = 0.75)
> head(nfactor1,3)
 normFactors
Sun071.PG1 1655
Sun027.BF2 1068
Sun066.PF1 1145
```

```
> # Normalize the samples
> normFactors(MRobj) <- rnorm(ncol(MRobj))
> head(normFactors(MRobj))
Sun071.PG1 Sun027.BF2 Sun066.PF1 Sun029.BF4 Sun068.PF3 Sun026.BF1
 -0.9120 1.5436 1.6109 0.3158 0.6128 -0.2461
```

12.4.7 Access the Library Sizes Using the libSize()

We can use the **libSize()** function to access the library sizes (sequencing depths):

```
> # Access the library sizes (sequencing depths)
> head(libSize(MRobj),3)
 [,1]
Sun071.PG1 61841
Sun027.BF2 42729
Sun066.PF1 54025
```

12.4.8   Save the Normalized Counts Using the exportMat()

The **exportMat()** function exports the normalized MRexperiment dataset as a matrix.

The syntax is given as.

```
exportMat(obj, log = TRUE, norm = TRUE, sep = "\t", file = "matrix.tsv")
```

where **obj** is a MRexperiment object or count matrix, **log** is used to specify whether or not to log transform the counts in the case of MRexperiment object, **norm** is used to specify whether or not to normalize the counts in the case of MRexperiment object, **sep** is the separator for writing out the count matrix, and **file** is used to name the output file. This function allows us to output the dataset to the workspace as a tab-delimited file for convenience of downstream analysis:

```
> exportMat(MRobj, norm = TRUE, sep = "\t", file = "QtRNA_norm.tsv")
```

12.4.9   Save the Sample Statistics Using the exportStats()

The **exportStats** () function exports a matrix of values for each sample including various sample statistics: sample ids, the sample scaling factor, quantile value, the number identified features, and library size (depth of coverage):

```
> exportStats(MRobj, file="QtRNA_norm.tsv")
Default value being used.
> head(read.csv(file= "QtRNA_norm.tsv", sep="\t"),3)
 Subject Scaling.factor Quantile.value Number.of.identified.features
Library.size
1 Sun071.PG1 -0.912 17 130 61841
2 Sun027.BF2 1.544 5 101 42729
3 Sun066.PF1 1.611 10 125 54025
```

12.4.10   Find Unique OTUs or Features

We can use the function **uniqueFeatures** () to find features absent from any number of classes. The returned table lists the feature ids, the number of positive features, and reads for each group. Thresholding for the number of positive samples or reads required are options:

```
> class = pData(obj_Post)[["Group"]]
> uniqueFeatures(obj_Post, class, nsamples = 5, nreads = 10)
 featureIndices Samp. in 0 Samp. in 1 Reads in 0 Reads in 1
```

```
OTU_323 41 0 10 0 527
OTU_420 81 5 0 109 0
OTU_429 86 6 0 90 0
OTU_510 94 0 5 0 285
```

12.4.11   Aggregate Taxa

Normalization is recommended at the OTU level. However, the count matrix (either normalized or not) can be aggregated based on a particular user-defined level of taxonomy. Currently, there are two functions called **aggregateByTaxonomy()** and **aggTax()** in the metagenomicsSeq package for using the feature-data information in the MRexperiment object to aggregate taxa. The aggregateByTaxonomy() and aggTax() are used to aggregate a MRexperiment object or count matrix to a particular level. We first call the **aggregateByTaxonomy()** function on the MRexperiment object **obj_Post** to aggregate the "Genus" level (a featureData column) counts using the default **colSums ()** function:

```
> colnames(fData(obj_Post))
[1] "Kingdom" "Phylum" "Class" "Order" "Family" "Genus" "Species"


> aggregateByTaxonomy(obj_Post,lvl="Genus", norm=TRUE,
aggfun=colSums)
Default value being used.
MRexperiment (storageMode: environment)
assayData: 54 features, 20 samples
 element names: counts
protocolData: none
phenoData
 sampleNames: Sun071.PG1 Sun066.PF1 ... Sun062.PE2 (20 total)
 varLabels: MouseID Genotype ... Total.Read (6 total)
 varMetadata: labelDescription
featureData
 featureNames: D_5__[Eubacterium] coprostanoligenes group D_5__
[Eubacterium] xylanophilum group ... Other (54 total)
 fvarLabels: Kingdom Phylum ... Genus (6 total)
 fvarMetadata: labelDescription
experimentData: use 'experimentData(object)'
Annotation:
```

The **colMedians()** can also be used:

```
> aggregateByTaxonomy(obj_Post, lvl="Genus", norm=TRUE,
aggfun=matrixStats::colMedians)
```

We then call the **aggTax ()** function on the MRexperiment object **obj_Post** to aggregate the "Phylum" level (a featureData column) counts using the default **colSums ()** function:

```
> aggTax(obj_Post, lvl='Phylum',norm=FALSE, alternate = FALSE, log =
FALSE,
+ aggfun = colSums, sl = 1000, featureOrder = NULL,
+ returnFullHierarchy = TRUE, out = "MRexperiment")
MRexperiment (storageMode: environment)
assayData: 8 features, 20 samples
 element names: counts
protocolData: none
phenoData
 sampleNames: Sun071.PG1 Sun066.PF1 ... Sun062.PE2 (20 total)
 varLabels: MouseID Genotype ... Total.Read (6 total)
 varMetadata: labelDescription
featureData
 featureNames: D_1__Actinobacteria D_1__Bacteroidetes ...
D_1__Tenericutes (8 total)
 fvarLabels: Kingdom Phylum
 fvarMetadata: labelDescription
experimentData: use 'experimentData(object)'
Annotation:
```

The **colMedians()** can also be used:

```
> aggTax(obj_Post, lvl='Phylum',norm=FALSE, alternate = FALSE, log =
FALSE,
+ aggfun = matrixStats::colMedians, sl = 1000, featureOrder = NULL,
+ returnFullHierarchy = TRUE, out = "MRexperiment")
```

12.4.12   Aggregate Samples

Samples can also be aggregated using the phenoData information in the MRexperiment object. Currently, there are two functions called **aggregateBySample()** and **aggSamp()** in the metagenomicsSeq package for using the phenoData information in the MRexperimen to aggregate a MRexperiment object or count matrix to by a factor.

In the following commands, we first call the **aggregateBySample()** function on the MRexperiment object **obj_Post** to use the Group (a phenoData column name) to aggregate counts with the default rowMeans () aggregating function:

```
> aggregateBySample(obj_Post, fct="Group",aggfun=rowSums)
MRexperiment (storageMode: environment)
assayData: 103 features, 2 samples
 element names: counts
protocolData: none
phenoData
 sampleNames: 0 1
 varLabels: phenoData
 varMetadata: labelDescription
```

```
featureData
featureNames: OTU_22 OTU_57 ... OTU_618 (103 total)
fvarLabels: Kingdom Phylum ... Species (7 total)
fvarMetadata: labelDescription
experimentData: use 'experimentData(object)'
Annotation:
```

The **rowMedians ()** can also be used:

```
> aggregateBySample(obj_Post, fct="Group",aggfun= matrixStats::
rowMedians)
```

We then call the **aggSamp()** function on the MRexperiment object **obj_Post** to aggregate the sample counts using the default **rowMeans ()** function:

```
> aggSamp(obj_Post, fct="Group",aggfun= rowMeans, out =
"MRexperiment")
MRexperiment (storageMode: environment)
assayData: 103 features, 2 samples
 element names: counts
protocolData: none
phenoData
 sampleNames: 0 1
 varLabels: phenoData
 varMetadata: labelDescription
featureData
 featureNames: OTU_22 OTU_57 ... OTU_618 (103 total)
 fvarLabels: Kingdom Phylum ... Species (7 total)
 fvarMetadata: labelDescription
experimentData: use 'experimentData(object)'
Annotation:
```

The **rowMaxs ()** can also be used:

```
> aggSamp(obj_Post, fct="Group",aggfun=matrixStats::rowMaxs)
```

In microbiome data analysis, the data is often required to be aggregated. The functions **aggregateByTaxonomy()**, **aggregateBySample()**, **aggTax()**, and **aggSamp()** are very flexible. They can facilitate data in either a matrix with a vector of labels or a MRexperiment object with a vector of labels or featureData column name. All these functions can output either a matrix or MRexperiment object.

12.5 Remarks on CSS Normalization, ZIG, and ZILN

CSS method divides raw counts into the cumulative sum of counts up to a percentile determined using a data-driven approach, which is an adaptive extension of the quantile normalization method (Bullard et al. 2010). CSS normalization may have the advantages, including the following: (1) CSS may have superior performance for weighted metrics, although it is arguable (Paulson et al. 2013, 2014; Costea et al. 2014; Weiss et al. 2017). (2) CSS is similar to log upper quartile (LUQ), but it is flexible than LUQ because it allows a dependent threshold to determine each sample's quantile divisor (Weiss et al. 2017). (3) It was shown that CSS along with GMPR is overall more robust to sample-specific outlier OTUs than TSS and RNA-seq normalization methods (Chen et al. 2018). For CSS, it is crucial to ensure that chosen quantile is appropriate so that the normalization approach does not introduce normalization-related artifacts in the data, in which the normalization method assumes that all the count distributions of samples are roughly equivalent and independent of each other up to this quantile. However, the challenge is that the percentiles that CSS is based on could not be determined for microbiome datasets with high variable counts (Chen et al. 2018).

ZIG and ZILN are implemented via the metagenomeSeq Bioconductor package. The metagenomeSeq methods have the advantages, including:

1. It was shown that metagenomeSeq outperforms other widely used statistical methods in the field including Metastats (White et al. 2009), LEfSe's Kruskal-Wallis test (Segata et al. 2011), and representative methods for RNA-seq analysis, edgeR (Robinson et al. 2010) and DESeq (Anders and Huber 2010) in terms of the area under the curve (AUC) scores (Paulson et al. 2013). Others also showed that metagenomeSeq has slightly higher powers than most of the other DA methods (Lin and Peddada 2020b). (2) It yields a more precise biological interpretation of microbiome data (Paulson et al. 2013). (3) ZIG performs well when there is an adequate number of biological replicates (McMurdie and Holmes 2014).

   However, the disadvantages of ZIG include a higher false-positive rate (FDR) (McMurdie and Holmes 2014; Mandal et al. 2015; Weiss et al. 2017), and the problem of FDR inflation gets worse when sample size or the effect size (i.e., fold change of mean absolute abundances) increases (Weiss et al. 2017; Lin and Peddada 2020a). This disadvantage is mostly criticized in microbiome literature; currently it was shown that ZIG is subject to inflate FDR even with the normalized observed abundances by its built-in scaling method (CSS) (Lin and Peddada 2020b).

   However, we should point out that some evaluated disadvantages of ZIG and ZILN were based on compositional approach. In microbiome literature, the compositional approach is also arguable. Additionally, ZIG was shown increasing FDR that was assessed, using rarefied data (McMurdie and Holmes 2014; Weiss et al. 2017). This also highlights the difference between count-based (e.g., ZIG) and compositional approach. ZIG was developed for 16S count data, and thus it is reasonable that it is not suitable for compositional or proportion data. ZIG and its CSS require original library size to capture the zero proportions. Rarefying biological count data omits available valid data, and hence it is statistically inadmissible (McMurdie and Holmes 2014). Therefore, criticizing ZIG based on rarefying is not a strong argument because the use of rarefication itself is also arguable.

2. ZIG was developed with the log transformation on the read counts using an empirical Bayes procedure to estimate the moderated variances. This was reviewed it is not clear how to extend the empirical Bayes method in longitudinal setting and is not trivial to simply include certain random effects by extending these methods (Chen and Li 2016).

3. MetagenomeSeq is specifically developed for handling the high number of zero observations encountered in metagenomic data. However, inference is done after transformation using a log transformation ($\log_2(y_{ij} + 1)$) followed by correction for zero inflation based on a Gaussian mixture model (Paulson et al. 2013). Thus, actually metagenomeSeq moderates the gene-specific variance estimates (Smyth 2004) by some kind of removing zeros.

So far, most performances of metagenomeSeq have been evaluated based on ZIG; recently, one study demonstrated that ZILN mixture model successfully controls the FDR under 5% in all simulations, but suffers a severe loss of power (Lin and Peddada 2020a). Additional evaluations based on the ZILN are needed for providing further information on the metagenomeSeq methods.

12.6 Summary

In this chapter, we focused on introducing the metagenomeSeq methods including ZIG and ZILN mixture models as well as additional statistical tests and useful functions in metagenomeSeq. We first introduced TSS and CSS normalization methods. CSS is one critical component of the metagenomeSeq methods. We then described the ZIG and ZILN models. Next, we illustrated how to implement ZILN via the metagenomeSeq package. Currently ZILN or fitFeatureModel() function is recommended for use over ZIG functions or fitZig() function in the metagenomeSeq manual. Followed that, we illustrated some additional statistical tests that are available in metagenomeSeq: log-normal permutation test of abundance, presence-absence testing of the proportion/odds, discovery odds ratio testing of the proportion of observed counts, as well as performing feature correlations. To facilitate analysis of microbiome using the metagenomeSeq methods, we also illustrated some useful functions in metagenomeSeq to access the MRexperiment object, subset the MRexperiment object, filter the MRexperiment object or count matrix, merge the MRexperiment object, call the normalized counts using the cumNormMat() and MRcounts(), calculate the normalization factors using the calcNormFactors(), access the library sizes using the libSize(), save the normalized counts using the exportMat(), save the sample statistics using the exportStats(), find unique OTUs or features, and aggregate taxa and samples. Finally, we provided some remarks on CSS, ZIG, and ZILN.

In Chap. 13, we will introduce zero-inflated beta models for microbiome data.

References

Anders, S., and W. Huber. 2010. Differential Expression Analysis for Sequence Count Data. *Genome Biology* 11 (10): R106. https://doi.org/10.1186/gb-2010-11-10-r106.

Bullard, James H., Elizabeth Purdom, Kasper D. Hansen, and Sandrine Dudoit. 2010. Evaluation of Statistical Methods for Normalization and Differential Expression in mRNA-Seq Experiments. *BMC Bioinformatics* 11 (1): 94. https://doi.org/10.1186/1471-2105-11-94.

Chen, E.Z., and H. Li. 2016. A Two-Part Mixed-Effects Model for Analyzing Longitudinal Microbiome Compositional Data. *Bioinformatics* 32 (17): 2611–2617. https://doi.org/10.1093/bioinformatics/btw308.

Chen, L., J. Reeve, L. Zhang, S. Huang, X. Wang, and J. Chen. 2018. GMPR: A Robust Normalization Method for Zero-Inflated Count Data with Application to Microbiome Sequencing Data. *PeerJ* 6: e4600. https://doi.org/10.7717/peerj.4600.

Costea, Paul I., Georg Zeller, Shinichi Sunagawa, and Peer Bork. 2014. A Fair Comparison. *Nature Methods* 11 (4): 359–359.

Dillies, Marie-Agnès, Andrea Rau, Julie Aubert, Christelle Hennequet-Antier, Marine Jeanmougin, Nicolas Servant, Céline Keime, Guillemette Marot, David Castel, Jordi Estelle, Gregory Guernec, Bernd Jagla, Luc Jouneau, Denis Laloë, Caroline Le Gall, Brigitte Schaëffer, Stéphane Le Crom, Mickaël Guedj, Florence Jaffrézic, and on behalf of The French StatOmique Consortium. 2012. A comprehensive evaluation of normalization methods for Illumina high-throughput RNA sequencing data analysis. *Briefings in Bioinformatics* 14 (6): 671–683. https://doi.org/10.1093/bib/bbs046.

Friedman, J., and E.J. Alm. 2012. Inferring correlation networks from genomic survey data. *PLoS Computational Biology* 8 (9): e1002687. https://doi.org/10.1371/journal.pcbi.1002687.

Gentleman, R., V. Carey, M. Morgan, and S. Falcon. 2020. *Biobase: Base functions for bioconductor, version 2.46.0*.

Lin, Huang, and Shyamal Das Peddada. 2020a. Analysis of compositions of microbiomes with bias correction. *Nature Communications* 11 (1): 3514. https://doi.org/10.1038/s41467-020-17041-7.

———. 2020b. Analysis of microbial compositions: A review of normalization and differential abundance analysis. *npj Biofilms and Microbiomes* 6 (1): 60. https://doi.org/10.1038/s41522-020-00160-w.

Mandal, Siddhartha, Will Van Treuren, Richard A. White, Merete Eggesbø, Rob Knight, and Shyamal D. Peddada. 2015. Analysis of composition of microbiomes: A novel method for studying microbial composition. *Microbial Ecology in Health and Disease* 26 (1): 27663.

McMurdie, Paul J., and Susan Holmes. 2014. Waste not, want not: Why rarefying microbiome data is inadmissible. *PLoS Computational Biology* 10 (4): e1003531. https://doi.org/10.1371/journal.pcbi.1003531.

Morton, James T., Jon Sanders, Robert A. Quinn, Daniel McDonald, Antonio Gonzalez, Yoshiki Vázquez-Baeza, Jose A. Navas-Molina, Se Jin Song, Jessica L. Metcalf, and Embriette R. Hyde. 2017. Balance trees reveal microbial niche differentiation. *MSystems* 2 (1).

Paulson, Joseph N., Hector Corrada Bravo, and Mihai Pop. 2014. A fair comparison reply. *Nature Methods* 11 (4): 359–360.

Paulson, Joseph N., Mihai Pop, and Hector Corrada Bravo. 2011. Metastats: An improved statistical method for analysis of metagenomic data. *Genome Biology* 12 (Suppl 1): P17–P17. https://doi.org/10.1186/gb-2011-12-s1-p17. https://www.ncbi.nlm.nih.gov/pmc/articles/PMC3439073/.

Paulson, Joseph N., O. Colin Stine, Héctor Corrada Bravo, and Mihai Pop. 2013. Differential abundance analysis for microbial marker-gene surveys. *Nature Methods* 10 (12): 1200–1202. https://doi.org/10.1038/nmeth.2658. https://pubmed.ncbi.nlm.nih.gov/24076764, https://www.ncbi.nlm.nih.gov/pmc/articles/PMC4010126/.

Paulson, Joseph Nathaniel, Modi ed: October 4, 2016. Compiled: February 3, 2020, 2020, *metagenomeSeq: Statistical analysis for sparse high-throughput sequencing.*

Robinson, Mark D., Davis J. McCarthy, and Gordon K. Smyth. 2010. edgeR: A bioconductor package for differential expression analysis of digital gene expression data. *Bioinformatics* 26 (1): 139–140.

Segata, Nicola, Jacques Izard, Levi Waldron, Dirk Gevers, Larisa Miropolsky, Wendy S. Garrett, and Curtis Huttenhower. 2011. Metagenomic biomarker discovery and explanation. *Genome Biology* 12 (6): 1–18.

Smyth, Gordon K. 2004. Linear models and empirical Bayes methods for assessing differential expression in microarray experiments. *Statistical Applications in Genetics and Molecular Biology* 3 (1).

———. 2005. Limma: Linear models for microarray data. In *Bioinformatics and computational biology solutions using R and bioconductor*, 397–420. Springer.

Soneson, Charlotte, and Mauro Delorenzi. 2013. A comparison of methods for differential expression analysis of RNA-seq data. *BMC Bioinformatics* 14 (1): 91. https://doi.org/10.1186/1471-2105-14-91.

Storey, John D., and Robert Tibshirani. 2003. Statistical significance for genomewide studies. *Proceedings of the National Academy of Sciences* 100 (16): 9440–9445. https://doi.org/10.1073/pnas.1530509100. https://www.pnas.org/content/pnas/100/16/9440.full.pdf.

Tsilimigras, Matthew C.B., and Anthony A. Fodor. 2016. Compositional Data Analysis of the Microbiome: Fundamentals, Tools, and Challenges. *Annals of Epidemiology* 26 (5): 330–335.

Weiss, Sophie, Xu Zhenjiang Zech, Shyamal Peddada, Amnon Amir, Kyle Bittinger, Antonio Gonzalez, Catherine Lozupone, Jesse R. Zaneveld, Yoshiki Vázquez-Baeza, Amanda Birmingham, Embriette R. Hyde, and Rob Knight. 2017. Normalization and microbial differential abundance strategies depend upon data characteristics. *Microbiome* 5 (1): 27. https://doi.org/10.1186/s40168-017-0237-y.

White, James Robert, Niranjan Nagarajan, and Mihai Pop. 2009. Statistical methods for detecting differentially abundant features in clinical metagenomic samples. *PLoS Computational Biology* 5 (4): e1000352. https://doi.org/10.1371/journal.pcbi.1000352.

Zhang, Jilei, Lu Rong, Yongguo Zhang, Żaneta Matuszek, Wen Zhang, Yinglin Xia, Tao Pan, and Jun Sun. 2020. tRNA Queuosine modification enzyme modulates the growth and microbiome recruitment to breast tumors. *Cancers* 12 (3): 628. https://doi.org/10.3390/cancers12030628. https://pubmed.ncbi.nlm.nih.gov/32182756, https://www.ncbi.nlm.nih.gov/pmc/articles/PMC7139606/.

Chapter 13
Zero-Inflated Beta Models
for Microbiome Data

Abstract This chapter introduces two specifically designed zero-inflated beta regression models for analyzing zero-inflated count microbiome data. First, it briefly introduces the zero-inflated beta modeling microbiome data. Then, it introduces the zero-inflated beta regression (ZIBSeq). Next, it introduces the zero-inflated beta-binomial model (ZIBB).

Keywords Zero-inflated beta · Zero-inflated beta regression (ZIBSeq) · Zero-inflated beta-binomial model (ZIBB) · Semi-continuous · Zero-inflated continuous · Zero-inflated beta regression model with random effects (ZIBR) · Marginalized two-part beta regression (MTPBR) · Likelihood ratio tests (LRT) · ZIBBSeqDiscovery · fitZIBB()

This chapter introduces two specifically designed zero-inflated beta regression models for analyzing zero-inflated count microbiome data. First we briefly introduce the zero-inflated beta modeling microbiome data (Sect. 13.1). Then we introduce the zero-inflated beta regression (ZIBSeq) and its implementation (Sect. 13.2). Next, we introduce the zero-inflated beta-binomial model (ZIBB) and its implementation (Sect. 13.3). Finally, we briefly summarize the zero-inflated beta regression in microbiome study (Sect. 13.4).

13.1 Zero-Inflated Beta Modeling Microbiome Data

One trend of statistical modeling microbiome data is to use two-part zero-inflated beta regression to detect differential abundance of microbes across clinical conditions or different treatments. The two-part zero-inflated beta regression is attractive because it takes the unique characteristics of microbiome data into account; that is,

Supplementary Information The online version contains supplementary material available at https://doi.org/10.1007/978-3-031-21391-5_13.

microbiome data are compositional (quantified by relative abundance), highly skewed, and bounded in [0, 1) and often have many zeros.

A "semi-continuous" or "zero-inflated continuous" data describes a kind of continuous data with a point mass at zero and a right-skewed continuous distribution with a positive values. The raw sequencing reads of microbiome data are absolute abundant counts. However, if the absolute abundant counts are transformed or normalized into relative abundance, then microbiome data can be characterized as a "semi-continuous" or "zero-inflated continuous" data; in which the zero values indicate that either the certain microbes are absent in the sample (structural zeros) or the rare microbes are present but missed due to under-sampling (sampling-zeros), while the continuous distribution of positive values describes the levels of relative abundance among the present microbes (Chai et al. 2018).

Typically, a two-part zero-inflated beta model consists of two sub-models: Part I uses a logistic regression to model the probability of a feature (taxon) having zero values, while Part II uses a beta regression to model the distribution of positive values, i.e., the relative abundance of feature (or taxon) conditionally presented.

So far several two-part beta regression models have been developed for modeling relative abundance of microbiome data, including zero-inflated beta regression (ZIBSeq) (Peng et al. 2016), zero-inflated beta-binomial model (ZIBB) (Hu et al. 2018), zero-inflated beta regression model with random effects (ZIBR) (Chen and Li 2016), and marginalized two-part beta regression (MTPBR) (Chai et al. 2018). Chai et al. (2018) thought that the regression coefficients in Part II in previous two-part zero-inflated beta models is not able to marginally (unconditionally) interpret the covariate effects on the microbial abundance, which is often the interest of microbiome researches. Thus, unlike BESeq and other two-part zero-inflated beta models, MTPBR was proposed to examine covariate effects on the overall marginal (unconditional) mean, while capturing the zero-inflation and skewness of microbiome data (Chai et al. 2018). MTPBR model was derived starting with the conventional two-part model with a beta component in Part II (Chen and Li 2016; Ospina and Ferrari 2012; Peng et al. 2016). Then, in order to obtain the impact of covariates on the overall marginal mean, MTPBR model reparameterizes the likelihood of the conventional two-part model and provides the different interpretation of covariate effects. The distinctive feature of MTPBR is its ability of modeling covariate effects on the marginal mean of the outcome. It was showed that in terms of controlling the type I error, and power as well as robustness, MTPBR has satisfactory performance and outperformed the conventional two-part (CTP) model using the likelihood ratio tests (LRT) for both simulated real data analyses (Chai et al. 2018).

In summary, all the proposed zero-inflated beta models use zero-inflated beta regression to fit relative abundant microbiome data; thus, they aim to take into account the compositional and zero-inflation nature of the microbiome data. However, all these models have their own advantages and limitations.

MTPBR is implemented via SAS NLMIXED procedure, and ZIBR is a longitudinal model that we have introduced in our previous book (2018) (Xia et al. 2018a); thus in this chapter we focus on introducing ZIBSeq and ZIBB.

13.2 Zero-Inflated Beta Regression (ZIBSeq)

ZIBSeq is a zero-inflated beta regression approach for differential abundance analysis of metagenomics data. It is an extension of the generalized linear model (GLM).

13.2.1 Introduction to ZIBseq

ZIBSeq (Peng et al. 2016) was developed to identify differentially abundant features between multiple clinical conditions while addressing the unique characteristics of metagenomics data with small sample size, high dimensionality, sparsity, often with a large number of zeros and skewed distribution under the compositions (proportions) setting. ZIBSeq directly handles the proportion data, assuming that proportional dependent variable can be characterized by the beta distribution.

13.2.1.1 ZIBseq Model

Assume x follows a beta distribution: $x \sim \text{Beta}(\mu, \phi)$, then the density of beta distribution (Ferrari and Cribari-Neto 2004) can be described as a function of u and ϕ:

$$f(\xi, \mu, \phi) = \frac{\Gamma(\phi)}{\Gamma(\mu\phi)\Gamma((1-\mu)\phi)} \xi^{\mu\phi-1}(1-t)^{(1-\mu)\phi-1}, 0 < \xi < 1 \qquad (13.1)$$

The mean and variance can be parameterized as:

$$\begin{cases} E(x) = \mu \\ \text{Var}(x) = \mu(1-\mu)/(\phi+1) \end{cases}, \qquad (13.2)$$

where $\Gamma(\cdot)$ is the gamma function, $\mu(0 < \mu < 1)$ is the mean, and $\phi(\phi > 0)$ is a precision parameter.

Beta distribution has a wide range of shapes depending on the values of mean and precision parameters. Thus, beta regression models are very suitable to model the continuous response variables of rates and proportions when their values are restricted to the interval $(0,1)$. To model the variable containing zero or one, Ospina and Ferrari in 2012 (Ospina and Ferrari 2012) proposed a more general class of zero-or-one inflated beta regression models for continuous proportions. Given that microbiome data often contains zero instead of one, ZIBSeq was developed based on the zero-inflated beta regression, assuming the response variable has a mixed continuous-discrete distribution with probability mass at zero.

The zero-inflated beta distribution adds a new parameter α to account for the probability of observations at zero. The subsequent mixture density is given as below:

$$Bi(x; \alpha, \mu, \phi) = \begin{cases} \alpha & \text{if } x = 0 \\ (1 - \alpha)f(x; \mu, \phi) & \text{if } 0 < x < 1 \end{cases}, \tag{13.3}$$

where $f(x; \mu, \phi)$ is the beta density in (13.1). The zero-inflated beta and hence ZIBSeq directly model the proportion of the response variables. Thus, the feature abundances are first normalized into the proportion of features. Let $x_i^{(j)}$ denote the normalized feature abundance, i.e., the proportion of feature j reads in sample i, ZIBSeq simply normalizes the feature abundance by dividing read counts of each given feature by the total mapped read counts in the sample: $x_i^{(j)} = c_{ij}/T_i$. Thus, the obtained $x_1^{(j)}, x_2^{(j)}, \ldots, x_n^{(j)}$ are proportions of feature $j, j = 1, \ldots, m$ on n samples. By assuming that each $x_i^{(j)}$ has a probability density function (13.3) with parameters $\alpha = \alpha_i^{(j)}$, $\mu = \mu_i^{(j)}$, and $\phi = \phi_i^{(j)}$. To fit the parameters in mixture distribution (13.3) for feature j, the ZIBSeq model is defined as below:

$$\text{logit}\left(\alpha_i^{(j)}\right) = \rho_0^{(j)}, \text{logit}\left(\mu_i^{(j)}\right) = \beta_0^{(j)} + \beta_1^{(j)} y_i, \text{ and } \phi_i^{(j)} = T_1 - 1. \tag{13.4}$$

In the above equations, $\rho_0^{(j)}, \beta_0^{(j)}$, and $\beta_1^{(j)}$ are unknown regression parameters to be estimated, while y_i is an outcome measurement indicating the class of sample i, while T_i is the ith sample depth. The dispersion parameter ϕ in (13.4) is approximated by a binomial distribution and a beta distribution under parameterization in (13.1). First, assume the number of reads on a particular feature c follows a binomial distribution with the sample depth T and the probability that the sample reads in this feature μ. Second, let $x = \frac{c}{T} - N\left(\mu, \frac{\mu(1-\mu)}{T}\right)$, then, the variance of proportion x will be $\text{Var}(x) = \frac{\mu(1-\mu)}{T}$. Third, in the case, the proportion x follows a beta distribution under parameterization in (13.1), the variance will be $\text{Var}(x) = \frac{\mu(1-\mu)}{\phi+1}$. An approximation of the dispersion parameter ϕ: $\phi \approx T - 1$ is derived from these two forms of variance.

13.2.1.2   Statistical Hypothesis Testing of Microbiome Composition and Outcome

The statistical hypothesis testing of no significant difference in relative abundance of taxa between class membership or experimental conditions is given as: $\beta_1^{(j)} = 0$. For large samples, the true null distributions can be approximated by a chi-squared distribution. ZIBSeq takes numerical approach to conduct the statistical hypothesis testing of null distributions to find the maximum likelihood estimates in (13.4). The t-statistic and corresponding P-value are obtained by implementing R package GAMLSS (Stasinopoulos and Rigby 2007).

13.2.2 Implement ZIBseq

The microbiome data structure that ZIBseq works on is a $n \times m$ matrix (see Table 11.1), with each of n rows presenting samples collected from two or more classes, and each of m columns presenting expression levels of genes or abundance of taxa measured for each sample. Each entry c_{ij} denotes the number of reads from sample i that mapped to feature (i.e., taxon) j. $T_i = \sum_{j=1}^{m} c_{ij}$ denotes the total number of reads of sample i. Each count c_{ij} not only depends on the reads of taxon j but also varies from the sequencing depth: the total number of reads of sample i. In Table 13.1, y_i, $i = 1, \ldots, n$ denotes the outcome variable associated with sample i, and it could be class membership of the sample.

ZIBseq is implemented via the function **ZIBseq()**. The syntax of function **ZIBseq** () is given below:

```
ZIBseq(data, outcome, transform = F, alpha = 0.05).
```

where the argument:

- **data** is a matrix recording count data, which could be extracted from a data frame with samples (or cases) as rows and taxa (or variables) as columns format recording the sample (or meta) and count data.
- **outcome** is a categorical vector of a clinical condition or treatment.
- **transform** is logical value indicating the square-root transform of the compositional matrix.
- **alpha** is used for customized threshold for calculating Q-values.

After running this function, the significant feature ("sigFeature"), the features that have been concerned ("useFeature"), and both Q-value ("qvalue") and P-value ("pvalue") will be provided in the output.

Example 13.1: Vitamin D Receptor (VDR) Knockout Mice

Data is from our study (2015) "Lack of Vitamin D Receptor Causes Dysbiosis and Changes the Functions of the Murine Intestinal Microbiome" (Jin et al. 2015). This dataset has been used in our previous book Statistical Analysis of Microbiome Data with R (Xia et al. 2018b).

The "otu_table_fecal_genus_vdr.csv" is a csv file containing 16S rRNA sequencing data matrix for the study samples. The VDR mice abundance data provides the

Table 13.1 Microbiome data structure for ZIBseq model

	$Feature_1$	$Feature_2$	...	$Feature_m$	Total	Outcome
$Sample_1$	c_{11}	c_{12}	...	c_{1m}	T_1	y_1
$Sample_2$	c_{21}	c_{22}	...	c_{2m}	T_2	y_2
...			...	...	...	...
$Sample_n$	c_{n1}	c_{n2}	...	c_{nm}	T_n	y_n

absolute sequence read counts for a total of 240 taxa at the genus rank level (rows) in 8 samples (columns), preprocessed as described in the main article. Among these eight samples, five samples with VDR were knockout and the remaining three samples were wild types. We are interested in whether dysbiosis and functions of intestinal microbiome are associated with the VDR knockout. Thus, the main metadata is the VDR condition status. We can provide additional metadata, which could be a CSV file containing the sample metadata in the data matrix to run the ZIBSeq. Here, we use commands to extract the grouping information from the sequencing data matrix.

First, set up the working directory and install the latest version of the ZIBseq package (latest version 1.2, June, 2017) in R or RStudio.

```
> setwd("~/Documents/QIIME2R/Ch13_ZeroInflatedBeta")
> install.packages("ZIBseq")
```

Then, we can perform ZIBSeq using the following four steps.

Step 1: Load the dataset.

```
> otu_tab=read.csv("otu_table_fecal_genus_vdr.csv", row.names=1,
check.names=FALSE)
> head(otu_tab,3)
        5_15_drySt-28F 1_11_drySt-28F 2_12_drySt-28F 3_13_drySt-28F
Tannerella          476            549            578            996
Lactococcus         326           2297            548           2378
Lactobacillus        94            434            719            322
        4_14_drySt-28F 7_22_drySt-28F 8_23_drySt-28F 9_24_drySt-28F
Tannerella          404            319            526            424
Lactococcus         471            882           1973           2308
Lactobacillus       205            644           2340           1000
```

Step 2: Create grouping variable.

We can use the sample ids to create grouping variable.

```
> otu_tab_t<-t(otu_tab)
> head(otu_tab_t,3)
        Tannerella Lactococcus Lactobacillus Lactobacillus::
Lactococcus
5_15_drySt-28F        476         326            94                 1
1_11_drySt-28F        549        2297           434                25
2_12_drySt-28F        578         548           719                 5


> grouping<-data.frame(row.names=rownames(otu_tab_t),t(as.data.
frame(strsplit(rownames(otu_tab_t),"_"))))
> grouping
                X1 X2     X3
```

```
5_15_dryySt-28F 5 15 dryySt-28F
1_11_dryySt-28F 1 11 dryySt-28F
2_12_dryySt-28F 2 12 dryySt-28F
3_13_dryySt-28F 3 13 dryySt-28F
4_14_dryySt-28F 4 14 dryySt-28F
7_22_dryySt-28F 7 22 dryySt-28F
8_23_dryySt-28F 8 23 dryySt-28F
9_24_dryySt-28F 9 24 dryySt-28F
```

```
> grouping$Group <- with(grouping,ifelse(X2%in% c(11,12,13,14,15),c
("Vdr-/-"), c("WT")))
> grouping
                X1 X2 X3 Group
5_15_dryySt-28F 5 15 dryySt-28F Vdr-/-
1_11_dryySt-28F 1 11 dryySt-28F Vdr-/-
2_12_dryySt-28F 2 12 dryySt-28F Vdr-/-
3_13_dryySt-28F 3 13 dryySt-28F Vdr-/-
4_14_dryySt-28F 4 14 dryySt-28F Vdr-/-
7_22_dryySt-28F 7 22 dryySt-28F WT
8_23_dryySt-28F 8 23 dryySt-28F WT
9_24_dryySt-28F 9 24 dryySt-28F WT
> grouping
[1] "Vdr-/-" "Vdr-/-" "Vdr-/-" "Vdr-/-" "Vdr-/-" "WT" "WT" "WT"
```

Step 3: Merge grouping and abundance taxa datasets to create a data frame.

```
> df_G <-as.data.frame(cbind(grouping,otu_tab_t))
> head(df_G,3)        grouping Tannerella Lactococcus Lactobacillus
5_15_dryySt-28F  Vdr-/-    476       326          94
1_11_dryySt-28F  Vdr-/-    549       2297         434
2_12_dryySt-28F  Vdr-/-    578       548          719
```

```
> class(df_G)
[1] "data.frame"
> row.names(df_G)
[1] "5_15_dryySt-28F" "1_11_dryySt-28F" "2_12_dryySt-28F" "3_13_dryySt-
28F"
[5] "4_14_dryySt-28F" "7_22_dryySt-28F" "8_23_dryySt-28F" "9_24_dryySt-
28F"
```

```
> ncol(df_G)
[1] 249
> dim(df_G)
[1] 8 249
> head(df_G,3)
```

The data frame has 249 columns, the first column is grouping variable, the remaining 248 columns are taxa.

Of course, we can create a meta dataset a priori and then load it into R and merge it to taxa abundance dataset.

Then, we can perform ZIBSeq analysis. The following steps including screening features, normalizing and transforming data as well as correcting multiple hypothesis testing are processed automatically by the **ZIBSeq()** function or through specifying such as to specify transform = TRUE or T will do a square root or cube root transformation. We explicitly list them below for reminding us what are really the function **ZIBSeq()** working on.

Step 4: Filter low abundant features.

Because taxa with a small number of reads are not reliable, so the function first remove any features with total counts less than 2 time large of the sample size to reduce the effects of noises and measurement errors.

Step 5: Normalize and transform data.

ZIBSeq uses a simple normalization procedure to convert the raw abundance measure to a proportion by dividing each feature read count by the total feature read counts in the sample, which results in relative abundance measure ranging [0,1]. After normalization, a square root or cube root transformations is performed to ensure the proportion data are better fitting a beta distribution if the distributions of the proportion are extremely left skewed.

Step 6: Define testing features, construct design matrix and run ZIBseq.

Perform zero-inflated beta regression (Ospina and Ferrari 2012) to predict each normalized feature (response variable) with outcome (explanatory variable); the *P*-value of the regression coefficient in each regression is obtained.

```
> library(gamlss)
> library(ZIBseq)
> yy=df_G[,2:249] # specify the 248 taxa
> for (i in 1:ncol(yy)){yy[,i]=as.numeric(as.character(yy[,i]))}
> grp=df_G[,1] # specify the group variable
> grp
[1] "Vdr-/-" "Vdr-/-" "Vdr-/-" "Vdr-/-" "Vdr-/-" "WT" "WT" "WT"
> grp1 <- gsub("Vdr-/-", "1",grp)
> grp2 <- gsub("WT", "0", grp1)
> grp3=as.numeric(grp2)
> grp3
[1] 1 1 1 1 1 0 0 0
> result=ZIBseq(data=yy,outcome=grp3,transform = T, alpha = 0.05)
```

We can use the **summary()** function to retrieve the summary information.

```
> summary(result)
 Length Class Mode
sigFeature 2 -none- character
useFeature 1 -none- numeric
qvalues 38 -none- numeric
pvalues 38 -none- numeric
```

We can also obtain the names of significant features and how many numbers of features have been used in this analysis.

```
> result$sigFeature
[1] "Lactobacillus" "Acinetobacter"
> result$useFeature
[1] 38
```

Step 7: Correct multiple hypothesis testing.

We can retrieve the *P*-values.

```
> result$pvalues
 [1] 1.447e-01 1.982e-01 5.668e-03 1.425e-01 6.503e-01 3.321e-01
1.108e-01
 [8] 1.843e-02 4.919e-01 1.495e-01 1.081e-01 7.490e-02 4.471e-01
2.722e-02
[15] 6.060e-01 1.080e-01 5.808e-02 1.603e-01 9.405e-01 5.989e-01
2.484e-01
[22] 3.193e-01 5.835e-01 1.411e-05 7.196e-01 7.299e-02 7.859e-01
2.423e-01
[29] 7.379e-01 1.334e-01 1.693e-01 3.883e-01 9.207e-01 3.591e-01
2.404e-01
[36] 5.300e-01 8.546e-02 6.135e-01
```

Use the FDR algorithm proposed by Storey and Tibshirani (2003) to estimate a conservative *Q*-value based on *P*-values obtained under the assumption that *P*-values are uniformly distributed.

```
> result$qvalues
 [1] 0.1258969 0.1392339 0.0358341 0.1258969 0.2491919 0.1825697
0.1258969
 [8] 0.0776776 0.2303659 0.1258969 0.1258969 0.1258969 0.2174525
0.0860320
[15] 0.2424364 0.1258969 0.1258969 0.1258969 0.3129562 0.2424364
0.1495700
[22] 0.1825697 0.2424364 0.0001784 0.2665825 0.1258969 0.2760442
0.1495700
[29] 0.2665825 0.1258969 0.1258969 0.1964160 0.3129562 0.1891691
0.1495700
[36] 0.2393545 0.1258969 0.2424364
```

Table 13.2 *P*-values and *Q*-values from ZIBSeq model

	pvalues	qvalues
1	0.144664728056042	0.125896940419006
2	0.198205269193142	0.139233854649929
3	0.00566793498498382	0.0358341428141313
4	0.142450703458304	0.125896940419006
5	0.650347961806409	0.249191944808199
6	0.332088919006491	0.18256970103783
7	0.110812381135047	0.125896940419006
8	0.018429552251035	0.0776775797080376
9	0.491903363519228	0.230365895238663
10	0.149481602689104	0.125896940419006
11	0.108146886935789	0.125896940419006
12	0.0748980599117641	0.125896940419006
13	0.447131785785204	0.217452473513295
14	0.0272155845567047	0.0860319628188012
15	0.606042815146284	0.242436357692026
16	0.107965948361024	0.125896940419006
17	0.0580814636233615	0.125896940419006
18	0.16028183039337	0.125896940419006
19	0.940513377708146	0.312956183240814
20	0.598904276700515	0.242436357692026
21	0.248405777366409	0.149570020619097
22	0.319334489842356	0.18256970103783
23	0.583495023804528	0.242436357692026
24	1.4105582565407e-05	0.000178358242098557
25	0.719627307693503	0.26658254635945
26	0.0729887026803861	0.125896940419006
27	0.785921203458262	0.27604422186671
28	0.242279775107805	0.149570020619097
29	0.737900153014585	0.26658254635945
30	0.133376612225457	0.125896940419006
31	0.169262963895601	0.125896940419006
32	0.388342311067966	0.196415998030281
33	0.920725205848764	0.312956183240814
34	0.3590536291346	0.189169116242343
35	0.240428856352592	0.149570020619097
36	0.530026294775602	0.239354468760951
37	0.085457319818132	0.125896940419006
38	0.613543801741082	0.242436357692026

Step 8: Write the results of *P*-values and *Q*-values as tables (Table 13.2).

```
> # Write results table
> # Make the table
> pvalues <- result$pvalues
> qvalues <- result$qvalues
> pq<- data.frame(cbind(pvalues,qvalues))

> library(xtable)
> table <- xtable(pq,caption = "Table of significant taxa",
lable="sig_taxa_table")
> print.xtable(table,type="html",file = "ZIBSeq_Table_VDR.html")
> write.csv(pq,file = paste("Results_ZIBSeq_Table_VDR.csv",sep = ""))
```

13.2.3 Remarks on ZIBSeq

When ZIBSeq (Peng et al. 2016) was proposed in 2015, it was evaluated with various simulated data and compared to the zero-inflated Gaussian (ZIG) model or metagenomeSeq (Paulson et al. 2013), which was proposed in 2013. It has been shown (Peng et al. 2016) that:

First, ZIBSeq has better performance in terms of large AUC values in identifying differential features based on ROC analysis via the R package ROCR (Sing et al. 2009) as well as can identify biologically important taxa in a real microbiome data application.

Second, ZIBSeq outperformed ZIG model based on two arguments: (1) compared to ZIG, ZIBSeq was more effective on the simulated multinomial distribution (MN) and binomial distribution (BI) data, although ZIG method performed slightly better than ZIBSeq with larger AUC values at different sample sizes with simulated data of zero-inflated Poisson (ZIP) and zero-inflated negative binomial (ZINB) distributions. (2) ZIBSeq method is more stable, whereas ZIG method is more likely to obtain negative *Q*-values because it tends to produce more small *P*-values. Thus, ZIG method may violate the assumption of uniform distribution in the *Q*-value calculation algorithm proposed by Storey and Tibshirani (2003).

Third, it almost had the same performance whether the response variable (features) is transformed with square root or not transformed in ZIBSeq.

ZIBSeq directly models the proportion after using the Total Sum Scaling (TSS) normalization; in contrast, ZIG models the zero-inflated count data by using the Cumulative Sum Scaling (CSS) normalization (see Sect. 12.1 of Chap. 12); thus, given different approaches of modeling and normalizations used in ZIBSeq and ZIG methods, it is no surprise that ZIBSeq method had better performance for simulated MN, and BI data, while ZIG method had better performance with simulated ZIP and ZINB data.

However, ZIBSeq takes the approach of analyzing relative abundance of individual taxa one by one at a time with a multiple testing correction procedure to control for type I error rate.

- This kind of approach was criticized as being not able to incorporate the inter-taxa correlation (Li et al. 2018; Liu and Lin 2018) and cannot provide *P*-values and correct statistical inferences for the selected taxa (Liu and Lin 2018).
- This kind of two-part beta regression model was also criticized that cannot provide a straightforward interpretation of covariate effects on the overall marginal (unconditional) mean (Chai et al. 2018).
- Cross-sectional two-part beta regression models including ZIBSeq method can handle zero-inflated proportion data; they cannot deal with repeatedly measured proportion data or longitudinal data (Chen and Li 2016).
- ZIBSeq uses GAMLSS package to conduct statistical hypothesis testing the significance, which generates *t*-statistic and corresponding *P*-value. Currently the tests are limited for two classes of samples or conditions. For more than two classes, multiple binary recoding is needed before apply zero-inflated beta regression to the recoded outcomes.

13.3   Zero-Inflated Beta-Binomial Model (ZIBB)

ZIBB is another specifically developed model for microbiome data based on zero-inflated beta regression.

13.3.1   Introduction to ZIBB

ZIBB (Hu et al. 2018) was proposed based on zero-inflated beta-binomial model to detect the association of microbiome with a continuous or categorical phenotype of interest. ZIBB model assumes that the count data distribution consists of a mixture of a point mass probability at zero (zero model) and a beta-binomial distribution (count model). Thus, ZIBB model has two components: (1) a zero model accounting for excess zeros and (2) a count model (the remaining component) allowing for over-dispersion effects through beta-binomial regression, which may have an appreciable additional mass at zero. The proposed ZIBB method is an extension of the beta-binomial model of BBSeq (Zhou et al. 2011) that was developed for the analysis of RNA sequence count data. Employing zero-inflated modeling and considering a constrained approach to estimate the over-dispersion parameters have been considered as two major improvements in the proposed ZIBB framework (Hu et al. 2018).

13.3.1.1   Zero and Count Models in ZIBB

ZIBB was proposed for analysis of 16S rRNA sequence count data, a $m \times n$ sequence count matrix with columns representing samples and rows representing OTUs (ASVs/Taxa). Let $Y = (y_{ij}) \in Z^{m \times n}$ be the count matrix, and each element y_{ij} represents the count of OTU i in sample j, where $(i = 1, \ldots, m, j = 1, \ldots n)$. Let $s_j = \sum_{i=1}^{m} y_{ij}$ be the library size for sample j. Then, the distribution of microbiome counts data is modeled by a two-stage canonical beta-binomial model, which assuming y_{ij} follows a binomial distribution $\text{Bin}(s_j, \mu_{ij})$, and $\mu_{ij} \sim \text{Beta}(\alpha_{1ij}, \alpha_{2ij})$. The probability mass function of the beta-binomial distribution with parameter $(s_j, \alpha_{1ij}, \alpha_{2ij})$ is written as $f(\cdot | \alpha_{1ij}, \alpha_{2ij})$ or

$$f\left(y_{ij} | \alpha_{1ij}, \alpha_{2ij}\right) = \binom{s_j}{y_{ij}} \frac{B\left(y_{ij} + \alpha_{1ij}, s_j - y_{ij} + \alpha_{2ij}\right)}{B\left(\alpha_{1ij}, \alpha_{2ij}\right)} \qquad (13.5)$$

In the above formulation, the probability is impliedly dependent on s_j. Since $\mu_{ij} \sim \text{Beta}(\alpha_{1ij}, \alpha_{2ij})$, then the mean and variance of $\text{Beta}(\alpha_{1ij}, \alpha_{2ij})$ distribution can be written as:

$E(\mu_{ij}) = \alpha_{1ij}/(\alpha_{1ij}, \alpha_{2ij})$ and $\phi_i E(\mu_{ij})(1 - E(\mu_{ij}))$. And it is trivial to solve that $\alpha_{1ij} = E(-\mu_{ij})(1 - \phi_i)/\phi_i$ and $\alpha_{2ij} = (1 - E(\mu_{ij}))(1 - \phi_i)/\phi_i$.

Through reparameterizing the parameters $(\alpha_{1ij}, \alpha_{2ij})$ of the beta distribution to $(E(\mu_{ij}), \phi_i)$, $\phi_i \geq 0$ in the reparameterized form can be clearly interpreted as the over-dispersion parameter. ZIBB model is a mixture of a point mass at zero (zero model) and a beta-binomial distribution (count model). The density of count y_{ij} is written as:

$$f\left(y_{ij} | \alpha_{1ij}, \alpha_{2ij}, \pi_{ij}\right) = \pi_{ij}\big|_{y_{ij}=0} + \left(1 - \pi_{ij}\right) f\left(y_{ij} | \alpha_{1ij}, \alpha_{2ij}\right), \qquad (13.6)$$

where π_{ij} is the point mass at zero and $f(\cdot)$ is the probability mass function of the canonical beta-binomial distribution as in (13.5). In the above ZIBB model, zero inflation has been taken accounted for by π_{ij}. Two link functions are used to include the effects of phenotype and covariates in the modeling: one for the zero model and another for the count model. The link function of the zero model is given below:

$$\text{logit}\left(\pi_{ij}\right) = \log\left(\frac{\pi_{ij}}{1 - \pi_{ij}}\right) = z_j^T \eta_i, \qquad (13.7)$$

where $z = (z_1, \ldots, z_n)^T \in R^{n \times q}$ is referred to as the design matrix for the zero model. $z = (z_{0, j}, \ldots, z_{q-1, j})^T \in R^q$ is the vector of zero-inflation related covariates for sample j (the $z_{0, j} = 1$ denotes the intercept) and $\eta_i = (\eta_{0, i}, \ldots, \eta_{q-1, i})^T \in R^q$ is the vector of corresponding coefficients for OTU i. When $q = 2$, the zero inflation-related covariates are $z_j^T = (1, \log s_j)$. The link function of the count model is given below:

$$\mathrm{logit}\left(E\left(\mu_{ij}\right)\right) = \log\left(\frac{E\left(\mu_{ij}\right)}{1 - E\left(\mu_{ij}\right)}\right) = X_j^T \beta_i, \tag{13.8}$$

where the $X = (X_1, \ldots X_n)^T \in R^{n \times P}$ is referred to as the design matrix for the count model. $X_j = (X_{0,j}, \ldots, X_{p-1,j})^T \in R^P$ is the vector of phenotypes of interest ($X_{0,j} = 1$ is the intercept of design matrix) for sample j and $\beta_i = (\beta_{0,i}, \ldots, \beta_{p-1,i})^T \in R^P$ is the vector of corresponding coefficients for OTU i.

To model the relationship between the mean and the variance, as well as between the mean and the over-dispersion. ZIBB uses the polynomial fit:

$$\mathrm{logit}(\phi_i) = \sum_{k=0}^{k} \gamma_k \{\mathrm{mean}(X\beta_i)\}^k \tag{13.9}$$

as the constraint between over-dispersion parameter ϕ_i and coefficients β_i in the count model.

The strategy of considering a constrained approach to model the over-dispersion as a polynomial function of the systematic (mean) component of the generalized linear model was adopted from BBSeq (Zhou et al. 2011). Based on the authors of ZIBB, $K = 3$ is typically sufficient to effectively model the relationship.

The distinctive feature of ZIBB lies on its modeling the mean-over-dispersion relationship. Thus, the authors of ZIBB called their model and the associated estimation approach as the constrained model and name the model that estimates ϕ_i separately for each OTU i as the free model.

13.3.1.2   Statistical Hypothesis Testing in ZIBB

ZIBB conducts parameter estimation by two steps: First, estimates the parameters in the free model by the maximum likelihood, and then uses the estimates of parameters in the constrained model to fit the polynomial model according to (13.9) and uses least squares' solutions to estimate the parameters in γ_k's. The main purpose of ZIBB is to test associations of OTUs with an experimental phenotypes of interest. Thus, statistical hypothesis testing is to test the statistical significance of β_{1i} for each OTU i, $i = 1, \ldots, m$. The null hypothesis is given as: $H_0 : \beta_{1i} = 0$. With $P = 2$, $\beta_i = (\beta_{0i}, \beta_{1i})^T$.

The Wald testing statistic is $\dfrac{\widehat{\beta_{1i}}}{SE(\widehat{\beta_{1i}})}$. For discrete (e.g., group indicator) phenotype of microbiome count data sets, a t distribution with degrees of freedom $n - 2$ is used to approximate the distribution of the Wald statistic under the null hypothesis, while for a continuous (e.g., body mass index values) of phenotype, a standard normal is used to approximate the Wald statistic's null distribution.

13.3.2   Implement ZIBB

ZIBB for the discovery analysis is implemented via the R package **ZIBBSeqDiscovery**. ZIBBSeqDiscovery uses zero-inflated beta-binomial model to analyze microbiome count data to detect the associate between phenotype of interest and the composition of the counts. The interested covariates can be adjusted via the link functions of zero and count models. The moment corrected correlation (MCC) approach is applied to adjust the *P-values*.

ZIBBSeqDiscovery takes two approaches to fit the ZIBB model: (1) The free approach treats the over-dispersion parameters for OTUs as independent, and (2) the constrained approach uses a mean-over-dispersion relationship to the count data.

ZIBBSeqDiscovery requires at least three data matrices: a *count data* matrix, *predictor (design)* matrix, and *zero inflation related* matrix. Count data matrix Y ($m \times n$) contains microbiome counts data with m rows and n columns representing m OTUs and n samples. Predictor matrix (design matrix) X ($n \times p$) contains n rows and p column. By default, the first column is the intercept term and the second column denotes the phenotype/covariate we are interested in. Thus, ZIBBSeqDiscovery tests the hypothesis that $\beta_{1i} = 0$. Zero inflation related matrix Z ($n \times q$) contains n rows and q columns. By default, the first column represents for the intercept term. The main function to fit ZIBB model is **fitZIBB()**. The syntax is given as:

```
fitZIBB(dataMatrix, X, ziMatrix, mode = "free", gn = 3, betastart =
matrix(NA, 0, 0), psi.start = vector(mode = "numeric", length = 0), eta.
start = matrix(NA, 0, 0))
```

where the argument:

- **dataMatrix** is the count matrix (m by n, m is the number of OTUs and n is the number of samples).
- **X** is the design matrix (n by p, p is the number of covariates) for the count model (e.g., beta-binomial), and intercept is included. The second column is assumed to be the covariate of interest.
- **ziMatrix** is the design matrix (n by q, q is the number of covariates) for the zero model, and intercept is included.
- **mode** is used to specify either "free" or "constrained" approach is used to estimate over-dispersion parameters.
- **gn** is used to specify a polynomial with degree of freedom **gn** to fit the mean-over-dispersion relationship. The default value for **gn** is 3. It is only used in constrained approach, thus **gn** is only valid when mode = "constrained".
- **betastart** is used for specifying the initial values for beta estimation matrix (p by m). Where beta are the effects (or coefficients) for the count model.
- **psi.start** is used to specify the initial values for the logit of over-dispersion parameters vector (with length m).

- **eta.start** is used to specify the initial values for eta estimation matrix (q by m), where eta are the effects (or coefficients) for the zero model.
- The arguments **psi. betastart**, **psi.start**, and **eta.start** are required only in constrained approach; they should be assigned as the beta estimation matrix, the psi estimation vector, and the eta estimation matrix, respectively, from free approach.

In the **ZIBBSeqDiscovery** package, the covariate of interest is referred to as phenotype. Current version (latest version 1.0, March 2018) of this package only considers one phenotype. In the design matrix X, the first column is the intercept, and the phenotype of interest is assumed to corresponding to the second column.

The **fitZIBB()** function returns several result values, including:

- **betahat** for the estimation matrix of beta (p by m) for count model.
- **bvar** for the estimation matrix of the variance of estimated betahat (p by m).
- **p** for the vector (with length of m) of P-values corresponding to the phenotype, in which the ith P-value corresponds to the hypothesis test of $H_0 : \beta_{1i} = 0$ for OTU i.
- **psi** for the estimation vector of the logit of the over-dispersion parameters (with length m).
- **zeroCoef** for the estimation matrix of eta (q by m) for zero model.
- **gamma** for the estimation vector of the coefficients in the mean-over-dispersion relationship in constrained approach (with length gn+1), which is only available when mode = "constrained" is specified.

Example 13.2: Microbiome and Colorectal Carcinoma Data
Dataset is from Kostic et al. (2012) "Genomic analysis identifies association of Fusobacterium with colorectal carcinoma" (Kostic et al. 2012). Sample collection and preparation, amplification, and processing of 16S sequence data were described as in the main article. The data matrix "otu_table_kostic.csv" is a csv file containing 16S rRNA sequencing data which has 2505 OTUs and 185 samples (90 tumor vs. 95 normal colon microbiota). The metadata matrix "meta_table_kostic.csv" is a csv file containing the health status for the samples in otu-table. The interested phenotype here is the health status for samples.

Since the authors of ZIBB (Hu et al. 2018) showed that constrained modeling of ZIBB is better than its free modeling of ZIBB in terms of the type I errors and power, thus we here only report the constrained version of ZIBB using this dataset.

First, install the package ZIBBSeqDiscovery from R CRAN. After the package was installed, we can take seven steps to perform the **fitZIBB ()** function via the ZIBBSeqDiscovery package.

Step 1: Load datasets.

```
> rm(list = ls())
> setwd("~/Documents/QIIME2R/Ch13_ZeroInflatedBeta")
```

```
> otu_tab=read.csv("otu_table_kostic.csv",row.names=1,check.
names=FALSE)
> meta_tab=read.csv("meta_table_kostic.csv",row.names=1,check.
names=FALSE)
```

```
> dim(otu_tab)
[1] 2505 185
> head(otu_tab,3)
       C0333.N.518126 C0333.T.518046 X38U4VAHB.518100 XZ33PN7O.518030
304309            40             4                1                2
469478             0             0                0                0
208196             0             0                0                0
```

We can see that "otu_tab" is a $m \times n$ count matrix: m is the number of OTUs (rows) and n is the number of samples(columns). Here there are 2505 taxa and 185 samples.

```
> dim(meta_tab)
[1] 185 1
> head(meta_tab,3)
 Group
1 Healthy
2 Tumor
3 Tumor
```

Step 2: Filter out OTUs and remove those OTUs that have zero count and few non-zero counts across all samples.

The ZIBB methods may fail to converge for a small proportion of OTUs. This typically occurs when OTUs have few (e.g., four or fewer) non-zero counts and for such OTUs. In such case, the power to detect association with experimental variables is small; thus before modeling using ZIBB methods, those OTUs need to be filtered out. For the remaining small number of OTUs that fail to converge (usually 2% or fewer), the ZIBB methods substitute *P*-values from the MCC package.

In below analysis, we removed all OTUs with zero counts across all samples. Based on the practice, for the discrete case, OTUs with zero counts across any one of groups were also removed.

```
> # Remove OTUs that have zero count across all samples
> otu_tab_kostic <- otu_tab[which(rowSums(otu_tab)>0),]
> dim(otu_tab_kostic)
[1] 2490 185
> head(otu_tab_kostic,3)
       C0333.N.518126 C0333.T.518046 X38U4VAHB.518100 XZ33PN7O.518030
304309            40             4                1                2
469478             0             0                0                0
208196             0             0                0                0
```

Step 3: Construct the design matrix and zero inflation related matrix.

We use below commands to construct the design matrix and zero inflation related matrix.

```
> # Construct the design matrix and zero inflation related matrix
> meta_tab_kostic <- cbind(1, meta_tab=="Tumor")
> zero_tab_kostic <- cbind(1, log(colSums(otu_tab_kostic)))
> head(meta_tab_kostic,3)
  Group
1 1   0
2 1   1
3 1   1
> head(zero_tab_kostic,3)
  [,1] [,2]
C0333.N.518126 1 8.640
C0333.T.518046 1 7.159
X38U4VAHB.518100 1 8.787


> class(otu_tab_kostic)
[1] "data.frame"
> class(zero_tab_kostic)
[1] "matrix" "array"
> class(meta_tab_kostic)
[1] "matrix" "array"


> otu_tab_kostic<-as.matrix(otu_tab_kostic)
> class(otu_tab_kostic)
[1] "matrix" "array"


> dim(otu_tab_kostic)
[1] 2490 185
> dim(meta_tab_kostic)
[1] 185 2
> dim(zero_tab_kostic)
[1] 185 2
```

Step 4: Fit the ZIBB model with free approach.

First, we use below commands to fit the ZIBB model with free approach.

```
> # Fit the ZIBB model with free approach
free_fit <- fitZIBB(otu_tab_kostic, meta_tab_kostic, zero_tab_kostic,
mode="free")
```

Step 5: Fit the ZIBB model with constrained approach.

Then we fit the constrained ZIBB model using the estimation from free approach as initial values.

```
> # Fit the constrained ZIBB model using the estimation from free
approach as initial values
> const_fit <- fitZIBB(otu_tab_kostic, meta_tab_kostic,
zero_tab_kostic, mode="constrained",
+                gn=3, betastart=free_fit$betahat,
+                psi.start=free_fit$psi, eta.start=free_fit$zeroCoef)
```

Depending on the configuration of your computer, the **fitZIBB()** function may take several minutes to complete the implementation.

Step 6: Use MCC method to adjust the *P*-values.

Finally, we can employ Moment Corrected Correlation (MCC) method to replace the NAs in the *P*-values. After running the **mcc.adj()** function, we will obtain the corrected *P*-values.

```
> free_mcc <- mcc.adj(free_fit,otu_tab_kostic, meta_tab_kostic,
zero_tab_kostic, K=4)
> const_mcc <- mcc.adj(const_fit,otu_tab_kostic, meta_tab_kostic,
zero_tab_kostic, K=4)
```

NA mostly appears when the OTU counts are zero in most of the samples. In this case, there are 185 samples, thus we want to check the cases such that OTU has 180, 181, 182, 183, and 184 zero counts across the 185 samples, respectively. The following commands can be used to check the effects of MCC adjustment. The NA counts in the reported *P*-values before and after MCC adjustment will be printed out.

```
> otu_tab_kostic_0 <- rowSums(otu_tab_kostic==0)
> for (i in 1:5) {
+   idx <- otu_tab_kostic_0 == (185-i)
+   if (i==1) {
+     df_sum <- data.frame(zero.counts = 185-i, N = sum(idx),
+              Num_NA_Infree = sum(is.na(free_fit$p[idx])),
+              Num_NA_Inconst = sum(is.na(const_fit$p[idx])),
+              Num_NA_Infree_MCC = sum(is.na(free_mcc$p[idx])),
+              Num_NA_Inconst_MCC = sum(is.na(const_mcc$p[idx])))
+   } else {
+     df_sum <- rbind(df_sum, c(185-i, sum(idx), sum(is.na(free_fit$p
[idx])),
+                sum(is.na(const_fit$p[idx])),
+                sum(is.na(free_mcc$p[idx])),
+                sum(is.na(const_mcc$p[idx]))))
+   }
+ }
```

```
> print.data.frame(df_sum, right = FALSE)
 zero.counts N  Num_NA_Infree Num_NA_Inconst Num_NA_Infree_MCC
1 184      731 599        535             236
2 183      334 36         107               0
3 182      214 29          33               0
4 181      146 21          18               0
5 180      113 18           6               0
 Num_NA_Inconst_MCC
1 227
2 0
3 0
4 0
5 0
```

Step 7: Write out the corrected *P*-values adjusted by the MCC method.

```
> const_mcc$p
   304309    469478    208196    358030     16076     35786    296165
7.073e-08 9.246e-01 9.306e-03        NA 7.102e-01 4.097e-02 9.816e-01
   174920    117676    326792     11380    527323    181344    561483
7.271e-01 9.158e-02 7.718e-01 7.295e-02 3.080e-138 1.000e+00
5.420e-02
   184450    272955    241674    301062    220782    177005    579750
7.230e-01 9.327e-03 2.110e-01 1.955e-01 5.811e-01 4.682e-01 7.667e-
01
   469778    149335    249661    110317    345556    348377    205119
2.364e-01 9.912e-01 5.594e-01 3.357e-01 9.739e-127 9.126e-01
5.710e-02
   555547    344111    230936    266445    527741    553728    180368
1.124e-01 1.000e+00 9.725e-01 0.000e+00        NA 7.923e-01 2.925e-01
    60136    181016    253953    297708    241520    148303    215231
4.473e-01 2.303e-01 4.434e-02 0.000e+00 1.000e+00        NA 1.062e-01
> mccPvalues <- const_mcc$p
> write.csv(mccPvalues,file = paste
("Results_ZIBB_Table_MCCadjusted_Pvalues_Kostic.csv",sep = ""))
```

13.3.3 Remarks on ZIBB

Overall, this study (Hu et al. 2018) showed that both zero inflation and the proposed constraint approaches are vital for accurate and powerful analysis of microbiome count data and ZIBB modeling with the constrained approach is preferred among the competing methods, including BBSeq, edgeR, and ZINB. Especially, it was demonstrated (Hu et al. 2018) that in terms of Type I errors and power for both discrete and continuous cases: (1) ZIBB has higher or comparable performance compared to competing methods. (2) Both the approaches of free and the constrained ZIBB modeling can provide accurate standard error estimates. And (3) the constrained modeling of ZIBB outperformed free modeling of ZIBB. However, the ZIBB methods may fail to converge for a small proportion of OTUs. This converge problem typically occurs when OTUs have four or fewer non-zero counts (Hu et al. 2018).

13.4 Summary

In this chapter we first briefly introduced zero-inflated beta modeling microbiome data and then focused on covering two specifically designed two-part zero-inflated beta regression models for microbiome data: ZIBSeq and ZIBB. We introduced their methodological developments and illustrated their applications with real study data. We also commented on their advantages and limitations. In Chap. 14, we will introduce compositional analysis of microbiome data.

References

Chai, Haitao, Hongmei Jiang, Lu Lin, and Lei Liu. 2018. A marginalized two-part Beta regression model for microbiome compositional data. *PLoS Computational Biology* 14 (7): e1006329. https://doi.org/10.1371/journal.pcbi.1006329.

Chen, Eric Z., and Hongzhe Li. 2016. A two-part mixed-effects model for analyzing longitudinal microbiome compositional data. *Bioinformatics* 32 (17): 2611–2617. https://doi.org/10.1093/bioinformatics/btw308.

Ferrari, Silvia, and Francisco Cribari-Neto. 2004. Beta regression for modelling rates and proportions. *Journal of Applied Statistics* 31 (7): 799–815.

Hu, Tao, Paul Gallins, and Yi-Hui Zhou. 2018. A zero-inflated beta-binomial model for microbiome data analysis. *Stat* 7 (1): e185. https://doi.org/10.1002/sta4.185. https://onlinelibrary.wiley.com/doi/abs/10.1002/sta4.185.

Jin, Dapeng, Wu Shaoping, Yong-guo Zhang, Lu Rong, Yinglin Xia, Hui Dong, and Jun Sun. 2015. Lack of vitamin D receptor causes dysbiosis and changes the functions of the murine intestinal microbiome. *Clinical Therapeutics* 37 (5): 996–1009.e7. https://doi.org/10.1016/j.clinthera. 2015.04.004. http://www.sciencedirect.com/science/article/pii/S0149291815002283.

Kostic, Aleksandar D., Dirk Gevers, Chandra Sekhar Pedamallu, Monia Michaud, Fujiko Duke, Ashlee M. Earl, Akinyemi I. Ojesina, Joonil Jung, Adam J. Bass, Josep Tabernero, José Baselga, Chen Liu, Ramesh A. Shivdasani, Shuji Ogino, Bruce W. Birren, Curtis Huttenhower, Wendy S. Garrett, and Matthew Meyerson. 2012. Genomic analysis identifies association of Fusobacterium with colorectal carcinoma. *Genome Research* 22 (2): 292–298. https://doi.org/10.1101/gr.126573.111. https://pubmed.ncbi.nlm.nih.gov/22009990. https://www.ncbi.nlm.nih.gov/pmc/articles/PMC3266036/.

Li, Z., K. Lee, M. R. Karagas, J. C. Madan, A. G. Hoen, A. J. O'Malley, and H. Li. 2018. Conditional Regression Based on a Multivariate Zero-Inflated Logistic-Normal Model for Microbiome Relative Abundance Data. *Stat Biosciences* 10 (3): 587–608. https://doi.org/10.1007/s12561-018-9219-2

Liu, Zhenqiu, and Shili Lin. 2018. Sparse Treatment-Effect Model for Taxon Identification with High-Dimensional Metagenomic Data. In *Microbiome Analysis: Methods and Protocols*, edited by Robert G. Beiko, Will Hsiao and John Parkinson, 309–318. New York, NY: Springer New York

Ospina, Raydonal, and Silvia L.P. Ferrari. 2012. A general class of zero-or-one inflated beta regression models. *Computational Statistics & Data Analysis* 56 (6): 1609–1623.

Paulson, Joseph N., O. Colin Stine, Héctor Corrada Bravo, and Mihai Pop. 2013. Robust methods for differential abundance analysis in marker gene surveys. *Nature Methods* 10 (12): 1200–1202. https://doi.org/10.1038/nmeth.2658.

Peng, X., G. Li, and Z. Liu. 2016. Zero-inflated beta regression for differential abundance analysis with metagenomics data. *Journal of Computational Biology* 23. https://doi.org/10.1089/cmb.2015.0157

Sing, Tobias, Oliver Sander, Niko Beerenwinkel, and Thomas Lengauer. 2009. ROCR: Visualizing the performance of scoring classifiers. *R Package Version* 1, no. 7

Stasinopoulos, D. Mikis, and Robert A. Rigby. 2007. Generalized additive models for location scale and shape (GAMLSS) in R. *Journal of Statistical Software* 23 (7): 1–46.

Storey, John D., and Robert Tibshirani. 2003. Statistical significance for genomewide studies. *Proceedings of the National Academy of Sciences* 100 (16): 9440–9445.

Xia, Yinglin, Jun Sun, and Ding-Geng Chen. 2018a. Modeling zero-inflated microbiome data. In *Statistical analysis of microbiome data with R*, ed. Yinglin Xia, Jun Sun, and Ding-Geng Chen, 453–496. Singapore: Springer Singapore.

———. 2018b. What are microbiome data? In *Statistical analysis of microbiome data with R*, ed. Yinglin Xia, Jun Sun, and Ding-Geng Chen, 29–41. Singapore: Springer Singapore.

Zhou, Yi-Hui, Kai Xia, and Fred A. Wright. 2011. A powerful and flexible approach to the analysis of RNA sequence count data. *Bioinformatics* 27 (19): 2672–2678.

Chapter 14
Compositional Analysis of Microbiome Data

Abstract This chapter focuses on compositional data analysis (CoDA). First, it describes overall compositional data, the rationale that microbiome data can be treated as compositional, Aitchison simplex, challenges of analysis of compositional data, fundamental principles of CoDA, and the family of log-ratio transformations. Then, it introduces three methods/models of CoDA: ANOVA-like compositional differential abundance analysis (ALDEx2), analysis of composition of microbiomes (ANCOM), analysis of composition of microbiomes-bias correction (ANCOM-BC). Next, it provides some remarks on CoDA approach.

Keywords Compositional data · Aitchison simplex · Log-ratio transformations · Additive log-ratio (alr) · Centered log-ratio (clr) · Isometric log-ratio (ilr) · Inter-quartile log-ratio (iqlr) · ALDEx2 · Analysis of composition of microbiomes (ANCOM) · Analysis of composition of microbiomes-bias correction (ANCOM-BC) · ANCOMBC package · Scaling invariance · Subcompositional coherence · Permutation invariance · Perturbation invariance · Subcompositional dominance · Bland-Altman plot · Difference plot · Tukey mean-difference plot · Effect size · xtable() · xtable package · edgeR · DESeq · q2-composition · qiime composition ancom · False positive rate (FDR) · Holm · Bonferroni · Bonferroni correction method · Benjamini · Hochberg (BH) procedure

This chapter focuses on compositional data analysis (CoDA). It is organized this way: First, we describe overall compositional data, the reasons that microbiome data can be treated as compositional, Aitchison simplex, challenges of analysis of compositional data, some fundamental principles of CoDA, and the family of log-ratio transformations (Sect. 14.1). Then, we introduce three methods or models of CoDA: ANOVA-like compositional differential abundance analysis (ALDEx2) (Sect. 14.2), analysis of composition of microbiomes (ANCOM) (Sect. 14.3), and analysis of composition of microbiomes-bias correction (ANCOM-BC) (Sect. 14.4).

Supplementary Information The online version contains supplementary material available at https://doi.org/10.1007/978-3-031-21391-5_14.

Next, we make some remarks on CoDA approach (Sect. 14.5). Finally, we complete this chapter with a brief summary of CoDA in Sect. 14.6.

14.1 Introduction to Compositional Data

14.1.1 What Are Compositional Data?

Composition is "the act of putting together parts or elements to form a whole" and is "the way in which such parts are combined or related: constitution" (Webster's II New College Dictionary, 2005, p.236) [also see (Xia et al. 2018a)]. As described in Aitchison's 1986 seminar work (Aitchison 1986b), a compositional dataset has the following four characteristics: (1) each row presents an observational unit; (2) each column presents an composition of whole; (3) each entry is non-negative; and (4) the sum of all the entries in each row equals to 1 or 100%. That is, compositional data quantitatively measure each element as a composition or describe the parts of the whole, a vector of non-zero positive values (i.e., components or parts) carrying only relative information between their components or parts (Pawlowsky-Glahn et al. 2015; Hron et al. 2010; Egozcue and Pawlowsky-Glahn 2011; Aitchison 1986a). Thus, compositional data exist as the proportions/fractions of a whole or the portions of a total (van den Boogaart and Tolosana-Delgado 2013a) and have the following unique properties: (1) The elements of the composition are non-negative and sum to unity (Bacon-Shone 2011; Xia et al. 2018a). (2) The total sum of all component values (i.e., the library size) is an artifact of the sampling procedure (Quinn et al. 2018a, b; van den Boogaart and Tolosana-Delgado 2008). (3) The difference between component values is only meaningful proportionally (Quinn et al. 2018a, b; van den Boogaart and Tolosana-Delgado 2008). The unique properties, especially adding up all the percentages of compositions necessarily to 100, introduces effects on correlations (Krumbein 1962).

14.1.2 Microbiome Data Are Treated as Compositional

Compositional data analysis is really only interested in relative frequencies rather than the absolute amount of data. Thus, compositional data analysis frequently arises in various research fields, including genomics, population genetics, demography, ecology, biology, chemistry, geology, petrology, sedimentology, geochemistry, planetology, psychology, marketing, survey analysis, economics, probability, and statistics.

Microbiome data are really count data. However, when microbiome study arises, some researchers consider or treat microbiome data as compositional mainly based on the following three reasons (see the details in (Xia et al. 2018a)):

(1) The high-throughput sequencing technology itself predefines or constrains the sequencing data including microbiome data to some constants (each sample read counts are constrained by an arbitrary total sum, i.e., library size) when sequencing platforms were used to generate the data (Quinn et al. 2018a, b), resulting in the total values of the data meaningless. Regardless the datasets are generated via 16S rRNA gene fragments sequencing or shotgun metagenomic sequencing, the observed number of reads (sequencing depth) is determined by the capacity of the sequencing platform used and the number of samples that are multiplexed in the run (Fernandes et al. 2014). Thus, the total reads mapped from the high-throughput sequencing methods are finite although they are large.

(2) Sample preparation and DNA/RNA extraction process cause each entry of microbiome datasets to carry only relative information in the measurements of feature (OTU or taxa abundance) (Lovell et al. 2011). RNA sequencing begins with extraction of the tissue of DNA/RNA samples. However, the tissue weight or volume is fixed. Thus, the number of sequence fragment reads obtained from a fixed volumes of total RNA is finite.

(3) In practice, to reduce experimental biases due to sampling depth and the biases of sample preparation and DNA/RNA extraction, the abundance read counts are typically divided by the total sum of counts in the sample(total library sizes) to normalize the data before analysis. All these result in microbiome dataset having compositional data structure and the abundance of each component (e.g., taxon/OTU) carries only relative information and hence is only coherently interpretable relative to other components within that sample (Quinn et al. 2018a, b).

14.1.3 Aitchison Simplex

Mathematically, a data is defined as compositional, if it contains D multiple parts of nonnegative numbers whose sum is 1 (Aitchison 1986a, p. 25) or any constant-sum constraint (Pawlowsky-Glahn et al. 2015, p. 10). It can be formally stated as:

$$S^D = \left\{ X = [x_1, x_2, \ldots, x_D] \middle| x_i > 0, i = 1, 2, \ldots, D; \sum_{i=1}^{D} x_i = \kappa \right\}. \quad (14.1)$$

That is, compositional data can be represented by constant sum real vectors with positive components. Where κ is arbitrary. Depending on the units of measurement or rescaling, frequent values are 1 (per unit, proportions), 100 (percent, %), (ppm, parts per million), and (ppb, parts per billion). The equation of 14.1 defines the sample space of compositional data as a hyperplane, called the simplex (Aitchison 1986a, p. 27). Also see (Mateu-Figueras et al. 2011; van den Boogaart and Tolosana-Delgado 2013a, p. 37) and (Pawlowsky-Glahn et al. 2015, p. 10). Compositional data do not exist in real Euclidean space, but rather in the simplex (a sub-space) (Aitchison 1986a).

14.1.4 Challenges of Analyzing Compositional Data

Standard methods (e.g., correlation analysis) rely on the assumption of Euclidean geometry in real space (i.e., P) (Eaton 1983) and assume that the differences between the tested variables are linear or additive. However, the simplex has one dimension less than real space (i.e., P-1), which represent special properties of the sample space (Aitchison 1986a). Therefore, in compositional data each component (or part) of the whole is dependent on other components. Because the sum of all proportions of each component equals to 1 (100%), at least two components are negatively correlated. This results in the dependency problem of the measured variables (e.g., taxa/OTUs/ASVs). In the statistical and especially in current microbiome literatures, the issue of dealing with proportions is referred as to compositionality. Microbial abundances are typically interpreted as proportions (e.g., relative abundance) to account for differences in sequencing depth (see Sect. 14.1.2 for details). Because proportions add to one, the change of a single microbial taxon will also change the proportions of the remaining microbial taxa. Therefore, it poses a challenge to infer exactly which microbial taxon is changing with treatments or conditions.

Compositionality makes most standard statistical methods and tests invalid. The challenges of analyzing compositional data have been described in our previous books (Xia 2020; Xia and Sun 2022a, b). Here, we summarize the main point as below.

- It was shown that two-sample t-tests and Wilcoxon rank sum tests have a higher FDR and low power to detect differential abundance because of ignoring the compositionality or dependency effect (microbial taxa in the same niche grow dependently) (Hawinkel et al. 2017).
- Analyzing compositional (or relative) data using Pearson and Spearman correlations will lead to the problem of "spurious correlation" between unrelated variables (Pearson 1897). Compositional data are not linear or monotonic; instead they exhibit dependence between components. Pearson and Spearman correlation analysis methods were originally proposed for absolute values. The dependence of each pair of components of compositional data violates the assumption that the paired data are randomly selected (independent) by linear (i.e., for Pearson) or rank (i.e., for Spearman) correlation.
- Visualizing or presenting compositional data using standard graphical tools (e.g., scatter plot, QQ plot) will result in a distorted graph.
- Analyzing compositional data using multivariate parametric models (e.g., MANOVA) violates the assumption of multivariate parametric analysis. Standard multivariate parametric models, such as MANOVA and multivariate linear regression, assume that the response variables are multivariate and normally distributed and have a linear relationship between response variables and predictors. However, compositional data are not multivariate normally distributed. Thus, using MANOVA (or ANOVA) and multivariate (or univariate) linear regression to test hypotheses on the response variable is meaningless due to dependence of the compositions.

14.1.5   Fundamental Principles of Compositional Data Analysis

Aitchison proposed and suggested (Aitchison 1982, 1986a) three fundamental principles for the analysis of compositional data that we should adhere to when analyzing compositional data. These fundamental principles are all rooted in the definition of compositional data: only ratios of components carry information and have been reformulated several times according to new theoretical developments (Barceló-Vidal et al. 2001; Martín-Fernández et al. 2003; Aitchison and Egozcue 2005; Egozcue 2009; Egozcue and Pawlowsky-Glahn 2011). The three fundamental principles are: (1) scaling invariance; (2) subcompositional coherence; and (3) permutation invariance.

In 2008, Aitchison summarized basic principle of compositional data analysis. For its formal expression, we cannot do better than to quote his definition (Aitchison 2008):

> Any meaningful function of a composition can be expressed in terms of ratios of the components of the composition. Perhaps equally important is that any function of a composition not expressible in terms of ratios of the components is meaningless.

Scaling invariance states that vectors with proportional positive components must be treated (analyzed) as representing the same composition (Lovell et al. 2015). That is, statistical inferences about compositional data should not depend on the scale used; i.e., a composition is multiplied by a constant k will not change the results (van den Boogaart and Tolosana-Delgado 2013b). Thus, the vector of per-units and the vector of percentages convey exactly the same information (Egozcue and Pawlowsky-Glahn 2011). We should obtain exactly the same results from analyzing proportions and percentages. For example, the vectors $a = [11, 2, 5]$, $b = [110, 20, 50]$, and $c = [1100, 200, 500]$ represent all the same composition because the relative importance (the *ratios*) between their components is the same (van den Boogaart and Tolosana-Delgado 2013b).

Subcompositional coherence states that analyses should depend only on data about components (or parts) within that subset, not depend on other non-involved components (or parts) (Egozcue and Pawlowsky-Glahn 2011); and statistical inferences about subcompositions (a particular subset of components) should be consistent, regardless of whether the inference is based on the subcomposition or the full composition (Lovell et al. 2015).

Permutation invariance states that the conclusions of a compositional analysis should not depend on the order (the sequence) of the components (the parts) (Egozcue and Pawlowsky-Glahn 2011; van den Boogaart and Tolosana-Delgado 2013b; Lovell et al. 2015). That is, in compositional analysis, the information from the order of the different components plays no role (i.e., changing the order of the components within a composition will not change the results) (van den Boogaart and Tolosana-Delgado 2013b; Quinn et al. 2018a, b). For example, it does not matter that we choose which component to be the "first," which component to be the "second," and which one to be the "last."

Other important properties of CoDa include perturbation invariance and subcompositional dominance. **Perturbation invariance** states that converting a composition between equivalent units will not change the results (van den Boogaart and Tolosana-Delgado 2013b; Quinn et al. 2018a, b).

Subcompositional dominance states that using a subset of a complete composition carries less information than using the whole (van den Boogaart and Tolosana-Delgado 2013b; Quinn et al. 2018a, b).

14.1.6 The Family of Log-Ratio Transformations

Compositional data exist in the Aitchison simplex (Aitchison 1986a), while the most statistical methods or models are valid in Euclidean space (real space). Aitchison (1981, 1982, 1983, 1984, 1986a; Aitchison and Egozcue 2005) approved that compositional data could be mapped into Euclidean space by using the log-ratio transformation and then can be analyzed using standard statistical methods or models such as an Euclidean distance metric (Aitchison et al. 2000). Thus, usually most compositional data analyses start with a log-ratio transformation (Xia et al. 2018a).

The algorithm behind the log-ratio transformation principle is that compositional vectors and associated log-ratio vectors exist a one-to-one correspondence so that any statement about compositions can be reformed in terms of log-ratios and vice versa (Pawlowsky-Glahn et al. 2015). Thus, log-ratio transformations solve the problem of a constrained sample space by projecting compositional data in the simplex into multivariate real space. Therefore, open up all available standard multivariate techniques (Pawlowsky-Glahn et al. 2015).

In order to transform the simplex to the real space, Aitchison in his seminal work (1986a, b) developed a set of fundamental principles, a variety of methods, operations, and tools, including the additive log-ratio (alr) for compositional data analysis.

The log-ratio transformation methodology has been accepted in various fields such as geology and ecology (Pawlowsky-Glahn and Buccianti 2011; van den Boogaart and Tolosana-Delgado 2013a; Aitchison 1982; Pawlowsky-Glahn et al. 2015). Here, we introduce the family of log-ratio transformations including the additive log-ratio (alr) (Aitchison 1986a, p. 135), the centered log-ratio (clr) (Aitchison 2003), the inter-quartile log-ratio (iqlr) (Fernandes et al. 2013, 2014), and isometric log-ratio (ilr) (Egozcue et al. 2003) transformations.

14.1.6.1 Additive Log-Ratio (alr) Transformation

The original approach of compositional data analysis proposed in Aitchison (1986a, b) was based on the alr-transformation. It is defined as:

$$\text{alr}(x) = \left[\ln\left(\frac{x_1}{x_D}\right), \ldots, \ln\left(\frac{x_i}{x_D}\right), \ldots, \ln\left(\frac{x_{D-1}}{x_D}\right) \right].$$ (14.2)

This formula maps a composition in the D-part Aitchison simplex none isometrically to a D-1 dimensional Euclidean vector. The alr-transformation chooses one component as a reference and takes the logarithm of each measurement within a composition vector (i.e., in the microbiome case, each sample vector containing relative abundances) after divided by a reference taxon (usually the taxon with index D, with D being the total number of taxon is chosen). Sum of all X_j to unity, but we can replace the components with any observed counts, which do not change the expression due to the library sizes cancelation. After the alr-transformation, any separation between the groups revealed by the ratios can be analyzed by standard statistical tools (Thomas and Aitchison 2006).

14.1.6.2 Centered Log-Ratio (clr) Transformation

The clr-transformation is defined as the logarithm of the components after dividing by the geometric mean of x:

$$\text{clr}(x) = \left[\ln\left(\frac{x_1}{g_m(x)}\right), \ldots, \ln\left(\frac{x_i}{g_m(x)}\right), \ldots, \ln\left(\frac{x_D}{g_m(x)}\right) \right],$$ (14.3)

with $g_m(x) = \sqrt[D]{x_1 \cdot x_2 \cdots x_D}$ ensuring that the sum of the elements of clr(x) is zero. Where $x = (x_1, \ldots, x_i, \ldots, x_D)$ represents the composition. The clr-transformation maps a composition in the D-part Aitchison simplex isometrically to a D-1 dimensional Euclidean vector. Unlike the alr-transformation in which a specific taxon is used as reference, the clr-transformation uses the geometric mean of the composition (i.e., sample vector) in place of x_D. Like the alr, after performing the clr-transformation, the standard unconstrained statistical methods can be used for analyzing compositional data (Aitchison 2003).

The clr-transformation algorithm has been adopted by some software (van den Boogaart and Tolosana-Delgado 2013b; Fernandes et al. 2013), and it was shown it could be used to analyze microbiome data and RNA-seq data and other next-generation sequencing data (Fernandes et al. 2014).

14.1.6.3 Isometric Log-Ratio (ilr) Transformation

The ilr-transformation is defined as:

$$y = \text{ilr}(x) = (y_i, \ldots, y_{D-1}) \in R^{D-1}.$$ (14.4)

where, $y_i = \frac{1}{\sqrt{i(i+1)}} \ln\left[\frac{\prod_{j=1}^{i} x_j}{(x_{i+1})i}\right].$

Like the clr, the ilr-transformation maps a composition in the D-part Aitchison simplex isometrically to a D-1 dimensional Euclidian vector. This ilr-transformation is an orthonormal isometry. It is the product of the clr and the transpose of a matrix which consists of elements. The elements are clr-transformed components of an orthonormal basis. The ilr-transformation transforms the data regarding an orthonormal coordinate system that performed from sequential binary partitions of taxa (van den Boogaart and Tolosana-Delgado 2013b).

Like alr and clr, ilr-transformation was developed to transform compositional data from the simplex into real space where standard statistical tools can be applied (Egozcue and Pawlowsky-Glahn 2005; Egozcue et al. 2003). That is, the ilr-transformed data can be analyzed using the standard statistical methods.

14.1.6.4   Inter-quartile Log-Ratio (iqlr) Transformation

The inter-quartile log-ratio (iqlr) transformation introduced in the ALDEx2 package (Fernandes et al. 2013, 2014) is defined as including only taxa that fall within the inter-quartile range of total variance in the geometric mean calculation. That is, the iqlr-transformation uses the geometric mean of a taxon subset as the reference.

14.1.7   Remarks on Log-Ratio Transformations

The difference of the three log-ratio (alr, clr, and ilr) transformations lies on choosing the divisor. Each transformation has its own strengths and weaknesses.

(1) alr-transformation

The strengths of alr-transformation are:

- It is the simplest transformation in the log-ratio transformation family and hence is relatively simple to interpret the results.
- The relation to the original D-1 first parts is preserved and it is still in wide use.

The weaknesses of alr-transformation are:

- It is not an isometric transformation from the Aitchison simplex metric into the real alr-space with the ordinary Euclidean metric. Although this weakness could be solved using an appropriate metric with oblique coordinates in real additive log-ratio (alr) space. However, it is not a standard practice (Aitchison and Egozcue 2005).
- Theoretically, we cannot use the standard statistical methods such as ANOVA and t-test to analyze the alr-transformed data. Although this weakness is a conceptual rather than practical problem (Aitchison et al. 2000) and this transformation was used in Aitchison (1986a, b) and further developed in Aitchison et al. (2000), by definition, the alr-transformation is asymmetric in

the parts of the composition (Egozcue et al. 2003); thus, the distances between points in the transformed space are not the same for different divisors (Bacon-Shone 2011).

- The practical problem is not always easy to choose an obvious reference and the choice of reference taxon is somewhat arbitrary (Li 2015), and results may vary substantially when the difference references are chosen (Tsilimigras and Fodor 2016). This may explain that the alr-transformation was an optional transformation approach but not default approach in Analyzing Compositional Data with R (van den Boogaart and Tolosana-Delgado 2013a, b).

(2) clr-transformation

The strengths of clr-transformation are:

- It avoids the alr-transformation problem of choosing a divisor (e.g., using one reference taxon) because the clr-transformation uses the geometric mean as the divisor.
- It is an isometric transformation of the simplex with the Aitchison metric, onto a subspace of real space with the ordinary Euclidean metric (Egozcue et al. 2003).
- Thus, it is most often used transformation in literature of compositional data analysis (CoDA) (Xia et al. 2018a).

The weaknesses of clr-transformation are:

- The clr covariance matrix is singular, thus it is difficult to use without adaption some standard statistical procedures (Bacon-Shone 2011).
- Its prominent weakness is not straightforward to obtain the orthogonal references in its subspace (Egozcue et al. 2003).

(3) ilr-transformation

The strengths of ilr-transformation are:

- It has significant conceptual advantages (Bacon-Shone 2011).
- It avoids the arbitrariness of alr and the singularity of clr; thus, it addresses certain difficulties of alr and clr and the ilr-transformed data can be analyzed using all the standard statistical methods.

However, the ilr-transformation has the weaknesses and been criticized because of the following difficulties:

- It has a major difficulty to naturally model the practical compositional situations in terms of a sequence of orthogonal logcontrasts, which is in contrast to the practical use of othogonality when using the simplicial singular value decomposition (Aitchison 2008). In other words, the ilr-transformation approach violates the practical use of the principal component analysis or principal logcontrast analysis in CoDA (Aitchison 1983, 1986a).
- Ensuring isometry has little to do with this compositional problem, and actually the coordinates in any ilr-transformation necessarily require a set of

orthogonal logcontrasts, which in clinical practice lacks of interpretability (Aitchison 2008) or its interpretability is dependent on the selection of its basis (Egozcue et al. 2003).

- There is no one-to-one relation between the original components and the transformed variables.

Due to these limitations and specifically the difficulty to interpret the results. In practice, ilr has somewhat limited its adoption or application (Egozcue et al. 2003; Xia et al. 2018a).

However, compared to alr-transformation that is not isometric and clr-transformation that is not an isomorphism, the isometric log-ratio transformation (ilr) is both an isomorphism and an isometry (so it is also known as balance). Thus, orthonormal coordinates can be defined using this transformation. Recently the ilr-transformation and the concept of groups of parts and their balances in CoDA (Egozcue and Pawlowsky-Glahn 2005) have been seen for the possible interesting applications in microbiome study to transform a rooted phylogeny and build the balances of taxa in inter-group analysis (Washburne et al. 2017; Rivera-Pinto et al. 2018). With the ilr-transformation, a sequential binary partition is used to construct a new set of coordinates, in application of microbiome, the sequential binary partition of the phylogeny is conducted in microbiome datasets until produces D-1 coordinates, x_i (called "balances") (Rivera-Pinto et al. 2018). Thus the statistical analysis will be performed on the "balances" (Rivera-Pinto et al. 2018) rather than the relative abundances, y_i, of D different OTUs/Taxa.

(4) **iqlr-transformation**

Compared to other three log-ratio transformations, the iqlr-transformation is not getting widely applied.

14.2   ANOVA-Like Compositional Differential Abundance Analysis

Previously, most existing tools for compositional data analysis have been used in the other fields, such as geology and ecology. **A**NOVA-**L**ike **D**ifferential **E**xpression (ALDEx) analysis is one of early statistical methods that were developed under a framework of compositional analysis for mixed population RNA-seq experiment.

14.2.1   Introduction to ALDEx2

ALDEx (Fernandes et al. 2013) and ALDEx2 (Fernandes et al. 2014) were initially developed for analyzing differential expression of mixed population RNA sequencing (RNA-seq) data, but it has been showed that this approach is essentially generalizable to nearly any type of high-throughput sequencing data, including three completely different experimental designs: the traditional RNA-seq, 16S rRNA gene amplicon-sequencing, and selective growth-type (SELEX) experiments (Fernandes et al. 2014; Gloor and Reid 2016; Gloor et al. 2016; Urbaniak et al. 2016). The R packages called ALDEx and ALDEx2 have been used to analyze unified high-throughput sequencing datasets such as RNA-seq, chromatin immunoprecipitation sequencing (ChIP-seq), 16S rRNA gene sequencing fragments, metagenomic sequencing, and selective growth experiments (Fernandes et al. 2014).

High-throughput sequencing (e.g., microbiome) data have several sources of variance including sampling replication, technical replication, variability within biological conditions, and variability between biological conditions. ALDEx and ALDEx2 were developed in a traditional ANOVA-like framework that decompose sample-to-sample variation into four parts: (1) within-condition variation, (2) between-condition variation, (3) sampling variation, and (4) general (unexplained) error.

Fernandes et al. (2013) highlighted the importance for partitioning and comparing biological between-condition and within-condition differences (variation), and hence they developed ALDEx, an ANOVA-like differential expression procedure, to identify genes (in the case, taxa/OTUs) with greater between- to within-condition differences via the parametric Welch's t-test or a non-parametric testing such as Wilcoxon rank sum test or Kruskal-Wallis test (Fernandes et al. 2013, 2014).

ALDEx was developed using the equivalency between Poisson and multinomial processes to infer proportions from counts and model the data as "compositional" or "proportional." In other words, ALDEx2 uses log-ratio transformation rather than effective library size normalization in count-based differential expression studies. However, when sample sizes are small, the marginal proportions have the large variance and extremely not normally distributed. Thus, ALDEx performs all inferences based on the compositionally valid Dirichlet distribution, i.e., using the full posterior distribution of probabilities drawn from the Dirichlet distribution through Bayesian techniques.

We can summarize the ALDEx2 methods into five fundamental parts (procedures) as below. For the details, the readers can refer to the original publications (Fernandes et al. 2013, 2014).

Part 1: Convert the observed abundances into relative abundances by Monte Carlo (MC) sampling.

Like other CoDa approaches, ALDEx is not interested in the total number of reads, instead of inferring proportions from counts. Denote n_i present the number of counts observed in taxon i, and assume that each taxon's read count was sampled from a Poisson process with rate μ_i, i.e., $n_i \sim \text{Poisson}(\mu_i)$ with $n = \sum_i n_i$. Then, the set of joint counts with given total read counts has a multinomial distribution, i.e., $\{[n_1, n_2, \ldots] | n\} \sim \text{Multinomial}(p_1, p_2, \ldots | n)$ where each $p_i = \mu_i / \sum_k \mu_k$ based on the equivalency between Poisson and multinomial processes.

Since the Poisson process is equivalent to the multinomial process, traditional methods use n_i to estimate μ_i and then use the set of μ_i to estimate p_i. These methods ignore that most datasets of this type contain large numbers of taxa with zero or small read counts; thus the maximum-likelihood estimate of p_i this way is often exponentially inaccurate. Therefore, ALDEx estimate the set of proportions p_i directly from the set of counts n_i. ALDEx uses standard Bayesian techniques to infer the posterior distribution of $[p_1, p_2, \ldots]$ as the product of the multinomial likelihood with a Dirichlet $(\frac{1}{2}, \frac{1}{2}, \ldots)$ prior. Considering the large variance and extreme non-normality of the marginal distributions p_i when the associated n_i are small, ALDEx does not summarize the posterior of p_i using point-estimates. Instead, it performs all inferences using the full posterior distribution of probabilities drawn from the Dirichlet distribution such that $[p_i, p_2, \ldots] \sim$ Dirichlet $\left([n_1, n_2, \ldots] + \frac{1}{2}\right)$. Adding 0.5 to the Dirichlet distribution, the multivariate distribution avoids the zero problem for the inferred proportions even if the associated count is zero and conserves the probability, i.e., $\sum_k p_k = 1$.

Because mathematically log-proportions are easily manipulated, ALDEx takes the component-wise logarithms and subtracts the constant $\frac{1}{m} \sum_k \log(p_k)$ from each log-proportion component for a set of m proportions $[p_i, p_2, \ldots, p_m]$. This results in the values of the relative abundances $q_i = \log(p_i) - \frac{1}{m} \sum_{k-1}^{m} \log(p_k)$ where $\sum_k q_k$ is always zero. Most important, it projects q onto a $m - 1$ dimensional Euclidean vector space with linearly independent components. Thus, a traditional ANOVA-like framework can be formed to analyze the q values $[q_i, q_2, \ldots, q_m]$.

Because it is cumbersome to directly compute q distribution, the software estimates the distribution of q from multiple Monte Carlo realizations of p given $[n_1, n_2, \ldots n_m]$. MC sampling is based on the compositionally valid Dirichlet distribution with the addition of a uniform prior. That is, it repeats the MC sampling for K times ($K = 128$ times by default) and consequently creating randomized instances dataset, in which for each taxon i in sample j, the observed abundance O_{ij} is represented by a vector of MC samples of relative abundances $\left(r_{ij}^{(1)}, \ldots, r_{ij}^{(k)}\right)^T$. This renders the data free of zeros.

Part 2: Perform log-ratio transformation on each of the so-called Monte Carlo (MC) instances usually most through clr- or iqlr-transformation

After obtaining the multivariate Dirichlet proportional distributions, to make a meaningful comparison between-sample values from proportional distributions, ALDEx uses the procedures developed by Aitchison, Egozcue, and others (Aitchison and Egozcue 2005; Egozcue and Pawlowsky-Glahn 2005; Egozcue et al. 2003) to transform component proportions into linearly independent components. That is, it performs the relative abundance vector $\left(r_{ij}^{(1)}, \ldots, r_{ij}^{(k)}\right)^T$ for each sample j and each MC Dirichlet realization k, $k = 1, \ldots, K$. The distributions are estimated from multiple independent Monte Carlo realizations of their underlying Dirichlet-distributed proportions for all taxa (genes) i simultaneously.

Part 3: Perform statistical hypothesis testing on each MC instance, i.e., each taxon in the vector of clr or iqlr transformed values, to generate *P*-values (P) for each transcript

The hypothesis testing is performed using the classical statistical tests, i.e., Welch's t and Wilcoxon rank sum tests for two groups, and glm and Kruskal-Wallis for two or more groups. Since there is a total of K MC Dirichlet samples, each taxon will have K *P*-values.

Let $i = \{1, 2, \ldots, I\}$ index taxa (genes), $j = \{1, 2, \ldots, J\}$ index the groups (conditions), and $k = \{1, 2, \ldots, K_j\}$ index the replicate of a given group(condition), using the framework of random-effect ANOVA models, the ALDEx model is given as below.

$$q_{ijk} = \mu_{ij} + \nu_{ijk} + \tau_{ijk} + \varepsilon_{ijk}, \tag{14.5}$$

where q_{ijk} is the adjusted log-abundance (expression); μ_{ij} is the expected abundance (expression) of taxon i (gene i) within each group (condition) j; ν_{ijk} is the sample-specific abundance (expression) change for replicate k; τ_{ijk} is the sampling variation from inferring abundance (expression) from read counts; and ε_{ijk} is the remaining nonspecific error.

Here, ν_{ijk} is assumed to be approximately normal as in the usual ANOVA models. The sampling error τ_{ijk} is given by the adjusted log-marginal distributions of the Dirichlet posterior and its distribution is very Gaussian-like. To ensure it is more appropriate for the analysis of high-throughput sequencing datasets, ALDEx does not assume that within-group (condition) sample-to-sample variation is small and essentially negligible.

Under the ANOVA framework, ALDEx conducts a null hypothesis to test the difference between groups (conditions) j and j', for all taxa i (genes i). $H_0 : \mu_{ij} = \mu_{ij'}$

Part 4: Average these *P*-values across all MC instances to yield expected *P*-values for each taxon.

The expected *P*-values are Benjamini-Hochberg adjusted *P*-values (BH) by using Benjamini-Hochberg correction procedure (Benjamini and Hochberg 1995).

Part 5: Develop an estimated effect size metric to compare the predicted between-group differences to within-group differences.

The authors of this model (Fernandes et al. 2013) emphasize the statistical significance by this hypothesis testing in Part 3 above does not imply that the groups (conditions) j and j' are meaningfully different. Instead, such meaning can be inferred through an estimated effect size that compares predicted between group (condition) differences to within-group(condition) differences.

Given the set of random variables p_{ijk} and q_{ijk}, the within-group (condition) distribution is:

$$W(i,j) = \sum_{k=1}^{K_j} q_{ijk},\qquad(14.6)$$

the absolute fold difference between-group (condition) distribution is:

$$\Delta_A(i,j,j') = W(i,j) - W(i,j'),\qquad(14.7)$$

the between sample, within-group(condition) difference is:

$$\Delta_W(i,j) = \max_{k \neq k'} |q_{ijk} - q_{ijk'}|,\qquad(14.8)$$

and the relative effect size:

$$\Delta_R(i,j,j') = \Delta_A(i,j,j') / \max\{\Delta_W(i,j), \Delta_W(i,j')\}.\qquad(14.9)$$

The effect size metric in ALDEx2 is another statistical method that was developed specifically for this package and is recommended for use over the P-values.

For normally distributed data, usually effect sizes measure the standardized mean difference between groups, such as Cohen's d (1988) measures the standardized difference between two means. Cohen's d has been widely used in behavioral sciences accompanying reporting of t-test and ANOVA results and meta-analysis. However, in high-throughput sequencing data, the data are often not normally distributed. For non-normality of data, it is very difficult to measure effect sizes and interpret them.

ALDEx2 effect size measure estimates the median standardized difference between groups. It is a standardized distributional effect size metric. To obtain this effect size, take the median of differences (about -2), which provides a non-parametric measure of the between-group difference. Then, scale (normalize) the median of differences by the dispersion: effect = median (difference / dispersion). Where dispersion = maximum (vector a-randomly permutated vector a, vector b- randomly permutated vector b). In the aldex.effect module output, these measures are denoted as: diff.btw = median(diff), diff.win = median(disp), and effect = eff. ALDEx2 also provides a 95% confidence interval of the effect size estimate if in the aldex.effect() function, CI=TRUE is specified (see Sect. 14.2.2).

It was shown that ALDEx2 effect size is somewhat robust and is approximately 1.4 times that of Cohen's d, as expected for a well-behaved non-parametric estimator. It is equally valid for normal, random uniform, and Cauchy distributions (Fernandes et al. 2018).

14.2.2 Implement ALDEx2 Using R

ALDEx2 takes the compositional data analysis approach that uses Bayesian methods to infer technical and statistical error (Fernandes et al. 2014). ALDEx2 incorporates a Bayesian estimate of the posterior probability of taxon abundance into a compositional framework; that is, it uses a Dirichlet distribution to transform the observed data and then estimates the distribution of taxon abundance by random sampling instances of the transformed data. There are two essential procedures: ALDEx2 first takes the original input data and generates a distribution of posterior probabilities of observing each taxon; then it uses the centered log-ratio transformation to transform this distribution. After centered log-ratio transforming the distribution, a parametric or nonparametric *t*-test or ANOVA can be performed for the univariate statistical tests and the *P-values* and Benjamini-Hochberg adjusted *P*-values are returned.

The ALDEx2 methods are available from the ALDEx2 package (current version 1.27.0, October 2021). There are three approaches that can be used to implement ALDEx2 methods: (1) Bioconductor ALDEx2 modular, (2) ALDEx2 wrapper, and (3) the aldex.glm module. Bioconductor version of ALDEx2 is modular, which is achieved by exposing the underlying center log-ratio transformed Dirichlet Monte-Carlo replicate values and hence is flexible for adding the specific R commands by the users based on their experimental design. Thus, it is suitable for the comparison of many different experimental designs. Currently, the ALDEx2 wrapper is limited to a two-sample *t*-test (and calculation of effect sizes) and one-way ANOVA design. The aldex.glm module was developed to implement the probabilistic compositional approach for complex study designs.

We illustrated this package in our previous book (Xia et al. 2018a). In this section, we illustrate these three approaches in turn with a new microbiome dataset.

Example 14.1: Breast cancer QTRT1 Mouse Gut Microbiome
This example dataset was introduced in Chap. 9 (Example 9.1). In this dataset, the mouse fecal samples were collected at two time points: pre-treatment and post-treatment. Here we are interested in comparison of the microbiome difference between Genotype (breast cancer cell lines MCF-7 cell (WT) and MCF-7 knockout) at post-treatment.

There are two ways to install the ALDEx2 package. The most recent version of ALDEx2 is available from github.com/ggloor/ALDEx_bioc. It is recommended to run the most up-to-date R and Bioconductor version of ALDEx2. Here we install this stable version of ALDEx2 from Bioconductor.

```
# Install stable version from Bioconductor
if (!requireNamespace("BiocManager", quietly = TRUE))
 install.packages("BiocManager")
BiocManager::install("ALDEx2")
```

Run the ALDEX Modular Step-by-Step
The aldex modular offers the users to specify their own tests, and then the ALDEx2 modular exposes the underlying intermediate data. To simplify, the ALDEx2

modular approach is just to call aldex.clr, aldex.ttest, and aldex.effect modules (functions) in turn and then merge the data into one data frame. We use the following seven steps to perform ALDEx2 by using the modular approach.

Step 1: Load OTU-table and sample metadata.

```
> setwd("~/Documents/QIIME2R/Ch14_Compositional/ALDEx2")
> otu_tab<-read.csv("otu_table_L7_MCF7_Post.csv",row.names = 1)
> meta_tab<-read.csv("metadata_QtRNA_Post.csv",row.names = 1)
```

Here, we want to compare post treatment between genotype (WT and MCF knockout). The above datasets are the subsets of the post treatment groups.

```
> Genotype<-meta_tab[, c('Genotype')]
> Genotype
 [1] "KO" "WT" "WT" "KO" "KO" "WT" "WT" "WT" "WT" "KO" "KO" "KO" "WT" "KO"
"WT"
[16] "KO" "WT" "KO" "KO" "WT"

> Group<-meta_tab[, c('Group')]
> Group
 [1] 1 0 0 1 1 0 0 0 0 1 1 1 0 1 0 1 0 1 1 0
```

Step 2: Run the aldex.clr module to generate the random centered log-ratio transformed values.

```
> library(ALDEx2)
> aldex_clr <- aldex.clr(otu_tab, Genotype, mc.samples=128 ,
denom="all", verbose=F)
```

where the argument:

- **counts table**(OTU abundance table), **a vector of groups**, and **the number of Monte-Carlo** are the three required inputs.
- **denom** is used to specify a string for indicating if iqlr, zero, or all features are used as the denominator.
- **verbosity** is used to specify the level of **verbosity** (TRUE or FALSE).

ALDEx2 recommends 128 or more mc.samples for the t-test, 1000 for a rigorous effect size calculation, and at least 16 for ANOVA.

Step 3: Run the aldex.ttest module to perform the Welch's t and Wilcoxon rank sum test.

Welch's t-test and Wilcoxon rank sum tests are used to conduct statistical testing of two conditions or groups.

```
> aldex_tt <- aldex.ttest(aldex_clr, Genotype, paired.test=FALSE,
verbose=FALSE)
```

where the argument:

- **aldex_clr** is the aldex object from aldex.clr module.
- **paired.test** is used to specify whether a paired test should be conducted or not (TRUE or FALSE).

The aldex.ttest() returns the values of we.ep (expected *P*-value of Welch's t test), we.eBH (expected Benjamini-Hochberg corrected *P*-value of Welch's *t* test), wi.ep (expected *P*-value of Wilcoxon rank test), and wi.eBH (expected Benjamini-Hochberg corrected *P*-value of Wilcoxon test).

As an alternative method of the *t*-test, we can run the aldex.kw module to perform the Kruskal-Wallis and glm tests for one-way ANOVA, which compares two or more groups.

```
> aldex_kw <- aldex.kw(aldex_clr)
```

It returns the values of kw.ep (expected *P*-value of Kruskal-Wallis test), kw.eBH (expected Benjamini-Hochberg corrected *P*-value of Kruskal-Wallis test), glm.ep (expected *P*-value of glm test), and glm.eBH (expected Benjamini-Hochberg corrected *P*-value of glm test). This module is slow and has not been evaluated for this version used.

Step 4: Run the aldex.effect module to estimate effect size and the within and between group values.

```
> aldex_effect <- aldex.effect(aldex_clr, CI=T, verbose=FALSE)
```

where the argument:

- **aldex_clr** is the aldex object from aldex.clr module.
- **CI** is used to indicate whether to include the 95% confidence interval information for the effect size estimate (TRUE or FALSE).
- **verbose** is used to specify the level of verbosity.

This step is required for plotting and is performed in the case of two groups. The aldex.effect module returns all the values:

- **rab.all** (median clr value for all samples in the feature).
- **rab.win.KO** (median clr value for the KO group of samples).
- **rab.win.WT** (median clr value for the WT group of samples).
- **dif.btw** (median difference in clr values between KO and WT groups).
- **dif.win** (median of the largest difference in clr values within KO and WT groups).
- **effect** (median effect size: diff.btw / max(diff.win) and **effect.low** and **effect.high** for all instances.

- **overlap** (proportion of effect size that overlaps between the Bayesian distribution of groups KO and WT; i.e., if the overlap is 0: no effect).

Step 5: Merge all data into one object and make a data frame for result viewing and downstream analysis.

The following commands merge the *t*-test and effect data into one object and name it as `aldex_all`.

```
> aldex_all <- data.frame(aldex_tt, aldex_effect)
```

The following head () function examines the first three lines of data.

```
> head(aldex_all,3)
 we.ep we.eBH wi.ep wi.eBH rab.all rab.win.KO rab.win.WT
OTU_2 0.5406304 0.8283945 0.5739729 0.8502291 -1.328501 -1.536681
-1.1439323
OTU_3 0.4959608 0.8083920 0.5036120 0.8216989 -1.150645 -1.359854
-0.8576377
OTU_4 0.4926840 0.7968770 0.5281530 0.8168477 -1.243212 -1.427288
-1.0419705
 diff.btw diff.win effect effect.low effect.high overlap
OTU_2 0.4165297 4.331159 0.08838384 -3.629842 4.652759 0.4515625
OTU_3 0.3566450 4.210935 0.07741749 -3.017144 5.117261 0.4531250
OTU_4 0.3932742 4.087157 0.07952276 -3.702736 4.159538 0.4578125
```

Step 6: Run the aldex.plot module to generate the MA and MW (effect) plots.

Bland-Altman plot (difference plot, or Tukey mean-difference plot), named after J. Martin Bland and Douglas G. Altman, is a data plotting method for analyzing the agreement between two different measures (Altman and Bland 1983; Martin Bland and Altman 1986; Bland and Altman 1999). Bland-Altman method states that any two methods designing to measure the same property or parameter are not merely highly correlated but also should have agreed sufficiently closely. ALDEx2 provides a **Bland-Altman (MA)** style plot to graphically compare the degree of agreement of measures between median $\log_2$ between-condition difference and median $\log_2$ relative abundance.

Effect size in ALDEx2 is defined as a measure of the mean ratio of the difference between groups (diff.btw) and the maximum difference within groups (diff.win or variance). The effect size can be obtained by the aldex.effect modular or by specifying effect=TRUE argument using the aldex wrapper. The aldex.plot modular plots median between-group difference versus median within-group difference to visualize differential abundance of the sample data, which are referred to as "**effect size**" plots in ALDEx2.

The following commands generate the MA and MW (effect) plots (Fig. 14.1).

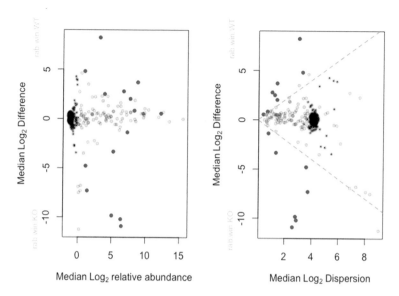

Fig. 14.1 MA and MW (Effect) plots of ALDEx2 output from aldex.plot() function. The left panel is a Bland-Altman or MA plot that shows the relationship between (relative) abundance and difference. The right panel is an MW (effect) effect plot that shows the relationship between difference and dispersion. In both plots, red represents the statistically significant features that are differentially abundant with $Q = 0.05$; gray are abundant, but not differentially abundant; black are rare, but not differentially abundant. This function uses the combined output from the aldex.ttest () and aldex.effect() functions. The Log-ratio abundance axis is the clr value for the feature

```
> # Figure 14.1
> par(mfrow=c(1,2))
> # Bland-Altman (MA) plot
> aldex.plot(aldex_all, type="MA", test="welch", cutoff=0.05, all.
cex=0.7, called.cex=1.1,
+      rare.col="black", called.col="red")
> aldex.plot(aldex_all, type="MW", test="welch",cutoff=0.05, all.
cex=0.7, called.cex=1.1,
+      rare.col="black", called.col="red")
```

ALDEx2 generates a posterior distribution of the probability of observing the count given the data collected. Importantly this approach generates the 95% CI of the effect size. **ALDEx2** uses a standardized effect size, similar to the Cohen's d metric. It was shown the effect size in **ALDEx2** is more robust and more conservative (being approximately 0.7 Cohen's d when the data are normally distributed based on Greg Gloor's note in his **AlDEx2**_vignette of **ANOVA**-Like Differential Expression tool for high throughput sequencing data, October 27, 2021).

In general, *P*-value is less robust than effect size. Thus, more researchers prefer to report effect size than to *P*-value. If sample size is *sufficiently large*, an effect size of 0.5 or greater is considered more likely corresponding to biological relevance.

A few features have an expected *Q*-value that is statistically significantly different. They are both relatively rare and have a relatively small difference. For those
features, it can be misleading even when identifying them based on the expected
effect size. ALDEx2 finds that the safest approach to identify those features is to find
where the 95% CI of the effect size does not cross 0. The 95% CI metric behaves
exactly in line with intuition: the precision of estimation of rare features is poor. To
identify rare features with more confidence, more deep sequencing is required. The
authors of ALDEx2 think this 95% CI approach can identify the biological variation
in the data as received (i.e., the experimental design is always as given). This is the
approach that was used in Macklaim et al. (2013), and it was independently validated
to be very robust (Nelson et al. 2015). In summary, this approach is not inferring any
additional biological variation but is identifying those features where simple random
sampling of the library would be expected to give the same result every time.

ALDEx2 considers an effect size cutoff of 1.5–2 and an overlap cutoff of 0.01 as
more appropriate to identify differential taxa of interest (Fernandes et al. 2013).
Below we illustrate two more effect size plots: (1) plot the effect size versus the *P*-
value, and (2) a volcano plot shows the difference between groups versus the *P*-value
(Fig. 14.2).

```
> # Figure 14.2
> # Effect size and volcano plots
> par(mfrow=c(1,2))
> plot(aldex_all$effect, aldex_all$wi.ep, log="y",pch=19, cex=0.7,
col=rgb(0,0,1,0.2),
```

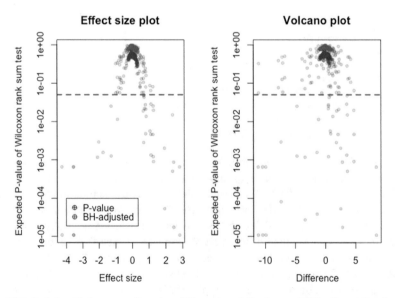

Fig. 14.2 Relationship between effect size, difference, and *P*-values and BH-adjusted *P*-values in
the tested dataset. This plot shows that the effect size has a much closer relationship to the *P*-value
than does the raw difference

```
+ xlab="Effect size", ylab="Expected P-value of Wilcoxon rank sum
test", main="Effect size plot")
> points(aldex_all$effect, aldex_all$wi.eBH,pch=19, cex=0.7,
col=rgb(1,0,0,0.2))
> abline(h=0.05, lty=2,lwd=2, col="red")
> legend(-4,0.0001, legend=c("P-value", "BH-adjusted"), pch=10,
col=c("blue", "red"))
> plot(aldex_all$diff.btw, aldex_all$wi.ep, log="y", pch=19,
cex=0.7, col=rgb(0,0,1,0.2),
+ xlab="Difference", ylab="Expected P-value of Wilcoxon rank sum
test", main="Volcano plot")
> points(aldex_all$diff.btw, aldex_all$wi.eBH, pch=19, cex=0.7,
col=rgb(1,0,0,0.2))
> abline(h=0.05, lty=2, lwd=2,col="red")
```

Step 7: Identify significant features by both Welch's t-test and Wilcoxon rank sum test.

```
> sig_by_both <- which(aldex_all$we.ep < 0.05 & aldex_all$wi.ep < 0.05)
> sig_by_both
 [1] 50 53 66 68 74 75 78 81 82 196 202 215 218 224 227 230 232 238 239 243
[21] 245 259 261 263 266 278 288
> sig_by_both_fdr <- which(aldex_all$we.eBH < 0.05 & aldex_all$wi.eBH <
0.05)
> sig_by_both_fdr
[1] 50 66 68 74 75 78 82 202 230 245 261 263
```

Twenty seven taxa are identified as significant by both Welch's *t*-test and Wilcoxon rank sum test, and twelve of these reach significance when the *P*-values are adjusted for multiple testing corrections using the Benjamini-Hochberg's method.

The following R commands use the **xtable()** function from **xtable** package to make a result table. The xtable package is used to create export tables, converting an R object to an xtable object, which can then be printed as a LaTeX or HTML table. Here, the print.xtable() function is used to export to HTML file. If you want to export the LaTeX file, then use type="latex", file="ALDEx2_Table_Coef_QtRNA. tex" instead.

```
> # Table 14.1
> # Write results table
> # Make the table
> library(xtable)
> table <-xtable(
+ aldex_all[sig_by_both,c(8:12,1,3,2,4)], caption="Table of
significant OTUs",lable="Coef_OTUs_table", digits=3,
+ label="sig.table", align=c("l",rep("r",9) )
+ )
```

```
> print.xtable(table,type="html",file = "ALDEx2_Table_Coef_QtRNA.
html")
> write.csv(table,file = paste("ALDEx2_Table_Coef_QtRNA.csv",sep =
""))
```

where the argument:

- **aldex_all**[sig_by_both,c(8:12,1,3,2,4)] is a R object; the element of the object "sig_by_both" is the row of output matrix, the element of the object "c(8: 12,1,3,2,4)] is the column of output matrix with the order of columns you want to be in export table.
- **caption** is used to specify the table's caption or title.
- **label** is used to specify the LaTeX label or HTML anchor.
- **align** indicates the alignment of the corresponding columns and is character vector with the length equal to the number of columns of the resulting table; the resulting table has 9 columns, so the number is 9.
- If the R object is a data.frame, the length of align is specified to be 1 + ncol (x) because the row names are printed in the first column.
- The left, right, and center alignment of each column are denoted by "l," "r," and "c," respectively. In this table, align=c("l",rep("r",9) indicates that first column is aligned left, and the remaining 9 columns are aligned right. The digits argument is used to specify the number of digits to display in the corresponding columns (Table 14.1).

Only those significant taxa detected in both Welch's t-test and Wilcoxon rank sum tests are printed in Table 14.1. We can interpret the table this way, for the OTU_98, the absolute difference between KO and WT groups can be up to -4.698, implying that the absolute fold change in the ratio between OTU_98 and all other taxa between KO and WT groups for this organism is on average $(1/2)^{-4.698} = 25.96$ fold across samples. The difference within the groups of 3.539 is roughly equivalent to the standard deviation, giving an effect size of $-4.698/3.539 = -1.327$ (here exactly: -1.269 [$-9.010, 0.347$]).

Run the ALDEX Wrapper

When running the aldex wrapper, it will link the modular elements together to emulate ALDEx2 prior to the modular approach. In the simplest case, the aldex wrapper performs a two-sample t-test and calculates effect sizes.

```
> aldex_t <- aldex(otu_tab, Genotype, mc.samples=128, test="t",
effect=TRUE,
+ include.sample.summary=FALSE, denom="all", verbose=FALSE)
```

Here the test group 'Genotype' has two levels (KO vs. WT). So this is two-sample t-test. We specify test = "t," and then the effect should be set to TRUE. The "t" option evaluates the data as a two-factor experiment using both Welch's t-test and the Wilcoxon rank sum test. Like other tests, t-tests also include a Benjamini-

Table 14.1 The significant features identified by both Welch's t-test and Wilcoxon rank sum test with *P*-values and BH-adjusted *P*-values in the tested dataset

	diff.btw	diff. win	effect	effect. low	effect. high	we. ep	wi.ep	we. eBH	wi. eBH
OUT_98	−4.698	3.539	−1.269	−9.010	0.347	0.002	0.001	0.033	0.025
OTU_102	1.253	1.333	0.854	−0.993	7.335	0.010	0.014	0.144	0.160
OTU_116	−10.435	2.831	−3.563	−23.145	−0.797	0.000	0.000	0.000	0.001
OTU_119	−3.402	1.443	−2.071	−16.377	−0.173	0.000	0.000	0.002	0.001
OTU_125	−9.904	2.786	−3.558	−26.823	−0.717	0.000	0.000	0.001	0.001
OTU_126	−11.171	2.684	−4.260	−33.527	−0.944	0.000	0.000	0.000	0.001
OTU_131	3.671	1.419	2.414	−0.834	18.509	0.000	0.001	0.002	0.015
OTU_134	3.560	2.865	1.137	−0.848	12.261	0.012	0.001	0.113	0.029
OTU_135	2.528	1.235	1.938	0.198	12.070	0.000	0.000	0.003	0.002
OTU_316	4.763	3.928	1.124	−1.458	10.936	0.015	0.005	0.119	0.064
OTU_323	−7.343	3.637	−2.004	−19.881	−0.072	0.000	0.000	0.003	0.003
OTU_342	0.639	0.830	0.725	−1.801	6.394	0.041	0.043	0.332	0.322
OTU_345	0.819	0.730	1.077	−0.969	7.081	0.004	0.006	0.072	0.086
OTU_354	1.882	2.367	0.676	−1.596	10.338	0.036	0.019	0.301	0.192
OTU_359	2.107	1.420	1.244	−1.761	12.310	0.002	0.005	0.043	0.081
OTU_362	2.818	1.069	2.474	0.351	14.778	0.000	0.000	0.000	0.001
OTU_365	4.726	3.273	1.280	−0.690	11.492	0.006	0.001	0.066	0.020
OTU_371	0.889	0.909	0.820	−0.746	11.161	0.013	0.005	0.177	0.080
OTU_373	2.223	2.327	0.831	−1.384	10.228	0.021	0.021	0.220	0.201
OTU_380	1.832	2.369	0.700	−0.694	16.919	0.031	0.003	0.294	0.057
OTU_385	0.513	0.397	1.265	−0.371	7.901	0.001	0.001	0.019	0.025
OTU_405	0.505	0.653	0.753	−1.231	7.099	0.019	0.022	0.206	0.206
OTU_408	8.446	3.040	2.826	0.644	25.680	0.000	0.000	0.001	0.001
OTU_410	−1.393	0.776	−1.742	−12.694	−0.161	0.000	0.000	0.002	0.002
OTU_413	0.518	0.636	0.808	−1.278	6.340	0.022	0.027	0.229	0.243
OTU_428	0.808	1.049	0.719	−1.418	7.025	0.043	0.049	0.34	0.344
OTU_448	−1.122	1.311	−0.887	−7.396	1.385	0.010	0.011	0.144	0.142

Table of significant OTUs

Hochberg correction of the raw *P*-values. The data also can be plotted onto Bland-Altman (MA) or effect (MW) plots (Fig. 14.3).

The inter-quartile log-ratio (iqlr) transformation was introduced in the ALDEx2 package (see Sect. 14.1.6.4). The following commands use iqlr transformation and generate MA and MW plots.

```
> aldex_t_iqlr <- aldex(otu_tab, Genotype, mc.samples=128, test="t",
effect=TRUE,
+        include.sample.summary=FALSE, denom="iqlr", verbose=FALSE)
> head(aldex_t_iqlr,3)
        rab.all rab.win.KO rab.win.WT diff.btw diff.win effect overlap
we.ep
OTU_2 -0.7688690 -0.9230157 -0.6292115 0.4898984 4.143746 0.09669726
```

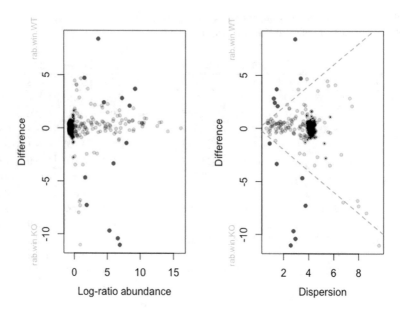

Fig. 14.3 MA and MW (effect) plots from t-tests in ALDEx2 output

```
0.4531250 0.5048162
OTU_3 -0.3587615 -0.5131136 -0.2026998 0.2097475 4.087856 0.04867811
0.4664587 0.5095202
OTU_4 -0.7754322 -0.9341565 -0.6243442 0.1912022 4.060661 0.03592633
0.4804992 0.4813148
        we.eBH  wi.ep  wi.eBH
OTU_2 0.8170884 0.5300930 0.8402394
OTU_3 0.8178020 0.5329214 0.8449501
OTU_4 0.8107567 0.5222389 0.8325645
```

```
> # Figure 14.3
> par(mfrow=c(1,2))
> aldex.plot(aldex_t_iqlr, type="MA", test="welch", cutoff=0.05,
all.cex=0.7, called.cex=1.1,
+       rare.col="black", called.col="red", xlab="Log-ratio
abundance",
+       ylab="Difference")
> aldex.plot(aldex_t_iqlr, type="MW", test="welch",cutoff=0.05,
all.cex=0.7, called.cex=1.1,
+       rare.col="black", called.col="red", xlab="Dispersion",
+       ylab="Difference")
```

The left panel is an Bland-Altman (MA) plot that shows the relationship between (relative) abundance and difference. The right panel is an MW (effect) plot that shows the relationship between difference and dispersion. In both plots, red dots indicate that the features are statistically significant and gray or black dots indicate the features are not significant. The log-ratio abundance axis is the clr value for the feature.

Run the ALDEX.GLM Module Using Complex Study Designs

ALDEx2 also has the aldex.glm module that can be used to implement the proba-bilistic compositional approach for complex study designs. This module is substan-tially slower compared to the above two-comparison tests; however, the users can implement their own study designs. Essentially, this approach is same as the above modular approach but requires the users to provide a model matrix and covariates to the **glm()** function in R.

Example 14.2: Breast Cancer QTRT1 Mouse Gut Microbiome, Example 14.1, Cont.

Here, we use the full dataset from Example 14.1 to illustrate the aldex.glm module. In this dataset, there are two main effect variables Group and Time. We are interested in comparison of the microbiome difference between Group (coded as WT and KO), Time (coded as Before and Post) and their interaction term.

Step 1: Load the count table and metadata.

```
> setwd("~/Documents/QIIME2R/Ch14_Compositional/ALDEx2")
> otu_table<-read.csv("otu_table_L7_MCF7_phyloseq.csv",
row.names = 1)
> meta_table<-read.csv("metadata_QtRNA.csv",row.names = 1)
> head(meta_table,3)
       MouseID Genotype Group  Time   Group4 Total.Read
Sun071.PG1    PG1     KO    1  Post  KO_POST    61851
Sun027.BF2    BF2     WT    0 Before WT_BEFORE 42738
Sun066.PF1    PF1     WT    0  Post  WT_POST    54043
```

Step 2: Set model matrix and covariates.

```
> Group<-meta_table[, c('Group')]
> Time<-meta_table[, c('Time')]
> TotalRead <- meta_table[, c('Total.Read')]

> covariates <- data.frame('Group', 'Time','TotalRead')
> mod <- model.matrix(~ Group*Time, covariates)
```

Step 3: Run the aldex.glm() function.

```
> aldex_clr_glm <- aldex.clr(otu_table, mod, mc.samples=128,
denom="all", verbose=F)
> aldex_glm <- aldex.glm(aldex_clr_glm, mod)
```

Fig. 14.4 Plot of effect sizes for Group by Time interaction at post-treatment in breast cancer QTRT1 mouse gut microbiome

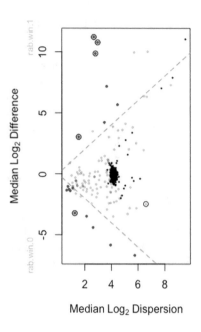

Step 4: Run the `aldex.glm.effect` () function.

```
> aldex_glm_effect <- aldex.glm.effect(aldex_clr_glm)
```
The `aldex.glm.effect` () function calculates the effect size for each binary predictor in the matrix and outputs to a named list.

Step 5: Plot the effect sizes.

Figure 14.4 can be generated using the following R commands, which plots the BH-corrected glm values for the actual test case vs. the effect size for the binary predictor.

```
> # Figure 14.4
> aldex.plot(aldex_glm_effect [["Group:TimePost"]], test="effect",
cutoff=2)
> sig <- aldex_glm [,20]<0.05
> points(aldex_glm_effect [["Group:TimePost"]]$diff.win[sig],
+   aldex_glm_effect [["Group:TimePost"]]$diff.btw[sig],
col="blue")
> sig <- aldex_glm [,20]<0.2
> points(aldex_glm_effect [["Group:TimePost"]]$diff.win[sig],
+   aldex_glm_effect [["Group:TimePost"]]$diff.btw[sig],
col="blue")
```

Table 14.2 The significant features identified by glm test with BH-adjusted *P*-values in the tested dataset

	Group. TimePost. diff.btw	Group. TimePost. diff.win	Group. TimePost. effect	Group. TimePost. overlap	model.Group. TimePost. Pr...t...BH
OUT_116	10.770	2.999	3.678	0.000	0.027
OTU_119	3.022	1.552	1.787	0.031	0.012
OTU_125	9.862	2.830	3.480	0.000	0.016
OTU_126	11.211	2.707	4.076	0.000	0.047
OTU_362	−3.225	1.255	−2.480	0.002	0.013

Table of significant OTUs

Step 6: Write out the glm test and effect table.

First, merge glm and glm effect tests into one object and make a data frame.

The following commands merge the aldex_glm and aldex_glm_effect data into one object and name it as `aldex_all_glm` (Table 14.2).

```
aldex_all_glm <- data.frame(aldex_glm, aldex_glm_effect)
> sig_glm <- which(aldex_all_glm[,20] < 0.05) # BH-corrected P-values of
Group.TimePost
```

```
> # Table 14.2
> # Write effect table
> # Make the table
> library(xtable)
> table <-xtable(
+ aldex_all_glm [sig_glm,c(38:41,20)], caption="Table of significant
OTUs", lable="Coef_OTUs_table", digits=3,
+ label="sig.table", align=c("l",rep("r",5) ))
> print.xtable(table,type="html",file =
"ALDEx2_Table2_Coef_QtRNA_glm.html")
> write.csv(table,file = paste("ALDEx2_Table2_Coef_QtRNA_glm.csv",
sep = ""))
```

14.2.3 Remarks on ALDEx2

ALDEx2(ALDEx) was designed to identify differential expression of genes, relative to the geometric mean abundance, between two or more groups. It was shown (Fernandes et al. 2013) that the ALDEx2 methods are robust comparing to existing representative methods in RNA-seq such as edgeR (Robinson et al. 2010) and DESeq (Anders and Huber 2010). It was also demonstrated that ALDEx2 has very high precision in identifying differentially expressed genes (and transcripts) for 16S rRNA data and high recall too given sufficient sample sizes (Quinn et al. 2018a, b). Although ALDEx2 was shown having the potential to be generalized using any type of high-throughput sequencing data and has often used for analyzing the differential abundant taxa (OTUs) among meta-genomics researchers (Urbaniak et al. 2016), ALDEx2 has several limitations, for example:

- ALDEx2 cannot address zero-inflation problem, instead replaces zero read counts using valid Dirichlet distribution (Fernandes et al. 2013).
- ALDEx2 statistical tests are hard to interpret until the log-ratio transformation sufficiently approximate an unchanged reference (Quinn et al. 2018a, b). The performance in this point is dependent on transformations; it was showed that the inter-quartile range log-ratio (iqlr) transformation outperform the centered log-ratio (clr) transformation and was recommended using as the default setting for ALDEx2 (Quinn et al. 2018a, b).
- ALDEx2 requires a large number of samples because it performs statistical hypothesis testing using non-parametric methods such as Wilcoxon rank sum test for comparisons of two groups and Kruskal-Wallis test for comparisons of more than two groups. Non-parametric differential expression methods have been suggested reducing statistical power and hence require large number of samples (Quinn et al. 2018a, b; Seyednasrollah et al. 2013; Tarazona et al. 2015; Williams et al. 2017; Quinn et al. 2018a, b). It was reviewed partially due to its non-parametric nature and weaknesses of ALDEx2 methods, the ALDEx2 package has not been widely adopted in the analysis of RNA-seq data (Quinn et al. 2018a, b).
- ALDEx2 interprets the log-ratio transformation as a normalization. Thus, ALDEx2 is majorly limited to this normalization. It is hard to interpret the results of the statistical tests even in the setting of large sample sizes when the log-ratio transformation does not sufficiently approximate an unchanged reference (Quinn et al. 2018a, b).

ALDEx2 was reported having difficulty to control FDR (Lin and Peddada 2020b) and to maintain statistical power compared to competing differential abundance methods (e.g., ANCOM, ANCOM-BC, edgeR, and DESeq2) (Lin and Peddada 2020b; Morton et al. 2019).

14.3   Analysis of Composition of Microbiomes (ANCOM)

Like ALDEx2, ANCOM was developed under statistical framework of ANOVA using a non-parametric testing to analyze relative abundance through its log-ratio-transformation of observed counts.

14.3.1   Introduction to ANCOM

ANCOM (Mandal et al. 2015), an alr (additive log-ratio)-based method, was proposed based on alr-transformation to account for the compositional structure of microbiome data. ANCOM repeatedly uses alr-transformation to choose each of the taxa in the dataset as a reference taxon at a time. Thus, given a total of m taxa, ANCOM will choose each of the m taxa to be a reference taxon one time and repeatedly perform the alr-transformation for each taxon and $m - 1$ regressions. Therefore, total $m(m - 1)$ regression models will be fitted.

Like an ANOVA model, ANCOM models the log-ratios of OTU abundances in each sample with a linear model, but, unlike typical ANOVA, ANCOM accommodates dependencies and correlations among the relative abundances of the OTUs by using of log-ratios. That is, ANCOM lies on framework of ANOVA, while accounts for the compositional structure of microbiome data. The dependencies and hence compositional structure arise in microbiome data because the relative abundances of OTUs sum to 1 or 100% in each sample and because relative abundances of different OTUs may be positively or negatively correlated.

ANCOM lies on the following two assumptions: (1) within the ecosystem, the mean log absolute abundance of two taxa are not different, and (2) within the ecosystem, the mean log absolute abundance of all taxa does not change by the same amount between two (or more) study groups (Mandal et al. 2015). Putting (1) and (2) together, ANCOM assumes that the abundant taxa are not differentiated with inter-taxa ratios transformation. We can summarize the methods of ANCOM into three main parts as below.

Part 1: Develop a statistical model under the framework of ANOVA to perform compositional differential analysis of each taxon.

Under these two assumptions, for the ith taxon and j^{th} sample, ANCOM developed a statistical model using a standard ANOVA model formulation to perform all possible DA analyses by successively using each taxon as a reference taxon.

$$\log\left(\frac{r_{ij}^{(g)}}{r_{i'j}^{(g)}}\right) = \alpha_{ii'} + \beta_{ii'}^{(g)} + \sum_{k} X_{jk}\beta_{ii'k} + \varepsilon_{ii'j}^{(g)}, \qquad (14.10)$$

where i' is the reference taxon, $i' \neq i = 1, 2, \ldots, m$, and $g = 1, 2, \ldots, G$ is the number of study groups.

Part 2: Conduct a null hypothesis regarding mean log absolute abundance in a unit volume of an ecosystem using relative abundances.

It was shown (Mandal et al. 2015) that by combining these two assumptions, it can conduct a null hypothesis regarding mean log absolute abundance in a unit volume of an ecosystem using relative abundances. And to test whether a taxon i is differentially abundant regarding a factor of interest with G levels is equivalent to test the null hypothesis and the alternative hypothesis (Lin and Peddada 2020b):

$$H_{0(ii')} : \beta_{ii'}^{(1)} = \cdots = \beta_{ii'}^{(G)} = 0, \text{ and } H_{a(ii')}$$

$$: \text{ not all } \beta_{ii'}^{(g)} = 0, \ \beta_{ii'}^{(g)} = 0, \text{ for every } i \neq i'.$$

Part 3: Adjust the *P*-values from the tests of taxon.

There are $\frac{m(m-1)}{2}$ *P*-values from the $\frac{m(m-1)}{2}$ tests of taxon. ANCOM uses Benjamini-Hochberg (BH) procedure (Benjamini and Hochberg 1995) or Bonferroni correction procedure (Dunn 1958, 1961) to adjust the *P*-values for a multiple testing correction. For each taxon, the number of rejections W_i is based on the empirical distribution of $\{W_1, W_2, \ldots, W_m\}$, which determines the cutoff value of significant taxon (Mandal et al. 2015). The large value of W_i indicates that taxon i is more likely differentially abundant. However, the decision rule of choosing cutoff value of W_i is kind of arbitrary in ANCOM, although 70th percentile of the W distribution is recommended. The users can select different threshold of cutoff value such as the 60th to 90th to output the statistical testing results.

14.3.2   Implement ANCOM Using QIIME 2

Example 14.3: Mouse Gut Microbiome Data, Example 9.2 Cont.
These mouse gut microbiome datasets were generated by DADA2 within QIIME 2 from raw 16S rRNA sequencing data and have been used for illustrating statistical analyses of alpha and beta diversities, as well other analyses and plots via QIIME 2 in Chaps. 9, 10, and 11. Here, we continue to use these datasets to illustrate implementation of ALDEx2 using QIIME 2. We showed in Chap. 10 via **emperor plot** that a lot of features were changed in abundance over time (Early and Late times). In this section, we illustrate how to test differential abundances with ANCOM using the same mouse gut data in Example 9.2.

ANCOM can be implemented via R and QIIME 2. Currently, ANCOM does not have an R package yet, to run ANCOM in R, we need to download the ANCOM code (ancom_v2.1.R) and then upload and store under the working directory, run the following code: source("ancom_v2.1.R"). Here, we implemented ANCOM via the **q2-composition** plugin. To perform ANCOM to identify features that are differentially abundant across groups, we need call **qiime composition ancom** command. To implement this command, two data files are required: (1) an artifact feature table (here, FeatureTableMiSeq_SOP.qza) and (2) a sample metadata (here, SampleMetadataMiSeq_SOP.tsv). The artifact feature table is used for ANCOM computation, while in the sample metadata file, a categorical sample metadata column is required to specify for the test of differential abundance across groups. A visualization output file name is also required.

Step 1: Use the filter-samples command to filter feature table.

ANCOM assumes that less than about 25% of the features are changing between groups. If this assumption is violated, then ANCOM will be more possible to increase both Type I and Type II errors. As we showed in Chap. 10 via **emperor plot** that a lot of features were changed in abundance over time (Early and Late times), here we will filter the full feature table to only contain late mouse gut samples and then perform ANCOM to identify which sequence variants and taxa are differentially abundant across the male and female mouse gut samples.

```
source activate qiime2-2022.2
cd QIIME2R-Biostatistics
cd ANCOM
# Filter out the samples with less than 1153 reads (see details in
Chapter 9)
qiime feature-table filter-samples\
  --i-table FeatureTableMiSeq_SOP.qza\
  --m-metadata-file SampleMetadataMiSeq_SOP.tsv\
  --p-where "[Time]='Late'"\
  --p-min-frequency 1153\
  --o-filtered-table LateMouseGutTable.qza
```

```
Saved FeatureTable[Frequency] to: LateMouseGutTable.qza
```

```
# Filter features that only appear in a single sample
qiime feature-table filter-features \
  --i-table LateMouseGutTable.qza \
  --p-min-samples 2 \
  --o-filtered-table LateMouseGutTable2.qza
```

```
Saved FeatureTable[Frequency] to: LateMouseGutTable2.qza
```

```
# Filter features that only appear 10 times or fewer across all samples
qiime feature-table filter-features \
  --i-table LateMouseGutTable.qza \
  --p-min-frequency 10 \
  --o-filtered-table LateMouseGutTable3.qza
```

```
Saved FeatureTable[Frequency] to: LateMouseGutTable3.qza
```

Step 2: Use taxa collapse command to collapse feature table

Support we want to analyze this data at genus level (Level 6), then we can apply following commands to collapse the feature table at genus level**.**

```
# Collapse feature table at genus level
qiime taxa collapse \
  --i-table LateMouseGutTable3.qza \
  --i-taxonomy TaxonomyMiSeq_SOP.qza \
  --p-level 6 \
  --o-collapsed-table LateMouseGutTableL6.qza
```

Saved FeatureTable[Frequency] to: LateMouseGutTableL6.qza

If we want to analyze this data at species level (Level 7), then we can apply the following commands to collapse the feature table.

```
# Collapse feature table at species level
qiime taxa collapse \
  --i-table LateMouseGutTable3.qza \
  --i-taxonomy TaxonomyMiSeq_SOP.qza \
  --p-level 7 \
  --o-collapsed-table LateMouseGutTableL7.qza
```

Saved FeatureTable[Frequency] to: LateMouseGutTableL7.qza

The following intermediate files can be removed from the directory by **rm** command, such as rm LateMouseGutTable.qza. However, we keep them there for later review.

```
LateMouseGutTable.qza
LateMouseGutTable2.qza
LateMouseGutTable3.qza
```

Step 3: Use the add-pseudocount command to add a small pseudocount to produce the compositional feature table artifact.

As a compositional method, ANCOM cannot address zero issue because frequencies of zero are not defined when taking log or log-ratio transformation. ANCOM operates on a FeatureTable[Composition] QIIME 2 artifact, which is based on frequencies of features on a per-sample basis. To build the composition artifact (a FeatureTable[Composition] artifact), a small pseudocount must be added to the FeatureTable[Frequency] artifact typically via an imputation method.

```
# Produce the compositional feature table artifact
qiime composition add-pseudocount \
  --i-table LateMouseGutTableL6.qza \
  --o-composition-table CompLateMouseGutTableL6.qza
```

Saved FeatureTable[Composition] to: CompLateMouseGutTableL6.qza

```
# Produce the compositional feature table artifact
qiime composition add-pseudocount \
  --i-table LateMouseGutTableL7.qza \
  --o-composition-table CompLateMouseGutTableL7.qza
```

Saved FeatureTable[Composition] to: CompLateMouseGutTableL7.qza

Step 4: Perform ANCOM to identity differential features across the mouse gut sex groups.

```
# Perform ANCOM to identity differential features
qiime composition ancom \
  --i-table CompLateMouseGutTableL6.qza \
  --m-metadata-file SampleMetadataMiSeq_SOP.tsv \
  --m-metadata-column Sex \
  --o-visualization AncomSexMisSeq_SOPL6.qzv
```

Saved Visualization to: AncomSexMisSeq_SOPL6.qzv

```
# Perform ANCOM to identity differential features
qiime composition ancom \
  --i-table CompLateMouseGutTableL7.qza \
  --m-metadata-file SampleMetadataMiSeq_SOP.tsv \
  --m-metadata-column Sex \
  --o-visualization AncomSexMisSeq_SOPL7.qzv
```

Saved Visualization to: AncomSexMisSeq_SOPL7.qzv

We can view the plots using the `qiime tools view` command:

`qiime tools view` AncomSexMisSeq_SOPL7.qzv

This view command displays three results: (1) ANCOM Volcano Plot, (2) ANCOM statistical results, and (3) percentile abundances of features by group. All these results can be exported, downloaded, and saved.

Table 14.3 is a copy of ANCOM statistical results. The first five taxa were identified as abundant taxa.

If we do not want to collapse feature table, then the following analyses are used to add pseudocount and perform ANCOM.

```
# Produce the compositional feature table artifact
qiime composition add-pseudocount \
  --i-table LateMouseGutTable3.qza \
  --o-composition-table CompLateMouseGutTable.qza
```

Saved FeatureTable[Composition] to: CompLateMouseGutTable.qza

Table 14.3 Abundant and not abundant taxa identified by ANCOM statistical testing

Kingdom	Phylum	Class	Order	Family	Genus	Species	W	Reject null hypothesis
k_Bacteria	p_Firmicutes	c_Bacilli	o_Lactobacillales	f_Lactobacillaceae	g_Lactobacillus	s_	48	TRUE
k_Bacteria	p_Firmicutes	c_Clostridia	o_Clostridiales	f_Clostridiaceae			48	TRUE
k_Bacteria	p_Firmicutes	c_Clostridia	o_Clostridiales	f_Peptococcaceae	g_rc4-4	s_	47	TRUE
k_Bacteria	p_Actinobacteria	c_Actinobacteria	o_Bifidobacteriales	f_Bifidobacteriaceae	g_Bifidobacterium	s_pseudolongum	44	TRUE
k_Bacteria	p_Verrucomicrobia	c_Verrucomicrobiae	o_Verrucomicrobiales	f_Verrucomicrobiaceae	g_Akkermansia	s_muciniphila	44	TRUE
k_Bacteria	p_TM7	c_TM7-3	o_CW040	f_F16	g_	s_	38	FALSE
k_Bacteria	p_Firmicutes	c_Bacilli	o_Lactobacillales	f_Lactobacillaceae	g_Lactobacillus		35	FALSE
k_Bacteria	p_Firmicutes	c_Bacilli	o_Turicibacterales	f_Turicibacteraceae	g_Turicibacter	s_	35	FALSE
k_Bacteria	p_Proteobacteria	c_Gammaproteobacteria	o_Pseudomonadales	f_Pseudomonadaceae	g_Pseudomonas	s_veronii	25	FALSE
k_Bacteria	p_Firmicutes	c_Clostridia	o_Clostridiales				23	FALSE
k_Bacteria	p_Firmicutes	c_Clostridia	o_Clostridiales	f_[Mogibacteriaceae]	g_	s_	18	FALSE
k_Bacteria	p_Tenericutes	c_Mollicutes	o_RF39	f_	g_	s_	17	FALSE
k_Bacteria	p_Firmicutes	c_Clostridia	o_Clostridiales	f_Lachnospiraceae	g_Dorea	s_	16	FALSE
k_Bacteria	p_Firmicutes	c_Clostridia	o_Clostridiales	f_Ruminococcaceae	g_Oscillospira	s_	14	FALSE
k_Bacteria	p_Firmicutes	c_Clostridia	o_Clostridiales	f_Lachnospiraceae	g_	s_	14	FALSE
k_Bacteria	p_Firmicutes	c_Clostridia	o_Clostridiales	f_	g_	s_	14	FALSE
k_Bacteria	p_Firmicutes	c_Clostridia	o_Clostridiales	f_Ruminococcaceae	g_Butyricicoccus	s_pullicaecorum	13	FALSE
k_Bacteria	p_Firmicutes	c_Clostridia	o_Clostridiales	f_Dehalobacteriaceae	g_Dehalobacterium	s_	13	FALSE
k_Bacteria	p_Firmicutes	c_Clostridia	o_Clostridiales	f_Lachnospiraceae	g_Clostridium	s_colinum	12	FALSE
k_Bacteria	p_Firmicutes	c_Clostridia	o_Clostridiales	f_Ruminococcaceae	g_Clostridium	s_methylpentosum	12	FALSE
k_Bacteria	p_Firmicutes	c_Erysipelotrichi	o_Erysipelotrichales	f_Erysipelotrichaceae	g_Allobaculum	s_	12	FALSE
k_Bacteria	p_Firmicutes	c_Clostridia	o_Clostridiales	f_Ruminococcaceae	g_	s_	11	FALSE
k_Bacteria	p_Firmicutes	c_Clostridia	o_Clostridiales	f_Ruminococcaceae			11	FALSE
k_Bacteria	p_Tenericutes	c_Mollicutes	o_Anaeroplasmatales	f_Anaeroplasmataceae	g_Anaeroplasma	s_	10	FALSE
k_Bacteria	p_Firmicutes	c_Clostridia	o_Clostridiales	f_Lachnospiraceae			10	FALSE
k_Bacteria	p_Firmicutes	c_Clostridia	o_Clostridiales	f_Lachnospiraceae	g_Roseburia	s_	10	FALSE

k_	p_	c_	o_	f_	g_	s_		
k_Bacteria	p_Firmicutes	c_Clostridia	o_Clostridiales	f_Christensenellaceae	g_	s_	10	FALSE
k_Bacteria	p_[Thermi]	c_Deinococci	o_Deinococcales	f_Deinococcaceae	g_Deinococcus	s_	9	FALSE
k_Bacteria	p_Firmicutes	c_Clostridia	o_Clostridiales	f_Clostridiaceae	g_Clostridium	s_butyricum	9	FALSE
k_Bacteria	p_Firmicutes	c_Erysipelotrichi	o_Erysipelotrichales	f_Erysipelotrichaceae	g_Coprobacillus	s_	9	FALSE
k_Bacteria	p_Firmicutes	c_Bacilli	o_Bacillales	f_Staphylococcaceae	g_Staphylococcus	_	9	FALSE
k_Bacteria	p_Proteobacteria	c_Betaproteobacteria	o_Neisseriales	f_Neisseriaceae	g_Neisseria	s_cinerea	9	FALSE
k_Bacteria	p_Firmicutes	c_Clostridia	o_Clostridiales	f_Lachnospiraceae	g_Blautia	_	9	FALSE
k_Bacteria	p_Proteobacteria	c_Gammaproteobacteria	o_Pseudomonadales	f_Moraxellaceae	g_Acinetobacter	s_guillouiae	9	FALSE
k_Bacteria	p_Firmicutes	c_Bacilli	o_Bacillales	f_Staphylococcaceae	g_Jeotgalicoccus	s_psychrophilus	9	FALSE
k_Bacteria	p_Proteobacteria	c_Alphaproteobacteria	o_Rickettsiales	f_mitochondria	_	_	9	FALSE
k_Bacteria	p_Proteobacteria	c_Gammaproteobacteria	o_Enterobacteriales	f_Enterobacteriaceae	_	_	9	FALSE
k_Bacteria	p_Bacteroidetes	c_Bacteroidia	o_Bacteroidales	f_Bacteroidaceae	g_Bacteroides	s_ovatus	9	FALSE
k_Bacteria	p_Actinobacteria	c_Coriobacteriia	o_Coriobacteriales	f_Coriobacteriaceae	_	_	8	FALSE
k_Bacteria	p_Firmicutes	c_Bacilli	o_Lactobacillales	f_Streptococcaceae	g_Streptococcus	s_	8	FALSE
k_Bacteria	p_Firmicutes	c_Clostridia	o_Clostridiales	f_Clostridiaceae	g_Candidatus Arthromitus	s_	8	FALSE
k_Bacteria	p_Firmicutes	c_Bacilli	o_Lactobacillales	f_Lactobacillaceae	g_Lactobacillus	s_reuteri	7	FALSE
k_Bacteria	p_Firmicutes	c_Clostridia	o_Clostridiales	f_Lachnospiraceae	g_[Ruminococcus]	s_gnavus	7	FALSE
k_Bacteria	p_Firmicutes	c_Clostridia	o_Clostridiales	f_Ruminococcaceae	g_Ruminococcus	s_	7	FALSE
k_Bacteria	p_Firmicutes	c_Clostridia	o_Clostridiales	f_Lachnospiraceae	g_Coprococcus	s_	7	FALSE
k_Bacteria	p_Actinobacteria	c_Coriobacteriia	o_Coriobacteriales	f_Coriobacteriaceae	g_	s_	6	FALSE
k_Bacteria	p_Cyanobacteria	c_Chloroplast	o_Streptophyta	f_	g_	s_	6	FALSE
k_Bacteria	p_Actinobacteria	c_Coriobacteriia	o_Coriobacteriales	f_Coriobacteriaceae	g_Adlercreutzia	s_	5	FALSE
k_Bacteria	p_Bacteroidetes	c_Bacteroidia	o_Bacteroidales	f_Rikenellaceae	g_	s_	5	FALSE
k_Bacteria	p_Bacteroidetes	c_Bacteroidia	o_Bacteroidales	f_S24-7	g_	s_	5	FALSE

```
# Perform ANCOM to identity differential features
qiime composition ancom \
  --i-table CompLateMouseGutTable.qza \
  --m-metadata-file SampleMetadataMiSeq_SOP.tsv \
  --m-metadata-column Sex \
  --o-visualization AncomSexMisSeq_SOP.qzv
  --verbose
```

Saved Visualization to: AncomSexMisSeq_SOP.qzv

ANCOM provides two optional parameters --p-transform-function and --p-difference-function to perform identity differential features. Both parameters are TEXT choices [('sqrt', 'log', 'clr'), ('mean_difference', 'f_statistic') for --p-transform-function and --p-difference-function, respectively]. The transform-function is used to specify the method to transform feature values before generating volcano plots. The default is "clr" (centered log ratio-transformation). Other two are square root ("sqrt") and log ("log") transformations. The difference-function is used to specify the method to visualize fold difference in feature abundances across groups for volcano plots. One command is provided below.

```
# Perform ANCOM to identity differential features
qiime composition ancom \
  --i-table CompLateMouseGutTableL7.qza \
  --m-metadata-file SampleMetadataMiSeq_SOP.tsv \
  --m-metadata-column Sex \
  --o-visualization AncomSexMisSeq_SOPL7.qzv
  --verbose
  --p-transform-function Choices'clr'\
  --p-difference-function Choices 'mean_difference'\
```

Saved Visualization to: AncomSexMisSeq_SOPL7.qzv

14.3.3 Remarks on ANCOM

It was shown that ANCOM has well controlled the FDR while maintaining power comparable with other methods (Mandal et al. 2015; Lin and Peddada 2020a). However, like ALDEx2, ANCOM actually uses log-ratio transformations as a kind of normalization (log-ratio "normalizations"). Thus, ANCOM suffers from some similar limitations as ALDEx2:

- Its usefulness mainly depends on interpreting the log-ratio transformation as a normalization. In other words, the statistical tests can be appropriately interpreted only when the log-ratio transformation sufficiently approximates an unchanged reference (Quinn et al. 2018a, b). In contrast, it was reviewed that the methods that do not require using log-ratio transformations as a kind of normalization are more appropriate (Quinn et al. 2018a, b).

- Cannot address zero-inflation problem, instead adds an arbitrary small pseudocount value such as 0.001, 0.5, or 1 to all read counts. This indicates compositional analysis approach fails in the presence of zero values (Xia et al. 2018a, p. 389). It was shown that due to improper handling of the zero counts, ANCOM inflated the number of false positives instead of controlled the false-positive rate (FDR) (Brill et al. 2019).
- It is underpowered and requires a large number of samples due to using non-parametric testing (Quinn et al. 2018a, b) as well as has decreased sensitivity on small datasets (e.g., less than 20 samples per group) partially because of its non-parametric nature (i.e., Mann-Whitney test) (Weiss et al. 2017; Quinn et al. 2018a, b).

Specifically, ANCOM has other weaknesses, including:

- ANCOM takes the approach of alr-transformation and transforms the observed abundances of each taxon to log-ratios of the observed abundance relative to a pre-specified reference taxon. Because when performing differential abundance analysis, typically, the comparing samples have thousands of taxa in microbiome data sets. Thus, it is computationally intensive to repeatedly apply alr-transformation to each taxon in the dataset as a reference taxon. The choice of this reference taxon is also a challenge when the number of taxa is large.
- Whether ANCOM can control FDR well or not, different studies have reported inconsistent results: some different simulation studies reported that ANCOM can control FDR reasonably well under various scenarios (Lin and Peddada 2020a; Weiss et al. 2017), whereas other studies reported that ANCOM could generate a potential false-positive result when a cutoff of 0.6 was used for the W statistic (Morton et al. 2019). Thus, when ANCOM is used in differential abundance analysis, a stricter cutoff value for the W statistic is recommended to reduce the chance of false positives.
- ANCOM uses the quantile of its test statistic W instead of P-values to conduct statistical testing for significance. This not only makes the analysis results difficult to interpret (Lin and Peddada 2020b) but also does not make ANCOM to improve its performance by filtering taxa before analysis. ANCOM was reported having reduced the number of detected differential abundant taxa. This is most likely related to the way that W statistics are calculated and used for significance in ANCOM (Wallen 2021).
- ANCOM does not provide P-value for individual taxon and cannot provide standard errors or confidence intervals of DA for each taxon (Lin and Peddada 2020a).
- It is difficult to interpret the testing differential abundance results because ANCOM uses presumed invariant features to guide the log-ratio transformation (Quinn et al. 2018a, b).

14.4   Analysis of Composition of Microbiomes-Bias Correction (ANCOM-BC)

ANCOM-BC is a bias correction version of ANCOM (analysis of compositions of microbiomes).

14.4.1   Introduction to ANCOM-BC

ANCOM-BC (Lin and Peddada 2020a) was proposed for differential abundance (DA) analysis of microbiome data with bias correction to ANCOM. ANCOM-BC was developed under the assumptions that (1) the observed abundance in a feature table is expected to be proportional to the unobservable absolute abundance of a taxon in a unit volume of the ecosystem. (2) The sampling fraction varies from sample to sample and hence inducing the estimation bias. Thus, to address the problem of unequal sampling fractions, ANCOM-BC uses a sample-specific offset term to serve as the bias correction (Lin and Peddada 2020a). The offset term is usually used in the generalized linear models such as Poisson, zero-inflated models to adjust for the sampling population (Xia et al. 2018b). Here ANCOM-BC uses a sample-specific offset term in a linear regression framework.

We can summarize the methods of ANCOM-BC into the following six parts.

Part 1: Define the sample-specific sampling fraction.

Unlike some other DA studies, which defines relative abundances of taxa as frequencies of these taxa in a sample, in ANCOM-BC, relative abundance of a taxon in the sample refers to the fraction of the taxon observed in the feature table relative to the sum of all observed taxa corresponding to the sample in the feature table (Lin and Peddada 2020a, b). Actually, it is the proportion of this taxon relative to the sum of all taxa in the sample with range of (0, 1). In ANCOM-BC, absolute abundance is defined as unobservable actual abundance of a taxon in a unit volume of an ecosystem, while observed abundance refers to the observed counts of features (OTUs or ASVs) in the feature table (Lin and Peddada 2020b).

Let O_{ij} denote the observed abundance of ith taxon in j^{th} sample, A_{ij} the unobserved abundance of ith taxon in the ecosystem of ith sample, and then the sample-specific sampling fraction c_j is defined as:

$$c_j = \frac{E(O_{ij}|A_{ij})}{A_{ij}}, \tag{14.11}$$

where c_j is the ratio of the expected abundance of taxon i in the jth sample to its absolute abundance in a unit volume of an ecosystem such as gut where the sample was derived from, which could be empirically estimated by the ratio of library size to the microbial load.

As we described in Sect. 14.1.2, both the library size and the microbial loads could drive microbiome data to be compositional. Thus, ANCOM-BC not only normalizes the library size to effect library size across samples but also takes the differences of the microbial loads into account. Li and Peddada (2020a) thought the normalized data using the sampling fractions is better than the normalized data obtained from the normalization methods that rely purely on the library sizes. The normalization method that accounts for differences in sampling fractions can avoid the bias in differential abundance analysis due to the increased false-positive and false-negative rates.

Part 2: Describe two model assumptions of ANCOM-BC.

Assumption 1 is given below.

$$E\left(O_{ijk}|A_{ijk}\right) = c_{ik}A_{ijk}$$
$$\mathrm{Var}\left(O_{ijk}|A_{ijk}\right) = \sigma^2_{w,ijk}, \tag{14.12}$$

where $\sigma^2_{w,ijk}$ is the variability between specimens within the kth sample from the jth group, which characterizes the within-sample variability. Usually the within-sample variability is not estimated since typically at a given time only one specimen is available in most microbiome studies. The assumption in (14.12) states that the absolute abundance of a taxon in a random sample is expected to be in constant proportion to the absolute abundance in the ecosystem of the sample. That is, the expected relative abundance of each taxon in a random sample equals to the relative abundance of the taxon in the ecosystem of the sample (Lin and Peddada 2020a).

Assumption 2 is given below. For each taxon i, $A_{ijk}, j = 1, \ldots, g, k = 1, \ldots, n_j$, are independently distributed with

$$E\left(A_{ijk}|\theta_{ij}\right) = \theta_{ij}$$
$$\mathrm{Var}\left(A_{ijk}|\theta_{ij}\right) = \sigma^2_{b,ij}, \tag{14.13}$$

where θ_{ij} is a fixed parameter rather than a random variable; $\sigma^2_{b,ij}$ is the between-sample variation within group j for the ith taxon. The assumption in (14.13) states that for a given taxon, all subjects within and between groups are independent.

Combining Assumption 1 in (14.12) and Assumption 2 in (14.13), the expected and variance of O_{ijk} are defined as:

$$E\left(O_{ijk}\right) = c_{jk}\theta_{ij}$$
$$\mathrm{Var}\left(O_{ijk}\right) = f\left(\sigma^2_{w,ijk}, \sigma^2_{b,ij}\right) := \sigma^2_{t,ijk}. \tag{14.14}$$

Part 3: Introduce a linear regression model framework for log-transformed OTU counts data to include the sample specific bias due to sampling fractions.

Under the above setting, the linear model framework for log-transformed OTU counts data is written as below.

$$y_{ijk} = d_{jk} + \mu_{ij} + \varepsilon_{ijk}, \tag{14.15}$$

where

$$\begin{aligned} E(\varepsilon_{ijk}) &= & 0, \\ E(y_{ijk}) &= & d_{jk} + \mu_{ij}, \\ \text{Var}(y_{ijk}) &= & \text{Var}(\varepsilon_{ijk}) := \sigma_{ijk}^2. \end{aligned} \tag{14.16}$$

Note that d in the above equation simply reflects the effect of c. It is not exactly log(c) due to Jensen's inequality. The above model formulation is distinctive from the standard one-way ANOVA: it shows that the sample-specific bias due to sampling fractions is introduced in the modeling. For details, the readers are referred to the original publication (Lin and Peddada 2020a).

Part 4: Develop a linear regression model that estimate the sample-specific bias and ensure that the estimator and the test statistic are asymptotically centered at zero under the null hypothesis.

Since the sample-specific bias is introduced because of the differential sampling fraction by each sample, thus, the goal of ANCOM-BC is to eliminate this bias. Given a large number of taxa on each subject, to estimate this bias, ANCOM-BC borrows information across taxa in its methodology.

The framework of least squares was used to develop bias and variance of bias estimation under the null hypothesis, which are estimated as follows.

$$\begin{aligned} \widehat{d}_{jk} &= \bar{y}_{.jk} - \bar{y}_{.j.}, k = 1, \ldots, n_j, j = 1, 2, \ldots, g, \\ \widehat{\mu}_{ij} &= \bar{y}_{ij.} - \widehat{\bar{d}}_{j.} = \bar{y}_{ij.}, i = 1, \cdots, m. \end{aligned} \tag{14.17}$$

where note that $E(\widehat{\mu}_{ij}) = E(\bar{y}_{ij.}) = \mu_{ij} + \bar{d}_{j.}$. Thus, for each $j = 1, 2, \ldots, g$, Lin and Peddada (2020a) approve that $\widehat{\mu}_{ij}$ is a biased estimator and $E(\widehat{\mu}_{i1} - \widehat{\mu}_{i2}) = (\mu_{i1} - \mu_{i2}) + (\bar{d}_{1.} - \bar{d}_{2.})$. For two experimental groups with balanced design (i.e., $g = 2$ and $n_1 = n_2 = n$) and given two ecosystems, for each taxon i, $i = 1, \cdots, m$, the test hypotheses are $\begin{aligned} H_0 &: \mu_{i1} = \mu_{i2}, \\ H_a &: \mu_{i1} \neq \mu_{i2}. \end{aligned}$

By denoting $\delta = (\bar{d}_{1.} - \bar{d}_{2.})$, under the null hypothesis, $E(\widehat{\mu}_{i1} - \widehat{\mu}_{i2}) = \delta \neq 0$ and hence is biased. From (12.15) and Lyapunov central limit theorem, Lin and Peddada (2020a) show that

$$\frac{\widehat{\mu}_{ij} - (\mu_{ij} + \bar{d}_{j.})}{\sigma_{ij}} \to_d N(0, 1) \quad \text{as } n \to \infty, \tag{14.18}$$

where $\sigma_{ij}^2 = \mathrm{Var}\left(\widehat{\mu}_{ij}\right) = \mathrm{Var}\left(\overline{y}_{ij.}\right) = \frac{1}{n^2} \sum_{k=1}^{n} \sigma_{ijk}^2$. Lin and Peddada (2020a) show that the taxa can be modeled using a Gaussian mixtures model and the expectation-maximization (EM) algorithm (i.e., $\widehat{\delta}_{\mathrm{EM}}$, which denotes the resulting estimator of δ). They also show that that $\widehat{\delta}_{\mathrm{EM}}$ and $\widehat{\delta}_{\mathrm{WLS}}$ (denoting the weighted least squares (WLS) estimator of δ) are highly correlated, are approximately unbiased, and thus can use $\widehat{\delta}_{\mathrm{WLS}}$ to approximate for $\widehat{\delta}_{\mathrm{EM}}$.

Therefore, under some further assumptions and developments, for hypothesis testing for two-group comparison: $\begin{aligned} H_0 &: \mu_{i1} = \mu_{i2} \\ H_a &: \mu_{i1} \neq \mu_{i2}, \end{aligned}$ for taxon i, the following test statistic is approximately centered at zero under the null hypothesis:

$$W_i = \frac{\widehat{\mu}_{i1} - \widehat{\mu}_{ii2} - \widehat{\delta}_{\mathrm{EM}}}{\sqrt{\sigma_{i1}^2 + \sigma_{i2}^2}}. \tag{14.19}$$

From Slutsky's theorem,

$$W_i \to_d N(0, 1) \text{ as } m, n \to \infty. \tag{14.20}$$

The above test statistic is modified as follows if the sample size is not very large and/or the number of non-null taxa is very large:

$$W_i^* = \frac{\widehat{\mu}_{i1} - \widehat{\mu}_{ii2} - \widehat{\delta}_{\mathrm{WLS}}}{\sqrt{\widehat{\sigma}_{i1}^2 + \widehat{\sigma}_{i2}^2 + \widehat{\mathrm{V}}\mathrm{ar}\left(\widehat{\delta}_{\mathrm{WLS}}\right) + 2\sqrt{\left(\widehat{\sigma}_{i1}^2 + \widehat{\sigma}_{i2}^2\right)\widehat{\mathrm{V}}\mathrm{ar}\left(\widehat{\delta}_{\mathrm{WLS}}\right)}}}. \tag{14.21}$$

Part 5: Adjust the *P*-values from the tests of taxon.

To control the FDR for multiple comparisons, ANCOM-BC recommends using the Holm-Bonferroni method (Holm 1979) or Bonferroni (Dunn 1958, 1961) correction instead of the Benjamini-Hochberg (BH) procedure (Benjamini and Hochberg 1995) to adjust the raw *P*-values. This is considering the findings in literatures that it is more appropriate to control the FDR using the Holm-Bonferroni and methods when *P*-values were not accurate (Lim et al. 2013) and the BH procedure controls the FDR if the data have either independence or some special correlation structures (e.g., perhaps positive regression dependence among taxa) (Benjamini and Hochberg 1995; Benjamini and Yekutieli 2001). Compared to the Bonferroni correction, ANCOM-BC results adjusted by BH procedure has larger power.

Part 6: Develop the test statistic for hypothesis testing for multigroup comparison to conduct the global test.

The test statistic for multigroup comparison was developed in the same way to the two-group comparison. That is, first getting the initial estimates of $\widehat{\mu}_{ij}$ and $\widehat{d}_{ij}$, then setting the reference group r (e.g., $r = 1$), and obtaining the estimator of the bias term $\widehat{\delta}_{rj}$ through E-M algorithm, the final estimator of mean absolute abundance of the ecosystem (in log scale) are obtained by transforming $\widehat{\mu}_{ij}$ of (14.17) into:

$$\widehat{\mu}_{ij}^{*} := \begin{cases} \widehat{\mu}_{ir}, & j = r \\ \widehat{\mu}_{ij} + \widehat{\delta}_{rj}, & j \neq r \in 1, \ldots, g \end{cases}. \tag{14.22}$$

The test statistic for pairwise comparison is defined as:

$$W_{i,jj'} = \frac{\widehat{\mu}_{ij}^{*} - \widehat{\mu}_{ij'}^{*}}{\sqrt{\widehat{\sigma}_{ij}^{2} + \widehat{\sigma}_{ij'}^{2}}}, i = 1, \ldots, m, j \neq j' \in \{1, \ldots, g\}. \tag{14.23}$$

For computational simplicity, the global test statistic is reformulated based on William's type of test (Williams 1971, 1977; Peddada et al. 2001; Farnan et al. 2014) as follows:

$$W_i = \max_{j \neq j' \in \{1, \ldots, g\}} | W_{i,jj'} |, i = 1, \ldots, m. \tag{14.24}$$

Under null, $W_{i,jj'} \to_d N(0, 1)$, thus for each specific taxon i, the null distribution of W_i can be constructed by simulations. Finally, P-value is calculated as:

$$p_i = \frac{1}{B} \sum_{b=1}^{B} I\left(W_i^{(b)} > W_i\right), i = 1, \ldots, m, \tag{14.25}$$

and the Bonferroni correction is applied to control the FDR, where B is the times of simulations (e.g., $B = 1000$).

14.4.2 Implement ANCOM-BC Using the ANCOMBC Package

ANCOM-BC method is implemented via the **ancombc** () function in the ANCOMBC package. This package was developed to identify those taxa that their absolute abundances per unit volume in the ecosystem (e.g., gut, mouth or vagina) are differentially abundant between treatment groups/conditions while allowing adjustment of other covariates of interest. In current version (ANCOM-BC (v.1.2.2), October 2021), ANCOM-BC is performed for cross-sectional data.

One syntax is given as:

```
ancombc(phyloseq, formula, p_adj_method = "holm", zero_cut = 0.9,
lib_cut = 1000, group = NULL, struc_zero = FALSE, neg_lb = FALSE, tol =
1e-05, max_iter = 100, conserve = FALSE, alpha = 0.05, global = TRUE)
```

where the argument:

- **phyloseq** is a phyloseq-class object, consisting of a feature (OUT/ASV) table (in ANCOM-BC, refers to microbial observed abundance table), a sample meta-data, a taxonomy table (optional), and a phylogenetic tree (optional). As we described in Chap. 2, to create a phyloseq-class object, the row names of the metadata must match the sample names of the feature table, and the row names of the taxonomy table must match the taxon (feature) names of the feature table.
- **formula** is a the character string expression. It is used to specify how the microbial absolute abundances for each taxon (response variable) depend on the variables in metadata (predictor variables or covariates) such as we can specify that formula = "Age + Gender + BMI + Group".
- **p_adj_method** is used to specify the method for adjusting *P*-values. The default method is "holm" (Holm 1979). The optional methods include "holm", "hochberg" (Hochberg 1988), "hommel" (Hochberg 1988), "bonferroni" (Bonferroni 1936), "BH" or its alias "fdr" (Benjamini and Hochberg 1995), "BY" (Benjamini and Yekutieli 2001), and "none." Tukey's method is not available in this package. Type **?p.adjust()** to check the options and references in R. If you do not want to adjust the *P*-value, use the pass-through option ("none").
- **zero_cut** is a numerical fraction with range of (0, 1). It is used to specify the proportion of zero values for taxa greater than the cutoff values will be excluded in the analysis. Default is 0.90.
- **lib_cut** is used to specify a numerical threshold for filtering samples based on library sizes. Samples with library sizes less than this cutoff values will be excluded in the analysis. Default is 0, without filtering any sample.
- **group** is the group variable in metadata. It is required for detecting structural zeros and performing global test.
- **struc_zero** is used to specify whether to detect structural zeros. Default is FALSE.
- **neg_lb** is used to specify to classify a taxon as a structural zero in the corresponding study group using its asymptotic lower bound. Default is FALSE.
- **tol** is the iteration convergence tolerance for the E-M algorithm. Default is 1e-05.
- **max_iter** is used to specify the maximum number of iterations for the E-M algorithm. Default is 100.
- **conserve** is used to specify a conservative variance estimate of the test statistic. Default is FALSE. Because ANCOM-BC may not perform well when the sample sizes are very small (e.g., $n \leq 5$ per group) or when the proportion of differentially abundant taxa is too large (e.g., $>75\%$) (Lin and Peddada 2020a), it is recommended to use in the case of small sample size and/or large number of differentially abundant taxa.

- **alpha** is used to specify the level of significance. Default is 0.05.
- **global** is used to specify a global test. Default is FALSE.

Example 14.4: Breast Cancer QTRT1 Mouse Gut Microbiome, Example 14.2, Cont.

In this dataset, the mouse fecal samples were collected for two genotype groups (KO and WT) at two time points: pretreatment and posttreatment. We first use ANCOMBC primary analysis to obtain primary results of genotype, time effects, and their interaction effects. Then we use ANCOMBC global test to obtain the global test results of genotype and time interaction.

First, type the following commands to download, install, and load the ANCOMBC package.

```
> if (!requireNamespace("BiocManager", quietly = TRUE))
+ install.packages("BiocManager")
> BiocManager::install("ANCOMBC")
> library(ANCOMBC)
```

Then, load OTU-table, taxonomy table, and sample metadata.

```
> setwd("~/Documents/QIIME2R/Ch14_Compositional/ANCOMBC")
> otu<-read.csv("otu_table_L7_MCF7_phyloseq.csv",row.names = 1)
> tax<-read.csv("tax_table_L7_MCF7_phyloseq.csv",row.names = 1)
> sam<-read.csv("metadata_QtRNA.csv",row.names = 1)
```

Next, create a phyloseq-class object.

```
> otumat<-as.matrix(otu)
> taxmat<-as.matrix(tax)
> class(otumat)
[1] "matrix" "array"
> class(taxmat)
[1] "matrix" "array"
> class(sam)
[1] "data.frame"
```

```
> library("phyloseq")
> # Merge otu table, taxa table and meta table to create a phyloseq object
> otu_tab = otu_table(otumat, taxa_are_rows = TRUE)
> tax_tab = tax_table(taxmat)
> meta_tab = sample_data(sam)
```

```
> physeq = phyloseq(otu_tab, tax_tab, meta_tab)
> physeq
```

```
phyloseq-class experiment-level object
otu_table() OTU Table: [ 635 taxa and 40 samples ]
sample_data() Sample Data: [ 40 samples by 6 sample variables ]
tax_table() Taxonomy Table: [ 635 taxa by 7 taxonomic ranks ]
```

Finally, implement the ancombc() function to perform primary analysis and global test.

We can specify global = TRUE to request both global test and primary analysis.

```
> library(ANCOMBC)
> output = ancombc(phyloseq = physeq, formula = "Group*Time",
+         p_adj_method = "holm", zero_cut = 0.90, lib_cut = 1000,
+       group = "Genotype", struc_zero = TRUE, neg_lb = TRUE, tol = 1e-5,
+          max_iter = 100, conserve = TRUE, alpha = 0.05, global = TRUE)
> rest_primary = output$res
> rest_global = output$res_global
```

ANCOMBC Primary Analysis.

The following commands request only the primary analysis.

```
> library(ANCOMBC)
> output = ancombc(phyloseq = physeq, formula = "Group*Time",
+         p_adj_method = "holm", zero_cut = 0.90, lib_cut = 1000,
+       group = "Genotype", struc_zero = TRUE, neg_lb = TRUE, tol = 1e-5,
+          max_iter = 100, conserve = TRUE, alpha = 0.05, global = FALSE)
```

The **ancombc()** function uses raw counts as input data. We specify the p_adj_method to be "holm," which is default. All other options were also left as default. This calling returns the following six primary results from the ANCOM-BC log-linear model to determine taxa that are differentially abundant based on the specified covariates:

1) **res$beta** (coefficients: a data.frame of coefficients).
2) **res$se** (standard errors: a data.frame of standard errors (SEs) of beta).
3) **res$W** (test statistics: a data.frame of test statistics. W = beta/se).
4) **res$p_val** (P-values: a data.frame of P-values obtained from two-sided Z-test using the test statistic W).
5) **res$q_val** (adjusted P-values: a data.frame of adjusted P-values obtained by applying the p_adj_method to p_val).
6) **res$diff_abn** (differentially abundant taxon indicators: a logical data.frame. TRUE if the taxon has q_val less than alpha otherwise FALSE).

Other returned results include:

- **res$feature_table**: a data.frame of pre-processed (based on zero_cut and lib_cut) microbial observed abundance table).

- **res\$zero_ind**: a logical `matrix` with TRUE indicating the taxon is identified as a structural zero for the specified `group` variable.
- **res\$samp_frac**: a numeric vector of estimated sampling fractions in log scale (natural log)).
- **res\$resid**: a `matrix` of residuals from the ANCOM-BC log-linear (natural log) model. Rows are taxa and columns are samples.
- **res\$delta_em**: estimated bias terms through E-M algorithm.
- **res\$delta_wls**: estimated bias terms through weighted least squares (WLS) algorithm.

Microbiome data may exist three kinds of zeros: structural zeros, sampling zeros, and rounded zeros. Their definitions can be found in Xia et al. (2018c) and the strategies of dealing with them in compositional microbiome data have been described in Xia et al. (2018a).

Briefly, structural zeros are referred to as zero accounts in experimental groups where the taxon is not expected to be present. ANCOM-BC uses the following two criteria defined in Kaul et al. (2017) to detect structural zeros. Let p_{ij} represent the proportion of the ith taxon from non-zero samples in the jth group, and let $\hat{p}_{ij} = \frac{1}{n_j}\sum_{k=1}^{n_j} I(O_{ijk} \neq 0)$ denote the estimate of p_{ij}. Then the ith taxon is declared to exist structural zeros in the jth group if either of the following is true.

1. $\hat{p}_{ij} = 0$.

2. $\hat{p}_{ij} - 1.96\sqrt{\dfrac{\hat{p}_{ij}(1-\hat{p}_{ij})}{n_j}} \leq 0$.

If a taxon is declared to be a structural zero in an experimental group, then, for that specific ecosystem, ANCOM-BC does not include this taxon in the analysis (Lin and Peddada 2020a). To apply both criteria stated in above definitions of structural zeros, set `neg_lb = TRUE`; otherwise, ANCOM-BC uses only Equation 1 to declare structural zeros. Setting `neg_lb = TRUE` is recommended when the sample size per group is relatively large (e.g., >30).

We create the object "rest_primary" to present the primary results(output\$res) for operation.

> rest_primary = output\$res

We use the following seven steps to obtain ANCOMBC primary results in which genotype, time, and their interaction effects are analyzed by the **ancombc()** function (Table 14.4).

Step 1: Obtain the estimated coefficients of `beta`.

```
> tab_coef = rest_primary$beta
> options(digits=2)
> tax_tab= physeq@tax_table[rownames(tab_coef), ]
> options(max.print=999999)
> coef_rest <- as.data.frame(cbind(tax_tab,tab_coef))
```

Table 14.4 The estimated coefficients of beta in QtRNA microbiome data

	Kingdom	Phylum	Class	Order	Family	Genus	Species	Group	TimePost	Group:TimePost
OTU_56	D_0_Bacteria	D_1_Actinobacteria	D_2_Coriobacteria	D_3_Coriobacteriales	D_4_Eggerthellaceae	D_5_Enterorhabdus	D_6_Enterorhabdus caecimuris B7	0.000463	0.049312	0.012773
OTU_57	D_0_Bacteria	D_1_Actinobacteria	D_2_Coriobacteriia	D_3_Coriobacteriales	D_4_Eggerthellaceae	D_5_Enterorhabdus	D_6_Enterorhabdus muris	0.856548	0.604616	– 0.83636
OTU_58	D_0_Bacteria	D_1_Actinobacteria	D_2_Coriobacteria	D_3_Coriobacteriales	D_4_Eggerthellaceae	D_5_Enterorhabdus	D_6_mouse gut metagenome	– 0.35507	1.136966	-0.0586
OTU_77	D_0_Bacteria	D_1_Bacteroidetes	D_2_Bacteroidia	D_3_Bacteroidales	D_4_Bacteroidaceae	D_5_Bacteroides	D_6_Bacteroides acidifaciens JCM 10556	– 0.06885	0.389432	– 0.32735
OTU_101	D_0_Bacteria	D_1_Bacteroidetes	D_2_Bacteroidia	D_3_Bacteroidales	D_4_Muribaculaceae	D_5_Muribaculum	D_6_Parabacteroides sp. YL27	0.000463	0.118627	3.06081
OTU_131	D_0_Bacteria	D_1_Bacteroidetes	D_2_Bacteroidia	D_3_Bacteroidales	D_4_Tannerellaceae	D_5_Parabacteroides	D_6_Parabacteroides goldsteinii CL02T12C30	-0.8368	1.42599	1.49273
OTU_221	D_0_Bacteria	D_1_Cyanobacteria	D_2_Oxyphotobacteria	D_3_Nostocales	D_4_Nostocaceae	Other	Other	0.375805	0.326571	-0.5012
OTU_263	D_0_Bacteria	D_1_Firmicutes	D_2_Bacilli	D_3_Bacillales	D_4_Staphylococcaceae	D_5_Staphylococcus	D_6_Staphylococcus saprophyticus subsp. saprophyticus	– 0.13817	0.47644	– 0.220717
OTU_264	D_0_Bacteria	D_1_Firmicutes	D_2_Bacilli	D_3_Bacillales	D_4_Staphylococcaceae	D_5_Staphylococcus	D_6_Staphylococcus sp. UAsDu23	– 0.93095	-1.7192	1.013497
OTU_276	D_0_Bacteria	D_1_Firmicutes	D_2_Bacilli	D_3_Lactobacillales	D_4_Lactobacillaceae	D_5_Lactobacillus	D_6_Lactobacillus murinus	0.139093	0.257256	-0.3338
OTU_279	D_0_Bacteria	D_1_Firmicutes	D_2_Bacilli	D_3_Lactobacillales	D_4_Lactobacillaceae	D_5_Lactobacillus	D_6_gut metagenome	0.103425	0.106848	-0.3338
OTU_287	D_0_Bacteria	D_1_Firmicutes	D_2_Bacilli	D_3_Lactobacillales	D_4_Streptococcaceae	D_5_Streptococcus	D_6_Streptococcus danieliae	– 0.25306	0.840292	0.151861
OTU_323	D_0_Bacteria	D_1_Firmicutes	D_2_Clostridia	D_3_Clostridiales	D_4_Lachnospiraceae	D_5_ASF356	D_6_Clostridium sp. ASF356	0.000463	-0.02	3.469499
OTU_325	D_0_Bacteria	D_1_Firmicutes	D_2_Clostridia	D_3_Clostridiales	D_4_Lachnospiraceae	D_5_Acetatifactor	D_6_Clostridiales bacterium CIEAF 015	0.156511	2.28364	– 0.26737
OTU_330	D_0_Bacteria	D_1_Firmicutes	D_2_Clostridia	D_3_Clostridiales	D_4_Lachnospiraceae	D_5_Anaerostipes	D_6_Clostridiales bacterium VE202–09	0.11611	0.50183	-0.3425
OTU_351	D_0_Bacteria	D_1_Firmicutes	D_2_Clostridia	D_3_Clostridiales	D_4_Lachnospiraceae	D_5_Lachnospiraceae NK4A136 group	D_6_Clostridiales bacterium CIEAF 020	0.748644	0.79693	– 1.77908
OTU_353	D_0_Bacteria	D_1_Firmicutes	D_2_Clostridia	D_3_Clostridiales	D_4_Lachnospiraceae	D_5_Lachnospiraceae NK4A136 group	D_6_Lachnospiraceae bacterium A4	-0.0377	0.42619	– 0.44784

(continued)

Table 14.4 (continued)

	Kingdom	Phylum	Class	Order	Family	Genus	Species	Group	TimePost	Group: TimePost
OTU_354	D_0_Bacteria	D_1_Firmicutes	D_2_Clostridia	D_3_Clostridiales	D_4_Lachnospiraceae	D_5_Lachnospiraceae NK4A136 group	D_6_Lachnospiraceae bacterium COE1	1.068315	0.69781	–
OTU_361	D_0_Bacteria	D_1_Firmicutes	D_2_Clostridia	D_3_Clostridiales	D_4_Lachnospiraceae	D_5_Lachnospiraceae UCG-006	D_6_Clostridium sp. ASF502	– 0.30308	– 0.63509	2.21009 0.45926
OTU_366	D_0_Bacteria	D_1_Firmicutes	D_2_Clostridia	D_3_Clostridiales	D_4_Lachnospiraceae	D_5_Marvinbryantia	D_6_Clostridiales bacterium CIEAF 012	0.03092	0.015579	– 0.63559
OTU_369	D_0_Bacteria	D_1_Firmicutes	D_2_Clostridia	D_3_Clostridiales	D_4_Lachnospiraceae	D_5_Roseburia	D_6_Clostridium sp. Clone-44	0.248954	–0.02	– 0.30503
OTU_370	D_0_Bacteria	D_1_Firmicutes	D_2_Clostridia	D_3_Clostridiales	D_4_Lachnospiraceae	D_5_Roseburia	D_6_Eubacterium sp. 14-2	– 1.37716	– 2.75447	1.538724
OTU_395	D_0_Bacteria	D_1_Firmicutes	D_2_Clostridia	D_3_Clostridiales	D_4_Ruminococcaceae	D_5_Anaerotruncus	D_6_Anaerotruncus sp. G3(2012)	0.133685	– 1.46779	0.071386
OTU_408	D_0_Bacteria	D_1_Firmicutes	D_2_Clostridia	D_3_Clostridiales	D_4_Ruminococcaceae	D_5_Ruminiclostridium 5	D_6_Clostridium sp. Culture Jar-8	0.000463	4.359209	– 4.29712
OTU_409	D_0_Bacteria	D_1_Firmicutes	D_2_Clostridia	D_3_Clostridiales	D_4_Ruminococcaceae	D_5_Ruminiclostridium 5	D_6_Ruminiclostridium sp. KB18	0.564388	2.346951	–1.2909
OTU_412	D_0_Bacteria	D_1_Firmicutes	D_2_Clostridia	D_3_Clostridiales	D_4_Ruminococcaceae	D_5_Ruminiclostridium 9	D_6_bacterium enrichment culture clone M244	– 0.34294	1.211262	1.14918
OTU_516	D_0_Bacteria	D_1_Proteobacteria	D_2_Alphaproteobacteria	D_3_Rickettsiales	D_4_Mitochondria	D_5_Triticum aestivum (bread wheat)	D_6_Triticum aestivum (bread wheat)	– 0.13817	– 0.15863	0.151402

```
> # Table 14.4
> # Write results table
> # Make the table
> library(xtable)
> table1 <- xtable(coef_rest,caption = "Table of Coefficients from the
Primary Results",lable="Coef_taxa_table")
> print.xtable(table1,type="html",file = "ANCOMBC_Table_Coef_QtRNA.
html")
> write.csv(coef_rest ,file = paste("ANCOMBC_Table_Coef_QtRNA.csv",
sep = ""))
```

The above table is extracted from the ANCOMBC_Table_Coef_QtRNA.

Step 2: Obtain the estimated standard errors of `beta`.

```
> tab_se = rest_primary$se
> options(max.print=999999)
> se_rest <- as.data.frame(cbind(tax_tab,tab_se))


> library(xtable)
> table2 <- xtable(se_rest,caption = "Table of Standard Erros from the
Primary Results",lable="SE_taxa_table")
> print.xtable(table2,type="html",file = "ANCOMBC_Table_SE_QtRNA.
html")
> write.csv(se_rest,file = paste("ANCOMBC_Table_SE_QtRNA.csv",sep =
""))
```

Step 3: Obtain the test statistics.

```
> tab_w = rest_primary$W
> options(max.print=999999)
> w_rest <- as.data.frame(cbind(tax_tab,tab_w))
> library(xtable)
> table3 <- xtable(w_rest,caption = "Table of Test Statistics from the
Primary Results",lable="Statistics_taxa_table")
> print.xtable(table3,type="html",file =
"ANCOMBC_Table_Statistics_QtRNA.html")
> write.csv(w_rest,file = paste("ANCOMBC_Table_Statistics_QtRNA.
csv",sep = ""))
```

Step 4: Obtain the *P*-values from the two-sided Z-test using the test statistics.

```
> tab_p = rest_primary$p_val
> options(max.print=999999)
> p_rest <- as.data.frame(cbind(tax_tab,tab_p))
> library(xtable)
```

```
> table4 <- xtable(p_rest,caption = "Table of P-values from the Primary
Results",lable="P-values_taxa_table")
> print.xtable(table4,type="html",file = "ANCOMBC_Table_P-
values_QtRNA.html")
> write.csv(p_rest,file = paste("ANCOMBC_Table_P-values_QtRNA.csv",
sep = ""))
```

Step 5: Obtain the Adjusted *P*-values by applying the `p_adj_method()` function.

As summarized in above Part 5 of ANCOM-BC methods, it was shown (Lim et al. 2013) that Holm-Bonferroni (Holm 1979) or Bonferroni (Bonferroni 1936; Dunn 1958, 1961) correction method is more appropriate to control the FDR when *P*-values were not accurate. The Benjamini-Hochberg (BH) procedure controls the FDR if the taxa have either independent structures or some positive correlation structures(Benjamini and Hochberg 1995; Benjamini and Yekutieli 2001). Thus, to control the FDR due to multiple comparisons, ANCOM-BC recommends applying the Holm-Bonferroni or Bonferroni correction method instead of the BH procedure to adjust the raw *P*-values. It was shown in ANCOM-BC (Lin and Peddada 2020a) that the *P*-values adjusted by Bonferroni correction is more conservative, while BH procedure results in FDR around the nominal level (5%), maintaining larger power (Table 14.5).

```
> tab_q = rest_primary$q
> options(max.print=999999)
> q_rest <- as.data.frame(cbind(tax_tab,tab_q))
> # Table 14.5
> # Write results table
> # Make the table
> library(xtable)
> table5 <- xtable(q_rest, caption = "Table of Adjusted p-values from
the Primary Results",lable="Adjusted p-values_taxa_table")
> print.xtable(table5,type="html",file = "ANCOMBC_Table_Adjusted
p-values_QtRNA.html")
> write.csv(q_rest,file = paste("ANCOMBC_Table_Adjusted
p-values_QtRNA.csv",sep = ""))
```

Step 6: Identify the differentially abundant taxa (Table 14.6).

```
> tab_diff = rest_primary$diff_abn
> options(max.print=999999)
> diff_rest <- as.data.frame(cbind(tax_tab,tab_diff))
> # Table 14.6
> # Write results table
> # Make the table
> library(xtable)
> table6 <- xtable(diff_rest, caption = "Table of Differentially
Abundant Taxa from the Primary Results",lable="DA_taxa_table")
```

Table 14.5 The adjusted *P*-values of differentially abundant analysis of taxa in QtRNA microbiome data

	Kingdom	Phylum	Class	Order	Family	Genus	Species	Group	TimePost	Group: TimePost
OTU_56	D_0_Bacteria	D_1_Actinobacteria	D_2_Coriobacteriia	D_3_Coriobacteriales	D_4_Eggerthellaceae	D_5_Enterorhabdus	D_6_Enterorhabdus caecimuris B7	1	1	1
OTU_57	D_0_Bacteria	D_1_Actinobacteria	D_2_Coriobacteriia	D_3_Coriobacteriales	D_4_Eggerthellaceae	D_5_Enterorhabdus	D_6_Enterorhabdus muris	1	1	1
OTU_58	D_0_Bacteria	D_1_Actinobacteria	D_2_Coriobacteriia	D_3_Coriobacteriales	D_4_Eggerthellaceae	D_5_Enterorhabdus	D_6_mouse gut metagenome	1	0.012024	1
OTU_77	D_0_Bacteria	D_1_Bacteroidetes	D_2_Bacteroidia	D_3_Bacteroidales	D_4_Bacteroidaceae	D_5_Bacteroides	D_6_Bacteroides acidifaciens JCM 10556	0	1	0
OTU_101	D_0_Bacteria	D_1_Bacteroidetes	D_2_Bacteroidia	D_3_Bacteroidales	D_4_Muribaculaceae	D_5_Muribaculum	D_6_Parabacteroides sp. YL27	1	1	0.409359
OTU_131	D_0_Bacteria	D_1_Bacteroidetes	D_2_Bacteroidia	D_3_Bacteroidales	D_4_Tannerellaceae	D_5_Parabacteroides	D_6_Parabacteroides goldsteinii CL02T12C30	1	0.522427	1
OTU_231	D_0_Bacteria	D_1_Deferribacteres	D_2_Deferribacteres	D_3_Deferribacterales	D_4_Deferribacteraceae	D_5_Mucispirillum	D_6_Mucispirillum schaedleri ASF457	1	1	1
OTU_263	D_0_Bacteria	D_1_Firmicutes	D_2_Bacilli	D_3_Bacillales	D_4_Staphylococcaceae	D_5_Staphylococcus	D_6_Staphylococcus saprophyticus subsp. saprophyticus	1	0.54287	1
OTU_264	D_0_Bacteria	D_1_Firmicutes	D_2_Bacilli	D_3_Bacillales	D_4_Staphylococcaceae	D_5_Staphylococcus	D_6_Staphylococcus sp. UAsDu23	1	4.80E-06	1
OTU_276	D_0_Bacteria	D_1_Firmicutes	D_2_Bacilli	D_3_Lactobacillales	D_4_Lactobacillaceae	D_5_Lactobacillus	D_6_Lactobacillus murinus	0	1	0
OTU_279	D_0_Bacteria	D_1_Firmicutes	D_2_Bacilli	D_3_Lactobacillales	D_4_Lactobacillaceae	D_5_Lactobacillus	D_6_gut metagenome	1	1	1
OTU_287	D_0_Bacteria	D_1_Firmicutes	D_2_Bacilli	D_3_Lactobacillales	D_4_Streptococcaceae	D_5_Streptococcus	D_6_Streptococcus danieliae	1	1	1
OTU_323	D_0_Bacteria	D_1_Firmicutes	D_2_Clostridia	D_3_Clostridiales	D_4_Lachnospiraceae	D_5_ASF356	D_6_Clostridium sp. ASF356	0	1	0
OTU_325	D_0_Bacteria	D_1_Firmicutes	D_2_Clostridia	D_3_Clostridiales	D_4_Lachnospiraceae	D_5_Acetatifactor	D_6_Clostridiales bacterium CIEAF 015	1	0.008759	1
OTU_330	D_0_Bacteria	D_1_Firmicutes	D_2_Clostridia	D_3_Clostridiales	D_4_Lachnospiraceae	D_5_Anaerostipes	D_6_Clostridiales bacterium VE202-09	1	1	1
OTU_351	D_0_Bacteria	D_1_Firmicutes	D_2_Clostridia	D_3_Clostridiales	D_4_Lachnospiraceae	D_5_Lachnospiraceae NK4A136 group	D_6_Clostridiales bacterium CIEAF 020	1	1	1
OTU_353	D_0_Bacteria	D_1_Firmicutes	D_2_Clostridia	D_3_Clostridiales	D_4_Lachnospiraceae			1	1	1

(continued)

Table 14.5 (continued)

	Kingdom	Phylum	Class	Order	Family	Genus	Species	Group	TimePost	Group: TimePost
OTU_354	D_0_Bacteria	D_1_Firmicutes	D_2_Clostridia	D_3_Clostridiales	D_4_Lachnospiraceae	D_5_Lachnospiraceae NK4A136 group	D_6_Lachnospiraceae bacterium A4	1	1	0.169092
OTU_361	D_0_Bacteria	D_1_Firmicutes	D_2_Clostridia	D_3_Clostridiales	D_4_Lachnospiraceae	D_5_Lachnospiraceae NK4A136 group	D_6_Lachnospiraceae bacterium COE1	1	1	1
OTU_366	D_0_Bacteria	D_1_Firmicutes	D_2_Clostridia	D_3_Clostridiales	D_4_Lachnospiraceae	D_5_Lachnospiraceae UCG-006	D_6_Clostridium sp. ASF502	1	1	1
OTU_369	D_0_Bacteria	D_1_Firmicutes	D_2_Clostridia	D_3_Clostridiales	D_4_Lachnospiraceae	D_5_Marvinbryantia	D_6_Clostridiales bacterium CIEAF 012	0	1	0
OTU_370	D_0_Bacteria	D_1_Firmicutes	D_2_Clostridia	D_3_Clostridiales	D_4_Lachnospiraceae	D_5_Roseburia	D_6_Clostridium sp. Clone-44	1	0.00047	1
OTU_395	D_0_Bacteria	D_1_Firmicutes	D_2_Clostridia	D_3_Clostridiales	D_4_Ruminococcaceae	D_5_Anaerotruncus	D_6_Anaerotruncus sp. G3(2012)	1	0.000105	1
OTU_408	D_0_Bacteria	D_1_Firmicutes	D_2_Clostridia	D_3_Clostridiales	D_4_Ruminococcaceae	D_5_Ruminiclostridium 5	D_6_Clostridium sp. Culture Jar-8	0	2.38E-47	0
OTU_409	D_0_Bacteria	D_1_Firmicutes	D_2_Clostridia	D_3_Clostridiales	D_4_Ruminococcaceae	D_5_Ruminiclostridium 5	D_6_Ruminiclostridium sp. KB18	1	0.636737	1
OTU_412	D_0_Bacteria	D_1_Firmicutes	D_2_Clostridia	D_3_Clostridiales	D_4_Ruminococcaceae	D_5_Ruminiclostridium 9	D_6_bacterium enrichment culture clone M244	0	1	0
OTU_516	D_0_Bacteria	D_1_Proteobacteria	D_2_Alphaproteobacteria	D_3_Rickettsiales	D_4_Mitochondria	D_5_Triticum aestivum (bread wheat)	D_6_Triticum aestivum (bread wheat)	0	1	0

Table 14.6 The identified differentially abundant taxa in QtRNA microbiome data

	Kingdom	Phylum	Class	Order	Family	Genus	Species	Group	TimePost	Group: TimePost
OTU_56	D_0__Bacteria	D_1__Actinobacteria	D_2__Coriobacteriia	D_3__Coriobacteriales	D_4__Eggerthellaceae	D_5__Enterorhabdus	D_6__Enterorhabdus caecimuris B7	FALSE	FALSE	FALSE
OTU_57	D_0__Bacteria	D_1__Actinobacteria	D_2__Coriobacteriia	D_3__Coriobacteriales	D_4__Eggerthellaceae	D_5__Enterorhabdus	D_6__Enterorhabdus muris	FALSE	FALSE	FALSE
OTU_58	D_0__Bacteria	D_1__Actinobacteria	D_2__Coriobacteriia	D_3__Coriobacteriales	D_4__Eggerthellaceae	D_5__Enterorhabdus	D_6__mouse gut metagenome	FALSE	TRUE	FALSE
OTU_77	D_0__Bacteria	D_1__Bacteroidetes	D_2__Bacteroidia	D_3__Bacteroidales	D_4__Bacteroidaceae	D_5__Bacteroides	D_6__Bacteroides acidifaciens JCM 10556	TRUE	FALSE	TRUE
OTU_101	D_0__Bacteria	D_1__Bacteroidetes	D_2__Bacteroidia	D_3__Bacteroidales	D_4__Muribaculaceae	D_5__Muribaculum	D_6__Parabacteroides sp. YL27	FALSE	FALSE	FALSE
OTU_131	D_0__Bacteria	D_1__Bacteroidetes	D_2__Bacteroidia	D_3__Bacteroidales	D_4__Tannerellaceae	D_5__Parabacteroides	D_6__Parabacteroides goldsteinii CL02T12C30	FALSE	FALSE	FALSE
OTU_231	D_0__Bacteria	D_1__Deferribacteres	D_2__Deferribacteres	D_3__Deferribacterales	D_4__Deferribacteraceae	D_5__Mucispirillum	D_6__Mucispirillum schaedleri ASF457	FALSE	FALSE	FALSE
OTU_263	D_0__Bacteria	D_1__Firmicutes	D_2__Bacilli	D_3__Bacillales	D_4__Staphylococcaceae	D_5__Staphylococcus	D_6__Staphylococcus saprophyticus subsp. saprophyticus	FALSE	FALSE	FALSE
OTU_264	D_0__Bacteria	D_1__Firmicutes	D_2__Bacilli	D_3__Bacillales	D_4__Staphylococcaceae	D_5__Staphylococcus	D_6__Staphylococcus sp. UAsDu23	FALSE	TRUE	FALSE
OTU_276	D_0__Bacteria	D_1__Firmicutes	D_2__Bacilli	D_3__Lactobacillales	D_4__Lactobacillaceae	D_5__Lactobacillus	D_6__Lactobacillus murinus	TRUE	FALSE	TRUE
OTU_279	D_0__Bacteria	D_1__Firmicutes	D_2__Bacilli	D_3__Lactobacillales	D_4__Lactobacillaceae	D_5__Lactobacillus	D_6__gut metagenome	FALSE	FALSE	FALSE
OTU_287	D_0__Bacteria	D_1__Firmicutes	D_2__Bacilli	D_3__Lactobacillales	D_4__Streptococcaceae	D_5__Streptococcus	D_6__Streptococcus danieliae	FALSE	FALSE	FALSE
OTU_323	D_0__Bacteria	D_1__Firmicutes	D_2__Clostridia	D_3__Clostridiales	D_4__Lachnospiraceae	D_5__ASF356	D_6__Clostridium sp. ASF356	TRUE	FALSE	TRUE
OTU_325	D_0__Bacteria	D_1__Firmicutes	D_2__Clostridia	D_3__Clostridiales	D_4__Lachnospiraceae	D_5__Acetatifactor	D_6__Clostridiales bacterium CIEAF 015	FALSE	TRUE	FALSE
OTU_330	D_0__Bacteria	D_1__Firmicutes	D_2__Clostridia	D_3__Clostridiales	D_4__Lachnospiraceae	D_5__Anaerostipes	D_6__Clostridiales bacterium VE202-09	FALSE	FALSE	FALSE

(continued)

Table 14.6 (continued)

	Kingdom	Phylum	Class	Order	Family	Genus	Species	Group	TimePost	Group: TimePost
OTU_351	D_0_Bacteria	D_1_Firmicutes	D_2_Clostridia	D_3_Clostridiales	D_4_Lachnospiraceae	D_5_Lachnospiraceae NK4A136 group	D_6_Clostridiales bacterium CIEAF 020	FALSE	FALSE	FALSE
OTU_353	D_0_Bacteria	D_1_Firmicutes	D_2_Clostridia	D_3_Clostridiales	D_4_Lachnospiraceae	D_5_Lachnospiraceae NK4A136 group	D_6_Lachnospiraceae bacterium A4	FALSE	FALSE	FALSE
OTU_354	D_0_Bacteria	D_1_Firmicutes	D_2_Clostridia	D_3_Clostridiales	D_4_Lachnospiraceae	D_5_Lachnospiraceae NK4A136 group	D_6_Lachnospiraceae bacterium COE1	FALSE	FALSE	FALSE
OTU_361	D_0_Bacteria	D_1_Firmicutes	D_2_Clostridia	D_3_Clostridiales	D_4_Lachnospiraceae	D_5_Lachnospiraceae UCG-006	D_6_Clostridium sp. ASF502	FALSE	FALSE	FALSE
OTU_366	D_0_Bacteria	D_1_Firmicutes	D_2_Clostridia	D_3_Clostridiales	D_4_Lachnospiraceae	D_5_Marvinbryantia	D_6_Clostridiales bacterium CIEAF 012	FALSE	FALSE	FALSE
OTU_369	D_0_Bacteria	D_1_Firmicutes	D_2_Clostridia	D_3_Clostridiales	D_4_Lachnospiraceae	D_5_Roseburia	D_6_Clostridium sp. Clone-44	TRUE	FALSE	TRUE
OTU_370	D_0_Bacteria	D_1_Firmicutes	D_2_Clostridia	D_3_Clostridiales	D_4_Lachnospiraceae	D_5_Roseburia	D_6_Eubacterium sp. 14-2	FALSE	TRUE	FALSE
OTU_395	D_0_Bacteria	D_1_Firmicutes	D_2_Clostridia	D_3_Clostridiales	D_4_Ruminococcaceae	D_5_Anaerotruncus	D_6_Anaerotruncus sp. G3(2012)	FALSE	TRUE	FALSE
OTU_408	D_0_Bacteria	D_1_Firmicutes	D_2_Clostridia	D_3_Clostridiales	D_4_Ruminococcaceae	D_5_Ruminiclostridium 5	D_6_Clostridium sp. Culture Jar-8	TRUE	TRUE	TRUE
OTU_409	D_0_Bacteria	D_1_Firmicutes	D_2_Clostridia	D_3_Clostridiales	D_4_Ruminococcaceae	D_5_Ruminiclostridium 5	D_6_Ruminiclostridium sp. KB18	FALSE	FALSE	FALSE
OTU_412	D_0_Bacteria	D_1_Firmicutes	D_2_Clostridia	D_3_Clostridiales	D_4_Ruminococcaceae	D_5_Ruminiclostridium 9	D_6_bacterium enrichment culture clone M244	TRUE	FALSE	TRUE
OTU_516	D_0_Bacteria	D_1_Proteobacteria	D_2_Alphaproteobacteria	D_3_Rickettsiales	D_4_Mitochondria	D_5_Triticum aestivum (bread wheat)	D_6_Triticum aestivum (bread wheat)	TRUE	FALSE	TRUE

```
> print.xtable(table6,type="html",file =
"ANCOMBC_Table_Differentially Abundant Taxa_QtRNA.html")
> write.csv(diff_rest,file = paste("ANCOMBC_Table_Differentially
Abundant Taxa_QtRNA.csv",sep = ""))
```

The above table is extracted from the ANCOMBC_Table_Differentially Abundant Taxa_QtRNA.

Step 7: Estimate and adjust the sample-specific sampling fractions to obtain the bias-adjusted abundances.

First, estimate sample-specific sampling fractions by natural log scale. Since the sampling fraction is not estimable for each sample with the presence of missing values, the missing values (NA) for any variable specified in the formula will be replaced with 0.

```
> samp_frac = output$samp_frac
> # Replace NA with 0
> samp_frac[is.na(samp_frac)] = 0
```

Then, adjust the log observed abundances by subtracting the estimated sampling fraction from log observed abundances of each sample. Note that to avoid taking the log of 0, which is undefined, a pseudocount (here 1) is added to each estimated abundances before taking natural log.

```
> # Add pesudo-count (1) to avoid taking the log of 0
> library(microbiome)
> log_obs_abn = log(abundances(physeq) + 1)


> # Adjust the log observed abundances
> log_obs_abn_adj = t(t(log_obs_abn) - samp_frac)


> library(xtable)
> table7 <- xtable(round(log_obs_abn_adj, 2), caption = "Table of Bias-
adjusted Log Observed Abundances from the Primary Results",
lable="Bias-adjusted Log-OA_taxa_table")
> print.xtable(table7,type="html",file = "ANCOMBC_Table_Bias-
adjusted Log Observed Abundances_QtRNA.html")
> write.csv(table7,file = paste("ANCOMBC_Table_Bias-adjusted Log
Observed Abundances_QtRNA.csv",sep = ""))
```

ANCOMBC Global Test.

In the sample metadata, we created a four levels of categorical variable "Group4" (WT_BEFORE, WT_POST, KO_BEFORE, and KO_POST). We used the following commands to illustrate the ANCOM-BC global test using this variable.

```
> library(ANCOMBC)
> output = ancombc(phyloseq = physeq, formula = "Group4",
+ p_adj_method = "holm", zero_cut = 0.90, lib_cut = 1000,
+ group = "Group4", struc_zero = TRUE, neg_lb = TRUE, tol = 1e-5,
+ max_iter = 100, conserve = TRUE, alpha = 0.05, global = TRUE)

> rest_global = output$res_global
```

The global test results are contained in the output$res_global (a data.frame for the variable specified in group).

Each column contains the test statistics (W), the *P*-values (p_val) from the two-sided chi-square test using W, the adjusted *P*-values (q_val) that are obtained by applying the p_adj_method to p_val, and the logical vector (diff_abn) of differentially abundant taxa with TRUE if the taxon has q_val less than alpha, otherwise with FALSE. Like the primary analysis, we can obtain the obtain the ANCOMBC global test results using the following steps:

Step 1: Obtain the test global statistics.

```
> tab_w = rest_global$W
> options(max.print=999999)
> w_rest <- as.data.frame(cbind(tax_tab, tab_w))
> library(xtable)
> table8 <- xtable(w_rest, caption = "Table of Global Test Statistics
from the Global Results", lable="Global_ Statistics_taxa_table")
> print.xtable(table8, type="html", file = "ANCOMBC_Table_
Global_Statistics_QtRNA.html")
> write.csv(w_rest, file = paste("ANCOMBC_Table_
Global_Statistics_QtRNA.csv", sep = ""))
```

Step 2: Obtain the *P*-values from the two-sided Z-test using the global test statistics.

```
> tab_p = rest_global$p_val
> options(max.print=999999)
> p_rest <- as.data.frame(cbind(tax_tab, tab_p))
> library(xtable)
> table9 <- xtable(p_rest, caption = "Table of P-values from the Global
Results", lable=" Global_P-values_taxa_table")
> print.xtable(table9, type="html", file = "ANCOMBC_Table_ Global_P-
values_QtRNA.html")
> write.csv(p_rest, file = paste("ANCOMBC_Table_ Global_P-
values_QtRNA.csv", sep = ""))
```

Step 3: Adjust the *P*-values by applying the `p_adj_method()` function for the global test.

```
> tab_q = rest_global$q
> options(max.print=999999)
> q_rest <- as.data.frame(cbind(tax_tab,tab_q))
> library(xtable)
> table10 <- xtable(q_rest, caption = "Table of Adjusted p-values from
the Global Results", lable="Adjusted p-values_taxa_table")
> print.xtable(table10,type="html",file = "ANCOMBC_Table_
Global_Adjusted p-values_QtRNA.html")
> write.csv(q_rest,file = paste("ANCOMBC_Table_ Global_Adjusted
p-values_QtRNA.csv",sep = ""))
```

Step 4: Identify the differentially abundant taxa `for the global test`.

```
> tab_diff = rest_global$diff_abn
> options(max.print=999999)
> diff_rest <- as.data.frame(cbind(tax_tab,tab_diff))
> library(xtable)
> table11 <- xtable(diff_rest, caption = "Table of Differentially
Abundant Taxa from the Global Results", lable="DA_taxa_table")
> print.xtable(table11,type="html",file = "ANCOMBC_Table_
Global_Differentially Abundant Taxa_QtRNA.html")
> write.csv(diff_rest,file = paste("ANCOMBC_Table_
Global_Differentially Abundant Taxa_QtRNA.csv",sep = ""))
```

14.4.3   Remarks on ANCOM-BC

ANCOM-BC method has some unique characteristics (Lin and Peddada 2020a): (1) It explicitly tests hypothesis of differential absolute abundance of individual taxon and provides valid *P*-values and confidence intervals for each taxon. (2) It provides an approach to estimate the sampling fraction and performs DA analysis by correcting bias due to (unobservable) differential sampling fractions across samples. (3) It does not rely on strong parametric assumptions. ANCOM-BC has some strengths (Lin and Peddada 2020a):

- It can construct confidence intervals for DA of each taxon whereas ANCOM and other DA methods cannot. For example, we can construct a 95% simultaneous confidence intervals using Bonferroni method for the mean DA of each taxon in the two experimental groups, and hence an effect size that is associated with each taxon can be obtained in comparing two experimental groups.
- It was demonstrated that like ANCOM, ANCOM-BC is able to well control the FDR while maintaining adequate power compared to other often used DA

methods, including count-based methods edgeR and DESeq2 in RNA-seq studies, the zero-inflated Gaussian mixture model used in metagenomeSeq (ZIG) (Lin and Peddada 2020a).

ANCOM-BC also has some weaknesses, for example:

- ANCOM-BC cannot well control the FDR for small sample sizes (e.g., $n = 5$ per group). Ten samples are required for controlling FDR with adequate power (Lin and Peddada 2020a). This is probably because the estimator and the test statistic of ANCOM-BC are developed asymptotically.
- ANCOM-BC assumes that the compositional data on the covariates is linear. Like ANCOM, ANCOM-BC cannot control the FDR well when this linear assumption is violated (Wang 2021).
- ANCOM-BC adopted the methodology (Kaul et al. 2017) for dealing structural zeros, the outlier zeros, and sampling zeros. In practice, ANCOM-BC actually drops taxa in the analysis when the structural zeros associated with these taxa are detected by the criteria (Kaul et al. 2017).
- ANCOM-BC uses log-transformation of OTU counts data rather than log-ratio transformation, ANCOM-BC methodology is not a typical CoDa in the sense of Aitchison's approach (Aitchison 1986a). To solve the problem of compositionality, ANCOM-BC uses the sum of ecosystem as reference to offset the data. This kind of normalization is not the typical Aitchison's approach, while like using offset in the count-based methods.

ANCOM and ANCOM-BC share some common features: (1) assuming that the observed sample is an unknown fraction of a unit volume of the ecosystem, and the sampling fraction varies from sample to sample, and (2) using log-ratio transformation (or log scale) to serve as normalization.

However, ANCOM-BC is different from ANCOM in some ways:

- Unlike ANCOM, ANCOM-BC does not directly implement a log-ratio transformation, instead simplifies DA analysis by recasting the problem as a linear regression problem with an offset to account for the sampling fraction. That is, ANCOM-BC's bias correction is a sample-specific offset term that is introduced into a linear regression and accounts for sampling fraction in conjunction with performing linear regression in log scale serves the same purpose as a log-ratio transformation.
- Compared to ANCOM, ANCOM-BC is very fast because it avoids repeatedly using a taxon as reference that ANCOM uses. ANCOM needs $\frac{m(m-1)}{2}$ linear regressions to fit the models, whereas ANCOM-BC requires only m linear regressions to fit.

14.5 Remarks on Compositional Data Analysis Approach

In microbiome study, two kinds of statistical methods are available: count-based approach and compositional data analysis (CoDa) approach. The count-based approach, i.e., negative binomial methods (e.g., edgeR and DESeq2), and over-dispersed and zero-inflated models (e.g., metagenomeSeq) do not take data transformations to bring the data to normality before analysis, instead of using the offset to scale the data during the modeling.

Different from count-based approach, CoDa approach including ALDEx2, ANCOM, and ANCOM-BC typically need some kind of log or log-ratio transformations before analysis.

In some instances, the log-ratio transformation is technically equivalent to a normalization such as alr- and clr-transformations can formally serve as effective library size normalizations. Thus sometime the log-ratio transformation is called the log-ratio "normalization" (Quinn et al. 2018a, b). The parametric methods (ALDEx2 t-test, ANCOM-BC) in CoDa approach further assume a Gaussian distribution of the data in their statistical models. Since it is difficult to measure total microbial load or absolute number of microorganisms to accurately determine taxonomic shifts (Morton et al. 2019), there remains a lack of a gold standard to measure microbiome data. Thus, DA analysis is still controversial in microbiome research. One current effort tries to establish microbial composition measurement standards with reference frames and define the concept of "reference frames" for inferring changes in abundance in CoDa. The "reference frames" are used as a constraint for inferencing how microbial populations change relatively given by other microbial populations. The reference frame for inferring changes is determined by denominator in a log-ratio (Morton et al. 2019).

CoDa methods have their advantages, providing a statistical tool to move from the simplex into real space (Aitchison et al. 2000) and hence enable statistical methods in Euclidean distances. Thus, log-ratio "normalizations" or log-ratio transformations are useful even when they do not normalize the data. As we reviewed in this chapter, ALDEx2, ANCOM, and ANCOM-BC methods have been evaluated and compared to edgeR, DESeq2, metagenomeSeq, and other competing methods. Their robustness and privileges are claimed by the proposed authors.

However, we notice that when ANCOM-BC claimed that both ANCOM and ANCOM-BC are outperformed over edgeR, and DESeq2, and ZIG in terms of FDR control and higher power, their DA analyses in concept are different. For example, ANCOM-BC differentially analyzes absolute abundance rather than differentially analyzes relative abundance. When ANCOM-BC states that it takes into account the compositionality of the microbiome data, it actually uses log-transformation to transform the OTU data or use an offset to account for the sampling fraction. Additionally, both edgeR and DESeq2 were developed based on NB models to account for small sample sizes and individual variation of genes (taxa). Overall, the concept and approach of reference frames remain within the CoDa with log-ratio transformation. Thus, it still has the challenge to deal with zero-problem and in practice how to pick an appropriate reference frame and to understand statistical properties of these reference frames (Morton et al. 2019). We should consider the context when we explain the performance of model comparisons.

However, on the one hand, whether or not microbiome data are really compositional is debatable. Some researchers acknowledged that next-generation sequencing (NGS) abundance data differ slightly from the formally defined compositional data because they contain integer values only, such as NGS abundance data do not require the arbitrary sum to represent complete unity in Aitchison's definition of compositional data (Aitchison 1982, 1986a). Thus, many datasets such as possibly NGS abundance data lack the information of potential components and hence exist as "incomplete compositions" (Quinn et al. 2018a, b). But this kind of data (e.g., NGS abundance data) is incidentally constrained to an arbitrary sum, and thus it still has compositional properties and can be called "count compositional data" and treated as compositional data (Lovell et al. 2015; Quinn et al. 2017, 2018a, b). In contrast, other researchers do not think that count data is purely relative—the count pair (1, 2) carries different information than counts of (1000, 2000) even though the ratio of the two pair is same. Instead, they think that treating count data as compositional is treating "really discrete count data" as continuous (Bacon-Shone 2008).

Briefly, there are five reasons that drive those researchers to take count-based approach to analyze microbiome data and other NGS data instead using compositional data analysis (see details in Xia et al. 2018a; Xia 2020):

(1) Compositionality (e.g., the spurious correlation concern) occurs when the data dimension is relatively low (i.e., the number of components is relatively small), such as in ecology studies. However, microbiome data is high dimensional and usually having the large number of taxa. Even there is the compositional effect (e.g., the spurious correlation), its effect is mild when the samples have large diversity (Zhang et al. 2017; Friedman and Alm 2012) because as the number of taxa increases, the compositional effect is attenuating (Hu et al. 2018).

(2) The mild data compositionality can be corrected by technological developments including the estimation of absolute cellular abundances from microbiome sequence data in microbiome data science (Vandeputte et al. 2017).

(3) The count-based models do not require the data to be transformed before analysis. It was shown these models have more statistical power to detect differential expression (Robinson and Oshlack 2010; Zhang and Yi 2020). Thus, these researchers advised to treat the high-throughput sequencing datasets as count data (Anders et al. 2013; Kuczynski et al. 2011).

(4) Compositional data analysis may lack strong biological interpretation. The results from transformation-based analyses should be interpreted with respect to the chosen reference or the results should be translated back into compositional terms. This remains complicated. Otherwise it is difficult to interpret biologically because one purpose of microbiome study is to detect which taxa are associated with the outcome (Hu et al. 2018) rather than to detect the association between ratio of taxa and the outcome. Thus, CoDa uses the ratio

of taxa to interpret the results have been criticized because of no clinical-direct-relevance (Xia 2020).

(5) Compositional data analysis usually takes log-ratio transformation to transform the abundance count data before analysis. To ensure log of zeros be defined, arbitrary small constant counts are added to a zero value. Thus, how best to handle zeros remains a topic of ongoing research and poses the major challenge of CoDa (Xia 2020).

Both compositional and count-based models are based on abundance read counts. Actually, functional study indicates that the rare taxa with lower abundant read counts do not necessarily have lower functions related to the health and disease. Thus, the statistical models that are sensitive to detect these rare taxa are especially important.

14.6 Summary

This chapter focused on principles, methods, and techniques of compositional data analysis in microbiome data. First, we described overall compositional data analysis (CoDa) including definition of compositional data, description of the reasons that some researchers treat microbiome data as compositional, definition of Aitchison simplex, description of the challenges of analyzing compositional data, fundamental principles of CoDa, and the family of log-ratio transformations: additive log-ratio (alr), centered log-ratio (clr), isometric log-ratio (ilr), and inter-quartile log-ratio (iqlr). Next, we introduced and illustrated three CoDa, respectively, in the order of these methods proposed: (1) ANOVA-Like Compositional Differential Abundance Analysis (ALDEx2), (2) Analysis of Composition of Microbiomes (ANCOM), and (3) Analysis of Composition of Microbiomes-Bias Correction (ANCOM-BC). We implemented ALDEx2 and ANCOM-BC with R and ANCOM using QIIME 2. We also provided some general remarks on the approach of CoDa.

In Chap. 15, we will introduce how to analyze longitudinal microbiome data through linear mixed-effects models.

References

Aitchison, John. 1981. A new approach to null correlations of proportions. *Mathematical Geology* 13 (2): 175–189.

Aitchison, J. 1982. The statistical analysis of compositional data (with discussion). *Journal of the Royal Statistical Society, Series B (Statistical Methodology)* 44 (2): 139–177.

———. 1983. Principal component analysis of compositional data. *Biometrika* 70 (1): 57–65. https://doi.org/10.1093/biomet/70.1.57.

———. 1984. Reducing the dimensionality of compositional data sets. *Journal of the International Association for Mathematical Geology* 16 (6): 617–635.

————. 1986a. *The statistical analysis of compositional data*, Monographs on statistics and applied probability. London: Chapman and Hall Ltd. Reprinted in 2003 with additional material by The Blackburn Press.

————. 1986b. *The statistical analysis of compositional data*. Chapman & Hall; Reprinted in 2003, with additional material, by The Blackburn Press.

Aitchison, John. 2003. A concise guide to compositional data analysis. In: *2nd compositional data analysis workshop, Girona, Italy, 2003*.

————. 2008. The single principle of compositional data analysis, continuing fallacies, confusions and misunderstandings and some suggested remedies. In: *Proceedings of CoDaWork'08*.

Aitchison, J., and J.J. Egozcue. 2005. Compositional data analysis: Where are we and where should we be heading? *Mathematical Geology* 37 (7): 829–850. https://doi.org/10.1007/s11004-005-7383-7.

Aitchison, J., C. Barceló-Vidal, J.A. Martín-Fernández, and V. Pawlowsky-Glahn. 2000. Logratio analysis and compositional distance. *Mathematical Geology* 32 (3): 271–275.

Altman, D.G., and J.M. Bland. 1983. Measurement in medicine: The analysis of method comparison studies. *Journal of the Royal Statistical Society. Series D (The Statistician)* 32 (3): 307–317. https://doi.org/10.2307/2987937, http://www.jstor.org/stable/2987937.

Anders, Simon, and Wolfgang Huber. 2010. Differential expression analysis for sequence count data. *Genome Biology* 11 (10): R106. https://doi.org/10.1186/gb-2010-11-10-r106.

Anders, S., D.J. McCarthy, Y. Chen, M. Okoniewski, G.K. Smyth, W. Huber, and M.D. Robinson. 2013. Count-based 631 differential expression analysis of RNA sequencing data using R and Bioconductor. *Nature Protocols* 8. https://doi.org/10.1038/nprot.2013.099.

Bacon-Shone, J. 2008. Discrete and continuous compositions. In *Proceedings of CODAWORK'08, the 3rd compositional data analysis workshop*, ed. J. Daunis-i-Estadella and J. Martin-Fernández. Girona: University of Girona.

————. 2011. A short history of compositional data analysis. In *Compositional data analysis: Theory and applications*, ed. V. Pawlowsky-Glahn and A. Buccianti. Chichester: Wiley.

Barceló-Vidal, Carles, Josep A Martín-Fernández, and Vera Pawlowsky-Glahn. 2001. Mathematical foundations of compositional data analysis. In *Proceedings of IAMG*.

Benjamini, Y., and Y. Hochberg. 1995. Controlling the false discovery rate: A practical and powerful approach to multiple testing. *Journal of the Royal Statistical Society, Series B (Methodology)* 57: 289–300.

Benjamini, Y., and D. Yekutieli. 2001. The control of the false discovery rate in multiple testing under dependency. *Annals of Statistics* 29: 1165–1188.

Bland, J. Martin, and Douglas G. Altman. 1999. Measuring agreement in method comparison studies. *Statistical Methods in Medical Research* 8 (2): 135–160. https://doi.org/10.1177/096228029900800204.

Bonferroni, C. E. 1936. Teoria statistica delle classi e calcolo delle probabilità. Pubblicazioni del R Istituto Superiore di Scienze Economiche e Commerciali di Firenze.

Brill, Barak, Amnon Amir, and Ruth Heller. 2019. Testing for differential abundance in compositional counts data, with application to microbiome studies. *arXiv preprint arXiv*: 1904.08937.

Cohen, J. 1988. *Statistical power analysis for the behavioral sciences*. 2nd ed. Hillsdale: Erlbaum.

Dunn, Olive Jean. 1958. Estimation of the means of dependent variables. *The Annals of Mathematical Statistics* 29: 1095–1111.

————. 1961. Multiple comparisons among means. *Journal of the American statistical association* 56 (293): 52–64.

Eaton, Morris L. 1983. *Multivariate statistics: A vector space approach*, 512. New York: Wiley.

Egozcue, Juan José. 2009. Reply to "On the Harker Variation Diagrams; ..." by J.A. Cortés. *Mathematical Geosciences* 41 (7): 829–834. https://doi.org/10.1007/s11004-009-9238-0.

Egozcue, Juan José, and Vera Pawlowsky-Glahn. 2005. Groups of parts and their balances in compositional data analysis. *Mathematical Geology* 37 (7): 795–828.

Egozcue, J.J., and V. Pawlowsky-Glahn. 2011. Basic concepts and procedures. In *Compositional data analysis: Theory and applications*, ed. V. Pawlowsky-Glahn and A. Buccianti. Chichester: Wiley.

Egozcue, Juan José, Vera Pawlowsky-Glahn, Glòria Mateu-Figueras, and Carles Barcelo-Vidal. 2003. Isometric logratio transformations for compositional data analysis. *Mathematical Geology* 35 (3): 279–300.

Farnan, Laura, Anastasia Ivanova, and Shyamal D. Peddada. 2014. Linear mixed effects models under inequality constraints with applications. *PloS One* 9 (1): e84778.

Fernandes, A.D., J.M. Macklaim, T.G. Linn, G. Reid, and G.B. Gloor. 2013. ANOVA-like differential expression (ALDEx) analysis for mixed population RNA-seq. *PLoS One* 8. https://doi.org/10.1371/journal.pone.0067019.

Fernandes, Andrew D., Jennifer Ns Reid, Jean M. Macklaim, Thomas A. McMurrough, David R. Edgell, and Gregory B. Gloor. 2014. Unifying the analysis of high-throughput sequencing datasets: characterizing RNA-seq, 16S rRNA gene sequencing and selective growth experiments by compositional data analysis. *Microbiome* 2: 15. https://doi.org/10.1186/2049-2618-2-15, https://pubmed.ncbi.nlm.nih.gov/24910773, https://www.ncbi.nlm.nih.gov/pmc/articles/PMC4030730/.

Fernandes, Andrew D., Michael T.H.Q. Vu, Lisa-Monique Edward, Jean M. Macklaim, and Gregory B. Gloor. 2018. A reproducible effect size is more useful than an irreproducible hypothesis test to analyze high throughput sequencing datasets. *arXiv preprint arXiv*: 1809.02623.

Friedman, Jonathan, and Eric J. Alm. 2012. Inferring Correlation Networks from Genomic Survey Data. *PLOS Computational Biology* 8 (9): e1002687. https://doi.org/10.1371/journal.pcbi.1002687.

Gloor, Gregory B., and Gregor Reid. 2016. Compositional analysis: a valid approach to analyze microbiome high-throughput sequencing data. *Canadian Journal of Microbiology* 62 (8): 692–703. https://doi.org/10.1139/cjm-2015-0821.

Gloor, G.B., J.R. Wu, V. Pawlowsky-Glahn, and J.J. Egozcue. 2016. It's all relative: analyzing microbiome data as compositions. *Annals of Epidemiology* 26 (5): 322–329.

Hawinkel, Stijn, Federico Mattiello, Luc Bijnens, and Olivier Thas. 2017. A broken promise: microbiome differential abundance methods do not control the false discovery rate. *Briefings in Bioinformatics* 20 (1): 210–221. https://doi.org/10.1093/bib/bbx104.

Hochberg, Y. 1988. A sharper Bonferroni procedure for multiple tests of significance. *Biometrika* 75: 800–803.

Holm, S. 1979. A simple sequentially rejective multiple test procedure. *Scandinavian Journal of Statistics* 6: 65–70.

Hron, K., M. Templ, and P. Filzmoser. 2010. Exploratory compositional data analysis using the R-package robCompositions. In *Proceedings 9th international conference on computer data analysis and modeling*, ed. S. Aivazian, P. Filzmoser, and Yu. Kharin, 179–186. Minsk: Belarusian State University.

Hŭ, J., H. Koh, L. He, M. Liu, M.J. Blaser, and H. Li. 2018. A two-stage microbial association mapping framework with advanced FDR control. *Microbiome* 6 (1): 131. https://doi.org/10.1186/s40168-018-0517-1.

Kaul, Abhishek, Siddhartha Mandal, Ori Davidov, and Shyamal D. Peddada. 2017. Analysis of microbiome data in the presence of excess zeros. *Frontiers in Microbiology* 8 (2114). https://doi.org/10.3389/fmicb.2017.02114, https://www.frontiersin.org/article/10.3389/fmicb.2017.02114.

Krumbein, W.C. 1962. Open and closed number systems in stratigraphic mapping. *AAPG Bulletin* 46 (12): 2229–2245.

Kuczynski, J., C.L. Lauber, W.A. Walters, L.W. Parfrey, J.C. Clemente, D. Gevers, and R. Knight. 2011. Experimental and analytical tools for studying the human microbiome. *Nature Reviews Genetics* 13 (1): 47–58.

Li, Hongzhe. 2015. Microbiome, metagenomics, and high-dimensional compositional data analysis. *Annual Review of Statistics and Its Application* 2: 73–94.

Lim, Changwon, Pranab K. Sen, and Shyamal D. Peddada. 2013. Robust analysis of high throughput screening (HTS) assay data. *Technometrics* 55 (2): 150–160. https://doi.org/10.1080/00401706.2012.749166, https://pubmed.ncbi.nlm.nih.gov/23908557, https://www.ncbi.nlm.nih.gov/pmc/articles/PMC3727440/.

Lin, Huang, and Shyamal Das Peddada. 2020a. Analysis of compositions of microbiomes with bias correction. *Nature Communications* 11 (1): 3514. https://doi.org/10.1038/s41467-020-17041-7.

———. 2020b. Analysis of microbial compositions: a review of normalization and differential abundance analysis. *npj Biofilms and Microbiomes* 6 (1): 60. https://doi.org/10.1038/s41522-020-00160-w.

Lovell, D., W. Müller, J. Taylor, A. Zwart, and C. Helliwell. 2011. Proportions, percentages, PPM: Do the molecular biosciences treat compositional data right? In *Compositional data analysis: Theory and applications*, ed. V. Pawlowsky-Glahn and A. Buccianti. Chichester: Wiley.

Lovell, David, Vera Pawlowsky-Glahn, Juan José Egozcue, Samuel Marguerat, and Jürg Bähler. 2015. Proportionality: A valid alternative to correlation for relative data. *PLoS Computational Biology* 11 (3): e1004075. https://doi.org/10.1371/journal.pcbi.1004075, http://www.ncbi.nlm.nih.gov/pmc/articles/PMC4361748/.

Macklaim, Jean M., Andrew D. Fernandes, Julia M. Di Bella, Jo-Anne Hammond, Gregor Reid, and Gregory B. Gloor. 2013. Comparative meta-RNA-seq of the vaginal microbiota and differential expression by Lactobacillus iners in health and dysbiosis. *Microbiome* 1 (1): 12. https://doi.org/10.1186/2049-2618-1-12.

Mandal, Siddhartha, Will Van Treuren, Richard A. White, Merete Eggesbø, Rob Knight, and Shyamal D. Peddada. 2015. Analysis of composition of microbiomes: a novel method for studying microbial composition. *Microbial Ecology in Health and Disease* 26 (1): 27663. https://doi.org/10.3402/mehd.v26.27663.

Martin Bland, J., and Douglas G. Altman. 1986. Statistical methods for assessing agreement between two methods of clinical measurement. *The Lancet* 327 (8476): 307–310. https://doi.org/10.1016/S0140-6736(86)90837-8.

Martín-Fernández, J.A., C. Barceló-Vidal, and V. Pawlowsky-Glahn. 2003. Dealing with zeros and missing values in compositional data sets using nonparametric imputation. *Mathematical Geology* 35 (3): 253–278. https://doi.org/10.1023/A:1023866030544.

Mateu-Figueras, G., V. Pawlowsky-Glahn, and J.J. Egozcue. 2011. The principle of working on coordinates. In *Compositional data analysis: Theory and applications*, ed. V. Pawlowsky-Glahn and A. Buccianti. Chichester: Wiley.

Morton, James T., Clarisse Marotz, Alex Washburne, Justin Silverman, Livia S. Zaramela, Anna Edlund, Karsten Zengler, and Rob Knight. 2019. Establishing microbial composition measurement standards with reference frames. *Nature Communications* 10 (1): 2719. https://doi.org/10.1038/s41467-019-10656-5.

Nelson, T.M., J.L. Borgogna, R.M. Brotman, J. Ravel, S.T. Walk, and C.J. Yeoman. 2015. Vaginal biogenic amines: biomarkers of bacterial vaginosis or precursors to vaginal dysbiosis? *Frontiers in Physiology* 6: 253. https://doi.org/10.3389/fphys.2015.00253.

Pawlowsky-Glahn, V., and A. Buccianti. 2011. In *Compositional data analysis: Theory and applications*, ed. V. Pawlowsky-Glahn and A. Buccianti. Wiley, Chichester.

Pawlowsky-Glahn, V., J.J. Egozcue, and R. Tolosana-Delgado. 2015. *Modeling and analysis of compositional data*. London: Wiley.

Pearson, K. 1897. Mathematical contributions to the theory of evolution. On a form of spurious correlation which may arise when indices are used in the measurement of organs. *Proceedings of the Royal Society of London* LX: 489–502.

Peddada, Shyamal D., Katherine E. Prescott, and Mark Conaway. 2001. Tests for order restrictions in binary data. *Biometrics* 57 (4): 1219–1227.

Quinn, Thomas P., Mark F. Richardson, David Lovell, and Tamsyn M. Crowley. 2017. propr: An R-package for identifying proportionally abundant features using compositional data analysis. *Scientific Reports* 7 (1): 16252. https://doi.org/10.1038/s41598-017-16520-0.

Quinn, Thomas P., Ionas Erb, Mark F. Richardson, and Tamsyn M. Crowley. 2018a. Understanding sequencing data as compositions: an outlook and review. *Bioinformatics* 34 (16): 2870–2878. https://doi.org/10.1093/bioinformatics/bty175.

Quinn, Thomas P., Tamsyn M. Crowley, and Mark F. Richardson. 2018b. Benchmarking differential expression analysis tools for RNA-Seq: normalization-based vs. log-ratio transformation-based methods. *BMC Bioinformatics* 19 (1): 274. https://doi.org/10.1186/s12859-018-2261-8.

Rivera-Pinto, J., J.J. Egozcue, V. Pawlowsky-Glahn, R. Paredes, M. Noguera-Julian, and M.L. Calle. 2018. Balances: A new perspective for microbiome analysis. *mSystems* 3 (4): e00053–e00018. https://doi.org/10.1128/mSystems.00053-18. https://pubmed.ncbi.nlm.nih.gov/30035234, https://www.ncbi.nlm.nih.gov/pmc/articles/PMC6050633/.

Robinson, Mark D., and Alicia Oshlack. 2010. A scaling normalization method for differential expression analysis of RNA-seq data. *Genome Biology* 11 (3): R25. https://doi.org/10.1186/gb-2010-11-3-r25, http://www.ncbi.nlm.nih.gov/pmc/articles/PMC2864565/.

Robinson, M.D., D.J. McCarthy, and G.K. Smyth. 2010. edgeR: A bioconductor package for differential expression analysis of digital gene expression data. *Bioinformatics* 26 (1): 139–140. https://doi.org/10.1093/bioinformatics/btp616.

Seyednasrollah, Fatemeh, Asta Laiho, and Laura L. Elo. 2013. Comparison of software packages for detecting differential expression in RNA-seq studies. *Briefings in Bioinformatics* 16 (1): 59–70. https://doi.org/10.1093/bib/bbt086.

Tarazona, Sonia, Pedro Furió-Tarí, David Turrà, Antonio Di Pietro, María José Nueda, Alberto Ferrer, and Ana Conesa. 2015. Data quality aware analysis of differential expression in RNA-seq with NOISeq R/Bioc package. *Nucleic Acids Research* 43 (21): e140–e140. https://doi.org/10.1093/nar/gkv711. https://pubmed.ncbi.nlm.nih.gov/26184878, https://www.ncbi.nlm.nih.gov/pmc/articles/PMC4666377/.

Thomas, C.W., and John Aitchison. 2006. Log-ratios and geochemical discrimination of Scottish Dalradian limestones: A case study. *Geological Society, London, Special Publications* 264 (1): 25–41.

Tsilimigras, Matthew C.B., and Anthony A. Fodor. 2016. Compositional data analysis of the microbiome: Fundamentals, tools, and challenges. *Annals of Epidemiology* 26 (5): 330–335. https://doi.org/10.1016/j.annepidem.2016.03.002, http://www.sciencedirect.com/science/article/pii/S1047279716300722.

Urbaniak, Camilla, Michelle Angelini, Gregory B. Gloor, and Gregor Reid. 2016. Human milk microbiota profiles in relation to birthing method, gestation and infant gender. *Microbiome* 4 (1): 1. https://doi.org/10.1186/s40168-015-0145-y.

van den Boogaart, K. Gerald, and Raimon Tolosana-Delgado. 2008. "Compositions": A unified R package to analyze compositional data. *Computers & Geosciences* 34 (4): 320–338. https://doi.org/10.1016/j.cageo.2006.11.017, https://www.sciencedirect.com/science/article/pii/S0098300400701001X.

van den Boogaart, G.K., and R. Tolosana-Delgado. 2013a. *Analyzing compositional data with R*. Berlin/Heidelberg: Springer-Verlag.

van den Boogaart, K. Gerald, and Raimon Tolosana-Delgado. 2013b. Fundamental concepts of compositional data analysis. In *Analyzing compositional data with R*, 13–50. Springer.

Wallen, Zachary D. 2021. Comparison study of differential abundance testing methods using two large Parkinson disease gut microbiome datasets derived from 16S amplicon sequencing. *BMC Bioinformatics* 22 (1): 1–29.

Wang, Shulei. 2021. Robust differential abundance test in compositional data. *arXiv preprint arXiv*: 2101.08765.

Washburne, Alex D., Justin D. Silverman, Jonathan W. Leff, Dominic J. Bennett, John L. Darcy, Sayan Mukherjee, Noah Fierer, and Lawrence A. David. 2017. Phylogenetic factorization of compositional data yields lineage-level associations in microbiome datasets. *PeerJ* 5: e2969. https://doi.org/10.7717/peerj.2969.

Webster's New World Dictionary, ed. 2005. Webster's II new college dictionary. Houghton Mifflin Harcourt. p.236

Weiss, Sophie, Xu Zhenjiang Zech, Shyamal Peddada, Amnon Amir, Kyle Bittinger, Antonio Gonzalez, Catherine Lozupone, Jesse R. Zaneveld, Yoshiki Vázquez-Baeza, Amanda Birmingham, Embriette R. Hyde, and Rob Knight. 2017. Normalization and microbial differential abundance strategies depend upon data characteristics. *Microbiome* 5 (1): 27. https://doi.org/10.1186/s40168-017-0237-y.

Williams, D.A. 1971. A test for differences between treatment means when several dose levels are compared with a zero dose control. *Biometrics* 27: 103–117.

Williams, David A. 1977. Some inference procedures for monotonically ordered normal means. *Biometrika* 64 (1): 9–14.

Williams, Claire R., Alyssa Baccarella, Jay Z. Parrish, and Charles C. Kim. 2017. Empirical assessment of analysis workflows for differential expression analysis of human samples using RNA-Seq. *BMC Bioinformatics* 18 (1): 38. https://doi.org/10.1186/s12859-016-1457-z.

Xia, Y. 2020. Correlation and association analyses in microbiome study integrating multiomics in health and disease. *Progress in Molecular Biology and Translational Science* 171: 309–491. https://doi.org/10.1016/bs.pmbts.2020.04.003.

Xia, Y., and J. Sun. 2022a. *An integrated analysis of microbiomes and metabolomics*. American Chemical Society.

———. 2022b. *Statistical data analysis of microbiomes and metabolomics*. American Chemical Society.

Xia, Yinglin, Jun Sun, and Ding-Geng Chen. 2018a. Compositional analysis of microbiome data. In *Statistical analysis of microbiome data with R*, 331–393. Singapore: Springer Singapore.

———. 2018b. Modeling zero-inflated microbiome data. In *Statistical analysis of microbiome data with R*, ed. Yinglin Xia, Jun Sun, and Ding-Geng Chen, 453–496. Singapore: Springer Singapore.

———. 2018c. What are microbiome data? In *Statistical analysis of microbiome data with R*, ed. Yinglin Xia, Jun Sun, and Ding-Geng Chen, 29–41. Singapore: Springer Singapore.

Zhang, X., and N. Yi. 2020. Fast zero-inflated negative binomial mixed modeling approach for analyzing longitudinal metagenomics data. *Bioinformatics* 36 (8): 2345–2351. https://doi.org/10.1093/bioinformatics/btz973.

Zhang, Yilong, Sung Won Han, Laura M. Cox, and Huilin Li. 2017. A multivariate distance-based analytic framework for microbial interdependence association test in longitudinal study. *Genetic Epidemiology* 41 (8): 769–778. https://doi.org/10.1002/gepi.22065, https://www.ncbi.nlm.nih.gov/pubmed/28872698, https://www.ncbi.nlm.nih.gov/pmc/articles/PMC5696116/.

Chapter 15
Linear Mixed-Effects Models for Longitudinal Microbiome Data

Abstract Longitudinal microbiome data analysis can be categorized into two approaches: univariate longitudinal analysis and multivariate longitudinal analysis. Univariate analysis analyzes the change of one taxon or alpha diversity over time. Multivariate analysis directly analyzes the change of multiple taxa simultaneously or distance/dissimilarity (beta diversities) over time. This chapter introduces using the classical univariate linear mixed-effects models (LMMs) to analyze microbiome data. First, it describes linear mixed-effects models (LMMs), and then it introduces implementation of LMMs to identify the significant taxa using the **nlme** package, and model the diversity indices using the **lme4** and **LmerTest** packages, respectively. Next, it introduces how to fit LMMs using QIIME 2 and provides some remarks on longitudinal microbiome studies based on LMMs.

Keywords Linear mixed-effects models (LMMs) · Fixed effect · Random effect · Hypothesis tests · nlme package · lme4 package · LmerTest package · Maximum likelihood (ML) · AIC · BIC · Restricted maximum likelihood (REML) · Generalized linear model (GLM) · Autoregressive of order 1 [AR(1)] · KRmodcomp() · pbkrtest · qiime longitudinal linear-mixed-effects · Volatility analysis

We can categorize longitudinal microbiome data analysis into univariate and multivariate approaches. Univariate analysis analyzes one taxon per time or first summarizes microbiome abundances into alpha diversities and then analyzes these diversities one by one. Multivariate analysis either analyzes multiple taxa simultaneously or directly tests distance/dissimilarity (beta diversities). This chapter focuses on introducing how to use the classical univariate linear mixed-effects models (LMMs) to analyze microbiome data. First, we describe linear mixed-effects models (LMMs) (Sect. 15.1), then we introduce how to implement LMMs to identify the significant taxa using the **nlme** package (Sect. 15.2) and model the diversity indices using the **lme4** and **LmerTest** packages (Sect. 15.3), respectively. In Sect. 15.4, we

---

Supplementary Information The online version contains supplementary material available at https://doi.org/10.1007/978-3-031-21391-5_15.

introduce how to fit LMMs using QIIME 2. In Sect. 15.5, we provide some remarks on longitudinal microbiome studies based on LMMs. Finally, we summarize the topics of this chapter (Sect. 15.6).

15.1   Introduction to Linear Mixed-Effects Models (LMMs)

In this section, we introduce the advantages of LMMs, the development of fixed and random effects model and definition of LMMs, as well as how to conduct statistical hypothesis testing and fit LMMs.

15.1.1   Advantages and Disadvantages of LMMs

The distinctive characteristic of longitudinal study is that the subjects are measured repeatedly during the study, permitting directly assess the changes in response variable over time (Fitzmaurice et al. 2004; Diggle et al. 2002). Thus, a longitudinal study can capture between-individual differences (heterogeneity among individuals) and within-subject dynamics. Linear mixed-effects models (LMMs, aka multi-level modeling) are an important class of statistical models incorporating fixed and random effects. LMMs as a statistical method can be used to analyze correlated or non-independent data. Such data are often collected in the settings of longitudinal/ repeated measurements in which the same statistical units were repeatedly measured or clustered observations where the clusters of related statistical units were measured. Such data also arise from a multilevel/hierarchical structure.

Because the circumstances of the measurements often cannot be fully controlled, considerable variation among individuals in the number and timing of observations may exist, hence resulting in unbalanced datasets. Typically, such unbalanced datasets are challenge to be analyzed using a general multivariate model with unrestricted covariance structure (Laird and Ware 1982). Moreover, missing values often arise in longitudinal/repeated measurements. Traditional statistical methods such as repeated measure analysis of variance are difficult to deal with missing values.

Mixed effects models often have advantages in dealing with unbalanced datasets and missing values.

LMMs integrate two (hierarchical) levels of observations of longitudinal data, i.e., within-group (subject) and between-group (subject) in a single model (Harville 1977; Arnau et al. 2010) and allows a variety of variance/covariance structures or correlation patterns to be explicitly modeled, which provides the opportunity to accurately capture the individual profile information over time. Typically the fixed effects are used to model the systematic mean patterns (i.e., treatment conditions), while random effects are used to model two types of random components: the correlation patterns between repeated measures within subjects or heterogeneities between subjects or both (Lee et al. 2006). That is, random effects estimate the variance of the response variable within and among these groups to reduce the probability of false positives (Type I error rates) and false negatives (Type II error rates) (Crawley 2012; Harrison et al. 2018).

In summary, the desirable features of LMMs are that they are not required for balance in the data and allow explicitly modeling and analysis of between- and within-individual variation (Laird and Ware 1982). Thus, LMMs have the capability of modeling variance and covariance (random effects) which makes this method more preferred to the classical linear model (Xia and Sun 2021). However, LMMs have the major limitation compared to the general multivariate model: a special form of the covariance structure should be assumed (Laird and Ware 1982).

15.1.2   Fixed and Random Effects

The core feature of LMMs is that they incorporate fixed and random effects and hence can analyze the fixed and random effects simultaneously. A fixed effect refers to a parameter that does not vary, whereas a random effect represents a parameter that is itself a random variable. In other words, the distribution of these parameters (or "random effects") vary over individuals (Laird and Ware 1982).

LMMs have a long history of development. In 1918 Ronald Fisher introduced the random effects models to study the correlations of trait values between relatives such as the correlation of parent and child (Fisher 1918; Laird and Ware 1982). Since the 1930s, researchers have emphasized taking account of random effects as accurately as possible when estimating experimental treatment effects (fixed effects) (Thompson 2008). In the 1950s, Charles Roy Henderson provided best linear unbiased estimates (BLUE) of fixed effects and best linear unbiased predictions (BLUP) of random effects (Robinson 1991; Henderson et al. 1959; McLean et al. 1991). Subsequently, Goldberger (1962), Henderson (Henderson 1963, 1973, 1984; Henderson et al. 1959), and Harville (1976a, b) developed mixed model procedures. The mixed model equations (Henderson 1950; Henderson et al. 1959) introduced by Charles Roy Henderson offers the base for a methodology that provides flexibility of fitting models with various fixed and random elements with the possible assumption of correlation among random effects (McLean et al. 1991; Thompson 2008).

15.1.3   Definition of LMMs

In matrix form a linear mixed model can be specified as:

$$y = X\beta + Zu + \varepsilon, \ u \sim N_q(0, G), \ \varepsilon \sim N_n(0, R), \tag{15.1}$$

where y is a $n \times 1$ column vector, the response (outcome) variable; X is a $n \times p$ design matrix of fixed-effects parameters (i.e., for the p predictor variables); β is a $p \times 1$ column vector of the fixed-effects parameters representing regression coefficients (the βs); Z is the $n \times q$ design matrix for the q random effects; u is a $q \times 1$ vector of

q random effects (the random complement to the fixed β); and ε is a $n \times 1$ column vector of the residuals, which is part of y that is not explained by the model, $X\beta + Zu$. u and ε are independent and $R = \sigma^2 I$.

G is the variance-covariance matrix of the random effects. Given the fixed effects are directly estimated, the random effects are just deviations around the value in β, which is the mean. Thus it the variance that is left to estimate. Various structures of the variance-covariance matrix of the random effects can be assumed and specified. If only a random intercept is specified, then G is just a 1×1 matrix, which models the variance of the random intercept. If a random intercept and a random slope are assumed, then we can specify:

$$G = \begin{bmatrix} \sigma^2_{int} & \sigma^2_{int,slope} \\ \sigma^2_{int,slope} & \sigma^2_{slope} \end{bmatrix}.$$ If we assume that the random effects are *independent*,

then the true structure is $G = \begin{bmatrix} \sigma^2_{int} & 0 \\ 0 & \sigma^2_{slope} \end{bmatrix}$. In Chap. 16 (Sect. 16.2), we will

introduce generalized linear mixed models (GLMMs).

15.1.4 Statistical Hypothesis Tests

In the statistical model building for experimental or observational data, LMMs can be used to serve two purposes (Bates 2005): (1) to characterize the dependence of a response (outcome variable) on covariate(s), conditioning on the treatment status and the time under treatment, and (2) to characterize the "unexplained" variation in the response.

A mixed-effects model (or more simply a mixed model) incorporates both fixed-effects terms and random-effects terms, which use a different way to model repeatable covariates and non-repeatable covariates (Bates 2005; Bates et al. 2015). The fixed-effects terms are used for a repeatable covariate to characterize the change in the response between different levels such as typically the change of the response over time under treatment or the difference of a response between treatment and the control groups. In contrast, the random-effects terms are used to characterize the variation induced in the response by the different levels of the non-repeatable covariate (a grouping factor).

Hypothesis Testing of the Fixed-Effects

To test a hypothesis on the fixed effects β, we can use likelihood-ratio test (LRT). The LRT can be constructed by specifying a nested (smaller) model with the same error structure as model (15.1).

$$y = X_0 \beta_0 + Zu + \varepsilon. \tag{15.2}$$

The LRT statistic for the test of the hypothesis is given:

$$H_0 = \beta \in \Theta_{\beta_0},$$
$$H_a = \beta \in \Theta_{\beta},$$

where Θ_{β_0} is a subspace of the parameter space Θ_{β} of the fixed effects β. The log-likelihood ratio test statistic for the null hypothesisis is given by:

$$\lambda_{\text{LRT}} = 2(ll - ll_0),$$

where ll and ll_0 are the log-likelihoods of the models in Eqs. (15.1) and (15.2), respectively. Under the hull hypothesis, λ_{LRT} follows asymptotically a χ^2 distribution. LRT is often used, but it tends to produce "anticonservative" P-values and sometimes quite badly so (Pinheiro and Bates 2000) (p. 88). Thus, an F test (although overall it does not exactly follow an F distribution) of the null hypothesis about the fixed effects β: $H_0 : L\beta = 0$, is considered in the literature with L is a contrast matrix of $q = \text{rank}(L) > 1$. The test statistic for this null hypothesis is given:

$$F = \frac{\left(L\hat{\beta}\right)^T \left(L\hat{C}L^T\right)^{-1} \left(L\hat{\beta}\right)}{q}, \tag{15.3}$$

We will describe more details about the LRT and other tests in Chap. 16 (Sect. 16.5.4 for the LRT).

15.1.5 How to Fit LMMs

To fit a LMM, we typically should start with a full model, i.e., including all independent variables of interest and some interaction terms of main factors. Then we can evaluate parameters and compare sub-models. In literature, some protocols and procedures have been proposed and discussed to avoid common and particularly statistical problems related to mixed-effects modeling and multi-model inference in ecology (Zuur et al. 2010; Zuur and Ieno 2016; Harrison et al. 2018). Here, we describe some main points based on our experience using LMMs and the discussion in the literature.

First of all, exactly determining whether a variable is a random or fixed is controversial.

Specifying a particular variable as fixed or random effect or even both simultaneously largely depends on experimental design and context. Generally fixed effects refer to all factors whose levels are experimentally determined or whose interest is in the specific effects of each level, such as treatments, time, other covariates, as well as treatment and time (or covariates) interactions. In contrast, random effects refer to all factors that qualify as sampling from a population or whose interest is in the variation among the population rather than the specific effects of each level. Random effects typically represent some grouping variables in both subject and unit levels (Breslow and Clayton 1993), such as individuals in repeated measurements, field trials, plots, blocks, batches.

Random effect models have several desirable properties, but it is more challenging to use them. Harrison et al. (2018) very well described four considerations when fitting random effects. We summarized them here:

1. Fitting random effects requires at least five "levels" (groups) for a random intercept term to achieve robust estimates of variance (Gelman and Hill 2007; Harrison 2015). The mixed model may not be able to estimate the among-group variance accurately for less than 5 levels (Harrison et al. (2018) because if the random effect models have less than 5 levels, then the variance estimate will either collapse to zero, making the model become an ordinary GLM (Gelman and Hill 2007) or be non-zero which is incorrect if the small number of groups that were sampled are not representative of true distribution of means (Harrison 2015). Thus, in practice, as a rule of thumb, the factors with fewer than 5 levels are considered as fixed effects, while the factors with numerous levels are considered as random effects in order to increase accurately estimating variance.
2. Models and especially the random slope models will be unstable if sample sizes across groups are highly unbalanced (Grueber et al. 2011).
3. It is difficult to determine the significance or importance of variance among groups.
4. Mis-specification of random effects or incorrectly parameterizing the random effects in the model could results in the model estimates being unreliable as ignoring the need for random effects altogether (Harrison et al. 2018).

Second, start with a full model to include all potential independent variables in the fixed model using maximum likelihood (ML).

Depending on experimental design and hypothesis, include as many as possible predictors and their interactions in the fixed-effect model. A linear regression (ordinary least squares) model can be fitted to compare with the fixed effect only model. The fixed-effect model is fitted under the statistical framework of ML to find the parameters of a model that maximizes the probability of the observed data (the likelihood). It is suggested to use information criteria (e.g., AIC and/or BIC, see Chap. 16 for details) to compare the model fit and perform variable selection. Because LRT is unreliable for small to moderate sample sizes, it is not recommended for testing fixed effects in GLMMs (Pinheiro and Bates 2000). To test fixed effects or compare any models with different fixed effects using LRT, the model is required to be fitted with ML too. Based on the significant fixed effects using ML estimation, an optimal fixed model structure will be determined, including the important main effects and their interactions. For more details on model selection in GLMMs, please see Chaps. 16 and 17.

Third, fit the random model using restricted maximum likelihood (REML) to optimize the random structure.

The goal of this step is to find the optimal variance structure in terms of heterogeneity. Thus, it is required to fit the model with REML to estimate the random-effect parameters (i.e., standard deviations) over the averaged values of the fixed-effect parameters. Because REML assumes that the fixed effects are unknown, thus generally REML estimation is less biased compared to corresponding ML estimation (Veroniki et al. 2016; Zuur et al. 2009). An identical generalized linear model (GLM) with REML can be fitted to serve as a reference for the fitted LMMs. Based on the significant random effects using REML estimation, an optimal random (variance) model structure will be determined. That is, the random intercept, random slopes and crossed or nested effect can be specified as the random effects. An error structure can also be chosen.

Finally, refit the optimal fixed and random models using REML, and validate and interpret the modeling results.

After updating (refitting) optimal model with REML, we can check the differences in the main effects and effects in interaction terms caused by the introduction of random effects. A diagnosing model fit procedure can be conducted to validate the fitted models. The reader is referred to Chap. 17 for details. We then can interpret the results based on the *prior* hypotheses.

15.2   Identifying the Significant Taxa Using the nlme Package

In this section, we introduce LMMs in microbiome research and illustrate how to fit LMMs to identify significant microbial taxa.

15.2.1   Introduction to LMMs in Microbiome Research

The microbiome is inherently dynamic, driven by interactions with the host and the environment, and varies over time. Longitudinal microbiome data analysis can enhance our understanding of short- and long-term trends of microbiome by intervention, such as diet, and the development and persistence of chronic diseases caused by microbiome. Thus, longitudinal microbiome data analysis provides rich information on the profile of microbiome with host and environment interactions (Xia et al. 2018a).

In the literatures of longitudinal statistical analysis of physical, biological, psychological, social, and behavioral sciences, the linear mixed-effects models (LMMs) (Laird and Ware 1982) are often used. The classical LMM method was also one of the first models that were used to analyze longitudinal microbiome data (e.g., Rosa et al. 2014; Kostic et al. 2015; DiGiulio et al. 2015; Wang et al. 2015) and currently is still used in microbiome and multi-omics literature (Lloyd-Price et al. 2019). The reasons that LMMs have been chosen to analyze longitudinal microbiome data mainly because (1) LMMs equip with fixed- and random-effects components, which provides a standardized and flexible approach to model both fixed- and random-effects. (2) LMMs can jointly model measures over all the time points to account for time-dependent correlations in longitudinal microbiome study designs. Specifically because (3) LMM method was an established methodology to remove the effects of fixed- and random-effect confounding variables (Ernest et al. 2012; Xia 2020) in the areas of microarray, genome-wide association studies (GWAS), and metabolomics (Fabregat-Traver et al. 2014; Zhao et al. 2019), which motivates its application to microbiome data.

15.2.2   Longitudinal Microbiome Data Structure

Microbiome data (i.e., OTUs/ASVs table) generated by each sample through either the 16S rRNA gene sequencing or the shotgun metagenomics sequencing consist of read counts of taxa at certain taxonomic levels, such as kingdom, phylum, class, order, family, genus, and species. In longitudinal microbiome study, each subject are measured at multiple time points (i.e., samples). Assume that there are n subjects, and subject i is measured at n_i time points t_{ij}, $j = 1, \ldots, n_i$ and $i = 1, \ldots, n$. Denote C_{ijm} as read counts for subject i measured at time j for taxon m, $m = 1, \ldots, m$, and denote the total sequence read counts for subject i at time j as T_{ij}, which is referred to as library size or depths of coverage; then a longitudinal microbiome data structure can be summarized in Table 15.1.

In Table 15.1, X_i denotes clinical or environmental variables for each subject. Under this data structure and notations, we can formulate a longitudinal microbiome model to identify the associations between the microbiome counts and the host trait of interest (i.e., covariate variables) and characterize the time trends of microbiome abundance both within subjects and between subjects.

Table 15.1 Longitudinal microbiome data structure

Subject-ID	Taxon 1	Taxon 2	$\cdots$	Taxon m	Total reads	Covariate	Time
Subject 1	C_{111}	C_{112}	$\cdots$	C_{11m}	T_{11}	X_1	t_{11}
Subject 1	C_{121}	C_{122}	$\cdots$	C_{12m}	T_{12}	X_1	t_{12}
Subject 1	C_{131}	C_{132}	$\cdots$	C_{13m}	T_{13}	X_1	t_{13}
Subject 2	C_{211}	C_{212}	$\cdots$	C_{21m}	T_{21}	X_2	t_{21}
Subject 2	C_{221}	C_{222}	$\cdots$	C_{22m}	T_{22}	X_2	t_{22}
Subject 2	C_{231}	C_{232}	$\cdots$	C_{23m}	T_{23}	X_2	t_{23}
$\vdots$	$\vdots$	$\vdots$	$\vdots$	$\vdots$	$\vdots$	$\vdots$	$\vdots$
Subject n	C_{n11}	C_{n12}	$\cdots$	C_{n1m}	T_{n1}	X_n	T_{n1}

To analyze the associations between the microbiome and the host trait of interest (i.e., covariate variables), two approaches can be applied via LMMs: (1) fit LMMs using the read counts as the outcome and (2) fit LMMs using the diversity index as the outcome. The LMMs can be fitted using the **nlme** (current version 3.1-157, March 2022) and **lme4** (current version 1.1-29, April 2022) packages. Below we illustrate their uses through our dataset of breast cancer QTRT1 mouse gut microbiome.

15.2.3   Fit LMMs Using the Read Counts as the Outcome

When applying LMM to fit microbiome and multi-omics abundance data, raw microbiome abundances should be transformed or normalized by an appropriate method (Xia 2020): (1) arcsine square root transformation (Kostic et al. 2015; Rosa et al. 2014), (2) log-transformation (with pseudocount 1 for zero values) (Lloyd-Price et al. 2019), and (3) log-transformation with no pseudocount for expression ratios (non-finite values removed) (Lloyd-Price et al. 2019).

It is arguable whether or not these transformation/normalization methods are appropriate to each outcomes of interest. Especially adding a pseudocount 1 for zero values may even not make sense because it forces the microbiomes from "nothing" (absence) to "being"(presence).

Example 15.1: Breast Cancer QTRT1 Mouse Gut Microbiome, Example 9.1, Cont.
In this dataset, the mouse fecal samples were collected at two time points: pretreatment and posttreatment. Below we use LMMs to identify the significant change species over time.

```
> setwd("~/Documents/QIIME2R/Ch15_LMMs")
```

Step 1: Load taxa abundance data and meta data.

```
> otu_tab=read.csv("otu_table_L7_MCF7.csv",row.names=1,check.
names=FALSE)
> meta_tab <- read.csv("metadata_QtRNA.csv",row.names=1,check.names
= FALSE)
```

Step 2: Check taxa and meta data.

```
> # Check the numbers of taxa and samples
> nrow(otu_tab)#taxa
[1] 59
> ncol(otu_tab)#samples
[1] 40
```

```
> # Check the numbers of samples and variables
> nrow(meta_tab) #samples
[1] 40
> ncol(meta_tab) #variables
[1] 6

> head(otu_tab,3)
> head(meta_tab,3)
```

For the modeling of LMMs, the taxa need to be in the columns. We use the t() function to transform the otu_tab.

```
> otu_tab_t<-t(otu_tab)
> head(otu_tab_t,3)
> dim(otu_tab_t)
[1] 40 59
```

Step 3: Create analysis dataset-outcomes and covariates.

```
# Create outcome matrix or data frame
> colnames(otu_tab_t)
> colnames(meta_tab)
> class(otu_tab_t)
[1] "matrix" "array"
> class(meta_tab)
[1] "data.frame"

> outcome =otu_tab_t

> # Match sample IDs in outcome matrix or data frame and meta_tab data
frame
> outcome = outcome[rownames(meta_tab),]
> class(outcome)
[1] "matrix" "array"
> class(meta_tab)
[1] "data.frame"
> head(outcome,3)
> yy = as.matrix(outcome)
> yy = ifelse(is.na(yy), 0, yy) #Exclude missing outcome variables if
any
> dim(yy)
[1] 40 59
```

There are 59 taxa in otu_tab. To effectively model the outcomes, we first reduce the number of taxa in the analysis.

In the literature, there are no unique procedures to filter abundance taxa data before modeling. In general, the unclassified taxa were removed, then either read counts or relative abundance data were used. Some researchers renormalized the data after the unclassified taxa were removed, while others directly used the data but adjusted the total read counts during the modeling. Some researchers consider low read counts contribute less to outcome results (although this argument underestimates the functional analysis or the role of rare taxa) and hence removed the low read counts based on a cutoff value. Because too few zeros and too many zeros in the abundance data may cause the models be inaccurate, most models used the proportion of zeros to filter abundance taxa data. Some even used double functions (sum() and quantile() functions) (Xia et al. 2018b) to filter the data. Actually, such arbitrary data preprocessing is an unapproved practice in microbiome research. We should recognize that it is a limitation of current microbiome research.

Here, we use sum() and quantile() functions to double filter the taxa that results in 17 taxa to be included in the analysis.

```
> # Filter abundance taxa data
> # Here we use double filters
> filter1 <- apply(yy, 2, function(x){sum(x>0)>0.4*length(x)})
> filter2 <- apply(yy, 2, function(x){quantile(x,0.9)>1})

> filter12 <-yy[,filter1&filter2]

> cat('after filter:','samples','taxa',dim(filter12),'\n')
after filter: samples taxa 40 17
> #cat(colnames(filter12),'\n')
> colnames(filter12)
 [1] "D_6__Anaerotruncus sp. G3(2012)"
 [2] "D_6__Clostridiales bacterium CIEAF 012"
 [3] "D_6__Clostridiales bacterium CIEAF 015"
 [4] "D_6__Clostridiales bacterium CIEAF 020"
 [5] "D_6__Clostridiales bacterium VE202-09"
 [6] "D_6__Clostridium sp. ASF502"
 [7] "D_6__Enterorhabdus muris"
 [8] "D_6__Eubacterium sp. 14-2"
 [9] "D_6__gut metagenome"
[10] "D_6__Lachnospiraceae bacterium A4"
[11] "D_6__Lachnospiraceae bacterium COE1"
[12] "D_6__mouse gut metagenome"
[13] "D_6__Mucispirillum schaedleri ASF457"
[14] "D_6__Parabacteroides goldsteinii CL02T12C30"
[15] "D_6__Ruminiclostridium sp. KB18"
[16] "D_6__Staphylococcus sp. UAsDu23"
[17] "D_6__Streptococcus danieliae"
```

If the meta data are incomplete, we can use the **complete.cases()** function to subset the data and apply complete cases.

Now we can define outcome matrix or data frame and determine the final taxa in the analysis.

```
> # Match the number of taxa to the filter12 dataset in the analysis
> yy = filter12 [rownames (meta_tab), ]
> dim(yy)
[1] 40 17
> # Create a matrix to hold the taxa
> taxa_all <- colnames (filter12)
> taxa_all
```

The following commands are used to specify covariates for the model.

```
> N = meta_tab[, "Total.Read"] # total reads
> mean(N); sd(N)
[1] 55619
[1] 7524
> mean(log(N)); sd(log(N))
[1] 10.92
[1] 0.1417
> subject = meta_tab[, "MouseID"]
> group = meta_tab[, "Group"]
> time = meta_tab[, "Time"]
```

Step 4: Run the lme() function in the nlme package to implement LMMs.

The **lme** () function is available in the R package **nlme**, which was developed to fit linear mixed models of the form described in Pinheiro and Bates (2000) (Pinheiro and Bates 2000, 2006).

We can assume independence of correlation matrix by setting R matrix as an identity matrix, which is the most simplified structure. However, various correlation matrices can be specified (Pinheiro and Bates 2000, 2006) such as autoregressive of order 1 [AR(1)] or continuous-time AR(1). The **lme** () function allows for multiple and correlated group-specific (random) effects (the argument random) and various types of within-group correlation structures (the argument correlation) described by corStruct in the nlme package.

Below we fit LMMs with main host factors: group and time as well as their interaction as fixed effects and random intercept only (Group*Time, random = ~ 1| Subject). Other LMMs also can be fitted such as with main host factors: group, time, and their interaction as fixed effects and random intercept only and specifying correlation matrix (Group*Time, random = ~ 1|Subject, correlation = corAR1()).

First we need to create list() and matrix to hold the objects of modeling results, which could be used as further processing the results and model comparisons. Then we need to call library(nlme) and run the **lme()** function iteratively to fit multiple taxa (outcome variables).

```
> yy <- yy[, colSums(is.na(yy)) == 0]
> # Create list() and the objects to hold the modeling results
> mod= list()
> rest1 = rest2 = rest3 = matrix(NA, ncol(yy), 1)
```

```
> library(nlme)
> for (j in 1:ncol(yy)){
+ tryCatch({
+ #filter12= na.omit(filter12)
+ y = yy[, j]
+ yt = asin(sqrt(y/N))
+ df = data.frame(y=y, Group=group, Time=time, Subject=subject)
+ mod= lme(yt ~ Group*Time, random = ~ 1|Subject,data=df)
+ rest_tab = summary(mod)$tTable
+ rest1[j, ] = summary(mod)$tTable[2, 5]
+ rest2[j, ] = summary(mod)$tTable[3, 5]
+ rest3[j, ] = summary(mod)$tTable[4, 5]
+ }, error=function(e){cat("ERROR :",conditionMessage(e), "\n")})
+ }
```

We can check the last outcome results using the object rest_tab and summary (mod).

```
> rest_tab
 Value Std.Error DF t-value p-value
(Intercept) 0.0140953 0.002828 36 4.98473 1.574e-05
Group -0.0003452 0.003999 36 -0.08631 9.317e-01
TimePost 0.0043009 0.003999 36 1.07551 2.893e-01
Group:TimePost -0.0012754 0.005655 36 -0.22551 8.229e-01
> summary(mod)
```

Step 5: Adjust *P*-values.

```
> all_lmm <- as.data.frame(cbind(rest1, rest2,rest3))
> # Adjust p-values
> Group_lmm_adj <-round(p.adjust(rest1,'fdr'),4)
> TimePost_lmm_adj <-round(p.adjust(rest2,'fdr'),4)
> Group.TimePost_lmm_adj <-round(p.adjust(rest3,'fdr'),4)

> all_lmm_adj <- as.data.frame(cbind(taxa_all,Group_lmm_adj,
TimePost_lmm_adj,Group.TimePost_lmm_adj))
```

Step 6: Write table to contain the results (Table 15.2).

```
> # Table 15.2
> # Write results table
> # Make the table
> library(xtable)
> table <- xtable(all_lmm_adj,caption = "Table of significant taxa",
lable="sig_taxa_table")
> print.xtable(table,type="html",file = "LMMs_Table_Model_QtRNA.
html")
> write.csv(all_lmm_adj ,file = paste("LMMs_Table_Model_QtRNA.csv",
sep = ""))
```

Table 15.2 Modeling results from linear mixed effects model based on QtRNA data

	taxa_all	Group_Imm_adj	TimePost_Imm_adj	Group. TimePost_Imm_adj
1	D_6__Anaerotruncus sp. G3(2012)	0.5954	2E-04	0.8432
2	D_6__Clostridiales bacterium CIEAF 012	0.9589	0.6811	0.4371
3	D_6__Clostridiales bacterium CIEAF 015	0.4575	2E-04	0.3968
4	D_6__Clostridiales bacterium CIEAF 020	0.3038	0.1708	0.1857
5	D_6__Clostridiales bacterium VE202-09	0.9589	0.2281	0.4862
6	D_6__Clostridium sp. ASF502	0.8211	0.3766	0.5562
7	D_6__Enterorhabdus muris	0.1482	0.8121	0.3488
8	D_6__Eubacterium sp. 14-2	0.0654	6E-04	0.1857
9	D_6__gut metagenome	0.8211	0.8808	0.5562
10	D_6__Lachnospiraceae bacterium A4	0.8309	0.6208	0.4494
11	D_6__Lachnospiraceae bacterium COE1	0.1329	0.4969	0.0277
12	D_6__mouse gut metagenome	0.4186	3E-04	0.8432
13	D_6__Mucispirillum schaedleri ASF457	0.0644	0.7033	0.1857
14	D_6__Parabacteroides goldsteinii CL02T12C30	0.353	0.0026	0.9993
15	D_6__Ruminiclostridium sp. KB18	0.8211	0.0026	0.1857
16	D_6__Staphylococcus sp. UAsDu23	0.1482	6E-04	0.3249
17	D_6__Streptococcus danieliae	0.9589	0.4471	0.8743

15.3   Modeling the Diversity Indices Using the lme4 and LmerTest Packages

In this section, we introduce the lme4 and lmerTest packages and illustrate how to fit
LMMs to analyze the diversity indices.

15.3.1   *Introduction to the lme4 and lmerTest Packages*

The **lme4** package (Bates 2005; Bates et al. 2015) and its **lmer()** function were developed by Bates and his colleagues to provide more flexible fitting of linear mixed models and provide extensions to generalized linear mixed models as well. Compared to the **lme ()** function, the **lmer()** function fits a larger range of models and is more reliable and faster, but the model specification has been changed slightly. For the details on how to use the **lmer()** function to fit linear mixed models, and the differences between lmer() and lme(), the interested reader is referred to the cited references for description of the methods (Bates 2005; Bates et al. 2015).

The R package **lmerTest** (Kuznetsova et al. 2017) (latest version 3.1-3, October 2020) was developed to facilitate the functionality of the "lmerMod" class of the lme4 package, including:

- Obtain the *P*-values for the *F*-and *t*-tests for objects returned by the **lmer()** through overloading the anova() and summary() functions of fixed effects.
- Implement the Satterthwaite's method for approximating degrees of freedom for the *t*-and *F*-tests.
- Implement the construction of Type I-III ANOVA tables.
- Obtain the summary and the anova table using the Kenward-Roger approximation for denominator degrees of freedom by using the **KRmodcomp()** function in the **pbkrtest** package.
- Perform backward elimination (a step-down model building method) of both random and fixed non-significant effects.
- Calculate population means and perform multiple comparison tests as well as plot facilities.

The **lme4** package provides *F*-statistics and the *t*-statistics in anova() and summary() functions, respectively; however, the *P*-values for the corresponding *F*- and *t*-tests are not provided. The **lmerTest** package implements and wraps Satterthwaite's method (Giesbrecht and Burns 1985; Hrong-Tai Fai and Cornelius 1996) into anova () and summary() functions for the object returned by lmer(). Satterthwaite's method has been implemented in SAS software (SAS Institute Inc. 1978, 2013). The SAS users should be familiar with Satterthwaite's method. As an alternative, the Kenward-Roger approximation method is also available, which is integrated through the **KRmodcomp()** function of the **pbkrtest** package (Halekoh and Højsgaard 2014, 2021). For details, the interested reader is referred to the paper of lmerTest package (Kuznetsova et al. 2017).

15.3.2   Fit LMMs Using the Diversity Index as the Outcome

Example 15.2: Breast Cancer QTRT1 Mouse Gut Microbiome, Example 15.1 Cont.

In this subsection, we use the same dataset from Example 15.1 to assess the association between microbiome diversity and group status (Genotype) by setting the Chao 1 richness index as the outcome and the group and time as the fixed-effect terms.

We fit the LMMs using the R package **lme4** (Bates et al. 2015) and perform the corresponding statistical test using the R package **lmerTest** (Kuznetsova et al. 2017).

For illustration, we select the Chao 1 richness index as the response variable and the group, time, and their interaction as the fixed-effect terms. We also consider group and group by time interaction as random terms. We use the following five steps to perform LMMs and illustrate the functionality of **lme4** and **LmerTest** packages.

The first two steps have done in Chap. 9 (Example 9.1), we duplicate them here for the reader's convenience.

Step 1: Load taxa abundance data and meta data.

```
> setwd("~/Documents/QIIME2R/Ch15_LMMs")
> otu_tab <- read.csv("otu_table_L7_MCF7_amp.csv", check.names =
FALSE)
> meta_tab <- read.csv("metadata_QtRNA_amp.csv", check.names = FALSE)
```

Step 2: Calculate alpha-diversity indices.

Here we choose Chao 1 richness index as response variable. Similarly, other diversity indices can be applied. Below we use the **ampvis2** package to calculate Chao 1 richness index. It can be calculated using other R packages such as **vegan** and **microbiome** packages.

```
> # Combine otu-table and meta-table using the amp_load() function
> library(ggplot2)
> library(ampvis2)
> ds <- amp_load(otutable = otu_tab, metadata = meta_tab)
```

We can calculate all the alpha-diversity indices including Chao 1 and ACE by specifying richness=TRUE that are available in the ampvis2 package.

```
> options(width=110,digits=4)
> alpha_amp <- amp_alphadiv(ds,
+ measure =,
```

```
+          richness = TRUE,
+          rarefy = NULL
+  )
```

```
> head(alpha_amp,3)
           SampleID Sample MouseID Genotype Group  Time   Group4 Total_Read
Reads ObservedOTUs Shannon Simpson invSimpson Chao1  ACE
Sun040.BH5 Sun040.BH5    40   BH5     KO   1 Before KO_BEFORE    39824 39813
110  2.385 0.8270    5.779 141.0 156.1
Sun039.BH4 Sun039.BH4    39   BH4     KO   1 Before KO_BEFORE    42068 42056
112  2.378 0.8336    6.009 160.0 170.5
Sun027.BF2 Sun027.BF2    27   BF2     WT   0 Before WT_BEFORE    42738 42729
101  2.276 0.7784    4.512 153.8 162.9
```

**Step 3: Run LMMs using the ANOVA method for objects returned by lmer()
     function.**

```
> install.packages("lmerTest")
> library(lme4)
> library("lmerTest")
> chao1 <- lmer(Chao1 ~ Group*Time + (1 |Group), data = alpha_amp)
```

The lmerTest package provides Type I, II, and III ANOVA tables as defined in the SAS software (SAS Institute Inc. 1978). The Type II and III tables do not depend on the order in which the effects are entered in the model, whereas the Type I ANOVA table is order dependent which performs the sequential decomposition of the contributions of the fixed-effects. The Type I table is the one produced by the ANOVA method of the lme4 package (Bates et al. 2015; Kuznetsova et al. 2017).

For balanced cases, these three ANOVA tables give same results. By default the lmerTest package provides the Type III ANOVA table, while lme4 provides the sequential (Type I) ANOVA table.

```
> anova(chao1)
Type III Analysis of Variance Table with Satterthwaite's method
       Sum Sq Mean Sq NumDF DenDF F value Pr(>F)
Group       405    405    1   36   0.56 0.458
Time        592    592    1   36   0.82 0.370
Group:Time 2376   2376    1   36   3.30 0.077 .
---
Signif. codes:  0 '***' 0.001 '**' 0.01 '*' 0.05 '.' 0.1 ' ' 1
```

The columns DenDF and Pr(>F) refer to denominator degrees of freedom and P-values, which are calculated using the Satterthwaite's method of approximation. Based on the P-values, the Chao 1 richness index for Group and Time interaction is marginally significant ($P = 0.077$).

```
> anova(chao1, type = 1)
Type I Analysis of Variance Table with Satterthwaite's method
      Sum Sq Mean Sq NumDF DenDF F value Pr(>F)
Group      405    405   1   36   0.56 0.4579
Time      6916   6916   1   36   9.61 0.0037 **
Group:Time 2376   2376   1   36   3.30 0.0775 .
---
Signif. codes: 0 '***' 0.001 '**' 0.01 '*' 0.05 '.' 0.1 ' ' 1
```

The Time variable is statistically significant ($P = 0.0037$) and the Group by Time interaction is marginally statistically significant ($P = 0.0775$) from the Type I ANOVA table with Satterthwaite's approximation.

We can require another type of ANOVA by changing the type argument and Kenward-Roger's method for calculating the F-test. The following commands require the Type I ANOVA table with Kenward-Roger's approximation.

```
> anova(chao1, type = 1, ddf = "Kenward-Roger")
Type I Analysis of Variance Table with Kenward-Roger's method
      Sum Sq Mean Sq NumDF DenDF F value Pr(>F)
Group      405    405   1 60.8   0.56 0.4559
Time      6916   6916   1 36.0   9.61 0.0037 **
Group:Time 2376   2376   1 36.0   3.30 0.0775 .
---
Signif. codes: 0 '***' 0.001 '**' 0.01 '*' 0.05 '.' 0.1 ' ' 1
```

Kenward-Roger's method achieves the same results.

Step 4: Run LMMs using the summary method for objects returned by the lmer () function.

```
> summary(chao1)
Linear mixed model fit by REML. t-tests use Satterthwaite's method
['lmerModLmerTest']
Formula: Chao1 ~ Group * Time + (1 | Group)
Data: alpha_amp
```

REML criterion at convergence: 348.2
Scaled residuals:

```
Min 1Q Median 3Q Max
-1.758 -0.710 -0.276 0.749 2.235
```

Random effects:

```
Groups Name Variance Std.Dev.
Group (Intercept) 10.8 3.28
Residual 719.4 26.82
Number of obs: 40, groups: Group, 2
```

Fixed effects:

```
          Estimate Std. Error    df t value Pr(>|t|)
(Intercept)    170.23     9.09 36.00  18.72  <2e-16 ***
Group           -8.16    12.86 36.00  -0.63   0.530
TimePost        10.88    11.99 36.00   0.91   0.370
Group:TimePost  30.83    16.96 36.00   1.82   0.077 .
---
Signif. codes:  0 '***' 0.001 '**' 0.01 '*' 0.05 '.' 0.1 ' ' 1

Correlation of Fixed Effects:
        (Intr) Group TimPst
Group      -0.707
TimePost   -0.660 0.466
Group:TmPst 0.466 -0.660 -0.707
optimizer (nloptwrap) convergence code: 0 (OK)
unable to evaluate scaled gradient
Hessian is numerically singular: parameters are not uniquely
determined
```

In the output, additional columns of df and Pr(>|t|) in the fixed effects are added to the lme4 package. The df refers to degrees of freedom based on Satterthwaite's approximation and Pr(>|t|) is the P-value for the t-test with df as degrees of freedom.

The following commands require the Kenward-Roger's approximation.

```
> summary(chao1, dff="Kenward-Roger")
```

Step 5: Perform backward elimination using the step method for objects returned by the lmer() function.

To illustrate the step method for backward elimination, here we consider a full model including Group, Time, and their interaction as fixed-effect terms, as well as Group and Group by Time interaction as random-effect terms.

```
> chao1a <- lmer(Chao1 ~ Group*Time + (1 | Group) + (1 | Group:Time), data
= alpha_amp)
> b_step <- step(chao1a)
> b_step
```

Backward reduced random-effect table:

```
Eliminated npar logLik AIC LRT Df Pr(>Chisq)
<none> 6 -174 360
(1 | Group:Time) 1 5 -174 358 5.68e-14 1 1
```

Backward reduced fixed-effect table:

```
      Eliminated Df Sum of Sq  RSS AIC F value Pr(>F)
Group:Time     1 1    2376 28273 268   3.30 0.0775 .
Group          2 1     526 28800 267   0.69 0.4119
Time           0 1    6916 35716 274   9.13 0.0045 **
---
Signif. codes: 0 '***' 0.001 '**' 0.01 '*' 0.05 '.' 0.1 ' ' 1
```

Model found:

```
Chao1 ~ Time
```

The reduced model showed that Group and Time interaction in fixed-effect terms is marginally significant ($P = 0.0775$) and Time is significant ($P = 0.0045$).

15.4   Implement LMMs Using QIIME 2

Example 15.3: Mouse Gut Microbiome Data, Example 9.2, Cont.
In Chap. 9 (Sect. 9.5.4), we used this study to illustrate how to obtain core alpha diversities via QIIME 2 and then we used pairwise Kruskal-Wallis test to compare Sex by Time variables interaction. There are total 349 samples (208 for early and 141 for late times). The results showed that later time has larger Shannon diversity than early time with P-value of 0.003227 and Q-value of 0.003227. In Chap. 9, we commented that Kruskal-Wallis test does not consider the dependency of the individuals/subjects and a longitudinal model is more appropriate. Here we use QIIME 2 to implement LMMs to test whether Shannon diversity index changed over time and had gender difference in this study.

15.4.1   Introduction to the QIIME Longitudinal Linear-Mixed-Effects Command

Linear mixed-effects models (LMMs) are more flexible in modeling the association between a single response variable and one or more independent variables in a longitudinal or repeated-measures design. The **qiime longitudinal linear mixed-effects** perform LMMs to evaluate the covariates effects of "group_columns" and "random_effects" to a single dependent variable ("metric") and plot line plots of each group column. To implement LMMs, at least one numeric state-column (e.g., time variable) and one or more comma-separated group-columns need to be specified as the fixed effects, i.e., independent variables. These fixed effects may be categorical or numeric metadata columns. After implementing this model, regression plots of the response variable ("metric") are plotted as a function of the state column and each group column.

15.4.2 Fit LMMs in QIIME 2

First, we need activate QIIME 2. Open the terminal and type: source activate qiime2-2022.2, which is latest version of QIIME 2. When the terminal appears "(qiime2-2022.2)," it indicates qiime2-2022.2 version has been activated. We can work with this version now.

We already create a folder LMM in the path:QIIME2-Biostatistics/longitudinal/LMM, so we can access to this folder using cd QIIME2R-Biostatistics/longitudinal/LMM. If this folder did not exist in your computer, you can create one: mkdir QIIME2R-Biostatistics/longitudinal/LMM. The following commands perform LMMs by specifying Sex and Time as fixed-effects terms, but not requiring the random-effects (Fig. 15.1 for Sex effects and Fig. 15.2 for Time effects).

```
# Figures 15.1 and 15.2
qiime longitudinal linear-mixed-effects \
    --m-metadata-file SampleMetadataMiSeq_SOP.tsv\
    --m-metadata-file shannon_vector.qza\
    --p-metric shannon\
    --p-group-columns Sex,Time\
    --p-state-column DPW\
```

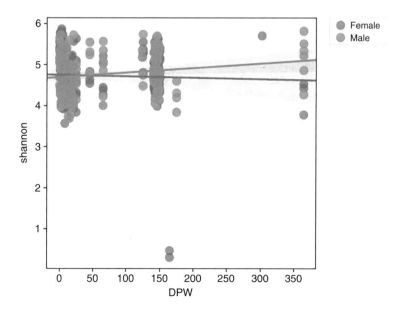

Fig. 15.1 Regression scatterplots of Shannon diversity by male and female mice

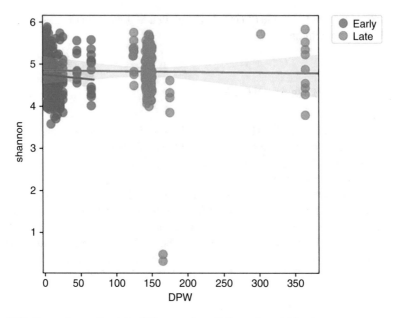

Fig. 15.2 Regression scatterplots of Shannon diversity by early and later times

```
--p-individual-id-column StudyID\
--o-visualization LMMEffectsMiSeq_SOP.qzv
```

Saved Visualization to: LMMEffectsMiSeq_SOP.qzv

where the four columns are required: (1) sample metadata (here, SampleMetadataMiSeq_SOP.tsv) which contains individual-id-column; (2) state-column parameter: metadata column containing state (time) variable (here, DPW); (3) individual-id-column parameter, metadata column containing study IDs for individual subjects (here, StudyID), which indicates the individual subject/site that was sampled repeatedly; and (4) the o-visualization output column.

The p-metric is used to specify the response (dependent) variable column name, which must be located in the metadata or feature table files if the feature table is provided as input data. The p-group columns is used to specify the metadata columns as the independent covariates that are used to determine mean structure of "metric." Several fixed-effects variables can be specified via a comma-separated list. The p-random-effects parameter can be added. The random-effects metadata columns are used to specify the independent covariates that are used to determine the variance and covariance structure (random effects) of "metric." A random intercept for each individual is set by default, while to specify a random slope, set the variable in the p-state-column, in which the state-column value is passed as input to the random-effects parameter. Same to fixed effects, to specify several random effects, a comma-separated list of random effects can be provided in the p-random-effects column. The o-visualization output is used to name VISUALIZATION outputs. We here named it as LMMEffectsMiSeq_SOP.qzv.

By reviewing LMMEffectsMiSeq_SOP.qzv via qiime2 view, we can see that there were no statistically significant effects of sex ($P = 0.088$) or sampling period (early vs. late, $P = 0.356$) on Shannon diversity index, which confirmed the results that were originally reported in Schloss et al. (2012).

The **qiime longitudinal linear-mixed-effects** command returns several results, including (1) the input parameters at the top of the visualization, (2) the **Model summary** of descriptive information about the LMMs, (3) the main **Model results**, which summarizes the effects of each fixed effect (and their interactions) on the dependent variable (here, Shannon diversity). This table shows parameter estimates, including standard errors, z scores, P-values (P>|z|), and 95% confidence interval upper and lower bounds for each parameter, (4) the **Regression scatterplots** that categorized by each "group column" at the bottom of the visualization, with linear regression lines and 95% confidence interval in gray for each group, and (5) **Projected Residuals,** the scatterplots of fit vs. residual plots. The plots show the relationship between metric predictions for each sample (on the x-axis), and the residual or observation error (prediction - actual value) for each sample (on the y-axis). They are used for diagnostics. The roughly zero-centered residuals suggest a well-fitted model.

Below we add random-effects parameters and specify DPW as a random effect. For details of how specifying random effects, the reader is referred to Sect. 15.1.5 How to Fit LMMs.

```
qiime longitudinal linear-mixed-effects \
   --m-metadata-file SampleMetadataMiSeq_SOP.tsv\
   --m-metadata-file shannon_vector.qza\
   --p-metric shannon\
   --p-group-columns Sex,Time\
   --p-random-effects DPW\
   --p-state-column DPW\
   --p-individual-id-column StudyID\
   --o-visualization LMMEffectsMiSeq_SOP_Random.qzv
```

Saved Visualization to: LMMEffectsMiSeq_SOP_Random.qzv

The above regression scatterplots in Figs. 15.1 and 15.2 just provide a quick summary information of the data. To interactively plot the longitudinal data, QIIME 2 recommends using the volatility plot.

15.4.3 *Perform Volatility Analysis*

Volatility analysis generates interactive line plots to visualize how a dependent variable is volatile over a continuous, independent variable (e.g., time) in group(s).

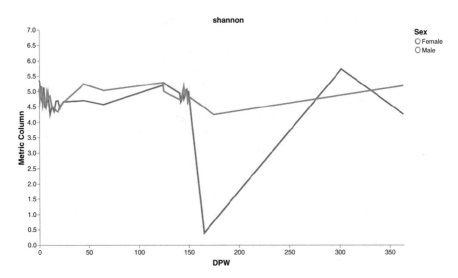

Fig. 15.3 Volatility plots of Shannon diversity by male and female mice

Thus, a volatility plot is a good qualitative tool to identify outliers that disproportionately drive the variance within individuals and groups.

The input data can be metadata files (e.g., alpha and beta diversity artifacts) and FeatureTable[RelativeFrequency] tables. Different dependent variables can be plotted on the y-axis. The following commands examine how variance in Shannon diversity and Sex changes over DPW (day post weaning) (specified in the state-column) (Fig. 15.3).

```
# Figure 15.3
qiime longitudinal volatility \
    --m-metadata-file SampleMetadataMiSeq_SOP.tsv\
    --m-metadata-file shannon_vector.qza\
    --p-default-metric shannon \
    --p-default-group-column Sex\
    --p-state-column DPW\
    --p-individual-id-column StudyID \
    --o-visualization VolatilitySexLMMEffectsMiSeq_SOP.qzv

Saved Visualization to: VolatilitySexLMMEffectsMiSeq_SOP.qzv
```

The following commands examine how variance in Shannon diversity and Time (early and later periods) changes over DPW (day post weaning) (specified in the state-column) (Fig. 15.4).

```
# Figure 15.4
qiime longitudinal volatility \
    --m-metadata-file SampleMetadataMiSeq_SOP.tsv\
    --m-metadata-file shannon_vector.qza\
```

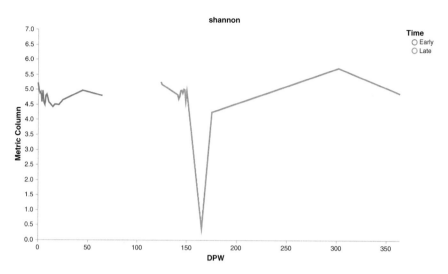

Fig. 15.4 Volatility plots of Shannon diversities by early and later times

```
--p-default-metric shannon \
--p-default-group-column Time\
--p-state-column DPW\
--p-individual-id-column StudyID \
--o-visualization VolatilityTimeLMMEffectsMiSeq_SOP.qzv
```

```
Saved Visualization to: VolatilityTimeLMMEffectsMiSeq_SOP.qzv
```

By reviewing the resulted output visualization using QIIME 2 view, the plot displays a line plot on the left-hand side of and a panel of "Plot Controls" to the right-hand side.

The "Plot Controls" panel is used to interactively adjust variables and parameters for determining how "groups" and "individuals" values change across the specified state-column (a single independent variable). For details on how use the interactive features, the reader is referred to QIIME 2 documents for further references or practicing the **output** visualization in QIIME 2 view.

15.5   Remarks on LMMs

The development of new methods for analysis of microbiome data often takes the advantage of LMM framework such as in Zhang and Yi (2020) and Hui et al. (2017). However, LMM is not specifically designed for microbiome data, and hence it does not address issues due to characteristics of microbiome data. LMM method is often criticized for using to analyze microbiome data because (1) it needs to transform absolute abundance read account into the relative abundance via a transformation method (e.g., arcsine square root transformation) before fitting the model, and most often treats transformed microbiome abundance as normally distributed responses (such as in Srinivas et al. 2013; Rosa et al. 2014; Leamy et al. 2014; Wang et al. 2015); (2) it does not explicitly handle the excess zeros in the data (Chen and Li 2016), and thus cannot fit the model with zero-inflation and over-dispersion to address the sparsity issue (Zhang and Yi 2020). Thus, directly applying LMM method to analyze microbiome data may be underpowered and have potential bias to identify the dynamic microbiome effects.

Several proposed longitudinal microbiome models have compared their methods to LMM method (Chen and Li 2016; Zhang et al. 2018; Zhang and Yi 2020).

15.6   Summary

This chapter focused on LMMs to microbiome data analysis. Section 15.1 covered the general topics of LMMs including the advantages and disadvantages of using LMMs, definitions of fixed and random effects, and its formulation, statistical hypothesis testing, and the procedures of fitting LMMs. Section 15.2 described how to identify the significant taxa with the nlme package. A general introduction to LMMs in microbiome research and longitudinal microbiome data structure were also described. Section 15.3 introduced the lme4 and LmerTest packages and how to use these two packages to analyze the diversity indices. Section 15.4 described how to implement LMMs in QIIME 2, including introduction to the QIIME longitudinal linear-mixed-effects command and volatility analysis. Section 15.5 provided some general remarks on LMMs in microbiome data.

Chapter 16 will describe the generalized linear mixed models (GLMMs), and Chap. 17 will introduce GLMMs and zero-inflated GLMMs for longitudinal microbiome data.

References

Arnau, Jaume, Roser Bono, Nekane Balluerka, and Arantxa Gorostiaga. 2010. General linear mixed model for analysing longitudinal data in developmental research. *Perceptual and Motor Skills* 110 (2): 547–566.

Bates, Douglas. 2005. Fitting linear mixed models in R. *R News* 5 (1): 27–30.

Bates, Douglas, Martin Mächler, Ben Bolker, and Steve Walker. 2015. Fitting linear mixed-effects models using lme4. *Journal of Statistical Software* 67 (1): 48. https://doi.org/10.18637/jss.v067.i01. https://www.jstatsoft.org/v067/i01.

Breslow, Norman E., and David G. Clayton. 1993. Approximate inference in generalized linear mixed models. *Journal of the American Statistical Association* 88 (421): 9–25.

Chen, E.Z., and H. Li. 2016. A two-part mixed-effects model for analyzing longitudinal microbiome compositional data. *Bioinformatics* 32 (17): 2611–2617. https://doi.org/10.1093/bioinformatics/btw308.

Crawley, Michael J. 2012. *The R book*. 2nd ed. Wiley.

Diggle, P.J., P. Heagerty, K.-Y. Liang, and S.L. Zeger. 2002. *Analysis of longitudinal data*. 2nd ed. Oxford: Oxford University Press.

DiGiulio, D.B., B.J. Callahan, P.J. McMurdie, E.K. Costello, D.J. Lyell, A. Robaczewska, C.L. Sun, D.S. Goltsman, R.J. Wong, G. Shaw, D.K. Stevenson, S.P. Holmes, and D.A. Relman. 2015. Temporal and spatial variation of the human microbiota during pregnancy. *Proceedings of the National Academy of Sciences of the United States of America* 112 (35): 11060–11065. https://doi.org/10.1073/pnas.1502875112.

Ernest, Ben, Jessica R. Gooding, Shawn R. Campagna, Arnold M. Saxton, and Brynn H. Voy. 2012. MetabR: An R script for linear model analysis of quantitative metabolomic data. *BMC Research Notes* 5: 596–596. https://doi.org/10.1186/1756-0500-5-596. https://www.ncbi.nlm.nih.gov/pubmed/23111096. https://www.ncbi.nlm.nih.gov/pmc/articles/PMC3532230/.

Fabregat-Traver, Diego, Sodbo Zh. Sharapov, Caroline Hayward, Igor Rudan, Harry Campbell, Yurii Aulchenko, and Paolo Bientinesi. 2014. High-performance mixed models based genome-wide association analysis with omicABEL software. *F1000Research* 3: 200–200. https://doi.org/10.12688/f1000research.4867.1. https://www.ncbi.nlm.nih.gov/pubmed/25717363. https://www.ncbi.nlm.nih.gov/pmc/articles/PMC4329600/.

Fisher, R.A. 1918. The correlation between relatives on the supposition of mendelian inheritance. *Transactions of the Royal Society of Edinburgh* 52 (02): 399–433.

Fitzmaurice, G.M., N.M. Laird, and J.H. Ware. 2004. *Applied longitudinal analysis*. Hoboken: Wiley.

Gelman, Andrew, and Jennifer Hill. 2007. *Data analysis using regression and hierarchical/multilevel models*. New York: Cambridge University Press.

Giesbrecht, F.G., and J.C. Burns. 1985. Two-stage analysis based on a mixed model: Large-sample asymptotic theory and small-sample simulation results. *Biometrics* 41 (2): 477–486. https://doi.org/10.2307/2530872. http://www.jstor.org.proxy.cc.uic.edu/stable/2530872.

Goldberger, Arthur S. 1962. Best linear unbiased prediction in the generalized linear regression model. *Journal of the American Statistical Association* 57 (298): 369–375.

Grueber, C.E., S. Nakagawa, R.J. Laws, and I.G. Jamieson. 2011. Multimodel inference in ecology and evolution: Challenges and solutions. *Journal of Evolutionary Biology* 24 (4): 699–711. https://doi.org/10.1111/j.1420-9101.2010.02210.x. https://onlinelibrary.wiley.com/doi/abs/10.1111/j.1420-9101.2010.02210.x.

Halekoh, Ulrich, and Søren Højsgaard. 2014. A kenward-roger approximation and parametric bootstrap methods for tests in linear mixed models–The R package pbkrtest. *Journal of Statistical Software* 59 (9): 1–30.

———. 2021. *pbkrtest: Parametric bootstrap, kenward-roger and satterthwaite based methods for test in mixed models*. Last Modified 2021-03-09. https://people.math.aau.dk/~sorenh/software/pbkrtest/.

Harrison, Xavier A. 2015. A comparison of observation-level random effect and Beta-Binomial models for modelling overdispersion in Binomial data in ecology & evolution. *PeerJ* 3: e1114. https://doi.org/10.7717/peerj.1114.

Harrison, Xavier A., Lynda Donaldson, Maria Eugenia Correa-Cano, Julian Evans, David N. Fisher, Cecily E.D. Goodwin, Beth S. Robinson, David J. Hodgson, and Richard Inger. 2018. A brief introduction to mixed effects modelling and multi-model inference in ecology. *PeerJ* 6: e4794–e4794. https://doi.org/10.7717/peerj.4794. https://pubmed.ncbi.nlm.nih.gov/29844961. https://www.ncbi.nlm.nih.gov/pmc/articles/PMC5970551/.

Harville, David. 1976a. Extension of the Gauss-Markov theorem to include the estimation of random effects. *The Annals of Statistics* 4: 384–395.

Harville, David A. 1976b. *Confidence intervals and sets for linear combinations of fixed and random effects*. Vol. 32, 403–407. *Biometrics*.

———. 1977. Maximum likelihood approaches to variance component estimation and to related problems. *Journal of the American Statistical Association* 72 (358): 320–338.

Henderson, Charles R. 1950. Estimation of genetic parameters. *Biometrics* 6 (2): 186–187.

————. 1963. Selection Index and Expected Genetic Advance. In: *Statistical Genetics and Plant Breeding*, National Academy of Sciences, No. 982, National Research Council Publication, Washington DC, 141–163.

————. 1973. Sire evaluation and genetic trends. *Journal of Animal Science* 1973 (Symposium): 10–41.

Henderson, C.R. 1984. *Applications of linear models in animal breedlinig*. Guelph: University of Guelph.

Henderson, C.R., S.R. Oscar Kempthorne, and Searle, and C. M. von Krosigk. 1959. The estimation of environmental and genetic trends from records subject to culling. *Biometrics* 15 (2): 192–218. https://doi.org/10.2307/2527669. http://www.jstor.org/stable/2527669.

Hrong-Tai Fai, Alex, and Paul L. Cornelius. 1996. Approximate F-tests of multiple degree of freedom hypotheses in generalized least squares analyses of unbalanced split-plot experiments. *Journal of Statistical Computation and Simulation* 54 (4): 363–378. https://doi.org/10.1080/00949659608811740.

Hui, Francis K.C., Samuel Müller, and A.H. Welsh. 2017. Joint selection in mixed models using regularized PQL. *Journal of the American Statistical Association* 112 (519): 1323–1333. https://doi.org/10.1080/01621459.2016.1215989.

Kostic, Aleksandar D., Dirk Gevers, Heli Siljander, Tommi Vatanen, Tuulia Hyötyläinen, Anu-Maaria Hämäläinen, Aleksandr Peet, Vallo Tillmann, Päivi Pöhö, Ismo Mattila, Harri Lähdesmäki, Eric A. Franzosa, Outi Vaarala, Marcus de Goffau, Hermie Harmsen, Jorma Ilonen, Suvi M. Virtanen, Clary B. Clish, Matej Orešič, Curtis Huttenhower, Mikael Knip, Diabimmune Study Group, and Ramnik J. Xavier. 2015. The dynamics of the human infant gut microbiome in development and in progression toward type 1 diabetes. *Cell Host & Microbe* 17 (2): 260–273. https://doi.org/10.1016/j.chom.2015.01.001. https://pubmed.ncbi.nlm.nih.gov/25662751. https://www.ncbi.nlm.nih.gov/pmc/articles/PMC4689191/.

Kuznetsova, Alexandra, Per B. Brockhoff, and Rune H.B. Christensen. 2017. lmerTest package: Tests in linear mixed effects models. *Journal of Statistical Software* 82 (13): 26. https://doi.org/10.18637/jss.v082.i13. https://www.jstatsoft.org/v082/i13.

Laird, N.M., and J.H. Ware. 1982. Random-effects models for longitudinal data. *Biometrics* 38 (4): 963–974.

Leamy, Larry J., Scott A. Kelly, Joseph Nietfeldt, Ryan M. Legge, Fangrui Ma, Kunjie Hua, Rohita Sinha, Daniel A. Peterson, Jens Walter, Andrew K. Benson, and Daniel Pomp. 2014. Host genetics and diet, but not immunoglobulin A expression, converge to shape compositional features of the gut microbiome in an advanced intercross population of mice. *Genome Biology* 15 (12): 552–552. https://doi.org/10.1186/s13059-014-0552-6. https://pubmed.ncbi.nlm.nih.gov/25516416. https://www.ncbi.nlm.nih.gov/pmc/articles/PMC4290092/.

Lee, Y., J.A. Nelder, and Y. Pawitan. 2006. *Generalised linear models with random effects*. London: Chapman and Hall.

Lloyd-Price, Jason, Cesar Arze, Ashwin N. Ananthakrishnan, Melanie Schirmer, Julian Avila-Pacheco, Tiffany W. Poon, Elizabeth Andrews, Nadim J. Ajami, Kevin S. Bonham, Colin J. Brislawn, David Casero, Holly Courtney, Antonio Gonzalez, Thomas G. Graeber, A. Brantley Hall, Kathleen Lake, Carol J. Landers, Himel Mallick, Damian R. Plichta, Mahadev Prasad, Gholamali Rahnavard, Jenny Sauk, Dmitry Shungin, Yoshiki Vázquez-Baeza, Richard A. White, Jason Bishai, Kevin Bullock, Amy Deik, Courtney Dennis, Jess L. Kaplan, Hamed Khalili, Lauren J. McIver, Christopher J. Moran, Long Nguyen, Kerry A. Pierce, Randall Schwager, Alexandra Sirota-Madi, Betsy W. Stevens, William Tan, Johanna J. ten Hoeve, George Weingart, Robin G. Wilson, Vijay Yajnik, Jonathan Braun, Lee A. Denson, Janet K. Jansson, Rob Knight, Subra Kugathasan, Dermot P.B. McGovern, Joseph F. Petrosino, Thaddeus S. Stappenbeck, Harland S. Winter, Clary B. Clish, Eric A. Franzosa, Hera Vlamakis, Ramnik J. Xavier, Curtis Huttenhower, and Ibdmdb Investigators. 2019. Multi-omics of the gut microbial ecosystem in inflammatory bowel diseases. *Nature* 569 (7758): 655–662. https://doi.org/10.1038/s41586-019-1237-9.

McLean, Robert A., William L. Sanders, and Walter W. Stroup. 1991. A unified approach to mixed linear models. *The American Statistician* 45 (1): 54–64. https://doi.org/10.2307/2685241. http://www.jstor.org/stable/2685241.

Pinheiro, J., and D. Bates. 2000. *Mixed-effects models in S and S-PLUS*. New York: Springer.

Pinheiro, José, and Douglas Bates. 2006. *Mixed-effects models in S and S-PLUS*. Springer.

Robinson, G.K. 1991. That BLUP is a good thing: The estimation of random effects. *Statistical Science* 6 (1): 15–32. http://www.jstor.org/stable/2245695.

Rosa, La, S. Patricio, Barbara B. Warner, Yanjiao Zhou, George M. Weinstock, Erica Sodergren, Carla M. Hall-Moore, Harold J. Stevens, William E. Bennett, Nurmohammad Shaikh, Laura A. Linneman, Julie A. Hoffmann, Aaron Hamvas, Elena Deych, Berkley A. Shands, William D. Shannon, and Phillip I. Tarr. 2014. Patterned progression of bacterial populations in the premature infant gut. *Proceedings of the National Academy of Sciences* 111 (34): 12522–12527. https://doi.org/10.1073/pnas.1409497111. https://www.pnas.org/content/pnas/111/34/12522.full.pdf.

SAS Institute Inc. 1978. Tests of hypotheses in fixed-effects linear models: SAS technical report R-101. SAS Institute Inc.

———. 2013. *The SAS system, version 9.4*. SAS Inc. http://www.sas.com/.

Schloss, Patrick D., Alyxandria M. Schubert, Joseph P. Zackular, Kathryn D. Iverson, Vincent B. Young, and Joseph F. Petrosino. 2012. Stabilization of the murine gut microbiome following weaning. *Gut Microbes* 3 (4): 383–393.

Srinivas, Girish, Steffen Möller, Jun Wang, Sven Künzel, Detlef Zillikens, John F. Baines, and Saleh M. Ibrahim. 2013. Genome-wide mapping of gene–microbiota interactions in susceptibility to autoimmune skin blistering. *Nature Communications* 4 (1): 2462. https://doi.org/10.1038/ncomms3462.

Thompson, Robin. 2008. Estimation of quantitative genetic parameters. *Proceedings of the Royal Society B: Biological Sciences* 275 (1635): 679–686.

Veroniki, Areti Angeliki, Dan Jackson, Wolfgang Viechtbauer, Ralf Bender, Jack Bowden, Guido Knapp, Oliver Kuss, Julian P.T. Higgins, Dean Langan, and Georgia Salanti. 2016. Methods to estimate the between-study variance and its uncertainty in meta-analysis. *Research Synthesis Methods* 7 (1): 55–79. https://doi.org/10.1002/jrsm.1164. https://pubmed.ncbi.nlm.nih.gov/26332144. https://www.ncbi.nlm.nih.gov/pmc/articles/PMC4950030/.

Wang, Jun, Shirin Kalyan, Natalie Steck, Leslie M. Turner, Bettina Harr, Sven Künzel, Marie Vallier, Robert Häsler, Andre Franke, Hans-Heinrich Oberg, Saleh M. Ibrahim, Guntram A. Grassl, Dieter Kabelitz, and John F. Baines. 2015. Analysis of intestinal microbiota in hybrid house mice reveals evolutionary divergence in a vertebrate hologenome. *Nature Communications* 6 (1): 6440. https://doi.org/10.1038/ncomms7440.

Xia, Y. 2020. Correlation and association analyses in microbiome study integrating multiomics in health and disease. *Progress in Molecular Biology and Translational Science* 171: 309–491. https://doi.org/10.1016/bs.pmbts.2020.04.003.

Xia, Yinglin, and Jun Sun. 2021. Longitudinal methods for analysis of microbiome data. In *Microbiome & Metabolomics: Statistical Data Analyses*, ACS in Focus. American Chemical Society.

Xia, Yinglin, Jun Sun, and Ding-Geng Chen. 2018a. Introductory overview of statistical analysis of microbiome data. *Statistical Analysis of Microbiome Data with R*: 43–75.

———. 2018b. Modeling zero-inflated microbiome data. In *Statistical analysis of microbiome data with R*, ed. Yinglin Xia, Jun Sun, and Ding-Geng Chen, 453–496. Singapore: Springer.

Zhang, X., and N. Yi. 2020. Fast zero-inflated negative binomial mixed modeling approach for analyzing longitudinal metagenomics data. *Bioinformatics*. https://doi.org/10.1093/bioinformatics/btz973.

Zhang, Xinyan, Yu-Fang Pei, Lei Zhang, Boyi Guo, Amanda H. Pendegraft, Wenzhuo Zhuang, and Nengjun Yi. 2018. Negative binomial mixed models for analyzing longitudinal microbiome data. *Frontiers in Microbiology* 9 (1683). https://doi.org/10.3389/fmicb.2018.01683. https://www.frontiersin.org/article/10.3389/fmicb.2018.01683.

Zhao, Xueheng, Liang Niu, Carlo Clerici, Roberta Russo, Melissa Byrd, and Kenneth D.R. Setchell. 2019. Data analysis of MS-based clinical lipidomics studies with crossover design: A tutorial mini-review of statistical methods. *Clinical Mass Spectrometry* 13: 5–17. https://doi.org/10.1016/j.clinms.2019.05.002. http://www.sciencedirect.com/science/article/pii/S2376999818300497.

Zuur, Alain F., and Elena N. Ieno. 2016. A protocol for conducting and presenting results of regression-type analyses. *Methods in Ecology and Evolution* 7 (6): 636–645.

Zuur, Alain F., Elena N. Ieno, Neil J. Walker, Anatoly A. Saveliev, and Graham M. Smith. 2009. *Mixed effects models and extensions in ecology with R.* Vol. 574. New York: Springer.

Zuur, Alain F., Elena N. Ieno, and Chris S. Elphick. 2010. A protocol for data exploration to avoid common statistical problems. *Methods in Ecology and Evolution* 1 (1): 3–14.

Chapter 16
Introduction to Generalized Linear Mixed Models

Abstract Chapter 15 investigated linear mixed-effects models (LMMs). This chapter introduces generalized linear mixed models (GLMMs), which can be considered as an extension of linear mixed models to allow response variables from different distributions, such as binary responses. First, it reviews the brief history of generalized linear models (GLMs) and generalized nonlinear models (GNLMs). Then it describes the generalized linear mixed models (GLMMs). Next, it introduces model estimation in GLMMs and investigates the algorithms for parameter estimation in GLMMs and particularly the parameter estimation algorithms for specifically developed GLMMs for microbiome research. Finally, it describes the statistical hypothesis testing and modeling in GLMMs.

Keywords GLMs · GNLMs · Generalized linear mixed models (GLMMs) · Penalized quasi-likelihood-based methods · Taylor-series linearization · Likelihood-based methods · Numerical integration · Laplace approximation · Gauss-Hermite quadrature (GHQ) · Adaptive Gauss-Hermite quadrature (AGQ) · Markov chain Monte Carlo-based integration · IWLS (iterative weighted least squares) algorithm · EM-IWLS algorithm · NBMMs · Zero-inflated negative binomial mixed models (ZINBMMs) · Machine Learning · Information criteria · AIC (Akaike's information criterion) · AIC_c (finite-sample corrected AIC) · QAIC and $QAIC_c$ (Quasi Akaike information criterion and corrected quasi-AIC) · BIC (Bayesian information criterion) · BC (bridge criterion) · DIC (deviance information criterion) · GIC_λ (generalized information criterion) · Vuong test · Pseudo-likelihood (PL) · Penalized quasi-likelihood (PQL)

In Chap. 15, we focused on linear mixed-effects models (LMMs), one of most widely used univariate longitudinal models in classical statistical literature and has recently been applied into microbiome data analysis. In this chapter, we introduce generalized linear mixed models (GLMMs), which can be considered as an extension of linear mixed models to allow response variables from different distributions, such as binary responses. Section 16.1 review the brief history of generalized linear models (GLMs) and generalized nonlinear models (GNLMs). Section 16.2 describes the generalized linear mixed models (GLMMs). Section 16.3 introduces model

© Springer Nature Switzerland AG 2023
Y. Xia, J. Sun, *Bioinformatic and Statistical Analysis of Microbiome Data*,
https://doi.org/10.1007/978-3-031-21391-5_16

estimation in GLMMs Sect. 16.4 investigates the algorithms for parameter estimation in GLMMs and the parameter estimation algorithms for specifically developed GLMMs for microbiome research. Section 16.5 describes the statistical hypothesis testing and modeling in GLMMs.. Finally, we summarize this chapter in Sect. 16.6.

16.1 Generalized Linear Models (GLMs) and Generalized Nonlinear Models (GNLMs)

In 1972, Nelder and Wedderburn (1972) introduced a class of generalized linear models (GLMs) in univariate setting that extends the family of Gaussian-based linear model to the exponential family of distributions (i.e., assuming errors from the exponential family), in which the predicted values are determined by discrete and continuous predictor variables and by the link function (e.g., logistic regression, Poisson regression). By using the quasi-likelihood estimation approach, GLMs provide a unified approach to model either discrete or continuous data.

To account for correlated response data, Liang and Zeger in 1986 and 1988 (Liang and Zeger 1986; Zeger et al. 1988) extended the likelihood and quasi-likelihood methods to the repeated measures and longitudinal settings using a non-likelihood generalized estimating equation (GEE) via a "working" correlation matrix (Zhang et al. 2011b).

In 1988, Prentice (1988) proposed a correlated binary regression to model covariates to binary response, and in 1991, Prentice and Zhao (1991) proposed a class of quadratic exponential family that extended the exponential family to model multivariate discrete and continuous responses. These works introduced a class of generalized nonlinear models (GNLMs). In literature, the GEE approach of Liang and Zeger is a non-likelihood semiparametric estimation technique for correlated data, which is referred to GEE approach, while the approach of Prentice (1988) and Prentice and Zhao (1991) is referred to GEE2, in which the estimation and inference are based on the second-order generalized equations.

GLMs and GNLMs share one common characteristic: their inferences are based the parameters that model an underlying marginal distribution. In other words, both GLMs and GNLMs are used to estimate the regression parameters that predict the average response in a population of individuals and model the effects of different covariates on the mean response of a given population (Vonesh 2012) (p. 205).

16.2 Generalized Linear Mixed Models (GLMMs)

In contrast to the population-averaged inference of GLMs and GNLMs, GLMMs focus on the subject-specific inference, in which the regression parameters estimate the "typical" or "average" individual's mean response instead of the average response for a population of individuals (Vonesh 2012). GLMMs combine statistical frameworks of linear mixed models (LMMs) and generalized linear models (GLMs): extending the properties of (1) LMMs via incorporating random effects and (2) GLMs by using link functions and exponential family distributions to allow response variables to handle non-normal data (e.g., Poisson or binomial). By combining the statistical frameworks of both LMMs and GLMs, GLMMs extend the capability of GLMs for handling non-normal data (e.g., logistic regression) (McCullagh and Nelder 2019) by using both fixed and random effects in the linear predictor (hence become mixed models) (Breslow and Clayton 1993; Stroup 2012; Jiang and Nguyen 2021).

The extension of the GLM of Nelder and Wedderburn (1972) to GLMMs through a number of publications in the mid-1980s and early 1990s such as mostly by Stiratelli et al. (1984), Zeger et al. (1988), Schall (1991), and Breslow and Clayton (1993). All the models proposed in these papers are limited to the conditionally independent models, i.e., they all considered that observations within clusters or subjects are conditionally independent given a set of random effects (Vonesh 2012). In 1993, Wolfinger and O'Connell (1993) extended the basic formulation by also allowing for within-cluster or within-subject correlation, and then a general equation of GLMMs is formulated (Vonesh 2012).

Assume y_i is a vector of grouped/clustered outcome for the ith sample unit $(i = 1, \ldots, n)$, and then GLMMs are generally defined as:

$$g[E(y_i|b_i)] = X_i\beta + Z_ib_i, \tag{16.1}$$

i.e., conditioned on the subject-specific random effects b_i, and the dependent variable y_i is distributed according to the exponential family with its expectation related to the linear predictor $X_i\beta + Z_ib_i$ via a monotonic link function $g(\cdot)$. Here, X_i is a design matrix for the fixed effects coefficients β and Z_i is a design matrix for the random effects coefficients b_i, respectively. In general, it assumes that the matrix Z_i is a subset of X_i and the random effects b_i in classical statistics is not estimated (like β, it would be a column vector if it had been estimated). Instead, we typically assume that b_i is distributed as normal with mean zero and variance: $b_i \sim N(0, \Psi(\theta_b))$ with $\Psi(\theta_b)$ representing a between-subject variance-covariance matrix.

Because the fixed effects (including the intercept of fixed effect) are directly estimated, what is left to estimate is the variance. In other words, the random effect complements are to model the deviations from the fixed effect, and the random effects are just deviations around the mean value in β and hence have mean zero. For a random intercept, the variance of the random intercept is just a 1×1 matrix. If a random intercept and a random slope are specified, then random effect complement is a variance-covariance matrix. This matrix has some properties such as square, symmetric, and positive semidefinite as well as redundant elements. For a $q \times q$ matrix, there are $q(q + 1)/2$ unique elements. Typically this matrix is estimated via θ (e.g., a triangular Cholesky factorization $G = LDL^T$) rather than is directly modeled.

In order to remove the redundant effects to simplify computation and ensure that the resulting estimate matrix is positive definite, usually, we can use various different ways to parameterize G and use various constraints to simplify the model. For example, we can (1) consider G representing the random effects and use the nonredundant elements such as taking the natural logarithm to ensure that the variances are positive and (2) assume the random effects have the simplest independent structure. Thus, the final element in the model is the variance-covariance matrix of the residuals, ε or the conditional covariance matrix of $y_i \mid X_i\beta + z_ib_i$.

If we assume a homogeneous residual variance for all (conditional) observations and that they are (conditionally) independent, then the most common residual covariance structure is $R = I\sigma_\varepsilon^2$, where I is the identity matrix (diagonal matrix of 1s) and σ_ε^2 is the residual variance. We can also assume other structures such as compound symmetry or autoregressive.

In the terminology of SAS, the variance-covariance matrix of the random effects is G matrix, while the residual covariance matrix of the random effects is R matrix. Thus, in GLMMs the final fixed elements are y_i, X_i, z_i, and ε. The final estimated elements are $\widehat{\beta}$, $\widehat{\theta}$, and $\widehat{R}$. The final model depends on the assumed distribution but generally has the form:

$$y_i \mid X_i\beta + z_ib_i \sim F(0, R). \tag{16.2}$$

In addition, the distribution may potentially have extra dispersion/shape parameters.

16.3   Model Estimation in GLMMs

GLMMs are relatively easy to perform. To fit the data, basically it is to just specify a distribution, link function, and structure of the random effects. Because the differences between groups can be modeled as a random effect, GLMMs provide a various models to fit the grouped data, including longitudinal data (Fitzmaurice et al. 2012).

GLMMs are fitted via maximum likelihood (ML), which involves integrating over the random effects. In general because the random effects in GLMMs are nonlinear, the marginal log-likelihood function or the marginal moments of GLMMs do not have closed form of expressions except in special cases (e.g., in linear mixed-effects model). In particular, the log-likelihood function of the observed y_i is given:

$$\ell(\theta) = \sum_{i=1}^{n} \log f(y_i; \theta) = \sum_{i=1}^{n} \log \int f(y_i|b_i; \theta) f(b_i; \theta) db_i, \tag{16.3}$$

where θ denotes the full parameter vector including the fixed effects, the extra potential dispersion/shape parameters ϕ, and the unique element of the covariance matrix G and $f(\cdot)$ denotes a probability density or probability mass function.

ML estimation requires maximization of the integral in (16.1), which in special cases (e.g., in linear mixed-effects model) can be evaluated analytically and maximized by standard methods (Raudenbush et al. 2000) such as the EM (Dempster et al. 1977, 1981), Fisher scoring (Goldstein 1986; Longford 1987), or Newton-Raphson (Lindstrom and Bates 1988). However, although the integrals and hence the log-likelihood function $\ell(\theta)$ seem simple in appearance and generally do not have a resolution of closed analytical form because to obtain the ML estimates, it needs to integrate likelihoods over all possible values of the random effects (Browne and Draper 2006; Lele 2006), i.e., a potentially high dimensional integration is needed when a large number of random effects (i.e., high dimension of b_i) is specified (Zhang et al. 2011a). Thus, the ML estimation usually requires numerical approximations.

16.4   Algorithms for Parameter Estimation in GLMMs

Various estimators have been proposed. The advantages and disadvantages of these algorithms and the software packages that implement these methods have been compared by the review articles such as Pinheiro and Chao (2006) and Bolker et al. (2009), as well as simulation studies such as Raudenbush et al. (2000), Zhang et al. (2011a), and Zhang and Yi (2020).

The estimation for GLMMs have been categorized into different basic approaches. For example:

- Zhang et al. (2011a) divided the estimation into (1) approximating the model using linearization and (2) approximating the log-likelihood function.
- Vonesh (2012) divided the estimation for generalized linear and nonlinear models into (1) a moment-based approach utilizing some form of Taylor-series linearization and (2) a likelihood-based approach using some form of numerical integration.

   Actually the approach of approximating the model and the moment-based approach stated a similar thing: using Taylor-series expansion to approximate the marginal moments (first- and second-order conditional moments) and hence approximate the model.
- Raudenbush et al. (2000) grouped the estimation as three prominent strategies: (1) quasi-likelihood inference; (2) Gauss-Hermite approximations; and (3) Monte Carlo integration.

In this chapter, we categorize four algorithms for estimation in GLMMs: (1) penalized quasi-likelihood-based methods using Taylor-series linearization; (2) likelihood-based methods using numerical integration; (3) Markov Chain Monte Carlo-based integration; and (4) IWLS (Iterative Weighted Least Squares) and EM-IWLS algorithms.

Among the various estimation techniques that have been proposed for estimating the parameters of GLMMs, linearization via Taylor series expansion and numerical integration are the most frequently used techniques. Taylor series linearization techniques are used to approximate either the marginal moments of an approximate marginal quasi-likelihood function (Zhang et al. 2011a; Vonesh 2012) or an integrated quasi-likelihood function corresponding to specified first-and second-order conditional moments (Vonesh 2012). Numerical integration techniques are used exclusively for estimation based on maximizing an integrated log-likelihood function. Particularly, the methods involving numerical quadrature (e.g., via Gauss-Hermite quadrature) or Markov chain Monte Carlo and Laplace approximation (Breslow and Clayton 1993) have increased in use due to increasing computing power and advances in methods in practice. Recently in microbiome field, IWLS algorithm and its extension of EM-IWLS algorithm have been increased in use.

16.4.1   Penalized Quasi-Likelihood-Based Methods Using Taylor-Series Linearization

This category can combine the approaches of pseudo-likelihood (PL) (Wolfinger and O'Connell 1993) and penalized quasi-likelihood (PQL) (Breslow and Clayton 1993). The PQL approach was termed by Breslow and Clayton (1993). Because the GLMM is nonlinear in the random effects, this approach approximates to the first- and second-order conditional moments of the model (Vonesh 2012) using linearization. Typically a doubly iteration algorithm is involved in linearization (Zhang et al. 2011a): First, the GLMM is approximated by a linear mixed-effects model based on current values of the covariance parameter estimates, and then the resulting linear mixed-effects model is fitted iteratively, and once convergence the linearization is updated using the new parameter estimates, resulting in a new linear mixed-effects model. The iterative process continues until the difference in parameter estimates between successive linear mixed model fits fall within a specified tolerance level.

The PQL approach has some **strengths** including:

- Can fit models with a large number of random effects, crossed random effects, multiple types of subjects, and even correlated response after conditioning on covariates in fixed effects and zero-inflated terms and random effects (Zhang et al. 2011a). Thus the PQL approach is flexible and widely implemented (Bolker et al. 2009).
- The linearized model has a relatively simple form that typically can be fitted based only on the mean and variance in the linearized form (Zhang et al. 2011a).

However, the PQL algorithm also has some **weaknesses**, for example:

- PQL computes a quasi-likelihood rather than a true likelihood. It is inappropriate to use quasi-likelihoods to make likelihood-based inference (Bolker et al. 2009; Pinheiro and Chao 2006). Thus, PQL algorithm only generates Wald-type test statistics and cannot perform likelihood-based tests such as the likelihood ratio statistic (Zhang et al. 2011a).

- PQL is inconsistent and is biased for estimates of regression coefficients. The bias is severely for sparse within-subject data (Breslow and Lin 1995; Lin and Breslow 1996), is very severe particularly for paired binary data (Breslow and Lin 1995; Breslow and Clayton 1993), and is most serious when the random effects have large variance and the binomial denominator is small (Raudenbush et al. 2000). Thus, although it was shown that the PQL estimator is consistent upon its order in terms of effect of cluster size, and even for discrete binary data (Vonesh et al. 2002) and is closely related to the Laplace-based ML estimator in some conditions (Vonesh 2012), in general, PQL estimator is biased for large variance or small means (Bolker et al. 2009).
- Particularly PQL estimator has poor performance for Poisson data when the mean number of counts per treatment is less than five, or for binomial (also binary) data if the expected numbers of successes and failures for each observation are both less than five (Breslow 2003).
- Even the algorithms that implement linearization can fail at both levels of the double iteration scheme (Zhang et al. 2011a).

16.4.2   Likelihood-Based Methods Using Numerical Integration

In contrast to the PQL approach using linearization, numerical integration approach solves the nonlinear issue of the random effects usually by approximating the marginal log likelihood (Vonesh 2012), which can be achieved via a singly iterative algorithm. **Laplace approximation** (Raudenbush et al. 2000; Breslow and Lin 1995; Lin and Breslow 1996), **Gauss-Hermite quadrature (GHQ)** (Pinheiro and Chao 2006; Anderson and Aitkin 1985; Hedeker and Gibbons 1994, 1996), and **Adaptive Gauss-Hermite quadrature (AGQ)** (Pinheiro and Bates 1995) can be categorized into this approach. We introduce them separately.

16.4.2.1   Laplace Approximation

Laplace approximation has a long history that used for approximating integrals and is often used in a Bayesian framework for approximating posterior moments, marginal densities (Vonesh 2012; Tierney and Kadane 1986). It is a special case of the GHQ with only one quadrature point (Bates et al. 2015).

Laplace approximation has some **strengths**, including:

- Approximates the true GLMM likelihood rather than a quasi-likelihood (Raudenbush et al. 2000).

- Is more accurate than PQL (Bolker et al. 2009). Particularly it was shown that sixth-order Laplace approximation was much faster than those required non-adaptive GHQ and AGQ, and produced comparable good results by the Gauss-Hermite method with 20 quadrature points and adaptive method with seven quadrature (Raudenbush et al. 2000).
- Like PQL estimator, Laplace-based ML estimator is consistent upon its order in terms of effect of cluster size and even for discrete binary data (Vonesh 1996).

However, Laplace approximation also has the **weaknesses**, for example:

- Produces estimates with unknown asymptotic properties due to not maximizing the underlying log-likelihood function, and hence it may not provide reliable and consistent estimates, except in the special linear mixed-effects model the maximum likelihood estimation (MLE) is really produced by this approach (Lindstrom and Bates 1988).
- Like PL and PQL, Laplace-based ML estimator is biased for estimates of regression coefficients. The bias is severe for sparse within-subject data (Breslow and Lin 1995; Lin and Breslow 1996) and is very severe particularly for paired binary data (Breslow and Lin 1995; Breslow and Clayton 1993).
- Is slower and less flexible than PQL.

16.4.2.2   Gauss-Hermite Quadrature (GHQ)

Compared to Laplace approximation, GHQ has the **strengths**:

- Is more accurate (Pinheiro and Chao 2006; Bolker et al. 2009) because GHQ yields estimates that approximate the MLE (Zhang et al. 2011a).
- GHQ overcomes the small sample bias associated with Laplace approximation and PQL (Pinheiro and Chao 2006), and thus GHQ is recommended for highly discrete data. In contrast, we should not use Laplace approximation and PQL unless the cluster sizes are larger or the random-effects dispersion is small (Vonesh 2012).

However, GHQ also has the **weaknesses**:

- Slower than Laplace approximation (Bolker et al. 2009).
- Limited to 2–3 random effects because the speed of GHQ decreases rapidly with increasing numbers of random effects (Bolker et al. 2009).

16.4.2.3   Adaptive Gauss-Hermite Quadrature (AGQ)

AGQ has been widely used. For example, SAS PROC NLMIXED by default uses AGQ estimates as described in Pinheiro and Bates (1995) to compute for nonlinear mixed regression models when carrying out the dual quasi-Newton optimization (SAS Institute Inc. 2015). AGQ has some **strengths**, including:

- AGQ approach produces more accurate results than its non-adaptive quadrature alternative does when the random effects have large dispersion (Raudenbush et al. 2000).
- AGQ approach provides the consistent, asymptotically normal estimates when using SAS NLMIXED (Zhang et al. 2011a).

However, AGQ also has some **weaknesses**, including:

- Is much more difficult as the number of correlated random effects per cluster increases (Raudenbush et al. 2000).
- It was shown that the R packages **lme4** and **glmmML** using the integral approach based on AGQ do not seem to provide more accurate estimates than its Laplace counterpart for these packages as in its SAS NLMIXED when fitting binary responses (Zhang et al. 2011a).
- Mean squared errors of AGQ with seven quadrature are larger than those of sixth-order Laplace approximation and the average run time for the AGQ analyses was very slower compared to sixth-order Laplace approximation (Raudenbush et al. 2000).

16.4.3   *Markov Chain Monte Carlo-Based Integration*

Markov chain Monte Carlo(MCMC) integration methods (Wei and Tanner 1990) generate random samples from the distributions of parameter values for fixed and random effects rather than explicitly integrate over random effects to compute the likelihood (Bolker et al. 2009). MCMC is usually used in a Bayesian framework (McCarthy 2007); it either incorporates prior information from previous knowledge about the parameters or specifies uninformative (weak) prior distributions to indicate without knowledge (McCarthy 2007). Thus in MCMC inference is based on summary statistics (i.e., mean, mode, quantiles, etc.) of the posterior distribution, which combines the prior distribution with the likelihood (McCarthy 2007).

The **strengths** of MCMC include:

- Is highly flexible, arbitrary number of random effects can be specified (Bolker et al. 2009; Gilks et al. 1995).
- Bayesian MCMC is similarly accurate to maximum-likelihood approaches when datasets are highly informative and little prior knowledge is assumed. There are two reasons that MCMC does not need assumption (Bolker et al. 2009): First, MCMC provides confidence intervals on GLMM parameters by naturally averaging over the uncertainty in both the fixed and random-effect parameters and hence avoiding most difficult approximations used in frequentist hypothesis testing. Second, Bayesian techniques define posterior model probabilities that automatically penalize more complex models, hence providing a way to select or average over models.

However, MCMC also has some **weaknesses**:

- It is very slow, involving potentially technically challenging (Bolker et al. 2009), such as requiring the statistical model is well posed, choosing appropriate priors (Berger 2006), and efficient algorithms for large problems (Carlin 2006), and assessing when chains have run long enough for reliable estimation (Cowles and Carlin 1996; Brooks and Gelman 1998; Paap 2002).
- It is computationally intensive and is convergent stochastically rather than numerically, which can be difficult to assess (Raudenbush et al. 2000).
- Some users do not favor the Bayesian framework (Bolker et al. 2009).

16.4.4   IWLS and EM-IWLS Algorithms

IWLS (iterative weighted least squares) algorithm is used to find the maximum likelihood estimates of a generalized linear model (Nelder and Wedderburn 1972; McCullagh and Nelder 1989). The standard IWLS algorithm is equivalent to PQL procedure. In 2017, Zhang et al. (2017) extended the commonly used IWLS algorithms for fitting generalized linear models (GLMs) and generalized linear mixed models (GLMMs) for fitting negative binomial mixed models (NBMMs) to analyze microbiome data with the independent within-subject errors in the linear mixed model.

In 2018, the same research group (Zhang et al. 2018) applied the IWLS algorithms to fit the NBMMs for longitudinal microbiome data to account for special within-subject correlation structures. In 2020, they (Zhang and Yi 2020) applied the IWLS algorithms again to fit zero-inflated negative binomial mixed models (ZINBMMs) for modeling longitudinal microbiome data to account for excess zeros in the logit part.

16.4.4.1   Extension of IWLS Algorithm for Fitting NBMMs

The extended IWLS algorithm for fitting NBMMs (Zhang et al. 2017) was developed based on the standard IWLS (equivalently PQL). Basically, the extension of the IWLS algorithm to NBMMs is used to iteratively approximate the negative binomial mixed model by a linear mixed model via iteratively updating the parameters of the fixed and random effects, within-subject correlation and fixed shape. Like quasi-GLMs (McCullagh and Nelder 1989) and GLMMs (Venables and Ripley 2002; Schall 1991; Breslow and Clayton 1993), the extended IWLS algorithm for fitting the NBMMs introduces a dispersion parameter to correct for over-dispersion to some extent even if shape parameter is not well estimated.

The iterative approximation is processed via two steps of updating:

1. Updating the parameters of the fixed and random effects, within-subject correlation by extending the IWLS algorithm (or equivalently the PQL procedure) for fitting GLMMs (Breslow and Clayton 1993; Schall 1991; Searle and McCulloch

2001; Venables and Ripley 2002; McCulloch and Searle 2004). The IWLS algorithm first uses a weighted normal likelihood to approximate the generalized linear model likelihood and then updates the parameters from the weighted normal model (McCullagh and Nelder 1989; Gelman et al. 2013), i.e., in the NBMMs case, the negative binomial (NB) likelihood can be approximated by the weighted normal likelihood conditional on the shape parameter, the fixed, and the random effects (Zhang et al. 2017).

2. Updating the shape parameter by maximizing the NB likelihood using the standard Newton-Raphson algorithm conditional on the fixed and random effects (Venables and Ripley 2002).

The extended IWLS algorithm to NBMMs has several **strengths**, including:

- Using the standard procedure for fitting the linear mixed models (Zhang et al. 2017) because it was developed based on the commonly used algorithms for fitting GLMs and GLMMs.
- Flexibility, stability and efficiency (Zhang et al. 2017).
- Being robust and efficient to deal with over-dispersed microbiome count data (Zhang et al. 2017) because it overcomes the issues (Saha and Paul 2005) of the lack of robustness, being severely biased and especially failure to converge for small sample size with the ML estimator of the shape parameter in NB models by adding a dispersion parameter to the framework of GLMMs,

Similar to other algorithms, this extended IWLS algorithm also has the **weaknesses**:

- Only the simple random effect with the independent within-subject errors can be fitted and thus ignores special within-subject correlation structures (Zhang et al. 2017, 2018).
- Not particularly designed to deal with zero-inflation (Zhang and Yi 2020; Zhang et al. 2017).

16.4.4.2 Extension of IWLS Algorithm for Fitting Longitudinal NBMMs

The extended IWLS algorithm for fitting longitudinal NBMMs (Zhang et al. 2018) was developed to extend the IWLS algorithm in Zhang et al. (2017) by (1) adding a correlation matrix (R matrix) to describe dependence among observations and (2) incorporating several choices of correlation matrix as described in Pinheiro and Bates (2000), including the commonly used autoregressive of order 1, AR(1) or continuous-time AR(1) into the NBMMs. This extension of IWLS algorithm for fitting longitudinal data has the **strengths**, including:

- Not only can address over-dispersion and varied total reads in microbiome count data but also account for correlation among the observations (non-constant variances or special within-subject correlation structures) (Zhang et al. 2018).

- Is stable and efficient (Zhang et al. 2017, 2018) and flexible to handle complex structured longitudinal data, allowing for incorporating any types of random effects and within-subject correlation structures as in Pinheiro and Bates (2000) and McCulloch and Searle (2001).

However, with IWLS algorithm being extended for fitting longitudinal NBMMs, the algorithm still has the **weakness:**

- It is not particularly designed to deal with zero-inflation (Zhang and Yi 2020; Zhang et al. 2017).

16.4.4.3 EM-IWLS Algorithm for Fitting Longitudinal ZINBMMs

The EM-IWLS algorithm for fitting longitudinal ZINBMMs (Zhang and Yi 2020) was developed (1) to distinguish the logit part for excess zeros and the NB distribution via a latent indicator variable; (2) to iteratively approximate the logistic likelihood using a weighted logistic mixed model via the standard IWLS algorithm (equivalently PQL procedure) for fitting GLMMs (Breslow and Clayton 1993; Schall 1991; Searle and McCulloch 2001; Venables and Ripley 2002); and (3) to update the parameters in the means of the NB part via the extended IWLS algorithm (Zhang et al. 2017).

This extension of IWLS algorithm for fitting longitudinal ZINBMMs has the **strengths**, including:

- Is able to model over-dispersed and zero-inflated longitudinal microbiome count data (Zhang and Yi 2020).
- Like other ZINBMM methods (Zhang et al. 2017, 2018), FZINBMM method takes advantage of the standard procedure of fitting LMMs to facilitate handling various types of fixed and random effects and within subject correlation structures (Zhang and Yi 2020).
- Particularly, it was demonstrated that the EM-IWLS algorithm has remarkable computational efficiency and statistically comparable with the adaptive Gaussian quadrature algorithm in GLMMadaptive package and the Laplace approximation in the glmmTMB package in terms of empirical power and false positive rates (Zhang and Yi 2020). Both GLMMadaptive and glmmTMB packages use numerical integration to fit ZINBMMs.
- Thus, FZINBMM method is fast and stable because of using model-fitting EM-IWLS algorithm.

Various algorithms for parameter estimation have been evaluated based on speed and accuracy. As the techniques advance, nowadays computer is running much fast, the difference of running time for different algorithms is within minutes, thus accuracy is much more important than speed.

16.5 Statistical Hypothesis Testing and Modeling in GLMMs

In this section, we review and describe some problems for performing statistical hypothesis testing and modeling in GLMMs including GLMs and ZIGLMM.

16.5.1 Model Selection in Statistics

Model selection is a crucial step in big data science including microbiome data analysis for reliable and reproducible statistical inference or prediction. The goal of model selection is to select an optimal statistical model from a set of candidate models based on a given dataset.

In statistics, model selection is defined from the perspective of "procedure," in which model selection can be defined as a pluralism of different inferential paradigms by using either statistical null hypothesis testing approach (Stephens et al. 2005) or the information-theoretic approach (Burnham and Anderson 2002). The null hypothesis testing method is to test the simpler nested models against more complex models (Stephens et al. 2005), while the information-theoretic method attempts to identify the (likely) best model, orders the models from the best to worst, and weights the evidence of each model to see if it really is the best as an inference (Burnham and Anderson 2002) (p. ix).

These two approaches may result in substantially different models being selected. Although the information-theoretic approach has been considered to offer powerful and compelling advantages over null hypothesis testing, the null hypothesis testing approach has their own benefits (Stephens et al. 2005). Both approaches actually have their own strengths and weaknesses depending on what kinds of data are used for the model selection.

16.5.2 Model Selection in Machine Learning

In machine learning, model selection is defined from the perspective of "objective," in which model selection can be defined as two directions and two roles they play, respectively: model selection for inference and model selection for prediction (Ding et al. 2018b). The goal of the first direction is to identify the best model for the data, thus for scientific interpretation the selected model is required not too sensitive to the sample size, which accordingly requires the evaluating methods can select the candidate model most robustly and consistently (Ding et al. 2018b). The goal of the second direction is to choose a model as machinery to offer excellent predictive performance. Thus, the selected model may simply be the lucky winner among a few close competitors, but the selected model should have the predictive performance as best as possible (Ding et al. 2018b).

When model selection is defined in two directions, it emphasizes the distinction of statistical inference and prediction. For example, Ding et al. (2018b) thought that (1) better fitting does not imply better predictive performance and (2) the predictive performance is optimal at a candidate model that typically depends on both the sample size and the unknown data-generating process. Thus, practically an appropriate model selection technique in parametric framework should be able to strongly select the best model for inference and prediction, while in nonparametric framework, it should be able to strike a good balance between the goodness of fit and model complexity on the observed data to facilitate optimal prediction.

Model selection defined from statistics and machine learning share a common objective: statistical inference. Whether model selection is defined based on the "procedure" or the "objective," the key is to select the (likely) best model. We summarize the important criteria of model selection used in the literature and put our thoughts on model selection in general and model selection specifically for microbiome data.

16.5.3   Information Criteria for Model Selection

Here we introduce some information criteria for model selection in GLMMs, including GLMs and ZIGLMMs whose asymptotic performances are well understood. Generally **Information Criteria** refer to model selection methods for parametric models using likelihood functions. A number of model selection criteria have been developed and investigated based on estimating Kullback's directed or symmetric divergence (information theory and statistics) (1968) (Kullback 1997). In general, information criteria is to quantify the Kullback-Leibler distance to measure the relative amount of information lost when a given model approximates the true data-generating process (Harrison et al. 2018).

16.5.3.1   AIC (Akaike's Information Criterion)

AIC (Akaike 1973, 1974; Burnham and Anderson 2004; Aho et al. 2014) was proposed to justify for a large-sample estimator of Kullback's directed divergence between the generating model and a fitted candidate model (Kim et al. 2014). AIC was developed based on the asymptotic properties of ML estimators and hence was considered by Akaike as an extension of R. A. Fisher's likelihood theory (Burnham and Anderson 2002) (p. 3). AIC is defined as:

$$\text{AIC} = -2 \log(L) + 2k, \tag{16.4}$$

where L is the likelihood of the estimated model and k is the number of parameters (the dimension of model). The first term in (16.4) is essentially the deviance and the second is a penalty for the number of parameters. AIC has the following characteristics:

- AIC can be used for comparing non-nested models.
- The smaller the AIC value, the better the model. Thus, the model with the lowest AIC indicates the least information lost and hence represents the best model in that it optimizes the trade-off between fit and complexity (Richards 2008).
- This statistic takes into consideration model parsimony penalizing for the number of predictors in the model. Thus, more complex models (with larger k) will suffer from larger penalties.
- AIC (a minimax optimal criterion) is safer for use than BIC when model selection is for prediction because the minimax consideration gives more protection in the worst case (Ding et al. 2018b).

However, as a model selection criterion, AIC has the limitations, for example:

- When the number of case (the sample size) N is small and k is relatively large (e.g., $k \simeq N/2$), AIC severely underestimates Kullback's directed divergence and results in inappropriately favoring unnecessarily large (or high dimensional candidate) models (Hurvich and Tsai 1989; Kim et al. 2014).
- Specifically for GLMMs, the use of AIC has two main concerns (Bolker et al. 2009): one is the boundary effects (Greven 2008) and another is the estimation of degrees of freedom for random effects (Vaida and Blanchard 2005).

16.5.3.2   AIC$_c$ (Finite-Sample Corrected AIC)

AIC$_c$ was originally proposed by Sugiura (1978) in the framework of linear regression models with normal errors. Then it was extended into various frameworks (e.g., normal and nonlinear regression) (Hurvich and Tsai 1989; Hurvich et al. 1990), which was proposed as an improved estimator of Kullback-Leibler information to correct the AIC for small samples. AIC$_c$ is defined as:

$$\text{AIC}_c = \text{AIC} + \frac{2(k+1)(k+2)}{N-k-2}, \tag{16.5}$$

where N is the number of case and k is the number of parameters (the dimension of model). AIC$_c$ has the following characteristics:

- It selects the model that minimizes AIC$_c$.
- There is little difference between AIC$_c$ and AIC unless the number of case N is small compared with model dimension.
- It is less generally applicable compared to AIC because the justification of AIC$_c$ is depending on the structure of the candidate modeling framework.

16.5.3.3 QAIC and QAIC$_c$ (Quasi Akaike Information Criterion and Corrected Quasi-AIC)

For over-dispersed count data, the simple modified versions of AIC and AIC$_c$ have been increasingly utilized (Burnham and Anderson 2002; Kim et al. 2014). Specifically, the quasi Akaike information criterion (quasi-AIC, QAIC) and its corrected version, QAIC$_c$ (Lebreton et al. 1992), have been proposed as the quasi-likelihood analogues of AIC and AIC$_c$ for modeling over-dispersed count or binary data. They are defined as below:

$$\text{QAIC} = - \left[\frac{2\log(L)}{\widehat{c}}\right] + 2k, \tag{16.6}$$

and

$$\begin{aligned}\text{QAIC}_c &= - \left[\frac{2\log(L)}{\widehat{c}}\right] + 2k + \frac{2k(k+1)}{N-k-1} \\ &= \text{QAIC} + \frac{2k(k+1)}{N-k-1},\end{aligned} \tag{16.7}$$

respectively.

Here, c is a single variance inflation factor used for approximation in modeling of count data (Cox and Snell 1989). The formulas for QAIC and QAIC$_c$ will reduce to AIC and AIC$_c$, respectively, when no over-dispersion exists (i.e., $c = 1$). The variance inflation factor c should be estimated from the global model, i.e., the candidate model that contains all covariates of interest and hence subsuming all of the models in the candidate family (Burnham and Anderson 2002; Kim et al. 2014). In practice, it can be estimated from the goodness-of-fit chi-square statistic (χ^2) of the global model and its degrees of freedom, $\widehat{c} = \frac{\chi^2}{df}$.

The reason that QAIC and QAIC$_c$ were developed because when count data are over-dispersed and $c > 1$, the proper likelihood is $\log(L)/c$ instead of just $\log(L)$, i.e., the quasi-likelihood methods are appropriate. Researchers have found that QAIC and QAIC$_c$ performed well for analyzing different levels of over-dispersion in product multinomial models (Anderson et al. 1994) because the quasi-likelihood criteria tend to outperform the ordinary likelihood criteria when dealing with over-dispersed count data (Kim et al. 2014). Because "quasi-likelihood methods of variance inflation are most appropriate only after a reasonable structural adequacy of the model has been achieved" (Burnham and Anderson 2002) in characterizing the mean, thus c should be computed only for the global model. However, we should note that although the estimation of c based on the global model is recommended, some standard software packages (e.g., SAS GENMOD) do not estimate the single variance inflation factor c using the global model (Burnham and Anderson 2002).

16.5.3.4   BIC (Bayesian Information Criterion)

BIC (Schwarz 1978) is defined as:

$$BIC = -2 \log(L) + \text{Log}(N) \times k, \qquad (16.8)$$

where L is the likelihood of the estimated model, N is the number of case, and k is the number of parameters (the dimension of model). BIC has the following characteristics:

- Like AIC, the smaller the BIC value, the better the model.
- Also like AIC, the penalties are there to reduce the effects of overfitting.
- The penalty is stronger for BIC than AIC for any reasonable sample size.
- BIC often yields oversimplified models because BIC imposes a harsher penalty for the estimation of each additional covariate.
- BIC performs better than AIC when the data have large heterogeneity due to the stronger penalty (Brewer et al. 2016).
- As a variant of AIC, BIC is useful when one wants to identify the number of parameters in a "true" model (Burnham and Anderson 2002).
- Both BIC and AIC_c perform penalty based on total sample size; compared to AIC_c, BIC penalize more severe for moderate sample sizes and penalize less severe for very low sample size (Brewer et al. 2016).

16.5.3.5   BC (Bridge Criterion)

BC is an adaptively asymptotic efficient method (Ding et al. 2018a). It is defined as:

$$BC = -2 \log(L) + C_n \left(1 + 2^{-1} + \cdots + k^{-1}\right), \qquad (16.9)$$

where L is the likelihood of the estimated model, $C_n = n^{2/3}$ is suggested, and k is the dimension of model. The model is chosen that minimizes BC over all of the candidate models whose dimensions are no larger than the dimension of the model selected by AIC (k_{AIC}). BC is recommended when it is not clear if a (practically) parametric framework is suitable (Ding et al. 2018b). BC has the following characteristics:

- BC aims to bridge the advantages of both AIC and BIC in the asymptotic regime (Ding et al. 2018a).
- The penalty is approximately $C_n \log(k)$, but it is written as a harmonic number for its nice interpretation (Ding et al. 2018b).
- BC performs similarly to AIC in a nonparametric framework and similarly to BIC in a parametric framework (Ding et al. 2018a, b).

- BC was shown achieving both consistency when the model class is well-specified and asymptotic efficiency when the model class is mis-specified under mild conditions (Ding et al. 2018a).

16.5.3.6 DIC (Deviance Information Criterion)

DIC (Spiegelhalter et al. 2002) is defined as:

$$\mathrm{DIC} = D\big(E_{\theta|z}(\theta)\big) + 2P_D, \tag{16.10}$$

DIC selects the model that minimizes the equation. DIC is defined with a relevant concept, the deviance under model: $D(\theta) = -2 \log P(y|\theta) + C$ and the effective number of parameters of the model: $P_D = E_{\theta \mid z}D(\theta) - D(E_{\theta \mid z}(\theta))$.

where C does not depend on the model being compared and $E_{\theta \mid z}(\cdot)$ is the expectation taken over θ conditional on all of the observed data z under model. DIC has the following characteristics:

- DIC was derived as a measure of Bayesian model complexity and can be considered as a Bayesian counterpart of AIC (Ding et al. 2018b).
- DIC makes weaker assumptions and automatically estimates a penalty for model complexity that is automatically calculated by the WinBUGS program (Spiegelhalter et al. 2003; Crainiceanu et al. 2005; Kéry and Schaub 2011).
- Although its properties are uncertain (Spiegelhalter et al. 2002), DIC is rapidly gaining popularity in ecology (Bolker et al. 2009). This may be due to the following three reasons: (1) DIC in concept is readily connected to AIC (Akaike 1974; Ding et al. 2018b); in DIC the MLE and model dimension in AIC are replaced with the posterior mean and effective number of parameters, respectively. (2) DIC has some computational advantages for comparing complex models whose likelihood functions may not even be in analytic forms compared to AIC (Ding et al. 2018b). (3) In Bayesian settings, we can use MCMC tools to simulate posterior distributions of each candidate model, which can be further used to efficiently compute DIC (Ding et al. 2018b). DIC is not yet widely used in microbiome field. We are looking forward to seeing more DIC applications in microbiome study.

16.5.3.7 GIC_λ (Generalized Information Criterion)

GIC_λ (Shao 1997; Nishii 1984) represents a wide class of criteria with penalties being linear in model dimension. It is defined as:

$$\text{GIC}_\lambda = \widehat{e} + \frac{\lambda \widehat{\sigma}^2 k}{N}, \tag{16.11}$$

where $\widehat{\sigma}^2$ is an estimator of σ^2 (the variance of the noise), and $\widehat{e} = \frac{\|y - \widehat{y}\|_2^2}{N}$ is the mean square error between the observations and least-squares estimates under the regression model. λ is a deterministic sequence of N for controlling the trade-off between the model fitting and model complexity. GIC_λ has the following characteristics:

- The regression model that minimizes GIC_λ will be selected.
- It was shown (Shao 1997; Ding et al. 2018b) under mild conditions that minimizing GIC_λ is equivalent to minimizing $\text{Log}(\widehat{e}) + \frac{\lambda k}{N}$ if replacing $\widehat{\sigma}^2$ with $\frac{N\widehat{e}}{(N-K)}$.
- In the case, when $\lambda = 2$, the formula corresponds to AIC, and $\lambda = \log N$ corresponds to BIC, while Mallows's C_p method (Mallows 1973) is a special case of GIC with $\sigma^2 \triangleq \frac{N\widehat{e_m}}{(N-k)}$ and $\lambda = 2$, where $\overline{m}$ indexes the largest model that includes all of the covariates.

16.5.4   Likelihood-Ratio Test

Likelihood-ratio test (LRT) (Felsenstein 1981; Huelsenbeck and Crandall 1997; Huelsenbeck and Rannala 1997) is a class of tests that compare two competing statistical models based on the ratio of their likelihoods via assessing the goodness of fit. LRT is usually used to compare the nested (null hypothesis) and reference (complex) models defining a hypothesis being tested. LRT statistic is defined as:

$$\lambda_{\text{LRT}} = -2\ln\left[\frac{L_s\left(\widehat{\theta}\right)}{L_g\left(\widehat{\theta}\right)}\right], \tag{16.12}$$

where s and g denote the simpler and the general model, respectively; and the simpler model (s) has fewer parameters than the general (g) model. The quantity inside the brackets is the ratio of two likelihood functions and its value is bounded between 0 and 1.

The log-likelihoods version of LRT, which is expressed as a difference between the log-likelihoods, is mathematically handled and most often used because they can be expressed in terms of deviance:

$$\begin{aligned} \lambda_{\text{LRT}} &= -2\left(\ln(L_s) - \ln(L_g)\right) \\ &= -2\ln(L_s) + 2\ln(L_g) \\ &= \text{Deviance}_s - \text{Deviance}_g. \end{aligned} \tag{16.13}$$

LRT has the following characteristics:

- This LRT statistic approximately (asymptotically) follows a χ^2 distribution, with degrees of freedom equal to the difference in the number of parameters between the two models.
- LRT tests whether likelihood-ratio is significantly different from 1, or equivalently whether its natural logarithm is significantly different from 0.
- LRT can be used to test for fixed-effect parameters and covariance parameters.
- To test the fixed-effect parameters: (1) Maximum likelihood (ML) estimation is used by setting the same set of covariance parameters but assuming different sets of fixed-effect parameters between the nested and reference models. (2) The difference of log-likelihood for the reference model from that for the nested model. (3) Restricted maximum likelihood (REML) test statistic is obtained by subtracting the -2 ML estimation is not appropriate in testing fixed-effect parameters (Morrell 1998; Pinheiro and Bates 2000; Verbeke and Molenberghs 2000; West et al. 2007).
- To test the covariance parameters: (1) REML estimation should be used by setting the same set of fixed-effect parameters but assuming different sets of covariance parameters. (2) The test statistic for covariance parameters is obtained by subtracting the -2 REML log-likelihood value for the reference model from that for the nested model. (3) REML estimation has been shown to reduce the bias inherent in ML estimates of covariance parameters (West et al. 2007; Morrell 1998).

However, the hypothesis testing via LRT has some challenging issues, such as:

- Testing the covariance parameters had to deal with the null hypothesis values lie on the boundary of the parameter space. Because standard deviations (σ) must be larger or equal to 0, the null hypothesis for random effects ($\sigma = 0$) violates the assumption that the parameters are not on the boundary (Molenberghs and Verbeke 2007; Bolker et al. 2009).
- The way that LRT chooses the better of a pair of nested models via pairwise comparisons has been criticized as abusing the hypothesis testing (Burnham and Anderson 2002; Whittingham et al. 2006; Bolker et al. 2009).
- LRT compares the change in deviance between nested and reference models with random-effect terms. This is conservative and hence increasing the risk of type II errors (Bolker et al. 2009).
- LRT is associated with stepwise deletion procedures, which have been criticized being prone to overestimate the effect size of significant predictors (Whittingham et al. 2006; Forstmeier and Schielzeth 2011; Burnham et al. 2011); and because stepwise deletion is prone to bias effect sizes, presenting means and SEs of parameters from the reference (complex) model should be more robust, especially when the ratio of data points (observations) (n) to estimated parameters (k) is low (Forstmeier and Schielzeth 2011; Harrison et al. 2018).
- It seems that achieving a "minimal adequate model" by using stepwise deletion with LRT only requires a single a priori hypothesis. But in fact it requires multiple significance tests (Whittingham et al. 2006; Forstmeier and Schielzeth 2011); and

such cryptic multiple testing can lead to hugely inflated Type I errors (Forstmeier and Schielzeth 2011; Harrison et al. 2018).

In summary, LRT is widely used throughout statistics, but it has some critical weaknesses such as tending to produce "anticonservative" P-values especially in unbalanced data or when the ratio of data points (n) to estimated parameters (k) is low (Pinheiro and Bates 2000) (p. 88) (see Chap. 15 for details), causing biased effect sizes and inflated Type I errors. Thus, LRT is only appropriate for testing fixed effects for both large ratio of the total sample size to the number of fixed effect levels being tested (Pinheiro and Bates 2000) and large number of random effect levels (Agresti 2002; Demidenko 2013); otherwise LRT can be unreliable for fixed effects in GLMMs. Since it is unfortunate that these two conditions are often for most ecological datasets, LRT is not recommended for testing fixed effects with both small total sample size and numbers of random effect levels in ecology (Bolker et al. 2009). LRT has been reviewed (Bolker et al. 2009) in general appropriate for inference on random effects (Scheipl et al. 2008).

16.5.5 Vuong Test

Vuong test (Long 1997; Vuong 1989) was proposed for a general model selection to test whether the competing models are nested, overlapping, or non-nested or whether the models are correctly specified. Particularly, in the case of zero-inflated and zero-hurdle models, Vuong test is used to test for over-dispersion and zero-inflation. The test statistic is defined as:

$$V = \frac{\sqrt{n}\,\overline{m}}{S_m},\qquad\qquad (16.14)$$

where n is the sample size, S_m is the standard error of the test statistic, $\overline{m} = \left(\frac{1}{n}\right)\sum_{i=1}^{n} m_i$ and $S_m^2 = \left(\frac{1}{n-1}\right)\sum_{i=1}^{n}(m_i - \overline{m})^2$, and $m_i = \log\left[\frac{f_1(y_i)}{f_2(y_i)}\right]$, f_1 and f_2 are two competing probability models. Vuong test has the following characteristics:

- Vuong's statistic is the average log-likelihood ratio suitably normalized and thus has an asymptotically standard normal distribution.
- The test is directional, with a large positive (negative) value favoring f_1 (f_2), and a value close to zero indicating that neither model fits the data well (Vuong 1989; Long 1997; Xia et al. 2012).
- In practice, we need to choose a critical value, c, for a significance level. When $V > c$, the statistic favors the model in the numerator, when $V < -c$, the statistic favors the model in the denominator, and when $V \in (-c, c)$ neither model is favored. It is often to set c as 1.96 for being consistent with standard convention $\alpha = 0.05$.

16.6   Summary

In this chapter we overviewed the generalized linear mixed models (GLMMs). Section 16.1 briefly reviewed the history of generalized linear models (GLMs) and generalized nonlinear models (GNLMs). Sections 16.2 and 16.3 introduced GLMMs and their model estimations, respectively.

In Sect. 16.4, we comprehensively compared four algorithms for parameter estimation in GLMMs, including (1) penalized quasi-likelihood-based methods using Taylor-series linearization; (2) likelihood-based methods using numerical integration (Laplace approximation, Gauss-Hermite quadrature (GHQ), adaptive Gauss-Hermite quadrature (AGQ)); (3) Markov chain Monte Carlo (MCMC)-based integration; and (4) IWLS (iterative weighted least squares) and EM-IWLS algorithms. Particularly, we introduced the extensions of IWLS algorithm for fitting GLMMs in microbiome data. In Sect. 16.5, we investigated the statistical hypothesis testing and modeling in GLMMs. In this section, we first briefly introduced model selection in statistics and in machine learning; we then described and discussed the three important kinds of criteria/tests for model selection: information criteria, likelihood-ratio test (LRT), and Vuong test. For information criteria, we introduced AIC(Akaike's information criterion), AIC_c (finite-sample corrected AIC), QAIC and $QAIC_c$ (quasi Akaike information criterion and corrected quasi-AIC), BIC (Bayesian information criterion), BC (bridge criterion), DIC (deviance information criterion), and GIC_λ (generalized information criterion).

Chapter 17 will introduce generalized linear mixed models for longitudinal microbiome data.

References

Agresti, Alan. 2002. *Categorical data analysis*. Hoboken: Wiley.

Aho, Ken, DeWayne Derryberry, and Teri Peterson. 2014. Model selection for ecologists: The worldviews of AIC and BIC. *Ecology* 95 (3): 631–636. https://doi.org/10.1890/13-1452.1.

Akaike, H. 1973. Information theory and an extension of the maximum likelihood principle. In *2nd international symposium on information theory*. Budapest: Akademiai Kiado.

———. 1974. A new look at the statistical model identification. *IEEE Transactions on Automatic Control* 19 (6): 716–723. https://doi.org/10.1109/tac.1974.1100705.

Anderson, Dorothy A., and Murray Aitkin. 1985. Variance component models with binary response: Interviewer variability. *Journal of the Royal Statistical Society. Series B (Methodological)* 47 (2): 203–210. http://www.jstor.org/stable/2345561.

Anderson, D.R., K.P. Burnham, and G.C. White. 1994. AIC model selection in overdispersed capture-recapture data. *Ecology* 75 (6): 1780–1793. https://doi.org/10.2307/1939637. http://www.jstor.org/stable/1939637.

Bates, Douglas, Martin Mächler, Ben Bolker, and Steve Walker. 2015. Fitting linear mixed-effects models using lme4. *Journal of Statistical Software* 67 (1): 48. https://doi.org/10.18637/jss.v067.i01. https://www.jstatsoft.org/v067/i01.

Berger, James. 2006. The case for objective Bayesian analysis. *Bayesian Analysis* 1 (3): 385–402.

Bolker, Benjamin M., Mollie E. Brooks, Connie J. Clark, Shane W. Geange, John R. Poulsen, M. Henry, H. Stevens, and Jada-Simone S. White. 2009. Generalized linear mixed models: A practical guide for ecology and evolution. *Trends in Ecology & Evolution* 24 (3): 127–135. https://doi.org/10.1016/j.tree.2008.10.008.

Breslow, Norm. 2003. Whither PQL? UW Biostatistics Working Paper Series. Working Paper 192. January 2003. http://biostats.bepress.com/uwbiostat/paper192.

Breslow, N.E., and D.G. Clayton. 1993. Approximate inference in generalized linear mixed models. *Journal of the American Statistical Association* 88 (421): 9–25. https://doi.org/10.2307/2290687. http://www.jstor.org/stable/2290687.

Breslow, Norman E., and Xihong Lin. 1995. Bias correction in generalised linear mixed models with a single component of dispersion. *Biometrika* 82 (1): 81–91.

Brewer, Mark J., Adam Butler, and Susan L. Cooksley. 2016. The relative performance of AIC, AICC and BIC in the presence of unobserved heterogeneity. *Methods in Ecology and Evolution* 7 (6): 679–692. https://doi.org/10.1111/2041-210x.12541.

Brooks, Stephen P., and Andrew Gelman. 1998. General methods for monitoring convergence of iterative simulations. *Journal of Computational and Graphical Statistics* 7 (4): 434–455.

Browne, William J., and David Draper. 2006. A comparison of Bayesian and likelihood-based methods for fitting multilevel models. *Bayesian Analysis* 1 (3): 473–514.

Burnham, K.P., and D.R. Anderson. 2002. *Model selection and multimodel inference: A practical information-theoretic approach.* New York: Springer.

Burnham, Kenneth P., and David R. Anderson. 2004. Multimodel inference: Understanding AIC and BIC in model selection. *Sociological Methods & Research* 33 (2): 261–304. https://doi.org/10.1177/0049124104268644.

Burnham, Kenneth P., David R. Anderson, and Kathryn P. Huyvaert. 2011. AIC model selection and multimodel inference in behavioral ecology: Some background, observations, and comparisons. *Behavioral Ecology and Sociobiology* 65 (1): 23–35. https://doi.org/10.1007/s00265-010-1029-6.

Carlin, B.P. 2006. Elements of hierarchical Bayesian inference. In *Hierarchical modelling for the environmental sciences*, ed. J. Clark and A.E. Gelfand, 3–24. Oxford: Oxford University Press.

Cowles, Mary Kathryn, and Bradley P. Carlin. 1996. Markov chain Monte Carlo convergence diagnostics: A comparative review. *Journal of the American Statistical Association* 91 (434): 883–904.

Cox, D.R., and E.J. Snell. 1989. *Analysis of binary data.* 2nd ed. New York: Chapman and Hall.

Crainiceanu, Ciprian M., David Ruppert, and Matthew P. Wand. 2005. Bayesian Analysis for Penalized Spline Regression Using WinBUGS. Journal of Statistical Software 14 (14): 1–24. https://doi.org/10.18637/jss.v014.i14. https://www.jstatsoft.org/index.php/jss/article/view/v014i1

Demidenko, Eugene. 2013. *Mixed models: Theory and applications with R.* Hoboken: Wiley.

Dempster, Arthur P., Nan M. Laird, and Donald B. Rubin. 1977. Maximum likelihood from incomplete data via the EM algorithm. *Journal of the Royal Statistical Society: Series B (Methodological)* 39 (1): 1–22.

Dempster, Arthur P., Donald B. Rubin, and Robert K. Tsutakawa. 1981. Estimation in covariance components models. *Journal of the American Statistical Association* 76 (374): 341–353.

Ding, J., V. Tarokh, and Y. Yang. 2018a. Bridging AIC and BIC: A new criterion for autoregression. *IEEE Transactions on Information Theory* 64 (6): 4024–4043. https://doi.org/10.1109/TIT.2017.2717599.

———. 2018b. Model selection techniques: An overview. *IEEE Signal Processing Magazine* 35 (6): 16–34. https://doi.org/10.1109/MSP.2018.2867638.

Felsenstein, J. 1981. Evolutionary trees from DNA sequences: A maximum likelihood approach. *Journal of Molecular Evolution* 17 (6): 368–376. https://doi.org/10.1007/bf01734359.

Fitzmaurice, Garrett M., Nan M. Laird, and James H. Ware. 2012. *Applied longitudinal analysis.* Vol. 998. Hoboken: Wiley.

Forstmeier, Wolfgang, and Holger Schielzeth. 2011. Cryptic multiple hypotheses testing in linear models: Overestimated effect sizes and the winner's curse. *Behavioral Ecology and Sociobiology* 65 (1): 47–55. https://doi.org/10.1007/s00265-010-1038-5.

Gelman, Andrew, John B. Carlin, Hal S. Stern, David B. Dunson, Aki Vehtari, and Donald B. Rubin. 2013. *Bayesian data analysis*. London: Chapman Hall.

Gilks, Walter R., Sylvia Richardson, and David Spiegelhalter. 1995. Introducing Markov chain Monte Carlo. In *Markov chain Monte Carlo in practice*, ed. Walter R. Gilks, 1–19. Boca Raton: Chapman and Hall/CRC Press.

Goldstein, Harvey. 1986. Multilevel mixed linear model analysis using iterative generalized least squares. *Biometrika* 73 (1): 43–56.

Greven, Sonja. 2008. *Non-standard problems in inference for additive and linear mixed models*. Göttingen: Cuvillier Verlag.

Harrison, Xavier A., Lynda Donaldson, Maria Eugenia Correa-Cano, Julian Evans, David N. Fisher, Cecily E.D. Goodwin, Beth S. Robinson, David J. Hodgson, and Richard Inger. 2018. A brief introduction to mixed effects modelling and multi-model inference in ecology. *PeerJ* 6: e4794.

Hedeker, Donald, and Robert D. Gibbons. 1994. A random-effects ordinal regression model for multilevel analysis. *Biometrics* 50 (4): 933–944. https://doi.org/10.2307/2533433. http://www.jstor.org/stable/2533433.

———. 1996. MIXOR: A computer program for mixed-effects ordinal regression analysis. *Computer Methods and Programs in Biomedicine* 49 (2): 157–176. https://doi.org/10.1016/0169-2607(96)01720-8. https://www.sciencedirect.com/science/article/pii/0169260796017208.

Huelsenbeck, John P., and Keith A. Crandall. 1997. Phylogeny estimation and hypothesis testing using maximum likelihood. *Annual Review of Ecology and Systematics* 28 (1): 437–466. https://doi.org/10.1146/annurev.ecolsys.28.1.437. https://www.annualreviews.org/doi/abs/10.1146/annurev.ecolsys.28.1.437.

Huelsenbeck, John P., and Bruce Rannala. 1997. Phylogenetic methods come of age: Testing hypotheses in an evolutionary context. *Science* 276 (5310): 227–232. https://doi.org/10.1126/science.276.5310.227. https://science.sciencemag.org/content/sci/276/5310/227.full.pdf.

Hurvich, Clifford M., and Chih-Ling Tsai. 1989. Regression and time series model selection in small samples. *Biometrika* 76 (2): 297–307.

Hurvich, Clifford M., Robert Shumway, and Chih-Ling Tsai. 1990. Improved estimators of Kullback–Leibler information for autoregressive model selection in small samples. *Biometrika* 77 (4): 709–719.

Jiang, Jiming, and Thuan Nguyen. 2021. *Linear and generalized linear mixed models and their applications*. New York: Springer Nature.

Kéry, Marc, and Michael Schaub. 2011. *Bayesian population analysis using WinBUGS: A hierarchical perspective*. Waltham: Academic.

Kim, Hyun-Joo, Joseph E. Cavanaugh, Tad A. Dallas, and Stephanie A. Foré. 2014. Model selection criteria for overdispersed data and their application to the characterization of a host-parasite relationship. *Environmental and Ecological Statistics* 21 (2): 329–350.

Kullback, Solomon. 1997. Information theory and statistics. Mineola, New York: Dover Publications, INC

Lebreton, Jean-Dominique, Kenneth P. Burnham, Jean Clobert, and David R. Anderson. 1992. Modeling survival and testing biological hypotheses using marked animals: A unified approach with case studies. *Ecological Monographs* 62 (1): 67–118. https://doi.org/10.2307/2937171. https://esajournals.onlinelibrary.wiley.com/doi/abs/10.2307/2937171.

Lele, Subhash R. 2006. Sampling variability and estimates of density dependence: A composite-likelihood approach. *Ecology* 87 (1): 189–202.

Liang, Kung-Yee, and Scott L. Zeger. 1986. Longitudinal data analysis using generalized linear models. *Biometrika* 73 (1): 13–22.

Lin, Xihong, and Norman E. Breslow. 1996. Bias correction in generalized linear mixed models with multiple components of dispersion. *Journal of the American Statistical Association* 91 (435): 1007–1016. https://doi.org/10.2307/2291720. http://www.jstor.org/stable/2291720.

Lindstrom, Mary J., and Douglas M. Bates. 1988. Newton—Raphson and EM algorithms for linear mixed-effects models for repeated-measures data. *Journal of the American Statistical Association* 83 (404): 1014–1022.

Long, J.S. 1997. *Regression models for categorical and limited dependent variables.* Thousand Oaks: Sage.

Longford, Nicholas T. 1987. A fast scoring algorithm for maximum likelihood estimation in unbalanced mixed models with nested random effects. *Biometrika* 74 (4): 817–827.

Mallows, C.L. 1973. Some comments on CP. *Technometrics* 15 (4): 661–675.

McCarthy, Michael A. 2007. *Bayesian methods for ecology.* Cambridge: Cambridge University Press.

McCullagh, P., and J.A. Nelder. 1989. *Generalized linear models.* London: Chapman and Hall.

McCullagh, Peter, and John A. Nelder. 2019. *Generalized linear models.* London: Routledge.

McCulloch, Charles E., and Shayle R. Searle. 2001. *Generalized, linear, and mixed models.* Hoboken: Wiley.

———. 2004. *Generalized, linear, and mixed models.* Boca Raton: Wiley.

Molenberghs, Geert, and Geert Verbeke. 2007. Likelihood ratio, score, and Wald tests in a constrained parameter space. *The American Statistician* 61 (1): 22–27.

Morrell, Christopher H. 1998. Likelihood ratio testing of variance components in the linear mixed-effects model using restricted maximum likelihood. *Biometrics* 54: 1560–1568.

Nelder, John Ashworth, and Robert W.M. Wedderburn. 1972. Generalized linear models. *Journal of the Royal Statistical Society: Series A (General)* 135 (3): 370–384.

Nishii, Ryuei. 1984. Asymptotic properties of criteria for selection of variables in multiple regression. *The Annals of Statistics* 12: 758–765.

Paap, Richard. 2002. What are the advantages of MCMC based inference in latent variable models? *Statistica Neerlandica* 56 (1): 2–22.

Pinheiro, José C., and Douglas M. Bates. 1995. Approximations to the log-likelihood function in the nonlinear mixed-effects model. *Journal of Computational and Graphical Statistics* 4 (1): 12–35.

Pinheiro, José, and Douglas Bates. 2000. *Mixed-effects models in S and S-PLUS.* New York: Springer.

Pinheiro, José C., and Edward C. Chao. 2006. Efficient Laplacian and adaptive Gaussian quadrature algorithms for multilevel generalized linear mixed models. *Journal of Computational and Graphical Statistics* 15 (1): 58–81. https://doi.org/10.1198/106186006X96962.

Prentice, Ross L. 1988. Correlated binary regression with covariates specific to each binary observation. *Biometrics* 44 (4): 1033–1048. https://doi.org/10.2307/2531733. http://www.jstor.org/stable/2531733.

Prentice, Ross L., and Lue Ping Zhao. 1991. Estimating equations for parameters in means and covariances of multivariate discrete and continuous responses. *Biometrics* 47 (3): 825–839. https://doi.org/10.2307/2532642. http://www.jstor.org/stable/2532642.

Raudenbush, Stephen W., Meng-Li Yang, and Matheos Yosef. 2000. Maximum likelihood for generalized linear models with nested random effects via high-order, multivariate Laplace approximation. *Journal of Computational and Graphical Statistics* 9 (1): 141–157.

Richards, Shane A. 2008. Dealing with overdispersed count data in applied ecology. *Journal of Applied Ecology* 45 (1): 218–227. https://doi.org/10.1111/j.1365-2664.2007.01377.x. https://besjournals.onlinelibrary.wiley.com/doi/abs/10.1111/j.1365-2664.2007.01377.x.

Saha, Krishna, and Sudhir Paul. 2005. Bias-corrected maximum likelihood estimator of the negative binomial dispersion parameter. *Biometrics* 61 (1): 179–185.

SAS Institute Inc. 2015. *SAS/STAT® 14.1 user's guide: The NLMIXED procedure.* Cary: SAS Institute Inc.

Schall, Robert. 1991. Estimation in generalized linear models with random effects. *Biometrika* 78 (4): 719–727.

Scheipl, Fabian, Sonja Greven, and Helmut Küchenhoff. 2008. Size and power of tests for a zero random effect variance or polynomial regression in additive and linear mixed models. *Computational Statistics & Data Analysis* 52 (7): 3283–3299.

Schwarz, Gideon. 1978. Estimating the dimension of a model. *The Annals of Statistics* 6 (2): 461–464. http://www.jstor.org/stable/2958889.

Searle, Shayle Robert, and Charles E. McCulloch. 2001. *Generalized, linear and mixed models*. Chichester: Wiley.

Shao, Jun. 1997. An asymptotic theory for linear model selection. *Statistica Sinica* 7: 221–242.

Spiegelhalter, David J., Nicola G. Best, Bradley P. Carlin, and Angelika Van Der Linde. 2002. Bayesian measures of model complexity and fit. *Journal of the Royal Statistical Society: Series B (Statistical Methodology)* 64 (4): 583–639.

Spiegelhalter, David J., Andrew Thomas, Nicky Best, and David Lunn. 2003. *WinBUGS version 1.4 user manual*. Cambridge: MRC Biostatistics Unit. http://www.mrc-bsu.cam.ac.uk/bugs.

Stephens, Philip A., Steven W. Buskirk, Gregory D. Hayward, and Carlos Martinez Del Rio. 2005. Information theory and hypothesis testing: A call for pluralism. *Journal of Applied Ecology* 42 (1): 4–12.

Stiratelli, R., N. Laird, and J.H. Ware. 1984. Random-effects models for serial observations with binary response. *Biometrics* 40 (4): 961–971.

Stroup, Walter W. 2012. *Generalized linear mixed models: Modern concepts, methods and applications*. Boca Raton: CRC Press.

Sugiura, Nariaki. 1978. Further analysts of the data by akaike's information criterion and the finite corrections: Further analysts of the data by akaike's. *Communications in Statistics-Theory and Methods* 7 (1): 13–26.

Tierney, Luke, and Joseph B. Kadane. 1986. Accurate approximations for posterior moments and marginal densities. *Journal of the American Statistical Association* 81 (393): 82–86.

Vaida, Florin, and Suzette Blanchard. 2005. Conditional akaike information for mixed-effects models. *Biometrika* 92 (2): 351–370.

Venables, W.N., and B.D. Ripley. 2002. *Modern applied statistics with S*. 4th ed. New York: Springer.

Verbeke, Geert, and Geert Molenberghs. 2000. *Linear mixed models for longitudinal data*. New York: Springer.

Vonesh, Edward F. 1996. A note on the use of Laplace's approximation for nonlinear mixed-effects models. *Biometrika* 83 (2): 447–452. http://www.jstor.org/stable/2337614.

———. 2012. *Generalized linear and nonlinear models for correlated data: Theory and applications using SAS*. Cary: SAS Institute.

Vonesh, Edward F., Hao Wang, Lei Nie, and Dibyen Majumdar. 2002. Conditional second-order generalized estimating equations for generalized linear and nonlinear mixed-effects models. *Journal of the American Statistical Association* 97 (457): 271–283. https://doi.org/10.1198/016214502753479400.

Vuong, Quang H. 1989. Likelihood ratio tests for model selection and non-nested hypotheses. *Econometrica* 57 (2): 307–333. https://doi.org/10.2307/1912557. http://www.jstor.org/stable/1912557.

Wei, Greg C.G., and Martin A. Tanner. 1990. A Monte Carlo implementation of the EM algorithm and the poor man's data augmentation algorithms. *Journal of the American Statistical Association* 85 (411): 699–704. https://doi.org/10.1080/01621459.1990.10474930. https://www.tandfonline.com/doi/abs/10.1080/01621459.1990.10474930.

West, Brady T., Kathleen B. Welch, and Andrzej T. Galecki. 2007. *Linear mixed models: A practical guide using statistical software*. Boca Raton: Chapman and Hall/CRC.

Whittingham, Mark J., Philip A. Stephens, Richard B. Bradbury, and Robert P. Freckleton. 2006. Why do we still use stepwise modelling in ecology and behaviour? *Journal of Animal Ecology*

75 (5): 1182–1189. https://doi.org/10.1111/j.1365-2656.2006.01141.x. https://besjournals. onlinelibrary.wiley.com/doi/abs/10.1111/j.1365-2656.2006.01141.x.

Wolfinger, Russ, and Michael O'Connell. 1993. Generalized linear mixed models a pseudo-likelihood approach. *Journal of Statistical Computation and Simulation* 48 (3–4): 233–243. https://doi.org/10.1080/00949659308811554.

Xia, Y., D. Morrison-Beedy, J. Ma, C. Feng, W. Cross, and X. Tu. 2012. Modeling count outcomes from HIV risk reduction interventions: A comparison of competing statistical models for count responses. *AIDS Research and Treatment* 2012: 593569. https://doi.org/10.1155/2012/593569. http://www.ncbi.nlm.nih.gov/pubmed/22536496.

Zeger, S., K. Liang, and P. Albert. 1988. Models for longitudinal data: A generalized estimating equation approach. *Biometrics* 44 (4): 1049–1060.

Zhang, Xinyan, and Nengjun Yi. 2020. Fast zero-inflated negative binomial mixed modeling approach for analyzing longitudinal metagenomics data. *Bioinformatics* 36 (8): 2345–2351.

Zhang, Hui, Lu Naiji, Changyong Feng, Sally W. Thurston, Yinglin Xia, Liang Zhu, and Xin M. Tu. 2011a. On fitting generalized linear mixed-effects models for binary responses using different statistical packages. *Statistics in Medicine* 30 (20): 2562–2572.

Zhang, Hui, Y. Xia, R. Chen, D. Gunzler, W. Tang, and Tu. Xin. 2011b. Modeling longitudinal binomial responses: Implications from two dueling paradigms. *Journal of Applied Statistics* 38 (11): 2373–2390.

Zhang, Xinyan, Himel Mallick, Zaixiang Tang, Lei Zhang, Xiangqin Cui, Andrew K. Benson, and Nengjun Yi. 2017. Negative binomial mixed models for analyzing microbiome count data. *BMC Bioinformatics* 18 (1): 4. https://doi.org/10.1186/s12859-016-1441-7.

Zhang, Xinyan, Yu-Fang Pei, Lei Zhang, Boyi Guo, Amanda H. Pendegraft, Wenzhuo Zhuang, and Nengjun Yi. 2018. Negative binomial mixed models for analyzing longitudinal microbiome data. *Frontiers in Microbiology* 9: 1683–1683. https://doi.org/10.3389/fmicb.2018.01683. https://pubmed.ncbi.nlm.nih.gov/30093893. https://www.ncbi.nlm.nih.gov/pmc/articles/PMC6070621/.

Chapter 17
Generalized Linear Mixed Models for Longitudinal Microbiome Data

Abstract Chapter 16 investigated some general topics of generalized linear mixed-effects models (GLMMs). This chapter focuses on some newly developed GLMMs that take account for correlated observations with random effects while considering over-dispersion and zero-inflation. First, it reviews and discusses some general issues of GLMMs in microbiome research. Then, it introduces three GLMMs that model over-dispersed and zero-inflated longitudinal microbiome data through using (1) glmmTMB package; (2) GLMMadaptive package; and (3) NBZIMM package for fast zero-inflated negative binomial mixed modeling (FZINBMM), which was specifically designed for analyzing longitudinal microbiome data. Finally, it provides some remarks on fitting GLMMs.

Keywords Generalized linear mixed models (GLMMs) · glmmTMB package · GLMMadaptive package · Fast zero-inflated negative binomial mixed modeling (FZINBMM) · NBZIMM package · Over-dispersed · Zero-inflated · pscl package · Zero-hurdle · Negative binomial (NB) model · Over-dispersion · Zero-inflated negative binomial (ZINB) · Zero-hurdle negative binomial (ZHNB) · Poisson · Zero-inflated Poisson (ZIP) · Zero-hurdle Poisson (ZHP) · Zero-inflated generalized linear mixed models (ZIGLMMs) · BIC · Wald t test · F test · Satterthwaite approximation · Kenward-Roger (KR) approximation · TMB (template model builder) package · AICtab () · BICtab() · bbmle package · DHARMa package · plotQQunif () · plotResiduals () · testDispersion() · testZeroinflation() · Conway-Maxwell-Poisson distribution · zicmp · marginal_coefs () · effectPlotData ().devtools package · QQ-plot · glmm.zinb() · lme() · nlme package · stats package · glmPQL() · MASS package

Generalized linear mixed-effects models (GLMMs) and the generalized estimating equations (GEEs) are the two most appealing paradigms in a longitudinal setting. Thus, GEEs and GLMMs not only have been adopted for the analysis of longitudinal

Supplementary Information The online version contains supplementary material available at https://doi.org/10.1007/978-3-031-21391-5_17.

microbiome data at the beginning stage of microbiome research, but also have been used as a statistical foundation to build statistical models that specifically target longitudinal microbiome data in recent years. In Chap. 16, we investigated some general topics of GLMMs. Microbiome count data are often over-dispersed and zero-inflated, containing excess zeros than would be expected from the typical error distributions. In this chapter, we introduce GLMMs, with focusing on some newly developed GLMMs that take account for correlated observations with random effects while considering over-dispersion and zero-inflation. In Sect. 17.1, we review and discuss some general issues of GLMMs in microbiome research. Then, we introduce three GLMMs that can be used for modeling over-dispersed and zero-inflated longitudinal microbiome data. They are generalized linear mixed modeling (1) using glmmTMB package (Sect. 17.2); (2) using GLMMadaptive package (Sect. 17.3); and (3) fast zero-inflated negative binomial mixed modeling (FZINBMM) using the NBZIMM package (Sect. 17.4), which was specifically designed for analyzing longitudinal microbiome data. In Sect. 17.5, we provide some remarks on fitting GLMMs. Finally, we summarize this chapter in Sect. 17.6.

17.1 Generalized Linear Mixed Models (GLMMs) in Microbiome Research

In longitudinal study or repeated measurements, the observations on the same individual are often correlated, which is usually modeled with a random effect in generalized linear mixed models (GLMMs). Count data are typically modeled with GLMs and GLMMs using either Poisson or negative binomial distributions. However, microbiome count data are over-dispersed and sparse and often contain many zeros. Thus, Poisson model is not able to analyze microbiome count data due to its assumption of equality of the variance and the mean. The negative binomial model and its extensions are usually used to analyze the over-dispersed microbiome data. To model zero-inflated microbiome data, the zero-inflated or zero-hurdle models with the negative binomial distribution (a mixture of Poisson distributions with Gamma-distributed rates) can be applied. Xia et al. (2018a, b) introduced over-dispersed and zero-inflated modeling microbiome data.

In our 2018 book, we mainly adopted two widely used negative binomial (NB) models edgeR and DESeq2 in RNA-seq literature to model over-dispersed microbiome data as well as the widely-used **pscl** package to fit zero-inflated and zero-hurdle GLMs using maximum likelihood estimation (MLE) (Zeileis et al. 2008) along with various model comparisons.

A negative binomial (NB) model can model over-dispersion due to clustering and but cannot model over-dispersion, sparsity caused by zero-inflation, which could result in biased parameter estimates (Xia et al. 2018a; Xia 2020). We concluded that the zero-inflated and zero-hurdle GLMs with NB regression, i.e., zero-inflated negative binomial (ZINB) and zero-hurdle negative binomial (ZHNB) are the best fitted models to over-dispersed and zero-inflated microbiome data among the competing models including Poisson, NB, zero-inflated Poisson(ZIP), and zero-hurdle Poisson (ZHP) (Xia et al. 2018b).

However, **pscl** is not equipped with the component of random effects and thus is limited to cross-sectional data and cannot model longitudinal data or repeated measurements. Without modeling random effects and thereby ignoring correlation could lead the statistical inference to be anti-conservative (Bolker et al. 2009; Bolker 2015). Here, we introduce other two R packages **glmmTMB** and **GLMMadaptive** that can be used to model over-dispersed and zero-inflated longitudinal microbiome data. After the introduction of glmmTMB and GLMMadaptive, we will move on modeling microbiome data from GLMs, GLMMs, to zero-inflated generalized linear mixed models (ZIGLMMs). Before we move on to introducing the glmmTMB and GLMMadaptive packages, in this section, we review and introduce some specific issues in modeling microbiome data, including data transformation, challenges of model selection, and statistical hypothesis testing in microbiome data when using GLMMs.

17.1.1   Data Transformation Versus Using GLMMs

Typically three kinds of zeros exist in microbiome data: structural zero, sampling zero, and rounded zero. Excess zeros not only causes the data to be zero-inflated but also to be over-dispersed (Xia et al. 2018c). Such kinds of zeros cannot be simply treated using a data processing step such as replacing with a small value or using transformation to make them be normally distributed because even the data steps process successfully, the resulting data might violate statistical assumptions or limit the scope of inference (e.g., the estimates of fixed effects cannot be extrapolated to new groups) (Bolker et al. 2009). For example, in practice, log transformation is often employed. Log transformation make multiplicative models additive to fascinate the data analysis and can perfectly remove heteroscedasticity if the relative standard deviation is constant (Kvalheim et al. 1994). However, log transformation has two main drawbacks: (1) it is unable to deal with the zero values because log zero is undefined; (2) the effect of log transformation on values with a large relative standard deviation is limited, which usually is the case in microbiome data with the low level of taxonomy ranks (e.g., genus and species). It was shown by simulation and real data (Feng et al. 2013, 2014) that the log transformation is often misused:

- Log transformation usually only can remove or reduce skewness of the original data that follows a log-normal distribution or approximately so, which in some cases comes at the cost of actually making the distribution more skewed than the original data.
- It is not generally true that the log transformation can reduce variability of data especially if the data includes outliers. In fact, whether or not log transformation can reduce variability depends on the magnitude of the mean of the observations — the larger the mean, the smaller the variability.
- It is difficult to interpret model estimates from log transformed data because the results obtained from standard statistical tests on log-transformed data are often

not relevant to the original, non-transformed data. In practice, the obtained model estimates from fitting the transformed data are usually required to translate back to the original scale through exponentiation to give a straightforward biological interpretation (Bland and Altman 1996). However, since no inverse function can map back exp(E(log X)) to the original scale in a meaningful fashion, it was advised that all interpretations should focus on the transformed scale once data are log-transformed (Feng et al. 2013).

- Fundamentally, statistical hypothesis testing of equality of (arithmetic) means of two samples is different from testing equality of (geometric) means of two samples after log transformation of right-skewed data. These two hypothesis tests are equivalent if and only if the two samples have equivalent standard deviations.

The underlying assumption of the log transformation is that the transformed data have a distribution equal or close to the normal distribution (Feng et al. 2013). Since this assumption is often violated, therefore, log transformation with adding the shift parameter on the one hand cannot help reduce the variability and on the another hand could be quite problematic to test the equality of means of two samples when there are values close to zero in the samples. Actually even nonparametric models still need assumptions (e.g., homogeneity of variance across groups). Thus, the over-dispersed and zero-inflated count models are recommended (Xia et al. 2018a, b).

17.1.2   Model Selection in Microbiome Data

In general, nested models are compared using likelihood or score test, for example, to compare ZIP vs. ZINB (ZIP is nested within ZINB) and ZHP vs. ZHNB (ZHP nested within ZHNB), while non-nested models are evaluated using AIC and/or the Vuong test (Long 1997; Vuong 1989).

In statistical theory, AIC (Akaike 1973, 1998) and BIC (Schwarz 1978; Wit et al. 2012) have served as a common criterion (or the golden rules) for model selection since they were proposed (Aho et al. 2014).

In GLMMs, estimates of AIC have been developed based on certain exponential family distributions (Saefken et al. 2014). As we described in Chap. 16 (Sect. 16.5.3 of Information Criteria for Model Selection), AIC and BIC have several theoretical properties: consistency in selection, asymptotic efficiency, and minimax-rate optimality. For example, it was shown that AIC is asymptotically efficient for the nonparametric framework and is also minimax optimal (Barron et al. 1999) and BIC is consistent and asymptotically efficient for the parametric framework. However, AIC and BIC also have their own drawbacks. For example, AIC is inconsistent in a parametric framework (at least exist two correct candidate models), and hence AIC is not asymptotically efficient in such a framework. Actually it is difficult to evaluate analytically the properties of AIC for small samples, and thus its statistical optimality could not be demonstrated (Shibata 1976). BIC, on the other hand, is not minimax optimal and asymptotically efficient in a nonparametric framework (Shao 1997; Foster and George 1994). For the details on the good properties and drawbacks of AIC and BIC, the reader is referred to Ding et al. (2018a, b) for general references on these topics. In order to leverage the strengths of AIC and BIC and overcome their weaknesses, hybrid or adaptive approaches of AIC and BIC have been proposed such as in these cited references (Ing 2007; Yang 2007; Liu and Yang 2011; van Erven et al. 2012; Zhang and Yang 2015; Ding et al. 2018a).

Generally, AIC and BIC are sufficient to be used to choose better models. Various variants of AIC and BIC may not improve the decision too much. However, relying solely on AIC may misinterpret the importance of the variables in the set of candidate models; thus, combining AIC and BIC (BIC is more appropriate for heterogeneous data due to using more penalties) is recommended in modeling microbiome data in GLMMs.

Microbiome data and ecology data share some common characteristics. However, microbiome data are more over-dispersed and zero-inflated compared to ecology data. There are two general characteristics in model selection for microbiome data.

1. Over-dispersed and zero-inflated models are usually more appropriate than their nested (simple) models. Since generally it is obvious that NB models have advantages compared to their nested Poisson models, similarly, zero-inflated NB and zero-hurdle NB models have advantages than their nested zero-inflated Poisson and zero-hurdle Poisson models; thus, the information-theoretic approach is much more important than the null hypothesis testing approach.
2. Most often the information-theoretic approach and the null hypothesis testing approach can provide consistent results for choosing the best models. Compared to ecology, model selection is relatively simple in microbiome study. Based on our experience, usually the test results from AIC, BIC, Vuong test, and LRT are consistent in microbiome data (Xia et al. 2018b). Considering them jointly should enable to choose a better model.

For the improvements of various variants of AIC and BIC, the readers can read the description of their properties and characteristics in Chap. 16 (Sect. 16.5.3).

17.1.3 Statistical Hypothesis Testing in Microbiome Data

In this subsection, we describe some challenges when conducting statistical hypothesis testing in microbiome data. Generally, when we use GLMMs, we need to consider two statistical approaches: (1) ML versus REML and (2) Wald test versus LRT. For details of consideration, see Chaps. 15 (Sect. 15.1.5) and 16 (Sect. 16.5). Particularly, the statistical hypothesis testing microbiome data using GLMMs is even challenging. We summarize some challenges as follows.

First, the statistical methods that are appropriate for testing over-dispersion and zero-inflation are limited.

In general, the null hypothesis of statistical Wald Z, χ^2, t, and F tests commonly measures a quantity (i.e., the resulting test statistic) of zero (null) or no difference between certain characteristics of a population (or data-generating process or models). However, Wald Z and χ^2 tests cannot be used for GLMMs with over-dispersion. When over-dispersion occurs, Wald t and F tests are appropriate for GLMMs because they account for the uncertainty in the estimates of over-dispersion (Littell et al. 2006; Moscatelli et al. 2012). Thus, for over-dispersed microbiome data, Wald t and F tests are required. Similarly, for zero-inflated microbiome data, the zero-inflated GLMMs are more appropriate. However, currently both over-dispersed and zero-inflated GLMMs or GLMMs that are capable of modeling both over-dispersed and zero-inflated microbiome data are still limited.

Second, testing the null hypothesis on the boundary poses more challenge on most statistical tests.

Molenberghs and Verbeke (2007) brought the boundary effects to attention: Most tests require that the standard deviations must be ≥ 0, the null hypothesis for random effects ($\sigma = 0$) violates the assumption that the null values of the parameters are not on the boundary of their allowable ranges. Bolker et al. (2009) discussed the boundary effects when using GLMMs for ecology and evolution data. Microbiome data exist many zeros and therefore is prone to be boundary on zeros. Additionally, the count microbiome data are often normalized into proportions for analysis, which makes the data potentially to be boundary at 0 and 1. Thus, for microbiome data, boundary issue is more challenging and urging to be solved.

Third, choosing an appropriate method to calculate the degrees of freedom is difficult.

Wald t or F tests or AIC (AICc) needs to calculate the degrees of freedom (df), which is between 1 and the number of random-effect levels $N - 1$. Several approaches are available for calculating df and different software packages use different ones. But there is no consensus on which one is appropriate. For example, the simplest approach is the minimum number of df, which is contributed by random effects that affect the term being tested. SAS uses this approach as default (Littell et al. 2006). While the more complicated approach uses Satterthwaite and Kenward-Roger (KR) approximations of the degrees of freedom to adjust for the standard errors (Littell et al. 2006; Schaalje et al. 2001). KR (only available in SAS) was reviewed having overall best performance at least for LMMs (Schaalje et al. 2002; Bolker et al. 2009). Additionally, the residual df can be estimated by the sample size n minus the trace t (i.e., the sum of the diagonal elements) of the hat matrix (Burnham 1998; Spiegelhalter et al. 2002). Due to the complexity, the calculating degrees of freedom and boundary effects have been reviewed as the two particular challenges to perform statistical testing the results of GLMMs (Bolker et al. 2009).

17.2 Generalized Linear Mixed Modeling Using the glmmTMB Package

In this section, we introduce the glmmTMB package and illustrate its use with microbiome data.

17.2.1 Introduction to glmmTMB

The R package **glmmTMB** (Brooks et al. 2017) was developed to estimate GLMs and GLMMs and to extend the GLMMs by including zero-inflated and hurdle GLMMs using ML. Thus, glmmTMB can handle a various range of statistical distributions including Gaussian, Poisson, binomial, negative binomial, and beta. Because glmmTMB extended GLMMs, it can not only model zero-inflation, heteroscedasticity, and autocorrelation but also handle various types of within-subject correlation structures.

The glmmTMB package has wrapped all these capabilities of estimates in GLMs, GLMMs, and their extensions including zero-inflated and hurdle GLMMs using ML. Currently glmmTMB focuses on zero-inflated counts although this package can be used to fit continuous distributions too (Brooks et al. 2017).

The zero-inflated (more broadly zero-altered) models wrapped in glmmTMB **was** developed under the framework of GLMs and GLMMs, which allow us to model count data using a mixture of a Poisson, NB. Especially glmmTMB uses the Conway-Maxwell-Poisson distribution (Huang 2017), which consists of both mean and dispersion parameters. It is a generalized version of the Poisson distribution with the Bernoulli and geometric distributions as special cases (Sellers and Shmueli 2010). Depending on the dispersion, Conway-Maxwell-Poisson distribution can have either longer or shorter upper tail than that of the Poisson (Sellers and Shmueli 2010). Thus, it can model either over- or under-dispersed count data (Shmueli et al. 2005; Lynch et al. 2014; Barriga and Louzada 2014; Brooks et al. 2017). Conway-Maxwell-Poisson distribution can flexibly model dispersion and skewness through the Sichel and Delaporte distributions (Stasinopoulos et al. 2017).

The glmmTMB package uses the interface and formula syntax of the **lme4** package (Bates et al. 2015) and performs estimation via the **TMB** (Template Model Builder) package (Kristensen et al. 2016). Like lme4, glmmTMB uses MLE and the Laplace approximation to integrate over random effects. Currently the restricted maximum likelihood (REML) is not available in glmmTMB, and the random effects are not integrated using the Gauss-Hermite quadrature, although they are available in lme4 (Brooks et al. 2017). However, a fundamental difference underlying implementation between glmmTMB and lme4 is that glmmTMB uses TMB to take the advantage of fast estimating non-Gaussian models and provides more flexibility to fit the classes of distributions that it can fit (Brooks et al. 2017).

A **glmmTMB** model consists of four main components: (1) a conditional model formula, (2) a distribution for the conditional model, (3) a dispersion model formula, and (4) a zero-inflation model formula. Both fixed and random effects models can be specified in the conditional and zero-inflated components of the model. For the dispersion parameter, only the fixed effects models can be specified.

One example syntax for full zero-inflated negative binomial GLMMs is given below:

```
zinb = glmmTMB(Count ~ Group* Time + (1|Subject), zi= ~ Group* Time,
data=ds, family = nbinom2)
```

One example syntax for full hurdle negative binomial GLMMs is given below:

```
hnb = glmmTMB(Count ~ Group* Time + (1|Subject), zi= ~ Group* Time, data
= ds, family = truncated_nbinom2)
```

17.2.1.1   Conditional Model Formula

glmmTMB is very flexible to fit various GLMs and GLMMs. For example, we can fit simple GLMs and GLMMs, just specify the conditional model while leaving the zero-inflation and dispersion formulas at their default values. To specify a mean of the conditional model, we can use a two-sided formula (the syntax is same as **lme4**): specify the response variable on the left and predictors and potentially include random effects and offsets on the right. One example syntax of conditional model formula is given as below:

```
Count ~ Taxon + (1 | Subject)
```

This formula specifies a conditional model for the dependence of mean count on taxon:

counts vary by taxon and vary randomly by subject.

One zero-inflated and dispersed GLMMs that can be modeled in glmmTMB is given as follows.

$$y = E(y_i|b_i, NSZ) = \exp(\beta_0 + \beta_1 + b_i),$$
$$b_i \sim N\left(0, \sigma_{b_i}^2\right),$$
$$\sigma^2 = Var(y_i|b_i, NSZ) = y(1 + y/\theta), \tag{17.1}$$
$$\text{logit}(p) = \beta_0^{(zi)} + \beta_1^{(zi)},$$
$$\log(\theta) = \beta_0^{(disp)} + \beta_2^{(disp)},$$

where y is the read abundance count, b_i is the subject specific random effect, NSZ is the probability of "non-structural zero," $p = 1 - \Pr(NSZ)$ is the probability of zero inflation, and β's are regression coefficients with subscript denoting covariate/level (with 0 denoting intercept).

The above model formulation in (17.1) allows the conditional mean to depend on whether or not a subject was from covariate 1 (group variable, e.g., treatment vs. control) and to vary randomly by subject. It also allows the number of structural (i.e., extra) zeros to depend on covariate 1 (group variable). Additionally, it allows the dispersion parameter to depend on covariate 2 (e.g., time).

17.2.1.2 Distribution for the Conditional Model

We can specify the distribution of the mean of the conditional model via the family argument. For the count data, the family of the distribution includes Poisson (family = poisson), negative binomial (family = nbinom1 or family = nbinom2), and Conway-Maxwell-Poisson to fit over- and under-dispersed data (family = compois). By default, the link function of Poisson, Conway-Maxwell-Poisson, and negative binomial distributions is log. We can specify other links using the family argument such as family = poisson(link = "identity").

glmmTMB provides two parameterizations of the negative binomial. They are different in the dependence of the variance (σ^2) on the mean (μ). The argument **family = nbinom1** is used to indicate that the variance increases linearly with the mean, i.e., $\sigma^2 = \mu(1 + \alpha)$, with $\alpha > 0$; while the argument **family = nbinom2** is used to indicate that the variance increases quadratically with the mean, i.e., $\sigma^2 = \mu(1 + \mu/\theta)$, with $\theta > 0$ (Hardin et al. 2007). The variance of the Conway-Maxwell-Poisson distribution does not have a closed form equation (Huang 2017).

17.2.1.3 Dispersion Model Formula

In **glmmTMB**, the dispersion parameter can be specified as either identical dispersion or covariate-dependent dispersion. For example, the default dispersion model (dispformula = ~1) treats the dispersion parameter (e.g., α or θ in the NB model) as identical for each observation; while the dispersion parameter can also be treated as varying with fixed effects, in which the dispersion model uses a log link. We can also use the dispersion model to account for heteroskedasticity. For example, to account for more variation (relative to the mean) of the response to a predictor variable, we can specify the one-sided formula dispformula = ~ this predictor variable in a NB distribution. Although when the conditional and dispersion models include the same variables, the mean-variance relationship can be manipulated; however, a potential non-convergence issue could happen (Brooks et al. 2017). To see the description of the dispersion parameter for each distribution, type?sigma.glmmTMB in R or RStudio.

17.2.1.4   Zero-Inflation Model Formula

The overall distribution of zero-inflated generalized models is a mixture of the conditional model and zero-inflation model (Lambert 1992). The zero-inflation model describes the probability of observing an extra (i.e., structural) zero that is not generated by the conditional model. Zero-inflation model creates an extra point mass of zeros in the response distribution, where the zero-inflation probability is bounded between zero and one by using a logit link.

To assume that the probability of producing a structural zero for all observations is equal, we can specify an intercept model as ziformula = ~1 to account for the structural zero. To model the structural zero of the response due to a specific predictor variable, we can specify ziformula = ~ this predictor variable in the zero-inflation model. **glmmTMB** also allows to include random effects in both conditional and zero-inflation models, but not allowing to specify a term of random effects in the dispersion model.

17.2.2   Implement GLMMs via glmmTMB

Example 17.1: Breast Cancer QTRT1 Mouse Gut Microbiome
This dataset was used in previous chapters. Here we continue to use this dataset to illustrate glmmTMB. The 16S rRNA microbiome data were measured at pretreatment and posttreatment times for 10 samples at each time. We are interested in the differences between the genotype factors or groups. We use the example data to illustrate how to fit and compare GLMMs, zero-inflated GLMMs, and hurdle GLMMs using glmmTMB and how to extract results from a model.

Here, we illustrate how to perform model selection to select the models that are better fitted to this abundance data in detecting the genotype differences. The compared models include Poisson, Conway-Maxwell-Poisson, negative binomial, zero-inflated, and zero-hurdle models. We use the species "D_6__Ruminiclostridium.sp.KB18" for this illusttration.

To run the **glmmTMB()** function to perform GLMMs, we first need to install the **glmmTMB** package, which is available from the Comprehensive R Archive Network (CRAN) (current version 1.1.2.3, February 2022) and from GitHub (development versions). We can install glmmTMB via CRAN using the command install. packages("glmmTMB") or via GitHub using devtools (Wickham and Chang 2017): devtools::install_github("glmmTMB/glmmTMB/glmmTMB").

Step 1: Load taxa abundance data and meta data.

```
> setwd("~/Documents/QIIME2R/Ch17_GLMMs2")
> otu_tab=read.csv("otu_table_L7_MCF7.csv",row.names=1,check.
names=FALSE)
> meta_tab <- read.csv("metadata_QtRNA.csv",row.names=1,check.names
= FALSE)
```

Step 2: Check taxa and meta data.

```
> nrow(otu_tab)#taxa
> ncol(otu_tab)#samples
> nrow(meta_tab)#samples
> ncol(meta_tab)#variables

> head(otu_tab,3)
> head(meta_tab,3)

> otu_tab_t<-t(otu_tab)
> head(otu_tab_t,3)
> dim(otu_tab_t)
```

Step 3: Create analysis dataset including outcomes and covariates.

```
> colnames(otu_tab_t)
> colnames(meta_tab)
> class(otu_tab_t)
> class(meta_tab)

> outcome =otu_tab_t
> # Match sample IDs in outcome matrix or data frame and meta_tab data
frame
> outcome = outcome[rownames(meta_tab),]
> class(outcome)
> head(outcome,3)
> yy = as.matrix(outcome)
> yy = ifelse(is.na(yy), 0, yy) #Exclude missing outcome variables if
any
> dim(yy)
> # Filter abundance taxa data
> # Here we use double filters
> filter1 <- apply(yy, 2, function(x){sum(x>0)>0.4*length(x)})
> filter2 <- apply(yy, 2, function(x){quantile(x,0.9)>1})

> filter12 <-yy[,filter1&filter2]
> cat('after filter:','samples','taxa',dim(filter12),'\n')
> colnames(filter12)

> yy = filter12[rownames(meta_tab),]
> ds = data.frame(meta_tab,yy)
> head(ds,3)
```

Step 4: Explore the distributions of the outcomes (taxa).

First, we summarize the distribution of the species "D_6__Ruminiclostridium.sp. KB18," including sample size, mean, sd, min, median, and percentage of zeros for each group at each time point.

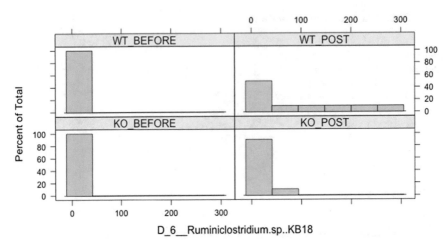

Fig. 17.1 Histogram of Ruminiclostridium.sp.KB18 over before and posttreatment in wild-type (WT) and knockout (KO) groups

```
> options(digits=4)
> library(FSA)
> # Summarize taxonomy by groups
> Summarize(D_6__Ruminiclostridium.sp..KB18 ~ Group4, data = ds)
 Group4 n mean sd min Q1 median Q3 max percZero
1 KO_BEFORE 10 4.1 7.838 0 0.00 0.0 6.00 24 70
2 KO_POST 10 11.8 13.951 0 0.75 7.5 17.75 44 30
3 WT_BEFORE 10 0.3 0.483 0 0.00 0.0 0.75 1 70
4 WT_POST 10 92.9 111.588 0 0.00 46.5 171.75 294 50
```

The species "D_6__Ruminiclostridium.sp.KB18" has 70% of zeros in KO_BEFORE and WT_BEFORE, 30% of zeros in KO_POST, and 50% of zeros in WT_POST. The variances are greater than their means. The distributions suggest that this species may be over-dispered and zero-inflated. See below plots (Figs. 17.1 and 17.2).

```
> # Figure 17.1
> # Individual plots in panel of 2 columns and 2 rows
> library(lattice)
> histogram(~ D_6__Ruminiclostridium.sp..KB18|Group4, data=ds,
layout=c(2,2))


> # Figure 17.2
> # Check outcome distribution and zeros
> par(mfrow=c(1,2))
> plot(table(ds$D_6__Ruminiclostridium.sp..KB18),
ylab="Frequencies",main="D_6__Eubacterium.sp..14.2",
+ xlab="Observed read values")
> plot(sort(ds$D_6__Ruminiclostridium.sp..KB18),
```

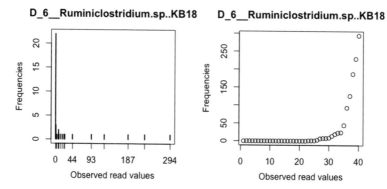

Fig. 17.2 Distribution of Ruminiclostridium.sp.KB18

```
ylab="Frequencies",main="D_6__Eubacterium.sp..14.2",
+ xlab="Observed read values")
```

Step 5: Run the glmmTMB() function to implement glmmTMB.

For each outcome taxonomy variable (here, D_6__Ruminiclostridium.sp.KB18), we fit fixed effects of two main factors Group, Time, and their interaction term. For random effects, we specify that taxonomy counts vary randomly by mouse.

Poisson Model(family = poisson)
The syntax for fitting GLMMs with glmmTMB is quite similar to using glmer. A log link is the standard link function for the Poisson distribution.

```
> library(glmmTMB)
# Using offset
> poi = glmmTMB(D_6__Ruminiclostridium.sp..KB18 ~ Group*Time + offset
(log(Total.Read))+(1|MouseID), data=ds, family=poisson)
# Without using offset
> poi = glmmTMB(D_6__Ruminiclostridium.sp..KB18 ~ Group*Time +(1|
MouseID), data=ds, family=poisson)
```

Conway-Maxwell-Poisson Model(family = compois)
Conway-Maxwell-Poisson model is fitted by using family = compois.

```
# Using offset
> cmpoi = glmmTMB(D_6__Ruminiclostridium.sp..KB18 ~ Group*Time +
offset(log(Total.Read)) + (1|MouseID), data=ds, family=compois)

# Without using offset
> cmpoi = glmmTMB(D_6__Ruminiclostridium.sp..KB18 ~ Group*Time + (1|
MouseID), data=ds, family=compois)
```

Negative Binomial Models
(family = nbinom1)

The following commands specify negative binomial model.

```
# Using offset
> nb1 = glmmTMB(D_6__Ruminiclostridium.sp..KB18 ~ Group*Time + offset
(log(Total.Read)) + (1|MouseID), data=ds, family=nbinom1)
# Without using offset
> nb1 = glmmTMB(D_6__Ruminiclostridium.sp..KB18 ~ Group*Time + (1|
MouseID), data=ds, family=nbinom1)
```

(family = nbinom2)

```
# Using offset
> nb2 = glmmTMB(D_6__Ruminiclostridium.sp..KB18 ~ Group*Time + offset
(log(Total.Read)) + (1|MouseID), data=ds, family=nbinom2)
# Without using offset
> nb2 = glmmTMB(D_6__Ruminiclostridium.sp..KB18 ~ Group*Time + (1|
MouseID), data=ds, family=nbinom2)
```

(family = nbinom1 and disp = ~Time)

Poisson distribution only has mean parameter, while the negative binomial distribution has both mean and dispersion parameters. The following commands specify that the taxonomy counts to become more dispersed (relative to the mean) over time using disp = ~ Time.

```
# Using offset
> nbdisp1 = glmmTMB(D_6__Ruminiclostridium.sp..KB18 ~ Group*Time +
offset(log(Total.Read)) + (1|MouseID), disp=~Time, data=ds,
family=nbinom1)
# Without using offset
> nbdisp1 = glmmTMB(D_6__Ruminiclostridium.sp..KB18 ~ Group*Time + (1|
MouseID), disp=~Time, data=ds, family=nbinom1)
```

(family = nbinom2 and disp = ~Time)

```
# Using offset
> nbdisp2 = glmmTMB(D_6__Ruminiclostridium.sp..KB18 ~ Group*Time +
offset(log(Total.Read)) + (1|MouseID), disp=~Time, data=ds,
family=nbinom2)
# Without using offset
> nbdisp2 = glmmTMB(D_6__Ruminiclostridium.sp..KB18 ~ Group*Time + (1|
MouseID), disp=~Time, data=ds,family=nbinom2)
```

Zero-Inflated Models

Unlike Poisson and negative binomial distributions, zero-inflated models have the capability to model how the probability of an extra zero (i.e., structural zero) will vary with predictors, which is fitted by using the ziformula argument or zi sub-model in glmmTMB. The formula of sub-model zi only has a right side because the left side of this formula has been specified in the first formula, which always models the probability of having a structural zero in the response. The probability of zero-inflation is always modeled with a logit link to ensure its value between 0 and 1. In the following commands, we specify the same two main factors Group, Time, and their interaction term as in conditional model. This specification assumes that absences will vary not only by group or time as well as by group and time interaction. Zero-inflation can be used with any of the distributions in glmmTMB, including zero-inflated Poisson, Conway-Maxwell-Poisson, and negative binomial distributions.

ZIP(family= poisson and zi = ~ Group)

The following commands specify a zero-inflated Poisson distribution with zero-inflated by group, time, and their interaction.

```
# Using offset
> zip0 = glmmTMB(D_6__Ruminiclostridium.sp..KB18 ~ Group*Time + offset
(log(Total.Read)) + (1|MouseID), zi=~ Group*Time, data=ds, family=
poisson)
Warning message:
In (function (start, objective, gradient = NULL, hessian = NULL, :
NA/NaN function evaluation
# Without using offset
> zip0 = glmmTMB(D_6__Ruminiclostridium.sp..KB18 ~ Group*Time + (1|
MouseID), zi=~ Group*Time, data=ds, family= poisson)
Warning messages:
1: In fitTMB(TMBStruc) :
Model convergence problem; non-positive-definite Hessian matrix. See
vignette('troubleshooting')
2: In fitTMB(TMBStruc) :
Model convergence problem; singular convergence (7). See vignette
('troubleshooting')
```

Warning messages tell us that the zero-inflated Poisson model specified in the object zip0 does not converge. Failed convergence are most likely caused by one or more following problems:

- When a model is overparameterized, then the data does not contain enough information to estimate the parameters reliably.
- When a random-effect variance is estimated to be zero, or random-effect terms are estimated to be perfectly correlated ("singular fit").
- When zero-inflation is estimated to be near zero (a strongly negative zero-inflation parameter).
- When dispersion is estimated to be near zero.
- When *complete separation* occurs in a binomial model: some categories in the model contain proportions that are either all 0 or all 1.

Commonly these problems are typically caused by trying to estimate parameters for which the data do not contain information, such as a zero-inflated and over-dispersed model is specified for the data that does not have zero-inflation and/or over-dispersion, or a random effect is assumed for varying not necessary predictors. The general model convergence issues are discussed in vignette ("troubleshooting"). The readers are referred to this vignette to obtain advice for troubleshooting convergence issues that have been developed ("troubleshooting", package = "glmmTMB"). For the detailed estimation challenges in GLMMs, The readers are referred to Sect. 17.1.3.

We should not consider the models that do not converge in model comparison. Since specifying zi = ~ Group*Time causes the model not to converge, we instead specify zi = ~Group below.

```
# Using offset
> zip = glmmTMB(D_6__Ruminiclostridium.sp..KB18 ~ Group*Time + offset
(log(Total.Read)) + (1|MouseID), zi=~ Group, data=ds, family=
poisson)
# Without using offset
> zip = glmmTMB(D_6__Ruminiclostridium.sp..KB18 ~ Group*Time + (1|
MouseID), zi=~ Group, data=ds, family= poisson)
```

ZICMP(family = compois and zi = ~ Group as well as zi = ~ Group*Time)

The following commands specify a zero-inflated Conway-Maxwell-Poisson distribution.

```
# Using offset
> zicmp0 = glmmTMB(D_6__Ruminiclostridium.sp..KB18 ~ Group*Time +
offset(log(Total.Read)) + (1|MouseID), zi=~ Group*Time, data=ds,
family=compois)
# Without using offset
> zicmp0 = glmmTMB(D_6__Ruminiclostridium.sp..KB18 ~ Group*Time + (1|
MouseID), zi=~ Group*Time, data=ds, family=compois)
```

```
# Using offset
> zicmp = glmmTMB(D_6__Ruminiclostridium.sp..KB18 ~ Group*Time +
offset(log(Total.Read)) + (1|MouseID), zi=~ Group, data=ds,
family=compois)
# Without using offset
> zicmp = glmmTMB(D_6__Ruminiclostridium.sp..KB18 ~ Group*Time + (1|
MouseID), zi=~ Group, data=ds, family=compois)
```

ZINB(family = nbinom1 and zi = ~ Group*Time)

The following commands specify a zero-inflated negative binomial distribution.

```
# Using offset
> zinb1 = glmmTMB(D_6__Ruminiclostridium.sp..KB18 ~ Group*Time +
offset(log(Total.Read)) + (1|MouseID), zi=~ Group*Time, data=ds,
family=nbinom1)
```

```
Warning messages:
1: In fitTMB(TMBStruc) :
 Model convergence problem; non-positive-definite Hessian matrix. See
vignette('troubleshooting')
2: In fitTMB(TMBStruc) :
 Model convergence problem; false convergence (8). See vignette
('troubleshooting')
# Without using offset
>zinb1 = glmmTMB(D_6__Ruminiclostridium.sp..KB18 ~ Group*Time + (1|
MouseID), zi=~ Group*Time, data=ds, family=nbinom1)

# Using offset
> zinb1a = glmmTMB(D_6__Ruminiclostridium.sp..KB18 ~ Group*Time +
offset(log(Total.Read)) + (1|MouseID), zi=~ Group, data=ds,
family=nbinom1)
Warning messages:
1: In fitTMB(TMBStruc) :
 Model convergence problem; non-positive-definite Hessian matrix. See
vignette('troubleshooting')
2: In fitTMB(TMBStruc) :
 Model convergence problem; false convergence (8). See vignette
('troubleshooting')

# Without using offset
>zinb1a = glmmTMB(D_6__Ruminiclostridium.sp..KB18 ~ Group*Time + (1|
MouseID), zi=~ Group, data=ds, family=nbinom1)
Error in eigen(h) : infinite or missing values in 'x'
In addition: Warning message:
In (function (start, objective, gradient = NULL, hessian = NULL, :
NA/NaN function evaluation
```

ZINB(family = nbinom2 and zi = ~ Group)

```
# Using offset
> zinb2a = glmmTMB(D_6__Ruminiclostridium.sp..KB18 ~ Group*Time +
offset(log(Total.Read)) + (1|MouseID), zi=~ Group*Time, data=ds,
family=nbinom2)
Warning message:
In fitTMB(TMBStruc) :
 Model convergence problem; singular convergence (7). See vignette
('troubleshooting')

# Without using offset
>zinb2a = glmmTMB(D_6__Ruminiclostridium.sp..KB18 ~ Group*Time + (1|
MouseID), zi=~ Group*Time, data=ds, family=nbinom2)
Warning messages:
1: In fitTMB(TMBStruc) :
 Model convergence problem; non-positive-definite Hessian matrix. See
vignette('troubleshooting')
2: In fitTMB(TMBStruc) :
```

```
Model convergence problem; singular convergence (7). See vignette
('troubleshooting')
```

Since zinb2a does not converge, we specify zi = ~ Group in below zinb2 instead.

```
# Using offset
> zinb2 = glmmTMB(D_6__Ruminiclostridium.sp..KB18 ~ Group*Time +
offset(log(Total.Read)) + (1|MouseID), zi=~ Group, data=ds,
family=nbinom2)
# Without using offset
> zinb2 = glmmTMB(D_6__Ruminiclostridium.sp..KB18 ~ Group*Time + (1|
MouseID), zi=~ Group, data=ds, family=nbinom2)
```

Hurdle Models

We fit zero-hurdle Poisson model by using a truncated Poisson distribution and specify the same predictors in both the conditional model and the zero-inflation model.

ZHP(family = truncated_poisson and zi = ~ Group*Time)

```
# Using offset
> zhp = glmmTMB(D_6__Ruminiclostridium.sp..KB18 ~ Group*Time + offset
(log(Total.Read)) + (1|MouseID), zi=~ Group*Time, data=ds,
family=truncated_poisson)

# Without using offset
> zhp = glmmTMB(D_6__Ruminiclostridium.sp..KB18 ~ Group*Time + (1|
MouseID), zi=~ Group*Time, data=ds, family=truncated_poisson)
```

ZHNB (family = truncated_nbinom1/truncated_nbinom2 and zi = ~ Group*Time)

The following commands fit a zero-hurdle negative binomial model by using a truncated negative binomial distribution and specify the same predictors in both the conditional model and the zero-inflation model.

```
# Using offset
> zhnb1 = glmmTMB(D_6__Ruminiclostridium.sp..KB18 ~ Group*Time +
offset(log(Total.Read)) + (1|MouseID), zi=~ Group*Time, data=ds,
family= truncated_nbinom1)
Warning messages:
1: In fitTMB(TMBStruc) :
 Model convergence problem; non-positive-definite Hessian matrix. See
vignette('troubleshooting')
2: In fitTMB(TMBStruc) :
 Model convergence problem; singular convergence (7). See vignette
('troubleshooting')
# Without using offset
```

```
> zhnb1 = glmmTMB(D_6__Ruminiclostridium.sp..KB18 ~ Group*Time + (1|
MouseID), zi=~ Group*Time, data=ds, family= truncated_nbinom1)
```

```
# Using offset
> zhnb2 = glmmTMB(D_6__Ruminiclostridium.sp..KB18 ~ Group*Time +
offset(log(Total.Read))
+ (1|MouseID), zi=~ Group*Time, data=ds, family= truncated_nbinom2)
Warning message:
In fitTMB(TMBStruc) :
 Model convergence problem; non-positive-definite Hessian matrix. See
vignette('troubleshooting')
# Without using offset
> zhnb2 = glmmTMB(D_6__Ruminiclostridium.sp..KB18 ~ Group*Time + (1|
MouseID), zi=~ Group*Time, data=ds, family= truncated_nbinom2)
```

Step 6: Perform model comparison and model selection using AIC and BIC.

We can use the functions **AICtab ()** and **BICtab()** (standing for a table of AIC and BIC model comparisons, respectively) from the **bbmle** package to compare all the GLMMs, including zero-inflated and hurdle models (K.P. Burnham and Anderson 2002). To test the effects of using offset in the models, we separately specify the models using offset to adjust for the total count reads and without using offset.

Model comparison using AIC and BIC and offset.

```
#install.packages("bbmle")
> library(bbmle)
# Using offset
> AICtab(poi, cmpoi,nb1,nb2, nbdisp1, nbdisp2, zip, zip0, zicmp,
zicmp0, zinb1,zinb2,zhp,zhnb1, zinb1a, zhnb2,zinb2a)
 dAIC df
zhp 0.0 9
zinb2a 0.7 10
zicmp0 0.8 10
zinb2 1.1 8
zicmp 1.5 8
zip0 2.4 9
zip 2.5 7
nb2 17.5 6
nbdisp2 19.4 7
nbdisp1 20.9 7
nb1 21.9 6
cmpoi 28.2 6
poi 95.1 5
zinb1 NA 10
zhnb1 NA 10
zinb1a NA 8
zhnb2 NA 10

> # Using offset
>BICtab(poi, cmpoi,nb1,nb2, nbdisp1, nbdisp2, zip,zip0, zicmp,zicmp0,
```

```
zinb1,zinb2,zhp,zhnb1, zinb1a, zhnb2,zinb2a)
 dBIC df
zip 0.0 7
zinb2 0.3 8
zicmp 0.7 8
zhp 0.9 9
zinb2a 3.3 10
zip0 3.3 9
zicmp0 3.4 10
nb2 13.3 6
nbdisp2 16.9 7
nb1 17.7 6
nbdisp1 18.5 7
cmpoi 24.0 6
poi 89.2 5
zinb1 NA 10
zhnb1 NA 10
zinb1a NA 8
zhnb2 NA 10
```

Model comparison using AIC and BIC but without offset.

```
# Without using offset
> AICtab(poi, cmpoi,nb1,nb2, nbdisp1, nbdisp2, zip, zip0,zicmp,
zicmp0, zinb1,zinb2,zhp,zhnb1,zhnb2,zinb2a)
 dAIC df
zhnb2 0.0 10
zhp 0.8 9
zicmp0 1.2 10
zinb2 1.7 8
zicmp 1.8 8
zhnb1 2.3 10
zinb1 4.1 10
nb2 19.3 6
nbdisp2 21.2 7
nbdisp1 22.6 7
nb1 23.2 6
zip 28.5 7
cmpoi 29.7 6
poi 94.6 5
zip0 NA 9
zinb1a NA 8
zinb2a NA 10

# Without using offset
> BICtab(poi, cmpoi,nb1,nb2, nbdisp1, nbdisp2, zip,zip0, zicmp,
zicmp0, zinb1,zinb2,zhp,zhnb1,zhnb2,zinb2a)
 dBIC df
zinb2 0.0 8
zicmp 0.1 8
```

```
zhp 0.7 9
zhnb2 1.6 10
zicmp0 2.9 10
zhnb1 3.9 10
zinb1 5.8 10
nb2 14.2 6
nbdisp2 17.8 7
nb1 18.1 6
nbdisp1 19.2 7
cmpoi 24.5 6
zip 25.1 7
poi 87.8 5
zip0 NA 9
zinb1a NA 8
zinb2a NA 10
```

Both AICtab () and BICtab()functions report the log-likelihood of the unconverged models as NA at the end of the AIC and BIC tables.

We print AIC and BIC values from each model in Table 17.1 to compare models and check the effects of using offset.

From Table 17.1, we can see using offset does not improve model fit for this specific outcome variable: either there are more non-converged models using offset or more models have slightly worse model fits. Using offsets only zero-inflated models have better model fits than those without using offsets, but some warning messages are generated, so the model evaluation is suspicious.

Based on above goodness of fit values evaluated by AIC and BIC, zero-hurdle NB models, zero-hurdle Poisson model, and the zero-inflated NB models, as well as zero-inflated Conway-Maxwell-Poisson models have better model fits compared to

Table 17.1 AIC and BIC values in compared models from glmmTMB

Model	No offset		Using offset	
	AIC	BIC	AIC	BIC
zhnb2	201.3	218.2	NA	NA
zhp	202.1	217.3	203.2	218.4
zicmp0	202.5	219.4	204.0	220.9
zinb2	203.1	216.6	204.3	217.9
zicmp	203.1	216.6	204.7	218.2
zhnb1	203.6	220.5	NA	NA
zinb1	205.4	222.3	NA	NA
nb2	220.6	230.8	220.7	230.9
nbdisp2	222.5	234.4	222.6	234.5
nbdisp1	223.9	235.7	224.2	236.0
nb1	224.5	234.6	225.1	235.2
zip	229.9	241.7	205.7	217.5
cmpoi	231.0	241.1	231.4	241.5
poi	295.9	304.3	298.3	306.7
zip0	NA	NA	205.6	220.8
zinb1a	NA	NA	NA	NA

other competing models. The one parametric Poisson model has the worst performance. The performance of negative binomial models is in the middle. The performance of zero-inflated Poisson (ZIP) is unstable and has some unreliable properties such as underestimating small counts and overestimating intermediate counts (Xia et al. 2012). Additionally, although ZIP has larger power than NB to estimate a larger probability of zeros when they have the same mean values and variances; however, ZIP has less capability to model over-dispersion than NB if they have the same mean values and probabilities of zeros (Feng et al. 2015).

Step 7: Perform residual diagnostics for GLMMs using DHARMa.

We can evaluate the goodness-of-fit of fitted GLMMs using the procedures described in the **DHARMa** package (version 0.4.5, 2022-01-16). The R package DHARMa (or strictly speaking, DHARM) (Hartig 2022) stands for "Diagnostics for HierArchical Regression Models." The name DHARMa is used to avoid confusion with the term Darm which in German means intestines and also taking the meaning of DHARMa in Hinduism which signifies behaviors should be in accord with rta, the order that makes life and universe possible, including duties, rights, laws, conduct, virtues, and "right way of living." The meaning of DHARMa in Hinduism is purposed to play for this package, i.e., to test whether fitted model is in harmony with the data tested.

DHARMa uses a simulation-based approach to create readily interpretable scaled (quantile) residuals for fitted (generalized) linear mixed models. Currently the linear and generalized linear (mixed) models supported by DHARMa are from "lme4" (classes "lmerMod," "glmerMod"), "glmmTMB," "GLMMadaptive" and "spam," generalized additive models ("gam" from "mgcv"), "glm" (including "negbin" from "MASS," but not quasi-distributions), and "lm" model classes.

DHARMa standardizes the resulting residuals to values between 0 and 1. Thus, the residuals can be intuitively interpreted as residuals from a linear regression. There are two basic steps behind the residual method: (1) Simulate new response data from the fitted model for each observation. (2) For each observation, calculate the empirical cumulative density function for the simulated observations, which describes the possible values and their probability given the observed predictors and assuming the fitted model is correct. The residual generated this way represents the value of the empirical density function at the value of the observed data. Thus, a residual of 0 means that all simulated values are larger than the observed value, and a residual of 0.5 means that half of the simulated values are larger than the observed value (Dunn and Smyth 1996; Gelman and Hill 2006).

The package provides a number of plot and test functions for diagnosing typical model misspecification problems, including over−/under-dispersion and zero-inflation.

DHARMa can be installed either by running the following command in R or RStudio:

```
install.packages("DHARMa")
```

Or through https://github.com/florianhartig/DHARMa to install a development version.

```
> library(DHARMa)
```

Most functions in DHARMa can be implemented directly on the fitted model object. For example, the following command tests for dispersion problems:

```
> testDispersion(fittedModel)
```

However, doing in this way, every test or plot needs to repeat residual calculation. Thus, DHARMa highly recommends first calculating the residuals once using the function simulateResiduals(). The syntax is given below:

```
> SimulatedOutput <- simulateResiduals(fittedModel = fittedModel, plot
= F)
```

The randomized quantile residuals are calculated by this function call. The function returns an object of class DHARMa, containing the simulations and the scaled residuals. We can pass them on to all other plots and test functions. If we specify the optional argument plot = T, then the standard DHARMa residual plot is displayed directly. We also can plot and test the calculated (scaled) residuals via a number of DHARMa functions, or access directly via calling the residuals () function:

```
> residuals(SimulatedOutput)
```

Here, we illustrate how to evaluate the goodness-of-fit of two fitted GLMMs (the worst model: Poisson, and one of the best models: zinb2) using the procedures described in the **DHARMa** package.

Step 7a: Simulate and plot the scaled residuals from the fitted model using the functions simulateResiduals() and plot.DHARMa().

Here, the following commands simulate and plot the scaled residuals from the fitted Poisson model (Fig. 17.3).

```
> # Figure 17.3
> SimulatedOutput_poi <- simulateResiduals(fittedModel = poi)
> plot(SimulatedOutput_poi)
```

The QQ-plot of the observed versus expected residuals is created by the plotQQunif () function to detect overall deviations from the expected distribution. The scaled residuals have a uniform distribution in the interval (0,1) for a well-specified model. The *P*-value reported in the plot is obtained from the Kolmogorov-

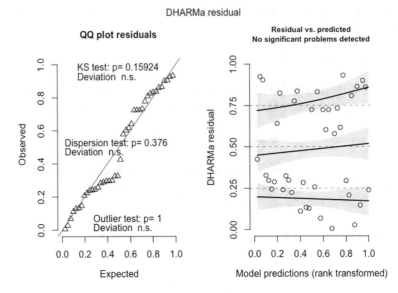

Fig. 17.3 QQ-plot residuals (left panel) and the scatterplot of the residuals against the predicted (fitted) values plot (right panel) in fitted Poisson model by the glmmTMB package

Smirnov test for testing uniformity. By default, the three generated tests are KS test (for correct distribution), dispersion test (for detecting dispersion), and outlier test (for detecting outliers). The plot generated by the **plotResiduals** () by default is the residuals against the predicted values, but the plot of the residuals against specific predictors is highly recommended. To visualize the effects in detecting deviations from uniformity in y-direction, the plot function also calculates a quantile regression to compare the empirical 0.25, 0.5, and 0.75 quantile lines across the plots in y-direction. These lines should be straight, horizontal, and at y-values of the theoretical 0.25, 0.5, and 0.75 quantiles.

The above plots can be created separately.

```
> plotQQunif(SimulatedOutput_poi) # Left plot in plot.DHARMa()
> plotResiduals(SimulatedOutput_poi) # Right plot in plot.DHARMa()
```

Step 7b: Perform goodness-of-fit tests on the simulated scaled residuals.

The following commands first use the function **plotResiduals()** to estimate and test the within group residuals. Then use the function **testDispersion()** to test if the simulated dispersion is equal to the observed dispersion, and the function **testZeroinflation()** to test if there are more zeros in the data than expected from the simulations. The generated three plots are represented in a panel with 1 row by 3 columns (Fig. 17.4).

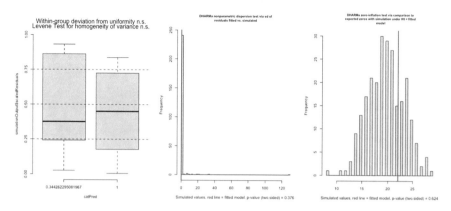

Fig. 17.4 Levene test for homogeneity of variance (left panel), nonparametric dispersion test (middle panel), and zero-inflation test (right panel) for fitted Poisson model by the glmmTMB package

```
> # Figure 17.4
> par(mfrow = c(1,3))
> plotResiduals(SimulatedOutput_poi, ds$Group)
> testDispersion(SimulatedOutput_poi)

DHARMa nonparametric dispersion test via sd of residuals fitted vs.
 simulated

data: simulationOutput
dispersion = 2.8e-05, p-value = 0.4
alternative hypothesis: two.sided

> testZeroInflation(SimulatedOutput_poi)

DHARMa zero-inflation test via comparison to expected zeros with
 simulation under H0 = fitted model

data: simulationOutput
ratioObsSim = 1.1, p-value = 0.6
alternative hypothesis: two.sided
```

In this case, both dispersion and zero-inflation tests are not statistically significant.

In DHARMa, several overdispersion tests are available for comparing the dispersion of simulated residuals to the observed residuals. The syntax is given:

```
> testDispersion(SimulatedOutput, alternative = c("two.sided",
"greater",
less"), plot = T, type = c("DHARMa", "PearsonChisq"))
```

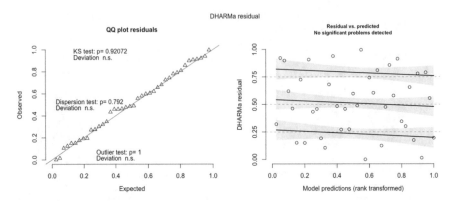

Fig. 17.5 QQ-plot residuals (left panel) and Residuals vs. predicted values plot (right panel) in fitted zinb2 model by the glmmTMB package

The default type is DHARMa, which is a non-parametric test to compare the variance of the simulated residuals to the observed residuals. Alternatively, it is *PearsonChisq*, which implements the Pearson-χ^2 test. While if residual simulations are done via *refit*, DHARMa will compare the Pearson residuals of the re-fitted simulations to the original Pearson residuals. This is essentially a nonparametric test. The author of the DHARMa package showed that the default test is fast, nearly unbiased for testing both under and over-dispersion, as well as reliably powerful.

One important resource of over-dispersion is zero-inflation. In DHARMa, zero-inflation can be formally tested using the function **testZeroInflation()**, such as here testZeroInflation(SimulatedOutput_poi), to compare the distribution of expected zeros in the data against the observed zeros.

Based on AIC and BIC model comparison results, zinb2 is one of the best models. In order to confirm this model comparison results and check if zinb2 has better model fit than Poisson model for this data, we simulate the scaled residuals from the fitted zinb2 model and perform goodness-of-fit tests on the simulated scaled residuals below (Figs. 17.5 and 17.6).

```
> # Figure 17.5
> rest_zinb2 <- simulateResiduals(fittedModel = zinb2, plot = T)
Or
> rest_zinb2 <- simulateResiduals(zinb2, plot = T)
```

```
> # Figure 17.6
par(mfrow = c(1,3))
> plotResiduals(rest_zinb2, ds$Group)
> testDispersion(rest_zinb2)
```

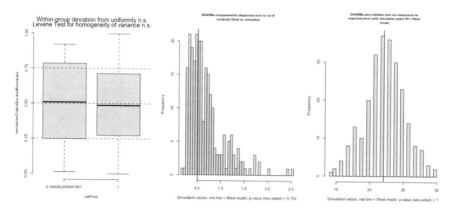

Fig. 17.6 Levene test for homogeneity of variance (left panel), nonparametric dispersion test (middle panel), and zero-inflation test (right panel) for fitted zinb2 model by the glmmTMB package

```
DHARMa nonparametric dispersion test via sd of residuals fitted vs.
simulated

data: simulationOutput
dispersion = 0.75, p-value = 0.8
alternative hypothesis: two.sided

> testZeroInflation(rest_zinb2)

DHARMa zero-inflation test via comparison to expected zeros with
simulation under H0 = fitted model

data: simulationOutput
ratioObsSim = 0.98, p-value = 1
alternative hypothesis: two.sided
```

It is obvious that the fitted zinb2 model outperforms the fitted Poisson model for the outcome variable D_6__Ruminiclostridium.sp.KB18.

Step 8: Summarize the modeling results and extract the outputs of selected models.

The following commands implement ZINB(family = nbinom2) and summarize the modeling results (Table 17.2).

```
>zinb2 = glmmTMB(D_6__Ruminiclostridium.sp..KB18 ~ Group*Time +
offset(log(Total.Read))
+ (1|MouseID), zi=~ Group, data=ds, family=nbinom2)
> summary(zinb2)
Family: nbinom2 ( log )
Formula:
```

Table 17.2 Modeling results from the models that use offset via glmmTMB

| Model | Conditional model | | | | |
	Interaction term	Estimate	Std. Error	z value	P-value
poi	Group:TimePost	−4.939	0.606	−8.15	3.6e-16
cmpoi	Group:TimePost	−4.347	1.032	−4.21	2.6e-05
nb1	Group:TimePost	−2.44	1.20	−2.03	0.0425
nb2	Group:TimePost	−4.905	1.512	−3.24	0.0012
nbdisp1	Group:TimePost	−1.669	1.107	−1.51	0.13
nbdisp2	Group:TimePost	−4.903	1.542	−3.18	0.0015
zip0	Group:TimePost	−6.879	0.645	−10.67	< 2e-16
zip	Group:TimePost	−6.291	0.732	−8.59	<2e-16
zicmp	Group:TimePost	−5.944	0.975	−6.10	1.1e-09
zicmp0	Group:TimePost	−6.423	0.909	−7.07	1.6e-12
zinb1	Group:TimePost	−5.827	0.939	−6.21	5.4e-10
zinb2	Group:TimePost	−5.936	0.858	−6.92	4.6e-12
zhp	Group:TimePost	−24.8	11825.8	0	1

```
D_6__Ruminiclostridium.sp..KB18 ~ Group * Time + offset(log(Total.
Read)) +
(1 | MouseID)
Zero inflation: ~Group
Data: ds

AIC BIC logLik deviance df.resid
204.3 217.9 -94.2 188.3 32

Random effects:

Conditional model:
Groups Name Variance Std.Dev.
MouseID (Intercept) 1.9e-08 0.000138
Number of obs: 40, groups: MouseID, 20

Dispersion parameter for nbinom2 family (): 3.09

Conditional model:
Estimate Std. Error z value Pr(>|z|)
(Intercept) -11.595 0.693 -16.73 < 2e-16 ***
Group 3.438 0.785 4.38 1.2e-05 ***
TimePost 5.861 0.739 7.93 2.2e-15 ***
Group:TimePost -5.936 0.858 -6.92 4.6e-12 ***
---
Signif. codes: 0 '***' 0.001 '**' 0.01 '*' 0.05 '.' 0.1 ' ' 1

Zero-inflation model:
Estimate Std. Error z value Pr(>|z|)
(Intercept) -0.212 0.580 -0.36 0.71
Group 0.201 0.734 0.27 0.78
```

The following commands implement ZHP(family = truncated_poisson) and summarize the modeling results.

```
> zhp = glmmTMB(D_6__Ruminiclostridium.sp..KB18 ~ Group*Time + offset
(log(Total.Read))
+ + (1|MouseID), zi=~ Group*Time, data=ds, family=truncated_poisson)
> summary(zhp)
 Family: truncated_poisson ( log )
Formula:
D_6__Ruminiclostridium.sp..KB18 ~ Group * Time + offset(log(Total.
Read)) +
 (1 | MouseID)
Zero inflation: ~Group * Time
Data: ds

AIC BIC logLik deviance df.resid
203.2 218.4 -92.6 185.2 31

Random effects:

Conditional model:
 Groups Name Variance Std.Dev.
 MouseID (Intercept) 0.366 0.605
Number of obs: 40, groups: MouseID, 20

Conditional model:
 Estimate Std. Error z value Pr(>|z|)
(Intercept) -30.0 11825.8 0 1
Group 22.1 11825.8 0 1
TimePost 24.2 11825.8 0 1
Group:TimePost -24.8 11825.8 0 1

Zero-inflation model:
 Estimate Std. Error z value Pr(>|z|)
(Intercept) 8.47e-01 6.90e-01 1.23 0.22
Group 5.23e-07 9.76e-01 0.00 1.00
TimePost -8.47e-01 9.36e-01 -0.90 0.37
Group:TimePost -8.47e-01 1.35e+00 -0.63 0.53
```

Significantly different Group by TimePost interaction in conditional model is identified in nbinom2 model; by contrast, this term is not significant difference in truncated Poisson model (Table 17.3).

Based on AIC, the best model is the zero-hurdle negative binomial (zhnb2) that assume the variance increases quadratically with the mean. We print the output below.

```
> zhnb2 = glmmTMB(D_6__Ruminiclostridium.sp..KB18 ~ Group*Time + (1|
MouseID), zi=~ Group*Time, data=ds, family= truncated_nbinom2)
> summary(zhnb2)
```

Table 17.3 Modeling results from the models that do not use offset via glmmTMB

| Model | Conditional model | | | | |
	Interaction term	Estimate	Std. Error	z value	P-value
poi	Group:TimePost	−4.678	0.606	- 7.72	1.2e-14
cmpoi	Group:TimePost	−4.135	1.026	−4.03	5.5e-05
nb1	Group:TimePost	−2.230	1.130	−1.97	0.04847
nb2	Group:TimePost	−4.68	1.51	- 3.10	0.0019
nbdisp1	Group:TimePost	−1.478	1.114	−1.33	0.18
nbdisp2	Group:TimePost	−4.68	1.54	−3.04	0.0024
zip	Group:TimePost	−6.037	0.627	−9.63	<2e-16
zicmp	Group:TimePost	−5.732	0.948	−6.05	1.5e-09
zicmp0	Group:TimePost	−6.271	0.857	−7.32	2.5e-13
zinb1	Group:TimePost	−6.247	0.937	−6.67	2.6e-11
zinb2	Group:TimePost	−5.711	0.840	−6.80	1.1e-11
zhp	Group:TimePost	−23.9	8656.4	0	1
zhnb1	Group:TimePost	−23.9	10093.7	0	1
zhnb2	Group:TimePost	−23.8	8597.3	0	1

```
Family: truncated_nbinom2 ( log )
Formula:
D_6__Ruminiclostridium.sp..KB18 ~ Group * Time + (1 | MouseID)
Zero inflation: ~Group * Time
Data: ds

AIC BIC logLik deviance df.resid
201.3 218.2 -90.7 181.3 30

Random effects:

Conditional model:
 Groups Name Variance Std.Dev.
 MouseID (Intercept) 0.0314 0.177
Number of obs: 40, groups: MouseID, 20

Dispersion parameter for truncated_nbinom2 family (): 3.64

Conditional model:
 Estimate Std. Error z value Pr(>|z|)
(Intercept) -18.8 8597.3 0 1
Group 21.4 8597.3 0 1
TimePost 24.0 8597.3 0 1
Group:TimePost -23.8 8597.3 0 1

Zero-inflation model:
 Estimate Std. Error z value Pr(>|z|)
(Intercept) 8.47e-01 6.90e-01 1.23 0.22
Group 4.26e-05 9.76e-01 0.00 1.00
```

```
TimePost -8.47e-01 9.36e-01 -0.90 0.37
Group:TimePost -8.47e-01 1.35e+00 -0.63 0.53
```

Group by TimePost interaction in conditional model is not statistically significant identified by truncated nbinom2 model.

Based on BIC, the best model is the zero-inflated negative binomial (zinb2) that assumes the variance increases quadratically with the mean. We print the output below.

```
> zinb2 = glmmTMB(D_6__Ruminiclostridium.sp..KB18 ~ Group*Time + (1|
MouseID), zi=~ Group, data=ds, family=nbinom2)
> summary(zinb2)
 Family: nbinom2 ( log )
Formula:
D_6__Ruminiclostridium.sp..KB18 ~ Group * Time + (1 | MouseID)
Zero inflation: ~Group
Data: ds

 AIC BIC logLik deviance df.resid
 203.1 216.6 -93.5 187.1 32

Random effects:

Conditional model:
 Groups Name Variance Std.Dev.
 MouseID (Intercept) 1.19e-08 0.000109
Number of obs: 40, groups: MouseID, 20

Dispersion parameter for nbinom2 family (): 3.39

Conditional model:
 Estimate Std. Error z value Pr(>|z|)
(Intercept) -0.705 0.686 -1.03 0.3
Group 3.310 0.772 4.29 1.8e-05 ***
TimePost 5.930 0.728 8.14 3.9e-16 ***
Group:TimePost -5.711 0.840 -6.80 1.1e-11 ***
---
Signif. codes: 0 '***' 0.001 '**' 0.01 '*' 0.05 '.' 0.1 ' ' 1

Zero-inflation model:
 Estimate Std. Error z value Pr(>|z|)
(Intercept) -0.215 0.580 -0.37 0.71
Group 0.208 0.733 0.28 0.78
```

Group by TimePost interaction in conditional model is statistically significant identified by zero-inflated nbinom2 model.

17.2.3 Remarks on glmmTMB

In this section, we demonstrated that glmmTMB can be easily used to fit complicated models, although maximally complex model might not be necessary and might not converge. The **glmmTMB** function has the similar interface to the **lme4** package. Overall, glmmTMB is a very flexible package for modeling zero-inflated and over-dispersed count data. Although it is not time efficient, glmmTMB is able to estimate the Conway-Maxwell-Poisson distribution parameterized by the mean (Brooks et al. 2017). Fitting Conway-Maxwell-Poisson is one unique feature among packages that fit zero-inflated and over-dispersed mixed models. The benefits of using this package (Brooks et al. 2017) include:

- Various models including GLMs, GLMMs, zero-inflated GLMMs, and hurdle models can be quickly fitted in a single package.
- The information criteria provided in the modeling results facilitates the model comparisons and model selection via using likelihood-based methods to compare the model fitting of the estimated models.
- At least one simulation study in microbiome literature (Xinyan Zhang and Yi 2020b) confirmed that **glmmTMB** controlled the false-positive rates close to the significance level over different sample sizes, while remaining reasonable empirical power.

However, glmmTMB also has some disadvantages, such as:

- Estimating Conway-Maxwell-Poisson distribution is time expensive when the model is over-parameterized.
- Convergence is an issue when the model is over specified with more than necessary predictors.

Based on AIC and BIC and conditional model in detecting significant interesting variables, we recommend using zinb (zinb1 and zinb2) and zicmp rather than zhp and zhnb (zhnb1 and zhnb2). They are all robust in modeling estimations and sensitive to identify significant interesting variables. The concept adjustment for using zero-hurdle and zero-inflated models, the readers are referred to Xia et al. (2018b).

The benefit of using offset has not been approved in microbiome research, and like the practice of sequencing depth rarefaction (rarefying sequencing to same depth), which is adopted from macro-ecology, the argument of using offset has not been solidly validated and is difficult to validate. In this study, we compared same GLMMs using and without using offset and found that the goodness-of-fit not necessarily improved by using offset based on AIC and BIC. Actually several advanced/completed GLMMs had model convergence problems. The non-improvement of using offset may be due to two reasons: (1) For univariate approach (using one taxon per time as the response variable), it is not appropriate or not necessary to adjust for sample total reads. (2) For advanced/completed GLMMs, it is a burden to add additional parameter to estimate.

17.3 Generalized Linear Mixed Modeling Using the GLMMadaptive Package

In this section, we introduce another R package GLMMadaptive and illustrate its use with microbiome data.

17.3.1 Introduction to GLMMadaptive

GLMMadaptive was developed by Dimitris Rizopoulos (2022a) to fit GLMMs for a single grouping factor under maximum likelihood approximating the integrals over the random effects. Unlike the lme4 and glmmTMB packages, this package uses an adaptive Gauss-Hermite quadrature (AGQ) rule (Pinheiro and Bates 1995). GLMMadaptive provides functions for fitting and post-processing mixed effects models for grouped/repeated measurements/clustered outcomes for which the integral over the random effects in the definition of the marginal likelihood cannot be solved analytically. This package fits mixed effects models for grouped/repeated measurements data which have a non-normal distribution, while allowing for multiple correlated random effects.

In GLMMadaptive, the AGQ rule is efficiently implemented to allow for specifying multiple correlated random effects terms for the grouping factor including random intercepts, random linear, and random quadratic slopes. It also offers several utility functions for extracting useful information from the fitted mixed effects models. Additionally, GLMMadaptive provides a hybrid optimization procedure: starting with implementing an EM algorithm, treating the random effects as "missing data," and then performing a direct optimization procedure with a quasi-Newton algorithm.

17.3.2 Implement GLMMs via GLMMadaptive

GLMMadaptive implement GLMMs via a single model-fitting function **mixed_model()** with four required arguments. One example syntax is given below.

mixed_model(fixed $= y \sim x_1 + x_2$, random $= \sim 1 \mid g$, data $=$ df, family $=$ zi.negative. binomial()).

where the argument:

- **fixed** is a formula that is used for specifying the fixed effects including the response (outcome) variable.
- **random** is a formula that is used for specifying the random effects.

Table 17.4 Family objects and models that can be fitted in GLMMadaptive

Family object	Model to fit
negative.binomial()	Negative binomial
zi.poisson()	Zero-inflated Poisson
zi.negative.binomial()	Zero-inflated negative binomial
hurdle.poisson()	Hurdle Poisson
hurdle.negative.binomial()	Hurdle negative binomial
hurdle.lognormal()	Two-part/hurdle mixed models for semi-continuous normal data
censored.normal()	Mixed models for censored normal data
cr_setup() and cr_marg_probs ()	Continuation ratio mixed models for ordinal data
beta.fam()	Beta
hurdle.beta.fam()	Hurdle beta
Gamma() or Gamma.fam()	Gamma mixed effects models
censored.normal()	Linear mixed effects models with right and left censored data

- **family** is a family object that is used for specifying the type of the repeatedly measured response variable (e.g., binomial() or poisson()).
- **data** is a data frame containing the variables required in fixed and random.

In the mixed effects model, y denotes a grouped/clustered outcome, x_1 and x_2 denote the covariates, and g denotes the grouping factor. The above example syntax of mixed effects model specifies y as outcome and x_1 and x_2 as fixed effects and random intercepts. The mixed effects model is fitted by zero-inflated negative binomial model. If we also want to specify both intercepts and x_1 as the random effects, then update the random = $\sim x_1$ | g (e.g., $\sim$ time | id) in the call to **mixed_model**()function.

GLMMadaptive provides several standard methods to access the returned modeling results: summary(), anova(), coef(), fixef(), ranef(), vcov(), logLik(), confint(), fitted(), residuals(), predict(), and simulate(). GLMMadaptive can fit various GLMMs. We summarize the family objects and the models that can be fitted by GLMMadaptive in Table 17.4.

For the repeated measurements response variable, we may specify our own log-density function and use the internal algorithms for the optimization. The marginalized coefficients can be calculated via the **marginal_coefs** () using the approach of Hedeker et al. (2018) to calculate marginal coefficients from mixed models with nonlinear link functions. The GLMMadaptive package also provides predictions and their confidence intervals for constructing effects plots via the **effectPlotData** ().

Example 17.2: Breast Cancer QTRT1 Mouse G Microbiome, Example 17.1 Cont.

Here, we first use the same example data in Example 17.1 to illustrate how to fit negative binomial, zero-inflated Poisson, zero-inflated negative binomial, hurdle

Poisson, and hurdle negative binomial models using the GLMMadaptive package. We then illustrate how to perform model comparison and model selection to choose the best models. As we described in Example 17.1, the species "D_6__Ruminiclostridium.sp.KB18" has 30%, 50%, and 70% of zeros in genotype and wild-type (WT) groups before and posttreatments. We use this species to illustrate implmenting GLMMs using GLMMadaptive through the following six steps.

Step 1: Install package and load data.

To install the GLMMadaptive package (version 0.8-5, 2022-02-07), type the following commands in R or RStudio.

```
install.packages("GLMMadaptive")
```

The development version of the GLMMadaptive package and its vignettes are available from GitHub, which can be installed using the **devtools** package:

```
devtools::install_github("drizopoulos/GLMMadaptive")
```

We can install the package with vignettes using the **devtools** package:

```
devtools::install_github("drizopoulos/GLMMadaptive",
build_opts=NULL)

library(GLMMadaptive)
help(GLMMadaptive)
```

```
> setwd("~/Documents/QIIME2R/Ch17_GLMMs2")
> otu_tab=read.csv("otu_table_L7_MCF7.csv",row.names=1,check.
names=FALSE)
> meta_tab <- read.csv("metadata_QtRNA.csv",row.names=1,check.names
= FALSE)
```

Step 2: Create analysis dataset including outcomes and covariates.

In Sect. 17.2 (Example 17.1), we double filtered and created an analysis dataset in order to effectively implement the models. We repeat the core commands below.

```
> otu_tab_t<-t(otu_tab)
> outcome =otu_tab_t

> # Match sample IDs in outcome matrix or data frame and meta_tab data
frame
> outcome = outcome[rownames(meta_tab),]
> yy = as.matrix(outcome)
> yy = ifelse(is.na(yy), 0, yy) #Exclude missing outcome variables if
```

```
any
> # Filter abundance taxa data
> # Here we use double filters
> filter1 <- apply(yy, 2, function(x){sum(x>0)>0.4*length(x)})
> filter2 <- apply(yy, 2, function(x){quantile(x,0.9)>1})
> filter12 <-yy[,filter1&filter2]

> ## Match the number of taxa to the filter12 dataset in the analysis
> yy = filter12[rownames(meta_tab),]

> ds = data.frame(meta_tab,yy)
```

Step 3: Fit candidated mixed models.

```
> library(GLMMadaptive)
```

Poisson Mixed Model(family = poisson())

The following commands call the **mixed_model** () function to fit Poisson model
with fixed effects of group, time, and their interaction term, and randomly vary over
time.

```
> poi <- mixed_model(fixed = D_6__Ruminiclostridium.sp..KB18 ~
Group*Time, random = ~ Time | MouseID, data = ds, family = poisson())
```

Negative Binomial Mixed Model(family = negative.binomial())

```
>  nb  <-  mixed_model(fixed  =  D_6__Ruminiclostridium.sp..KB18  ~
Group*Time, random = ~ Time | MouseID, data = ds, family = negative.
binomial())
```

Zero-Inflated Poisson Mixed Effects Model(family = zi.poisson())

The following commands fit the zero-inflated Poisson mixed model to improve the
simple Poisson model, allowing for excess zeros. We specify intercepts as fixed
effects in the zero-inflated fixed model and assume that intercepts vary over mouse.

```
> zip <- mixed_model(fixed = D_6__Ruminiclostridium.sp..KB18 ~
Group*Time, random = ~ Time | MouseID, data = ds, family = zi.poisson(),
zi_fixed = ~ 1, zi_random = ~ 1| MouseID)
```

Zero-Inflated Negative Binomial Mixed Effects Model(family = zi.negative. binomial())

The following commands fit a zero-inflated negative binomial mixed effects model
with intercepts as the zero-inflated fixed and random effects.

```
> zinb <- mixed_model(fixed = D_6__Ruminiclostridium.sp..KB18 ~
Group*Time, random = ~ Time | MouseID, data = ds, family = zi.negative.
binomial(), zi_fixed = ~1, zi_random = ~ 1| MouseID)
```

Hurdle Poisson Mixed Effects Model(family = hurdle.poisson())

Hurdle Poisson mixed models can be fitted by mixed_model() using the family objects hurdle.poisson(). For hurdle Poisson typically a truncated at zero is used.

```
> zhp<- mixed_model(fixed = D_6__Ruminiclostridium.sp..KB18 ~
Group*Time, random = ~ Time | MouseID, data = ds, family = hurdle.poisson
(), zi_fixed = ~ 1, iter_EM = 0)
```

Hurdle Negative Binomial Mixed Effects Model (family = hurdle.negative.binomial())

Hurdle negative binomial mixed models can be fitted by mixed_model() using the family objects hurdle.negative.binomial(). For hurdle negative binomial mixed model typically a truncated at zero is used.

```
> zhnb<- mixed_model(fixed = D_6__Ruminiclostridium.sp..KB18 ~
Group*Time, random = ~ Time | MouseID, data = ds, family = hurdle.
negative.binomial(), zi_fixed = ~ 1, iter_EM = 0, iter_qN_outer = 0)
```

Step 4: Perform model comparisons and model selection.

Model Comparisons Using Likelihood Ratio Test

```
> # Use a likelihood ratio test to compare NB with Poisson
> anova(poi, nb)
 AIC BIC log.Lik LRT df p.value
poi 231.4 238.4 -108.7
nb 224.7 232.6 -104.3 8.79 1 0.003
```

```
> # Use a likelihood ratio test to compare ZINB with ZIP
> anova(zip, zinb)
 AIC BIC log.Lik LRT df p.value
zip 203.7 214.7 -90.86
zinb 205.8 217.7 -90.88 0.05 1 0.828
```

```
> # Use a likelihood ratio test to compare ZHNB with ZHP
> anova(zhp, zhnb)
 AIC BIC log.Lik LRT df p.value
zhp 201.1 209.1 -92.57
zhnb 226.2 235.2 -104.10 23.06 1 <0.0001
Warning messages:
1: In anova.MixMod(zhp, zhnb) : it seems that 'zhnb' has not converged.
2: In anova.MixMod(zhp, zhnb) :
 it seems that the two objects represent model with different families;
are the models nested? If not, you should set 'test' to FALSE.
```

Model Comparisons Using the bbmle Package

```
> library(bbmle)
> AICtab(poi,nb,zip,zinb,zhp,zhnb)
 dAIC df
zhp 0.0 8
zip 2.6 11
zinb 4.6 12
```

Table 17.5 AIC and BIC values in compared models from GLMMadaptive

Model	AIC	BIC
Zhp	201.1	209.1
Zip	203.7	214.7
Zinb	205.8	217.7
Nb	224.6	232.6
Zhnb	226.2	235.2
Poi	231.4	238.4

```
nb 23.5 8
zhnb 25.1 9
poi 30.3 7

> BICtab(poi,nb,zip, zinb,zhp, zhnb)
 dBIC df
zhp 0.0 8
zip 5.6 11
zinb 8.6 12
nb 23.5 8
zhnb 26.1 9
poi 29.3 7
```

In this case the zero-truncated hurdle (zero-inflated) Poisson models are a little bit better fitted to the data than zero-inflated negative binomial model and better than zero-inflated (negative binomial) models, suggesting that the variances of outcome variable "D_6__Ruminiclostridium.sp.KB18" is more due to zero-inflation than over-dispersion (see Example 17.1: there are 30%, 50%, and 70% of zeros in the samples before filtering) (Table 17.5).

Step 5: Evaluate goodness of fit using simulated residuals via the DHARMa **package.**

We can evaluate the goodness-of-fit of mixed models fitted by the mixed_model() function using the procedures described in the **DHARMa** package (version 0.4.5, 2022-01-16). Currently **DHARMa** does not support objects of class MixMod yet, to enable the use of the procedures of this package, a wrapper function for scaled simulated residuals has been developed by Dimitris Rizopoulos (2022b).

```
# Function for Scaled Simulated Residuals
# resids_plot:
 resids_plot <- function (object, y, nsim = 1000,
 type = c("subject_specific", "mean_subject"),
 integerResponse = NULL) {
 if (!inherits(object, "MixMod"))
 stop("this function works for 'MixMod' objects.\n")
 type <- match.arg(type)
 if (is.null(integerResponse)) {
 integer_families <- c("binomial", "poisson", "negative binomial",
 "zero-inflated poisson", "zero-inflated negative binomial",
 "hurdle poisson", "hurdle negative binomial")
```

```
numeric_families <- c("hurdle log-normal", "beta", "hurdle beta",
"Gamma")
if (object$family$family %in% integer_families) {
integerResponse <- TRUE
} else if (object$family$family %in% numeric_families) {
integerResponse <- FALSE
} else {
stop("non build-in family object; you need to specify the
'integerResponse',\n",
"\targument indicating whether the outcome variable is integer or not.
\n")
}
}
sims <- simulate(object, nsim = nsim, type = type)
fits <- fitted(object, type = type)
dharmaRes <- DHARMa::createDHARMa(simulatedResponse = sims,
observedResponse = y,
fittedPredictedResponse = fits,
integerResponse = integerResponse)
DHARMa:::plot.DHARMa(dharmaRes, quantreg = FALSE)
}
```

In above function, the argument:

- **object** is an object inheriting from class MixMod.
- **y** is the observed response vector.
- **nsim** is used to specify an integer number of simulated datasets(defaults is 1000).
- **type** is what type of fitted values and data to simulate; it is used for including the random effects or setting them to zero.
- **integerResponse** is a logical argument; it sets to TRUE for discrete grouped/ cluster outcome data. Based on the chosen family for fitting the model, the integerResponse is automatically determined by this function. But for user-specified family objects, this argument needs to be defined by the user. Here, we evaluate the goodness-of-fit of the chosen family, so the integerResponse argument is automatically determined.

We can either store the "resids_plot.R" wrapper function in R Script panel and call it directly or store it in Source File Location (here, "~/Documents/QIIME2R/ Ch17_GLMMs") and call it using the following command.

```
> source('resids_plot.R')
```

The following commands call the function **resids_plot()** and via the package DHARMa to generate two residual plots: QQ plot residuals on the left panel and residuals vs. predicted values (ranks transformed) on the right panel.

In the residual plot of residuals vs. predicted values, in order to visualize in detecting deviations from uniformity in y-direction, an (optional default) quantile regression is calculated to compare the empirical 0.25, 0.5, and 0.75 quantiles in y direction (red solid lines) with the theoretical 0.25, 0.5 and 0.75 quantiles (dashed

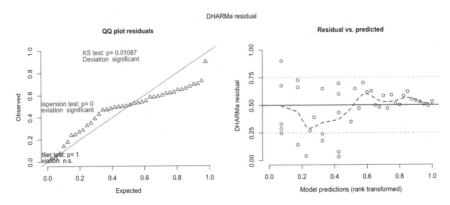

Fig. 17.7 QQ-plot residuals (left panel) and residuals vs. predicted values plot(right panel) in fitted Poisson model by the GLMMadaptive package

black line) and to provide a *P*-value for the deviation from the expected quantile. The significance of the deviation to the expected quantiles is tested and displayed visually and can be additionally extracted with the testQuantiles function (Fig. 17.7).

```
> # Figure 17.7
> library(DHARMa)
> source('resids_plot.R')
> resids_plot(poi, ds$D_6__Ruminiclostridium.sp..KB18)
```

Unlike in the glmmTMB, it has obvious evidence from these two plots that the Poisson model does not well fit the data satisfactorily (Fig. 17.8).

```
> # Figure 17.8
> testQuantiles(poi,ds$D_6__Ruminiclostridium.sp..KB18)
 Test for location of quantiles via qgam
data: simulationOutput
p-value = 3e-09
alternative hypothesis: both
```

First, let's check the simulated residuals from the fitted zero-truncated Poisson distribution (Figs. 17.9 and 17.10).

```
  > # Figure 17.9
  > resids_plot(zhp, ds$D_6__Ruminiclostridium.sp..KB18)
```

```
> # Figure 17.10
> testQuantiles(zhp,ds$D_6__Ruminiclostridium.sp..KB18)
 Test for location of quantiles via qgam
data: simulationOutput
p-value <2e-16
alternative hypothesis: both
```

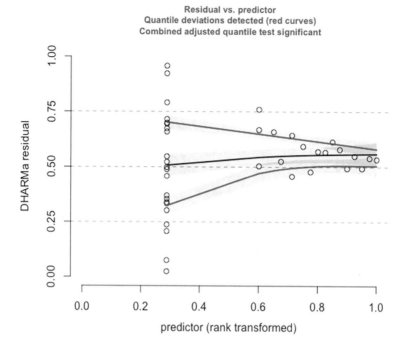

Fig. 17.8 The significance of the deviation of the empirical quantiles to the expected quantiles in fitted Poisson model is tested and visualized with the testQuantiles () function

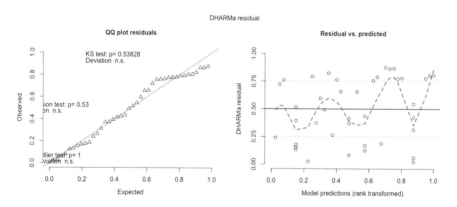

Fig. 17.9 QQ-plot residuals (left panel) and residuals vs. predicted values plot (right panel) in fitted zero-truncated Poisson model by the GLMMadaptive package

QQ-plot showed that the zero-truncated Poisson model has improved the model fit compared to the Poisson model. It seems to be acceptable. But based on the testing of the significance of the deviation of the empirical quantiles to the expected quantiles, the zero-truncated Poisson is still not well fitted.

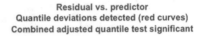

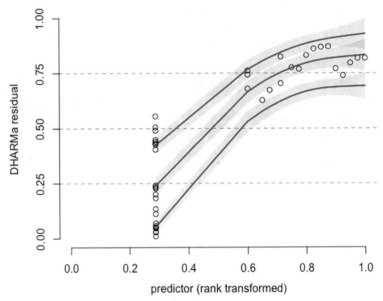

Fig. 17.10 The significance of the deviation of the empirical quantiles to the expected quantiles in fitted zero-truncated Poisson model is tested and visualized with the testQuantiles () function

The following commands check the simulated residuals from the fitted zero-inflated negative binomial distribution (Figs. 17.11 and 17.12).

```
> # Figure 17.11
> resids_plot(zinb, ds$D_6__Ruminiclostridium.sp..KB18)
```

```
> # Figure 17.12
> testQuantiles(zinb,ds$D_6__Ruminiclostridium.sp..KB18)
  Test for location of quantiles via qgam
data: simulationOutput
p-value = 3e-06
alternative hypothesis: both
```

QQ-plot showed that the model fit of zero-inflated negative binomial model is comparable to the model fit of the zero-truncated Poisson model. But the deviation between the empirical quantiles and the expected quantiles is still significant.

Similarly, the simulated residuals from other fitted models can be checked using the following commands.

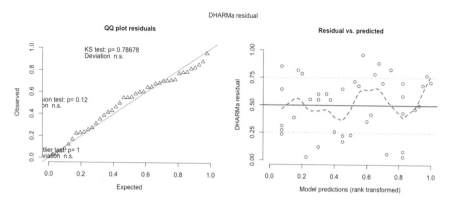

Fig. 17.11 QQ-plot residuals (left panel) and residuals vs. predicted values plot (right panel) in fitted zero-inflated negative binomial model by the GLMMadaptive package

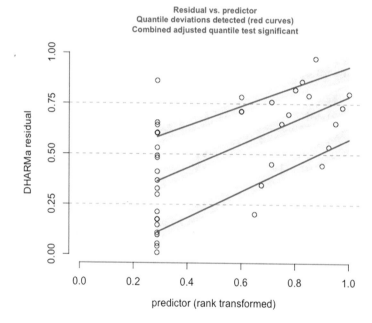

Fig. 17.12 The significance of the deviation of the empirical quantiles to the expected quantiles in fitted zero-inflated negative binomial model is tested and visualized with the testQuantiles () function

```
> resids_plot(nb, ds$D_6__Ruminiclostridium.sp..KB18)
>resids_plot(zip, ds$D_6__Ruminiclostridium.sp..KB18)
> resids_plot(zhnb, ds$D_6__Ruminiclostridium.sp..KB18)
```

Step 6: Summarize the modeling results and extract the outputs of selected models.

The partial output from ZINB is printed below.

```
> zinb <- mixed_model(fixed = D_6__Ruminiclostridium.sp..KB18 ~
Group*Time, random = ~ Time | MouseID, data = ds, family = zi.negative.
binomial(), zi_fixed = ~1, zi_random = ~ 1| MouseID)

Model:
 family: zero-inflated negative binomial
 link: log

Fit statistics:
 log.Lik AIC BIC
 -90.88 205.8 217.7

Random effects covariance matrix:
 StdDev Corr
(Intercept) 0.7268 (Intr) TimPst
TimePost 1.3550 -0.9886
zi_(Intercept) 1.9091 -0.9382 0.9060

Fixed effects:
 Estimate Std.Err z-value p-value
(Intercept) -1.125 0.9867 -1.140 0.25422
Group 2.984 0.7790 3.830 0.00013
TimePost 6.673 1.0771 6.196 < 1e-04
Group:TimePost -5.716 0.9889 -5.780 < 1e-04
```

The partial output from ZHP is printed below.

```
> zhp<- mixed_model(fixed = D_6__Ruminiclostridium.sp..KB18 ~
Group*Time, random = ~ Time | MouseID, data = ds, family = hurdle.poisson
(), zi_fixed = ~ 1, iter_EM = 0)
> summary(zhp)

Model:
 family: hurdle poisson
 link: log

Fit statistics:
 log.Lik AIC BIC
 -92.57 201.1 209.1

Random effects covariance matrix:
 StdDev Corr
```

Table 17.6 The modeling results of Group by Time Interaction from the selected models

| Model | Fixed effects | | | | |
	Interaction term	Estimate	Std. Error	z value	P-value
Poi	Group:TimePost	−3.539	1.654	−2.139	0.0324
Nb	Group:TimePost	−4.678	1.5099	−3.098	0.0019
Zip	Group:TimePost	−5.697	0.9800	−5.814	< 1e-04
Zinb	Group:TimePost	−5.716	0.9889	−5.780	< 1e-04
Zhp	Group:TimePost	−10.538	13.05	−0.8077	0.42
Zhnb	Group:TimePost	−4.678	1.957	−2.391	0.01682

```
(Intercept) 0.5510
TimePost 1.0677 -0.9375

Fixed effects:
 Estimate Std.Err z-value p-value
(Intercept) -5.699 13.03 -0.4374 0.66
Group 7.992 13.03 0.6132 0.54
TimePost 10.841 13.04 0.8317 0.41
Group:TimePost -10.538 13.05 -0.8077 0.42
```

Significantly different fixed effects of Group by TimePost interaction is identified in zero-inflated negative binomial model; by contrast, this term is not significant difference in zero-hurdle Poisson model (Table 17.6).

17.3.3 Remarks on GLMMadaptive

Various numerical approximating methods or Markov chain Monte Carlo (MCMC) have increased in use to estimate GLLMs in practice. For example, MCMC and Hamiltonian Monte Carlo can provide accurate evaluation of the integrals under the Bayesian paradigm or approaches. However, none has good properties for all possible models and data. As reviewed in Chap. 16 (Sects. 16.4.1 and 16.4.3), the numerical approximating methods or MCMC approach are particularly problematic for binary/dichotomous data and count data with small counts and few repeated measurements because in these conditions the accuracy of this approximation is rather low.

In contrast, the numerical integration algorithm that both **GLMMadaptive** and **glmmTMB** are used is considered as more accurate than the numerical approximating methods. Especially AGQ algorithm is commonly considered as the gold standard numerical approximation method (Pinheiro and Bates 1995). GLMMadaptive has adopted this method and focuses on maximum likelihood estimation. Along with glmmTMB, GLMMadaptive approach has been confirmed controlling the false-positive rates close to the significance level over different sample sizes, while remaining reasonable empirical power (Zhang and Yi 2020b). However GLMMadaptive also has the weaknesses, including:

- The AGQ rule is more computationally intensive compared to the numerical approximating methods or MCMC approach.
- The current version (0.8.5, February 2022) only allows a single grouping factor; i.e., no nested or crossed random effects designs are provided.

Based on the model fit evaluated by AIC and BIC and the capability of the fixed effects model in detecting significant variables of interest, we recommend using ZINB to analyze microbiome data when using GLMMadaptive.

17.4   Fast Zero-Inflated Negative Binomial Mixed Modeling (FZINBMM)

FZINBMM (Zhang and Yi 2020b) was specifically designed for longitudinal microbiome data. Similar to glmmTMB and GLMMadaptive, FZINBMM belongs to univariate approach of longitudinal microbiome data analysis. However, different from glmmTMB and GLMMadaptive, which use the Laplace approximation to integrate over random effects and an adaptive Gauss-Hermite quadrature (AGQ) rule, respectively, FZINBMM employs the EM-IWLS algorithm.

17.4.1   Introduction to FZINBMM

FZINBMM (Zhang and Yi 2020b) was developed to analyze high-dimensional longitudinal metagenomic count data, including both 16S rRNA and whole-metagenome shotgun sequencing data. The goal of FZINBMM is to simultaneously address the main challenges of longitudinal metagenomics data, including high-dimensionality, dependence among samples and zero-inflation of observed counts. FZINBMM takes two advantages: (1) built on zero-inflated negative binomial mixed models (ZINBMMs), FZINBMM is able to analyze over-dispersed and zero-inflated longitudinal metagenomic count data; (2) using a fast and stable EM-iterative weighted least squares (IWLS) model-fitting algorithm to fit the ZINBMMs, which takes advantage of fitting linear mixed models (LMMs). Thus, FZINBMM can handle various types of fixed and random effects and within-subject correlation structures and analyze many taxa fast.

17.4.1.1   Zero-Inflated Negative Binomial Mixed Models (ZINBMMs)

Let c_{ijm} denote the observed count for the mth taxon from the jth sample of the ith subject in a longitudinal metagenomics data setting and T_{ij} as the total sequence read, where $y_{ij} = c_{ijm}, j = 1, \ldots, n_i, i = 1, \ldots, n$, n is the total number of subjects, and n_i is the total number of samples for each subject. For any given taxon m, assume that y_{ij} is distributed as the zero-inflated negative binomial (ZINB). ZINB is a two-part model with Part I (a logistic model) being used for predicting excess zeros and Part II (a NB model) being used for modeling over-dispersed counts. The ZINB distribution is written as:

$$y_{ij} \sim \begin{cases} 0 & p_{ij} \\ NB(y_{ij}|\mu_{ij}, \theta) & 1 - p_{ij} \end{cases}, \qquad (17.2)$$

where P_{ij} denotes the unknown probability of y_{ij} being from the excess zero state and μ_{ij} and θ are the means and the dispersion parameter of NB distribution, respectively. The relationships between the probabilities and the potential covariates are linked through the logit link functions. There are two logit link functions for logistic model. The logit link function in (17.3) only includes an intercept in Z_{ij}, indicating that the model assumes the same probability of belonging to the excess zeros for all observed zeros.

$$\text{logit}(p_{ij}) = Z_{ij}\alpha. \qquad (17.3)$$

To include subject-specific effects (random effects), we can extend the above logistic model as follow:

$$\text{logit}(p_{ij}) = Z_{ij}\alpha + G_{ij}\alpha_i. \qquad (17.4)$$

In (17.3) and (17.4), Z_{ij} denotes the potential covariates associated with the excess zeros, and a is the vector of effects. a_i is the subject-specific effects which are usually assumed to follow a multivariate normal distribution (McCulloch and Searle 2001; Searle and McCulloch 2001; Pinheiro and Bates 2000):

$$\alpha_i \sim N(0, \Psi_\alpha), \qquad (17.5)$$

where Ψ_α is the variance-covariance matrix for the random effects. In the NB model, the means μ_{ij} are assumed to link with the covariates via a logarithmic link function:

$$\log(\mu_{ij}) = \log(T_{ij}) + X_{ij}\beta + G_{ij}b_i, \qquad (17.6)$$

where $\log(T_{ij})$ is the offset for accounting for the sequencing depth of the total sequence reads; β is the vector of fixed effects (i.e., population-level effects);

b_i denotes the random effects (i.e., the vector of subject-specific effects). X_{ij} is the vector of covariates for the jth sample of the ith subject. In longitudinal studies, the jth sample of the ith subject is collected in different time t_{ij}. Thus, X_{ij} could be $(1 - X_i)$, $(1 - X_i, t_{ij})$, or $\left(1 - X_i, t_{ij}, X_i^s t_{ij}\right)$, where X_i^s is an indicator variable of interest in X_i such as an indicator variable for the case-control group. G_{ij} denotes the vector of group-level covariates with 1 indicating only a subject-specific random intercept being specified in the model, with $(1, t_{ij})$ indicating a subject-specific random intercept and random time effect are included in the model.

Similar to a_i in (17.4), the subject-specific effects b_i (random effects) in (17.6) are assumed to follow a multivariate normal distribution (McCulloch and Searle 2001; Searle and McCulloch 2001; J. Pinheiro and Bates 2000):

$$b_i \sim N(0, \Psi_b), \tag{17.7}$$

where Ψ_b is the variance-covariance matrix for the random effects b_i. The matrices, Ψ_a and Ψ_b are the general positive-definite matrices to model the correlation structure among the random covariates. In the simplest case, they are restricted to be diagonal (see details on introduction to GLMMs in Sect. 16.2).

17.4.1.2   EM-IWLS Algorithm for Fitting the ZINBMMs

The above ZINBMMs are fitted via a proposed fast and stable EM-IWLS algorithm. A vector of latent indicator variables $\xi = \left(\xi_{i1}, \ldots, \xi_{in_j}\right)$ is used to distinguish the logit part for excess zeros and the NB distribution with $\xi_{ij} = 1$ indicating that y_{ij} is from the excess zero and $\xi_{ij} = 0$ indicating that y_{ij} is from the NB distribution. The log-likelihood of the complete data (y, ξ) is written as:

$$
\begin{aligned}
L(\Phi; y, \xi) &= \sum_{i=1}^{n} \sum_{j=1}^{n_i} \log\left[p_{ij}^{\xi_{ij}}\left(1 - p_{ij}\right)^{1-\xi_{ij}}\right] \\
&+ \sum_{i=1}^{n} \sum_{j=1}^{n_i} \log\left[(1 - \xi_{ij}) \log\left(NB\left(y_{ij}|\mu_{ij}, \phi\right)\right)\right),
\end{aligned}
\tag{17.8}
$$

where Φ denotes all the parameters of fixed and random effects in the ZINBMMs. The E-step in the EM-IWLS algorithm replaces the indicator variables ξ_{ij} by their conditional expectations $\widehat{\xi}_{ij}$. The $\widehat{\xi}_{ij}$ can be estimated by the conditional function of ξ_{ij} given all the parameters Φ and data y_{ij}:

$$
\begin{aligned}
\widehat{\xi}_{ij} &= p\left(\xi_{ij} = 1|\Phi, y_{ij}\right) \\
&= \frac{p\left(y_{ij}|\mu_{ij}, \phi, \xi_{ij} = 1\right)p\left(\xi_{ij} = 1|p_{ij}\right)}{p\left(y_{ij}|\mu_{ij}, \phi, \xi_{ij} = 0\right)p\left(\xi_{ij} = 0|p_{ij}\right) + p\left(y_{ij}|\mu_{ij}, \phi, \xi_{ij} = 1\right)p\left(\xi_{ij} = 1|p_{ij}\right)},
\end{aligned}
\tag{17.9}
$$

If $y_{ij} \neq 0$, we have $p(y_{ij} \mid \mu_{ij}, \phi, \xi_{ij} = 1) = 0$, and thus $\widehat{\xi}_{ij} = 0$.
If $y_{ij} = 0$, we have

$$
\begin{aligned}
\widehat{\xi}_{ij} &= \left[\frac{p(\xi_{ij}=0)|p_{ij}}{p(\xi_{ij}=1)|p_{ij}} p(y_{ij} = 0 | \mu_{ij}, \phi, \xi_{ij} = 0) + 1 \right]^{-1} \\
&= \left[\frac{1 - p_{ij}}{p_{ij}} NB(y_{ij} = 0 | \mu_{ij}, \phi) + 1 \right]^{-1},
\end{aligned}
\tag{17.10}
$$

In the M-step, the parameters in the two parts are updated separately by maximizing $L\left(\Phi; y, \widehat{\xi}\right)$.

The parameters in the logit part for excess zeros are updated by running a logistic regression with $\widehat{\xi}_{ij}$ as response in (17.11) without including a random-effect term:

$$
\widehat{\xi}_{ij} \sim Bin(1, p_{ij}), \quad \mathrm{logit}(p_{ij}) = Z_{ij}\alpha,
\tag{17.11}
$$

If a random-effect term is included in the logit part, then the parameters in that part are updated by fitting the logistic mixed model as response in (17.12):

$$
\widehat{\xi}_{ij} \sim Bin(1, p_{ij}), \quad \mathrm{logit}(p_{ij}) = Z_{ij}\alpha + G_{ij}\alpha_i, \alpha_i \sim N(0, \Psi_a).
\tag{17.12}
$$

Both the IWLS algorithm and the penalized quasi-likelihood (PQL) procedure can equivalently
approximate the logistic regression or logistic mixed model. The PQL procedure iteratively approximates the logistic likelihood by a weighted LMM, which has been used for fitting GLMMs (Breslow and Clayton 1993; McCulloch and Searle 2001; Schall 1991; Venables and Ripley 2002).

FZINBMM updates the parameters in the means of the NB part through using an extended IWLS algorithm, which iteratively approximates the weighted NB likelihood (i.e., the second part of (17.8)) by the weighted LMM:

$$
\begin{aligned}
z_{ij} &= \log(T_{ij}) + X_{ij}\beta + G_{ij}b_i + \left(1 - \widehat{\xi}_{ij}\right)^{-1/2} w_{ij}^{-1/2} e_{ij} \\
b_i &\sim N(0, \Psi), e_{ij} = (e_{i1}, \ldots, e_{in_i})' \sim N(0, \sigma^2 R_i),
\end{aligned}
\tag{17.13}
$$

The extended IWLS algorithm was developed based on the standard IWLS (equivalently, PQL) by Zhang et al. (2017). In (17.13), z_{ij} and w_{ij} are the pseudo-responses and the pseudo-weights, respectively, and R_i is a correlation matrix accounting for within subject correlation structures. Here the z_{ij} and w_{ij} are calculated in the same way as in Zhang et al. (2017), where the IWLS algorithm is used for fitting NB mixed models.

Like in Pinheiro and Bates (2000), in FZINBMM framework various correlation matrices can be specified, e.g., autoregressive of order 1 [AR(1)] or continuous-time

AR(1) except for assuming the simplest independence structure of correlation matrix by setting R_i as an identity matrix.

Finally, FZINBMM uses the standard Newton-Raphson algorithm to update the dispersion parameter θ by maximizing the NB likelihood:

$$L(\phi) = \sum\sum\left(1-\widehat{\xi}_{ij}\right)\log NB\left(y_{ij}|\widehat{\mu}_{ij},\phi\right), \qquad (17.14)$$

where $\widehat{\mu}_{ij} = \exp\left(\log\left(T_{ij}\right) + X_{ij}\widehat{\beta} + G_{ij}\widehat{b}_i\right)$. The convergence algorithm is based the criterion:

$$\sum_{i=1}^{n}\sum_{j=1}^{n_i}\left[\left(\eta_{ij}^{(t)} - \eta_{ij}^{(t-1)}\right)^2 + \left(\gamma_{ij}^{(t)} - \gamma_{ij}^{(t-1)}\right)^2\right]$$
$$< \varepsilon\left(\sum_{i=1}^{n}\sum_{j=1}^{n_i}\left[\left(\eta_{ij}^{(t)}\right)^2 + \left(\gamma_{ij}^{(t)}\right)^2\right]\right), \qquad (17.15)$$

where $\eta_{ij}^{(t)} = \log\left(T_{ij}\right) + X_{ij}\beta^{(t)} + G_{ij}b_i^{(t)}$, $\gamma_{ij}^{(t)} = Z_{ij}\alpha^{(t)} + G_{ij}\alpha_i^{(t)}$, and ε is a small value (say 10^5). All the estimated steps are repeated until convergence.

After the model is convergent, FZINBMM provides the maximum likelihood estimates of the fixed effects β_k and the associated standard deviations and the estimates of the random effects α_k and the associated standard deviations.

We can perform two null statistical hypotheses tests, respectively, in FZINBMM:

$$H_0 : \beta_k = 0 \text{ and } H_0 : \alpha_k = 0.$$

17.4.2 Implement FZINBMM Using the NBZIMM Package

FZINBMM was developed under the LMM framework and is implemented via the function **glmm.zinb()** in the package **NBZIMM**. So it can deal with any types of random effects and within subject correlation structures as the function **lme()**. The function **glmm.zinb()** works by repeatedly calling the function **lme()** of the package **nlme** to fit the weighted LMM and GLM in the **stats** package or **glmPQL()** in the **MASS** package to fit the logistic regression or logistic mixed model.

The package NBZIMM is freely available from the public GitHub repository (Yi 2021) (current version 1.0, April 2022). One example syntax of the **glmm.zinb()** function is given below:

glmm.zinb(fixed, random, data, correlation, zi_fixed = ~1, zi_random = NULL, niter = 30, epsilon = 1e-05, verbose = TRUE, ...)

where the arguments:

- **fixed and random** are used to specify the formulas for the fixed-effects (including the count outcome) and the random-effects parts of the negative binomial model, respectively. They are the same as in the function **lme**() in the package **nlme**.
- **random** only contains the right-hand side part, e.g., ~ time | id, where time is a variable, and id is the grouping factor.
- **data** is a data.frame containing all the variables that the model uses.
- **correlation** is used to specify an optional correlation structure. It is the same as in the function **lme**() in the package **nlme**.
- **zi_fixed** and **zi_random** are used to specify the formulas for the fixed and random effects of the zero inflated part, respectively. They only contain the right-hand side part.
- **niter** denotes the maximum number of iterations.
- **epsilon** denotes the positive convergence tolerance.
- **verbose** is a logical argument, with verbose = TRUE printing out the number of iterations and computational time.
- ... denotes the further arguments for **lme**.

After implementing the **glmm.zinb** () function, the package NBZIMM returns both fitted model objects of the count distribution part and the zero-inflation part. The fitted model object is the class **lme**, which can be summarized by functions in the package **nlme**. The object for the zero-inflation part contains additional components, including the estimate of the dispersion parameter (**theta**), the zero-state probabilities (**zero.prob**), the conditional expectations of the zero indicators (**zero. indicator**), and the fitted logistic (mixed) model for the zero-inflation part (**fit.zero**).

Example 17.3: Breast Cancer QTRT1 Mouse Gut Microbiome, Example 17.1 Cont.

In Chap. 15, we used this dataset to illustrate LMMs to identify the significant change species over time and to assess the association of the Chao 1 richness index with group status over time.

In Examples 17.1 and 17.2, we used this dataset to illustrate glmmTMB and GLMMadaptive, respectively. Here, we use it to illustrate how to identify the significant change species over time by performing FZINBMM.

The data processing steps including loading and checking data and creating analysis dataset are same as in previous examples. For the reader's convenience, we copy the core commands here and omit the details of interpretation.

Step 1: Process data to create analysis dataset consisting of outcomes and covariates.

```
> setwd("~/Documents/QIIME2R/Ch17_GLMMs2")
> otu_tab=read.csv("otu_table_L7_MCF7.csv",row.names=1,check.
names=FALSE)
> meta_tab <- read.csv("metadata_QtRNA.csv",row.names=1,check.names
= FALSE)
```

```
> otu_tab_t<-t(otu_tab)
> colnames(otu_tab_t)

> outcome =otu_tab_t
> outcome = outcome[rownames(meta_tab),]
> yy = as.matrix(outcome)
> yy = ifelse(is.na(yy), 0, yy) #Exclude missing outcome variables if
any

> filter1 <- apply(yy, 2, function(x){sum(x>0)>0.4*length(x)})
> filter2 <- apply(yy, 2, function(x){quantile(x,0.9)>1})
> filter12 <-yy[,filter1&filter2]

> yy = filter12[rownames(meta_tab),]
> taxa_all <- colnames(filter12)

> N = meta_tab[, "Total.Read"] # total reads
> mean(N); sd(N)
> mean(log(N)); sd(log(N))
> subject = meta_tab[, "MouseID"]
> group = meta_tab[, "Group"]
> time = meta_tab[, "Time"]
```

Step 2: Run the glmm.zinb() function to implement FZINBMM.

Type the following commands to install the NBZIMM package.

```
> install.packages("remotes")
> remotes::install_github("nyiuab/NBZIMM")
```

Like in running the **lme()** function in Chap. 14, we first create list() and matrix to hold modeling results.

```
> yy <- yy[, colSums(is.na(yy)) == 0]
> # Create list() and matrix to hold the modeling results
> mod = list()
> rest1 = rest2 = rest3 = matrix(NA, ncol(yy), 1)
```

Then we call library(NBZIMM) and run the **glmm.zinb()** function iteratively to fit multiple taxa (outcome variables).

```
> library(NBZIMM)
> for (j in 1:ncol(yy)){
+ tryCatch({
+ y = as.numeric(yy[, j])
+ df = data.frame(y=y, Group=group, Time = time, N=N, Subject=subject)
+ mod = glmm.zinb(y ~ Group*Time + offset(log(N)),
+ random = ~ 1|Subject, data = df)
+ rest_tab = summary(mod)$tTable
```

```
+ rest1[j, ] = summary(mod)$tTable[2, 5]
+ rest2[j, ] = summary(mod)$tTable[3, 5]
+ rest3[j, ] = summary(mod)$tTable[4, 5]
+ }, error=function(e){cat("ERROR :",conditionMessage(e), "\n")})
+ }

> rest_tab
 Value Std.Error DF t-value p-value
(Intercept) -7.7431 0.1837 18 -42.1423 1.920e-19
Group 0.2473 0.2687 18 0.9204 3.695e-01
TimePost -0.2203 0.2418 18 -0.9111 3.743e-01
Group:TimePost -0.4221 0.3492 18 -1.2086 2.424e-01
```

Like in LMMs of Chap. 15, we include two main factors group, time, and their interaction term as fixed effects and include intercept only as random term. Other models can also be fitted. For example, we can fit the same fixed effects with group, time, and their interaction term but fit both intercept and time as random effects or fit random intercept and the within-subject correlation of AR(1). We can also include other covariates in the above models to test the hypothesis of interest.

For LMMS, we first perform an arcsine transformation (**asin(sqrt())**) to transform the read counts of each taxon, then model the transformed values of each taxon. Unlike LMMs, here we model the read counts of each taxon directly and use log of total read counts as offset.

We can obtain the information of random and fixed effects and zero model for this NBZIMM using the summary() function as implementing in the **nlme()** function.

```
> summary(mod)
> fixed(mod)
> random.effects(mod)
> summary(mod$fit.zero)
```

Step 3: Adjust p-values.

```
> all_zinb <- as.data.frame(cbind(rest1, rest2,rest3))
> # Adjust p-values
> group_zinb_adj <-round(p.adjust(rest1,'fdr'),4)
> TimePost_zinb_adj <-round(p.adjust(rest2,'fdr'),4)
> Group.TimePost_zinb_adj <-round(p.adjust(rest3,'fdr'),4)
> options(width = 110)
> all_zinb_adj <- as.data.frame(cbind(taxa_all,Group_zinb_adj,
TimePost_zinb_adj,Group.TimePost_zinb_adj))
```

Step 4: Write table to contain the results.

Finally we can write out the modeling results into Table 17.7.

Table 17.7 Modeling results from fast zero-inflated negative binomial mixed model in QtRNA data

	taxa_all	Group_zinb_adj	Time Post_zinb_adj	Group. TimePost_zinb_adj
1	D_6__Anaerotruncus sp. G3 (2012)	NA	NA	NA
2	D_6__Clostridiales bacterium CIEAF 012	0.8725	0.3586	0.4475
3	D_6__Clostridiales bacterium CIEAF 015	0.5081	0.0035	0.0263
4	D_6__Clostridiales bacterium CIEAF 020	NA	NA	NA
5	D_6__Clostridiales bacterium VE202–09	0.2278	0.0242	0.1321
6	D_6__Clostridium sp. ASF502	NA	NA	NA
7	D_6__Enterorhabdus muris	0.4301	0.0111	0.2015
8	D_6__Eubacterium sp. 14–2	0.0048	0	0.0263
9	D_6__gut metagenome	0.8539	0.4978	0.3914
10	D_6__Lachnospiraceae bacterium A4	NA	NA	NA
11	D_6__Lachnospiraceae bacterium COE1	0.8725	0.236	0.0263
12	D_6__mouse gut metagenome	NA	NA	NA
13	D_6__Mucispirillum schaedleri ASF457	0.0034	0.1926	0.0011
14	D_6__Parabacteroides goldsteinii CL02T12C30	NA	NA	NA
15	D_6__Ruminiclostridium sp. KB18	0.0023	0	0
16	D_6__Staphylococcus sp. UAsDu23	0.4301	0.009	0.4475
17	D_6__Streptococcus danieliae	0.5061	0.4117	0.3334

```
> # Write results table
> # Make the table
> library(xtable)
> table <- xtable(all_zinb_adj,caption = "Table of significant taxa",
lable="sig_taxa_table")
> print.xtable(table,type="html",file = "FZINBMM_Table_Model_QtRNA.
html")
> write.csv(all_zinb_adj,file = paste("FZINBMM_Table_Model_QtRNA.
csv",sep = ""))
```

17.4.3   Remarks on FZINBMM

It was demonstrated (Zhang and Yi 2020a) that like the two ZINBMMs that use numerical integration algorithm, glmmTMB (Brooks et al. 2017) and GLMMadaptive (Rizopoulos 2019), FZINBMM controlled the false-positive rates well, and all three methods had comparable performances in terms of empirical power, whereas FZINBMM outperformed glmmTMB and GLMMadaptive in terms of computational efficiency.

Zhang and Yi (2020a, b) demonstrated that:

- FZINBMM outperformed linear mixed models (LMMs), negative binomial mixed models (NBMMs) (Zhang et al. 2017, 2018), and zero-inflated Gaussian mixed models (ZIGMMs).
- FZINBMM detected a higher proportion of associated taxa than LMMs, NBMMs, and zero-inflated Gaussian mixed models (ZIGMMs) methods.
- Generally transforming the count data could decrease the power in detecting significant taxa. Thus, the methods that directly analyzed the counts (i.e., NBMMs and FZINBMM) performed better than those methods that analyzed the transformed data (i.e., LMMs and ZIGMMs).
- The models that are able to address the zero-inflation problem could have higher the power in detecting the significant microbiome taxa and their dynamic associations between the outcome and the microbiome composition. For example, they found that ZIGMMs and FZINBMM both worked better than LMMs and NBMMs in this cited study.
- FZINBMM performed similarly to ZIGMMs and NBMMs when the microbiome data were not highly sparse. Whereas FZINBMM has better performance when the microbiome data is sparse.

However, FZINBMM also has several limitations, such as:

- As an univariate method, FZINBMM analyzes one taxon at a time and then adjusts for multiple comparisons.
- Assuming subject-specific effects (random effects) are followed as a multivariate normal distribution is difficult to validate.
- FZINBMM also shares most other limitations of ZIBR (zero-inflated beta random effect model) as described in Yinglin Xia (2020).

17.5   Remarks on Fitting GLMMs

In modeling cross-sectional microbiome count data with GLMMs, zero-inflated GLMMs, and hurdle models, we have concluded that generally zero-inflated negative binomial and two-parts hurdle negative binomial models are the two best models for analyzing zero-inflated and over-dispersed microbiome data (Xia et al. 2018b). For longitudinal zero-inflated and over-dispersed microbiome count data, although the findings are more complicated because more complicated models are specified to be estimated, the overall conclusion remains the same: zero-inflated and zero-hurdle negative binomial models are the best models among the competing models. Conway-Maxwell-Poisson model is also a good alternative for this kind of microbiome data; however, it is time-consuming to run this model when the specified models do not match the data. In specific case, zero-hurdle Poisson is also a choice; however, this model cannot model over-dispersion if the non-zero parts of data are over-dispersed. Therefore, we would like to recommend using zero-inflated and zero-hurdle negative binomial models for analyzing microbiome data.

We noticed that zero-hurdle Poisson and negative binomial models are not stable cross different software and different estimation algorithms in longitudinal microbiome data and sometimes fail to detect significant taxa compared to zero-inflated negative binomial models (ZINB). Thus, we prefer to recommend using ZINB for longitudinal microbiome data.

We also recommend performing model comparisons to choose an appropriate model for applying into microbiome data.

- It is not necessary to specify a maximally complex model to overestimate parameters because complicated models not only might not converge but also might not improve much over simplified models.
- Model selection is very important. The choice of better models should be based on model fitting criteria rather than based on whether more significant taxa are identified by the models (Xia 2020). We emphasize here again it is misleading to say one model outperforms other models because it can identify more significant taxa. The more significant taxa identified could be due to higher false-positive rates of this model (Xia et al. 2012).
- We diagnose the model fits from glmmTMB, GLMMadaptive through the DHARMa package, which takes the approach of simulating, scaling, and plotting residuals. The readers can use this approach as a reference, but model evaluation should not fully be determined based on its results because the simulation approach itself still need to be further validated.

17.6   Summary

This chapter introduced the GLMMs for analysis of longitudinal microbiome data. We organized this chapter into five sections. Section 17.1 provided an overview of GLMMs in microbiome research. We compared and discussed two approaches of analysis of microbiome data (data transformation versus using GLMMs directly) and particularly model selection as well as the challenges of statistical hypothesis testing in microbiome research. Next three sections (Sects. 17.2, 17.3, and 17.4) focused on investigating three GLMMs or packages and illustrated them with microbiome data. The first two packages (glmmTMB in Sect. 17.2 and GLMMadaptive in Sect. 17.3) were developed based on numerical integration algorithms, while the third package ZINBMMs was developed based on EM-IWLS algorithm (Sect. 17.4). For each model or package, we commended on their advantages and disadvantages and illustrated their model fits. The modeling results from glmmTMB and GLMMadaptive were evaluated by information criteria AIC and BIC and/or LRT as well as diagnosed by the DHARMa package. In Sect. 17.5, we provided some general remarks on fitting GLMMs, including the zero-inflation and over-dispersion issues and count-based model versus transformation-based approaches.

LMMs in Chap. 15, glmmTMB, GLMMadaptive, and FZINBMM in this chapter take the univariate approach to analyze longitudinal microbiome data. In Chap. 18, the last chapter of this book, we will introduce a multivariate longitudinal microbiome data analysis: non-parametric microbial interdependence test (NMIT).

References

Aho, K., D. Derryberry, and T. Peterson. 2014. Model selection for ecologists: The worldviews of AIC and BIC. *Ecology* 95 (3): 631–636. https://doi.org/10.1890/13-1452.1.

Akaike, Hirotugu. 1973. *Information theory and an extension of the maximum likelihood principle*, [w:]. Proceedings of the 2nd international symposium on information, bn petrow, f. Czaki, Akademiai Kiado, Budapest.

Akaike, Hirotogu. 1998. Information theory and an extension of the maximum likelihood principle. In *Selected papers of hirotugu akaike*, 199–213. Springer.

Barriga, Gladys D.C., and Francisco Louzada. 2014. The zero-inflated Conway–Maxwell–Poisson distribution: Bayesian inference, regression modeling and influence diagnostic. *Statistical Methodology* 21: 23–34. https://doi.org/10.1016/j.stamet.2013.11.003. https://www.sciencedirect.com/science/article/pii/S1572312714000148.

Barron, Andrew, Lucien Birgé, and Pascal Massart. 1999. Risk bounds for model selection via penalization. *Probability Theory and Related Fields* 113 (3): 301–413.

Bates, Douglas, Martin Mächler, Ben Bolker, and Steve Walker. 2015. Fitting linear mixed-effects models using lme4. *Journal of Statistical Software* 67 (1): 48. https://doi.org/10.18637/jss.v067.i01. https://www.jstatsoft.org/v067/i01.

Bland, J. Martin, and Douglas G. Altman. 1996. Transformations, means, and confidence intervals. *BMJ: British Medical Journal* 312 (7038): 1079.

Bolker, Benjamin M. 2015. Linear and generalized linear mixed models. In *Ecological Statistics: Contemporary theory and application*, ed. Gordon A. Fox et al., 309–333. New York: Oxford University Press.

Bolker, Benjamin M., Mollie E. Brooks, Connie J. Clark, Shane W. Geange, John R. Poulsen, M. Henry, H. Stevens, and Jada-Simone S. White. 2009. Generalized linear mixed models: A practical guide for ecology and evolution. *Trends in Ecology & Evolution* 24 (3): 127–135. https://doi.org/10.1016/j.tree.2008.10.008.

Breslow, Norman E., and David G. Clayton. 1993. Approximate inference in generalized linear mixed models. *Journal of the American Statistical Association* 88 (421): 9–25.

Brooks, Mollie, Kasper Kristensen, Koen van Benthem, C.W. Arni Magnusson, Anders Nielsen Berg, Hans Skaug, Martin Mächler, and Benjamin Bolker. 2017. glmmTMB balances speed and flexibility among packages for zero-inflated generalized linear mixed modeling. *The R Journal* 9: 378–400.

Burnham, Kenneth P. 1998. *Model selection and multimodel inference. A practical information-theoretic approach*. New York: Springer.

Burnham, K.P., and D.R. Anderson. 2002. *Model selection and multimodel inference: A practical information-theoretic approach*. New York: Springer.

Ding, J., V. Tarokh, and Y. Yang. 2018a. Bridging AIC and BIC: A new criterion for autoregression. *IEEE Transactions on Information Theory* 64 (6): 4024–4043. https://doi.org/10.1109/TIT.2017.2717599.

———. 2018b. Model selection techniques: An overview. *IEEE Signal Processing Magazine* 35 (6): 16–34. https://doi.org/10.1109/MSP.2018.2867638.

Dunn, Peter K., and Gordon K. Smyth. 1996. Randomized quantile residuals. *Journal of Computational and Graphical Statistics* 5 (3): 236–244.

Feng, Changyong, Hongyue Wang, Lu Naiji, and Xin M. Tu. 2013. Log transformation: Application and interpretation in biomedical research. *Statistics in Medicine* 32 (2): 230–239. https://doi.org/10.1002/sim.5486. https://onlinelibrary.wiley.com/doi/abs/10.1002/sim.5486.

Feng, Changyong, Hongyue Wang, Lu Naiji, Tian Chen, Hua He, Ying Lu, and Xin M. Tu. 2014. Log-transformation and its implications for data analysis. *Shanghai Archives of Psychiatry* 26 (2): 105–109. https://doi.org/10.3969/j.issn.1002-0829.2014.02.009. https://pubmed.ncbi.nlm.nih.gov/25092958; https://www.ncbi.nlm.nih.gov/pmc/articles/PMC4120293/.

Feng, Changyong, Hongyue Wang, Yu Han, Yinglin Xia, Naiji Lu, and Xin M. Tu. 2015. Some theoretical comparisons of negative binomial and zero-inflated poisson distributions. *Communications in Statistics – Theory and Methods* 44 (15): 3266–3277. https://doi.org/10.1080/03610926.2013.823203.

Foster, Dean P., and Edward I. George. 1994. The risk inflation criterion for multiple regression. *The Annals of Statistics* 22 (4): 1947–1975.

Gelman, Andrew, and Jennifer Hill. 2006. *Data analysis using regression and multilevel/hierarchical models*. Cambridge, MA: Cambridge University Press.

Hardin, James W., James William Hardin, Joseph M. Hilbe, and Joseph Hilbe. 2007. *Generalized linear models and extensions*. College Station: Stata press.

Hartig, Florian. 2022. *DHARMa: Residual diagnostics for hierarchical (multi-level/mixed) regression models*. R package version 0.4.5. https://CRAN.R-project.org/package=DHARMa.

Hedeker, Donald, Stephen H.C. du Toit, Hakan Demirtas, and Robert D. Gibbons. 2018. A note on marginalization of regression parameters from mixed models of binary outcomes. *Biometrics* 74 (1): 354–361.

Huang, Alan. 2017. Mean-parametrized Conway–Maxwell–Poisson regression models for dispersed counts. *Statistical Modelling* 17 (6): 359–380. https://doi.org/10.1177/1471082X17697749.

Ing, Ching-Kang. 2007. Accumulated prediction errors, information criteria and optimal forecasting for autoregressive time series. *The Annals of Statistics* 35 (3): 1238–1277.

Kristensen, Kasper, Anders Nielsen, Casper W. Berg, Hans Skaug, and Bradley M. Bell. 2016. TMB: Automatic differentiation and laplace approximation. *Journal of Statistical Software* 70 (5): 21. https://doi.org/10.18637/jss.v070.i05. https://www.jstatsoft.org/v070/i05.

Kvalheim, Olav M., Frode Brakstad, and Yizeng Liang. 1994. Preprocessing of analytical profiles in the presence of homoscedastic or heteroscedastic noise. *Analytical Chemistry* 66 (1): 43–51.

Lambert, Diane. 1992. Zero-inflated Poisson regression, with an application to defects in manufacturing. *Technometrics* 34 (1): 1–14. https://doi.org/10.2307/1269547. http://www.jstor.org/stable/1269547.

Littell, R.C., G.A. Milliken, W.W. Stroup, R.D. Wolfinger, and O. Schabenberger. 2006. *SAS for mixed models*. Cary: SAS Institute Inc.

Liu, Wei, and Yuhong Yang. 2011. Parametric or nonparametric? A parametricness index for model selection. *The Annals of Statistics* 39 (4): 2074–2102.

Long, J.S. 1997. *Regression models for categorical and limited dependent variables*. Thousand Oaks: Sage.

Lynch, Heather J., James T. Thorson, and Andrew Olaf Shelton. 2014. Dealing with under- and over-dispersed count data in life history, spatial, and community ecology. *Ecology* 95 (11): 3173–3180. https://doi.org/10.1890/13-1912.1. https://esajournals.onlinelibrary.wiley.com/doi/abs/10.1890/13-1912.1.

McCulloch, Charles E., and Shayle R. Searle. 2001. *Generalized, linear, and mixed models*, Wiley series in probability and statistics. New York: Wiley.

Molenberghs, Geert, and Geert Verbeke. 2007. Likelihood ratio, score, and Wald tests in a constrained parameter space. *The American Statistician* 61 (1): 22–27.

Moscatelli, Alessandro, Maura Mezzetti, and Francesco Lacquaniti. 2012. Modeling psychophysical data at the population-level: The generalized linear mixed model. *Journal of Vision* 12 (11): 26–26. https://doi.org/10.1167/12.11.26.

Pinheiro, José C., and Douglas M. Bates. 1995. Approximations to the log-likelihood function in the nonlinear mixed-effects model. *Journal of Computational and Graphical Statistics* 4 (1): 12–35.

Pinheiro, J., and D. Bates. 2000. *Mixed-effects models in S and S-PLUS*. New York: Springer.

Rizopoulos, Dimitris. 2019. *GLMMadaptive: Generalized linear mixed models using adaptive Gaussian quadrature*. R package version 0.6-0. https://drizopoulos.github.io/GLMMadaptive/ . Accessed 9 Jan, 2020.

———. 2022a. *Generalized linear mixed models using adaptive gaussian quadrature*. Last Modified 2022-02-07. https://github.com/drizopoulos/GLMMadaptive; https://drizopoulos. github.io/GLMMadaptive/. Accessed 11 Feb, 2022.

———. 2022b. *Goodness of fit for MixMod objects*. Last Modified 2022-02-08. https://drizopoulos. github.io/GLMMadaptive/articles/Goodness_of_Fit.html. Accessed 16 Feb, 2022.

Saefken, Benjamin, Thomas Kneib, Clara-Sophie van Waveren, and Sonja Greven. 2014. A unifying approach to the estimation of the conditional Akaike information in generalized linear mixed models. *Electronic Journal of Statistics* 8 (1): 201-225, 25. https://doi.org/10.1214/14-EJS881.

Schaalje, G. Bruce, Justin B. McBride, and Gilbert W. Fellingham. 2001. Approximations to distributions of test statistics in complex mixed linear models using SAS Proc MIXED. *SUGI (SAS User's Group International)* 26 (262): 1–5.

———. 2002. Adequacy of approximations to distributions of test statistics in complex mixed linear models. *Journal of Agricultural, Biological, and Environmental Statistics* 7 (4): 512–524.

Schall, Robert. 1991. Estimation in generalized linear models with random effects. *Biometrika* 78 (4): 719–727.

Schwarz, Gideon. 1978. Estimating the dimension of a model. *The Annals of Statistics* 6 (2): 461–464. https://doi.org/10.1214/aos/1176344136.

Searle, Shayle Robert, and Charles E. McCulloch. 2001. *Generalized, linear and mixed models*. New York: Wiley.

Sellers, Kimberly F., and Galit Shmueli. 2010. A flexible regression model for count data. *The Annals of Applied Statistics* 4 (2): 943–961., 19. https://doi.org/10.1214/09-AOAS306.

Shao, Jun. 1997. An asymptotic theory for linear model selection. *Statistica Sinica*: 221–242.

Shibata, Ritei. 1976. Selection of the order of an autoregressive model by Akaike's information criterion. *Biometrika* 63 (1): 117–126.

Shmueli, Galit, Thomas P. Minka, Joseph B. Kadane, Sharad Borle, and Peter Boatwright. 2005. A useful distribution for fitting discrete data: Revival of the Conway–Maxwell–Poisson distribution. *Journal of the Royal Statistical Society: Series C (Applied Statistics)* 54 (1): 127–142. https://doi.org/10.1111/j.1467-9876.2005.00474.x. https://rss.onlinelibrary.wiley.com/doi/abs/10.1111/j.1467-9876.2005.00474.x.

Spiegelhalter, David J., Nicola G. Best, Bradley P. Carlin, and Angelika Van Der Linde. 2002. Bayesian measures of model complexity and fit. *Journal of the Royal Statistical Society: Series B (Statistical Methodology)* 64 (4): 583–639.

Stasinopoulos, Mikis D., Robert A. Rigby, Gillian Z. Heller, Vlasios Voudouris, and Fernanda De Bastiani. 2017. *Flexible regression and smoothing: Using GAMLSS in R*. New York: CRC Press.

van Erven, Tim, Peter Grünwald, and Steven De Rooij. 2012. Catching up faster by switching sooner: A predictive approach to adaptive estimation with an application to the AIC–BIC dilemma. *Journal of the Royal Statistical Society: Series B (Statistical Methodology)* 74 (3): 361–417.

Venables, W.N., and B.D. Ripley. 2002. *Modern applied statistics with S*. New York: Springer.

Vuong, Quang H. 1989. Likelihood ratio tests for model selection and non-nested hypotheses. *Econometrica* 57 (2): 307–333. https://doi.org/10.2307/1912557. http://www.jstor.org/stable/1 912557.

Wickham, H., and W. Chang. 2017. *devtools: Tools to make developing R packages easier*. R package version 1.13.4. https://CRAN.R-project.org/package=devtools.

Wit, Ernst, Edwin van den Heuvel, and Jan-Willem Romeijn. 2012. 'All models are wrong . . .': An introduction to model uncertainty. *Statistica Neerlandica* 66 (3): 217–236. https://doi.org/10. 1111/j.1467-9574.2012.00530.x. https://onlinelibrary.wiley.com/doi/abs/10.1111/j.14 67-9574.2012.00530.x.

Xia, Yinglin. 2020. Chapter 11: Correlation and association analyses in microbiome study integrating multiomics in health and disease. In *Progress in molecular biology and translational science*, ed. Jun Sun, 309–491. New York: Academic Press.

Xia, Y., D. Morrison-Beedy, J. Ma, C. Feng, W. Cross, and X. Tu. 2012. Modeling count outcomes from HIV risk reduction interventions: A comparison of competing statistical models for count responses. *AIDS Research and Treatment* 2012: 593569. https://doi.org/10.1155/2012/593569. http://www.ncbi.nlm.nih.gov/pubmed/22536496.

Xia, Yinglin, Jun Sun, and Ding-Geng Chen. 2018a. Modeling over-dispersed microbiome data. In *Statistical analysis of microbiome data with R*, ed. Yinglin Xia, Jun Sun, and Ding-Geng Chen, 395–451. Singapore: Springer.

———. 2018b. Modeling zero-inflated microbiome data. In *Statistical analysis of microbiome data with R*, ed. Yinglin Xia, Jun Sun, and Ding-Geng Chen, 453–496. Singapore: Springer.

———. 2018c. What are microbiome data? In *Statistical analysis of microbiome data with R*, ed. Yinglin Xia, Jun Sun, and Ding-Geng Chen, 29–41. Singapore: Springer Singapore.

Yang, Yuhong. 2007. Prediction/estimation with simple linear models: Is it really that simple? *Econometric Theory* 23 (1): 1–36.

Yi, Nengjun. 2021. *NBZIMM: Negative binomial and zero-inflated mixed models, with applications to microbiome data analysis*. https://github.com/nyiuab/NBZIMM.

Zeileis, Achim, Christian Kleiber, and Simon Jackman. 2008. Regression models for count data in R. *Journal of Statistical Software* 27 (8): 1–25.

Zhang, Yongli, and Yuhong Yang. 2015. Cross-validation for selecting a model selection procedure. *Journal of Econometrics* 187 (1): 95–112. https://doi.org/10.1016/j.jeconom.2015.02.006. https://www.sciencedirect.com/science/article/pii/S0304407615000305.

Zhang, Xinyan, and Nengjun Yi. 2020a. Fast zero-inflated negative binomial mixed modeling approach for analyzing longitudinal metagenomics data. *Bioinformatics*. https://doi.org/10. 1093/bioinformatics/btz973.

———. 2020b. Fast zero-inflated negative binomial mixed modeling approach for analyzing longitudinal metagenomics data. *Bioinformatics* 36 (8): 2345–2351.

Zhang, X., H. Mallick, Z. Tang, L. Zhang, X. Cui, and A.K. Benson. 2017. Negative binomial mixed models for analyzing microbiome count data. *BMC Bioinformatics* 18. https://doi.org/10. 1186/s12859-016-1441-7.

Zhang, Xinyan, Yu-Fang Pei, Lei Zhang, Boyi Guo, Amanda H. Pendegraft, Wenzhuo Zhuang, and Nengjun Yi. 2018. Negative binomial mixed models for analyzing longitudinal microbiome data. *Frontiers in Microbiology* 9 (1683). https://doi.org/10.3389/fmicb.2018.01683. https://www.frontiersin.org/article/10.3389/fmicb.2018.01683.

Chapter 18
Multivariate Longitudinal Microbiome Models

Abstract Chapter 15 mainly introduced the glmmTMB, GLMMadaptive, and FZINBMM packages. However, all these three packages can only analyze univariate longitudinal microbiome data. This chapter first provides an *overview* of multivariate longitudinal microbiome analysis. Then it introduces the non-parametric microbial interdependence test (NMIT) and illustrates its implementation using R and QIIME 2. The large P small N problem is also discussed in this chapter.

Keywords Multivariate longitudinal microbiome analysis · Multivariate distance/ kernel-based Longitudinal models · Multi-omics methods · Univariate analysis · Multivariate analysis · Non-parametric microbial interdependence test (NMIT) · Correlated sequence kernel association test (cSKAT) · Correlation matrix · Distance matrix · Large P small N problem · GLMM-MiRKAT · aGLMM-MiRKAT · Frobenius norm · qiime longitudinal nmit · Kendall correlation method · Pearson correlation method · Spearman correlation method · p-corr method

In Chap. 15, we focused on linear mixed-effects models (LMMs), one of most widely univariate longitudinal model in classical statistical literature and has recently been applied into microbiome data analysis. In Chap. 16, we comprehensively introduced the estimation and modeling in the generalized linear mixed models (GLMMs), and in Chap. 17 we introduced and illustrated how to use GLMMs for modeling microbiome data via the glmmTMB, GLMMadaptive, and FZINBMM packages. However, all these three packages can only analyze univariate longitudinal microbiome data. In this chapter, we first provide an *overview* of multivariate longitudinal microbiome analysis (Sect. 18.1). We then introduce non-parametric microbial interdependence test (NMIT), one of newly developed multivariate longitudinal microbiome analysis methods, and illustrate how to implement NMIT using R and QIIME 2 (Sect. 18.2). In Sect. 18.3, we discuss the large P small N problem. Finally, we complete this chapter with a brief summary in Sect. 18.4.

Supplementary Information The online version contains supplementary material available at https://doi.org/10.1007/978-3-031-21391-5_18.

18.1 Overview of Multivariate Longitudinal Microbiome Analysis

Currently longitudinal models for discovering dynamic nature of the multivariate microbiome data are still rare. In this section, we review some newly developed multivariate longitudinal methods/approaches for modeling dynamic microbiomes.

18.1.1 Multivariate Distance/Kernel-Based Longitudinal Models

Recently the kernel machine regression framework has been extended to test the association between the outcomes and the microbiome community in longitudinal setting. Below, we introduce some longitudinal multivariate distance/kernel-based association tests of microbiome data.

Correlated sequence kernel association test (**cSKAT**) (Zhan et al. 2018) was proposed to directly test microbiome association with related outcomes either from longitudinal and family studies using the linear mixed-effects models (LMMs). It employs (1) random effects to account for the outcome correlations and the effect of covariates and (2) a small-sample adjusted kernel variance component score test called cSKAT to account for high dimensionality, address the problem of small samples but large number of association tests. cSKAT is flexible, allowing both longitudinal and clustered data to be analyzed. However, cSKAT is limited to a continuous outcome and to the item by-item use of the ecological distances (Koh et al. 2019). Because its test may be conservative, it is not a perfect exact test (Zhan et al. 2018). Actually, cSKAT is still not a multivariate longitudinal method because it is limited to a continuous outcome but cannot model multiple taxa (multiple outcomes) simultaneously.

GLMM-MiRKAT and **aGLMM-MiRKAT** (Koh et al. 2019) are a distance-based kernel association test and its data-driven adaptive test based on the generalized linear mixed model (GLMM) (Breslow and Clayton 1993). They were built on two statistical frameworks: (1) GLMM and (2) ecological distance/dissimilarity for microbial community analysis to model data dependency due to clusters (repeated measures) and to account for the within cluster correlation in responses. Because GLMM-MiRKAT uses the framework of GLMM, it not only can model Gaussian distributed host outcomes but also can handle non-Gaussian host outcomes such as binomial and Poisson regression models. In the sense of generalization, GLMM-MiRKAT is an extension of cSKAT (Zhan et al. 2018). Thus, GLMM-MiRKAT is able to analyze the association between microbial composition and various host outcomes adjusting for covariates. By using framework of ecological distances, e.g., Jaccard (Jaccard 1912) /Bray-Curtis dissimilarity (Bray and Curtis 1957), unique fraction distance, GLMM-MiRKAT can handle the multiple features or taxa in a multivariate approach. Furthermore, GLMM-MiRKAT uses the kernel trick

(Cristianini and Shawe-Taylor 2000) and performs a variance component test (Lin 1997) to deal with the large P and small N problem due to high dimensionality. The aGLMM-MiRKAT adapts the test statistic of the minimum P-value from multiple item-by-item GLMM-MiRKAT analyses so that it avoids the need to choose the optimal distance measures from various abundance-based and phylogeny-based distances. The aGLMM-MiRKAT is useful to detect diverse types of host outcomes with robust power and valid statistical inference (Koh et al. 2019). However, one drawback of aGLMM-MiRKAT is just due to its adaption approach: in practice, adaption of a minimum P-value ignores the natures of test statistics and the distance measures. GLMM-MiRKAT and aGLMM-MiRKAT ignore compositional nature of taxa and treat taxa or OTUs as independent. Additionally, the practice of its data processing is also arbitrary such as excluding measurements with low sequencing depth (i.e., <10,000 total reads) and removing OTUs with average relative abundance $<10^{-5}$.

18.1.2 Multivariate Integration of Multi-omics Methods

Lê Cao and her team (Bodein et al. 2019) propose a generic multivariate framework to integrate microbiome data with multi-omics datasets for longitudinal studies. This generic data-driven framework consists of data pre-processing, modeling, data clustering, and integration. The framework was mainly built on two statistical techniques: (1) a linear mixed model (LMM) including smoothing splines to model profiles across groups of samples and (2) sparse multivariate ordination methods to identify sets of variables that are highly associated across the data types and across time. Due to using linear mixed model splines (LMMS) with multivariate dimension reduction techniques, the proposed framework (Bodein et al. 2019) (1) is able to reduce the data dimension and to account for the individual variability and hence can identify the main patterns of longitudinal variation and (2) is also able to analyze data at different time points. However, it also have several limitations (Bodein et al. 2019), including (1) in this framework the data is actually fitted with simple linear regression models because a high individual variability between biological replicates limits the LMMS modeling step. (2) A large number of time points can result in noisy profiles and clusters are modeled which is often due to high individual variability. (3) The overall performance would be optimal for regularly spaced time points in the longitudinal omics experiments, although it was shown that clustering seems to not be impacted by the LMMS interpolation of missing time points. And (4) the issue of analyzing time-course compositional data have not been fully addressed.

The limitations of multivariate distance/kernel-based longitudinal models and multivariate integration of multi-omics methods really highlight the challenges in multivariate longitudinal microbiome data: (1) how to analyze multiple taxa simultaneously; (2) how to collect and integrate multiple omics data at different time points; (3) how to model multiple taxa as the response variables; and (4) how to deal with the large P small N problem (see Sect. 18.3 for details).

18.1.3   Univariate Analysis Versus Multivariate Analysis

Different from the univariate approach of microbiome data analysis, which analyzes each single microbial taxon separately or one-by-one, multivariate analysis analyzes multiple microbial taxa simultaneously. However, when there are too many microbial taxa of interest, it is difficult to directly analyze them simultaneously. In such cases, in general, two strategies could be taken: (1) reduce dimensions directly through dimension reduction techniques such as clustering and ordinations (e.g., PCA or PCoA, NMDS) and (2) use summary statistics such as distance/dissimilarity measures (e.g., Bray-Curtis dissimilarity) and/or correlation analysis (e.g., Kendall, Pearson, Spearman). For the details of using the approach of dimension reduction techniques, the readers are referred to Chap. 10. For the approach of using summary statistics, please see the following NMIT analysis.

In either way, the statistical hypothesis testing will be performed on the measured values, the summary statistics, or the reduced spaces (e.g., principal components or clusters) rather than directly on taxa (e.g., genes or species).

Both univariate and multivariate approaches have their own advantages and disadvantages. Multivariate methods can analyze all taxa simultaneously and deal with the simultaneous relationship among taxa. In general, univariate methods use the mean or median and the variance of a single taxon, while multivariate methods use covariances or correlations of multiple taxa. Microbiome data are multivariate. Thus, multivariate analysis methods are more appropriate to deal with a significant amount of collinearity among taxa in the data matrix.

18.2   Nonparametric Microbial Interdependence Test (NMIT)

Multivariate distance-based framework can be extended to test the association between the outcomes and the microbial community in longitudinal setting. NMIT is one of newly developed multivariate longitudinal distance-based method.

18.2.1   Introduction to NMIT

NMIT (Zhang et al. 2017) is a multivariate distance-based nonparametric longitudinal model and provides a nonparametric group comparison for microbial interdependence to adequately capture the dynamic nature of the microbiome data.

NMIT aims to identify differences in microbial interdependent relationships between groups over time. NMIT is used to compare temporal microbial interdependence structures between groups and test its association with covariates which are either binary outcome (e.g., disease status, gender, or case-control group indicator) or disease-associated continuous environmental factors or quantitative variables (e.g., age, BMI, blood pressure or biomarker measurement).

The NMIT framework consists of two major steps of statistical analysis or testing, one step of network analysis and one step of heatmap for result visualization. First, the core part of NMIT is to perform pairwise correlation analysis within each subject using the longitudinal microbial measurements to capture the individual microbial dependencies over time. Second, perform permutation MANOVA (Anderson 2001; McArdle and Anderson 2001; Tang et al. 2016) to test whether the correlation structure is different between groups or associated with an interested outcome or not. Third, use network analysis to visualize microbial dependency in different groups, and finally, use heatmap to visualize the results of the differences of temporal correlation structure between groups.

18.2.1.1   Calculate Interdependence Correlation Matrix

The goal of NMIT is to test the association between the temporal microbial interdependence profiles and the covariates of interest. NMIT assumes that the relative abundance of subjects for taxa over time points have been collected. Let $Y_{ij} = (y_{ij1}, \ldots, y_{ijm})^T$ denote the relative abundances of taxa M for the ith subject at time J, and let $X_i = (x_{i1}, \ldots, x_{ip})$ denote P covariates collected for subject i.

Under this temporal microbial interdependence setting, NMIT first calculates the $M \times M$ pair-wised taxa correlation matrix to capture the interdependence profile for each subject. Let $u_{i;\,m,m'}$ be the temporal correlation between taxa m and m', and then let $U_i = \{u_{i;m,m'}\}_{m,m'=1}^{M}$ denote the $M \times M$ temporal correlation matrix of subject i. The core part of the NMIT is to construct the interdependence correlation matrix for each subject to summarize their microbial interdependent correlation structures using the repeated microbiome measurements. NMIT has evaluated three correlation methods: (1) Pearson correlation for the linear association between two random variables; (2) the non-parametric Kendall's rank correlation for the nonlinear association among microbial species; and (3) the maximal information coefficient (MIC) (Reshef et al. 2011), measuring how much information two random variables share to capture both linear and nonlinear associations between two continuous variables.

18.2.1.2   Calculate Distance Matrix

NMIT then calculates the distance matrix $D = \{d_{ii'}\}_{i,i'=1}^{N}$, which measures the temporal correlation distance between subjects i and i'. NMIT has considered many distance metrics to calculate the distance of two correlation matrices. For example, one commonly used distance is Frobenius norm which is defined as:

$$d_{ii'} = M^{-1} \sqrt{\sum_{m=1}^{M} \sum_{m'}^{M} \left(u_{i;m,m'} - u_{i';m,m'}\right)^2}. \tag{18.1}$$

18.2.1.3   Perform Statistical Hypothesis Testing of Distance Matrix

Next, with any predetermined distance metric, NMIT performs permutation MANOVA to compare whether the distances between subjects within groups are shorter than the distances between subjects across groups to infer the group difference in terms of microbial interdependence profile. Specifically let $X = (X_1, \ldots, X_N)^T$ be the $N \times P$ covariate matrix, then the projection matrix is $H = X(X^TX)^{-1}X^T$ and the centroid matrix is $G = (I - 11^T/n)A(1 - 11^T/n)$, in which A is the adjacent matrix of D defined as $A = \{a_{ii'}\} = \{-d_{ii'}^2/2\}$, $i = 1, \ldots, N$, I is an $N \times N$ identity matrix, and 1 is an N by 1 vector with all elements being 1. The pseudo-F statistics can be constructed as:

$$F = \frac{tr(HGH)}{tr[(I-H)G(I-H)]}, \tag{18.2}$$

where $tr(\cdot)$ is the trace of a matrix.

Statistical significance testing the null distribution of the pseudo-F statistics is difficult. Typically permutation strategy is used to assess the statistical significance. In the literatures of metagenomics and genetics, permutation test has been widely adopted (Reiss et al. 2010), such as to analyze single time microbial location differences. Like in other studies of metagenomics and genetics, NMIT uses permutation test to perform statistical significance testing of distance matrix.

18.2.2   Implement NMIT Using R

The syntax of NMIT is given as:

NMIT(otu, id.var, cov.var, time.var, method = "kendall", dist.type = "F", heatmap = T, classify = F, fill.na = 0)

where the argument:

- **otu** is a matrix or data frame of OTU table.
- **id.var** is a vector of subjects.
- **cov.var** is a vector of covariates.
- **time.var** is a vector of time variable.
- **method** is used to specify the correlation method ("pearson", "kendall","spearman"). The default method is "kendall".
- **dist.type** is a character string, which is used to specify the type of matrix norm to be computed. The default is "F"or "f" which specifies the Frobenius norm (the Euclidean norm of x treated as if it were a vector). Specifying M″ or "m" is for the maximum modulus of all the elements in x. Specifying "O", "o" or "1" is for the one norm (maximum absolute column sum) and "I" or "i" is for the infinity norm (maximum absolute row sum).
- **heatmap** is a logical value indicating whether to draw heatmap, by default is TRUE.
- **classify** is a logical value indicating whether to draw classifier tree, by default is FALSE.
- **fill.na** is a number between 0 and 1 to fill the missing value; by default the missing value is filled with 0.

After implementing NMIT ()function, it returns a typical, but limited, output for analysis of variance (general linear models), including:

- **aov.tab** is a typical AOV table showing sources of variation, degrees of freedom, sequential sums of squares, mean squares, F statistics, partial R-squared, and P-values, based on N permutations.
- **coefficients** is a matrix of coefficients of the linear model, with rows representing sources of variation and columns representing taxa (i.e., OTU, genus, species); each column represents a fit of a taxon abundance to the linear model. These are what we typically obtain when we fit one taxon to the predictors. These are only available when the sample by taxa (i.e., site x species) matrix is provided, while they are NOT available when the distance matrix in the formula is supplied.
- **coef.sites** is a matrix of coefficients of the linear model, with rows representing sources of variation and columns representing samples or sites; each column represents a fit of sample or site distances (from all other sites) to the linear model. These are what we typically obtain when we fit distances of one site to the predictors.
- **f.perms** is an N by m matrix of the null F statistics for each source of variation based on N permutations of the data. The permutations can be inspected with permustats and its support functions.
- **model.matrix** is the model.matrix for the right hand side of the formula.
- **terms** is the terms component of the model.

Example 18.1: Longitudinal Pregnant Women's Vaginal Microbiota Study

Romero et al. (2014) conducted a longitudinal study using 16S rRNA sequencing to compare the different vaginal microbiota for the pregnant women who had a term delivery and those women who had a preterm delivery. The study collected

72 pregnant women who had a term delivery and 18 women who had a spontaneous preterm delivery. In this study, the interested variable is the binary case-control indicator (the group variable) for term versus preterm delivery.

The original study concluded that the healthy pregnant women had a higher abundance of vaginal microbiota *Lactobacillus* spp. than nonpregnant women. Here, we use this data to illustrate NMIT method to examine whether the pregnant women present a different microbial interdependence profile from those non-pregnant women.

First, we set up working directory and install the NMIT package (latest version 0.1, 2016-11-19).

```
> setwd("~/Documents/QIIME2R/Ch18_NMIT")

> library(devtools)
> install_github("elong0527/NMIT")
> library(NMIT)
```

Then we can use the functions **NMIT()** or **NMIT_phyloseq()** to perform the Nonparametric Microbial Interdependence Test (NMIT). By using **NMIT()**, we need a matrix of OTU table and sample/meta dataset as input data, while by using **NMIT_phyloseq()**, the input data need to be the phyloseq data structure (phyloseq object). Here we only illustrate the **NMIT()** function.

We can take five main steps to perform the **NMIT()** function.

Step 1: Load OTU table and sample dataset.

```
> otu_tab <- read.csv("RomeroOTU.csv", row.names = 1)
> meta_tab <- read.csv("RomeroSampleData.csv",row.names = 1

> head(otu_tab,3)
 Lactobacillus.iners Lactobacillus.crispatus Atopobium.vaginae
Lactobacillus
1 1050 2 0 791
2 2029 0 0 903
3 1425 0 0 672
> head(meta_tab,3)
 Subect_ID Sample_ID GA_Days Age Race Nugent.Score CST Total.Read.
Counts
1 N001 33604 19.29 19 1 0 II 4338
2 N001 35062 23.29 19 1 0 II 4610
3 N001 36790 27.71 19 1 0 II 3596
 pregnant
1 1
2 1
3 1
> colnames(meta_tab)
[1] "Subect_ID" "Sample_ID" "GA_Days"
[4] "Age" "Race" "Nugent.Score"
[7] "CST" "Total.Read.Counts" "pregnant"
```

```
> dim(meta_tab)
[1] 900 9
> dim(meta_tab)
[1] 900 9
```

Step 2: Filter data and obtain the subject level taxa relative abundance over time.

Originally, there are 900 samples, 143 taxa. We remove 3 NAs in GA_Days and filter prevalent taxa with relative abundance $>0.1\%$ in $\geq 20\%$ of all samples, thus resulting in 143 taxa in the 897 samples. Among these 897 samples, 139 were from pregnant women and 758 were from nonpregnant women.

```
> # Remove 3 NAs in GA_Days
> meta_tab = meta_tab[complete.cases(meta_tab[,3]), ]
> colnames(meta_tab)
[1] "Subect_ID" "Sample_ID" "GA_Days"
[4] "Age" "Race" "Nugent.Score"
[7] "CST" "Total.Read.Counts" "pregnant"
> dim(meta_tab)
[1] 897 9

> table(meta_tab$pregnant)
 0 1
758 139
```

Either counts or relative abundance data can be used in the otu_table. Here we use relative abundance data and illustrate two approaches to convert absolute read counts into relative abundance values. In the following commands, we use the **decostand ()** function from the **vegan** package to convert read counts into relative abundances.

```
> library(vegan)

> otu_tab_rel <- decostand(otu_tab, method = "total")
> head(otu_tab_rel,3)
 Lactobacillus.iners Lactobacillus.crispatus Atopobium.vaginae
Lactobacillus
1 0.2420 0.000461 0 0.1823
2 0.4401 0.000000 0 0.1959
3 0.3963 0.000000 0 0.1869
```

We can confirm whether or not this conversion is successful.

```
> head(apply(otu_tab_rel, 1, sum),3)
1 2 3
1 1 1
```

We can also use the **apply()** function to convert read counts into relative abundances below:

```
> otu_tab_rel1<-apply(otu_tab,1,function(X){X/sum(X)})
> head(otu_tab_rel1,3)
 1 2 3 4 5 6
Lactobacillus.iners 0.242047 0.4401 0.3963 0.4054 0.327488 0.5566161
 7 8 9 10 11 12
Lactobacillus.iners 0.3354928 0.4877345 0.06685 0.003204 0.002233
0.446091
 13 14 15 16 17
Lactobacillus.iners 0.185113 0.5519315 0.9528253 0.9089599 0.9395604
```

We can confirm whether or not this conversion is successful.

```
> head(apply(otu_tab_rel1, 2, sum),3)
1 2 3
1 1 1
```

By default, the function **NMIT()** in **NMIT** package uses the **OTUscreen()** function to keep the major taxa with a predefined error rate and percentage threshold. The syntax of this function is

OTUscreen (ana, error_rate = 0.1, pct_threshold = 20).

where the argument:

- **ana** is a matrix of OTU table or a phyloseq object with counts/relative abundance OTU table depending on using NMIT() or NMIT_phyloseq().
- **error_rate** is error rate percentage; the default is 0.1%.
- **pct_threshold** is the occurrence percentage threshold; the default percentage threshold is 20%.

By implementing the function **NMIT()**, a matrix with major taxa has been used in the analysis. We can check subsetting in the modeling results with the information of **coef.sites** or **model.matrix**.

Step 3: Compute the subject level taxa temporal correlation.

In this step, for each subject, a temporal correlation between any pair of taxa are computed. This step will return a list of temporal correlation matrix for each subject. The OTU temporal interdependence for each subjects is obtained by running the function **tscor()**. The syntax of **tscor()** is given:

tscor(ana, method='kendall', subject_var, fill.na = 0)

where the argument:

- **ana** is a matrix of OTU table or a phyloseq object of counts/relative abundance data.
- **method** is used to specify an option of the correlation method ("pearson", "kendall", "spearman"). In NMIT, three correlation methods are available: Pearson, Spearman, and Kendall. The default method is "spearman." The argument **subject_var** is a numeric vector of subject. The argument **fill.na** is a number between 0 and 1 to fill the missing value. The default value is 0.

Step 4: Calculate the distance between temporal correlation matrices.

This step is implemented in the **NMIT()** function. In this step, the distance between any two correlation matrices in Step 3 is to be calculated based on the specified **dist.type** (the distance measure). This must be one of "Euclidean," "maximum," "manhattan," "Canberra," "binary," or "minkowski." **Frobenius** norm is the default type.

Step 5: Perform permutation MANOVA to test the differences of distances between subjects obtained in Step 4.

The permutation MANOVA is implemented via the **adonis()** in vegan package, which is called in the NMIT() function. In the NMIT() function, a heatmap function is also developed through calling the **gplots::heatmap.2()** to show the within-group distances and the between-group distances. We can choose to draw heatmap by specifying the value is "TRUE" or "FALSE". The default value is TRUE.

```
> # Fit method = "spearman"
> set.seed(123)
> fit_s<-NMIT(
+ otu=otu_tab_rel,
+ id.var=meta_tab$Subect_ID,
+ cov.var= meta_tab$pregnant,
+ time.var= meta_tab$GA_Days,
+ method = "spearman",
+ dist.type = "F",
+ heatmap = F,
+ classify = F,
+ fill.na = 0
+ )
```

We can retrieve the ANOVA table below:

```
> fit_s$aov.tab
Permutation: free
Number of permutations: 999

Terms added sequentially (first to last)

Df SumsOfSqs MeanSqs F.Model R2 Pr(>F)
cov.var 1 711 711 1.6 0.03 0.007 **
Residuals 52 23151 445 0.97
```

```
Total 53 23862 1.00
---
Signif. codes: 0 '***' 0.001 '**' 0.01 '*' 0.05 '.' 0.1 ' ' 1
```

Other results can be obtained such as using the following commands to retrieve coef.sites, f.perms, and model.matrix:

```
> fit_s$coef.sites
> fit_s$f.perms
> fit_s$model.matrix

> # Fit method = "kendall"
> set.seed(123)
> fit_k<-NMIT(
+ otu=otu_tab,
+ id.var=meta_tab$Subect_ID,
+ cov.var= meta_tab$pregnant,
+ time.var= meta_tab$GA_Days,
+ method = "kendall",
+ dist.type = "F",
+ heatmap = F,
+ classify = F,
+ fill.na = 0
+ )

> fit_k$aov.tab
Permutation: free
Number of permutations: 999

Terms added sequentially (first to last)

 Df SumsOfSqs MeanSqs F.Model R2 Pr(>F)
cov.var 1 583 583 1.53 0.029 0.007 **
Residuals 52 19806 381 0.971
Total 53 20389 1.000
---
Signif. codes: 0 '***' 0.001 '**' 0.01 '*' 0.05 '.' 0.1 ' ' 1

> # Fit method = "pearson"
> set.seed(123)
> fit_p<-NMIT(
+ otu=otu_tab_rel,
+ id.var=meta_tab$Subect_ID,
+ cov.var= meta_tab$pregnant,
+ time.var= meta_tab$GA_Days,
+ method = "pearson",
+ dist.type = "F",
+ heatmap = F,
+ classify = F,
+ fill.na = 0
+ )
```

```
> fit_p$aov.tab
Permutation: free
Number of permutations: 999


Terms added sequentially (first to last)

 Df SumsOfSqs MeanSqs F.Model R2 Pr(>F)
cov.var 1 700 700 1.46 0.027 0.005 **
Residuals 52 24986 481 0.973
Total 53 25686 1.000
---
Signif. codes: 0 '***' 0.001 '**' 0.01 '*' 0.05 '.' 0.1 ' ' 1
```

The above permutation MANOVA test showed that the pregnancy can explain the distances between subjects with *P*-values of 0.007, 0.007, and 0.005 using Spearman, Kendall, and Pearson correlation methods, respectively. Thus, we can conclude that microbial interdependence profiles over the studied period are statistically different between the pregnant women and those nonpregnant women.

18.2.3 Implement NMIT Using QIIME 2

Example 18.2: Early Childhood Antibiotics and the Microbiome (ECAM)
ECAM (Bokulich et al. 2016) is a longitudinal human study. This study monitored the microbiome profiles of 43 children since their birth over the first 2 years. During the first year of life, stool samples were collected at each visit from these infants. Of the 43 children, 24 were vaginally delivered and 19 were cesarean-delivered. Thirty-one of infants were dominantly (>50% of feedings) breast-fed, and 12 of them were dominantly formula-fed for the first 3 months of life.

The data of this study have been used for testing the NMIT method in the original NMIT paper (Zhang et al. 2017) and QIIME 2 documentation. Both tested whether the microbial interdependence is interrupted during babies' first year of life between two delivery modes (vaginally and cesarean-delivered) with either adjusting their gender, diet type (breast feeding or on formula), and other covariates (in the original NMIT paper) or without adjusting covariates (in QIIME 2 documentation). Here, we use the same dataset to illustrate NMIT via QIIME 2 to evaluate whether the microbial interdependence is different between breast-dominant children and formula-dominant children during their first-year life. Each child had collected multiple stool samples during this period. In total, 505 stool samples had been collected; at average each child had collected 12 stool samples and minimally had 7 samples. The abundance data had 498 different taxa at the genus level. The 23 major taxa with their relative abundance >0.1% in >15% of the total samples were used for original NMIT paper and QIIME 2 analyses.

In this section, we implement NMIT via QIIME 2 using the data from this ECAM study. The datasets were downloaded from the QIIME 2 website, where the data were used to run NMIT to test delivery modes using the Pearson's method to calculate the temporal correlation matrix for each sample. Here, we use the data to illustrate the NMIT method via QIIME 2 to test diet types. We use the Kendall's method to calculate the temporal correlation matrix and the Frobenius norm to evaluate the distance between any two correlation matrices.

NMIT calculates the interdependence correlation matrix based on the repeated microbiome measurements within each subject. Thus, the different time points across subjects or missing observations for some subjects do not affect the value of interdependence correlation matrix (Zhang et al. 2017). Therefore, NMIT does not require that samples are collected at identical time points and hence is robust to missing samples. However, based on QIIME 2 documentation, this may have negative effects on data quality if subjects are highly under-sampled due to missing values or may not provide biologically meaningful results if subjects' sampling times do not overlap. QIIME 2 also suggests at least 5–6 samples at each time point per subject should be used. For a very large feature table, NMIT may take a long time to run, thus to improve runtime, in practice, the low-abundance features will be filtered and the feature tables on taxonomy will be collapsed such as into to genus level. Here, the illustrated feature tables on taxonomy are genus-level taxa with a relative abundance >0.1% in more than 15% of the total samples.

First, we need to activate QIIME 2. Open the terminal and type: source activate qiime2-2022.2, which is based on which version of QIIME 2 on your computer. When the terminal appears "(qiime2-2022.2)," it indicates qiime2-2022.2 version has been activated. We can work on this version now.

We already create a folder NMIT in the path:QIIME2-Biostatistics/longitudinal/NMIT, so we can use cd QIIME2R-Biostatistics/longitudinal/NMIT to this folder. If this folder did not exist in your computer, you can create one: mkdir QIIME2R-Biostatistics/longitudinal/NMIT.

Next, we can take four main steps to implement NMIT via QIIME 2's **qiime longitudinal nmit** command to evaluate longitudinal sample similarity as a function of temporal microbial composition. We illustrate NMIT using Kendall correlation method, the reader can easily use Pearson and Spearman methods just replacing "kendall" in the **p-corr-method** parameter column with either "pearson" or "spearman."

Step 1: Perform NMIT using qiime longitudinal nmit command.

```
source activate qiime2-2022.2
cd QIIME2R-Biostatistics/longitudinal/NMIT
qiime longitudinal nmit\
 --i-table FeatureRelativeTableEcam.qza\
 --m-metadata-file SampleMetadataEcam.tsv\
 --p-individual-id-column studyid\
 --p-corr-method kendall\
 --p-dist-method fro\
 --o-distance-matrix NmitDMKendallEcam.qza
```

where (1) the required input table is a FeatureTable[RelativeFrequency] artifact (here, FeatureRelativeTableEcam.qza). (2) The sample metadata file containing individual-id-column (here, SampleMetadataEcam.tsv) is also required, which is specified by the m-metadata-file. (3) The p-individual-id-column parameter is also required, which is used to specify the metadata column containing study IDs for individual subjects (here, studyid). (4) The p-corr-method parameter is used to specify the temporal correlation method to be applied. We can choose one of them from: "kendall"(default), "pearson," "spearman." This column is required too. (5) The p-dist-method parameter is optional. It is used to specify the temporal distance method either "fro" (**Frobenius** norm) or "nuc" (nuclear norm) with the default is the **Frobenius** norm. (6) Finally, we need to name and save the resulting distance matrix (i.e., NmitDMKendallEcam.qza) via the o-distance-matrix parameter, which is an artifact.

The saved output of the distance matrix artifact can be passed to other QIIME 2's commands for statistical testing of significance and visualization.

Step 2: Conduct statistical significance testing of distance matrix via qiime diversity beta-group-significance command.

```
qiime diversity beta-group-significance \
 --i-distance-matrix NmitDMKendallEcam.qza \
 --m-metadata-file SampleMetadataEcam.tsv \
 --m-metadata-column diet \
 --o-visualization NmitDietKendallEcam.qzv
```

This step actually performs PERMANOVA to evaluate whether between-group distances are larger than within-group distance. As we stated in Chap. 11 (Sect. 11. 3.2), it calls the function **adonis()** in vegan package. By QIIME2 view, we can see the distances within the dominantly breast-fed and dominantly formula-fed groups are smaller than the distance between these two groups. The group difference related to the microbial interdependence is statistically significant with P-value of 0.045 based on Kendall method. Thus, we concluded that the diet type did alter the microbial interdependence in early childhood.

Step 3: Compute principal coordinates via qiime diversity pcoa command.

```
qiime diversity pcoa \
 --i-distance-matrix NmitDMKendallEcam.qza \
 --o-pcoa NmitPCKendallEcam.qza
```

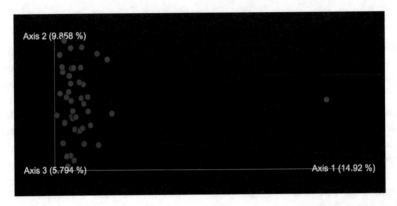

Fig. 18.1 Principal coordinates generated by PCoA and the similarities among subjects plotted by Emperor plot

Saved PCoAResults to: NmitPCKendallEcam.qza

Step 4: Use Emperor to visualize similarities among subjects via qiime emperor plot command.

Finally, we can also use Emperor plot to visualize similarities among **subjects** (Fig. 18.1).

```
# Figure 18.1
qiime emperor plot \
 --i-pcoa NmitPCKendallEcam.qza \
 --m-metadata-file SampleMetadataEcam.tsv \
 --o-visualization NmitEmperorKendallEcam.qzv
```

Saved Visualization to: NmitEmperorKendallEcam.qzv

Please note that the computed principal coordinates and similarities visualized by Emperor plot are among subjects but not among individual samples.

18.2.4   Remarks on NMIT

In the literature, microbiome data have been considered having the unique compositional structure. To address the spurious correlation problem with compositional data (Pearson 1896; Ranganathan and Borges 2011), a family of log-ratio transformations has been proposed (Xia et al. 2018a), which consists of the additive log-ratio (alr) transformation (Aitchison 1986) (p.113), the centered log-ratio (clr) transformation (Aitchison 2003), and the isometric log-ratio (ilr) transformation (Egozcue et al. 2003) (see details on Chap. 10 in this book).

The clr-representation of composition $x = (x_1, \ldots, x_i, \ldots, x_D)$ is defined as the logarithm of the components after dividing by the geometric mean of x (a vector of relative abundance for taxa), which is most often implemented to remove the possible compositional effect:

$$clr(x) = \left[\ln\left(\frac{x_1}{g_m(x)}\right), \cdots \ln\left(\frac{x_i}{g_m(x)}\right), \cdots \ln\left(\frac{x_D}{g_m(x)}\right) \right], \qquad (18.3)$$

with $g_m(x) = \sqrt[D]{x_1 \cdot x_2 \cdots x_D}$ ensuring that the sum of the elements of clr(x) is zero. The clr transforms the composition to the Euclidean sample space, and hence providing the possibility of using standard unconstrained statistical methods for analyzing compositional data (Aitchison 2003).

NMIT does not use the clr transformation in its test based on the following two reasons: (1) The correlation is calculated between any pair of taxa using the microbiome measurements at the multiple time points within each subject. The original correlation will be contaminated using the transformation because the geometric mean function varies across different time points and the transformation is performed at each time point independently. (2) The compositional effect (e.g., the spurious correlation) of the microbiome data is mild (Friedman and Alm 2012) compared to the ecology data because the number of taxa in the microbiome studies is usually larger than that in the ecology studies.

NMIT is a statistical method for discovering the microbiome data structure. This test is one of the first steps to understand the dynamic nature of microbiome data. However, the NMIT methods have the drawbacks, for example:

- As a distance-based testing method, NMIT provides an overall assessment of the group difference in terms of the interdependent relationship (microbial interdependence similarity) among taxa (Zhang et al. 2017). Further steps are required to identify the specific key taxa that are associated with outcomes or biological covariates.
- NMIT mainly uses the correlation tests (Pearson correlation, Kendall's rank correlation, and maximal information coefficient (MIC)). However, using correlation coefficients to detect dependencies of microbial taxa cannot avoid detecting spurious correlations due to compositionality (Chen and Li 2016) and could be severely underpowered owing to the relatively low number of samples (Layeghifard et al. 2018).
- The study demonstrated Kendall method is always comparable or has a slight power edge over Pearson method, while MIC method has less power than Pearson and Kendall methods (Zhang et al. 2017). These findings are not all consistent with the literature. Thus, these arguments need to be further validated.
- Current version of NMIT cannot handle time-varying covariate (Zhang et al. 2017).

Recently NMIT and linear mixed-effects models (LMMs) (Lindstrom and Bates 1988) were built in QIIME 2 as the two q2-longitudinal plugins to facilitate streamlined analysis and visualization of longitudinal and paired sample data sets (Bokulich et al. 2018). When the reader performs statistical analysis through QIIME 2, the user needs to know the advantages and disadvantages of the underlying methods.

18.3   The Large P Small N Problem

As we mentioned in Sect. 18.1, with many microbial taxa presented in a limited sample size, multivariate analysis of microbiome data can result in the large P small N problem (Xia and Sun 2022). The "large P small N" problem exist in both univariate and multivariate analyses.

In univariate analysis, the "large P small N" problem refers to the statistical modeling and inference problem in analyzing a response variable Y, with the larger number of covariates P in a set of vector $X = (X_1, X_2, \ldots, X_P)$ and the substantially smaller number of available sample size N (i.e., $N \ll P$) (Chakraborty et al. 2012). Here, P usually refers to the dimension of predictors (independent variables), and N refers to the sample size. That is, the design matrix (matrix of predictors) of $N \times P$ (Bernardo et al. 2003) is high-dimensional, so that it is challenging to directly estimate the parameters. The problem is more deteriorated if multicollinearity exists in the predictors.

In multivariate analysis, the "large P small N" problem means that there are more features than data points, where P refers to the dimension of response (dependent) variables. In terms of data matrix, P refers to the number of columns (graphically dimensional space), and N refers to the number of rows. The "large P small N" problem refers to small data points (samples: N), but each data point contains large P features (dimensional space).

Such situations usually arise in multiple-omics data. For example, in genome-wide association studies (GWAS), the "large P small N" problem could occur (Diao and Vidyashankar 2013) because the number of observations N is usually less hundreds or thousands, whereas the number of markers P is approximately hundreds of thousands (Mei and Wang 2016). One main characteristic of metabolomics datasets is that the number of metabolites (P) is greater than the number of observations (N) (Johnstone and Titterington 2009). Microbiome sequence data sets are also high dimensional with the number of features (ASV/OTU or taxa) being much higher than the number of samples (Xia et al. 2018b; Xia 2020; Xia and Sun 2022).

In statistical modeling and inference, a larger P needs a larger N. When the large P with small N, then it results in the problem of "curse of dimensionality." The problem of "large P small N" makes traditional linear regression methods infeasible because design matrix is singular, is underdetermined, and hence is no longer invertible, and no unique least-squares solution exists as well. When the multiple responses are correlated, the "large P small N" problem is more complicated.

In analysis of high dimensional omics data, the general strategy is either to reduce the number of variables to keep only significant variables in models or to project variables to lower dimension. Ideally, the multivariate analysis methods used are capable of dealing with significant amounts of collinearity in data matrix. Thus, some kinds of special statistical methods, such as variable selection, principal component related methods (of which PCA and PLS are prime examples), penalized likelihood, constraint or shrinkage methods, Bayesian methods, and ridge regressions are often used.

18.4 Summary

In the last chapter, we first provided an overview of multivariate longitudinal microbiome methods, including multivariate distance/kernel-based longitudinal models and multivariate integration of multi-omics methods, and briefly discussed the approaches of univariate analysis and multivariate analysis. We then specifically introduced the NMIT and illustrated its implements using R and QIIME 2 with real microbiome data. Finally, we discussed the large P small N problem in general statistics and especially in multiple-omics studies.

References

Aitchison, J. 1986. *The statistical analysis of compositional data. Monographs on statistics and applied probability.* London: Chapman & Hall Ltd.. (Reprinted in 2003 with additional material by The Blackburn Press). 416 p.

Aitchison, John. 2003. A concise guide to compositional data analysis. In *2nd compositional data analysis workshop. Girona, Italy, 2003.*

Anderson, Marti J. 2001. A new method for non-parametric multivariate analysis of variance. *Austral Ecology* 26 (1): 32–46. https://doi.org/10.1111/j.1442-9993.2001.01070.pp.x. https://onlinelibrary.wiley.com/doi/abs/10.1111/j.1442-9993.2001.01070.pp.x.

Bernardo, J.M., M.J. Bayarri, J.O. Berger, A.P. Dawid, D. Heckerman, A. Smith, and M. West. 2003. Bayesian factor regression models in the "large p, small n" paradigm. *Bayesian Statistics* 7: 733–742.

Bodein, Antoine, Olivier Chapleur, Arnaud Droit, and Kim-Anh Lê Cao. 2019. A generic multivariate framework for the integration of microbiome longitudinal studies with other data types. *Frontiers in Genetics* 10: 963–963. https://doi.org/10.3389/fgene.2019.00963. https://pubmed.ncbi.nlm.nih.gov/31803221; https://www.ncbi.nlm.nih.gov/pmc/articles/PMC6875829/.

Bokulich, Nicholas A., Jennifer Chung, Thomas Battaglia, Nora Henderson, Melanie Jay, Huilin Li, Arnon D. Lieber, Fen Wu, Guillermo I. Perez-Perez, and Yu Chen. 2016. Antibiotics, birth mode, and diet shape microbiome maturation during early life. *Science Translational Medicine* 8 (343): 343ra82-343ra82.

Bokulich, Nicholas A., Matthew R. Dillon, Yilong Zhang, Jai Ram Rideout, Evan Bolyen, Huilin Li, Paul S. Albert, and J. Gregory Caporaso. 2018. q2-longitudinal: Longitudinal and paired-sample analyses of microbiome data. *mSystems* 3 (6): e00219–18. https://doi.org/10.1128/mSystems.00219-18. https://msystems.asm.org/content/msys/3/6/e00219-18.full.pdf.

Bray, J. Roger, and J.T. Curtis. 1957. An ordination of the upland forest communities of Southern Wisconsin. *Ecological Monographs* 27 (4): 325–349. https://doi.org/10.2307/1942268. https://esajournals.onlinelibrary.wiley.com/doi/abs/10.2307/1942268.

Breslow, N.E., and D.G. Clayton. 1993. Approximate inference in generalized linear mixed models. *Journal of the American Statistical Association* 88 (421): 9–25. https://doi.org/10.2307/2290687. http://www.jstor.org/stable/2290687.

Chakraborty, Sounak, Malay Ghosh, and Bani K. Mallick. 2012. Bayesian nonlinear regression for large p small n problems. *Journal of Multivariate Analysis* 108: 28–40. https://doi.org/10.1016/j.jmva.2012.01.015. http://www.sciencedirect.com/science/article/pii/S0047259X12000164.

Chen, Eric Z., and Hongzhe Li. 2016. A two-part mixed-effects model for analyzing longitudinal microbiome compositional data. *Bioinformatics* 32 (17): 2611–2617. https://doi.org/10.1093/bioinformatics/btw308.

Cristianini, Nello, and John Shawe-Taylor. 2000. *An introduction to support vector machines and other kernel-based learning methods.* Cambridge: Cambridge University Press.

Diao, Guoqing, and Anand N. Vidyashankar. 2013. Assessing genome-wide statistical significance for large p small n problems. *Genetics* 194 (3): 781–783. https://doi.org/10.1534/genetics.113.150896. https://pubmed.ncbi.nlm.nih.gov/23666935; https://www.ncbi.nlm.nih.gov/pmc/articles/PMC3697980/.

Egozcue, Juan José, Vera Pawlowsky-Glahn, Glòria Mateu-Figueras, and Carles Barcelo-Vidal. 2003. Isometric logratio transformations for compositional data analysis. *Mathematical Geology* 35 (3): 279–300.

Friedman, Jonathan, and Eric J. Alm. 2012. Inferring correlation networks from genomic survey data. *PLoS Computational Biology* 8 (9): e1002687.

Jaccard, Paul. 1912. The distribution of the flora in the alpine zone.1. New Phytologist 11 (2): 37–50. https://doi.org/10.1111/j.1469-8137.1912.tb05611.x. https://nph.onlinelibrary.wiley.com/doi/abs/10.1111/j.1469-8137.1912.tb05611.x.

Johnstone, Iain M., and D. Michael Titterington. 2009. Statistical challenges of high-dimensional data. *Philosophical Transactions of the Royal Society A: Mathematical, Physical and Engineering Sciences* 367 (1906): 4237–4253. https://doi.org/10.1098/rsta.2009.0159. https://royalsocietypublishing.org/doi/abs/10.1098/rsta.2009.0159.

Koh, Hyunwook, Yutong Li, Xiang Zhan, Jun Chen, and Ni Zhao. 2019. A distance-based kernel association test based on the generalized linear mixed model for correlated microbiome studies. *Frontiers in Genetics* 10 (458). https://doi.org/10.3389/fgene.2019.00458. https://www.frontiersin.org/article/10.3389/fgene.2019.00458.

Layeghifard, Mehdi, David M. Hwang, and David S. Guttman. 2018. Constructing and analyzing microbiome networks in R. In *Microbiome analysis*, ed. Robert G. Beiko et al., 243–266. Springer.

Lin, Xihong. 1997. Variance component testing in generalised linear models with random effects. *Biometrika* 84 (2): 309–326. https://doi.org/10.1093/biomet/84.2.309.

Lindstrom, Mary J., and Douglas M. Bates. 1988. Newton-Raphson and EM algorithms for linear mixed-effects models for repeated-measures datam. *Journal of the American Statistical Association* 83 (404): 1014–1022. https://doi.org/10.2307/2290128. https://www.jstor.org/stable/2290128.

McArdle, Brian H., and Marti J. Anderson. 2001. Fitting multivariate models to community data: A comment on distance-based redundancy analysis. *Ecology* 82 (1): 290–297. https://doi.org/10.1890/0012-9658(2001)082[0290:Fmmtcd]2.0.Co;2. https://esajournals.onlinelibrary.wiley.com/doi/abs/10.1890/0012-9658%282001%29082%5B0290%3AFMMTCD%5D2.0.CO%3B2.

Mei, Bujun, and Zhihua Wang. 2016. An efficient method to handle the 'large p, small n'problem for genomewide association studies using Haseman–Elston regression. *Journal of Genetics* 95 (4): 847–852.

Pearson, Karl. 1896. Mathematical contributions to the theory of evolution – On a form of spurious correlation which may arise when indices are used in the measurement of organs. *Proceedings of the Royal Society of London* 60 (359–367): 489–498.

Ranganathan, Yuvaraj, and Renee M. Borges. 2011. To transform or not to transform: That is the dilemma in the statistical analysis of plant volatiles. *Plant Signaling & Behavior* 6 (1): 113–116.

Reiss, Philip T., M. Henry, H. Stevens, Zarrar Shehzad, Eva Petkova, and Michael P. Milham. 2010. On distance-based permutation tests for between-group comparisons. *Biometrics* 66 (2): 636–643.

Reshef, David N., Yakir A. Reshef, Hilary K. Finucane, Sharon R. Grossman, Gilean McVean, Peter J. Turnbaugh, Eric S. Lander, Michael Mitzenmacher, and Pardis C. Sabeti. 2011. Detecting novel associations in large data sets. *Science* 334 (6062): 1518–1524.

Romero, Roberto, Sonia S. Hassan, Pawel Gajer, Adi L. Tarca, Douglas W. Fadrosh, Janine Bieda, Piya Chaemsaithong, Jezid Miranda, Tinnakorn Chaiworapongsa, and Jacques Ravel. 2014. The vaginal microbiota of pregnant women who subsequently have spontaneous preterm labor and delivery and those with a normal delivery at term. *Microbiome* 2 (1): 18. https://doi.org/10.1186/2049-2618-2-18.

Tang, Zheng-Zheng, Guanhua Chen, and Alexander V. Alekseyenko. 2016. PERMANOVA-S: Association test for microbial community composition that accommodates confounders and multiple distances. *Bioinformatics* 32 (17): 2618–2625. https://doi.org/10.1093/bioinformatics/btw311.

Xia, Yinglin. 2020. Correlation and association analyses in microbiome study integrating multiomics in health and disease. *Progress in Molecular Biology and Translational Science* 171: 309–491.

Xia, Yinglin, and Jun Sun. 2022. *Statistical data analysis of microbiomes and metabolomics.* Washington, DC: American Chemical Society.

Xia, Yinglin, Jun Sun, and Ding-Geng Chen. 2018a. Compositional analysis of microbiome data. In *Statistical analysis of microbiome data with R,* ed. Yinglin Xia et al., 331–393. Springer.

———. 2018b. What are microbiome data? In *Statistical analysis of microbiome data with R,* ed. Yinglin Xia et al., 29–41. Springer.

Zhan, Xiang, Lingzhou Xue, Haotian Zheng, Anna Plantinga, Michael C. Wu, Daniel J. Schaid, Ni Zhao, and Jun Chen. 2018. A small-sample kernel association test for correlated data with application to microbiome association studies. *Genetic Epidemiology* 42 (8): 772–782. https://doi.org/10.1002/gepi.22160. https://onlinelibrary.wiley.com/doi/abs/10.1002/gepi.22160.

Zhang, Yilong, Sung Won Han, Laura M. Cox, and Huilin Li. 2017. A multivariate distance-based analytic framework for microbial interdependence association test in longitudinal study. *Genetic Epidemiology* 41 (8): 769–778. https://doi.org/10.1002/gepi.22065. https://www.ncbi.nlm.nih.gov/pubmed/28872698; https://www.ncbi.nlm.nih.gov/pmc/articles/PMC5696116/.

Correction to: Bioinformatic and Statistical Analysis of Microbiome Data

Yinglin Xia and Jun Sun

Correction to:
Y. Xia, J. Sun, *Bioinformatic and Statistical Analysis of Microbiome Data,*
https://doi.org/10.1007/978-3-031-21391-5

The original version of the book was inadvertently published without uploading relevant extra supplementary materials pertaining to Chapters 1, 2, 3, 4, 5, 6, 9, 10, 11, 12, 13, 14, 15, 17, and 18. The chapters have now been updated and supplementary materials have been uploaded to Springer website as downloadable files.

The updated original version of these chapters can be found at
https://doi.org/10.1007/978-3-031-21391-5_1
https://doi.org/10.1007/978-3-031-21391-5_2
https://doi.org/10.1007/978-3-031-21391-5_3
https://doi.org/10.1007/978-3-031-21391-5_4
https://doi.org/10.1007/978-3-031-21391-5_5
https://doi.org/10.1007/978-3-031-21391-5_6
https://doi.org/10.1007/978-3-031-21391-5_9
https://doi.org/10.1007/978-3-031-21391-5_10
https://doi.org/10.1007/978-3-031-21391-5_11
https://doi.org/10.1007/978-3-031-21391-5_12
https://doi.org/10.1007/978-3-031-21391-5_13
https://doi.org/10.1007/978-3-031-21391-5_14
https://doi.org/10.1007/978-3-031-21391-5_15
https://doi.org/10.1007/978-3-031-21391-5_17
https://doi.org/10.1007/978-3-031-21391-5_18

© Springer Nature Switzerland AG 2023
Y. Xia, J. Sun, *Bioinformatic and Statistical Analysis of Microbiome Data,*
https://doi.org/10.1007/978-3-031-21391-5_19

Index

A

Abundance based coverage estimator (ACE), 290–298, 330, 572, 573
Adaptive Gauss-Hermite quadrature (AGQ), 593–595, 608, 647, 659, 660
Additive log-ratio (alr), 496–498, 519, 551, 690
adonis(), 409, 410, 412, 418, 426, 685, 689
adonis2(), 409–414, 416
aGLMM-MiRKAT, 676, 677
AICtab (), 633, 635
Aitchison simplex, 491, 493, 496–498, 551
Akaike information criterion (AIC), 562, 575, 576, 600–605, 608, 618–620, 633–635, 640, 642–646, 651, 652, 658, 660, 670
ALDEx2, 491, 498, 500–518, 520, 526, 527, 549, 551
Alpha diversity, 2, 5, 47, 135, 136, 154, 156, 257, 272, 289–330, 336, 338, 557, 572, 576
Alpha-phylogenetic method, 326
ampvis2 package, 293–296, 300, 303, 308, 312, 319, 352–353, 355, 357, 388, 572
Analysis of composition of microbiomes (ANCOM), 491, 518–528, 532, 547–549, 551
Analysis of composition of microbiomes-bias correction (ANCOM-BC), 491, 518, 528–549, 551
Analysis of similarity (ANOSIM), 236, 341, 397–405, 409, 417, 418, 429, 431
ANCOMBC package, 532–547
Ape, 11, 29, 48–53, 75, 346, 352, 356
Artifacts, 2–5, 8, 65, 68, 69, 73, 76–79, 87–90, 92, 100–101, 104, 107, 110, 114, 116, 117, 127, 132, 133, 136–138, 149–151, 153, 154, 156, 230, 233, 323, 371, 382, 429, 465, 492, 520, 522, 523, 580, 689
Autoregressive of order 1[AR(1)], 568, 597, 663, 667
Average-linkage clustering, 162, 174, 177, 196–198, 214, 215, 218, 230, 238, 239, 259

B

Bayesian information criterion (BIC), 562, 601, 603, 605, 608, 618, 619, 633–635, 640, 642–646, 651, 652, 658, 660, 670
bbmle package, 633, 651
Benjamini–Hochberg (BH) procedure, 503, 511, 531, 540
betadisper(), 418–420
Beta diversity, 2, 43, 47, 152, 154, 209, 210, 261, 271, 289, 290, 322, 323, 330, 335–348, 371, 381–384, 388, 397–431, 441, 520, 557, 580
BICtab(), 633
Biological classifications, 174–176, 209, 218
BIOM format, 47, 54–56, 61, 74, 92, 134, 294
Bland-Altman plot, 508, 509, 514
Bonferroni correction method, 540
Box plots, 18–21, 102, 112, 113, 115
Bray-Curtis distance, 86, 87, 382, 383, 406, 409, 418, 429–430
Bray-Curtis index, 336–340
Bridge criterion (BC), 251, 532, 603–604, 608

© Springer Nature Switzerland AG 2023
Y. Xia, J. Sun, *Bioinformatic and Statistical Analysis of Microbiome Data*,
https://doi.org/10.1007/978-3-031-21391-5

C

calcNormFactors(), 460, 466

Canonical correspondence analysis (CCA), 351–353, 355–357, 359, 363, 368, 377–381, 385, 388, 398

Castor, 11, 48, 52–54, 75

Centered log-ratio (clr), 45, 344, 496, 497, 505, 506, 518, 551, 690

Chao 1, 290–298, 325, 330, 572, 573, 665

Characters, 12, 15, 16, 27, 49, 52, 53, 67, 69–71, 97, 137, 231, 236, 245–247, 267, 351, 360, 386, 443, 444, 512, 533, 681

Closed-reference clustering, 149–152, 154–156

Cluster-free filtering (CFF), 149–152, 154–157, 161, 254, 259–261, 269, 271, 273

Clustering, 2, 8, 103, 104, 119, 123, 125, 140, 147–157, 161, 162, 164, 176, 177, 191, 193–200, 202–204, 208–210, 214–218, 228–235, 237–241, 248–255, 257, 258, 260–269, 335, 345, 384–388, 398, 404, 449, 616, 677, 678

Clustering-based OTU methods, 161, 162, 209–218, 227–250, 252–254, 261, 273

Commonly clustering-based OTU methods, 161, 177, 192–207, 215, 216, 218, 273

Complete linkage clustering, 196–198, 200, 214, 229, 230, 259

Compositional data, 491–500, 505, 548–551, 677, 690, 691

Conway-Maxwell-Poisson distribution, 621, 623, 630, 646

Correlated sequence Kernel association test (cSKAT), 676

Correlation matrix, 140, 180, 188, 451, 568, 588, 597, 663, 664, 679, 680, 684, 685, 688

Correspondence analysis (CA), 351, 355, 366–372, 377, 387

CSV, 25, 46, 47, 97, 327, 430, 474

cumNormMat(), 459–460, 466

Cumulative sum scaling (CSS), 436, 438, 440, 441, 454, 465–467, 479

curatedMetagenomicData, 24, 48

D

DADA2, 2, 95, 103–108, 113, 114, 116–119, 124, 125, 128–132, 141, 148, 149, 156, 254, 256–259, 261–266, 268–270, 273, 296, 520

Deblur, 2, 95, 99, 103, 104, 108, 113–119, 148–150, 254, 256–259, 262, 264–265, 269, 270, 273

Demultiplexed, 71–73, 78, 95–108, 112–119, 149

Demultiplexed paired-end FASTQ data, 78, 95–108, 119

Denoising-based methods, 250, 254, 256–267, 273

De novo clustering, 153–156, 230, 234, 253

Density plots, 22

DESeq, 465, 517

Detrended correspondence analysis (DCA), 351, 352, 356, 371–374, 378, 385, 388

Deviance information criterion (DIC), 604, 608

devtools package, 649

DHARMa package, 640, 670

Difference plot, 508

Discovery odds ratio testing, 455, 466

Discriminant analysis, 193, 194, 203–207, 216–218

dist(), 401, 419

Distance matrix, 4, 52, 79, 86–87, 92, 135, 203, 209, 229, 235, 345, 347, 348, 351, 352, 360, 399, 407, 409, 424, 429, 680, 681, 689

Distribution-based clustering (DBC), 252–254, 260, 273

download.file(), 15

E

Ecological similarity, 236, 240, 260, 261

edgeR, 465, 488, 517, 518, 548, 549, 616

effectPlotData (), 648

Effect size, 168, 400, 465, 503–512, 516, 547, 606, 607

EM-IWLS algorithm, 592, 598, 660, 662–664, 670

Emperor plots, 382–384, 388, 428, 520, 521, 690

Entropy-based methods, 255

Eukaryote species, 242, 243

exportMat(), 461, 467

exportStats(), 461, 467

EzBioCloud, 124

F

Factor analysis, 162, 168, 180, 181, 193, 195, 201, 202, 209, 218, 349, 350

False positive rate (FDR), 140, 423, 437, 465, 466, 477, 494, 518, 526, 527, 531, 532, 540, 547–549

FASTA, 1, 3, 66–69, 71, 92, 128, 129

FASTQ, 66, 69–73, 78, 92, 95–109, 113–119

FastTree, 4, 135, 137, 138
Fast zero-inflated negative binomial mixed
    modeling (FZINBMM), 598, 616, 660–
    669, 671, 675
Feature correlations, 441, 455–456, 466
Feature table, 5, 8, 15, 25, 66, 73–76, 78–83,
    85–88, 90–92, 95–119, 123, 133, 134,
    147, 150, 151, 153, 269, 323, 324, 330,
    442, 520–523, 528, 533, 578, 688
Filter, 41, 58, 79–87, 104, 105, 116, 118, 241,
    294, 309, 353, 356, 448, 458, 466, 476,
    485, 521, 625, 650, 683
Finite-sample corrected AIC (AIC$_c$), 601–603,
    608, 620
fitZIBB(), 483, 484, 487
Fixed effect, 558–563, 568, 571, 573, 575–579,
    589, 590, 592, 606, 607, 617, 622, 623,
    627, 647, 648, 650, 658–661, 664,
    665, 667
Frobenius norm, 680, 681, 685, 688, 689
F test, 205, 209, 410, 571, 574, 620
Functional analysis, 271–273, 567

G
Gauss-Hermite quadrature (GHQ), 593, 594,
    608, 621
Generalized information criterion (GIC$_\lambda$),
    604, 608
Generalized linear mixed models (GLMMs),
    560, 562, 571, 582, 587–608, 615–622,
    624–659, 662, 663, 669–671, 675, 676
Generalized linear models (GLMs), 440, 471,
    482, 515, 528, 562, 563, 587–589, 596,
    597, 599, 600, 608, 616, 617, 621, 622,
    646, 664
Generalized nonlinear models (GNLMs), 587,
    588, 608
Generalized UniFrac, 135, 336, 344, 347
Genome Taxonomy Database (GTDB), 128, 130
ggpubr, 17–23, 313, 314, 316
GLMMadaptive package, 598, 616, 617,
    647–660
GLMM-MiRKAT, 676, 677
glmmTMB package, 598, 616, 621–647
glmm.zinb(), 664, 666
glmPQL(), 664
Greengenes, 116, 119, 124, 125, 127–129, 132,
    151, 155, 228, 256, 257

H
Heuristic clustering OTU methods,
    229–231, 273
Hierarchical clustering OTU methods, 229–230

Histogram plots, 22–23
HITdb, 130
Holm-Bonferroni, 531
Homogeneity, 102, 174, 177, 191, 198, 210,
    366, 397, 418–422, 431, 618, 639, 641
Hypothesis tests, 163, 208, 230, 397, 410, 484,
    560–561, 618

I
Information criteria, 562, 600–605, 608, 618,
    646, 670
Inter-quartile log-ratio (iqlr), 496, 498, 500,
    502, 503, 506, 513, 514, 518, 551
Inverse Simpson diversity, 303–304
Isometric log-ratio (ilr), 496–498, 500, 551, 691
Iterative weighted least squares (IWLS)
    algorithm, 591, 592, 596–598, 608,
    660, 663

J
Jaccard distance, 344, 409, 418, 429, 430
Jaccard index, 322, 336, 339–340

K
Keemei, 98, 109
Kendall correlation method, 688
Kenward-Roger (KR) approximation, 571
KRmodcomp(), 571
Kruskal-Wallis test, 318–321, 330, 465, 501,
    518, 576

L
Laplace approximation, 592–595, 598, 608,
    621, 660
Large P small N problem, 349, 675, 692–693
Library sizes, 437, 460, 461, 466, 481, 492,
    493, 497, 501, 528, 529, 533, 549, 564
libSize(), 460, 466
Likelihood-based methods, 591, 593–595,
    608, 646
Likelihood ratio test (LRT), 470, 560–562, 575,
    605–608, 619, 651, 670
Linear mixed-effects models (LMMs), 551,
    561, 565, 577, 581, 582, 587, 589–592,
    594, 598, 615, 620, 660, 663–665, 667,
    669, 671, 675–677, 692
list.files(), 96
lme(), 568, 571, 664–666
lme4 package, 571, 573, 575, 621, 646
LmerTest package, 557, 570–576, 582
load(), 14–15, 356, 572

Log-normal permutation test, 453–454
Log-ratio transformations, 344, 491, 496–502, 505, 518, 522, 526, 527, 548, 549, 551, 690

M
Machine learning, 3, 128, 599–600, 608
MAFFT, 136
marginal_coefs (), 648
Marginalized two-part beta regression (MTPBR), 470
Markov Chain Monte Carlo (MCMC)-based integration, 591, 592, 595, 596, 608
MASS package, 664
Maximum likelihood (ML), 52, 137, 472, 482, 562, 563, 590, 591, 593, 594, 596, 597, 600, 606, 616, 619, 621, 647, 659, 664
metagenomeSeq, 48, 74, 431, 436, 438, 440–466, 479, 548, 549
Microbiome package, 11, 34–36, 44, 46, 47, 61, 293, 296–298, 300, 303–305, 307, 312–316, 348, 377, 411, 426–429, 431, 572
Mothur, 2, 47, 74, 104, 124, 177, 216, 228, 230, 256–258, 263, 265, 294
MRcounts(), 456, 459–460, 466
MRexperiment Object, 441, 443–446, 448, 454–459, 461–464, 466
Multi-omics integration, 271, 273
Multi-omics methods, 677–678, 693
Multiplexed paired-end FASTQ data, 95, 108–112, 119
Multivariate analysis, 207, 244, 342, 347, 350, 351, 371, 384, 397, 398, 404, 405, 409, 417–419, 431, 557, 678, 692, 693
Multivariate distance/Kernel-based longitudinal models, 676–677, 693
Multivariate longitudinal microbiome analysis, 675–678

N
National Center for Biotechnology Information (NCBI), 124, 125, 130, 139, 256
Natural classification, 166, 169, 171–172, 175, 212, 215, 218
NBZIMM package, 616, 664–668
Negative binomial mixed models (NBMMs), 596–598, 650, 651, 669
Negative binomial (NB) model, 616
Newick tree format, 53
nifHdada2, 131
nlme package, 563–569, 582

Non-metric multidimensional scaling (NMDS), 202, 203, 351, 352, 355–357, 362–366, 385–388, 399, 678
Nonparametric MANOVA, 398, 405
Non-parametric microbial interdependence test (NMIT), 671, 675, 678–693
Normalization factors, 438, 439, 444, 445, 447, 460, 466
Normalized counts, 260, 437–439, 449, 456, 459–461, 466
Numerical integration, 591–595, 598, 608, 659, 668, 670
Numerical taxonomy, 140, 161–218, 227–231, 236, 239, 240, 244, 246, 260, 267, 349–351, 360, 385–387

O
Oligotyping, 254, 255, 260, 269
Open-reference clustering, 149, 154–157, 161
Operational taxonomic units (OTUs), 8, 25–30, 32, 36, 37, 40, 44, 45, 47, 54, 56, 57, 74, 91, 95, 103, 104, 107, 116, 118, 119, 123, 125, 126, 128, 132, 141, 147–157, 161–218, 227–241, 246–273, 291–295, 322, 325, 328, 345, 348, 350–356, 360, 364, 366, 368, 370, 378, 385, 386, 402, 408, 428, 435, 437–440, 442, 444, 447–449, 458, 461–462, 465, 467, 481–485, 487, 489, 493, 494, 500, 501, 506, 508, 511–514, 517–519, 528–530, 535, 548, 549, 564, 677, 681, 682, 684, 692
Ordination, 47, 154, 162, 176, 193, 194, 202, 209, 210, 216, 217, 335, 336, 339, 345, 349–353, 355, 357–379, 381–388, 397–399, 404–406, 431, 453, 678
Ordination methods, 177, 195, 202–203, 218, 330, 335, 349–381, 385–388, 398, 404, 677
Over-dispersed, 211, 387, 549, 597, 598, 602, 615–620, 629, 646, 660, 661, 669, 670
Over-dispersion, 379, 435, 437, 480–484, 582, 596, 597, 602, 607, 616, 620, 630, 636, 640, 652, 670, 671

P
p.adjust(), 423, 533
pairwise.perm.manova(), 423
pbkrtest, 571
p-corr-method, 688, 689
Pearson correlation method, 687

Penalized quasi-likelihood (PQL), 592–594, 596, 598, 663
Penalized quasi-likelihood-based methods, 591–593, 608
Permutational MANOVA (PERMANOVA), 341, 345, 397, 400, 405–420, 422–426, 428, 429, 431, 689
Permutation invariance, 495
Permutation tests, 208, 210, 398–400, 405, 409, 417, 419, 420, 423, 431, 448, 453–454, 466, 680
Perturbation invariance, 496
Phenetics, 162, 164–172, 174–178, 188, 194, 201, 202, 210–213, 215–218, 241, 243, 244, 246, 247, 350, 351, 385, 386
Phenetic taxonomy, 169–176, 202, 216
Phylogenetic diversity, 135, 136, 138, 249, 305, 306, 322, 326, 327, 330
Phylogenetic entropy, 305–307, 330
Phylogenetic quadratic entropy, 305, 307, 330
Phylogenetics, 2, 11, 25, 48–54, 61, 125, 136, 139, 140, 162–164, 170, 173–176, 212, 214, 232, 236, 237, 243, 245–249, 251, 261, 267, 290, 305–307, 322, 324, 326, 328, 335, 336, 342–348, 360
Phylogenetic trees, 4, 7, 8, 25, 29, 36, 41, 42, 48–53, 66, 74–77, 92, 119, 123–141, 234, 290, 305–307, 322, 326, 328, 343, 345, 356, 533
Phyloseq object, 24–31, 33, 34, 36, 46, 48, 58, 59, 88, 89, 297, 316, 345, 346, 411, 534
Phyloseq package, 24, 25, 30, 34, 35, 58, 74, 88, 179, 297, 298, 345
Physiological characteristics, 248–249
Phytools, 11, 48, 50–52
Pielou's evenness, 304–305, 326, 327, 330
pldist, 344–345
plotQQunif (), 637
plotResiduals (), 638
Plugins, 2–5, 8, 65, 80, 83, 85, 86, 95, 99, 103–112, 114–116, 119, 128, 133, 136, 148–150, 153, 155, 261, 322, 381, 428, 520, 692
Poisson, 168, 211, 260, 262, 437, 501, 528, 588, 589, 593, 676
pr2database, 131
Presence absence testing, 454–455, 466
Principal component analysis (PCA), 154, 202, 351–360, 363, 364, 367, 368, 370, 375, 378, 385–388, 452, 453, 499, 678, 693
Principal coordinate analysis (PCoA), 154, 202, 207, 323, 351, 352, 355–357, 359–364, 382–388, 421, 422, 428, 678, 689, 690
Prokaryote/bacterial species, 242

pscl package, 616
Pseudo-likelihood (PL), 592, 594
Pyrosequencing flowgrams, 258–259, 273

Q
q2-composition, 5, 520
q2-cutadapt, 5, 110, 111, 148
q2-data2, 2
q2-deblur plugin, 95, 113–119
q2-feature-classifier, 5, 126, 128–135, 141
q2-feature-table, 2, 5, 80, 86
QIIME, 1–4, 7, 65–67, 92, 100, 104, 124–127, 136, 138, 155, 156, 228, 231, 232, 234, 256, 257, 263, 270, 293, 323
QIIME 2, 1–8, 11, 15, 48, 65–92, 95–101, 103, 104, 106, 108–110, 113–116, 119, 124–128, 132, 133, 136, 137, 141, 147–149, 151, 153, 157, 256, 257, 265, 270, 290, 293, 294, 296, 306, 322–330, 335, 336, 381–384, 388, 397, 428–431, 520–526, 551, 557, 576–582, 675, 687–690, 692, 693
QIIME 2 archives, 8, 77–79, 92
qiime composition ancom, 520, 523, 526
qiime longitudinal linear-mixed-effects, 576–577, 579, 582
qiime longitudinal nmit, 688
qiime2R package, 15, 87–90, 92, 296
QIIME 2 view, 3, 87, 92, 323, 581
Qiime zipped artifacts (.qza), 3, 4, 7, 8, 65, 76, 77, 87, 88, 100–101, 110, 119, 132, 151, 296, 324
QQ-plot, 494, 637, 638, 640, 653–657
q-score, 114, 116, 149
q-score-joined, 116, 149
Q-technique, 167, 180–182
q2-types, 2, 5
Quality filter, 114, 116, 132, 149
Quality of the reads, 112–113
Quasi Akaike Information Criterion and Corrected Quasi-AIC (QAIC and QAICc), 602
q2-vsearch, 5, 115, 148–150, 157
.qza file, 88

R
Random effect, 470, 503, 558–564, 574, 575, 577–579, 582, 589–598, 601, 606, 607, 676
Rarefaction, 108, 154, 323, 328–330, 348, 437, 646

Raw sequence data, 1–4, 78, 96, 99,
    109–110, 119
read.csv(), 12, 57, 443
read.csv2(), 12, 443
read.delim(), 12, 442
readr, 16–17, 46, 60
readRDS(), 14
read.table(), 12, 443
Redundancy analysis (RDA), 351–353,
    355–357, 359, 363, 374–378, 385, 388
Reference databases, 123–129, 132, 135, 141,
    150–153, 155, 228, 256, 257, 293
Restricted maximum likelihood (REML), 563,
    574, 606, 619, 621
Ribosomal Database Project (RDP), 74, 124,
    128, 130, 228–230, 256
R-technique, 180–182
RVAideMemoire package, 423–426, 431

S
Sample metadata, 1, 25, 36, 56, 79, 86, 87,
    96–99, 109, 111, 112, 119, 133, 134,
    327, 328, 330, 351, 412, 429, 474, 506,
    520, 533, 534, 545, 578, 689
Sample size calculation, 166–168, 218
Satterthwaite approximation, 571,
    573–575, 620
save(), 13, 14
save.image(), 14
saveRDS(), 12–13
Scaling invariance, 495
SeekDeep, 254, 255, 259, 263–266, 268, 273
Semantic type, 4, 8, 68, 69, 73
Semi continuous, 469, 470, 648
seqkit, 68, 96
Sequence similarity, 228, 231, 236, 237, 240,
    241, 247, 248, 260, 261, 267, 271, 273
Sequencing error, 81, 95, 103, 118, 154, 230,
    235, 240–241, 251–253, 256, 259, 260,
    266, 329, 330
Shannon diversity, 43, 44, 298–300, 576–581
SILVA, 124, 127, 130, 151, 228, 257
Similarity coefficients, 162, 167, 169, 176,
    182–194, 196–198, 200, 208–210, 212,
    213, 216, 218, 336, 339, 340, 351, 354
Similarity/resemblance matrix, 193–194
Simpson diversity, 295, 298, 300–304
Simpson evenness, 303–304
Single linkage clustering, 162, 196–197, 200,
    214, 215, 234, 238, 252
Single-nucleotide resolution-based OTU
    methods, 250–254
Sørensen index, 336, 340–341

Spearman correlation method, 140, 494
Species and species-level analysis, 227,
    241–249, 273
16S rRNA method, 246–249
stats package, 664
Subcompositional coherence, 495
Subcompositional dominance, 496
Sub-OTU methods, 252, 269–271, 273
Swarm2, 252–254, 269, 273

T
Taxa, 5, 25, 84, 119, 134, 236, 342, 472, 494,
    566, 650, 677
Taxonomic classification, 36, 56, 103,
    123–125, 132, 133, 135, 141, 212, 213,
    236, 267
Taxonomic rank, 28, 124, 169–170, 213, 218,
    231, 236, 535
Taxonomic resemblance, 161, 178–192, 218
Taxonomic structure, 162, 164, 168, 176–207,
    209, 210, 214, 216, 218, 385, 387
Taxonomy, 2, 75, 119, 123, 228, 535, 626, 688
Taylor-series linearization, 591–593, 608
Template model builder (TMB) package, 621
testDispersion(), 638
testZeroinflation(), 638, 640
tidyverse, 23–24, 58–60
Total sum scaling (TSS), 436–438, 465,
    466, 479
.tsv file, 296, 324, 430
Tukey mean-difference plot, 508
TukeyHSD(), 418

U
UCLUST, 155, 228, 231, 232, 238, 253,
    265, 269
UNITE, 124, 127, 128, 151
Univariate analysis, 557, 678, 692, 693
UNOISE2, 254, 257, 259, 263–266, 268, 273
UNOISE3, 256–258, 263–264, 273
Unweighted UniFrac, 135, 136, 322, 336, 342,
    345, 347, 348, 409, 431
Unweighted UniFrac distance, 343, 344, 348,
    384, 409, 429, 430
USEARCH, 2, 74, 149, 155, 228, 231–234,
    257, 264–266, 293

V
vegan package, 24, 47, 179, 336, 339–342, 348,
    352, 356, 401–404, 409–419, 453, 683,
    685, 689

vegdist(), 193, 336, 338–341, 356, 401, 410, 418, 419
Violin plots, 20–21, 316–318, 330
Visualizations, 3–5, 7, 8, 17, 56, 61, 87, 101, 105, 106, 108, 109, 112, 119, 133, 134, 322, 323, 349, 353, 354, 381, 382, 428, 441, 449, 521, 579, 581, 679, 689, 692
Volatility analysis, 579–582
VSEARCH, 128, 149–151, 153, 157, 234
Vuong test, 607, 608, 618, 619

W
Wald *t* test, 619, 620
Weighted UniFrac, 135, 258, 322, 336, 343, 345, 347, 409, 429, 431
Weighted UniFrac distance, 136, 336, 343–345, 347, 429, 431
write.table(), 12

X
xtable(), 511
xtable package, 511

Z
Zero-hurdle, 607, 616, 619, 624, 635, 643, 646, 659, 669
Zero-hurdle negative binomial (ZHNB), 616, 618, 632, 643, 651, 669, 670

Zero-hurdle Poisson (ZHP), 616, 618, 619, 632, 635, 643, 651, 658, 659, 670
Zero-inflated, 211, 387, 436, 439, 440, 470, 479, 480, 548, 592, 598, 607, 616–622, 626, 628–630, 633, 635, 646, 660, 665, 669, 670
Zero-inflated beta, 467, 469–489
Zero-inflated beta-binomial model (ZIBB), 469, 470, 480–489
Zero-inflated beta regression (ZIBSeq), 469–480, 489
Zero-inflated beta regression model with random effects (ZIBR), 470
Zero-inflated continuous, 469, 470
Zero-inflated Gaussian (ZIG), 435–441, 465–467, 479, 480, 548, 549, 669
Zero-inflated generalized linear mixed models (ZIGLMMs), 599, 600, 617
Zero-inflated log-normal (ZILN), 435–453, 465–467
Zero-inflated negative binomial (ZINB), 479, 480, 488, 616, 618, 622, 630, 631, 641, 645, 648, 650–652, 656–661, 669, 670
Zero-inflated negative binomial mixed models (ZINBMMs), 596, 598, 660–664, 668, 670
Zero-inflated Poisson (ZIP), 479, 480, 616, 618, 619, 629, 636, 648, 650–652
ZIBBSeqDiscovery, 483, 484
zicmp, 630, 633–635, 642, 644, 646

Printed in the United States
by Baker & Taylor Publisher Services